Transparent Semiconducting Oxides

Transparent Semiconducting Oxides

Bulk Crystal Growth and Fundamental Properties

Zbigniew Galazka

JENNY STANFORD
PUBLISHING

Published by

Jenny Stanford Publishing Pte. Ltd.
Level 34, Centennial Tower
3 Temasek Avenue
Singapore 039190

Email: editorial@jennystanford.com
Web: www.jennystanford.com

British Library Cataloguing-in-Publication Data
A catalogue record for this book is available from the British Library.

Transparent Semiconducting Oxides: Bulk Crystal Growth and Fundamental Properties

ISBN 978-981-4800-94-5 (Hardcover)
ISBN 978-1-003-04520-5 (eBook)

To my loving wife Beata, daughter Marietta, and grandson Alek

Contents

Preface xvii
Acknowledgments xxi
Abbreviations and Symbols xxiii

1. Basics of Growth Techniques Used for Bulk Single Crystals of Transparent Semiconducting Oxides 1
1.1 Introduction 1
1.2 Gas Phase 4
1.2.1 Chemical Vapor Transport (CVT) 4
1.2.2 Physical Vapor Transport (PVT) 7
1.3 Solution 9
1.3.1 Hydrothermal 10
1.3.2 Flux 12
1.4 Melt 14
1.4.1 Verneuil 14
1.4.2 Optical Floating Zone (OFZ) 16
1.4.3 Cold Crucible 18
1.4.4 Bridgman–Stockbarger (BS) and Vertical Gradient Freeze (VGF) 20
1.4.5 Czochralski 22
1.4.6 Kyropoulos 25
1.4.7 Edge-Defined Film-Fed Growth (EFG) 26
1.4.8 Levitation-Assisted Self-Seeding Crystal Growth Method (LASSCGM) 28
1.5 Summary 30

2. Thermodynamics of Transparent Semiconducting Oxides 43
2.1 Introduction 43
2.2 Basics of Thermodynamics 44
2.3 Thermodynamics of TSOs 49
2.3.1 The Al–O_2 System 50
2.3.2 The Ga–O_2 System 50
2.3.3 The In–O_2 System 51
2.3.4 The Sn–O_2 System 52
2.3.5 The Sb–O_2 System 54

2.3.6 The $Cu–O_2$ System 55
2.3.7 The $Zn–O_2$ System 56
2.4 Crucible Materials 57
2.5 Growth Atmosphere 61
2.6 Summary 65

3. Zinc Oxide (ZnO) 69
3.1 Introduction 69
3.2 Crystal Structure 71
3.3 Band Structure and Native Point Defects 72
3.3.1 Band Structure 72
3.3.2 Native Point Defects 74
3.4 Bulk Crystal Growth 75
3.4.1 Growth from Gas Phase 76
3.4.1.1 Sublimating and oxidizing of elemental Zn 79
3.4.1.2 Use of ZnX_2 as a source material and an oxidizing agent 81
3.4.1.3 Use of ZnY as a source material and an oxidizing agent 86
3.4.1.4 Use of ZnO as the source material and a chemically active transport agent 88
3.4.1.5 Use of ZnO as a source material, a reducing agent, and an oxidizing agent 92
3.4.2 Flux Techniques 117
3.4.3 Hydrothermal 123
3.4.3.1 Autoclaves 125
3.4.3.2 Growth conditions/operating parameters 126
3.4.3.3 Impurities and intentional doping 132
3.4.3.4 Growth morphology and growth rate 135
3.4.3.5 Hydrothermally grown ZnO crystals and structural quality 139

3.4.3.6 Effect of annealing on the surface quality 140
3.4.3.7 Examples 142
3.4.4 Cold Crucible 157
3.4.5 Bridgman 159
3.5 Thermal and Mechanical Properties 165
3.5.1 Melting Point 165
3.5.2 Thermal Conductivity and Specific Heat Capacity 166
3.5.3 Thermal Expansion 167
3.5.4 Stiffness and Hardness 169
3.5.5 Conclusion 172
3.6 Basic Optical Properties 174
3.6.1 Transmittance and Optical Energy Gap 174
3.6.2 Refractive Index and Dielectric Constant 178
3.6.3 Luminescence 179
3.6.4 Phonon Modes 182
3.6.5 Conclusion 182
3.7 Basic Electrical Properties 184
3.7.1 Origin of Electrical Conductivity 184
3.7.2 Free Electron Concentration, Hall Mobility, and Resistivity/Conductivity at RT 188
3.7.3 Temperature-Dependent Electrical Properties and Ionization Energy of Shallow Donors 191
3.7.4 Effective Electron and Polaron Masses 194
3.7.5 Effect of Annealing on Electrical Properties 195
3.7.6 Effect of Intentional Doping on Electrical Properties 198
3.7.7 Deep Traps 199
3.7.8 Conclusion 201
3.8 Key Points 202
3.9 Summary 203

4. Gallium Oxide (β-Ga_2O_3) 223
4.1 Introduction 223
4.2 Polymorphism of Ga_2O_3 225

4.3 Crystal Structure of β-Ga_2O_3 227
4.4 Band Structure and Native Point Defects 228
4.4.1 Band Structure 228
4.4.2 Native Point Defects 230
4.5 Bulk Crystal Growth 231
4.5.1 Gas Phase 233
4.5.1.1 Physical vapor transport 233
4.5.1.2 Chemical vapor transport 234
4.5.2 Flux 242
4.5.2.1 Crystal growth 242
4.5.2.2 Examples 243
4.5.3 Verneuil 245
4.5.4 Optical Floating Zone (OFZ) 247
4.5.4.1 Crystal growth 247
4.5.4.2 Examples 249
4.5.5 Edge-Defined Film-Fed Growth (EFG) 254
4.5.6 Czochralski 259
4.5.6.1 Crystal growth 259
4.5.6.2 Pulling direction 260
4.5.6.3 Impact of the oxygen partial pressure 262
4.5.6.4 Free carriers and heat transport 272
4.5.6.5 Intentional doping and dopant segregation 276
4.5.6.6 Examples 290
4.5.7 Vertical Gradient Freeze (VGF) and Vertical Bridgman (VB) Methods 291
4.5.8 Conclusion 296
4.6 Surfaces 298
4.6.1 Surface Properties 298
4.6.2 Surface Preparation 300
4.6.3 Conclusion 303
4.7 Thermal and Mechanical Properties 304
4.7.1 Melting Point and Density 305
4.7.2 Thermal Conductivity, Diffusivity, and Specific Heat 305
4.7.3 Thermal Expansion 308
4.7.4 Stiffness and Hardness 309
4.7.5 Cleavage Planes 310

4.7.6 Conclusion 311
4.8 Basic Optical Properties 312
4.8.1 Transmittance, Absorption, Bandgap 313
4.8.2 Refractive Index and Dielectric Constant 316
4.8.3 Luminescence 318
4.8.4 Phonon Modes 322
4.8.5 Conclusion 323
4.9 Scintillation Properties 324
4.10 Basic Electrical Properties 328
4.10.1 Origin of Electrical Conductivity 329
4.10.2 Free Electron Concentration, Hall Mobility, and Resistivity/Conductivity at RT 332
4.10.3 Impact of Annealing on Electrical Properties 336
4.10.4 Temperature-Dependent Electrical Properties 340
4.10.5 Effective Electron Mass and Electron Mobility 342
4.10.6 Ionization Energy of Shallow Donors 344
4.10.7 Deep Traps 344
4.10.8 Conclusion 347
4.11 Key Points 349
4.12 Summary 349

5. Indium Oxide (In_2O_3) 371
5.1 Introduction 371
5.2 Crystal Structure 372
5.3 Band Structure and Native Point Defects 374
5.3.1 Band Structure 374
5.3.2 Native Point Defects 375
5.4 Bulk Crystal Growth 377
5.4.1 Stability Fields of In–O System 377
5.4.2 Historical Events 378
5.4.3 Vapor Phase 378
5.4.3.1 Thermal decomposition of In_2O_3 (PVT) 379
5.4.3.2 Use of a reducing agent 380
5.4.3.3 Use of chemically active

transport agent (CVT) 383
5.4.4 Solution Methods 388
5.4.4.1 Hydrothermal 388
5.4.4.2 Flux 389
5.4.4.3 Salt electrolysis 394
5.4.5 Levitation-Assisted Self-Seeding Crystal Growth Method (LASSCGM) 395
5.4.5.1 Thermodynamics 395
5.4.5.2 Concept of LASSCGM 397
5.4.5.3 Operation of LASSCGM 400
5.4.5.4 Furnace design 405
5.4.5.5 Growth conditions 407
5.4.5.6 Operating parameters 410
5.4.5.7 Bulk single crystals 410
5.4.4.8 Thermal stability of melt-grown bulk single crystals 415
5.4.6 Conclusion 417
5.5 Thermal and Mechanical Properties 419
5.5.1 Thermal Properties 419
5.5.2 Mechanical Properties 421
5.6 Optical Properties and Effect of Annealing 422
5.6.1 Transmittance, Absorption Edge, and Optical Bandgap 423
5.6.2 Dielectric Constant, Refractive Index, Reflectance 427
5.6.3 Luminescence 428
5.6.4 Phonon Modes 430
5.6.5 Conclusion 431
5.7 Electrical Properties 433
5.7.1 Free Electron Concentration, Hall Mobility, and Resistivity/ Conductivity at RT 434
5.7.2 Temperature-Dependent Electrical Properties 436
5.7.3 Impact of Annealing on Electrical Properties 438
5.7.4 Surface Electron Accumulation Layer 444
5.7.5 Effective Electron Mass and Electron Mobility 446

5.7.6 Ionization Energy of Shallow Donors 447
5.7.7 Conclusion 447
5.8 Key Points 449
5.9 Summary 450

6. Tin Oxide (SnO_2) 463
6.1 Introduction 463
6.2 Crystal Structure 465
6.3 Band Structure and Native Point Defects 465
6.3.1 Band Structure 465
6.3.2 Native Point Defects 466
6.3.3 Toward *p*-Type Conductivity 467
6.4 Bulk Crystal Growth 468
6.4.1 Stability Fields of the Sn–O System 468
6.4.2 Historical Events in Bulk SnO_2 Crystal Growth 470
6.4.3 Vapor Phase 470
6.4.3.1 Physical vapor transport (PVT) 471
6.4.3.2 Chemical vapor transport (CVT) 492
6.4.4 Solution 505
6.4.4.1 Hydrothermal 505
6.4.4.2 Flux 506
6.4.5 Conclusion 510
6.5 Thermal Stability of Bulk SnO_2 Crystals and Their Surfaces 511
6.5.1 Thermal Stability 511
6.5.2 Surfaces of SnO_2 and Their Stability 514
6.6 Thermal and Mechanical Properties 516
6.6.1 Thermal Properties 516
6.6.2 Mechanical Properties 518
6.6.3 Conclusion 519
6.7 Optical Properties and Effect of Annealing 520
6.7.1 Transmittance, Absorption Edge, and Bandgap 520
6.7.2 Dielectric Constant and Refractive Index 523
6.7.3 Luminescence 525
6.7.4 Phonon Modes 526

6.7.5 Conclusion 528
6.8 Electrical Properties and Effect of Annealing 529
6.8.1 Free Electron Concentration, Hall Mobility, and Resistivity/Conductivity at RT: As-Grown Crystals and Effect of Annealing 530
6.8.2 Temperature-Dependent Electrical Properties 535
6.8.3 Effective Electron Mass and Electron Mobility 538
6.8.4 Ionization Energy of Shallow Donors 542
6.8.5 Deep-Level Traps 543
6.8.6 Surface Band Bending 543
6.8.7 Conclusion 544
6.9 Key Points 546
6.10 Summary 547

7. Barium Stannate ($BaSnO_3$) 559
7.1 Introduction 559
7.2 Crystal Structure 560
7.3 Band Structure and Native Point Defects 562
7.3.1 Band Structure 562
7.3.2 Native Point Defects 563
7.4 Bulk Crystal Growth 565
7.4.1 Phase Diagram of the $BaO–SnO_2$ System 565
7.4.2 Historical Events 565
7.4.3 Flux 567
7.4.3.1 Crystal growth 567
7.4.3.2 Intentional doping 569
7.4.3.3 Examples 569
7.4.4 Growth from the Melt 571
7.4.4.1 Thermodynamics 571
7.4.4.2 Vertical gradient freeze (VGF) 575
7.4.4.3 Examples 585
7.4.5 Thermal Stability and Surfaces of Bulk Single Crystals 589
7.4.6 Conclusion 590
7.5 Thermal and Mechanical Properties 592
7.6 Optical Properties 596

7.6.1 Transmittance, Absorption Edge, and Optical Bandgap 596
7.6.2 Dielectric Constant and Refractive Index 599
7.6.3 Luminescence 600
7.6.4 Phonon Modes 601
7.6.5 Conclusion 602
7.7 Electrical Properties 603
7.7.1 Free Electron Concentration, Hall Mobility, and Resistivity/Conductivity at RT 604
7.7.2 Temperature-Dependent Electrical Properties 607
7.7.3 Effective Electron Mass and Electron Mobility 610
7.7.4 Two-Dimensional Electron Gas 617
7.7.5 Conclusion 619
7.8 Key Points 620
7.9 Summary 621

8. Ga-Based Spinels: $ZnGa_2O_4$, $MgGa_2O_4$, $(Zn,Mg)Ga_2O_4$, $CoGa_2O_4$ 629
8.1 Introduction 629
8.2 Crystal Structure 630
8.3 Band Structure and Native Point Defects 632
8.3.1 Band Structure 632
8.3.2 Intrinsic and Extrinsic Point Defects versus Electrical Conductivity 636
8.4 Bulk Crystal Growth 637
8.4.1 Phase Diagrams 637
8.4.2 Historical Events 639
8.4.3 Growth from the Gas Phase 640
8.4.4 Growth from the Flux 641
8.4.4.1 Crystal growth 641
8.4.4.2 Examples 642
8.4.5 Growth from the Melt 645
8.4.5.1 Thermodynamics 645
8.4.5.2 Laser heated pedestal (LHPD) growth 647

8.4.5.3 Optical floating zone (OFZ) 647
8.4.5.4 Vertical gradient freeze (VGF)/vertical Bridgman (VB) 648
8.4.5.5 Czochralski and Kyropoulos-like methods 652
8.4.5.6 Composition and structural quality of bulk Ga-based spinel crystals obtained from the melt 658
8.4.5.7 Thermal stability 662
8.4.6 Conclusion 668
8.5 Thermal and Mechanical Properties 670
8.5.1 Thermal Properties 670
8.5.2 Mechanical Properties 672
8.5.3 Conclusion 675
8.6 Optical Properties 678
8.6.1 Transmittance, Absorption Edge, and Optical Bandgap 678
8.6.2 Dielectric Constant and Refractive Index 681
8.6.3 Luminescence 683
8.6.4 Phonon Modes 686
8.6.5 Conclusion 687
8.7 Electrical Properties 689
8.7.1 Free Electron Concentration, Hall Mobility, and Resistivity at RT 690
8.7.2 Effective Electron Mass and Electron Mobility 693
8.7.3 Conclusion 695
8.8 Key Points 697
8.9 Summary 697

Index 709

Preface

Transparent semiconducting oxides (TSOs) constitute a class of materials that combine semiconducting properties and transparency in the visible spectrum, even down to deep ultraviolet spectral region. They are, therefore, regarded as wide or ultrawide bandgap semiconductors that are complementary to nitride semiconductors (GaN, AlN) or silicon carbide. The TSOs, as such, are sought to expand the functionality of existing devices or design new ones with unprecedented functionality. In fact, many TSO-based devices have already been demonstrated in a variety of technical areas, including electronics, optoelectronics, photovoltaics, and sensing devices. In a vast majority of cases, thin films/layers, nanostructures, or ceramics are used for both market devices and device demonstrations. Indeed, such material forms are relatively easily accessible worldwide at universities, research institutes, or companies, as a result of the wide spread of epitaxial techniques and/or preparation of amorphous layers/ceramics, which is easy and cost-effective. With bulk oxide single crystals, the situation is different, as typically they have high melting points, which require expensive growth stations (especially when large volume is in quest), high investments in consumables (noble metal crucibles, high-purity starting materials, thermal insulation, etc.), and unique know-how, typically created by experience. With bulk TSO single crystals, the situation is yet more complex, as all of them are thermally unstable at high temperatures, which highly limit the possibilities of obtaining bulk single crystals of a reasonable size and structural quality. However, each problem has a solution, and different concepts as well as novel technical approaches discussed in this book enable to obtain larger single crystals with improved structural quality.

The fundamental question is whether we really need bulk TSO single crystals. In the same way, one may ask why do we need bulk single crystals of classical or compound semiconductors? Bulk TSO single crystals can be differentiated in three ways based on their functions as (i) active device components (e.g., vertical FETs, gas detector, photodetector), (ii) substrates for epitaxy (both homo- and heteroepitaxy), and (iii) reference for properties. A device typically consists of both substrate and a structure formed "thereon," where the substrate can function as an active or a passive component. Moreover, native substrates significantly improve the quality and properties of the films (homoepitaxy) and resulting device functionality. To bring the device functionality to an acceptable or high level, bulk crystals are, in most cases, highly demanded. Moreover, properties of bulk single crystals are closer to the intrinsic properties of an "ideal" compound compared with thin films/layers or ceramics due to, in general, much lower defect density that significantly affects (among others) the electrical, optical, magnetic, mechanical, and thermal properties. If the defect density in a bulk single crystal becomes too high, it simply becomes a polycrystal, but a thin film/layer or ceramic will still remain a thin film/layer or ceramic. Bulk single crystals constitute a very good reference for properties that are targeted for thin films and reliably show the expected properties for novel compounds.

The bulk TSO single crystals, their preparation, and properties are, therefore, a foundation for further research toward a final device and prerequisite for industrial interest. The essence of this book is to discuss various aspects of different bulk TSO single crystals in terms of thermodynamics, bulk crystals growth with the use of different methods (from the gas phase, solution, and the melt), the resulting crystal size, appearance, and structural quality, as well as the basic properties that were gathered from bulk TSO single crystals. The properties of bulk TSO single crystals are briefly accompanied by the theory (band structure, native point defects) and occasionally by properties obtained for thin films. Although there are a number of review articles, book chapters, and books on some of the aspects of TSOs, this book is the only one so far that collects almost all available data on the growth and properties of bulk TSO single crystals. A reader may get a fast and direct access to a compound of interest as a chapter has been dedicated to each of the compounds (ZnO, β-Ga_2O_3, In_2O_3, SnO_2, $BaSnO_3$) or a group of compounds (Ga-based spinels), which has been arranged in the following order: introduction, structural properties, band structure and native point defects, detailed bulk crystal growth (all methods used for a given compound), thermal and mechanical properties, optical properties, and electrical properties. Occasionally, other features characteristic of a given compound have also been included. The content of the book relies not only on the published data, but also on new data that have not been published so far.

I believe, the book is prepared in a way that will be easily understandable by a wide community from different fields of activity, who seek condensed information on material preparation and specific properties of this class of materials. It may be useful for the research community in both academia and industry to get extended information about a particular TSO material, further explore the material's properties, and utilize the material in novel device concepts of unique functionality.

Zbigniew Galazka

Winter 2020

Acknowledgments

I would like to express my special thanks to Dr. Detlef Klimm (Leibniz-Institute für Kristallzüchtung (IKZ), Berlin, Germany) for contributing in chapter preparation on semiconducting oxide thermodynamics (Chapter 2) and overall discussion on the book content. I express my gratitude to Dr. Klaus Irmscher, Dr. Robert Schewski, Dr. Andreas Fiedler, Dr. Reinhard Uecker, Dr. Steffen Ganschow, Dr. Martin Albrecht, Dr. Rainer Bertram, Dr. Saud Bin Anooz, Dr. Andreas Popp, Dr. Günter Wagner, Dr. Detlev Schulz, Dr. Tobias Schulz, Mike Pietsch, and Albert Kwasniewski (all from IKZ) for crystal characterization and/or helpful discussions and brainstorming, as well as Dr. Uta Juda, Thomas Wurche (IKZ), and Knut Peters (CrysTec GmbH, Berlin, Germany) for crystal sample preparation. Many thanks to Isabelle M. Hanke, Mario Brützam, and Andreas Tauchert (IKZ) for technical support in crystal growth experiments. I thank very much Dr. Joel B. Varley (Lawrence Livermore National Laboratory, CA, USA) for his dedicated theoretical calculations included in this book and Prof. Winicjusz Drozdowski and colleagues from his group (Nicolaus Copernicus University, Toruń, Poland) for measurement of scintillation properties of some of the TSOs and helpful discussions. Dr. Oliver Bierwagen (Paul-Drude-Institut, Berlin, Germany), Prof. Recardo Manzke (Humboldt-Universität zu Berlin, Germany), Dr. Christoph Janowitz (Brandenburgische Technische Universität Cottbus/Humboldt-Universität zu Berlin, Germany), Dr. Mattia Mulazzi (Helmholtz-Zentrum Berlin, Germany), and Dr. Takahiro Nagata (National Institute for Materials Science, Tsukuba, Japan) are much appreciated for helpful discussions. I greatly appreciate Marietta Galazka for preparing a number of drawings included in the book. I would like to thank Prof. Roberto Fornari (Università di Parma, Italy), Prof. Günther Tränkle (Ferdinand-Braun-Institut, Berlin, Germany), and Profs. Matthias Bickermann and Thomas Schröder (IKZ) for creating a research environment at IKZ related to my activity on transparent semiconducting oxides, on which a large portion of this book is based.

Abbreviations and Symbols

Crystal structure

a, *b*, *c*	lattice parameters (constants)
(x y z)	plane orientation
{x y z}	family of planes
[x y z]	crystallographic direction vector
<x y z>	family of directions (by symmetry operations)

Growth methods for bulk crystals

CVT	chemical vapor transport
CWCVT	cold-wall chemical vapor transport
EFG	edge-defined film-fed growth
LASSCGM	levitation-assisted self-seeding crystal growth method
LHPG	laser heated pedestal growth
OFZ	optical floating zone
PVT	physical vapor transport
TSFZ	traveling-solvent floating zone
TSSG	top-seeded solution growth
VB	vertical Bridgman
VGF	vertical gradient freeze

Growth methods for layers

CSD	chemical solution deposition
CVD	chemical vapor deposition
HVPE	halide vapor phase epitaxy
LMBE	laser molecular beam epitaxy
MBE	molecular beam epitaxy
MOCVD	metal-organic chemical vapor deposition
MOVPE	metal-organic vapor phase epitaxy
PAMBE	plasma-assisted molecular beam epitaxy
PLD	pulsed laser deposition

Electronics devices

HEMT	high electron mobility transistor
MESFET	metal-semiconductor field-effect transistor
MOSFET	metal-oxide field-effect transistor
SBD	Schottky barrier diode
TFT	thin-film field-effect transistor

Theoretical calculations

B3LYP	hybrid functional of Becke
DFT	density functional theory
Gau-PBE	functional of Gau, Perdew, Burke, Ernzerhof
GGA	generalized gradient approximation
GW	Hedin approximation (Green function and Coulomb potential)
HSE06	hybrid functional of Heyd, Scuseria, and Ernzerhof
LCAO	linear combination of atomic orbitals approximation
LDA	local density approximation

Characterization techniques

AFM	atomic force microscopy
ARPES	angular-resolved photoemission spectroscopy
CL	cathodoluminescence
DLTS	deep-level transient spectroscopy
DLOS	deep-level optical spectroscopy
DSC	differential scanning calorimetry
EPR	electron paramagnetic resonance
FTIR	Fourier-transform infrared spectroscopy
GDMS	glow discharge mass spectroscopy
HXPS/HXPES	hard X-ray photoemission/photoelectron spectroscopy
ICP-OES	inductively coupled plasma optical emission spectroscopy
ICP-MS	inductively coupled plasma mass spectroscopy
LEED	low-energy electron diffraction
NDIC	Nomarski differential interference contrast
PICTS	photoinduced current transient spectroscopy

PL	photoluminescence
RL	radioluminescence
SIMS	secondary ion mass spectroscopy
STEM	scanning transmission electron microscopy
STM	scanning tunneling microscopy
SXPS/SXPES	soft X-ray photoemission/photoelectron spectroscopy
TDTR	time domain thermoreflectance
TEM	transmission electron microscopy
TG	thermogravimetry
TSC	thermally stimulated current
XPS/XPES	X-ray photoemission/photoelectron spectroscopy
XRD	X-ray powder diffraction
XRF	X-ray fluorescence

Thermodynamics

c_p	specific heat capacity
E_0, E_1	initial and final energy of a system
ΔE	energy difference
G	Gibbs free energy
ΔG	Gibbs free energy change
H, H_0	enthalpy
ΔH	enthalpy change
ΔH_f	latent heat of fusion
K	equilibrium constant
p	pressure
p_0	standard pressure (1 bar)
R	gas constant
S	entropy
ΔS	entropy change
T	temperature
T_f	fusion temperature (melting point)
T_0	standard temperature (25°C)

Crystal Growth

CATA	chemically active transport agent
C_S	dopant concentration in a crystal

C_0	initial dopant concentration in the melt
EP	explosion point
FF	fill fraction
g	crystallization fraction
k_{eff}	effective segregation coefficient
MP, T_m	melting point
OP	overheating point
p	pressure
$p(x)$	partial pressure of species x
SP	solidification point
T_D	dissolution temperature
T_F	final growth temperature
T_G	growth temperature
T_H	hole temperature
T_{HIGH}	high temperature
T_{MAX}	maximum temperature
TP	transition point
T_S	source temperature
ΔT	temperature difference
RF	radiofrequency
RT	room temperature

Thermal and mechanical properties

α	thermal expansion coefficient
B_0, G, E	bulk, shear, and Young's moduli
c_{ij}	elastic stiffness coefficients
c_p	heat capacity
D	thermal diffusivity
λ	thermal conductivity
ρ_s, ρ_l	solid and liquid density
θ_D	Debye temperature
ν	Poisson ratio
v_T, v_L	transverse and longitudinal sound velocity
v_{AV}	average sound velocity

Optical properties

α	absorption coefficient
E	energy

E_g	energy gap (bandgap)
$E_{g,\mathrm{dir}}$, $E_{g,\mathrm{indir}}$	direct, indirect bandgap
E_{ph}	photon energy
E_{phonon}	phonon energy
h	Planck constant
ε_{r}, ε_0	relative static dielectric constant
ε_{∞}	high-frequency dielectric constant
$\varepsilon_{\mathrm{AV}}$	average dielectric constant
n, n_0, n_{a}, n_{b}, n_{c},	refractive index
NIR	near-infrared
Δn	birefringence
UV	ultraviolet
ν	radiation frequency
R	reflectance
TR	transmittance

Electrical properties

2DEG	two-dimensional electron gas
BZ	Brillouin zone
CB	conduction band
CBM	conduction band minimum
CNL	charge neutrality level
e	electron charge
E1, E2...	deep level traps
E_{C}, E_{CBM}	energy level of CBM
E_{CD}	ionization energy of shallow donors
E_{V}, E_{VBM}	energy level of VBM
Γ	point at the Brillouin zone
μ	carrier mobility
m_0	free electron rest mass
m_{e}^*, m_{a}^*, m_{b}^*, m_{c}^*	effective electron mass
$m_{\mathrm{e}}^*{}_{\mathrm{AV}}$	average effective electron mass
m_{p}^*	effective hole mass
n_{e}	free electron concentration
N_{D}, N_{A}	donor, acceptor concentration
ρ	resistivity
SEAL	surface electron accumulation layer
STE	self-trapped exciton
STH	self-trapped hole

τ	electron scattering rate
VB	valence band
VBM	valence band maximum

Materials and treatment

CMP	chemical-mechanical polishing
EPD	etch pit density
FWHM	full width at half maximum of the rocking curve
ICP	inductively coupled plasma
RIE	reactive ion etching
RMS	root mean square
TSO	transparent semiconducting oxide
UHV	ultra-high vacuum
UWBG	ultra-wide bandgap

Chapter 1

Basics of Growth Techniques Used for Bulk Single Crystals of Transparent Semiconducting Oxides

1.1 Introduction

This chapter briefly discusses the basics of a diversity of growth techniques used so far to obtain bulk single crystals of transparent semiconducting oxides (TSOs), along with their advantages and drawbacks. As the crystal size, defect formation, purity, and optical/electrical properties depend on the growth environment, i.e., furnace design, furnace materials, growth temperature, temperature gradients in the furnace, atmosphere, pressure, time, and phase, it is important to introduce particularities of the growth techniques and their basic growth conditions. Here, it is not our intention to describe details of these techniques and their modifications, neither nucleation nor transport theories, which can be easily found in rich literature, but only basics, which may be helpful in understanding differences between bulk crystals of the same compound obtained by different techniques. We would like to stress that this description is made from the point of bulk TSO single crystals only. Details of growth environments and resulted bulk TSO crystals will be described and discussed in Chapters 3–8.

Transparent Semiconducting Oxides: Bulk Crystal Growth and Fundamental Properties
Zbigniew Galazka

ISBN 978-981-4800-94-5 (Hardcover), 978-1-003-04520-5 (eBook)
www.jennystanford.com

The growth techniques can be grouped into three categories (Fig. 1.1): gas phase, solution, and melt growth. Gas phase techniques include chemical vapor transport (CVT) and physical vapor transport (PVT). Solution growth techniques include flux and hydrothermal methods. Melt growth techniques cover a wide spectrum of methods, such as Verneuil, optical floating zone (OFZ), and cold crucible methods that constitute crucible-free techniques, and the Bridgman, vertical gradient freeze (VGF)/direct solidification, Czochralski, Kyropulous, edge-defined film-fed growth (EFG), and levitation-assisted self-seeding crystal growth method (LASSCGM) that utilize a metal crucible.

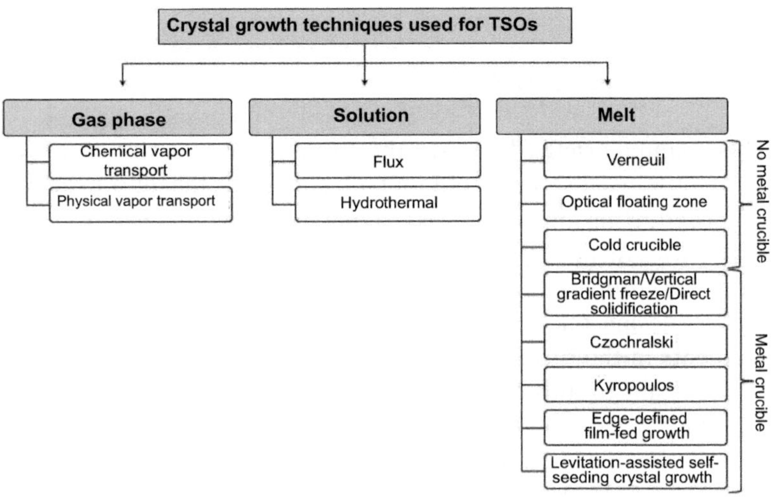

Figure 1.1 Techniques used for growing bulk TSO single crystals.

In terms of crystal growth from the gas phase, the CVT growth proceeds at temperature not only much below the melting point of the TSOs to be grown, but also below their thermal decomposition temperature. The transport of species to a growth zone proceeds with the use of a chemically active transport agent (CATA). On the other hand, the PVT method utilizes the thermal decomposition as the driving force for transporting volatile species to a growth region. It also proceeds below the melting point of the TSOs to be grown. The CVT method typically produces small TSO crystals, while the PVT may produce TSO crystals of sufficient size for wafer fabrication suitable for epitaxy.

The growth from solution also proceeds at temperatures below the melting and thermal decomposition points of the TSOs. This is achieved by using a solvent that can be either a low-melting-point oxide/fluoride (flux method) or aqueous alkaline solution under an overpressure (hydrothermal method). The solved TSO precipitates in the form of single crystals in the flux during slow cooling down, or on crystal seeds due to temperature difference and resulted solvent convection. The flux method produces very small TSO crystals, while the hydrothermal method may produce large crystals in large amount in one run.

The melt growth methods operate at high temperatures (roughly above 1800°C) at which the thermal decomposition becomes a big issue. Among these techniques, the crucible-free methods produce high-purity crystals but rather of small size and often of not so high crystal perfection. On the other hand, the melt growth methods utilizing the metal crucible are the most efficient techniques as they enable large crystal volume within a relatively short time and high crystal perfection. However, a common feature of the TSOs is their thermal instability at high temperatures, which in combination with high melting points leads to a strong decomposition. As a result, melt growth techniques are highly limited for this class of materials and often require further, non-standard modifications. Moreover, due to the thermal instability of TSOs, a reasonable oxygen partial pressure is required in a growth atmosphere, which limits crucible materials to iridium and platinum–rhodium alloys in some cases.

Small bulk crystals obtained by the CVT, flux, or OFZ methods are sufficient for fundamental studies of material properties and demonstration of functionality of test devices but are not sufficient for large-scale applications. For that purpose, the melt growth methods utilizing the metal crucible, the hydrothermal, and possibly the PVT method adapted for individual TSO compounds may satisfy the needs in a large scale.

Below we present the basics of different techniques utilized to obtain bulk TSO single crystals that highlight the principle or concept rather than describing in detail the design, physics, operation, and particularities of the growth methods. Some of these aspects, especially the technical design of growth setups and growth conditions, will be discussed in other chapters when discussing individual compounds. It is important to stress that each of the techniques have many modifications or variations in the technical design and growth conditions, but the fundamental concept remains the same. Although we touch the historical background to locate on time axis the origin of the particular method, it is rather selective, neither complete nor exhaustive. The historical development and application of each of the growth methods for bulk crystals can be easily found in a number of dedicated textbooks and review articles.

1.2 Gas Phase

This class of techniques first utilizes sublimation at a combustion zone, i.e., a direct solid → gas phase transition of a source (starting) material to be grown. Then the gas phase is transported to a growth zone, where it deposits in a reverse process, i.e., in a gas → solid phase transition. The deposition takes place on a crystal seed surface, or on another, typically amorphous surface, e.g., on an ampoule wall. The transport of the gas phase can be direct (in the case of volatile materials) or indirect by using a CATA (in the case of less or non-volatile materials at operating temperatures). The former technique is called the physical vapor transport (PVT) method. In the latter case, there is a chemical reaction between the source material and the transport agent, and additionally with oxygen in the growth zone. This technique is referred to the chemical vapor transport (CVT) method. There are different arrangements in setups and environments for the growth from the gas phase that are described in Chapters 3–8 for TSOs, and in general, e.g., by Wilke [1].

1.2.1 Chemical Vapor Transport (CVT)

The CVT method, also called vapor transport reaction method, is suitable for growing crystals of a compound of interest at temperature far below its melting point. It is a versatile method in terms of compounds to be grown and a furnace design. According to the historical review on the CVT method by Binnewies et al. [2], the first who used the vapor transport reaction method was R. Bunsen [3] in the nineteenth century, who carried out laboratory crystallization of Fe_2O_3 inspired by observing the formation of hematite in the presence of HCl during volcano eruption. Soon after other oxide materials were crystallized by this method, such as TiO_2, MnO, and MgO [4], and later elemental metals, such as W, Ti, Zr, Hf, and Th [5, 6]. A wider use of that method was possible after the work of Herald Schäfer [7–9], who in the 1950s quantitatively described transport reactions from thermodynamic point of view with a development of the CVT method. Nowadays, after over one and a half century of development, the CVT method is widely used for growing different oxides, including the TSOs such as ZnO [10–16] (and many other works), β-Ga_2O_3 [17–23], In_2O_3 [24–29], and SnO_2 [30–37]. The CVT-grown TSO crystals are used mainly for scientific purposes.

A growth furnace for the CVT method to grow oxide crystals, as schematically shown in Fig. 1.2, typically includes a sealed glass or quartz ampoule inserted into a horizontal resistance furnace consisting of a heater and thermal insulation. The ampoule is subjected to different temperatures at both ends, at a source temperature T_S at one end where the source material is placed, and at a growth temperature T_G at the opposite end. The ampoule is also filled up with a CATA and possibly with oxygen at predefined partial pressures (depending on the source material). The CATA reacts with the source material and transports it as vapor to the opposite end, where the transport agent is released and returns to the source material. On the other hand, the released source material reacts with oxygen at the opposite end and deposits on the ampoule wall or on a seed located in that end. The transport direction proceeds either from a hotter to a cooler end of the ampoule in the case of endothermic reaction ($T_S > T_G$), or from the cooler to the hotter end of the ampoule in the case of exothermic reaction ($T_S < T_G$) between the source material and the transport agent. The transport proceeds by convection and diffusion with a dominance of one of them depending on the total pressure and ampoule geometry. Typical CATAs for TSOs include halogens (Cl_2, Br_2, I_2), hydrogen halides (HCl, HBr), and ammonium halides (NH_4Cl, NH_4Br, and NH_4I).

In a simple system with a metal oxide as the source material, the chemical reactions can be defined as follows (in the downstream direction):

$$Me_xO_y + CATA \rightarrow Me_xCATA + \tfrac{1}{2}y\,O_2 \text{ at one end} \qquad (1.1)$$

$$Me_xCATA \rightarrow x Me + CATA \text{ at opposite end} \qquad (1.2)$$

$$x\text{Me} + \tfrac{1}{2}y\,\text{O}_2 \rightarrow \text{Me}_x\text{O}_y \text{ at opposite end} \tag{1.3}$$

Here Me and CATA indicate a metal and chemically active transport agent, respectively.

The source material can be elemental metal, metal oxide, or metal halide.

Operating temperatures usually do not exceed about 1100°C, although higher temperatures were also reported. A temperature difference between both ends of the ampoule $\Delta T = T_S - T_G$ could be even several hundred Kelvin, depending on the source material and reaction type. The volume of individual TSO single crystals is typically small, up to about 0.5 cm^3, although larger volumes are occasionally reported.

There are different technical designs of the furnaces, which may comprise a sealed ampoule, as shown in Fig. 1.2, or a sealed system with flowing gasses therethrough, but the principle remains the same.

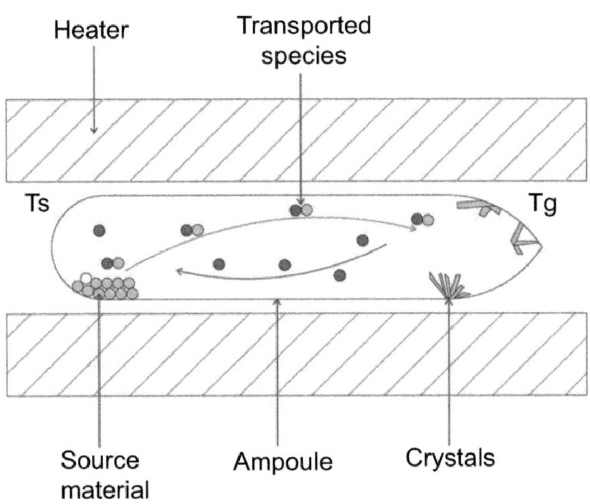

Figure 1.2 Schematic diagram describing a principle of the CVT method.

Most important growth conditions: Temperature of the source material, growth temperature, transport rate, type of the CATA, partial pressures of gaseous species, total pressure, type of the source material, furnace geometry, furnace materials, growth time.

TSOs materials: ZnO, β-Ga_2O_3, In_2O_3, SnO_2, $CoGa_2O_4$

Pros: Simple in operation, cheap, easily available furnace materials, low operating temperature, often high crystal perfection, typically low impurity level, suitable for most of the TSOs.

Cons: Small growth rate, very small crystal size, in many cases spontaneous nucleation, sensitive to a number of growth conditions.

1.2.2 Physical Vapor Transport (PVT)

It is rather difficult to track back the birth of the PVT method, but it was already used for growing oxide crystals in the 1950s, e.g., by Scharovsky [38], as well as by Bogner and Mollwo [39], who obtained ZnO single crystals by sublimation and oxidation of elemental Zn (can be referred to the PVT-like method), or in early 1960s by Weiher [40], and by Marley and MacAvoy [41], who obtained by this method In_2O_3 and SnO_2 crystals, respectively. Later, the PVT or PVT-like method was continuously used to grow ZnO [42, 43] (and other works), In_2O_3 [44–47], and in particular SnO_2 [48–57] single crystals. Fundamental background for the PVT growth can be found, e.g., in a book chapter by Kaldis and Piechotka [58].

The PVT is best suited for highly volatile compounds that are thermally unstable much below their melting points, which are typically very high (>1800°C), or the liquid phase cannot be easily accessed due to very high partial pressure of volatile species. This technique is industrially relevant as it is widely used for growing SiC [59, 60] and AlN [61] single crystals at large scale. It is also used to grow some of the oxides, however, not to a large extent as most of them can be grown by one of the melt growth techniques. The PVT method was basically derived from the CVT method, where the lack of the CATA could also lead to single crystal formation by a direct vapor transport of a compound in quest if it could sublimate at a reasonable temperature. To promote sublimation at not-too-high temperatures, sometimes a reducing agent is used, such as carbon or hydrogen. When using the reducing agent for enhancing the sublimation, the growth method is often referred to the CVT instead of the PVT method. Indeed, there is an initial reaction between the starting material and the reducing agent (e.g., $ZnO(s) + C(s) \leftrightharpoons Zn(g) + CO(g)$); however, the created volatile species (here Zn(g)) is transported by itself, without the transport agent. Therefore, it is rather the physical transport.

A simplified PVT growth furnace for oxides (Fig. 1.3) consists of a thermal insulation, a metal crucible containing a source (starting) material, and a seed on top of the crucible, on which the source material is grown in the form of a single crystal. This technique requires volatile species, which are transported from the source material at a high temperature T_S (source temperature) to the seed at lower temperature T_G (growth temperature) by gas convection and diffusion. The temperature difference $\Delta T = T_S - T_G$ spans in a wide range, roughly between 20 and 350 K. In operation, the crucible is heated up to a high temperature T_S at which the solid source material sublimates through thermal decomposition, and the volatile species are transported toward the seed located at the lower temperature T_G enabling deposition in the solid phase. The source material is typically a metal or metal oxide to be grown, while the growth atmosphere is typically oxidizing or neutral. In the latter case, the oxygen needed for forming a metal oxide is an oxygen from the metal oxide used as the starting material and residual oxygen in the neutral atmosphere (usually at a pressure of at least 10^{-5} bar).

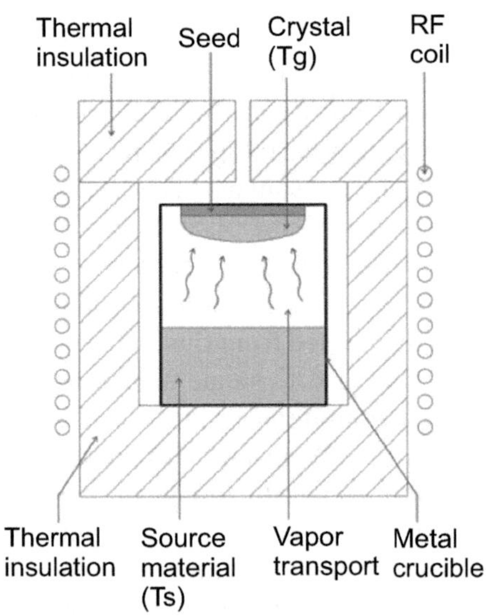

Figure 1.3 Schematic diagram of a growth furnace for the PVT method.

If the metal oxide is used as the source material, it decomposes into a metal suboxide and oxygen both in the gas phase:

$$Me_xO_y(s) \rightarrow xMeO(g) + \tfrac{1}{2}yO_2(g), \tag{1.4}$$

where $x = 1$ or 2, and $y = 2$ or 3
or into a metal and oxygen:

$$MeO(s) \rightarrow Me(g) + \tfrac{1}{2}O_2(g) \tag{1.5}$$

If the source material is an elemental metal, it is simply transferred from solid or liquid to a gas phase, and back to the solid phase after oxidation.

The operating temperatures of the PVT are typically higher than the operating temperatures of the CVT method, usually between 1400 and 1600°C especially when the metal oxide is used as the source material. Because of high operating temperatures, crucibles for the source materials are usually made of platinum (Pt) or iridium (Ir), which is typically heated inductively by a radiofrequency coil powered by a radiofrequency generator. The crystal size depends on the furnace design and growth conditions. In the case of the TSOs, the crystal volume can be as large as 5 cm^3.

Most important growth conditions: Temperatures of the source material and a seed, growth atmosphere and oxygen partial pressure, growth rate, furnace materials, source material, furnace geometry, growth time.

TSOs materials: ZnO, In_2O_3, SnO_2.

Pros: Relatively simple, low impurity level, different growth atmospheres can be used, possible large crystal volume after substantial development.

Cons: Usually small crystal volume, often not so high crystal perfection, limited to highly volatile compounds.

1.3 Solution

A solid starting material to be grown (solute) is first dissolved within a solvent, from which the starting material precipitates in the form of crystals either spontaneously or on a crystal seed. The phase transition can be described as solid → solid phase transition. In a classical solution growth, the operating temperature is low, just above room temperature. If the operating temperature is high at ambient pressure, usually above 1000°C, then the solvent is called a flux, which solidifies before reaching room temperature. The flux technique is also referred to high-temperature solution growth. In the case of an aqueous solution as the solvent, which remains in the liquid phase at room temperature, and high pressure, we have the hydrothermal method, which is also named high-pressure solution growth. Solution growth techniques enable to grow crystals at relatively low temperatures, much below their melting points, as well as below decomposition temperatures or phase transitions, if any.

1.3.1 Hydrothermal

Basing on the works of Byrappa [62], and Byrappa and Yoshimura [63], the origin of the hydrothermal method dates back to the nineteenth century and it was studied by geoscientists to explain the formation of different minerals. The first experiments on quartz growth by the hydrothermal method were conducted in 1851 by de Sénarmont [64] and later by Spezia (1904). That method became industrially relevant once high pressure could be applied and large autoclaves were constructed. This happened during and soon after World War II, with an intensive process development, largely contributed by the Hydrothermal Laboratory at the Institute of Crystallography in Moscow and in Bell Labs [65]. The method is suitable for growing crystals having low-temperature phase transition, like quartz, but also those that are thermally unstable at high temperatures (e.g., ZnO). This method is used to grow large-volume crystals at industrial scale.

Growth runs for oxides are conducted in so-called autoclaves as demonstrated in Fig. 1.4. They are made of steel with a liner typically made of Pt or other base-resistant metal, such as silver. The autoclave is filled up with a mineralizer constituting an aqueous alkaline solution. Examples of mineralizers used for oxides include aqua solutions of NaOH, LiOH, and KOH. The interior of the autoclave is divided by a baffle into two zones: a dissolution zone at the bottom part and a growth zone in the upper part of the autoclave. In the dissolution zone, a source material to be grown is dissolved, while in the growth zone, a plurality of seeds is mounted. This method uses overpressure from several to dozens of hundreds of atmospheres, while the operating temperature is typically between 200 and 500°C. In operation, a temperature difference of several to dozens of Kelvins between the dissolution and growth zones drives convection carrying the starting material to the growth zone, where it precipitates on the seeds. The growth rate is very small, about 1 mm/day, but the yield is enhanced by multiple crystals grown in one run. The hydrothermal method has intensively been used and developed for ZnO single crystals [66–72], which reached as large diameter as 2–3 inch and a thickness of about 1 cm. Also, small hydrothermally grown In_2O_3 [73–75] and SnO_2 [76, 77] crystals were demonstrated, however, with no further development.

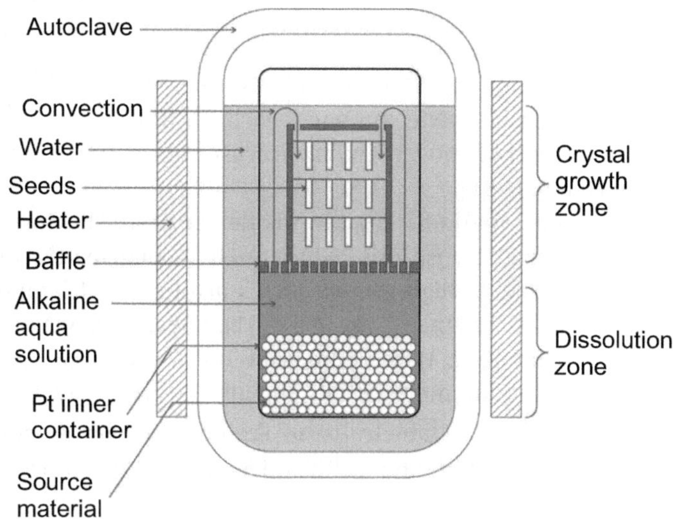

Figure 1.4 Schematic representation of an autoclave for the hydrothermal method.

Most important growth conditions: Type of alkaline solution and its composition, temperatures of the dissolution and growth zones, pressure, form of the starting material, autoclave design, growth time.

TSOs materials: ZnO, also In_2O_3 and SnO_2.

Pros: Large crystal volume per growth run, high crystal perfection, low operating temperature.

Cons: Expensive, long growth time, long time to establish a growth process, high pressure, use of chemically aggressive alkalines, possible contamination with mineralizer elements.

1.3.2 Flux

Fundamentally, the flux method, which is also called high-temperature solution growth, is a variant of the hydrothermal method where the solvent is not an aqua solution but is either a metal for non-metal compounds or an oxide for oxide compounds. The growth from the metal fluxes, according to Kanatzidis et al. [78], started at around 1900 by Henri Maissan and Paul Lebeau. From metal fluxes, phosphides, borides, carbides, and silicides, and recently nitrides (GaN and BN) were grown. The development of the flux method for oxides started in the 1950s, when Remeika [79] grew $BaTiO_3$, and soon later Nielson and Dearborn [80] grew magnetic garnets. In the early 1960s, TSOs single crystals grown from the flux were reported [81–84].

By the flux method, mainly oxide crystals having low-temperature phase transitions, which melt incongruently, or are thermally unstable at high temperatures are grown. Examples of such materials, as shown, e.g., by Wilke [1], Elwell and Scheel [85], Scheel [86], Brice [87], White [88], and Tolksdorf [89], who also discuss that method in more details, include $Y_3Fe_5O_{12}$ (YIG), $YFeO_3$, $BaTiO_3$, $GdAlO_3$, $KNbO_3$, $KTa_{1-x}Nb_xO_3$, Fe_2O_3, CeO_2, HfO_2, ZrO_2, TiO_2, Y_2O_3, MgO, and many others. The following bulk TSOs single crystals were obtained by the flux method: ZnO [90–95], β-Ga_2O_3 [96–100], In_2O_3 [101–107], SnO_2 [108–114], $BaSnO_3$ [115–123], $ZnGa_2O_4$ [124–127], $MgGa_2O_4$ [128], and $CoGa_2O_4$ [129].

Here, the source material (solute) is dissolved within the flux having much lower melting point than the source material to be grown. The flux setup for oxides, as depicted in Fig. 1.5, comprises a crucible, typically a metal crucible made of Pt, which is covered with a Pt lid. The crucible containing a mixture of the source material and the flux is placed within a furnace consisting of a thermal insulation and a heater. Because of the not-too-high operating temperature, usually between 1100 and 1300°C, a resistive heating is typically utilized. The flux is an oxide or fluoride having a relatively low melting point, such as Cu_2O, PbO, PbF_2, B_2O_3, Bi_2O_3, and V_2O_5. Important in flux selection are sufficient solubility of the component to be grown and low solubility of the flux in the crystal, low vapor pressure at operating temperatures, should not be aggressive to the crucible and furnace, and should be easily separated from the obtained crystals. Some of the flux compounds are used alone, but often in combination of at least two of them.

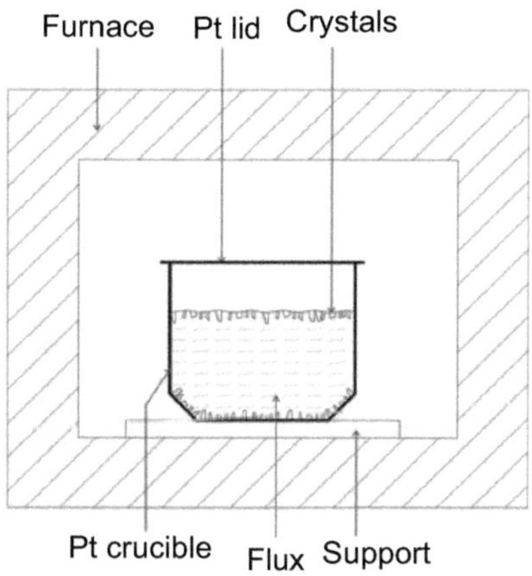

Figure 1.5 Schematic representation of a growth arrangement for the flux method.

In operation, the crucible is heated up above the melting point of the flux, then homogenized, and next very slowly cooled down at the rate of about 1–5 K/h. During slow cooling down, the source material spontaneously nucleates from the flux volume in the form of small crystals, e.g., at the crucible wall or bottom or at top part of the flux, depending on temperature gradients and densities of the material to be grown and the flux. Once cooled down to room temperature, the crystals are released from the flux with an acid that dissolves the flux, leaving the obtained crystals intact. The TSO crystal volume can reach the volume of about 1 cm^3.

A variant of the flux method is the top-seeded solution growth (TSSG), which differs from the simple flux method in growing a crystal on a rotating seed dipped into the flux from the top, which is next pulled up very slowly (about 1 mm/day). The seed is slowly enlarged to a predefined diameter, which is then separated from the flux after it reaches a predefined length and cooled down to room temperature. The TSSG is very similar to the Czochralski method (see Section 1.4.5), where the pure compound to be grown is replaced with a compound mixed with a flux material.

Yet another variant of the flux technique is the traveling-solvent floating zone (TSFZ), which is the OFZ method (see Section 1.4.2), where the feed rod is not a pure compound to be grown, but a mixture of a material to be grown and a flux material.

Most important growth conditions: Flux composition, molar ratio of the source material to flux, maximum temperature, cooling rate, growth time.

TSOs materials: ZnO, β-Ga_2O_3, In_2O_3, SnO_2, $BaSnO_3$

Pros: Simple, not so high operating temperature, most of the TSOs can be grown.

Cons: Low growth rate, very small crystal volume, in most cases spontaneous nucleation, not so good crystal perfection, flux impurities.

1.4 Melt

In melt growth techniques, a solid starting material to be grown is first converted into a liquid phase, which is then converted back into a solid phase in the form of a single crystal. A liquid–solid phase transition is driven by temperature gradients. That transition can be spontaneous, or in a majority of cases, it proceeds on a crystal seed with a defined crystallographic orientation. This type of techniques requires high operating temperatures, challenging insulation materials and crucible materials (if utilized). Materials having phase transitions or being highly unstable, rather, cannot be grown from the melt. In the case of the TSOs in quest, the phase transition is usually not an issue, but high thermal instability combined with high melting points is really a challenge for melt growth techniques.

1.4.1 Verneuil

The Verneuil method was described in 1902 by the French chemist Auguste Victor Louis Verneuil for growing synthetic gemstones, such as sapphire, ruby, TiO_2, and $SrTiO_3$. Later many other oxide crystals were grown with the use of the Verneuil method, such as ZrO_2, HfO_2, $MgAl_2O_4$, $Y_3Fe_5O_{12}$, and many other compounds [1]. This method is suitable for high melting point oxides and/or oxides showing features that limit the use melt growth techniques utilizing metal crucibles. Out of bulk TSO single crystals, only β-Ga_2O_3 was demonstrated by this technique [130]. The method uses combustion of oxyhydrogen at high temperatures, which melts a starting material to be grown. This is why the Verneuil method is also referred to as the flame fusion method. The advantage of this process is its simplicity and low production costs; however, it is occupied by lower crystal perfection.

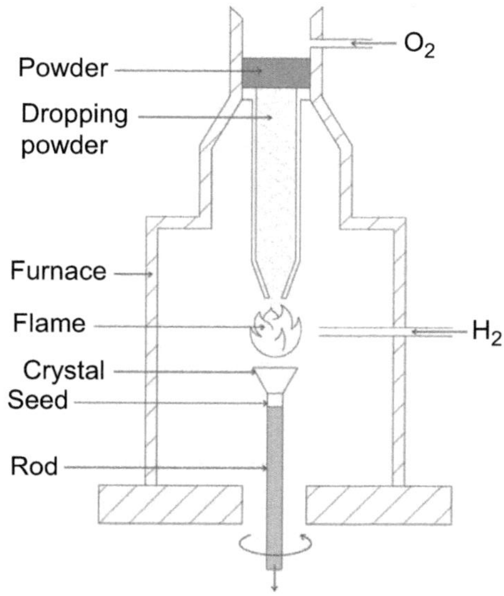

Figure 1.6 Schematic diagram of the Verneuil method.

A growth furnace of the Verneuil method is schematically shown in Fig. 1.6. It has in its upper part a container accommodating the starting material and connected to an oxygen supply, an outer passage introducing hydrogen, and a thermal insulation around the powder path, and the combustion and crystal growth zones. The starting material in the form of powder is slowly released from the container, which travels downstream to the combustion zone, where oxygen and hydrogen create high-temperature flame that melts the starting material. The molten materials in the form of droplets pass downward to the cooler crystal growth zone, where they solidify on a crystal seed, which is rotated and translated downward as the growth proceeds. The growth rate is usually several mm/h, while the rotation rate is up to about 60 rpm. Typically, the crystal diameter may reach about 30 mm (or even 3 inch in the case of sapphire), while the length may exceed 100 mm resulting in a volume of 70 cm^3. More details of the Verneuil method can be found, e.g., in a description by Wilke [1], Falckenberg [131], and Borodin et al. [132].

Most important growth conditions: Furnace design including the burner, paths and flow rates of hydrogen to oxygen, powder supply rate, growth rate, crystal rotation, cooling rate.

TSOs materials: β-Ga_2O_3.

Pros: Cheap, no metal crucible, oxidizing or reducing atmosphere can be used, relatively large crystal volume, short growth time, low impurity level.

Cons: Rather poor crystal perfection, limited scale-up.

1.4.2 Optical Floating Zone (OFZ)

The OFZ method is a derivative of the classical floating zone technique, where an inductive heat source for electrically conducting materials is replaced with an optical heat source, especially for dielectric materials. The ancestor of the floating zone method, following Bohm et al. [133], is Kapitza [134], who invented in 1928 a zone melting setup for metals. This technique was developed by Pfann [135] in 1952 for classical semiconductors (purification purpose) and later for oxides by Chapman and Clark (UO_2) [136]. In the floating zone technique, a feed rod (e.g., a polycrystal) passes through a narrow induction coil that produces a molten zone by the direct inductive heating of the feed rod, which crystalizes on a seed located just beneath the feed rod. The growing crystal contains much lower level of impurities having the segregation coefficient below unity (most cases). The first application of the optical source was demonstrated in 1961 for oxides by Kooy and Couwenberg [137] followed by Akashi et al. [138] who demonstrated an OFZ growth station with a halogen lamp and an elliptical reflector that could operate under an oxygen pressure of 30 atm. A higher overpressure OFZ growth station, up to 100 atm, was developed and applied for oxides by

Balbashov and Egarov [139]. Nowadays, the OFZ growth stations are commercially available in different designs and widely used to crystallize a large number of metals, alloys, and oxide compounds, as listed, e.g., by Balbashov and Egarov [139], Koohpayeh et al. [140], and by Dąbkowska and Dąbkowski [141]. Among bulk TSO crystal growth, only β-Ga_2O_3 [142–145], $MgGa_2O_4$ [146], and $CoGa_2O_4$ [147] were demonstrated.

In the OFZ method (Fig. 1.7), the heater is constructed from halogen or xenon lamps or lasers, and elliptical mirrors, which focus light onto the tip of the feed rod acting as a starting material in the form of a ceramic. The arrangement of the heat source and mirrors can be either horizontal (Fig. 1.7) or vertical (not shown). Just below the feed rod, a crystal seed is positioned. Light melts the tip of the feed rod, which is brought into contact with the crystal seed. Both the feed rod and the seed with the growing crystal are translated downward with rotation, usually in opposite directions. In this process, a ceramic feed rod is converted into a single crystal. Although the OFZ method is widely used, it is not large-volume industrial technique due to small crystal diameter, typically 5–10 mm, although larger diameters were demonstrated as well (β-Ga_2O_3 of 1 inch [144]).

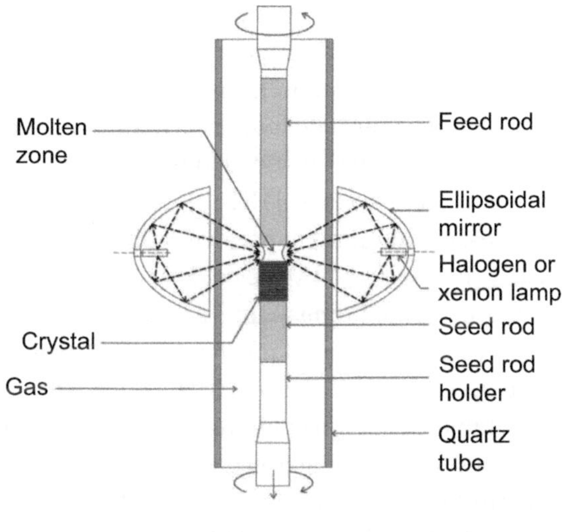

Figure 1.7 Schematic diagram of a growth furnace illustrating the principle of the optical floating zone method.

Most important growth conditions: Growth atmosphere, growth and rotation rates, feed rod preparation, pressure.

TSOs materials: β-Ga_2O_3, $MgGa_2O_4$, $CoGa_2O_4$.

Pros: No metal crucible, highly oxidizing atmosphere and overpressure can be used, short growth time, low impurity level.

Cons: Small and limited crystal size and scale-up, often not so good crystal perfection.

1.4.3 Cold Crucible

According to Mühlbauer [148], the history of inductive heating, being the starting point for the cold crucible method, originates from the work of many researchers, scientists, and engineers in the nineteenth–twentieth century, including William Siemens, Henri Moissan, Paul Heroult, Sebastian de Ferranti, F. A. Kjellin, E. A. Colby, A. Debuch, F. Jacoviello, Edwin F. Northrup, James R. Wyatt, Valentin P. Vologdin, and many others, who used and developed the inductive heating for melting and treating metals. Initial experiments with a direct inductive heating were done at Siemens & Halske in the 1920s to melt metals without a ceramic crucible, which typically contaminates the metals [149]. As described by Mühlbauer [150], the cold crucible technique has been used in the 1950s–60s for silicon processing (H. F. Sterling and R. W. Warren, J. Reboux, and others) as well as for metals and alloys processing. An extensive development of the cold crucible method, including growth stations, started in Russia (P. N. Lebedev Physical Institute of the Academy of Sciences) in the 1960s by Aleksandrov et al. [151] for a diversity of compounds (fluorides, halides, oxides), and in parallel in the USA by Monforte et al. [152] in 1961 (for ferrites). By the cold crucible method, different materials were grown, including metals and alloys (e.g., Ti-Al) [150], semiconductors (e.g., Si) [150], fluorides (e.g., LiF, CaF_2) [151], and oxides (e.g., Al_2O_3, MgO, Sc_2O_3, Yb_2O_3, Y_2O_3, ZrO_2, HfO_2) [151]. This technique is suitable particularly for very high melting point oxides (up to about 3000°C), for which crucible and thermal insulation materials constitute a big issue. This method found an application for growing bulk ZnO single crystals [153–155].

A cold crucible growth furnace for oxides is shown in Fig. 1.8. It consists of a container made of water-cooled copper fingers, and of a radiofrequency coil located around the container and connected to a radiofrequency generator. Both the container and the coil are disposed within a water-cooled growth chamber. The starting material to be grown is packed within the container, with an addition of pieces of electrically conducting material (e.g., a metal or graphite) for inductive coupling. The metal pieces are heated up inductively, which in turn heat up the surrounding starting material until its melting. Because the container is water cooled, the outer layer of the starting material of about 1–2 mm thick is not molten and constitutes a self-sustained crucible for the melt. Next the system is slowly cooled down, and the melt spontaneously solidifies in the form of multiple crystals. The operating frequency is between 0.1 and 4 MHz. The single crystal size of oxides may reach the diameter of 1 inch and length of several centimeters.

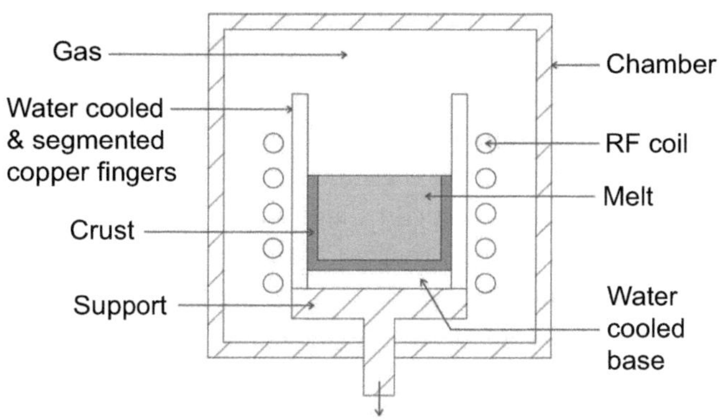

Figure 1.8 Schematic diagram of the cold crucible concept.

Most important growth conditions: Growth atmosphere, pressure, generator's frequency, the presence of the conducting material inside the starting material, furnace design.

TSOs materials: ZnO.

Pros: No metal crucible, highly oxidizing atmosphere can be used, acceptable crystal volume, fairly good crystal perfection, high-purity crystals, short growth time.

Cons: Uncontrolled crystallization, limited scale-up.

1.4.4 Bridgman–Stockbarger (BS) and Vertical Gradient Freeze (VGF)

This method was published in 1925 by Percy W. Bridgman to grow low-melting-point metals [156] with later modifications by Donald C. Stockbarger to grow LiF [157], although contribution of others (e.g., Tammann, Obreimow, and Schubnikov) is also acknowledged by Lawson and Nielsen [158], and Buckley [159]. It is a freezing growth technique, and often it is simply called the Bridgman method. In addition to metals (e.g., Sb, Bi, Cd, Zn, Sn) and alloys, this technique has widely been used to grow III–V (e.g., GaAs, GaP, InP) and II–VI (e.g., CdTe, ZnSe, CdSe) compounds, halides (e.g., CaF_2, MgF_2, BaF_2, NaI, CsI, $LaBr_3$, $LaCl_3$) and oxides (e.g., sapphire, $La_3Ga_5SiO_{14}$, $Bi_4Ge_3O_{12}$, $PbWO_4$, TeO_2). It may have either horizontal or vertical configuration. Also, several TSO single crystals were grown by the Bridgman/VGF method, including ZnO [160–164], β-Ga_2O_3 [165, 166], $BaSnO_3$ [167], and Ga-based spinels [168, 169].

The growth furnace in the vertical configuration for growing TSO single crystals, as shown in Fig. 1.9, consists of a metal crucible containing a starting material and supported by a translation mechanism, a lid covering the metal crucible, a thermal insulation surrounding the metal crucible, and an inductive coil surrounding the metal crucible and the thermal insulation. The inductive heating can potentially be replaced with resistive heating. The metal crucible has a conical bottom part, which acts as the nucleation center. Alternatively, at that conical part, a crystal seed is accommodated with a defined crystal orientation. The growth furnace by its design is subjected to a specific temperature profile along the crucible, which allows melting and crystallization of the starting material. To grow a crystal, the crucible is positioned at high temperature, T_f, enabling melting the starting material, except the seed portion (if a seed is used), which is subjected to a lower temperature $T_c < T_f$, that avoids seed melting. Then the metal crucible along with the furnace is slowly translated downward out of the coil (or heater in general) to the lower-temperature part of the temperature profile enabling crystallization of the melt on the crystal seed. The same applies for the horizontal configuration, although it has not been demonstrated so far for the TSOs.

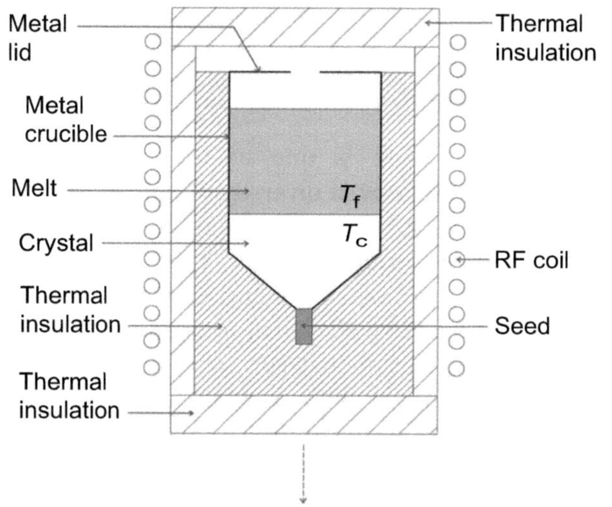

Figure 1.9 Schematic view of the Bridgman/VGF/direct solidification method.

In an alternative technical realization, the crucible is in the static position, but the overall temperature slowly decreases, and the crystallization follows the temperature gradients. Such modification is called the vertical gradient freeze (VGF) method.

Another variant of the VGF method can be called a direct solidification method, which involves fast rather than slow solidification rate to minimize the decomposition of a compound in quest. This method can be used for highly unstable multielement oxides with high decomposition rate, when other melt growth methods fail.

Most important growth conditions: Crucible shape, furnace geometry, temperature gradients (or temperature profile), crucible translation rate and/or cooling rate.

TSOs materials: ZnO, β-Ga_2O_3, $BaSnO_3$, $MgGa_2O_4$, $ZnGa_2O_4$.

Pros: Simple, short growth time, large crystal size, thermally unstable compounds can be grown, relatively good structural quality.

Cons: A metal crucible, extended defects due to a direct contact between the crystal and crucible wall, often not defined crystallographic orientation, not suitable for oxides having very high melting points (>2500°C) or having phase transition below their melting points, trace impurities from crucibles.

1.4.5 Czochralski

The Czochralski method is one of the most important and efficient techniques to grow a diversity of bulk crystals of classical semiconductors (e.g., Si, Ge), compound semiconductors (e.g., GaAs, GaP, InSb), oxides (e.g., sapphire, YVO_4, $Y_3Al_5O_{12}$, $Gd_3Ga_5O_{12}$, $YAlO_3$, $LiNbO_3$, $Bi_4Ge_3O_{12}$, Lu_2SiO_5, and many, many others), and halides (e.g., NaCl, KCl, KBr, CaF_2, BaF_2, LaF_3, $LiCaAlF_6$). The Czochralski method is one of the most industrially important growth techniques for large-volume single crystals of high crystal perfection.

This method was invented in 1916 by Polish scientist Jan Czochralski [170] in Allgemeine Elektricitäts-Gesellschaft (AEG), Berlin, Germany, during measurements of the crystallization velocity of Sn, Zn, and Pb metals. Basing on the detailed study on the Czochralski method by Uecker [171] and Tomaszewski [172], the concept of Czochralski with the use of a simple apparatus had soon after been continuously modified by numerous researches, including Gomperz [173] (Germany, 1922) who introduced a die and a gas coolant for a growing crystal, Grüneisen and Goens [174] (Germany, 1923) who introduced oriented crystal seeds, Hoyem and Tyndall [175] (USA, 1929) who introduced a seed necking, and Walther [176] (USA, 1937), who introduced a crystal rotation and control of the shape and diameter of the growing crystal, as well as redesigned a furnace for relatively high melting point materials and large diameter (2 cm) oriented crystals of NaCl. These developments, combined with the fundamental concept of Jan Czochralski to pull a crystal up from the melt, have established key aspects of the method already before World War II. After World War II, the Czochralski method was used to grow single crystals of elemental semiconductor Ge by Teal and Little [177] (USA, 1950), while introduction by Dash [178] a neck procedure in 1958 enabled obtaining of high-quality, dislocation-free Si crystals. This method spread rapidly in 50-ties of the twentieth century and enabled growing large-diameter (6 inch) Ge and Si single crystals, which led to their industrialization. Further modifications related to application of a magnetic field to damp the melt convection of conductive melts by Hoshikawa et al.

[179] (Japan, 1980), a continuous melt feeding developed by Petrov and Zemskov [180] (Russia, 1957), liquid encapsulation by Metz et al. [181] (USA, 1962), and introduction of electronic load cells for crystals weighting by Bardsley et al. [182] (UK, 1972) and for crucible weighting by Kyle and Zydzik [183] (USA, 1973), which enabled automatic diameter control of a growing crystal. The first use of the Czochralski method for growing oxide single crystals ($CaWO_4$) by the Czochralski method was done in 1960 by Nassau and Van Uitert [184] (USA).

The Czochralski method is nowadays used worldwide to grow different types of single crystals: metals, elemental semiconductors, compound semiconductors, halides, and oxides. As large as 450 mm diameter semiconductor and halide crystals and 200 mm diameter oxide crystals (sapphire, $Bi_4Ge_3O_{12}$) are grown nowadays by the Czochralski method. Large crystal diameter, high structural quality, and low production costs per volume unit make the Czochralski method one of the most frequently used technique at research and industrial levels for growing a diversity of single crystals. Among TSOs, the Czochralski method was successfully applied for β-Ga_2O_3 [185–189] and Ga-based spinel single crystals [168, 169].

Figure 1.10 schematically shows a growth furnace illustrating the Czochralski method for growing oxide single crystals, which is placed within a water-cooled growth chamber. The growth furnace includes a metal crucible containing a starting material of a compound to be grown. The crucible is surrounded from all sides by a thermal insulation (typically alumina and/or zirconia) to achieve, among other important aspects, acceptable temperature gradients to minimize cracking of a growing crystal and the formation of other defects, such as dislocations, twins, low-angle grain boundaries, etc. In the upper part of the insulation, i.e., above the crucible, there is a free space to accommodate the growing crystal. In that space, a metal afterheater may be placed to further decease temperature gradients. In the case of high melting point oxides, the metal crucible and metal afterheater (if used) are heated up inductively by a radiofrequency coil powered by a radiofrequency generator controlled by a controller. In the free space above the crucible, a crystal seed is axisymmetrically provided, which is mounted within a seed holder coupled via a pulling rod with a translation and rotation mechanisms. This allows the growing crystal to be pulled up while rotating. Both operating parameters, i.e., the growth rate and rotation rate depend mainly on the thermophysical properties of the melt and the growing crystal and typically range between 1–3 mm/h and 4–20 rpm, respectively. Another key operating parameter is an atmosphere in which the growth proceeds. They span from reducing through neutral to oxidizing atmosphere. However, to stabilize TSOs an oxidizing atmosphere is required, but this creates additional difficulties to the crucible material as the TSOs have high melting points.

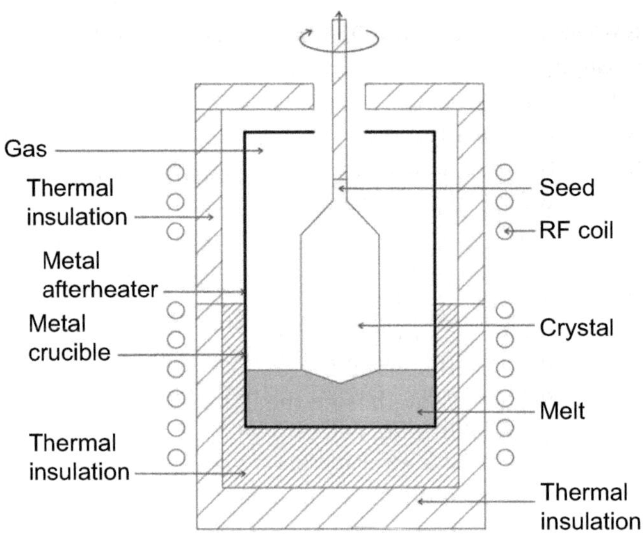

Figure 1.10 Schematic diagram of a growth furnace illustrating the Czochralski method.

Most important growth conditions: Furnace design, furnace materials, crucible material, temperature gradients, growth atmosphere, purity of the starting material, growth and rotation rates.

TSOs materials: β-Ga_2O_3, $MgGa_2O_4$, $ZnGa_2O_4$, $CoGa_2O_4$.

Pros: Large crystal volume, high crystal perfection, relatively short growth time, suitable for high melting point oxides, wide range of temperature gradients, growth rates and rotation rates, flexibility in a furnace design.

Cons: Metal crucible, not suitable for oxides having very high melting points (>2500°C) or having phase transition below their melting points, and for oxides being highly thermally unstable, trace impurities from crucibles.

1.4.6 Kyropoulos

The Kyropoulos method was developed by Spyro Kyropoulos in 1926 to grow halides [190]. This technique has successfully been used for growing III–V compounds (e.g., InP [191], and GaAs [192]) and oxides (e.g., sapphire [193]). By the Kyropoulus method, some of the Ga-based spinels were demonstrated [168].

In principle, this is a modification of the Czochralski method with a similar growth furnace concept, as demonstrated in Fig. 1.11. In the Kyropoulos method a crystal is grown on a seed by a slow cooling down of the melt substantially with no pulling up of the crystal, although very small pulling rates (smaller than that in the Czochralski method) and crystal rotation can also be applied. The growth proceeds inside the melt until substantially the whole melt is crystallized on the seed, after which the crystal is cooled down to room temperature. The Kyropoulos method requires very low temperature gradients in the melt and efficient heat dissipation, which is produced at the growth interface during liquid–solid phase transition. Technically, this is achieved by a furnace design with a possible cooled seed rod and crucible bottom, and heated crucible wall, typically inductively, although a resistance heating can be applied as well. It requires suitable adjustment of temperature gradients to control the growth rate and minimize fast solidification of the melt. The Kyropoulos method enables large crystal volume with the diameter close to a crucible diameter. Alternatively, the melt may freeze on the seed being in a contact with the crucible wall. Thermal insulation and crucible materials, as well as the growth atmosphere, are typically the same as those used for the Czochralski method, depending on the compound to be grown and its melting point.

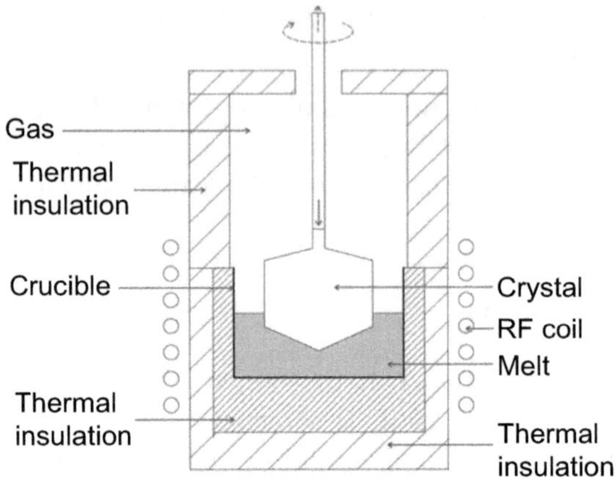

Figure 1.11 Schematic representation of the Kyropoulos method.

Most important growth conditions: Furnace design, furnace materials, crucible material, temperature gradients, growth atmosphere, purity of the starting material, growth rate.

TSOs materials: $MgGa_2O_4$, $CoGa_2O_4$.

Pros: Large crystal volume, good crystal perfection.

Cons: A metal crucible, difficult crystal diameter control, not suitable for oxides having very high melting points or having phase transition below their melting points, and for oxides being highly thermally unstable, trace impurities from crucibles.

1.4.7 Edge-Defined Film-Fed Growth (EFG)

The EFG method belongs to a class of growth techniques utilizing a shaper or a die for a crystal of defined shape, such as a rod, tube, plate, ribbon, and other shapes. The precursor of the shaped growth techniques is Gomperz [173], who introduced a die to the Czochralski method in 1922. In the 1950s, Alexandr Vasil'evitch Stepanov analyzed in detail the growth of shaped crystals defined by a die-shaped melt column [194]. The EFG method is a variant of the Stepanov method as proposed by LaBelle and Mlavsky [195] in 1967, which differs in that the crystal shape is defined by the edge of the die, not by the melt column shape characteristic for the Stepanov method. In the EFG method, the melt is transported to the die surface by capillary forces. A detailed description and analysis of the shaped melt growth techniques, including EFG, can be find, e.g., in the works of Tatarchenko [196–200], Antonov et al. [201, 202], Kurlov et al. [203], and Feigelson [204]. The shaped growth techniques, including the EFG, have been widely used for sapphire [205–207] but also for other oxides, fluorides, metals, and semiconductors [202]. Recently, the EFG method was successfully applied and developed for β-Ga_2O_3 [208, 209].

The EFG growth furnace, as depicted in Fig. 1.12, has a metal crucible containing a starting material, a die located within the crucible, a thermal insulation surrounding the crucible and a free space above a growing crystal, an inductive coil around the crucible and thermal insulation, and a pulling mechanism to which a seed holder with a crystal seed is attached. The die has a narrow vertical slit through which the melt is transported upward by the capillary forces and a shaped top surface defining the crystal shape. Although different shapes of the die can be applied, for wafer fabrication thin slabs of large surface area are desirable. The crucible may have either circular or rectangular cross section.

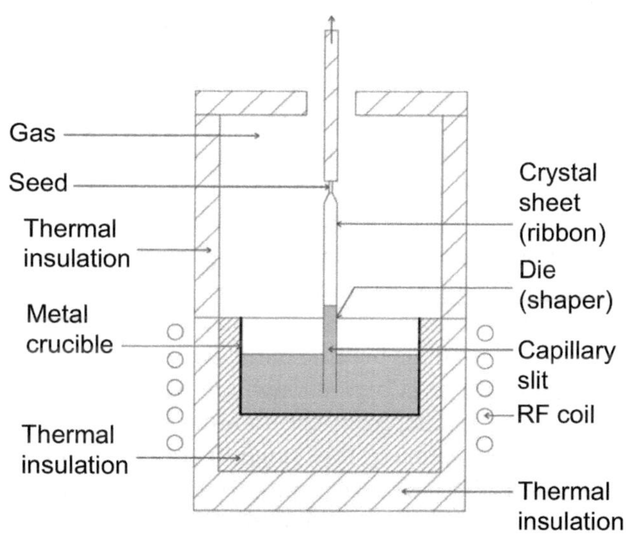

Figure 1.12 Schematic representation of the EFG method.

Most important growth conditions: Furnace and, particularly, die design, temperature gradients, crystal shape, growth atmosphere, growth rate.

TSOs materials: β-Ga_2O_3.

Pros: Relatively short growth time, good crystal quality, large one face enabling large wafers, fixed interface position, a plurality of crystals possible in one run.

Cons: Additional noble metal element (die), more complex cleaning of noble metal components after use, limited oxygen concentration at high temperatures, one crystal dimension much smaller than others.

1.4.8 Levitation-Assisted Self-Seeding Crystal Growth Method (LASSCGM)

This is a very recent growth technique defined and developed by Galazka et al. [210, 211] in 2012 (Germany). The fundamental idea was to convert an initially electrically insulating starting material into electrical conductor in the liquid phase through controlled decomposition of the starting material and couple the electrically conducting melt with an electromagnetic field from an inductive coil through a thin metal crucible wall. Upon such coupling, a top portion of the melt is levitated, more at the crucible wall and less at the central part thereof. This will lead to a formation of a liquid neck (or a liquid bridge) connecting the levitated top portion with a non-levitated bottom portion of the melt. The liquid neck will act as a crystal seed during solidification of such melt structure. (Fig. 1.13). This technique is suitable for oxide compounds that have high electrical conductivity in the liquid phase, particularly for those being thermally unstable at high temperatures. As the starting material within an Ir crucible is heated up, it becomes more and more electrically conducting as the result of the thermal decomposition. Around the melting point, the electrical conductivity of the starting material is high enough for effective coupling with the electromagnetic field from the inductive coil powered by a radiofrequency generator. In this stage, there is an additional (to conduction and radiation) heat source (Joule heat) that enhances melting of the starting material and an eddy current flowing through it creates its own electromagnetic field that counteracts the electromagnetic field from the coil causing levitation of a portion of the obtained melt. Under properly defined growth environment, levitation of a portion of the melt leads to a centro-axial formation of a liquid neck as the result of lower temperature in that region as compared with temperature along the crucible wall. In the liquid phase, the starting material (here In_2O_3) is highly unstable and decomposes, while the volatile products are removed by the openings in a lid covering the crucible. Cooling down the crucible with molten starting material leads to crystallization of the melt in such structure and formation of one or two single crystals on the opposite sides of the seed. The growth atmosphere is oxidizing but enabling a controlled decomposition that drives electrical conductivity of the starting material and consequently levitation. More details on that novel melt growth method are given in Chapter 5.

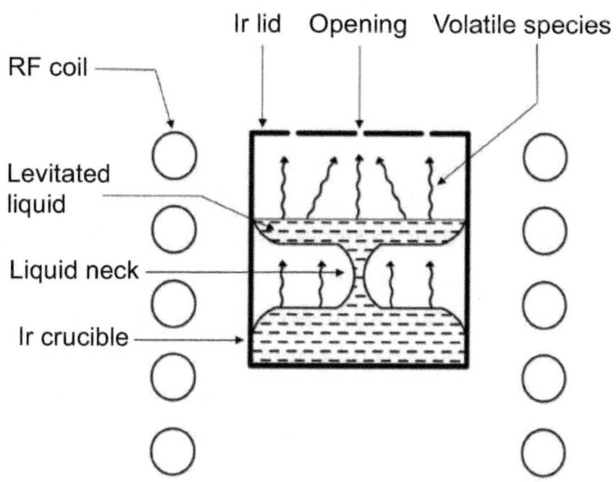

Figure 1.13 Schematic representation of the principle of the levitation-assisted self-seeding crystal growth method. Thermal insulation around the seed is not shown here (see Chapter 5 for more details).

Most important growth conditions: Design of particular furnace components (coil, crucible, lid), crucible–coil mutual position, temperature gradients, growth atmosphere, purity of the starting material, growth time.

TSOs materials: In_2O_3.

Pros: Short growth time, relatively large crystal size, suitable for highly electrically conducting oxide compounds that are thermally unstable at high temperatures.

Cons: Difficult to control, not defined crystal orientation.

1.5 Summary

The techniques described above have been applied so far for growing bulk TSO single crystals, but demand for large crystal size of high structural perfection as well as for new compounds encourages researches for modifying the existing methods or finding yet new techniques. Generally, all growth methods from gas phase, solution, and melt were successfully utilized for this class of materials, although not all for each of the compound in quest.

When considering the growth from the gas phase, the CVT method is versatile and was used to grow all binary TSO crystals (ZnO, β-Ga_2O_3, In_2O_3, SnO_2); however, it produces only very small crystals. This method operates at relatively low temperatures with no high demand for furnace design and furnace materials. It is widely used in the academia as the obtained crystals are sufficient for basic characterization. On the other hand, the PVT method could be utilized only for the most thermally unstable compounds, including ZnO, In_2O_3, and SnO_2. It enables much larger crystal size as compared with the CVT method. The PVT method is an industrially important technique for other materials (e.g., for SiC, AlN).

In the case of solution techniques, the flux method is also versatile and has found its application for growing most of the TSO crystals, i.e., ZnO, β-Ga_2O_3, In_2O_3, SnO_2, $BaSnO_3$. The flux-grown crystals are typically very small and often contaminated with flux elements, but the simplicity of the method and not-too-high operating temperatures spread that method in the academia for investigation of fundamental properties of the obtained material. In contrast to the flux method, the hydrothermal technique requires large, expensive, and complex autoclaves that withstand high pressure. It is an industrial method used for a number of crystals, especially for quartz. This method was successfully used for large ZnO single crystals (3-inch diameter) in large scale. Also, small crystals of In_2O_3 and SnO_2 were demonstrated by the hydrothermal method as well, but not developed.

Among melt growth techniques, the crucible-free methods were used for growing single crystals of β-Ga_2O_3 (Verneuil, OFZ) and ZnO (cold crucible). Although these techniques can provide high-purity crystals, their size and perfection are typically not as good as in the case of using melt growth methods utilizing a noble metal crucible. The Verneuil and cold crucible methods are used for growing some oxide crystals at relatively large scale, which cannot be readily obtained by other melt growth methods.

Melt growth methods involving the crucible (metal crucibles in the case of oxides) are relevant at the industrial scale (due to high yield) for growing a diversity of compounds, including metals, halides, semiconductors, compound semiconductors, and oxides. This is a result of large crystal volume combined with high structural perfection, and relatively short growth time. Freezing growth methods (Bridgman and VGF) were applied to grow ZnO, β-Ga_2O_3, $BaSnO_3$, $MgGa_2O_4$, and $ZnGa_2O_4$ with relatively large size and high structural quality suitable for wafer fabrication for epitaxial purposes. The Czochralski method is a very efficient growth technique, which is often the first choice for most oxide compounds. Applying a novel approach to the Czochralski methods enabled the growth of large (2-inch diameter and 4-inch long) single crystals of β-Ga_2O_3 with further scale-up capabilities. The Czochralski method was also demonstrated for Ga-based spinel single crystals ($MgGa_2O_4$, $ZnGa_2O_4$, $CoGa_2O_4$), although with smaller diameter, up to 1 inch. A technical variant of the Czochralski method, called the Kyropoulos method, was successfully tested for Ga-based spinel crystals ($MgGa_2O_4$, $CoGa_2O_4$). Another industrially important technique is the EFG method, which produces crystal sheets or slabs enabling large surface area. This method has successfully been utilized for growing β-Ga_2O_3 crystals. And finally, a novel melt growth technique emerged, called levitation-assisted self-seeding crystal growth method, which was defined and developed specifically for bulk In_2O_3 single crystals.

The development of bulk TSO single crystals has enormously been intensified in the last decade. Access to bulk crystals enables exploration of material properties (which is not always fully available for thin films due to their poor structural quality at an early stage of development) and use of the bulk crystals as substrates for homo- or heteroepitaxy. In particular, epitaxial layers grown on native substrates typically ensure much better structural quality of the layers as compared with the layers grown on foreign substrates and better electrical/optical properties. Therefore, bulk TSO crystals are substantial for understanding and designing devices based on oxide semiconductors.

References

1. K. Th. Wilke; "Kristallzüchtung", VEB Deutscher Verlag der Wissenschaften, Berlin, 1973.
2. M. Binnewies, R. Glaum, M. Schmidt, P. Schmidt; *Z. Anorg. Allg. Chem.* **639** (2013), 219–229.
3. R. Bunsen; *J. Prakt. Chem.* **56** (1852), 53–55.
4. H. Saint-Claire Deville; *Justus Liebigs Ann. Chem.* **120** (1861), 176.
5. A. E. van Arkel; *Physica* **3** (1923), 76–87.
6. A. E. van Arkel, J. H. de Boer; *Z. Anorg. Allg. Chem.* **148** (1925), 345–350.
7. H. Schäfer, H. Jacob, K. Etzel; *Z. Anorg. Allg. Chem.* **286** (1956), 27–41.
8. H. Schäfer; *Chemische Transportreaktionen*, VCH, Weinheim, 1962.
9. H. Schäfer; *Chemical Transport Reactions*, Academic Press, New York, London, 1964.
10. I. Kubo; *Jpn. J. Appl. Phys.* **4** (1965), 225–226.
11. M. Hirose, I. Kubo; *Jpn. J. Appl. Phys.* **8** (1969), 402.
12. M. Hirose, Y. Furuya; *Jpn. J. Appl. Phys.* **11** (1972), 423–424.
13. W. Piekarczyk, S. Gazda, T. Niemyski; *J. Cryst. Growth* **12** (1972), 272–276.
14. E. Ziegler, A. Heinrich, H. Oppermann, G. Stöver; *Phys. Status Solidi* **66** (1981), 635–648.
15. K. Matsumoto, G. Shimaoka; *J. Cryst. Growth* **86** (1988), 410–414.
16. K. Grasza, A. Mycielski, J. Domagala, V. Domukhovski, W. Paszkowicz, H. Sakowska, W. Hofman; *Phys. Status Solidi* (*c*) **3** (2006), 793–796.
17. T. Matsumoto, M. Aoki, A. Kinoshita, T. Aono; *Jpn. J. Appl. Phys.* **13** (1974), 737–738.
18. T. Matsumoto, M. Aoki, A. Kinoshita, T. Aono; *Jpn. J. Appl. Phys.* **13** (1974), 1578–1582.
19. Von U. Gerlach, H. Oppermann; *Z. Anorg. Allg. Chem.* **432** (1977), 17–25.
20. H. Juskowiak, A. Pajączkowska; *J. Mater. Sci.* **21** (1986), 3430–3434.
21. H. Juskowiak, A. Pajączkowska; *Mater. Res. Bull.* **23** (1988), 1071–1077.
22. A. Pajączkowska, H. Juskowiak; *J. Mater. Sci.* **21** (1986), 3435–3439.
23. A. Pajączkowska, H. Juskowiak; *J. Cryst. Growth* **79** (1986), 421–426.
24. J. H. W. De Wit; *J. Cryst. Growth* **12** (1972), 183–184.
25. M. Jozefowicz, W. Piekarczyk; *Mater. Res. Bull.* **22** (1987), 775–780.

26. J. Werner, G. Behr, W. Bieger, G. Krabbes; *J. Cryst. Growth* **165** (1996), 258–267.
27. G. Behr, J. Werner, S. Oswald, G. Krabbes, P. Dordor, D. Elefant, W. Pitschke; *Solid State Ion.* **101–103** (1997), 1183–1187.
28. G. Frank, L. Brock, H. D. Bausen; *J. Cryst. Growth* **36** (1976), 179–180.
29. V. Scherer, C. Janowitz, A. Krapf, H. Dwelk, D. Braun, R. Manzke; *Appl. Phys. Lett.* **100** (2012), 212108.
30. D. F. Crabtree, R. N. S. M. Mehdi, D. A. Wright; *Brit. J. Appl. Phys.* **2** (1969), 1503–1505.
31. D. F. Crabtree; *J. Phys. D: Appl. Phys.* **7** (1974), L17–L21.
32. D. F. Crabtree; *J. Phys. D: Appl. Phys.* **7** (1974), L22–L26.
33. M. Nagasawa, S. Shionoya, S. Makishima; *Jpn. J. Appl. Phys.* **4** (1965), 195–202.
34. M. Hirose, Y. Furuya; *Sci. Bull. Fac. Educ., Nagasaki Univ.* **32** (1981), 21–36.
35. Y. Furuya, M. Hirose, Y. Taneda; *Jpn. J. Appl. Phys.* **20** (1981), 785–786.
36. C. G. Fonstad, A. Linz, R. H. Rediker; *J. Electrochem. Soc.: Solid State Sci.* **116** (1969), 1269–1271.
37. K. Matsumoto, S. Kaneko, K. Takagi; *J. Cryst. Growth* **40** (1977), 291–297.
38. E. Scharovsky; *Zeitschrift für Physik* **135** (1953), 318–330.
39. G. Bogner, E. Mollwo; *J. Phys. Chem. Solids* **6** (1958), 136–143.
40. R. L. Weiher; *J. Appl. Phys.* **33** (1962), 2834–2839.
41. J. A. Marley, T. C. MacAvoy; *J. Appl. Phys.* **32** (1961), 2504–2505.
42. S. Ito, S. Shimada, A. Tsunashima, K. Kodaira, T. Matsushita; *Z. Anorg. Allg. Chem.* **439** (1978), 259–264.
43. J. L. Santailler, C. Audoin, G. Chichignoud, R. Obrecht, B. Kaouache, P. Marotel, D. Pelenc, S. Brochen, J. Merlin, I. Bisotto, C. Granier, G. Feuillet, F. Levy; *J. Cryst. Growth* **312** (2010), 3417–3424.
44. R. L. Weiher, R. P. Ley; *J. Appl. Phys.* **37** (1964), 299–302.
45. S. Shimada, O. Sato, A. Tsunashima, K. Kodaira, T. Matsushita; *J. Cryst. Growth* **80** (1987), 366–370.
46. S. Shimada, K. J. D. MacKenzie; *J. Cryst. Growth* **55** (1981), 453–456.
47. Z. Galazka, R. Uecker, K. Irmscher, D. Schulz, D. Klimm, M. Albrecht, M. Pietsch, S. Ganschow, A. Kwasniewski, R. Fornari; *J. Cryst. Growth* **362** (2013), 349–352.

48. J. A. Marley, R. C. Dockerty; *Phys. Rev.* **140** (1965), A304–A310.
49. M. de Murcia, J. P. Fillard; *Mat. Res. Bull.* **11** (1976), 189–196.
50. M. de Murcia, M. Egee, J. P. Fillard; *J. Phys. Chem. Solids* **39** (1978), 629–635.
51. M. de Murcia, J. Bonnafe, J. C. Manifacier, J. P. Fillard; *J. Appl. Phys.* **49** (1978), 1177–1187.
52. B. Thiel, R. Helbig; *J. Cryst. Growth*; **32** (1976), 259–264.
53. T. B. Reed, J. T. Roddy, A. N. Mariano; *J. Appl. Phys.* **33** (1962), 1014–1015.
54. D. F. Morgan, D. A. Wright; *Brit. J. Appl. Phys.* **17** (1966), 337–340.
55. D. B. Ghare; *J. Cryst. Growth* **23** (1974), 157–159.
56. T. Takizawa, T. Sakurai; *Jpn. J. Appl. Phys.* **12** (1973), 1323–1325.
57. Z. Galazka, R. Uecker, D. Klimm, K. Irmscher, M. Pietsch, R. Schewski, M. Albrecht, A. Kwasniewski, S. Ganschow, D. Schulz, C. Guguschev, R. Bertram, M. Bickermann, R. Fornari; *Phys. Status Solidi* (*a*) **211** (2014), 66–73.
58. E. Kaldis, M. Piechotka; *Handbook of Crystal Growth*, Elsevier Science B. V., 1994, Vol. 2a, Chapter 11, pp. 612–658.
59. S. A. Sakwe, M. Stockmeier, P. Hens, R. Müller, D. Queren, U. Kunecke, K. Konias, R. Hock, A. Magerl, M. Pons, A. Winnacker, P. Wellmann; *Silicon Carbide*, Vol. 1: *Growth, Defects, and Novel Applications*, Edited by Peter Friedrichs, Tsunenobu Kimoto, Lothar Ley, and Gerhard Pensl, WILEY-VCH Verlag GmbH & Co. KGaA, Weinheim, 2010.
60. P. J. Wellmann; *Z. Anorg. Allg. Chem.* **643** (2017), 1312–1322.
61. D. Ehrentraut, Z. Sitar; *MRS Bulletin* **34** (2009), 259–265.
62. K. Byrappa; *Handbook of Crystal Growth*, Elsevier Science B. V., 1994, Vol. 2a, Chapter 9, pp. 465–562.
63. K. Byrappa, M. Yoshimura; *Handbook of Hydrothermal Technology, A Technology for Crystal Growth and Materials Processing*, Chapter 2, 2001, pp. 53-81.
64. H. de Sénarmont; *Ann. Chim. Phys.* **32** (1851), 129–132.
65. A.C. Walker; *J. Am. Ceram. Soc.* **36** (1953), 250–256.
66. R. A. Laudise, E. D. Kolb, A. J. Caporaso; *J. Am. Ceram. Soc.* **47** (1964), 9–12.
67. D. F. Croxall, R. C. C. Ward, C. A. Wallace, R. C. Kell; *J. Cryst. Growth* **22** (1974), 117–124.
68. L. N. Dem'yanets, V. I. Lyutin; *J. Cryst. Growth* **310** (2008), 993–999.

69. L. N. Demianets, D. V. Kostamarov; *Ann. Chim. Sci. Mat.* **26** (2001), 193–198.

70. N. Sakagami; *J. Cryst. Growth* **99** (1990), 905–909.

71. K. Maeda, M. Sato, I. Niikura, T. Fukuda; *Semicond. Sci. Technol.* **20** (2005), S49–S54.

72. D. Ehrentraut, H. Sato, Y. Kagamitani, H. Sato, A. Yoshikawa, T. Fukuda; *Prog. Cryst. Growth Charact. Mater.* **52** (2006), 280–335.

73. R. Roy, M. W. Shafer; *J. Phys. Chem.* **58** (1954), 372.

74. A. N. Christensen, N. C. Broch, O. von Heidenstam, Å. Nilsson; *Acta Chem. Scand.* **21** (1967), 1046–1056.

75. B. N. Litvin, L. I. Ivanova; *Kristallographia* **14** (1969), 1122.

76. M. L. Harvil, R. Roy; *Am. Ceram. Soc. Bull.* **44** (1965), 297.

77. Y. Fujiki, Y. Suzuki; *J. Jpn. Assoc. Min. Petr. Econ. Geol.* **68** (1973), 277–283.

78. M. G. Kanatzidis, R. Pöttgen, W. Jeitschko; *Angew. Chem. Int. Ed.* **44** (2005), 6996–7023.

79. J. P. Remeika; *J. Amer. Chem. Soc.* **76** (1954), 940–941.

80. J. W. Nielsen, E. F. Dearborn; *J. Phys. Chem. Solids* **5** (1958), 202–207.

81. J. W. Nielsen, E. F. Dearborn; *J. Phys. Chem.* **64** (1960), 1762–1763.

82. J. P. Remeika; US 3075831 (1963).

83. J. P. Remeika, E. G. Spencer; *J. Appl. Phys.* **35** (1964), 2803.

84. H. F. Kunkle, E. E. Kohnke; *J. Appl. Phys.* **36** (1965), 1489.

85. D. Elwell, H. J. Scheel; *Crystal Growth from High temperature Solution*, Academic Press, London, 1975.

86. H. J. Scheel; *Prog. Cryst. Growth Charact.* **5** (1982), 277–290.

87. J. C. Brice; *Rep. Prog. Phys.* **40** (1977), 567–602.

88. E. A. D. White; *Br. J. Appl. Phys.* **16** (1965), 1415–1428.

89. W. Tolksdorf; *Handbook of Crystal Growth*, Elsevier Science B. V., 1994, Vol. 2a, Chapter 10, pp. 563–611.

90. J. W. Nielsen, E. F. Dearborn; *J. Phys. Chem.* **64** (1960), 1762–1763.

91. S. C. Kashyap; *J. Appl. Phys.* **44** (1973), 4381–4384.

92. K. Fischer, E. Sinn; *Cryst. Res. Technol.* **16** (1981), 689–894.

93. X. H. Li, J. Y. Xu, X. M. Li; *Appl. Phys. A* **82** (2006), 373–376.

94. X. Li, J. Xu, M. Jin, X. Wu, G. Qian, X. Li; *Cryst. Res. Technol.* **42** (2007), 221–226.

95. K. Oka, H. Shibata, S. Kashiwaya; *J. Cryst. Growth* **237–239** (2002), 509–513.
96. J. P. Remeika; US 3075831 (1963).
97. G. Katz, R. Roy; *J. Am. Ceram. Soc.* **49** (1966), 168–169.
98. G. Garton, S. H. Smith, B. M. Wanklyn; *J. Cryst. Growth* **13/14** (1972), 588–592.
99. M. Schieber; *J. Am. Ceram. Soc.* **47** (1964), 537–538.
100. V. I. Chani, K. Inoue, K. Shimamura, K. Sugiyama, T. Fukuda; *J. Cryst. Growth* **132** (1993), 335–336.
101. J. P. Remeika, E. G. Spencer; *J. Appl. Phys.* **35** (1964), 2803.
102. A. B. Chase, W. R. Wilcox; *J. Am. Ceram. Soc.* **50** (1967), 332–333.
103. A. B. Chase, H. H. Tippins; *J. Appl. Phys.* **38** (1967), 2469.
104. A. B. Chase; *J. Am. Ceram. Soc.* **51** (1968), 507–510.
105. A. B. Chase, R. Teviotdale; *J. Appl. Phys.* **39** (1968), 3574–3576.
106. S. J. Wen, G. Couturier, J. P. Chaminade, E. Marquestaut, J. Claverie, P. Hagenmuller; *J. Solid State Chem.* **101** (1992), 203–210.
107. D. R. Hagleitner, M. Menhart, P. Jacobsen, S. Blomberg, K. Schulte, E. Lundgren, M. Kubicek, J. Fleig, F. Kubel, C. Puls, A. Limbeck, H. Hutter, L. A. Boatner, M. Schmid, U. Diebold; *Phys. Rev. B* **85** (2012), 115441.
108. H. F. Kunkle, E. E. Kohnke; *J. Appl. Phys.* **36** (1965), 1489.
109. F. Kawamura, M. Kamei, I. Yasui, I. Sunagawa; *J. Am. Ceram. Soc.* **82** (1999), 774–776.
110. F. Kawamura, I. Yasui, M. Kamei, I. Sunagawa; *J. Am. Ceram. Soc.* **84** (2001), 1341–1346.
111. F. Kawamura, T. Takahashi, I. Yasui, I. Sunagawa; *J. Cryst. Growth* **233** (2001), 259–268.
112. S. Hidenori, U. Kenji; JP 60051699 A (1985).
113. S. Shimada, K. Kodaira, T. Matsushita; *Chem. Lett.* (1976), 235–236.
114. S. Shimada, K. Kodaira, T. Matsushita; *J. Cryst. Growth* **59** (1982), 662–664.
115. X. Luo, Y. S. Oh, A. Sirenko, P. Gao, T. A. Tyson, K. Char, S.-W. Cheong; *Appl. Phys. Lett.* **100** (2012), 172112.
116. H. J. Kim, U. Kim, T. H. Kim, J. Kim, H. M. Kim, B.-G. Jeon, W.-J. Lee, H. S. Mun, K. T. Hong, J. Yu, K. Char, K. H. Kim; *Phys. Rev. B* **86** (2012), 165205.
117. H. J. Kim, J. Kim, T. H. Kim, W.-J. Lee, B-G. Jeon, J.-Y. Park, W. S. Choi, D. W. Jeong, S. H. Lee, J. Yu, T. W. Noh, K. H. Kim; *Phys. Rev. B* **88** (2013), 125204.

118. T. N. Stanislavchuk, A. A. Sirenko, A. P. Litvinchuk, X. Luo, S.-W. Cheong; *J. Appl. Phys.* **112** (2012), 044108.

119. H. J. Kim, U. Kim, H. M. Kim, T. H. Kim, H. S. Mun, B.-G. Jeon, K. T. Hong, W.-J. Lee, C. Ju, K. H. Kim, K. Char; *Appl. Phys. Express* **5** (2012), 061102.

120. W.-J. Lee, H. J. Kim, E. Sohn, T. H. Kim, J.-Y. Park, W. Park, H. Jeong, T. Lee, J. H. Kim, K.-Y. Choi, K. H. Kim; *Appl. Phys. Lett.* **108** (2016), 082105.

121. W.-J. Lee, H. Lee, K.-T. Ko, J. Kang, H. J. Kim, T. Lee, J.-H. Park, K. H. Kim; *Appl. Phys. Lett.* **111** (2017), 231604.

122. W.-J. Lee, H. J. Kim, J. Kang, D. H. Jang, T. H. Kim, J. H. Lee, K. H. Kim; *Annu. Rev. Mater. Res.* **47** (2017), 391–423.

123. E. McCalla, D. Phelan, M. J. Krogstad, B. Dabrowski, C. Leighton; *Phys. Rev. Meter.* **2** (2018), 084601.

124. A. B. Chase, J. A. Osmer; *J. Am. Ceram. Soc.* **50** (1967), 325–328.

125. H. van den Boom, J. C. M. Henning, J. P. M. Damen; *Sol. State Comm.* **8** (1970), 717–719.

126. P. J. M. Van Der Straten, R. Metselaar, H. D. Jonker; *J. Cryst. Growth* **43** (1978), 270–272.

127. Z. Yan, H. Takei; *J. Cryst. Growth* **171** (1997), 131–135.

128. E. A. Giess; *J. Appl. Phys.* **33** (1962), 2143.

129. A. Nakatsuka, Y. Ikeda, N. Nakayama, T. Mizota; *Acta Crystallogr.* **E62** (2006), i109–i111.

130. A. B. Chase; *J. Am. Ceram. Soc.* **47** (1964), 470.

131. R. Falckenberg; *Crystal Growth Theory and Techniques*, edited by C. H. Goodman, Plenum Press, New York, 1978, Chapter 3, pp. 109–184.

132. V. A. Borodin, E. A. Brener, V. A. Tatarchenko; *Acta Physica Academiae Scientiarum Hungaricae, Tomus* **47** (1979), 151–158.

133. J. Bohm, A. Lüdge, W. Schröder; *Handbook of Crystal Growth*, Elsevier Science B. V., 1994, Vol. 2a, Chapter 4, pp. 214–257.

134. P. Kapitza; *Proc. R. Soc. A* **119** (1928), 358–443.

135. W. G. Pfann; *Trans. AIME* **194** (1952), 747–753.

136. A. T. Chapman, G. W. Clark; *J. Am. Ceram. Soc.* **48** (1965), 494–495.

137. C. Kooy, H. J. M. Couwenberg; *Philips Techn. Rundschau* **23** (1961/62), 143.

138. T. Akashi, K. Matumi, T. Okada, T. Mizutani; *IEEE Trans. Mag.* **5** (1969), 285–289.

139. A. M. Balbashov, S. K. Egorov; *J. Cryst. Growth* **52** (1981), 498–504.

140. S. M. Koohpayeh, D. Fort, J. S. Abell; *Prog. Cryst. Growth Charact. Mater.* **54** (2008), 121–137.

141. H. A. Dąbkowska, A. B. Dąbkowski; *Handbook of Crystal Growth Second Edition*, edited by T. Nishinga and P. Rudolph, Elsevier B. V., 2015, Vol. II, Part A, Chapter 8, pp. 281–312.

142. V. I. Vasyltsiv, Y. I. Rym, Y. Zakharko; *Phys. Status Solidi (b)* **195** (1996), 653–658.

143. N. Ueda, H. Hosono, R. Waseda, H. Kawazoe; *Appl. Phys. Lett.* **71** (1997), 933–935.

144. E. G. Villora, K. Shimamura, Y. Yoshikawa, K. Aoki, N. Ichinose; *J. Cryst. Growth* **270** (2004), 420–426.

145. E. G. Víllora, K. Shimamura, Y. Yoshikawa, T. Ujiie, K. Aoki; *Appl. Phys. Lett.* **92** (2008), 202120.

146. T. Suzuki, M. Hughes, Y. Ohishi; *J. Lumin.* **130** (2010), 121.

147. T. Suzuki, M. Hughes, Y. Ohishi; *J. Lumin.* **130** (2010), **121** (1985), 329–336.

148. A. Mühlbauer; *History of Inductive Heating and Melting*, Vulkan-Verlag GmbH, 2008.

149. Siemens & Halske, Berlin-Siemensstadt, DE518499C, (1931).

150. A. Mühlbauer; *History of Inductive Heating and Melting*, Vulkan-Verlag GmbH, 2008.

151. V. I. Aleksandrov, V. V. Osiko, A. M. Prokhorov, V. M. Tatarintsev; *Current Topics in Materials Science*, North-Holland Publishing Company, 1978, Vol. 1, Chapter 6, pp. 421–479.

152. F. R. Monforte, F. W. Swanekamp, L. G. Van Uitert; *J. Appl. Phys.* **32** (1961), 959–960.

153. G. Agarwal, J. E. Nause, D. N. Hill; *Mat. Res. Soc. Symp. Proc.* **512** (1988), 41–46.

154. J. E. Nause; *III-Vs Review* **12** (1999), 28–31.

155. J. Nause, B. Nemeth; *Semicond. Sci. Technol.* **20** (2005), S45–S48.

156. P. W. Bridgman; *Proc. Am. Acad. Arts Sci.* **60** (1925), 305–383.

157. D. C. Stockbarger; *Rev. Sci. Instrum.* **7** (1936), 133–136.

158. W. D. Lawson, S. Nielsen; *Preparation of Single Crystals*, Butterworths Scientific Publications, 1958, London.

159. H. E. Buckley; *Crystal Growth*, John Wiley & Sons Inc., 1951, New York, pp. 73–82.

160. D. Schulz, S. Ganschow, D. Klimm, K. Struve; *J. Cryst. Growth* **310** (2008), 1832–1835.

161. D. Klimm, S. Ganschow, D. Schulz, R. Fornari; *J. Cryst. Growth* **310** (2008), 3009–3013.

162. D. Schulz, S. Ganschow, D. Klimm; *Mater. Res. Soc. Symp. Proc.* **1201** (2010), 120-H06-10.

163. K. Jacobs, D. Schulz, D. Klimm, S. Ganschow; *Solid State Sci.* **12** (2010), 307–310.

164. R. Fornari, S. Ganschow, D. Klimm, M. Neubert, D. Schulz; DE 10 2007 006 731 (2011).

165. Z. Galazka, R. Uecker, D. Klimm, M. Bickermann; PCT/EP2015/079938 (WO/2016/110385) (2015).

166. K. Hoshikawa, E. Ohba, T. Kobayashi, J. Yanagisawa, C. Miyagawa, Y. Nakamura; *J. Cryst. Growth* **447** (2016), 36–41.

167. Z. Galazka, R. Uecker, K. Irmscher, D. Klimm, R. Bertram, A. Kwasniewski, M. Naumann, R. Schewski, M. Pietsch, U. Juda, A. Fiedler, M. Albrecht, S. Ganschow, T. Markurt, C. Guguschev, M. Bickermann; *J. Phys.: Condens. Matter.* **29** (2017), 075701.

168. Z. Galazka, D. Klimm, K. Irmscher, R. Uecker, M. Pietsch, R. Bertram, M. Naumann, M. Albrecht, A. Kwasniewski, R. Schewski, M. Bickermann; *Phys. Status Solidi (a)* **212** (2014), 1455–1460.

169. Z. Galazka, S. Ganschow, R. Schewski, K. Irmscher, D. Klimm, A. Kwasniewski, M. Pietsch, A. Fiedler, I. Schulze-Jonack, M. Albrecht, T. Schröder, M. Bickermann; *APL Mater.* **7** (2019), 022512.

170. J. Czochralski; *Z. Phys. Chem.* **92** (1918), 219–221.

171. R. Uecker; *J. Cryst. Growth* **401** (2014), 7–24.

172. P. E. Tomaszewski; *Jan Czochralski Restored*, Oficyna Wydawnicza ATUT, Wrocław, 2013.

173. E. V. Gomperz; *Z. Phys.* **8** (1922), 184–190.

174. E. Grüneisen, E. Goens; *Z. Phys.* **24** (1923), 506–510.

175. A. G. Hoyem, E. P. T. Tyndall; *Phys. Rev.* **33** (1929), 81–89.

176. H. Walther; *Rev. Sci. Instrum.* **8** (1937), 406–408.

177. J. B. Little, G. K. Teal; *Phys. Rev.* **78** (1950), 647.

178. W. C. Dash; *J. Appl. Phys.* **30** (1959), 459–474.

179. K. Hoshikawa, H. Konda, H. Hirata, H. Nakanishi; *Jpn. J. Appl. Phys.* **19** (1980), L33–36.

180. D. A. Petrov, V. S. Zemskov; *Rost Kristallov* **1** (1957), 262–272.

181. E. A. P. Metz, R. C. Miller, R. Mazelsky; *J. Appl. Phys.* **33** (1962), 2016–2017.

182. W. Bardsley, G. W. Green, C. H. Holliday, D. T. J. Hurle; *J. Cryst. Growth* **16** (1972), 277–279.

183. T. R. Kyle, G. Zydzik; *Mater. Res. Bull.* **8** (1973), 443–450.

184. K. Nassau, L. G. Van Uitert; *J. Appl. Phys.* **31** (1960), 1508.

185. Y. Tomm, P. Reiche, D. Klimm, T. Fukuda; *J. Cryst. Growth* **220** (2000), 510–514.

186. Z. Galazka, R. Uecker, K. Irmscher, M. Albrecht, D. Klimm, M. Pietsch, M. Brützam, R. Bertram, S. Ganschow, R. Fornari; *Cryst. Res. Technol.* **45** (2010), 1229–1236.

187. Z. Galazka, K. Irmscher, R. Uecker, R. Bertram, M. Pietsch, A. Kwasniewski, M. Naumann, T. Schulz, R. Schewski, D. Klimm, M. Bickermann; *J. Cryst. Growth* **404** (2014), 184–191.

188. Z. Galazka, R. Uecker, D. Klimm, K. Irmscher, M. Naumann, M. Pietsch, A. Kwasniewski, R. Bertram, S. Ganschow, M. Bickermann; *ECS J. Solid State Sci. Technol.* **6** (2017), Q3007–Q3011.

189. Z. Galazka, S. Ganschow, A. Fiedler, R. Bertram, D. Klimm, K. Irmscher, R. Schewski, M. Pietsch, M. Albrecht, M. Bickermann; *J. Cryst. Growth* **486** (2018), 82–90.

190. S. Kyropoulos; *Z. Anorg. Chem.* **154** (1926), 308–313.

191. D. F. Bliss, R. M. Hilton, S. Bachowski, J. A. Adamski; *J. Electron. Mater.* **20** (1991), 967–971.

192. G. Jacob; *J. Cryst. Growth* **58** (1982), 455–459.

193. V. N. Kurlov; "Sapphire: properties, growth, and applications," in *Reference Module in Materials Science and Materials Engineering*, edited by S. Hashmi, Elsevier 2016, pp. 1–11.

194. A. V. Stepanov; *Zh. Tekh. Fiz.* **29** (1959), 382–393.

195. H. E. LaBelle, A. I. Mlavsky; *Nature* **216** (1967), 574–575.

196. V. A. Tatarchenko; *J. Cryst. Growth* **37** (1977), 272–284.

197. V. A. Tatarchenko, G. A. Satunkin; *J. Cryst. Growth* **37** (1977), 285–288.

198. V. A. Tatarchenko; *J. Cryst. Growth* **82** (1987), 74–80.

199. V. A. Tatarchenko; *Handbook of Crystal Growth*, edited by D. T. J. Hurle, Elsevier Science B. V., 1994, , Vol. 2b, Chapter 16, pp. 1011–1110.

200. V. A. Tatarchenko, V. S. Uspenski, E. V. Tatarchenko, J. Ph. Nabot, T. Duffafr, B. Roux; *J. Cryst. Growth* **180** (1997), 615–626.

201. P. I. Antonov, V. N. Kurlov; *Prog. Cryst. Growth Charact. Mater.* **44** (2002), 63–122.

202. P. I. Antonov, S. P. Nikanorov, V. A. Tatarchenko; *J. Cryst. Growth* **42** (1977), 447–452.

203. V. N. Kurlov, S. N. Rossolenko, N. V. Abrosimov, K. Lebbou; "Shaped crystal growth," in *Crystal Growth Processes Based on Capillarity: Czochralski, Floating Zone, Shaping and Crucible Techniques*, edited by T. Duffar, John Wiley & Sons, Ltd., 2010, Chapter 5, pp. 277–354.

204. R. S. Feigelson; *Growth of Shaped Crystals in Crystal Growth in Science and Technology*, edited by H. Arend and J. Hulliger, Vol. 210, 1989, pp. 275–302.

205. V. A. Borodin, V. V. Sidorov, T. A. Steriopolo, V. A. Tatarchenko; *J. Cryst. Growth* **82** (1987), 89–94.

206. V. A. Borodin, V. V. Sidorov, S. N. Rossolenko, T. A. Steriopolo, V. A. Tatarchenko, T. N. Yalovets; *J. Cryst. Growth* **104** (1990), 69–76.

207. D. C. Harris; "A century of sapphire crystal growth: origin of the EFG method," *Proc. SPIE 7425, Optical Materials and Structures Technologies IV* (2009), 74250P.

208. H. Aida, K. Nishiguchi, H. Takeda, N. Aota, K. Sunakawa, Y. Yaguchi; *Jpn. J. Appl. Phys.* **47** (2008), 8506–8509.

209. A. Kuramata, K. Koshi, S. Watanabe, Y. Yamaoka, T. Masui, S. Yamakoshi; *Jpn. J. Appl. Phys.* **55** (2016), 1202A2.

210. Z. Galazka, R. Uecker, R. Fornari; PCT/EP2012/057447 (WO 2013/159808) (2012).

211. Z. Galazka, R. Uecker, R. Fornari; *J. Cryst. Growth* **388** (2014), 61–69.

Chapter 2

Thermodynamics of Transparent Semiconducting Oxides*

2.1 Introduction

Many crystal growth processes are performed in dense media (solids, liquids, gases near ambient pressure) and at elevated temperatures of, at least, several hundred degrees centigrade. Under such conditions, the thermal activation of atoms or ions as building blocks of the corresponding phases in the system is often sufficient to overcome kinetic barriers for chemical reactions or structural phase transitions, and the systems tend toward equilibrium within periods of time that are short compared to typical time constants of the growth process itself. In contrast, crystal growth processes that are performed from highly diluted media at low temperatures (such as molecular beam epitaxy) proceed often far from equilibrium.

Equilibrium states are described by the laws of chemical thermodynamics, and the term thermodynamics was introduced by Lord Kelvin as "the relation of heat to forces acting between contiguous parts of bodies" [1]. This early definition was given in the period of steam engines and paid special attention on the work that can be produced by engines, an aspect that is usually irrelevant for crystal growth processes. Nevertheless, the basic concepts that

*Contributed by Detlef Klimm

Transparent Semiconducting Oxides: Bulk Crystal Growth and Fundamental Properties
Zbigniew Galazka

ISBN 978-981-4800-94-5 (Hardcover), 978-1-003-04520-5 (eBook)
www.jennystanford.com

were developed by Lord Kelvin are valid for all processes close to thermodynamic equilibrium, and this holds for all processes that are described in the framework of this book, which includes growth from the melt (e.g., β-Ga_2O_3, In_2O_3, ZnO, and others) as well as growth from the gas phase (e.g., SnO_2).

2.2 Basics of Thermodynamics

The conservation of energy is one of the fundamental laws of thermodynamics. Besides one can assume that every system aspires to a state of minimum energy. The difference ΔE between the initial energy of the system E_0 and the final (equilibrium) energy E_1 is released during this process:

$$E_0 \rightarrow E_1 + \Delta E \tag{2.1}$$

All compounds are composed of atoms of the chemical elements; hence it is useful to use data of the pure elements and some defined standard conditions as basis. The US "National Bureau of Standards" defined $T_0 = 25°\text{C} = 298.15$ K and $p_0 = 1\times10^5$ Pa = 1 bar as "standard ambient temperature and pressure" [2]. The enthalpy of every chemical element in the state of aggregation that is stable under these standard conditions is per definition zero: $H_0 = 0$ J/mol. If the temperature of the element is changed to T by the introduction or removal of heat energy, the enthalpy change correlates with the specific heat capacity c_p according to the following relation:

$$H(T) = H_0 + \int_{T_0}^{T} c_\text{p}(T)dT \tag{2.2}$$

Figure 2.1 demonstrates for indium metal that the function $H(T)$ grows usually slowly and almost linearly. This can be expected from Eq. 2.2 because c_p does not change significantly far above the Debye temperature, which is around 111 K for In [3]. The jump at 430 K comes from the fusion, because the introduction of ΔH_f (latent heat of fusion) increases the enthalpy suddenly.

The enthalpy H corresponds to the amount of energy that is stored inside a solid, liquid, or gaseous phase, and the main portion of this energy is stored by elastic vibrational energy of the constituents (atoms, molecules, or the whole crystal lattice). Besides, the stored energy creates structural disorder or inharmonic processes, which is described by a growing entropy S of the phase. Only at absolute zero, the entropy is zero (at least for most substances), and the total energy content is expressed by H. Above absolute zero, the recoverable energy content is expressed by

$$G = H - TS \tag{2.3}$$

which is the "Gibbs free energy." The entropy S, which is a measure for disorder, starts with $S = 0$ at 0 K and, for higher T, can also be derived from c_p by

$$S(T) = H_0 + \int_{T_0}^{T} \frac{c_\mathrm{p}(T)}{T} \mathrm{d}T \tag{2.4}$$

For metallic indium, the function $G(T)$ is, together with $H(T)$, displayed in Fig. 2.1. Because S is always positive, $G(T)$ monotonously drops for nearly all substances. From Eq. 2.3, it is obvious that the negative slope of the curve equals S. Above the melting point T_f = 430 K, the liquid In phase has a higher disorder, hence entropy, than solid In. This is expressed by the somewhat steeper slope of the $G(T)$ curve. The Gibbs free energy G is the most important quantity for the description of thermodynamic equilibria, because under equilibrium conditions, G for the whole system has a minimum.

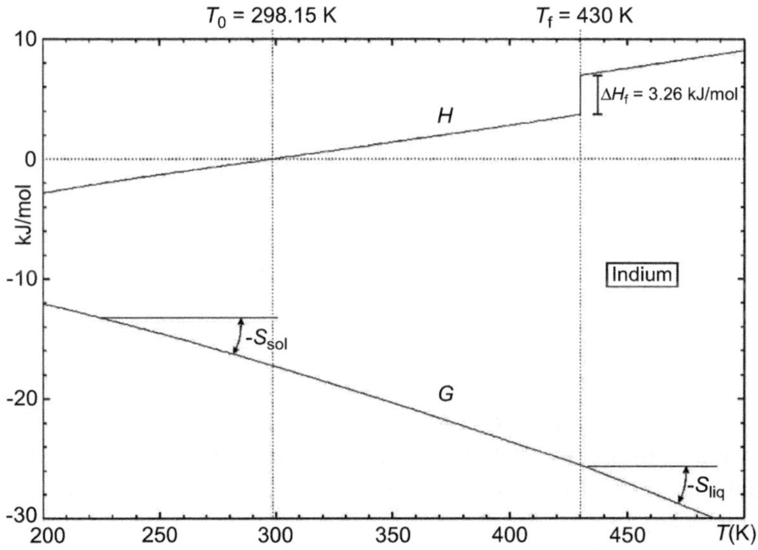

Figure 2.1 Enthalpy H and Gibbs free energy G of indium (p = 1 bar).

For all chemical elements, $H = 0$ can be defined for some arbitrary (e.g., "standard") conditions, because elements cannot interchange under thermodynamic (chemical) conditions. If two or more elements are reacting under the formation of a chemical compound, usually heat energy is released. The (negative) amount of energy is the "enthalpy of formation" ΔH of this compound. Usually ΔH is negative for stable compounds, because otherwise the compound would not be formed (see Table 2.1). ΔH is a function of temperature, but it is often given under the same standard conditions mentioned above. The stability of chemical compounds is especially high, if their ΔH is very negative. This explains why sapphire (α-Al_2O_3) is chemically extremely stable. In contrast, the energy gain during the formation of the other oxides in Table 2.1 is significantly lower, because the corresponding metals are somewhat "more noble." The formation enthalpies of silver and gold oxides are very low; this explains why Ag_2O and Au_2O_3 decompose to metal and O_2 quickly [4].

Table 2.1 Enthalpies of formation (p = 1 bar, 298.15 K) for some solid oxides from the elements

Binary compound	ΔH (kJ/mol)
α-Al_2O_3	−1675.70
β-Ga_2O_3	−1089.10
In_2O_3	−925.789
ZnO	−350.460
SnO_2	−577.630
Cu_2O	−170.707
Ag_2O	−31.0490
Au_2O_3	−3.347

Note: Data from FactSage [5]. ΔH is the formation energy (enthalpy of formation).

Thermodynamic data for simple oxides (binary systems) and other simple compounds such as halides or chalcogenides are often available in the original literature or in compilations [5, 6], whereas data for more complex substances such as oxides of two or more metals are scarce.

The stability of such complex compounds is basically determined by the stability of their constituents. For three examples, this is demonstrated in Fig. 2.2, where the enthalpy for the chemical reaction that forms the corresponding complex oxide is plotted versus temperature. Always these enthalpies are significantly smaller, compared to the formation enthalpies of the simple oxides that are given in Table 2.1. This means that for an approximation, it is often sufficient to decide if the constituent (simple) oxides are stable under the given process conditions. The additional influence of the formation enthalpy will usually not be very large. At least, this influence will be stabilizing, because the formation enthalpies of the complex oxides, such as shown in Fig. 2.2, are usually negative.

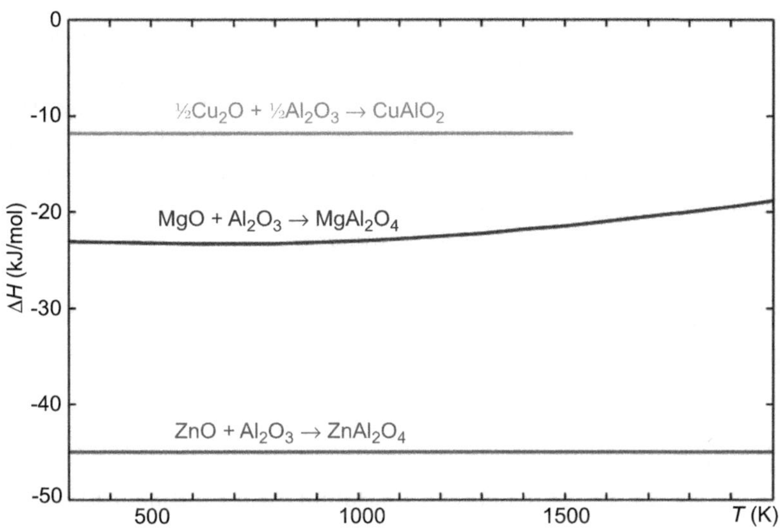

Figure 2.2 Formation enthalpies of three complex oxides from their element oxides.

Besides the formation of complex oxides, solution formation or other mainly entropic effects can influence G (Eq. 2.3) and hence the stability of phases. In the simplest case, one has an ideal mixture of two components with concentrations x and $1-x$, respectively. Then the ideal entropy of mixing

$$\Delta S_{\text{mix}} = -R[x \ln x + (1-x)\ln(1-x)] \tag{2.5}$$

(R is gas constant) reduces, for 1:1 mixtures ($x = 0.5$), the Gibbs energy (Eq. 2.3) of the ideal mixture by 5.76 kJ/mol at 1000 K and by 11.5 kJ/mol at 2000 K. Comparing these numbers with the data from Fig. 2.2 shows that the contribution of mixing to G can be in the order of compound formation. If the number of components is larger than two, the right-hand side of Eq. 2.5 has to be replaced by $-R\,\Sigma x_i \ln x_i$, where x_i is the concentration of the i-th component. Then ΔS_{mix} grows drastically, and such as for "high entropy (metallic) alloys," unexpected oxide phases can be stabilized at high temperature [7].

Redistribution of atoms over different lattice sites is another process that can increase the entropy and this way stabilize a phase. Spinels are a prominent example: These $A(B_2)X_4$ (X is often oxygen) compounds can crystallize as "normal spinels" where A occupies 1/8 of the tetrahedral sites of the cubic dense O packing and B occupies 1/2 of the octahedral sites. Depending on ionic radii and temperature, an "inverse spinel" structure $B(AB)X_4$ can be formed where all tetrahedral sites are occupied by B and each 1/4 of the octahedral sites by A and B. The inverse structure shows a higher degree of disorder, hence a larger entropy. At very high temperature, all ions tend to be distributed randomly over their lattice sites, and such "random spinels" have the highest entropy. If this configurational entropy is set to $S_c = 0$ for the normal spinel, it is 11.5 J/(mol K) for an "inverse" and 15.9 J/(mol K) for a "random" spinel [8].

One can summarize that the thermodynamic stability of oxides is mainly determined by the equilibria of the corresponding metals with oxygen. To some degree, the stability can be enhanced by the formation of intermediate compounds (complex oxides) due to their formation enthalpy. Other possible stabilization effects are possible through the formation of solid solutions (mixed crystals) through structural disorder inside one specific phase. The latter two effects are entropic and have no influence at low temperature but can result in significant stabilization at high temperature because the Gibbs free energy change scales with T (Eq. 2.3). In the following sections, a few metal oxides (binary systems) that are interesting as transparent semiconducting oxides (TSOs) or used in devices containing TSOs are discussed in more detail.

2.3 Thermodynamics of TSOs

Every chemical element Me (Me stands here for a metal or a non-metal) is with its oxides of valence state m in equilibrium via reactions of the type:

$$2MeO_{m/2} + \tfrac{1}{2}O_2 \rightleftarrows 2MeO_{(m+1)/2} \quad (2.6)$$

For $m = 0$, the pure metal is the first substance. If the fugacities of both metal oxides can be neglected with respect to the oxygen fugacity, the equilibrium constant K of Eq. 2.6 can be written as:

$$-RT \ln K = RT \ln p_{O2} = \Delta G^0 = \Delta H^0 - TS^0 \quad (2.7)$$

which results in linear plots of $\Delta G^0 = RT \ln p_{O2}$ versus temperature T [9]. Such plots are called Ellingham diagrams and are a valuable tool for the description of the chemical stability of metals and metal oxides under given thermodynamic conditions. Ellingham plots for several subsequent oxidation states of one specific element (e.g., $Fe \rightleftarrows FeO \rightleftarrows Fe_3O_4 \rightleftarrows Fe_2O_3$ [10]) are forming phase fields where only one oxide is stable. Often these phase fields are limited on the temperature axis by phase transitions between solid phases or between solid and melt. In the following, such Ellingham-type predominance diagrams will be provided for some Me–O_2 systems that are relevant for transparent conducting oxides. The order of diagrams starts with some main group metals of the periodic system, followed by some metals from subgroups. All diagrams were calculated for a total pressure of 1 bar with the FactSage software and database suite [5].

Ellingham diagrams can certainly be calculated also for conditions where the MeO_x fugacities cannot be neglected—just the $\Delta G^0(T)$ plots are then no longer linear. Moreover, it should be noted that fugacities of gases are activities and are almost identical to vapor pressures. Therefore, already in Eq. 2.7, the oxygen fugacity could be replaced by the oxygen partial pressure p_{O2} of a growth atmosphere. This means that the chemical equilibria (Eq. 2.6) between neighboring valence states of oxides can be manipulated with an appropriate choice of the process conditions T and p_{O2}.

2.3.1 The Al–O_2 System

Although Al_2O_3 itself is not considered to be a TSO, aluminum is a component of some complex TSOs, such as doping oxide or a gate in field effect transistors. The predominance diagram in Fig. 2.3 is ruled by the large phase field of α-Al_2O_3, which is very stable with respect to temperature and oxygen partial pressure. Only under extremely reducing conditions, the oxide is reduced to the metal, which melts at 933 K, indicated by the vertical line. The melting point of Al_2O_3 is just beyond the temperature scale of Fig. 2.3 at 2327 K. All fugacities at the triple point A are very low, and Al(gas) is the main species, followed by aluminum suboxide Al_2O(gas). It should be noted that the right scale is valid only for its position at $T = 2300$ K, because the rest of the diagram is scaled with the left abscissa $RT \ln p_{O2}$, where T is a factor.

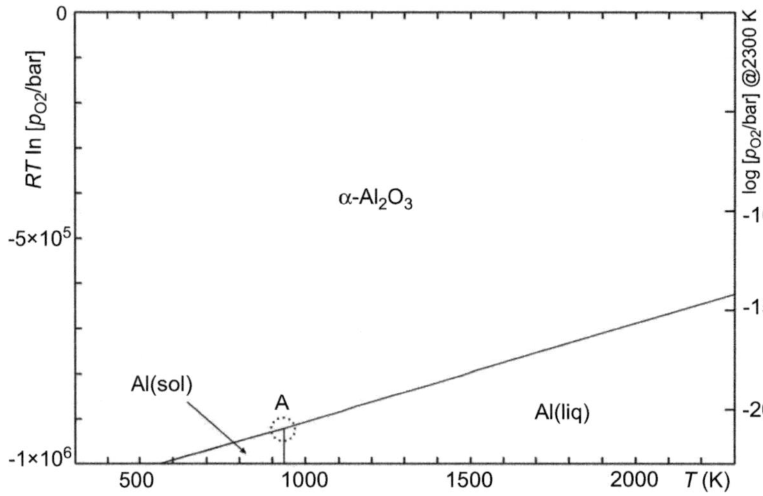

Figure 2.3 Predominance diagram for the Al–O_2 system. Gas phase fugacities (in bar) at point A: Al: 3.4×10^{-12}, Al_2O: 7.6×10^{-14}, Al_2: 2.2×10^{-19}, AlO: 6.3×10^{-26}, Al_2O_2: 1.2×10^{-28}, O: 2.7×10^{-37}, AlO_2: 2.3×10^{-48}, Al_2O_3: 4.0×10^{-50}, O_2: 2.3×10^{-52}.

2.3.2 The Ga–O_2 System

Monoclinic β-Ga_2O_3 is the only stable phase of this compound; it melts at 2073 K. However, Ga is significantly "more noble" compared to Al. Consequently, Ga_2O_3 undergoes decomposition to Ga(liq) or to gaseous suboxides (mainly Ga_2O) if the oxygen partial pressure in the atmosphere is too low. At the melting point, the lower p_{O2} limit (triple point "B" in Fig. 2.4) is in the same order of magnitude like rest oxygen impurities in commercial "inert gases" such as N_2 or Ar (see Section 2.5). This means some extra oxygen source must be added to the growth atmosphere. If p_{O2} is very high (the upper limit of Fig. 2.4 corresponds to pure oxygen, p_{O2} = 1 bar), no problem occurs and Ga_2O_3 can be molten without significant decomposition [11]. A detailed assessment of the system was performed by Zinkevich [12].

Metallic gallium has the largest liquid range of all chemical elements, ranging from 303 K (melting, almost at the low T limit of Fig. 2.4) to 2479 K (evaporation under a total pressure of 1 bar). Consequently, Ga(liq) is the only stable phase of gallium.

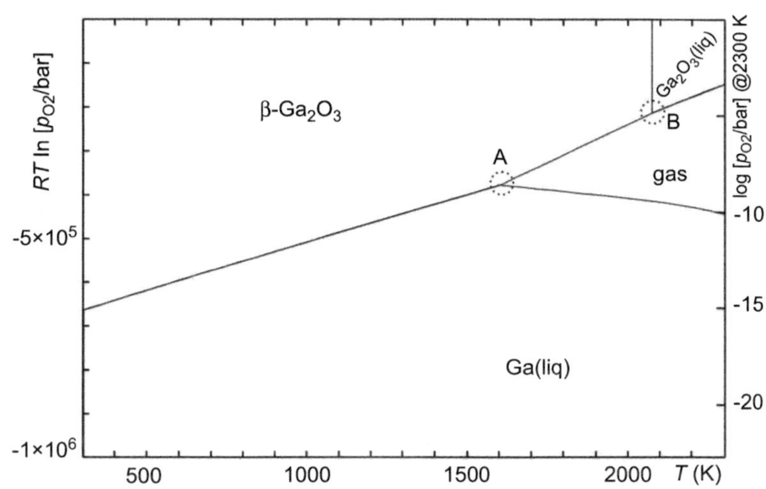

Figure 2.4 Predominance diagram for the Ga–O_2 system. Gas phase fugacities (in bar) at point A: Ga_2O: 0.99, Ga: 1.0×10^{-3}, GaO: 8.1×10^{-8}, O: 1.0×10^{-11}, O_2: 4.7×10^{-13}. B: Ga_2O: 0.99, Ga: 4.9×10^{-3}, GaO: 1.3×10^{-4}, O_2: 4.1×10^{-6}, O: 2.3×10^{-6}.

2.3.3 The In–O_2 System

The system is similar to Ga–O_2, but the lower phase boundaries of the In_2O_3(sol/liq) phase fields are shifted significantly to higher p_{O2}, which is a technological challenge. Already at the triple point

"C," one has an oxygen fugacity of 5.4 × 10^{-2} bar, which requires under ambient total pressure at least 5.4% oxygen in the growth atmosphere if growth shall be performed from the melt. To stabilize the melt and reduce evaporation of In_2O, an even higher oxygen concentration is desirable. The melting point of In_2O_3 was measured recently at 2223 ± 30 K [13], which is slightly higher than 2183 K that was reported previously [14]. Metallic indium melts at 430 K.

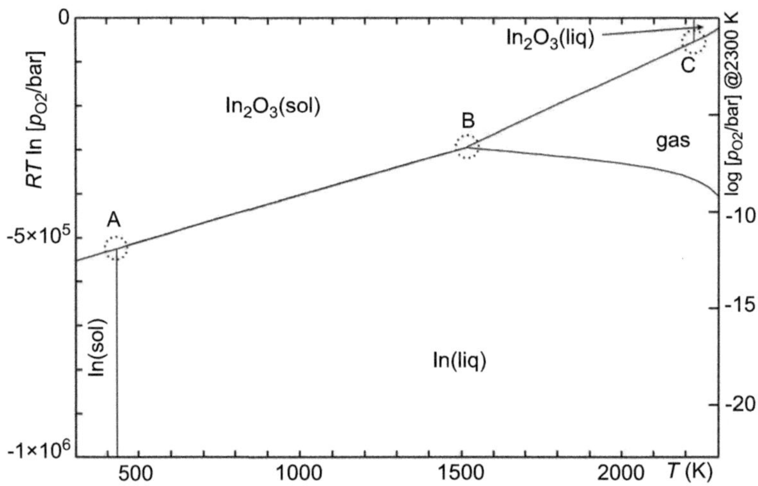

Figure 2.5 Predominance diagram for the In–O_2 system. Gas phase fugacities (in bar) at point A: In_2O: 1.4 × 10^{-20}, In: 2.1 × 10^{-24}, O: 7.1 × 10^{-60}, O_2: 1.2 × 10^{-64}. B: In_2O: 0.998, In: 1.5 × 10^{-3}, O_2: 7.2 × 10^{-11}, O: 4.3 × 10^{-11}, InO: 5.9 × 10^{-16}. C: In_2O: 0.92, O_2: 5.4 × 10^{-2}, In: 1.1 × 10^{-2}, O: 7.2 × 10^{-4}, InO: 3.1 × 10^{-9}.

2.3.4 The Sn–O_2 System

Figure 2.6 gives only an overview of the system, which is valid especially for high temperatures where SnO_2 is the only stable oxide phase. A thermodynamic assessment that takes into account especially the low temperature regime reports additionally tin(II) oxide SnO as stable phase below 543 ± 20 K, where it decomposes in a peritectoid reaction to Sn^+ another oxide Sn_3O_4, which is expected to decompose peritectically to SnO_2 and liquid tin near 723 K [15]. Sn_2O_3 as another possible intermediate phase is discussed elsewhere [16], but generally thermodynamic data about these tin oxides are scarce and partially contradictory. This is very unfortunate, because the tin(II) compounds SnO and especially $K_2Sn_2O_3$ could be prospective *p*-type TSO. SnO and its polymers are also major species in the gas phase (Fig. 2.6). It is surprising that even for SnO_2, wrong data prevail in the literature and in databases, and often a melting point around 1900 K is reported for this substance [5, 17, 18], which is obviously wrong. The upper panel in Fig. 2.6 shows DSC/TG curves that were measured by heating a pressed powder sample of pure SnO_2 in air [19]. Around 1600 K, the sample begins to evaporate, and this mass loss amounts to ca. 7% during the whole measurement. It is obviously the heat of evaporation that leads to the endothermal DSC effect there, which could be misinterpreted as the beginning of a melting peak. Optical inspection of the sample after this measurement revealed, however, that it was not molten. This agrees with bulk crystal growth experiments where melting below 2000°C could be excluded [20].

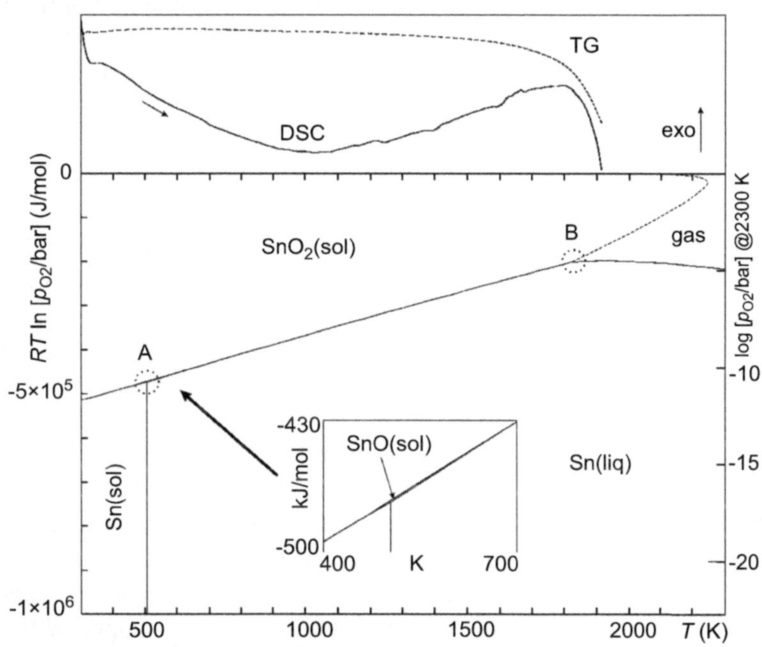

Figure 2.6 Bottom: Predominance diagram for the Sn–O_2 system. Gas phase fugacities (in bar) at point A: SnO: 2.3 × 10^{-23}, Sn_2O_2: 9.7 × 10^{-24}, Sn_4O_4: 6.9 × 10^{-24}, Sn_3O_3: 2.6 × 10^{-44}, Sn: 2.9 × 10^{-26}, O: 6.4 × 10^{-48}. B: Sn_2O_2: 0.430, SnO: 0.362, Sn_3O_3: 0.109, Sn_4O_4: 9.7 × 10^{-2}, Sn: 7.6 × 10^{-4}, O_2: 1.7 × 10^{-6}. Top: Simultaneous thermogravimetry (TG) and differential scanning calorimetry (DSC) of 82 mg SnO_2 in air up to T_{max} = 1913 K.

In the Ellingham-type predominance diagram of Fig. 2.6, calculated with FactSage data [5], the corresponding SnO_2(liq) phase field is missing due to incorrect melting point data of SnO_2 (1903 K), that was removed for calculations. This might evoke some uncertainty of the sublimation curve, which is dashed for this reason. A minor SnO(sol) phase field spans between SnO_2(sol) and Sn(sol/liq), which can hardly be seen in the scaling of Fig. 2.6 (insert). According to this phase diagram, SnO(sol) decomposes peritectically at ≈650 K to SnO_2(sol) + Sn(liq).

2.3.5 The Sb–O_2 System

Oxides of antimony are so far not very important as a pure TSO compound, but in concentrations of several percent, antimony is relevant especially as dopant for SnO_2 (antimony tin oxide, ATO) [21]. The affinity of antimony toward oxygen is smaller, compared to the metals that were described so far. This means that already at comparably mildly reducing conditions, they can be reduced even to the elemental metal. Sb and its oxides can exist in several crystal structures. For Sb itself, a rhombohedral structure is stable, which melts at 904 K and evaporates around 2000 K, depending on the pressure and composition of the atmosphere.

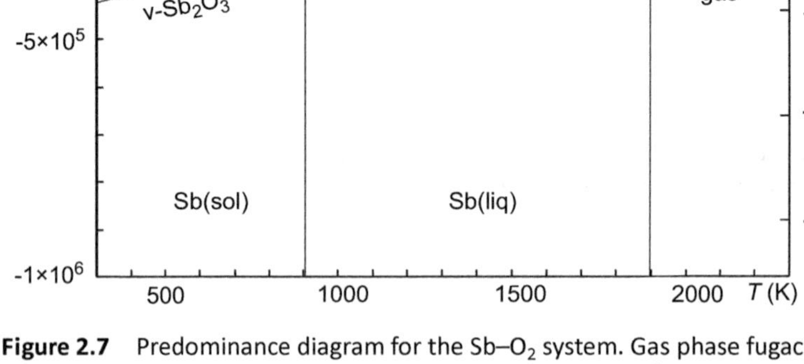

Figure 2.7 Predominance diagram for the Sb–O_2 system. Gas phase fugacities (main species, in bar) at point A: Sb_4O_6: 0.152, Sb_4: 1.1×10^{-4}, SbO: 6.7×10^{-11}, O_2: 1.4×10^{-19}. B: Sb_4O_6: 0.182, SbO: 2.2×10^{-12}, O_2: 1.4×10^{-13}. C: Sb_4O_6: 0.542, SbO: 2.4×10^{-11}, O_2: 2.5×10^{-12}. D: Sb_4O_6: 0.974, O_2: 2.6×10^{-2}, SbO: 2.7×10^{-6}.

A detailed report on the Sb–O system was published recently [22], and according to this paper, Sb_2O_3, Sb_4O_6, and Sb_6O_{13} are equilibrium phases in the system, whereas antimony pentoxide occurs always hydrated as $Sb_2O_5 \cdot xH_2O$. In contrast, Fig. 2.7 has Sb_2O_5 as the oxide of highest valence. Sb_2O_4 is not antimony(IV) oxide [23], because it contains only Sb^{3+} and Sb^{5+}. Both Sb_2O_4 and Sb_2O_3 show polymorphism, and their volatility is even higher than the volatility of the metal.

2.3.6 The Cu–O_2 System

Cuprite Cu_2O is one of the first oxides where semiconducting properties could be observed, and it belongs to the small group of *p*-type TSOs [24]. Moreover, copper(I) oxide is an important component of some delafossites such as $CuAlO_2$, which are also *p*-type semiconductors. Copper belongs to the first subgroup of the periodic system (Cu–Ag–Au), and all these metals are comparably noble (increasingly in this order, see Table 2.1). Accordingly, Cu_2O loses its oxygen easily, and already heating in a virtually "oxygen-free" atmosphere such as technical Ar or N_2 is sufficient to convert it to metallic copper. This can be dangerous if metallic crucibles such as Pt are used, because Cu can alloy and consequently destroy the crucible (see Section 2.4).

Cupric oxide CuO is so reactive that it can be used as oxygen source for oxidation reactions especially in organic chemistry, and also upon heating to ca. 1100 K, it loses oxygen quickly and converts to Cu_2O. It is practically impossible to melt CuO congruently: A recent thermodynamic assessment of the Cu–O system showed that congruent melting would be possible only under $p_{O2} \gg 100$ MPa [25, 26]. The intermediate Cu_4O_3 is not included in this assessment; it seems questionable if this compound is an equilibrium phase at all [27]. The switching of copper ion charge states (+1, +2, possibly +3) in $YBa_2Cu_3O_{7-\delta}$ high-T superconductors can be controlled by the oxygen fugacity in the atmosphere and plays a significant role in their superconducting properties [28, 29].

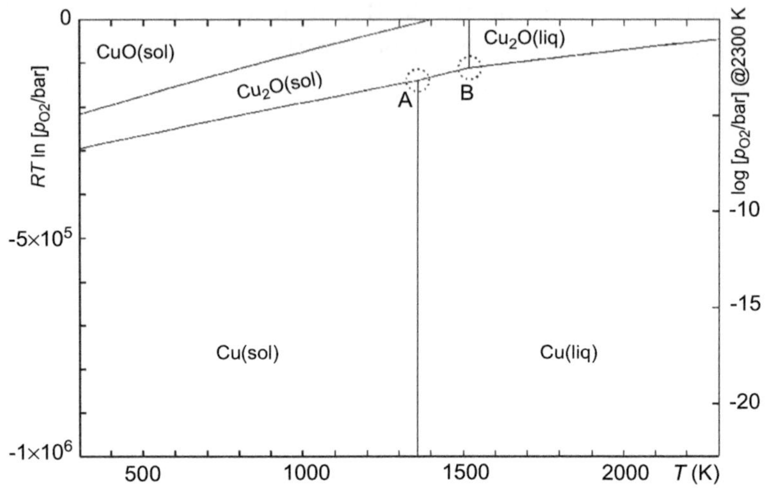

Figure 2.8 Predominance diagram for the Cu–O_2 system. Gas phase fugacities (in bar) at point A: O_2: 4.0×10^{-6}, Cu: 5.8×10^{-7}, O: 9.6×10^{-10}, CuO: 3.1×10^{-10}. B: O_2: 1.4×10^{-4}, Cu: 1.1×10^{-5}, O: 6.0×10^{-8}, CuO: 2.5×10^{-8}.

2.3.7 The Zn–O_2 System

The predominance diagram in Fig. 2.9 shows only ZnO as single oxide phase. Other oxides such as Zn_2O_3 or ZnO_2 are reported as nanostructures or are possibly metastable below 500 K [30, 31] and are discarded here. The chemical affinity of Zn is comparable to Ga (Fig. 2.4), but in contrast to gallium, zinc metal evaporates easily at 1081 K under ambient pressure. Figure 2.9 shows at point B a higher boiling point of 1254 K because this diagram is calculated for p = 2 bar, in contrast to all previous diagrams that were calculated for 1 bar. This was done to allow the formation of ZnO(liq), which has a vapor pressure of 1.06 bar at its melting point (2248 K) [5]. This requires mild overpressure to stabilize liquid ZnO but, nevertheless, is observed also at ambient pressure after quick heating [32]. Under suitable oxygen-rich conditions, ZnO crystals can be grown from the melt [33, 34] or by gas phase transport [35, 36].

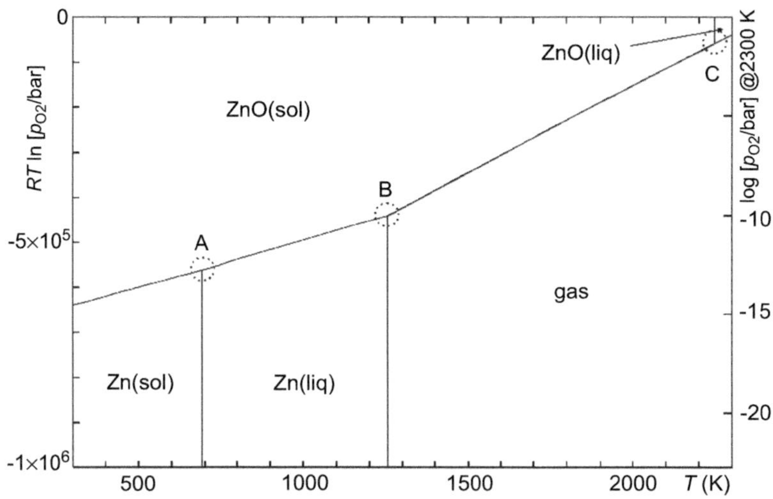

Figure 2.9 Predominance diagram for the Zn–O_2 system (total pressure $p = 2$ bar). Gas phase fugacities (in bar) at point A: Zn: 2.1×10^{-4}, O: 1.4×10^{-37}. B: Zn: 2.0, O: 4.7×10^{-17}. C: Zn: 1.95, O_2: 0.05.

2.4 Crucible Materials

Generally, it is desirable to use very noble crucible materials for the growth of crystals that require a high oxygen fugacity to maintain their stability. With the exception of Al_2O_3 (see Section 2.3.1), all the TSO compounds that were mentioned in the previous sections tend to lose oxygen under the formation of the corresponding metal or of oxides with lower valence (e.g., Ga_2O(gas), see Fig. 2.4). Platinum is a very popular crucible material, because it combines a high melting point with very good chemical resistivity. It is used as a crucible material for thermal analysis as well as crystal growth up to almost 1900 K [37, 38]. An Ellingham-type predominance diagram in the same scaling as used in Section 2.3 would show only Pt(sol) and Pt(liq) phase fields because the oxide PtO_2(sol) is never formed. With the formation reaction (Eq. 2.6) for this compound, one calculates a $\Delta G(T) > +160$ kJ/mol, with a minimum (160.9 kJ/mol) near 1500 K; the always positive ΔG prevents platinum from being oxidized in the bulk. Over a wide temperature range, gold is even more stable than platinum, and consequently Au would be an almost ideal crucible material if its maximum load temperature would not be so low. It should not be forgotten, however, that irrespective of the high resistivity of Pt and Au against O_2, some gaseous metal or oxide can be formed, leading especially to Pt loss if crucibles are used for a longer time at high *T*. Figure 2.10 shows that the fugacity of PtO_2 over Pt approaches 10^{-4} bar if one heats the crucibles in pure oxygen close to their load limit (gray area). The applicability of gold crucibles is very limited due to the low stability temperature threshold. Nevertheless, Au crucibles were, e.g., used for the growth of cuprate superconductors [39].

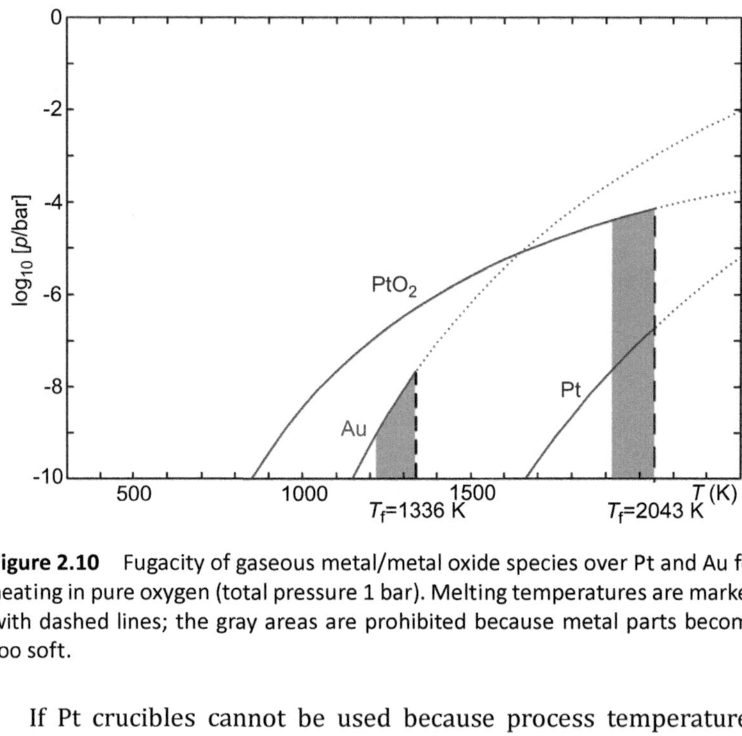

Figure 2.10 Fugacity of gaseous metal/metal oxide species over Pt and Au for heating in pure oxygen (total pressure 1 bar). Melting temperatures are marked with dashed lines; the gray areas are prohibited because metal parts become too soft.

If Pt crucibles cannot be used because process temperatures are too high, very often Ir crucibles are used. Figure 2.11 compares the stability of Ir with that of two other alternatives, Re and W. (Molybdenum is, from the chemical point of view, very similar to W but melts significantly lower at $T_f = 2896$ K. This means it combines high chemical sensitivity with a melting point just 180 K above Ir. The main advantage is just the lower price.) For the sake of simplicity, Fig. 2.11 just shows the upper phase boundary of the Me(sol) field for the three metals. Above this limit, the corresponding metal is destroyed either by the formation of solid or gaseous oxides.

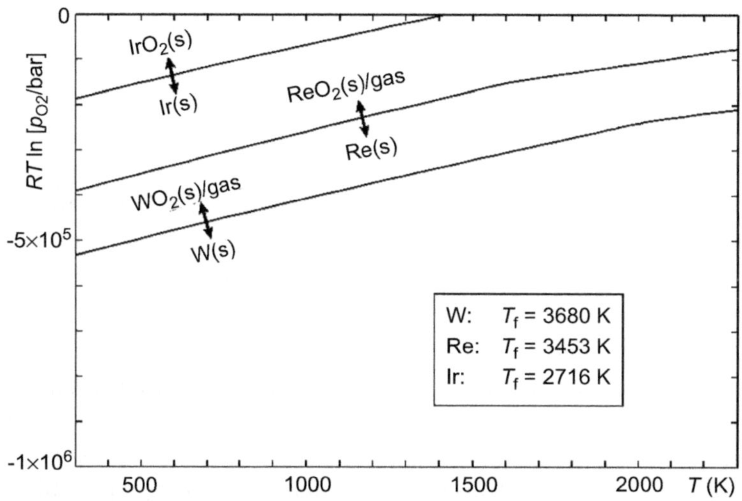

Figure 2.11 The stability of tungsten, rhenium, and iridium against oxidation grows in this order; but in the same order, their thermal stability (expressed by the melting point T_f) drops.

From Fig. 2.11, it is obvious that W and Re, the metals with highest T_f, can be used only in an atmosphere that is almost free of oxygen. Tungsten was successfully used, for e.g., for growth and thermal analysis of sapphire and some perovskites [40–42]. It should be noted that Fig. 2.3 (Al–O_2 system) is the only predominance diagram in Section 2.3 where the stability field of the oxide (Al_2O_3) is in agreement with the stability range of W(s) in Fig. 2.11. Rhenium is suitable even for the growth of rare-earth oxides with melting points near 2700 K [43].

Most TSOs have melting points beyond the stability limit of Pt, and then iridium crucibles are often the "working horse" for growing crystals from the melt [37]. The upper abscissa of Fig. 2.11 corresponds to pure O_2 under ambient pressure, which means that above ca. 1400 K, iridium crucibles should be stable even in a pure oxygen atmosphere. Often only virtually oxygen-free atmospheres such as Ar or N_2 are used together with Ir crucibles; sometimes a few percent O_2 is added to suppress the formation of suboxides such as Ga_2O [44]. It was shown recently, however, that Ir crucibles can be used under suitable process conditions in 100% O_2, which is in agreement with Fig. 2.11 [11].

Elemental metal crucibles that are used for growing oxide single crystals from the melt are listed in Table 2.2. Due to the requirement of significant oxygen concentration in a growth atmosphere during the growth of bulk TSO crystals from the melt, the crucibles made of Mo, Re, and W are excluded. As the TSOs have high melting point (above 1790°C for the compounds discussed in this book), the only elemental crucible that "accepts" some oxygen and withstands high temperatures is Ir. Pt cannot be used as a crucible material for growing the discussed TSO crystals directly from the melt, but it is used as a crucible material for growing TSOs from the flux, or as a lining material in autoclaves in the hydrothermal method. Au has too low melting point either for the melt growth or from the flux, but it is occasionally used as a lining material in the hydrothermal method. It should be noted that also alloyed crucibles are occasionally used, e.g. Pt–Rh or Pt–Ir, which have higher melting points than pure Pt and enable higher oxygen partial pressure, but they are more expensive, more difficult to recycle, and Rh usually contaminates a metal oxide melt, and thus a growing crystal.

Table 2.2 Metal crucibles used for growing oxide single crystals from the melt

Metal crucible	Melting point [°C]	Allowable oxygen	Applicable to TSOs
Au	1064	Yes	Yes[(a)]
Pt	1768	Yes	Yes[(b)]
Ir	2466	Yes[(c)]	Yes
Mo	2623	No	No
Re	3185	No	No
W	3422	No	No

[(a)]Only as lining material in autoclaves (hydrothermal method); [(b)]Only for TSOs with low melting point (<1500°C) or solution growth (flux (crucible) and hydrothermal (lining) methods); [(c)]Limited oxygen concentration (typically <2%)

2.5 Growth Atmosphere

Often "inert" gases such as N_2 or Ar are used for crystal growth. One can show that the term "inert" must be used with care. If one heats, e.g., iron(II) oxide FeO in argon gas with 5N (99.999%) purity above its melting point $T_f = 1644$ K, it is gradually oxidized to Fe_3O_4 [45]. This means that this gas cannot be considered inert under these experimental conditions. The explanation is given easily by a predominance diagram of the Fe–O_2 system where the FeO phase field is almost triangular with oxygen fugacities between $p_{O2} = 2.4 \times 10^{-6}$ bar ($T \approx 1700$ K) and 7.5×10^{-26} bar ($T \approx 850$ K). For higher p_{O2}, Fe_3O_4 is formed. A purity of 99.999% means that the gas contains 0.001% impurities, and typically air with 21% oxygen will be the main impurity. From this, one can conclude that a 5N gas has a constant background oxygen fugacity in the order of 2×10^{-6} bar.

Generally, the term "inert" should not be used simply for noble gases or other virtually oxygen-free atmospheres; rather, an atmosphere behaves inert for a given oxide material if the oxidation state of this oxide is not changed. This is the case if, during the corresponding process, which involves typically temperature changes under certain atmospheric conditions, the desired phase field of the oxide is not left. For the predominance diagrams in Section 2.3, with $RT \ln[p_{O2}/\text{bar}]$ as a coordinate, the top abscissa means $p_{O2} = 1$ bar. This top abscissa runs for oxides such as Al_2O_3, Ga_2O_3, In_2O_3 completely through the (solid or liquid) phase fields of these oxides, but not for the Sb–O_2 system (Fig. 2.7) where Sb_2O_5(sol) and two Sb_2O_4(sol) fields are touched, or for Cu–O_2 (Fig. 2.8) where CuO(sol) and Cu_2O(sol) are touched. This means that only for the former oxides from the third main group, pure oxygen behaves as an inert gas, whereas for the latter oxides, it leads to oxygen loss upon heating.

The Ellingham-type diagrams (with $RT \ln[p_{O2}/\text{bar}]$ as ordinate) have almost straight phase boundaries between subsequent oxidation levels. Alternatively, one could use $\log_{10}[p_{O2}/\text{bar}]$ as the coordinate. Then phase boundaries are bent, but the coordinate values correspond directly to the oxygen content in the atmosphere. Especially the constant oxygen fugacity, which is given by some "inert" gas with minor oxygen impurities (such as the 5N Ar mentioned above), is expressed by a horizontal line. From such $\log_{10}[p_{O2}/\text{bar}]$ (T) diagrams, it is immediately obvious that no gas mixture with

p_{O2} = constant is possible that maintains any of the oxidation levels FeO, Fe_3O_4, or Fe_2O_3 of iron stable over an extended T range [46]. For chromium, in contrast, Cr_2O_3 is stable at least up to 2800 K if the oxygen content of the atmosphere is ca. 1% or more [47].

Fortunately, not only for metals, but also for most nonmetals, the equilibria of the kind in Eq. 2.6 exist. Here the focus is especially on the equilibria of

$$CO_2 \rightleftarrows CO + \tfrac{1}{2}O_2 \tag{2.8}$$

$$CO \rightleftarrows C + \tfrac{1}{2}O_2 \tag{2.9}$$

which shift both to the product side with T. The temperature dependence $p_{O2}(T)$ for CO_2, CO, and for mixtures of both gases is similar to that of several metal oxides. Consequently, the desired oxidation states MeO_x can often be stabilized over an extended temperature range that is necessary, e.g., for crystal growth processes, if these processes are performed in a suitable CO_2/CO mixture. This was demonstrated with a mixture of 85% Ar/10% CO_2/5% CO with Fe^{2+} for wüstite FeO as well as olivines (Mg, Fe)$_2$SiO$_4$ [45]. The non-reactive carrier gas Ar was added in this case for security reasons, to keep the concentration of poisonous carbon monoxide as low as possible. CO_2/CO mixtures are very tolerant with respect to impurities resulting, e.g., from leaks, because they act like a "buffer." More examples are discussed elsewhere [47, 48].

Figure 2.12 gives an overview of which $RT \ln[p_{O2}]$ ranges are accessible by gas mixtures between pure CO_2 and pure CO. The diagram was calculated for a total pressure p = 1 bar. The curve of pure CO is, at the same time, the (Ellingham) phase boundary between C (graphite) and gas, which is obvious from Eq. 2.8. Especially mixtures containing only small amounts of CO_2 show negative slope at high T, because oxygen then occurs mainly as atoms rather than O_2 molecules, but only the latter contributes to the coordinate value. Besides, the growing T, which is a factor for the coordinate, contributes to the more negative coordinate value.

For practical reasons, it is sometimes more convenient to plot $\log[p_{O2}/\text{bar}]$ versus T, rather than $RT \ln[p_{O2}]$, because then the experimental parameter, i.e., the oxygen fugacity, can be read directly from the diagrams. However, straight lines from Ellingham-type predominance diagrams then become bent, as shown in Fig. 2.13 [49].

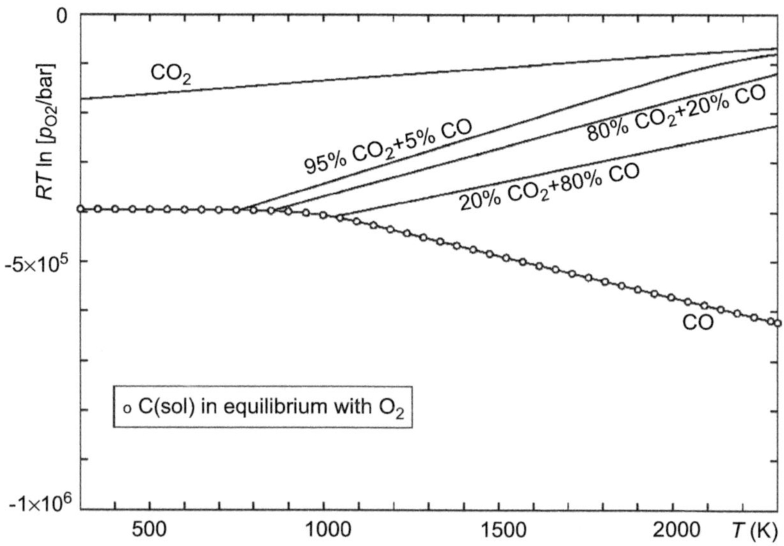

Figure 2.12 Carbon dioxide CO_2, carbon monoxide CO, and their mixtures can deliver T-dependent oxygen fugacities that agree partially with several Me–O_2 systems.

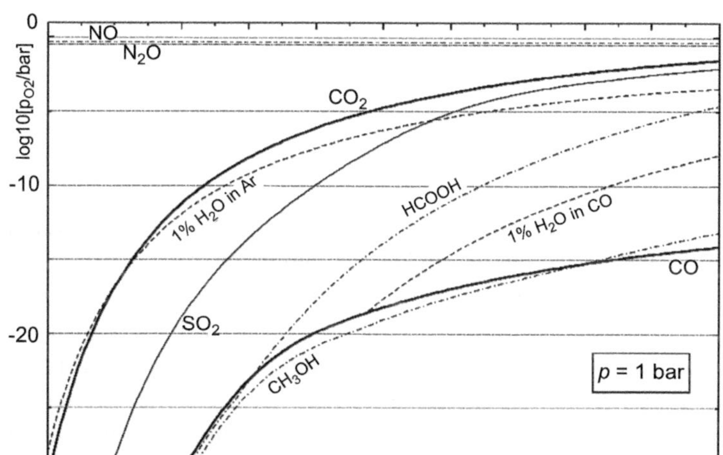

Figure 2.13 Oxygen fugacity p_{O2} of several gases and gas mixtures as a function of temperature.

There are not so many gases that can be used as an oxygen source in a manner described by Eqs. 2.7 and 2.8., and some options are demonstrated in Fig. 2.13. Carbon (with CO and CO_2) is the only useful element in the fourth main group with gaseous oxides. Nitrogen oxides (fifth main group) are endothermal compounds where the stability grows with *T*. At room temperature, they are under equilibrium conditions almost completely dissociated to N_2 and O_2, which is indicated by the double-dot dashed lines for NO and N_2O in the uppermost part of Fig. 2.13. (It is just a question of stoichiometry that NO delivers slightly more oxygen than N_2O.) SO_2 gives a $p_{O2}(T)$ function between CO_2 and CO, just a bit steeper. A similar function could be reached by CO_2/CO mixtures. It should be noted that nitrogen oxides and sulfur oxides are anhydrides of aggressive and toxic acids and consequently they should be handled with care! Moreover, it could happen that the "oxygen carriers" N or S themselves are incorporated in the crystals to be grown. Nitride N^{3-}, nitrate NO_3^-, nitrite NO_2^-, sulfide S^{2-}, sulfite SO_3^{2-}, and sulfate SO_4^{2-} are, together with suitable cations, partially quite stable anions. (Fortunately, most carbonates CO_3^{2-} of metals that are relevant for manipulation of oxidation state are comparably unstable. The same holds for most metal carbonyls with the [CO] group.)

Also, water releases oxygen at high temperatures via the reaction:

$$H_2O \rightleftarrows H_2 + \tfrac{1}{2}O_2 \qquad (2.10)$$

For silicon, for instance, the "wet oxidation process" that produces insulating SiO_2 layers is an industrial standard procedure [50] that can also be used to create defined oxygen fugacities for the stabilization of other oxides. The water is typically added by bubbling a carrier gas through water kept at some constant temperature. A higher bubbler temperature results in a higher water concentration in the wet gas. The upper dashed line in Fig. 2.13 (1% H_2O in Ar) shows that a carrier that is inert against water results in high p_{O2} levels, and the slope is similar to CO_2. With CO-rich carrier gases (lower dashed line: 1% H_2O in CO), the behavior is first dominated by CO, because H_2O does not dissociate sufficiently at low *T*. At some intermediate temperatures, an upward bend occurs.

Briefly, also oxygen-carrying organics shall be mentioned here, e.g., alcohols, aldehydes, and aliphatic acids. In this order, the oxygen content of these compounds becomes larger, and indeed for methanol CH_3OH (the simplest alcohol), the calculated $p_{O2}(T)$ function is lower than for formic acid HCOOH (the simplest aliphatic acid). (The simplest aldehyde HCHO is not shown in the figure as it is similar to CH_3OH.) However, these compounds should be handled with great care because many of them are poisonous, aggressive, or explosive in contact with air or oxygen.

2.6 Summary

Most crystal growth processes that are performed under not-too-low total pressure (typically above a few millibars) and at sufficiently high temperature (typically exceeding several hundred degrees centigrade) can be described at least approximately by equilibrium thermodynamics. For oxides, this means that every oxidation level from zero (the metal) to a maximum of eight (e.g., for OsO_4) has a limited (stability) phase field in the coordinates temperature T and oxygen fugacity (partial pressure) p_{O2}. If T is too high, and/or p_{O2} is too low, an oxide with lower valence (or the metal itself) is formed. If T is too low, and/or p_{O2} is too high, an oxide with higher valence is formed, if it exists at all.

This is uncritical if the corresponding oxide has a very wide stability field, which is the case, e.g., for Al_2O_3 (Fig. 2.3). In this case, it is impossible to obtain further oxidation, and reduction to the metal is only possible under extremely reducing conditions. (This is the reason why metallic aluminum was invented quite late in 1825, and why it is still produced by electrolytic reduction from its ores.)

Most of other oxides that are relevant as TSOs have substantially smaller stability fields. This means in phase diagrams of $\log[p_{O2}]$ versus T or $RT\ln[p_{O2}]$ versus T, respectively, they are stable only within a narrow corridor. An analogous behavior is shown by the metals that are used as crucible material for the growth of bulk TSO crystals from the liquid phase; here the only difference is that the metal itself must be stabilized at a sufficiently low p_{O2}. During the crystal growth process, crucible as well as oxide material must be kept stable and in equilibrium, which narrows the experimental corridor even more.

Often these requirements can be fulfilled by "reactive atmospheres"—these are gases or gas mixtures that supply an oxygen fugacity rising in an appropriate way with temperature [46]. Depending on the desired oxide material, gas mixtures based on CO_2, CO, and partially H_2O can fulfill this task quite often. Partial dilution of these gases with Ar or N_2 is useful to lower the concentration of poisonous or explosive components such as CO. "Pure" Ar or N_2, however, should not be considered to be always "inert" gases: They always contain some finite background of oxygen fugacity, typically in the order of several 10^{-6} bar, which can result in unpredictable redox processes.

References

1. W. Thomson; *Earth Envron. Sci. Trans. Royal Soc. Edinburgh*, **21** (1857), 123–171.
2. J. M. W. Chase, J. L. Curnutt, J. J. R. Downey, R. A. McDonald, A. N. Syverud, E. A. Valenzuela; *J. Phys. Chem. Ref. Data* **11** (1982), 695–940.
3. B. S. Chandrasekhar, J. A. Rayne; *Phys. Rev.* **124** (1961), 1011–1044.
4. J. F. Weaver, G. B. Hoflund; *Chem. Mater.* **10** (1994), 1693–1699.
5. www.factsage.com; FactSage 7.1, GTT Technologies Herzogenrath (2017).
6. I. Barin; *Thermodynamic Data of Pure Substances*, VCH, Weinheim, 1993.
7. C. M. Rost, E. Sachet, T. Borman, A. Moballegh, E. C. Dickey, D. Hou, J. L. Jones, S. Curtarolo, J. P. Maria; *Nat. Commun.* **6** (2015), 8485.
8. H. S. C. O'Neill, A. Navrotsky; *Am. Mineral.* **68** (1983), 181–194.
9. A. D. Pelton; "Thermodynamics and phase diagrams of materials" in *Material Science and Technology*, edited by R. W. Cahn, P. Haasen, E. J. Kramer, VCH, Weinheim, 1991, pp. 1–73.
10. D. Klimm, S. Ganschow; *J. Cryst. Growth.* **275** (2005), e849–e854.
11. Z. Galazka, R. Uecker, D. Klimm, K. Irmscher, M. Naumann, M. Pietsch, A. Kwasniewski, R. Bertram, S. Ganschow, M. Bickermann; *ECS J. Solid State Sci. Technol.* **6** (2017), Q3007–Q3011.
12. M. Zinkevich, F. Aldinger; *J. Am. Ceram. Soc.* **87** (2004), 683–691.
13. Z. Galazka, R. Uecker, K. Irmscher, D. Schulz, D. Klimm, M. Albrecht, M. Pietsch, S. Ganschow, A. Kwasniewski, R. Fornari; *J. Cryst. Growth.* **362** (2013), 349–352.
14. K. Sasaki, P. Bohac, L. J. Gauckler; *J. Am. Ceram. Soc.* **76** (1993), 689–698.

15. Cahen, N. David, J. M. Fiorani, A. Maître, M. Vilasi; *Thermochim. Acta.* **403** (2003), 275–285.

16. P. Sarker, M. N. Huda; *Comput. Mater. Sci.* **111** (2016), 359–365.

17. V. J. Barczak, R. H. Insley; *J. Amer. Ceram. Soc.* **45** (1962), 144.

18. ACerS-NIST Phase Equilibria Diagrams, V. 3.4.0, entry 4441.

19. D. Klimm; Leibniz-Institut für Kristallzüchtung, Berlin, Germany (2009), unpublished data.

20. Z. Galazka, R. Uecker, D. Klimm, K. Irmscher, M. Pietsch, R. Schewski, M. Albrecht, A. Kwasniewski, S. Ganschow, D. Schulz, C. Guguschev, R. Bertram, M. Bickermann, R. Fornari; *Phys. Status Solidi* (*a*) **211** (2014), 66–73.

21. J. Rockenberger, U. Zum Felde, M. Tischer, L. Troger, M. Haase, H. Weller; *J. Chem. Phys.* **112** (2000), 4296.

22. N. A. Asryan, A. S. Alikhanyan, G. D. Nipan; *Inorg. Mater.* **40** (2004), 720–725.

23. J. Amador, E. Gutierrez Puebla, M. A. Monge, I. Rasines, C. Ruiz Valero; *Inorg. Chem.* **27** (1988), 1367–1370.

24. W. H. Brattain; *Rev. Mod. Phys.* **23** (1951), 203–212.

25. L. Schramm, G. Behr, W. Löser, K. Wetzig; *J. Phase Equilib. Diff.* **26** (2005), 605–612.

26. G. Behr, W. Löser, M.-O. Apostu, W. Gruner, M. Hücker, L. Schramm, D. Souptel, A. Teresiak, J. Werner; *Cryst. Res. Technol.* **40** (2005), 21–25.

27. L. Debbichi, M. C. Marco de Lucas, P. Krüger; *Mater. Chem. Phys.* **148** (2014), 293–298.

28. L.T. Shi, K. N. Tu; *Appl. Phys. Lett.* **55** (1989), 1351–1353.

29. H. Verweij, W. H. M. Bruggink, R. A. Steeman, E. Frikkee, R. B. Helmholdt; *Physica C.* **166** (1990), 372–384.

30. T. Marković, Z. Pavlović, E. Nad; *Werkstoffe und Korrosion.* **17** (1966), 957–964 (in German).

31. W. Chen, Y. H. Lu, M. Wang, L. Kroner, H. Paul, H.-J. Fecht, J. Bednarcik, K. Stahl, Z. L. Zhang, U. Wiedwald, U. Kaiser, P. Ziemann, T. Kikegawa, C. D. Wu, J. Z. Jiang; *J. Phys. Chem. C* **113** (2009), 1320–1324

32. O. Fritsch; *Ann. D. Phys.* **22** (1935), 375–401 (in German).

33. D. Klimm, S. Ganschow, D. Schulz, R. Fornari; *J. Cryst. Growth.* **310** (2008), 3009–3013.

34. J. Nause, B. Nemeth; *Semicond. Sci. Technol.* **20** (2005), S45–S48.

35. R. Helbig; *J. Cryst. Growth.* **15** (1972), 25–31 (in German).
36. D. C. Look, D. C. Reynolds, J. R. Sizelove, R. L. Jones, C. W. Litton, G. Cantwell, W. C. Harsch; *Solid State Commun.* **105** (1998), 399–401.
37. D. Klimm, R. Bertram, Z. Galazka, S. Ganschow, D. Schulz, R. Uecker; *Cryst. Res. Technol.* **47** (2012), 247–252.
38. S. Zerfoss, L. R. Johnson, O. Imber; *Phys. Rev.* **75** (1949), 320.
39. Y. K. Tao, H. C. Chen, L. Martini, J. Bechtold, Z. J. Huang, P. H. Hor; *J. Cryst. Growth.* **114** (1991), 279–282.
40. P. Arhipov, S. Tkachenko, I. Gerasymov, O. Sidletskiy, K. Hubenko, S. Vasyukov, N. Shiran, V. Baumer, P. Mateychenko, A. Fedorchenko, Y. Zorenko, Y. Zhydachevskii, K. Lebbou, M. Korjik; *J. Cryst. Growth* **430** (2015), 116–121.
41. D. Klimm; *J. Cryst. Growth.* **312** (2010), 730–733.
42. R. E. Fahey, A. J. Strauss, A. C. Anderson; *J. Cryst. Growth* **128** (1993), 672–679.
43. R. Peters, C. Kränkel, K. Petermann, G. Huber; *J. Cryst. Growth* **310** (2008), 1934–1938.
44. W. Piekarczyk, A. Pajaczkowska; *J. Cryst. Growth* **46** (1979), 483–486.
45. D. Klimm, S. Ganschow; *J. Cryst. Growth* **275** (2005), e849–e854.
46. D. Klimm, S. Ganschow, Z. Galazka, R. Bertram, D. Schulz, R. Uecker; "Reactive atmospheres for oxide crystal growth" in *Advances in Solid Oxide Fuel Cells and Electronic Ceramics*, edited by N. P. Bansal, M. Kusnezoff, K. Shimamura, John Wiley & Sons, Inc., Hoboken, NJ, USA, Chapter 16, 2016: pp. 157–167.
47. S. Ganschow, D. Schulz, D. Klimm, R. Bertram, R. Uecker; *Cryst. Res. Technol.* **45** (2010), 1219–1224.
48. D. Klimm, S. Ganschow, D. Schulz, R. Bertram, R. Uecker, P. Reiche, R. Fornari; *J. Cryst. Growth.* **311** (2009), 534–536.
49. D. Klimm; "Thermodynamic and kinetic aspects of crystal growth" in *Handbook of Solid-State Chemistry*, edited by R. Dronskowski, S. Kikkawa, A. Stein, Wiley-VCH, Weinheim, 2017.
50. B. E. Deal, A. S. Grove; *J. Appl. Phys.* **36** (1965), 3770–3778.

Chapter 3

Zinc Oxide (ZnO)

3.1 Introduction

The substance ZnO occurs in nature as the rare mineral zincite. Its crystals are typically needles with hexagonal cross section. The color varies mainly from yellow to red, which is the result of iron or manganese impurities. Zincite is sometimes formed also in zinc smelters, because liquid zinc (melting point 419.6°C) evaporates easily (boiling point 908°C under atmospheric pressure) and reacts with oxygen if air is introduced through leaks. Such "artificial zincite" needles are sometimes more than 5 cm long.

ZnO is one of the firstly discovered semiconductors that belongs to a class of materials called transparent semiconducting oxides (TSOs). Semiconducting properties of ZnO were revealed early, and zincite crystals were among the active materials for the first "crystal radios" in the early 20th century. Varistors (voltage-dependent resistors) were developed in the 1960s and are still in use. These are ceramic components made from polycrystalline ZnO (intrinsically n-type), which is doped with additives such as Bi_2O_3 (intrinsically p-type). This way diodes are formed between the n-type grains and the p-type interfaces, and the varistor becomes a short circuit above some critical threshold voltage, allowing to save sensitive electronic parts from overvoltage.

Transparent Semiconducting Oxides: Bulk Crystal Growth and Fundamental Properties
Zbigniew Galazka

ISBN 978-981-4800-94-5 (Hardcover), 978-1-003-04520-5 (eBook)
www.jennystanford.com

Certainly, ZnO is the TCO material with the largest annual world production, of about 10^5 tones [1]. The largest share of this huge amount is used for the vulcanization of rubber. Other significant quantities are used in medicine and cosmetics. ZnO is even added to animal food because it raises the growth rate of farm animals.

In addition to semiconducting properties (both *n*-type and *p*-type) and wide bandgap (3.3 eV), ZnO also shows piezoelectric [2–4], pyroelectric [5, 6], ferromagnetic [7–9], and scintillation [10, 11] properties. Such a variety of properties puts ZnO among other TSOs in the frontline in material science, growth technology, and device fabrication of different functionality. Indeed, following varistors [12], also other sensing, electronic, and optoelectronic devices were demonstrated, such as surface acoustic wave devices [13], pyroelectric sensors [14, 15], scintillators [16–18], diode devices [19–23], thin-film transistors [24, 25], field-effect transistors [26–29], UV photodetectors [30–32], and light-emitting devices [25–36].

ZnO single crystals have no significant technical relevance so far; nevertheless, they are widely used for basic research and device demonstration. Single crystals are grown with a large variety of techniques, including (i) growth from the gas phase either by direct evaporation and precipitation of ZnO, by evaporation of Zn and subsequent oxidation to ZnO (like mentioned above in zinc smelters), or by chemical transport; (ii) grown from melt solutions consisting of different oxides and/or alkaline hydroxides; (iii) growth from aqueous or ammonia solutions, often under hydrothermal conditions, and (iv) grown from the melt either by indirect heating in an iridium crucible or by direct heating in the cold crucible technique.

In parallel to bulk single crystals, ZnO was also intensively grown as thin films and nanostructures either as pure ZnO or doped with a wide spectrum of elements. For example, thin films were grown by a diversity of techniques, including sol-gel [37–41], spray pyrolysis [42–44], sputtering [45–50], pulsed laser deposition (PLD) [51–56], chemical vapor deposition (CVD) [57–59], metal-organic chemical vapor deposition (MOCVD) [60–63], molecular beam epitaxy (MBE) [64–67], and many others.

An intensive research on ZnO is granted by a number of review papers, book chapters, and books, e.g. by Triboulet [68], Triboulet and Perriére [69], Pearton et al. [70], Ehrentraut et al. [71], Özgür et al. [72], C. Klingshirn [1, 73], Klingshrin et al. [74], Janotti and Van de Walle [75], Rodnyi and Khodyuk [76], Klimm et al. [77], Teherani and Litton (editors) [78], Reynolds and Reynolds [79], and Feng (editor) [80].

3.2 Crystal Structure

The crystal structure is among all TCO structures most closely related to that of the important semiconductors silicon and gallium arsenide: Like these, it consists crystals of stacked tetrahedra (Fig. 3.1). For Si (diamond structure) and GaAs (sphalerite structure), however, the tetrahedra have a stacking sequence A – B – C – A – B – C, which results in a cubic symmetry having three equivalent axes. For ZnO, the larger radius difference of its constituents (Zn^{2+}: 74 pm; 124 pm [81]) in conjunction with larger ionicity leads the alternative stacking sequence A – B – A – B, which is called wurtzite structure, results in hexagonal symmetry with space group $P6_3mc$. The energetic difference between both stackings is rather small, and it can be regarded as curiosity that the eponym minerals for both structures, sphalerite (= zincblende) and wurtzite, are both chemically ZnS. As a result of the similar lattice energy, stacking faults are often found for sphalerite- as well as wurtzite-type crystals, and also for ZnO [82].

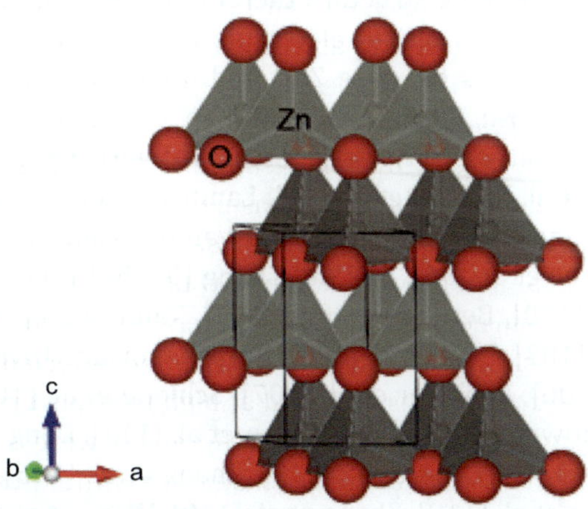

Figure 3.1 The wurtzite crystal structure of ZnO consists of A – B – A – B stacked tetrahedra.

From the lattice parameters a = 3.253(1) Å, c = 5.213(1) Å [83] follows an axis ratio c/a = 1.603, which represents a significant deviation from an ideal wurtzite structure. Such ideal wurtzite structure is built with perfect tetrahedra and has $c/a = (8/3)^{1/2}$ = 1.633. In wurtzite ZnO, the Zn–O bonds consequently have no identical length, because the tetrahedra are slightly compressed along the c-axis [0001] [84], and become disphenoids. The 6_3 screw axis is polar, and consequently ZnO is a piezoelectric and pyroelectric material.

Typical ZnO substrate orientations are "C-plane": (0001) or $(000\bar{1})$; "A-plane": $(11\bar{2}0)$; "M-plane": $(1\bar{1}00)$. Of these, only M-plane is not a polar face. Under high pressure p > 6 GPa ZnO transforms from wurtzite type (coordination number 4) to rocksalt type (coordination number 6) [85]. The stability of rocksalt-type ZnO can be extended to lower pressures by Mg doping, which is possible up to <7% in the bulk [86], or by epitaxial growth of $Zn_{1-x}Mg_xO$ epilayers [87].

3.3 Band Structure and Native Point Defects

3.3.1 Band Structure

ZnO is one of the most investigated TSO materials, including study of the electronic structure thereof. Intensive theoretical and experimental studies of the electronic structure are reflected in a number of reports for pure ZnO and doped with a diversity of elements (H, Li, Na, K, Au, Ag, Cu, Mn, Fe, Co, Ni, Mg, B, Al, Ga, Sc, Ti, V, Cr, Sr, Si, Sn, Zr, N, P, As, La, Ce, Pr, Nd, Eu), e.g., by Schröer et al. [88], Mi et al. [89], Ni et al. [90], Lambrecht et al. [91], Chien et al. [92], Hu and Pan [93], Janotti and Van de Walle [75, 94], Lyons et al. [95], Lee et al. [96], Lee and Chang [97, 98] Erhart et al. [99], Duan et al. [100], Berrezoug et al. [101], Tamura and Ozaki [102], Xiong et al. [103], Petit et al. [104], Karazhanov and Ulyashin [105], Guo et al. [106], Raebiger et al. [107], Schleife et al. [108], Dixit et al. [109], Chowdhury et al. [110], Han et al. [111], Deng et al. [112], Qu et al. [113, 114], Liu et al. [115], Zhang et al. [116], Sharma et al. [117], Lee et al. [118], Zhang et al. [119], Wang et al. [120], Peng et al. [121], Lorke et al. [122], Chen and Wu [123], Mahmood et al.

[124], Ataide et al. [125], Hammi et al. [126], Bashyal et al. [127]. The dopants aim to affect structural, electrical, optical, and magnetic (ferromagnetism) properties. Experimental valence-band structure with the use of angle-resolved photoemission spectroscopy (ARPES) and X-ray photoemission spectroscopy (XPS) was reported, e.g., by Sawada et al. [128], Ozawa et al. [129, 130], Preston et al. [131], and King et al. [132].

The calculated band structure of ZnO is depicted in Fig. 3.2 (Janotti and Van de Walle [75], DFT, hybrid HSE). At the Γ point (momentum k = 0) of the Brillouin zone (BZ), there is a global minimum and maximum in the conduction and valence band (CB and VB), respectively, indicating a direct bandgap. The lowest part of the CB is formed by empty 4s states of Zn^{2+} or the antibonding sp^3 hybrid states, while the top part of the VB originates from the occupied 2p orbitals of O^{2-} or from the bonding sp^3 orbitals (Klingshrin [73]).

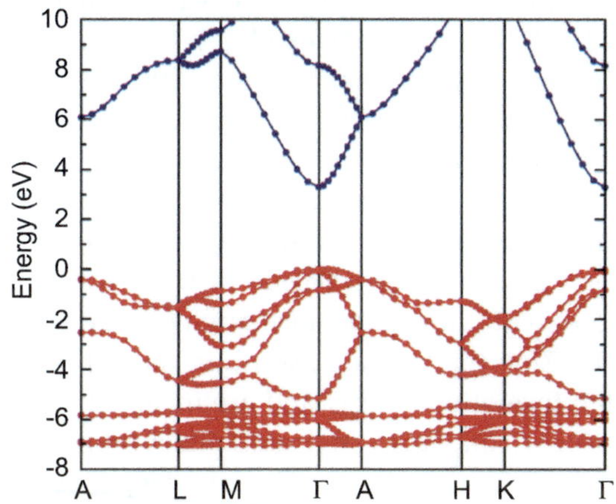

Figure 3.2 Calculated band structure of ZnO using the HSE hybrid functional. The energy of the valence-band maximum (VBM) was set to zero. Reprinted with permission from Ref. [95], Copyright 2009, American Physical Society.

Theoretical values of the bandgap spread in a wide spectrum depending on the calculation method, and often they are underestimated. Exemplary values of a theoretical direct bandgap (at Γ - Γ) of ZnO, which is close to experimental values, are 3.21 (DFT, hybrid HSE03 + GW) [108], 3.27 (DFT, GGA-PBE) [101], 3.4 (DFT + U, hybrid-HSE [127], and 3.49 (DFT, LDA+A-1/2) [125] eV.

A computed effective electron mass in ZnO was found to be almost isotropic, with a value of around $m_e*/m_0 = 0.28$ [73], 0.27 [120], and 0.30 [108], where m_0 is the rest electron mass. The effective hole mass was also found to be relatively isotropic, with a value of about $m_h*/m_{0h} = 0.59$ [73].

3.3.2 Native Point Defects

Native point defects in ZnO were also a subject of intensive theoretical studies, with some of them been experimentally confirmed. Examples of theoretical investigation of the native defects can be found in the works of Erhart et al. [99], Hu and Pan [93], Janotti and Van de Walle [75, 133, 134], and Sun and Wang [135].

Following Joanotti and Van de Walle [75, 133, 134], the native point defects in ZnO may include oxygen vacancies (V_O^0 and V_O^{2+}), oxygen interstitials (O_i^0 and O_i^{2-}), oxygen antisites (O_{Zn}), zinc interstitials (Zn_i^{2+}), zinc vacancies (V_{Zn}^{2-}), and zinc antisites (Zn_O). At early times of ZnO development, some of the native point defects, such as oxygen vacancies and zinc interstitials, were ascribed to *n*-type conductivity. However, oxygen vacancies (V_O) were found to be deep donors, have high formation energy, and can compensate *p*-type doping. Oxygen interstitials (O_i) can exist as electrically inactive interstitials or as deep acceptors. Oxygen antisites (O_{Zn}) are deep acceptors with the highest formation energy among all native point defects. Zinc interstitials (Zn_i) are, indeed, shallow donors but have high formation energy and are unstable as isolated point defects. Also zinc antisites (Zn_O) are shallow donors with a high formation energy. Zinc vacancies (V_{Zn}) are deep acceptors with low formation energy. They can compensate the *n*-type conductivity. Migration barriers of the native point defects span between 0.57 and 2.36 eV and can be annealed out at low and moderate temperatures of 219–909 K. As the result, the underlying *n*-type conductivity of ZnO originates rather from extrinsic defects, such as impurities and/or complexes of impurities and native point defects (for more details, see Section 3.4).

One particular extrinsic point defect that may induce the electrical conductivity in ZnO is hydrogen, as theoretically described, e.g., by Van de Walle [136, 137], Janotti and Van de Walle [75, 94, 133], Karazhanov and Ulyashin [105], Wardle et al. [138, 139], and Du and

Biswas [140]. The presence of hydrogen in any preparation process of ZnO is ubiquitous, and its incorporation cannot be avoided. Hydrogen was found to be a shallow donor when in an interstitial position (H_i^+) [75, 137, 140], substituting oxygen (H_O^+) [75, 94], as a molecule H_2 in the oxygen vacancy [140], or in complexes with 3d transition metals [139]. In particular, substitutional hydrogen (H_O^+) forms a multicenter configuration that is sensitive to the oxygen partial pressure [94].

3.4 Bulk Crystal Growth

ZnO, among other TSOs, has extensively been studied since several decades, together with enormous efforts undertaken to obtain bulk crystals for perspective applications.

The first attempts to obtain bulk ZnO crystals were performed by Fritsch in 1935 at the University of Erlangen [141]. Tiny ZnO crystals were obtained by sintering pressed ZnO powder into a solid body through which a current was next passed that caused partial melting of the sintered body. Next trials for bulk ZnO crystals were performed in the 1950s to 1960s of the twentieth century by using vapor phase and solution techniques. In 1953, Scharowsky [142] obtained ZnO crystals from the vapor phase. His method with modifications was followed by others for at least one decade. Walker [143], for the first time, mentioned the growth of bulk ZnO single crystals by the hydrothermal method, but without providing any details. Conditions for the hydrothermal growth were reported by Laudise and Ballman [144] in 1960. Flux-grown bulk ZnO crystals were reported for the first time by Nielsen and Dearborn [145] in 1960. The PVT-like method was studied for bulk ZnO crystals in 1972 by Helbig [146]. An important event in growing ZnO from the melt was the use of the cold crucible technique in 1998 by Agarwal et al. [147], and a few years later in 2006, the Bridgman method by Schulz et al. [148]. In the same year, large 3-inch-diameter ZnO crystals obtained by the hydrothermal method were reported by Ehrentraut et al. [71].

In summary, there was intensive activity in bulk ZnO growth, particularly in the 1950s and 1960s (Fig. 3.3), but due to difficulties in obtaining large enough crystals, failure in obtaining *p*-type conductivity, and possible domination of classical semiconductors, this activity declined for the next 25 years. New opportunities for TSOs at the turn of twentieth and twenty-first centuries restarted initiatives toward large ZnO crystals. At this time, melt-grown techniques and the hydrothermal method became leading technologies.

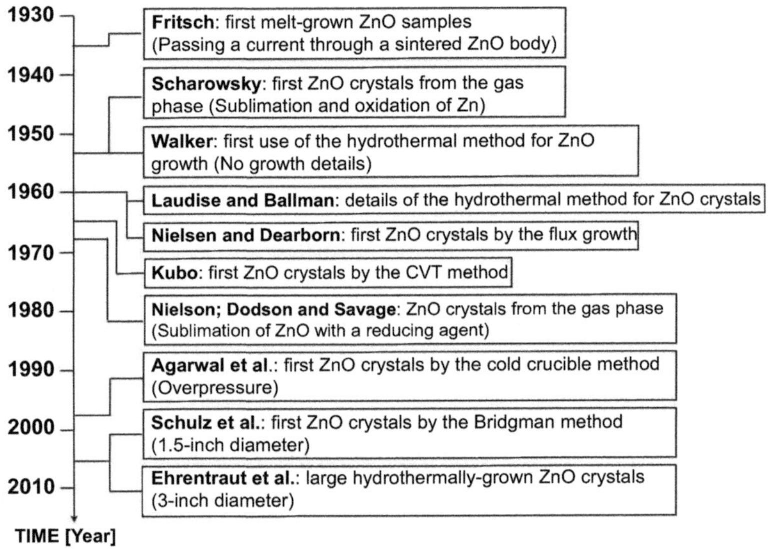

Figure 3.3 Main events in growth of bulk ZnO single crystals.

3.4.1 Growth from Gas Phase

Physical vapor transport (PVT) is a purely sublimation-controlled process in the presence of a non-reactive gas, while chemical vapor transport (CVT) is a reaction-controlled process in the presence of a reactive gas, i.e., the species is transported with a chemically active transport agent (reactive gas). Although the PVT and CVT methods can be differentiated from each other from the point of view of chemical reactions and mass transport, in the case of ZnO growth, the border between both methods is not fixed, as even a very small amount of reactive agents in a growth atmosphere may enhance the transport in the physical transport method. Reactive agents, such as O_2, H_2, and H_2O, are omnipresent, and residual traces thereof will always be present even in highly "purified" growth systems. This can be illustrated by partial pressures of gaseous Zn and O in the ZnO–Ar system (PVT method, Ar is chemically neutral) and ZnO–CO system (CVT method, CO is the reactive agent). As shown in Fig. 3.4, the same partial pressure of Zn requires much higher temperature in the PVT method as compared to the CVT method. For instance, Zn vapor partial pressure, p(Zn), of 10^{-2} atm can be obtained at about 800°C in the CVT method, while above 1500°C in the PVT method (see Fig. 3.4). However, in the PVT method, the partial pressures of Zn and O_2 are similar, while in the case of the CVT method, the partial pressure of O_2, $p(O_2)$, is much smaller than p(Zn), and it typically requires an additional oxygen source to obtain stoichiometric or near-stoichiometric ZnO crystals. In the example of the ZnO–CO system, Zn vapor is not transported by the chemically active transport agent as it is used to enhance sublimation only, not to transport Zn vapor. However, such systems are referred in the literature mainly to the CVT method.

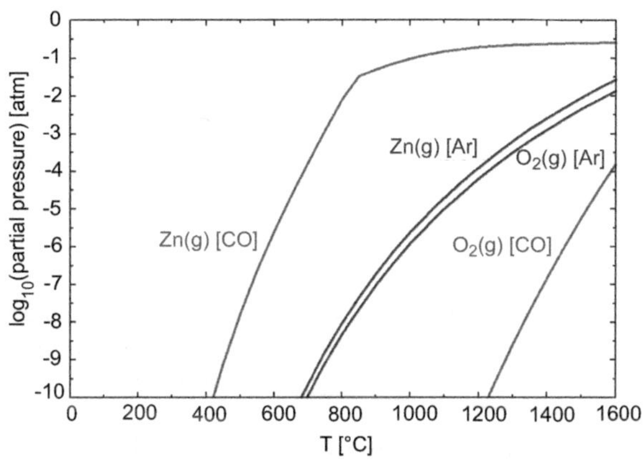

Figure 3.4 Partial pressure of gaseous Zn and O_2 as a function of temperature in the ZnO–Ar system simulating the PVT method and ZnO–CO system simulating the CVT method. For the calculations, the amount of ZnO, Ar, and CO was 1 mol and the total pressure was set to 1 bar.

The temperature and pressure for growth by the PVT method can be concluded from the p–T diagram of ZnO shown in Fig. 3.5. The PVT growth proceeds when the solid ZnO(s) and gaseous Zn(g) coexist. This is indicated by a phase boundary separating solid ZnO(s) and "gas" phase depending on the temperature. For example at 1600°C, p(Zn) is about 0.04 atm.

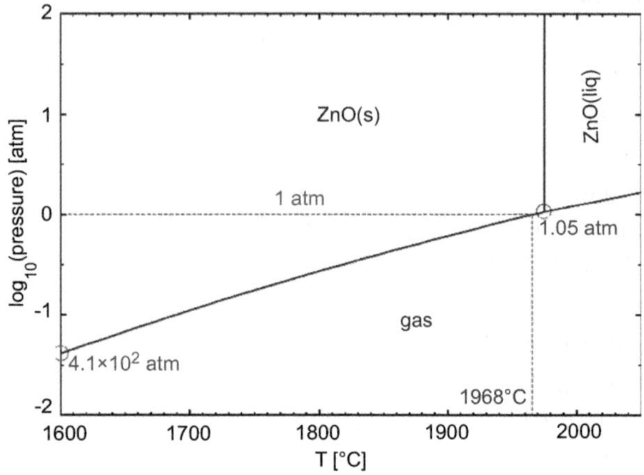

Figure 3.5 p–T equilibrium phase diagram of ZnO. PVT growth can be performed when a useful pressure of gaseous ZnO (>10^{-2} atm) is in equilibrium with solid ZnO.

As during ZnO growth from the gas phase, the border between the PVT and CVT methods is not clear, we will use different criteria to distinguish a diversity of growth concept, as listed below. Further, for PVT/CVT notation, we will use the original notation as it was used in the literature for a given concept.

A vapor of Zn-containing species to be oxidized to ZnO crystals is obtained and transported in a different way, which can be conducted by:

(i) Sublimating and oxidizing elemental Zn (can be referred to the PVT-like method);
(ii) Direct use of ZnX_2 (X = F, Cl, Br, I) as a source material and an oxidizing agent (CVT method);
(iii) Use of a ZnY solid (Y = Se, S) as a source material and an oxidizing agent;
(iv) Use of a ZnO solid as a source material and a chemically active transport agent, such as Cl_2, Br_2, I_2, HCl, NH_3, their mixtures, and other agents (CVT method);
(v) Use of ZnO solid as a source material, a reducing agent (such

as H_2, H_2O, C, CO, or their combinations) to produce a Zn vapor at lower temperatures as compared to the PVT method, and an oxidizing agent, which can be just a residual oxygen or water.

Each of these techniques requires a different furnace and/or ampoule design, which can be "open" or closed (better control of oxygen partial pressure), horizontal or vertical, with the use of different furnace/ampoule materials, gas compositions and their flow rates or partial pressures, total pressures, operating temperatures, and temperature gradients/differences.

3.4.1.1 Sublimating and oxidizing of elemental Zn

The melting point of metallic Zn is 420°C, and already slightly above that temperature, the partial pressure of Zn(g) is already high enough for an effective transport, for example at 600°C it is 0.015 atm. The concept of growing ZnO crystals simply by sublimation and oxidizing metallic Zn was experimentally carried out by Scharowsky [142]. He used a two-zone horizontal furnace consisting of evaporation and growth zones having different inner diameters (Fig. 3.6). The furnace was made of porcelain with a quartz tube extending through the evaporation zone toward the growth zone, where it is ended by a funnel acting as a surface for ZnO nucleation and growth. Inside the quartz tube in the evaporation zone, metallic Zn is placed, which is heated to 600°C, i.e., above its melting point. The growth is defined by a temperature of about 1150°C. Through the quartz tube, N_2 and H_2 flew at the rates of 1100 and 60 cm^3/min, respectively. A small amount of H_2 was needed to avoid oxidation of metallic Zn. For the oxidation of Zn(g) to ZnO(s), O_2 was supplied by air that diffused from the right-hand side into furnace II. Such growth conditions enabled to obtain needle-shaped ZnO crystals up to 4 cm long and below 1 mm in diameter with a hexagonal cross section, but also larger, leaf-shaped crystals. The crystals were generally colorless, some of them with a yellow tint. The crystals got more yellow after annealing in the Zn vapor at 800°C. If the partial pressure of Zn was too high, powdered ZnO instead of crystals was formed.

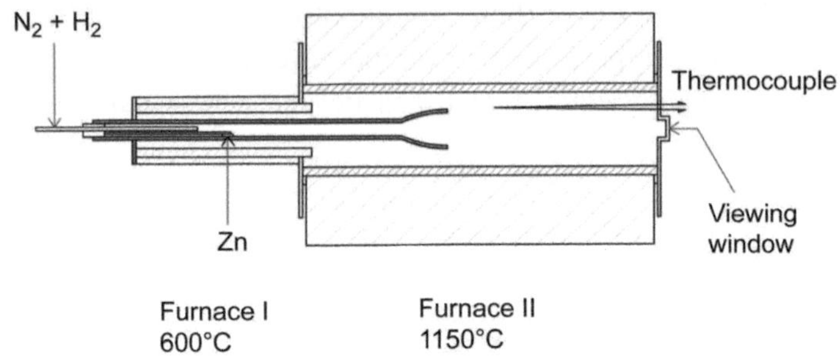

Figure 3.6 Furnace for growing ZnO crystals by evaporating elemental Zn. Redrawn from Ref. [142] with some modifications.

Another three-zone vertical furnace for growing ZnO crystals by evaporation and oxidation of elemental Zn was proposed by Bogner and Mollwo [149]. Metallic Zn was loaded into a quartz boat within a quartz tube in the evaporation zone heated to 720–760°C. Through that tube, N_2 and H_2 flew at the rates of 200 and 150 cm^3/min, respectively. The Zn vapor was transported upward through an intermediate zone to a growth zone held at 1330–1380°C. To the growth zone, O_2 was supplied with the flow rate of 5000 cm^3/min by a separate quartz tube. In these conditions, the rate of Zn evaporation was about 0.1 g/min. The growth duration was only 20–50 min. The obtained ZnO crystals were greenish hexagonal bars of 0.3–1 mm in diameter and 10–20 mm long or colorless plates of 0.3–0.5 mm wide, 10–20 mm long, and 0.05–0.15 mm thick, whether they deposited on porcelain material or alumina/quartz, respectively. The main identified impurities were Si, Al, Mg, and Cu at the concentration below 10^{-3} wt.%.

ZnO crystals were also doped during growth with Ag, Cu, and In. The dopants were loaded into quartz crucibles in the growth zone with a temperature profile covering a range of 800–1400°C. By hanging the crucibles, it was possible to place the dopant in a desired temperature. The dopants were evaporated at temperatures between 1000 and 1400°C. The obtained crystals were typically ribbons of very similar dimensions as undoped crystals, but high doping level of In changed the crystal shape to prisms. In the obtained ZnO crystals, no Ag could be detected, but Cu and In concentrations were ~10^{17}–10^{19} cm^{-3} and ~2×10^{18}–2×10^{20} cm^{-3}, respectively.

The In-doped crystals were light blue to dark blue with increasing In concentration.

3.4.1.2 Use of ZnX_2 as a source material and an oxidizing agent

In this technique, Zn is transported directly from the source to a growth zone by a compound with a general formula ZnX_2, where X is a halogen (F, Cl, Br, I). ZnX_2 transport is enhanced via a neutral transport gas, typically N_2 or Ar, to a growth zone (of higher temperature), where it reacts with the supplied oxygen to form ZnO crystals:

$$ZnX_2(g) + \tfrac{1}{2}O_2(g) \rightarrow ZnO(s) + X_2(g) \quad (3.1)$$

If H_2O is supplied instead of or in addition to O_2, there is also a reaction with H_2O:

$$ZnX_2(g) + H_2O(g) \rightarrow ZnO(s) + 2HX(g) \quad (3.2)$$

The effectiveness of the reactions 3.1 and 3.2 is defined by the equilibrium constant K_{eq}, which is shown in Fig. 3.7a,b as a function of temperature for different halogens.

We can, therefore, expect a very efficient reaction of ZnI_2 and $ZnBr_2$ with O_2 and of ZnF_2 with H_2O, as K_{eq} values reach very high values. In the case of $ZnCl_2$, its reaction with O_2 and H_2O is comparable. A decrease in K_{eq} with temperature establishes the mass transport from a cold to a hot zone, while its increase from the hot to the cold zone. Moreover, larger changes in K_{eq} lead to a more efficient mass transport. The use of different ZnX_2 source materials would require different oxidizing agents and different operating temperatures and parameters. These differences are, indeed, well reflected by experimental results.

Kubo [150] used ZnF_2 as the source material, which was loaded (10 g) into a Pt crucible resistively heated in a vertical furnace. Inside the furnace, there was a Pt pipe ended with a Pt plate located just above the crucible. The crucible with ZnF_2 was heated at 1100°C for 12 h. If dry nitrogen was used, no ZnO was obtained. However, when using wet N_2 (i.e., containing H_2O), yellowish, needle-shaped ZnO crystals were obtained on the Pt pipe above the plate at temperature of about 900°C. The largest crystals had the length of 10 mm and the diameter of 1 mm.

pit density (after etching in dilute HCl) was between 2.5 × 10^4 and 1.3 × 10^5 cm^{-2}. The impurity concentration (in wt. ppm) was as follows: Si = 5–50, Pb = 5–50, Cu = 5–50, and B = 0.3–3. In the growth setup (Fig. 3.9) according to Takahashi et al. [153], $ZnCl_2$ was held in a quartz vessel and heated to 530°C, corresponding to about 0.02 atm of saturated vapor pressure. In addition to $ZnCl_2$, also H_2O vapor was introduced to the growth zone through separate quartz tubes by carrier and reaction gases consisting of N_2 and O_2, respectively. The growth zone was kept at a temperature of 1300°C. Good results were obtained for the O_2 flow at the rates of 100–200 cm^3/min. The growth duration of 24 h produced plate-shaped ZnO crystals with (0001) orientation and a maximum size of 1 cm^2.

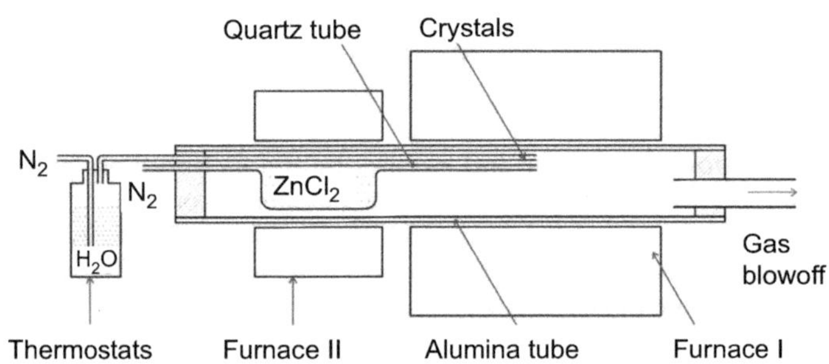

Figure 3.9 Schematic diagram of a furnace for growing ZnO single crystals by the CVT method with the use of $ZnCl_2$ as the source material. Redrawn from Ref. [153] with some modifications.

The use of ZnI_2 as the source material for growing ZnO crystals by the CVT method was studied extensively by Hirose et al. [154–157]. A two-zone muffle furnace was utilized, as shown in Fig. 3.10. It has an outer alumina/quartz tube (e.g., 50 mm diameter and 1000 mm long) containing therein two horizontal inner quartz tubs (40 and 20 mm in diameter) disposed at opposite ends of the outer tube and extending inward. Open ends of both inner quartz tubes face each other at a certain distance (about 15 mm) at the crystal growth zone. A powdered ZnI_2 starting material (20 g) is placed inside the smaller quartz tube (directly or in a combustion boat) in the vaporization zone that was heated to 420–430°C. Through the quartz tube with ZnI_2, a carrier gas of N_2 flew at the rate of 120 cm^3/min toward the growth zone, and through the opposite inner quartz tube at the opposite end of the furnace, reacting gases of either dry or wet O_2

or wet N_2 flew at the rate of 700 cm^3/min toward the growth zone, which was held at a temperature of 1150–1200°C. No growth was observed at temperatures below 1000°C. It came out that the wet N_2 was not suitable for the growth. ZnO crystals grew in the growth zone at the tip of the smaller quartz tube containing the ZnI_2 starting material.

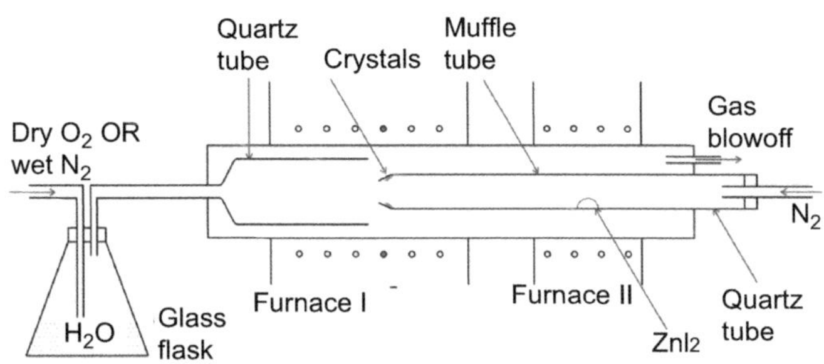

Figure 3.10 Schematic diagram of a furnace for growing ZnO crystals by the CVT method with the use of ZnI_2 as the source material. Redrawn from Ref. [154] with some modifications.

Adjusting temperature profiles in the growth zone enabled to define high temperature growth region (1150–1200°C) and low temperature growth region (970–1020°C) for oxidation and hydrolysis reactions, respectively. The crystals grew at the mouths of the opposite quartz tubes. At the high-temperature region, only growth with dry O_2 was observed, but not with wet N_2. At the low-temperature region, the situation was inverse, and the growth occurred only with wet N_2. At the high-temperature region, ZnO grew by oxidation defined by the reaction 3.1, while at the low-temperature region by hydrolysis defined by the reaction 3.2, likely aided by oxidation reaction, as some oxygen partial pressure will always exist in the carrier gas. Other experiments were conducted in parallel gas flow configuration to suppress gas collision from opposite ends of the inner tubes in the area of the growth zone. Here, there was one inner quartz tube extending from the vaporization to the growth zone. The carrier gas (N_2) flew through the inner quartz tube containing ZnI_2, while the reaction gas (O_2) was introduced to the outer tube parallel to the inner quartz tube, at the end of which O_2 and ZnI_2/N_2 flows met. The growth parameters were similar to those used in the configuration with counter flows. Generally, ZnO crystals obtained from the ZnI_2 source material had shape of facetted needles, ribbons, or plates, as schematically shown in Fig. 3.11.

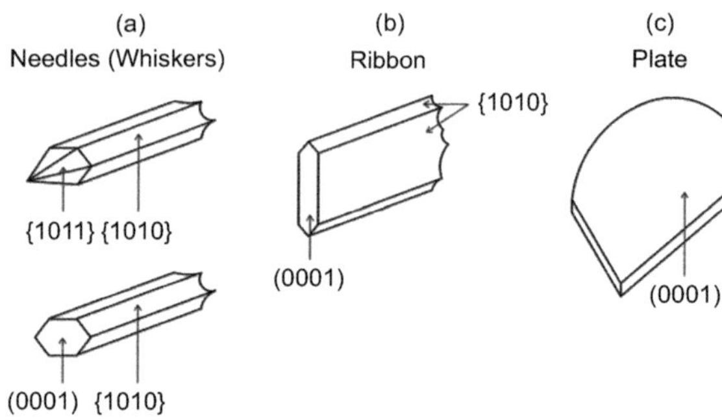

Figure 3.11 ZnO crystal shapes obtained by the CVT method from ZnI_2 as the starting material: needles or whiskers (a); ribbons (b); and plates (c). Redrawn from Ref. [156] with some modifications.

Thermodynamically, the growth of ZnO from $ZnBr_2$ as the source material should be similar to that of ZnI_2. Hirose and Furuya [158] used a similar furnace configuration with counter flows of carrier/reaction gases and growth conditions/operating parameters as for ZnI_2 and obtained very similar results in terms of the crystal shape and size, crystal perfection, growth habit, and crystal orientation. The flows of the carrier gas (N_2) and the reaction gas (O_2 or wet N_2) were 600 and about 700 cm^3/min, respectively, while $ZnBr_2$ was held in a combustion boat at temperature 450°C. The growth region was at the tips of both inner tubes having high- and low-temperature regions 1200–1250°C and 900–1000°C, respectively. Under O_2, ZnO crystals were obtained at the high-temperature region (20 h growth duration). They were yellow needles of 13 mm long and 1.3 mm diameter. Under wet N_2, ZnO crystals grew at the low-temperature region in the form of needles or rods. The growth rate was about 0.13 and 0.02 μ/s in the <0001> and <$000\bar{1}$> directions, respectively.

3.4.1.3 Use of ZnY as a source material and an oxidizing agent

The use of ZnY (Y = S, Se, and Te) as the source material for growing ZnO crystals was experimentally studied by Park and Reynolds [159]. ZnS or ZnSe powder is placed in a quartz boat, which in turn is located inside a quartz tube inserted into a mullite tube (Fig. 3.12). The quartz tube was about 50 mm in diameter, while the mullite tube was about 65 mm. During the heating up, only an argon flow was maintained at the rate of 4 cm^3/min, and when the starting material reached its sublimation temperature, oxygen flow was initiated at the rate of 0.8 cm^3/min. Such a procedure was to avoid the formation of an oxide layer on the source material during an early stage of heating. In the case of ZnS, the temperatures were 1375°C and 1300°C for the source material and growth zone, respectively. For ZnSe, the temperatures were about 100 K lower. The growth duration was 16 h.

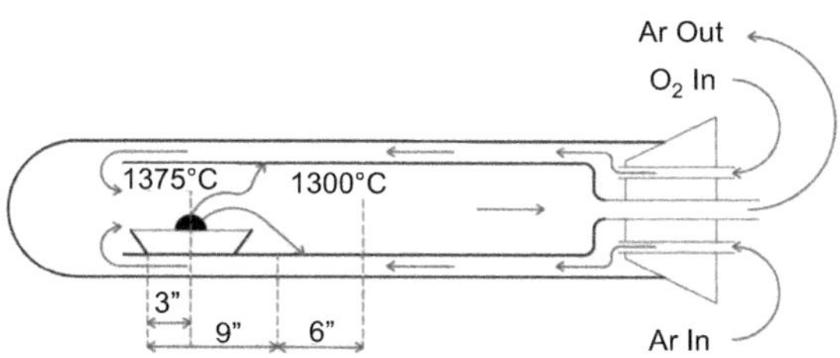

Figure 3.12 Furnace for growing ZnO crystals by the CVT method from ZnS/ZnSe as the source material. Here the temperature values are for the ZnS source material. Redrawn from Ref. [159] with some modifications.

From ZnS, prism-type ZnO crystals grew mainly in the quartz boat, while platelet-type ZnO crystals grew mainly in the quartz tube above the quartz boat and further downstream at lower temperatures. From ZnSe, all ZnO prisms and also some platelets occurred in the quartz boat and many smaller ZnO platelets in the quartz tube at lower temperatures. Dimensions of the largest crystals were 30 × 5 × 5 mm^3 and 10 × 6 × 1 mm^3 for the prisms and platelets, respectively. The ZnO prisms grown from ZnS had the hexagonal geometry and were bounded by the $(10\bar{1}0)$ and $(11\bar{2}0)$ planes. In the case of the plates, the *c*-axis was always perpendicular to the plane of the plates. On the other hand, the ZnO prisms grown from ZnSe were hollow and the plates had *c*-axis normal to the plane of the plates. The differences in the growth habit of the ZnO crystals from ZnS and ZnSe likely arise from a greater interaction of Se (Se has a higher boiling point than S) with a growth interface that alters the diffusion process. Growing ZnO crystals from ZnTe as the source material was not successful due to the high boiling point of Te (1390°C).

- **for Cl_2:** concentration of the chemically active transport agent = 2 – 4 × 10^{-5} mol/ml, temperature of the growth zone = 900–1000°C, and the temperature difference = 50°C.

The crystal growth habit was highly influenced by the transport in the case of Br_2, but less in the case of Cl_2. The obtained ZnO crystals were colorless and sometimes had a yellow–green tint. They were in the form of bars, prisms, or sometimes hexagonal plates. If the transport rate was higher than 30 mg/h, usually polycrystalline materials were obtained. The number of crystals obtained with Cl_2 was smaller than that obtained with Br_2, but they were 2–3 times larger. The crystal dimensions were 5 mm long and 2 mm^2 in cross section, and 5–15 mm long and a few mm^2 in cross section for Br_2 (growth duration 140–170 h) and Cl_2 (growth duration 250–350 h), respectively.

Widmer [161] used Br_2 as the chemically active transport agent within a sealed silica ampoule enabling a vacuum of 10^{-5} Torr. The ampoule consisted of two integrated ampoules one within other, wherein the source material was located within the inner ampoule having a capillary or conical section acting as a seeding section. To avoid a reaction of the source material with silica, the inner ampoule could be coated with alumina. At operating temperature of 990–1010°C, the obtained ZnO crystals were hexagonal plates of the size 4 × 3 × 0.5 mm^3.

Ziegler et al. [162], who performed ZnO growth experiments in closed quartz ampoules, pointed out Br_2 as an effective transport agent. With an optimum growth temperature between 900 and 1000°C, the obtained ZnO crystals were mainly rod shaped with a hexagonal cross section. The crystals were typically colorless and clear, but some of them were opaque as well.

Hydrogen halides (HCl, HBr) defined by reaction 3.4 are not effective chemically active transport agents according to several experimental works. Piekarczyk et al. [160] mentioned that experiments with HCl led to polycrystalline precipitates at similar growth conditions to those with Br_2 and Cl_2 that produced single crystals. Also, Ziegler et al. [162] and Shiloh and Gutman [163] stressed the lowest transport with HCl as compared with halogens or ammonium halides at similar growth temperatures. Reaction between ZnO and HBr is exothermic (ΔH = −13.3 kJ/mol at 900°C and 1 atm) and the transport proceeds from the cold to the hot zone. Ziegler et al. [162] performed growth experiments with HBr and extra elemental Zn in the ZnO source material. At growth temperature between 900 and 1000°C, the obtained ZnO crystals were orange to brown with an increase in the color brightness with growth temperature. In such a system, often Zn particles adhered to the crystals.

Better results were achieved with the use $HgCl_2$ as compared to HCl or HBr. This chemically active transport agent was experimentally tested by Shiloh and Gutman [163]. Growth experiments were performed in sealed quartz ampoules having an inner diameter of 11 mm and length of 150 mm. Before sealing the ampoules, ZnO was heated at 350°C in vacuum. The temperatures of the source material in the combustion zone and in the growth zone were 1295°C and 1000°C, respectively. Such high temperature difference (about 300 K) was achieved by separation of the combustion and growth zones with a channel for an effective vapor transport. The crystals grew on a ZnO seed fused to a quartz rod in the growth zone. ZnO crystals were yellowish in color and of 2 mm diameter and 8 mm length, wherein they were less colored when obtained with low initial pressure of the chemically active transport agent (i.e., growth at lower transport rate). The crystals grew typically along the *c*-axis.

Experimental studies of ammonium halides (NH_4Cl, NH_4Br, and NH_4I) as chemically active transport agents for growing ZnO crystals were conducted by Shiloh and Gutman [163], Ziegler et al. [162], and Matsumoto and Shimaoka [164]. The results of ZnO growth with NH_4X in terms of growth rate and crystal size were better as compared to the growth with the use of HX. Matsumoto and Shimaoka [164] used closed silica ampoules of 9.5 mm diameter and 90 mm length. The temperature of the source material was 1000°C, and the temperature difference between the combustion and growth zones was between 20 and 200 K. The transport rate was almost proportional to the temperature difference between the combustion and growth zones, and increased from 0.6 to 5 mg/cm^3 with the increase in the temperature difference from 20 to 200 K. The obtained crystals were typically rod shaped and of 1.5 mm in diameter and 8 mm in length or prismatic, if a small amount of the transport agent and large temperature difference were applied. The largest ZnO crystals were 1.5 mm in diameter and 8 mm in length. Some of the crystals were colorless, but others were brownish.

The crystals were brown if elemental Zn was used in addition to NH_4Cl as the transport agent. Ziegler et al. [162], who carried out growth experiments in closed quartz ampoules at 900–1000°C, noticed a relatively high transport rate when using NH_4Br and NH_4Cl as chemically active transport agents. The obtained crystals were typically rod shaped with orange to brown coloration when, additionally, an excess of elemental Zn was used in addition to ZnO as the source material.

Yet other combinations of the ZnO source material and chemically active transport agents can be used. For instance, Matsumoto et al. [165] used a $ZnCl_2$ transport agent and also elemental Zn in the powdered ZnO source material to increase the transport rate. The ZnO source material was first heated up in air at 600°C for 1 h and then elemental Zn was added. $ZnCl_2$ was added to the ZnO powder as a water solution, which was then dried at 100°C. The experiments were performed in a closed fused silica ampoule of 12.5 mm diameter and 180 mm length. The ampoule with the source material and the transport agent was evacuated to about 10^{-3} Pa at room temperature and then sealed off. The ampoule was inserted into a vertical Kanthal furnace in such a way that the source material was kept at the temperature of 1000°C while the growth zone at a temperature of 900°C. At such growth conditions, the transport rate was 10–20 mg/h. When using only $ZnCl_2$ as the transport agent, the obtained ZnO crystals were typically needle-shaped; however, after adding elemental Zn, prismatic and substantially colorless crystals of about 1.1 mm diameter and 4.5 mm length were obtained after 100 h growth duration. With a larger excess of elemental Zn, the crystals were yellow to orange.

3.4.1.5 Use of ZnO as a source material, a reducing agent, and an oxidizing agent

The PVT method requires relatively high operating temperatures (typically well above 1500°C) for thermal decomposition of ZnO enabling a high p(Zn) sufficient for physical transport. To increase the p(Zn) at lower temperatures, the ZnO decomposition is assisted by a chemical reaction with a reducing agent, and the Zn vapor re-oxidizes with an oxidizing agent in a growth zone of higher temperature to form ZnO crystals:

$$ZnO(s) + X(g, s) \leftrightharpoons Zn(g) + XO(g) \quad (3.7)$$

$$Zn(g) + \tfrac{1}{2}O_2(g) \leftrightharpoons ZnO(s) \quad (3.8)$$

where X is the reducing agent, which can be either in the gas (e.g., H_2, CO, CH_4) or solid phase (e.g., C).

Oxygen is typically supplied as O_2, H_2O, or CO_2.

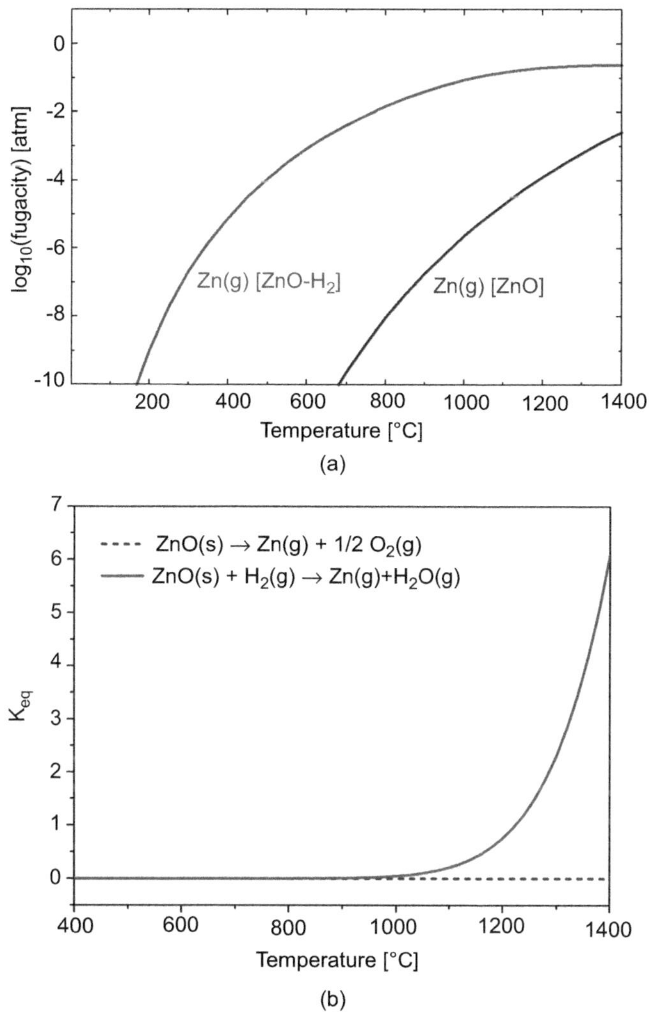

Figure 3.14 Partial pressure of Zn(g) during decomposition of ZnO with and without H_2 (a), and corresponding equilibrium constants of ZnO decomposition with and without H_2 (b).

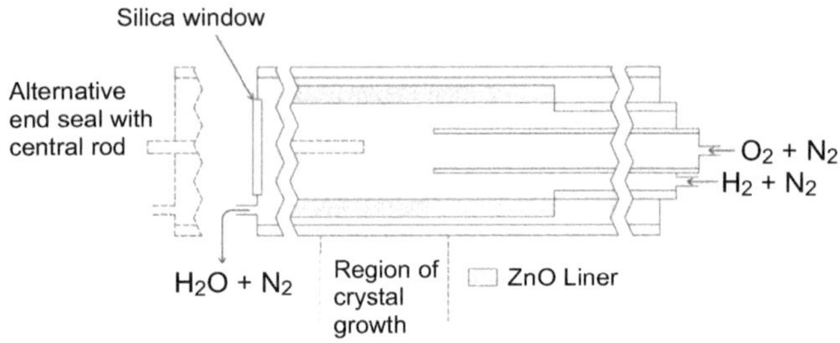

Figure 3.16 ZnO growth furnace by the CVT method with the reduction of ZnO source material by H_2 and oxidation of Zn vapor by O_2. Redrawn from Ref. [167] with some modifications.

It was found that the optimum total gas flow of $H_2 + N_2$ and $O_2 + N_2$ should be 500 cm^3/min, with H_2 and O_2 flows of 12–16 and 10–18 cm^3/min, respectively, growth temperature of 1280 to 1325°C, and the growth duration of 5–10 h. Lower growth temperatures or higher total gas flows led to ZnO whiskers or "wool," polycrystalline ZnO or ZnO powder. The obtained needle-type crystals were clear and colorless, 20 mm long, and 2–3 mm in diameter. The growth direction was parallel to the <0001> direction, and the tips of the needles were terminated with 6 faces $\{10\bar{1}1\}$ or $\{11\bar{2}1\}$ or both. The main impurities were Ce, Fe, and Sn between 1 and 2 wt. ppm. Also doping with Li and Cu was carried out, by soaking the ZnO liner with LiOH/LiF, or $CuSO_4$ solutions.

A horizontal growth furnace used by Helbig [146] was made of alumina (170 × 220 × 335 mm^3), with a smaller diameter tube (48 mm) extending therein, as illustrated in Fig. 3.17. Within that tube, a ZnO starting material (600–1000 g) was located and through which, an $N_2 + H_2$ gas mixture was introduced. At the end of that tube, a ceramic cone of Al_2O_3 or ZnO was situated, which acted as a seed for the vapor growth of ZnO. By a separate tube, O_2 was supplied to the growth zone.

The gas flows were H_2 = 2 l/h, N_2 = 80 l/h, and O_2 = 0.5–2 l/h. The temperature in the growth zone was 1300°C, and the growth duration was about 200 h. The undoped crystals were substantially colorless hexagonal bars with the length aligned with the *c*-axis. The length of the crystals was up to 40 mm and diameter up to 7 mm (up to 20 g). Also, pyramidal crystals were grown as well, with a surface area up to 2–3 cm^2. The main impurities were Al, Sn, Si, Fe, Cr, V, K, and Na, at a typical concentration of a few wt. ppm each. The crystals were also doped with different elements: Li, Na, Cu, Ga, In, and Mn.

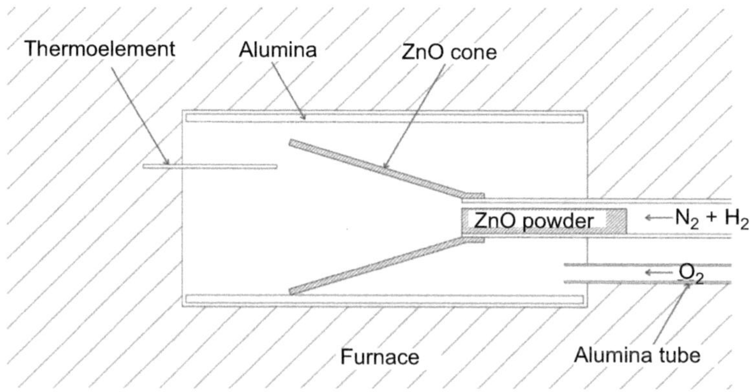

Figure 3.17 ZnO growth furnace by the CVT method with the reduction of ZnO source material by H_2 and oxidation of Zn vapor by O_2. Redrawn from Ref. [146] with some modifications.

Fischer [169] conducted ZnO growth experiments in an open horizontal growth furnace shown in Fig. 3.18. It consists of three concentric mullite or alumina tubes subjected to a predefined temperature profile. The ZnO source material in the form of spheres (25 mm in diameter), sintered in air at 1100°C, was placed in the first section of 12 cm length and cross section of 5 cm^2, where the temperature increases from about 950°C to 1275°C. Into this tube, H_2 with a flow rate of 0.3–1 l/h and N_2 with a flow rate of 60–80 l/h as a carrier gas were introduced. When the carrier gas entered the growth zone, its velocity decreased from about 4 to 0.2 cm/s due to an increase of a tube cross section from 5 to 110 cm^2. In the growth zone, a “substrate” was placed in the form of a wire or filament with a diameter of 0.1–3.5 mm and made of different materials, such as alumina, silica, Kanthal, mullite, or metals (e.g., Ni). Smaller seed diameter enabled smaller quantity but larger ZnO crystals. The growth duration was between 30 and 250 h. The growth velocity depends on supersaturation. In the case of high supersaturation, whiskers were formed, while at lower supersaturation at the same temperature, short but thick crystals were formed. The optimum growth temperature was found to be 1275–1280°C. The crystals grew mainly as hexagonal prisms in the <0001> direction. They were 100 mm long and 2–3 mm thick or 35 mm long and 8 mm thick. The number of crystals grown at the lower range of temperatures (1212–1275°C) was larger than at higher temperature range; they were clear and mostly hexagonal prisms. At a higher temperature range (1280–1347°C), the crystals were transparent grey and had irregular hexagonal shape.

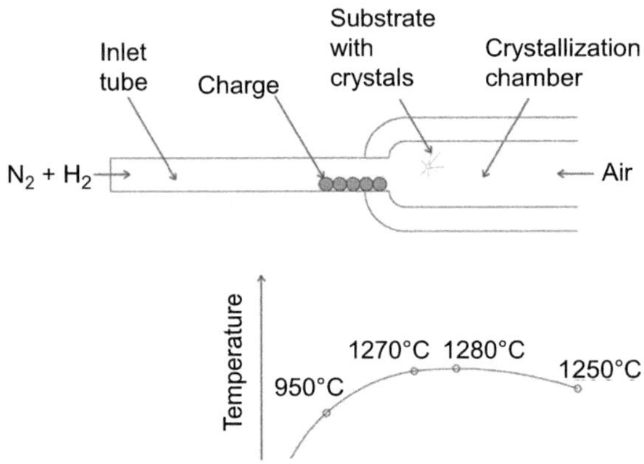

Figure 3.18 Open ZnO growth furnace with a ZnO source material, H_2 as the reducing agent, and air as the oxidizing agent. Redrawn from Ref. [169] with some modifications.

Growth experiments mentioned by Look et al. [170] were performed in a horizontal, nearly closed furnace. The hot zone was held at 1150°C, and the cold zone was kept at 1100°C. To the ZnO–H_2 system, a small amount of H_2O was added as an oxidizing agent. After 150–175 h of growth time, the obtained crystals were 2 inch in diameter and 1 cm thick.

Grasza et al. [171–174], Shalimov et al. [175], and Skupinski et al. [176] designed the so-called contactless chemical vapor transport (CCVT) method with H_2 as a reducing agent to increase the Zn partial pressure. According to this design concept, inside a silica glass ampoule (25 mm in diameter and 12–25 cm in length), a cylindrical sapphire pedestal is arranged with a diameter slightly smaller than the inner ampoule diameter. The pedestal is located at the opposite end of the ampoule to the ZnO source material. In addition to H_2, also H_2O as an oxidizing agent and inert gases (Ar or N_2) as carrier gases were used. To further increase the Zn partial pressure, an excess of elemental Zn was used, preferably separately at a conical tip of the ampoule behind the sapphire pedestal. For doping purposes with Cu and Cr, reactive gases ($CuBr_2$, $CrCl_2$) were used to transport the dopants. Their high partial pressures do not require to use additional carrier gases. The ampoule was arranged inside a quartz tube heated by heaters (Fig. 3.19a). Through the quartz tube containing a quartz ampoule, a constant hydrogen flow was applied to maintain hydrogen concentration in the ampoule. The driving force for the crystal growth is the supersaturation of the vapor components, which in turn is driven by a temperature difference between the source and growth zones. This was achieved by translation of the ampoule in a relatively steep parabolic temperature profile (Fig. 3.19b) with the rate of 1–5 mm/day. Optimum growth conditions were defined by a ZnO source temperature of 1050°C and growth temperature of 950–1000°C. The growth proceeds in this way that the source material closest to pedestal recrystallizes into a cone, the tip of which can be a single-crystal grain. That grain adheres to the sapphire pedestal and eventually separates from the source material. Then the growth proceeds on the sapphire pedestal with no contact with the ampoule wall, although a small Zn fraction may deposit in the cold zone behind the pedestal, due to transport through a slit between the ampoule wall and sapphire pedestal. The obtained ZnO material of 25 mm in diameter consisted of large single-crystal grains in the form of facetted pyramids (the facets were (110) and (001)) enabling 0.5 cm^2 wafers (Fig. 3.19c). The full width at half maximum (FWHM) of the wafers was about 60 arcsec in average, the lowest value being 29 arcsec. In the case of doping, the Cr and Cu concentrations in the crystals were uniform at the level of 10^{18} at. cm^{-3}.

The growth experiments at different ratios of partial pressures of Zn and H, $p(Zn)/p(H_2)$ = 0.15–2.5, revealed the stable growth for Zn/H = 2, when the crystallization front was smooth and slightly convex. However, depressions on the growth interface merged in the crystal and eventually they were overgrown, resulting in voids. Crystals grown at low and high p(Zn) were yellowish and red, respectively. With an increase in the temperature difference between the source and growth zones, the whole crystallization front becomes rough (constitutional supersaturation). At high Zn/H_2 ratio, most of the crystal surface was flat and smooth, but independent growth of separated grains was observed, making the obtained material mosaic. At $Zn/H_2 = 2.5$, there were also Zn droplets on the surface.

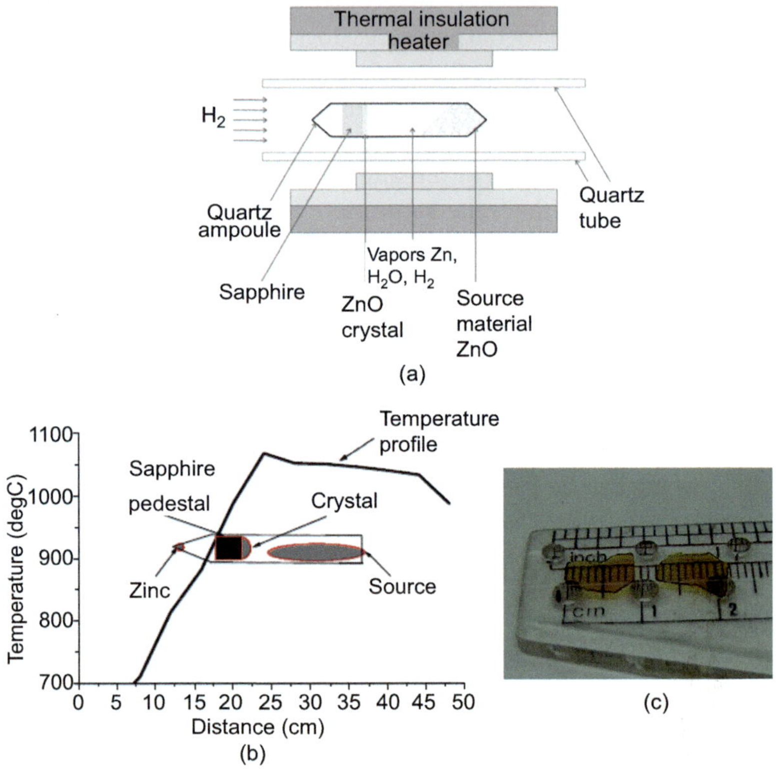

Figure 3.19 Contactless chemical vapor transport method: (a) growth furnace, redrawn from Ref. [174] with some modifications; (b) temperature profile with the ampoule location; and (c) wafers prepared from the obtained ZnO crystals. Figures (b) and (c) reprinted with permission from Ref. [172], Copyright 2006, John Wiley and Sons.

Instead of the sapphire pedestal, ZnO seeds were also used with the Zn-terminated (0001), O-terminated $(000\bar{1})$, and nonpolar $(1\bar{1}00)$ orientations. The obtained crystals maintained the seed orientation, but the structural quality deteriorated with the distance from the growth interface, namely, polycrystalline grains occurred 17, 6, and 2 mm beyond the crystal seed interface in the case of Zn-terminated, nonpolar, and O-terminated seed orientation, respectively. This is related to the growth on particular faces, which is the fastest and slowest for Zn- and O-terminated faces, respectively. The obtained crystals were of pink color as a result of deviation from stoichiometry caused by Zn excess. In the seeded growth, three kinds of defects were observed in the crystals independent of seed orientation: (i) flat voids with a size of about 200 μm, (ii) spherical voids (bubbles) of about 10 μm in diameter, and (iii) micropipes. The flat voids were likely formed in the free spaces between the seed and the ampoule wall. The vapor first condenses at the cooler ampoule wall and moves toward the seed, enclosing a free space within the crystal. A portion of the seed around the enclosed space subjected to a higher temperature may sublimate and condensate on another, cooler portion of the enclosed space. This effect would cause a motion of the flat voids through the seed and the growing crystal, the traces of which have been experimentally observed. The spherical voids are typically arranged in rows. They could be formed by splitting micropipes into spherical voids in order to decrease their surface energy, or by overgrowth of depressions at the growth interface. The micropipes could be formed by motion of the voids.

After annealing ZnO samples (from the crystals grown on the sapphire pedestal) in air, N_2, O_2, and Zn or As vapors at 750–1070°C for up to 160 h in quartz ampoules low index facets did not show big changes in morphology (except annealing in As vapor); however, there were changes in the morphology of the stepped slopes of the facets, which resulted in step bunching and an increase in the roughness. These changes were the slowest after annealing in Zn vapor. Annealing in air, O_2, and N_2 made the crystal samples transparent, but annealing in Zn vapor, made them yellow, orange, red, or brown depending on the time, temperature, and Zn partial pressure. After annealing in As vapor, the surface of ZnO samples was damaged even after annealing at low temperature range and short annealing times. The damage was irreversible. The surface was milky (which could arise from surface preparation), but As was not detected on the annealed sample surfaces.

Using carbon as the reducing agent for a ZnO source material was investigated by Matsushita et al. [177], Ito et al. [178], Hassani et al. [179], Tena-Zaera et al. [180, 181], Pfisterer et al. [182], Mycielski et al. [183], Polosz [184], Wei et al. [185], Mikami et al. [186–188], Hong et al. [189, 190], Munoz-Sanjosé [191], Abe et al. [192, 193], Santailler et al. [194], Jokela et al. [195], and Ntep et al. [196].

In the ZnO–C system, the following reactions occur:

$$ZnO(s) + C(s) \leftrightharpoons Zn(g) + CO(g) \quad (3.10)$$

$$ZnO(s) + \tfrac{1}{2}C(s) \leftrightharpoons Zn(g) + \tfrac{1}{2}CO_2(g) \quad (3.11)$$

$$Zn(g) + \tfrac{1}{2}O_2 \leftrightharpoons ZnO(s) \quad (3.12)$$

The change in Gibbs free energy for the above reducing and oxidizing reactions, as shown in Fig. 3.20, is negative over a wide temperature range, meaning that ZnO(s) can be reduced by carbon and Zn(g) can be oxidized by free oxygen.

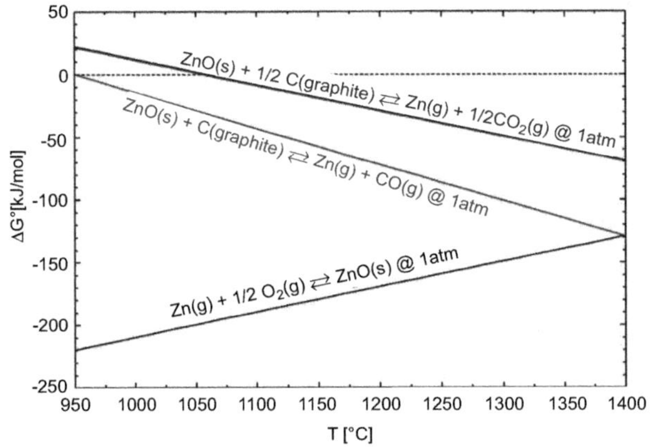

Figure 3.20 Gibbs free energy balance of reactions 3.10, 3.11, and 3.12.

Additionally, oxygen from ZnO decomposition [$ZnO(s) \leftrightharpoons Zn(g) + \tfrac{1}{2}O_2(g)$] is partly consumed by carbon:

$$C(s) + \tfrac{1}{2}O_2 \leftrightharpoons CO(g) \quad (3.13)$$

$$C(s) + O_2 \leftrightharpoons CO_2(g) \quad (3.14)$$

and leads to an excess of elemental Zn.

In the ZnO–C system, oxygen is transported by CO and CO_2, which also interact with the ZnO source material and C:

$$ZnO(s) + CO(g) \leftrightharpoons Zn(g) + CO_2(g) \quad (3.15)$$

$$CO_2(g) + C(s) \leftrightharpoons 2CO(g) \quad (3.16)$$

Gasification of carbon (Eq. 3.16) leads to much lower mass fluxes as compared to the ZnO–H_2 system according to Polosz [184] and requires higher operating temperatures.

Partial pressures of the most volatile species in the ZnO–C system, i.e., Zn, CO, and CO_2 at a total pressure of 1 atm, are shown in Fig. 3.21. The maximum partial pressure of Zn vapor is reached at about 900°C, which is then substantially constant because then the ZnO source is exhausted.

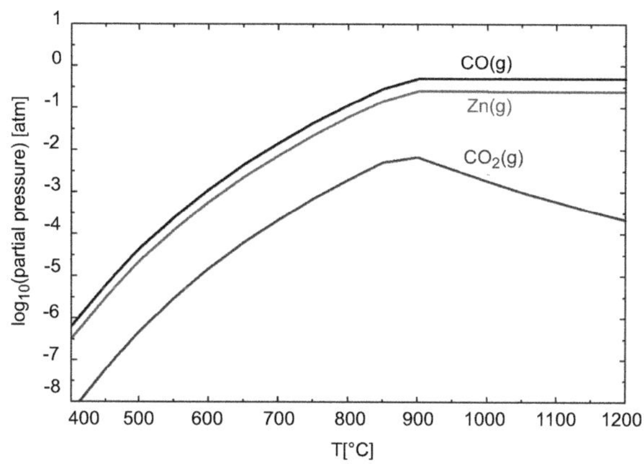

Figure 3.21 Partial pressures of gaseous Zn, CO, and CO_2 in the ZnO–C system. For the calculations, the amount of ZnO and C was 1 mol, and the pressure 1 bar.

Carbon can be used as solid C, gaseous CO, or a combination of both. They both have a strong and similar reducing action on ZnO, leading to a high $p(\mathrm{Zn})$ (Fig. 3.22a), but low $p(\mathrm{O_2})$ (Fig. 3.22b).

The partial pressures of Zn and oxygen can be further controlled by another C-containing species, CO_2, which has an oxidizing action on ZnO, as shown in Fig. 3.22a,b; however, a relatively high Zn partial pressure can be achieved only at high temperatures, above 1400°C. In the ZnO–C–CO system, ZnO is reduced directly by carbon (Eq. 3.10) and CO (Eq. 3.15), while in the ZnO–C–CO_2 system, directly by carbon and indirectly by CO, which is produced by CO_2 decomposition.

Instead or in addition to CO_2, also H_2O or O_2 can be used as an oxidizing agent in the ZnO–C system.

The presence of H_2O in the ZnO–C system produces other reducing species according to the reaction:

$$\mathrm{C(s) + H_2O(g) \leftrightharpoons CO(g) + H_2(g)} \tag{3.17}$$

wherein both H_2(g) and CO(g) will reduce ZnO according to Eqs. 3.9 and 3.15.

Using CO_2 in the ZnO–C system will lead to the formation of CO:

$$C(s) + CO_2(g) \leftrightharpoons 2CO(g) \tag{3.18}$$

which reduces ZnO according to Eq. 3.15.

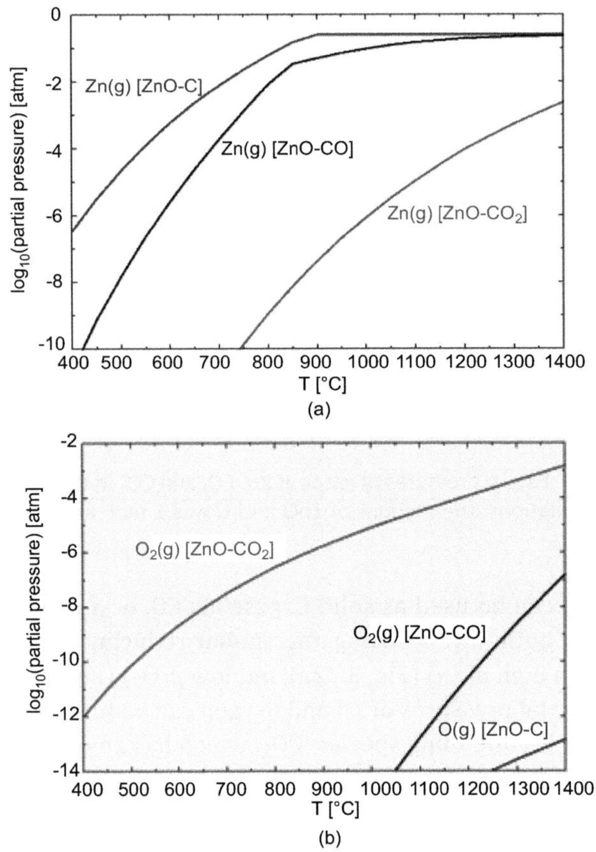

Figure 3.22 Partial pressure of Zn vapor (a) and oxygen (b) in the ZnO–C, ZnO–CO, and ZnO–CO_2 systems. The amount of ZnO, C, CO, and CO_2 used for calculations was 1 mol, and the pressure 1 bar.

CO will be produced also when using O_2 in the ZnO–C system:

$$C(s) + \tfrac{1}{2}O_2(g) \leftrightharpoons CO(g) \tag{3.19}$$

Ammonia (NH_3), in contact with C in the ZnO–C system, will generate hydrogen:

$$C(s) + 2NH_3(g) \leftrightharpoons CH_4(g) + N_2(g) + H_2(g) \tag{3.20}$$

Carbon can be mixed directly with ZnO source material, or it comes from a graphitized silica ampoule, which prevents reaction between ZnO and silica at high temperatures. Silica also creates unwanted nucleation centers for ZnO crystals. Graphitizing is typically performed with CH_4, e.g., at a pressure of 250 mbar, at about 1050°C for 30 min.

The ZnO starting material is typically dried prior use at several hundred degrees in vacuum for at least several hours to remove water, which has an impact on the growth process. Sealed off ampoules with the source material are typically maintained at inverse temperature profile (e.g., for 12 h) to thermally clean the crystallization zone and minimize or avoid the parasitic nucleation.

The possibility of using different C-containing species, which can be combined with other species in the ZnO–C system, a variety of growth conditions and a diversity of furnace/ampoule design are well described in the state of the art.

In a design according to Matsushita [177] (Fig. 3.23) through a furnace, a combustion tube extends. Inside the combustion tube, a mullite tube with an alumina boat containing powdered ZnO and carbon (4–10 wt.%) is arranged. N_2 and O_2 gases were introduced into the combustion tube with flow rates of 70–200 and 5–100 cm^3/min, respectively. The temperature in the growth furnace was between 1100 and 1200°C. For 3 h growth duration, the obtained crystals on the mullite tube and on the alumina boat were needles of 200 mm long and 0.1 mm diameter. The growth direction was along the *a*-axis lengthwise.

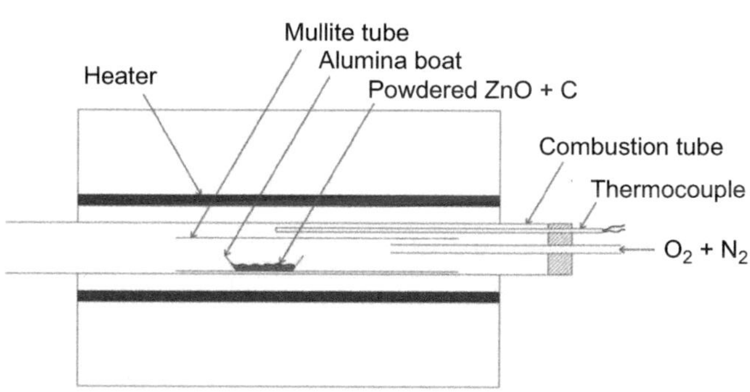

Figure 3.23 ZnO growth furnace with a ZnO source material, carbon as the reducing agent, and O_2 as the oxidizing agent. Redrawn from Ref. [177] with some modifications.

The needle-shaped crystal growth is due to the low partial pressure of Zn(g), which is controlled by the reduction rate of ZnO with C. If O_2 concentration was higher than 10% in the flowing gas, the crystals were small and polycrystalline due to the instantaneous oxidation of Zn vapor.

A growth furnace of Fig. 3.24 according to Ito et al. [178] consists of a porcelain reaction tube of 40 mm in diameter and another short porcelain tube located in the reaction tube near its inlet end. In the short porcelain tube, there is a porcelain boat with the source material constituting a mixture of ZnO powder and graphite (7–23.1 wt.%), which was pressed into pellets (0.2–0.6 g). At the inlet end of the short porcelain tube with the source material, an N_2 carrier gas was introduced. Separately, air mixed with N_2, to control oxygen flow or concentration, was introduced as the oxygen source to the reaction tube at the outlet end of the short porcelain tube.

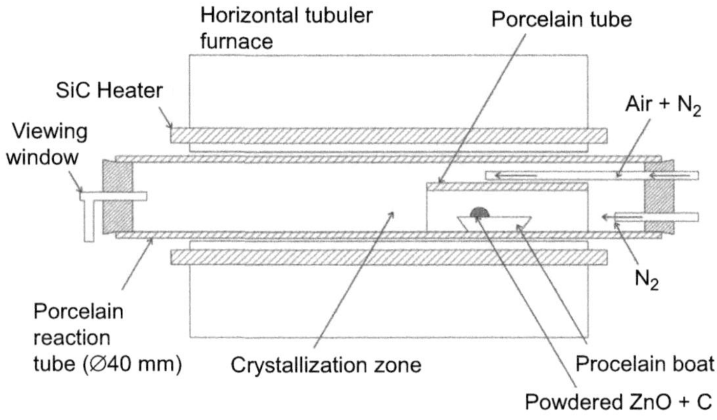

Figure 3.24 ZnO growth furnace with a ZnO source material, graphite as the reducing agent, and air as the oxidizing agent. Redrawn from Ref. [178] with some modifications.

Initially needle-shaped ZnO crystals were grown at the following most optimum conditions: N_2 flow rate = 200 ml/min; air + N_2 flow rate = 5 ml/min + 40 ml/min (O_2 flow rate = 4.3×10^{-5} mol/min) and growth temperature = 1200°C. After 5 h of growth at such conditions, the obtained needle-shaped crystals were up to 0.1 mm thick and up to 3 cm long. At lower O_2 flow rate, metallic Zn powder deposited at the end of the reaction tube, while at too high O_2 flow rate, ZnO powder precipitated. There was no growth at temperatures above 1200°C. When using the largest needle crystals as seeds at the same conditions, but for longer time (55 h), relatively large ZnO crystals in the form of platelets or prisms 2–4 mm wide and 22 mm long were obtained. They were gray and transparent. In the crystals, traces of Al, Si, and Mg were detected.

Hassani et al. [179] used a silica ampoule of 80 mm long and 15 mm in inner diameter (enabling a laminar flow at about 1000°C), which was placed in a multizone furnace. The ampoule had a conical end, which defined the growth zone that allowed to localize and control the onset of nucleation. The ZnO powdered source material was mixed with 3 to 10% C and heated to 150°C for 24 h under dynamic vacuum to remove water. Alternatively, instead of solid C, gaseous CH_4 was used as the reducing agent. Then the ampoule was sealed off at 10^{-7} Torr. In terms of the geometrical configuration only with cone angles below 40°, reasonable single crystals were obtained. With larger angles, there were several nucleation sites and consequently polycrystalline material was obtained. The obtained ZnO crystals were conical (up to 15 mm long) and orange (Fig. 3.25), which turned substantially colorless after annealing in the presence of molecular oxygen. The FWHM of the rocking curve was 28 arcsec.

Figure 3.25 ZnO crystal obtained with the use of C as the reducing agent. Reprinted with permission from Ref. [179], Copyright 2002, John Wiley and Sons.

Tena-Zaera et al. [180, 181] and Pfisterer et al. [182] approached a furnace design for growth experiments through numerical simulations of vapor transport in a closed ampoule, which were next verified experimentally. The simulation revealed that at low and moderate pressures (<0.1 atm.), the flow is laminar, while at about 1 atm, a convective cell is formed near the growth interface, which then occupies the whole ampoule. Experimentally, it was proved that the PVT transport at 1000°C was not observed even after 2 weeks. However, adding 0.25% Zn excess, growth was observed, but it was very low, below 0.05 g/h.

As a growth ampoule, a closed graphitized silica was used, which was of length 75 mm and of inner diameter about 15 mm. One end of the ampoule was conical (25–35°) to promote an initial nucleation. The graphitizing was performed by introducing into the ampoules CH_4 at about 60 mbar at room temperature, and then by heating at about 1050°C for 30 min. ZnO of 4N purity powder was used as a source material, which was initially heated at about 200°C for 24 h in vacuum to remove residual species, especially water. The temperature at the source zone was kept at 1000°C, and the temperature difference between the source and the growth zones was between 15 and 70 K. The temperature gradient in the growth zone was about 5 K/cm. Before each of the growth run, the ampoule was maintained in reverse temperature gradients for 12 h to clean the growth zone and remove parasitic nucleation sites. In graphitized ampoules, the partial pressure of oxygen can be controlled by temperature gradients between the source and growth zones. Large temperature gradients promote a fast transport rate and short contact period between O_2 and graphite, and consequently lower O_2 consumption (low Zn excess), while at low temperature gradients, the situation is reversed that leads to a higher O_2 consumption and larger Zn excess. Additionally, the O_2 partial pressure can be controlled by the graphite amount in the ampoule or by adding CO or CO_2 or Ar + CO_2 gases, which are introduced into ampoules under vacuum conditions. The difference in such growth conditions were well reflected in a crystal coloration, i.e., the crystals obtained at high transport rate of about 11×10^{-3} g/h (large temperature difference) were substantially colorless, and they became orange and darker with slower transport rate (low temperature difference) or by a larger amount of graphite in the ampoule. In the darker crystals, elemental Zn was identified. At moderate transport rates, the crystals were pale yellow. The obtained conical ZnO crystals were up to about 2 cm long and about 5–7 mm in diameter at the base. They typically consisted of multiple single-crystal grains. The full width of the rocking curve and the EPD were 50–200 arcsec and 10^5 cm^{-2}, respectively. Crystals annealed at 1000°C in O_2 for 20 h became colorless, while those annealed in Zn vapor at similar conditions became red.

Mycielski et al. [183] used a graphitized quartz ampoule and H_2O as an oxidizing agent. Additionally, N_2 or H_2 were used as a carrier or reducing agent, respectively. The ZnO starting material at a reaction zone was kept at a temperature of 1100–1150°C, while the crystallization took place at a growth zone having temperature lower than the reaction zone by about 50 K. The growth rate was about 1–2 mm/day. The growth direction was along the *c*-axis. The obtained ZnO material was a polycrystal of about 3 cm^3 consisted of single-crystal grains 5 × 6 × 7 mm^3 (0.2–0.5 cm^3). The color of the crystals was always red, independent of the chemical agents.

Polosz [184] compared the CVT growth of ZnO with the use of C (from graphitized ampoules) or CO as reducing agents. Silica glass ampoules of 12 cm length and 1.5 cm inner diameter were outgassed at 1000°C for at least 20 h prior use. When using CO, single crystals could be obtained below 1000°C, resulting from the relatively high mass transport. However, at similar temperatures, polycrystals were formed when C was used due to lower mass transport resulted from C gasification (Eq. 3.19). Therefore, the ZnO–C system required higher temperatures, about 1200°C.

Wei et al. [185] used the ZnO–C system in a seeded CVT method. The seeds were (0001) GaN and ZnO wafers of 25–45 mm in diameter. The obtained crystals were red due to Zn-rich growth conditions. After annealing at 900°C in air for 20 h, the crystals changed their color from red or pale yellow to colorless. The crystals contained Ga and Al residual impurities at the concentration of about 35 and 5 wt. ppm, respectively.

Mikami et al. [186] applied the standard CVT growth in vertical silica ampoules of 80 mm length and 15 mm inner diameter, where ZnO powder (2 g) as a source material was mixed with C powder (0.05 g). After loading the source material, the ampoules were evacuated for 6 h at 250°C (to remove traces of water from the source material), sealed under vacuum conditions, and placed in a furnace. Prior to growth, an inverse temperature gradient was applied to clean the nucleation zone from any contamination. The growth was performed with a source temperature of 950–1050°C and a temperature difference between the source and growth zones from 4 to 30 K. The growth time was 2 weeks. The largest crystals (often polycrystals) of 5 × 12 mm^2 cross section were grown at higher source temperature (1020°C) and low temperature difference of 9 K. The study showed that the deposition rate was approximately proportional to the temperature difference. All the crystals were orange or red (attributed to a stoichiometry shift toward Zn-rich composition), but annealing the crystals in oxygen for 3 h at 800°C removed the orange-red coloration. A subsequent annealing of the colorless crystal at 900°C for 48 h in Zn vapor made it orange red again. The color change was found to be reversible.

The authors concluded in the ZnO–C system the reaction between ZnO and C is so slow that the growth mode is rather reaction-controlled. To change the growth mode to diffusion-controlled, Mikami et al. [187, 188] and Hong et al. [189, 190] came with a different concept of an ampoule design (as shown in Fig. 3.26b,c) compared with a classical vertical configuration that is reaction-limited (Fig. 3.26a). In the reaction-controlled configuration, the vapor transport proceeds from the bottom of the ampoule held at high temperature (T_S), where a ZnO source material is loaded, toward the top part at lower temperature (T_G), where ZnO deposits. In the diffusion-controlled design, the ampoule consists of an outer silica tube and an inner silica tube communicating with the outer silica tube either via a side hole (Fig. 3.26b) or via a central capillary tube (Fig. 3.26c). In the reaction-controlled arrangement, the growth is driven by a temperature difference between the source and growth zones ($T_S - T_G$), while in the diffusion-controlled configuration by a temperature difference between the hole and source zone temperature ($T_H - T_S$) and between the source and growth zones temperature ($T_S - T_G$). For instance, in the design of Fig. 3.26b, the outer silica tube is 11 mm in diameter and 120 mm long and the inner silica tube is 10 mm in diameter and 90 mm long, while the hole diameter is 2 mm. The capillary tube of Fig. 3.26c is of 2 mm in inner diameter and 40 mm in length. The source material consisting of a polycrystalline ZnO (0.2–2.5 g) and C (0.1 g) was loaded into the inner tube. Instead of ZnO–C, also another system, ZnO–CO_2 + Zn, was used as well. At the bottom of the outer tube, a ZnO crystal seed was placed (e.g., of 1.5 × 1.5 × 0.5 mm^3). Prior growth, the ampoule was evacuated at 150°C for 3 h and then sealed off under vacuum higher than 7 × 10^{-3} Torr. If CO_2 was used, the ampoule was sealed off at 0.2 atm of CO_2 at room temperature.

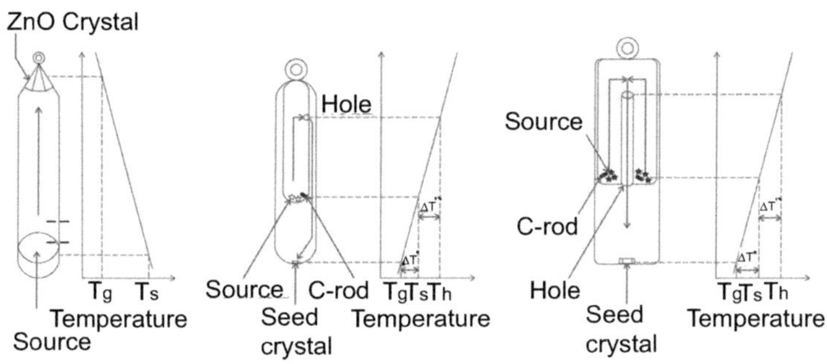

Figure 3.26 Ampoule design for a CVT growth in a vertical configuration: (a) reaction controlled, (b) and (c) transport controlled. Redrawn from Ref. [189] with some modifications.

The optimum temperature differences in the ampoule design with the side hole that enabled ZnO single crystals were $T_S - T_G = 2$ K and $T_H - T_S = 5$ K. At $T_H - T_S < 4$ K, parasitic nucleation was observed, which consequently led to polycrystalline ZnO. Likely that temperature difference defines a transition point between the reaction- and diffusion-controlled growth modes. After a week's growth, the obtained crystals were about 2 mm in size with the FWHM of the rocking curve of about 73 arcsec.

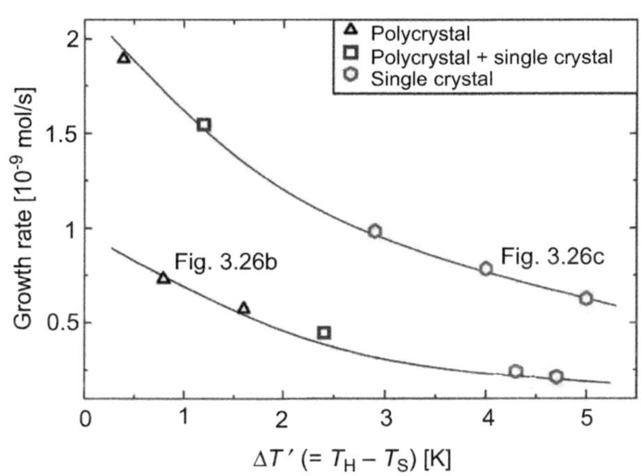

Figure 3.27 ZnO growth rate as a function of temperature difference $T_H - T_S$ at $T_G = 965°C$ and $T_S - T_G = 5$ K. Reprinted from Ref. [189], Copyright 2009, with permission from Elsevier.

However, the growth rate in such ampoule design was very small as compared to the standard ampoule. Higher growth rates were achieved in the ampoule design with a capillary tube, where the optimum temperature differences were the same as in the case with the side hole: $T_S - T_G = 5$ K, $T_H - T_S = 2$ K, at $T_S = 970$°C and $T_G = 965$°C. The comparison of the growth rates between the designs with a side hole and capillary tube is demonstrated in Fig. 3.27. At the same $T_H - T_S$, the growth rate of the design with the capillary tube was two–three times higher than that with the side hole.

After a growth time of 21 days, the obtained ZnO crystals (Fig. 3.28) were orange, facetted, and sized up to 5 × 5 × 5 mm^3. The FWHM of the rocking curve was 44 arcsec. C has not been detected by SIMS. After annealing at 1000°C for 5 h in oxygen, the orange crystals became colorless and transparent.

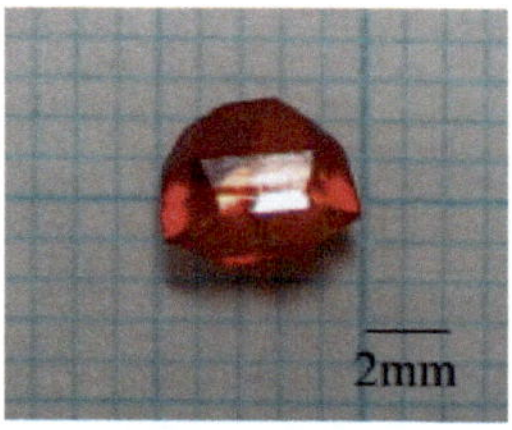

Figure 3.28 As-grown ZnO crystal obtained by the CVT with the use of the ZnO–C system. Reprinted from Ref. [189], Copyright 2009, with permission from Elsevier.

In the ZnO–CO_2 + Zn system, the obtained crystals were about 2 mm long and 1 mm in diameter, with the FWHM of the rocking curve of 25 arcsec. The crystal coloration changed from red to yellow to colorless when decreasing Zn vapor concentration from 1 to 0.3 mg/cm^3. At high Zn vapor concentrations, metallic Zn droplets were formed on the crystal surface.

Conventional CVT techniques provide low growth rates of ZnO as a result of Zn excess according to theoretical calculations of mass transport rates [180, 181, 184]. To reduce Zn excess, a cold-wall CVT (CWCVT) technique was proposed by Abe et al. [192, 193], in which Zn excess in vapor can be adsorbed on a reaction tube held at relatively low temperature. The concept of the CWCVT is presented in Fig. 3.29. The growth furnace utilizes an inductively heated graphite crucible located inside the reaction tube. The crucible of 30 mm in diameter and 22 mm in height is enclosed by a lid with a central opening to communicate with the atmosphere in the reaction tube. Also, the crucible wall has small openings. The crucible was loaded with 1.8 g of ZnO powder, while on the underside of the crucible lid, a ZnO crystal seed (Zn-polar) prepared from hydrothermally grown crystals was attached.

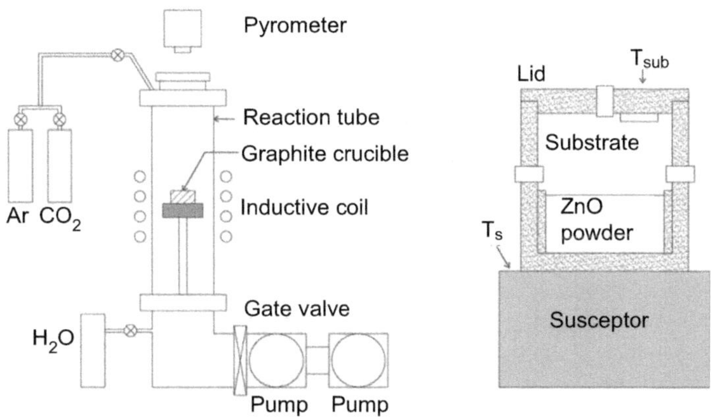

Figure 3.29 Schematic diagram of the CWCVT system and a graphite growth furnace. Redrawn from Ref. [192] with some modifications.

Before the growth, the reaction tube was evacuated to 2×10^{-5} Torr and then it was filled with a gas mixture of Ar, H_2O, and CO_2. With a total pressure of 610 Torr in the reaction tube, p(Ar), $p(H_2O)$, and $p(CO_2)$ were 355, 25, and 230 Torr, respectively. With no CO_2, p(Ar) and $p(H_2O)$ were 585 and 25 Torr, respectively. After 2 h of growth, the total pressure increased to 990 Torr. Temperatures of the source and substrate were at 1020 and 860°C, respectively.

When using a fully enclosed crucible, the growth rate was about 50 μm/h; however, with a small opening, the growth rate increased to 100–175 μm/h. The growth rate, however, decreases with $p(CO_2)$ as indicated in Fig. 3.30. At $p(CO_2)$ = 375 Torr, the growth rate decreased by a factor of about 2 in average as compared with the growth rate with no CO_2.

In the absence of CO_2, the O/Zn atomic ratio was between 0.94 and 0.96; however, the crystals grown with the use of CO_2 had a better stoichiometry defined by the O/Zn ratio between 0.97 and 1. The O/Zn ratio was normalized to hydrothermally grown ZnO used as the seed. A better ZnO stoichiometry as well as lower growth rates when using CO_2 is simply the result of lower Zn and higher O_2 partial pressures in the system, as demonstrated in Fig. 3.22.

The FWHM of the rocking curve of the crystals was around or below 100 arcsec.

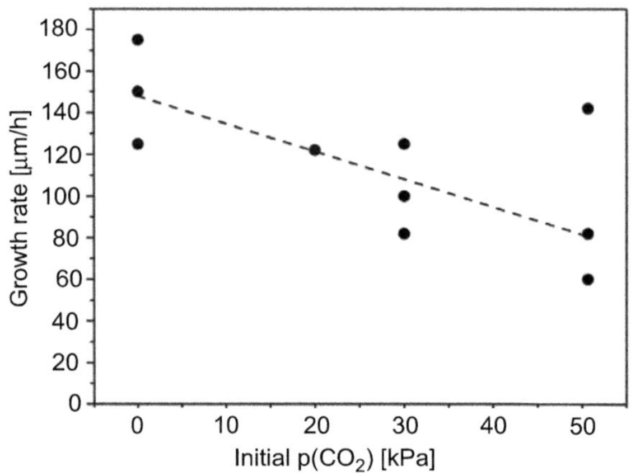

Figure 3.30 Growth rate of ZnO as a function of CO_2 partial pressure in the reaction tube of the CWCVT system. Reconstructed basing on data from Ref. [193], with some modifications and different visualization.

To obtain a good stoichiometry of ZnO crystals and high growth rate, Santailler [194] used in growth experiments reducing (CO), neutral (Ar, N_2), and oxidizing (O_2) gases at the same time. The proposed setup is shown in Fig. 3.31. At the bottom, a crucible contained a graphite chamber, through which O_2 flew to produce CO passing through a high-purity ZnO source material situated above the graphite chamber. Through the ZnO starting material, a separate flow of O_2 was applied. Above the crucible, a seed was mounted. The crucible was surrounded from the side and bottom by a susceptor, thermal insulation, and a silica tube, surrounded by an inductive coil. The space between the ZnO source material and the seed defined a sublimation chamber.

The growth experiments were performed in the following way. The furnace after assembling was pumped down and maintained at vacuum for several hours at 400°C. To avoid ZnO sublimation, Ar was introduced into the furnace. Next the furnace was heated up to sublimation temperature of 1100–1300°C and kept for 1 h to reach thermal equilibrium. After that the pressure was reduced to initiate ZnO sublimation. Decomposition species of ZnO were transported by thermal gradients to a seed kept at lower temperature of 900–1100°C. At this initial growth stage, PVT growth conditions were used. Then the CVT process was activated by flowing O_2 through the graphite chamber, which produced CO and CO_2 being transported through the ZnO starting material. The Zn concentration in the sublimation chamber has been significantly increased enabling ZnO crystallization on the seed at higher rate. The stoichiometry of the growing ZnO could be controlled by a separate O_2 flow through the ZnO starting material. The growth duration lasted for 24 to 36 h. Once the growth run was completed, the O_2 flow through the graphite chamber was switched off, the growth furnace was pumped down, and inert (N_2, Ar) or oxidizing (O_2) atmosphere was introduced therein. After that the furnace was slowly cooled down. The total growth run duration was about 64 h. At similar pressure, the sublimation yield was 10–20 (relative to the CVT method) and 80–100% for the PVT and CVT conditions, respectively. The sublimation yield of 2–6 g/h was possible in these experiments corresponding to a growth rate of 0.1–0.4 mm/h. The condensation yield depends on the ratio between the sublimation chamber and seed diameter (it is higher for larger seed diameter).

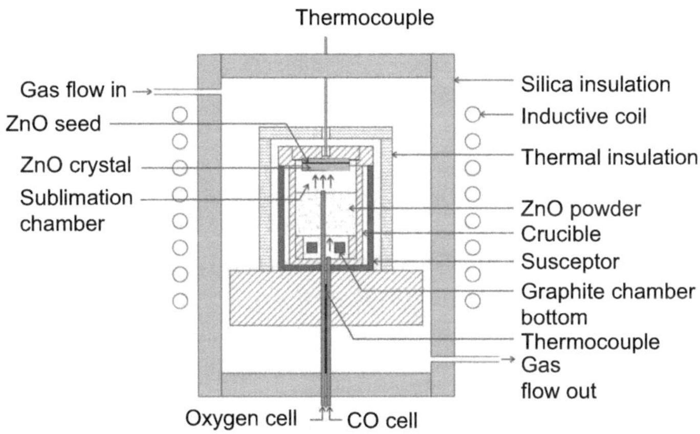

Figure 3.31 Schematic diagram of the CVT setup with the use of the reducing, neutral, and oxidizing gases. Redrawn from Ref. [194] with some modifications.

Three kinds of seeds were used: Al_2O_3 <0001>, heteroepitaxially grown ZnO <0001> on Al_2O_3 <0001> (template), and ZnO <0001>. In the case of ZnO seeds, both O-terminated and Zn-terminated faces were used. ZnO obtained on sapphire seeds was polycrystalline, while on ZnO seeds, it was a single crystal with the diameter of 17 or 46 mm (depending on seed diameter) and thickness of 10 or 8 mm as shown in Fig. 3.32. The FWHM of the rocking curve was 200–400 and 90–200 arcsec for the ZnO/Al_2O_3 and ZnO seeds, respectively. EPD for ZnO and ZnO/Al_2O_3 seeds was approximately the same as the EPD of the seeds (at the level of 10^4–10^5 cm^{-2} on the Zn-terminated face). The total content of impurities of CVT-grown ZnO was about 10 wt. ppm, while in the case of hydrothermally grown ZnO (comparison), it was about 100 wt. ppm (the major impurities are Li, Si, Mg, Fe, and Al).

Figure 3.32 ZnO single crystal grown on the ZnO <0001>/Al_2O_3 <0001> seed by the CVT method. Reprinted from Ref. [194], Copyright 2010, with permission from Elsevier.

Jokela [195] used C and anhydrous NH_3 as reducing agents for growing ZnO crystals by the CVT method. ZnO (1.5 g) and graphite (2 g) powders were placed at one end of a fused silica ampoule, while a crystal seed was mounted at the opposite end thereof. The ampoule was evacuated and filled with NH_3 several times, with the pressure of 0.5 atm in the last filling. After that the ampoule was sealed off and placed in a three-zone horizontal furnace. The ZnO starting material and graphite were kept at 1000°C and the seed at 880°C. The growth duration was 40 h.

In the ZnO–C system, Ntep et al. [196] used, additionally, Cl_2 as a transport gas. For growth experiments, a short cylindrical, graphitized silica ampoule of 2 cm diameter and a distance between the source and growth zones of 5 cm was used. Such geometry should enable at low temperature gradients laminar vapor flow with no convection cells or turbulence. The ampoule containing 3 g ZnO was sealed off under a residual Cl_2 pressure of about 100 mbar and was placed in a vertical furnace. The growth occurred at source temperature T_S = 950–1000°C and source-growth zone temperature difference $\Delta T = T_S - T_G$ = 10–100°C. The best crystals were obtained with T_S = 1000°C, ΔT = 30 K, and deposition time of 40 days. The crystals approached the size of about 1 cm, and the FWHM of the rocking curve was about 30 arcsec on the (0001) plane. Traces of Cl and C have been found in the crystals. If the source temperature was lower than 950°C, the crystals were less transparent with grayish or orange coloration.

The above survey clearly shows that the crystal coloration, crystal size, and structural quality strongly depend on the growth environment, such as type of growth technique, type of growth furnace, starting material, transport agents, reducing and/or oxidizing conditions, pressure, growth temperature, seed type (if any), and growth time. The basic outcome from the growth of bulk ZnO single crystals obtained from the gas phase (vapor) is summarized in Table 3.1.

3.4.2 Flux Techniques

The high melting point of ZnO (1975°C) and its thermal instability at high temperatures make the growth of ZnO crystals directly from melt rather difficult and challenging. To grow the crystals from the liquid phase at much lower temperatures than the melting point of ZnO at atmospheric pressure, it is necessary to use a solvent, which is called a flux. The flux for ZnO source material is an oxide, fluoride, or hydroxide with a low melting point, such as PbO, PbF_2, PbO + PbF_2, V_2O_5 + B_2O_3, V_2O_5 + MoO_3, KOH, or NaOH. The growth experiments are performed in Pt crucibles in an oxidizing growth atmosphere, typically air or oxygen. Instead of ZnO, also ZnF_2 can be used as the source material.

PbF_2 as the flux for growing ZnO crystals was used by Nielsen and Dearborn [145], Fischer and Sinn [197], and Li et al. [198–201]. According to the solubility curve of ZnO in PbF_2, this system has eutectic at about 8.8 mol.% of ZnO at 733°C, as illustrated in Fig. 3.33.

Table 3.1 Summary of ZnO growth from the gas phase. Here *D* is diameter, *L* is length, *T* is thickness, *W* is width, and *S* is surface area

Source material	Chemically active agent	Oxidizing agent	Crystal shape	Exemplary crystal dimensions [mm]	Crystal coloration	Reference
Zn	H_2	Air, O_2	Needles, leaf, plates	D=1, L=40; 0.15×0.5×20	Colorless, yellowish; greenish	142, 149
ZnX_2 (X=F, Cl, Br, I)	—	H_2O, O_2	Needles, ribbons, prisms, plates	D=1, L=40; D=2, L=20; W=4, T=0.3; W=1.7, L=5; S=4–8, T=0.2–0.4	Colorless, yellow, green, amber	150, 151, 152, 153, 154, 155, 156, 157, 158
ZnY (Y=Se, S)	—	O_2	Prisms, platelets	30×5×5, 10×6×1		159
ZnO	Cl_2, Br_2, I_2, HCl, HBr, NH_3, NH_4Cl, NH_4Br, NH_4I	—	Rods, prisms, plates	D=2, L=8; 4×3×0.5; S=few, L=5–15	Colorless, yellow, yellow–green, orange, brown	160, 161, 162, 163, 164, 165
ZnO	H_2, C, CO, CH_4	O_2, H_2O, CO_2	Needles, rods, prisms, pyramids, platelets, conical, disks	D=0.1, L=200; D=2–3, L=20; D=6–7, L=60–70; T=2–3, L=100; T=8, L=35; W=2–4, L=22; 5×6×7; 5×5×5; D=46, T=8	Colorless, gray, yellow, orange, red	146, 166, 167, 168, 169, 170, 171, 172, 173, 174, 175, 176, 177, 178, 179, 180, 181, 182, 183, 184, 185, 186, 187, 188, 189, 190, 191, 192, 193, 194, 195, 196

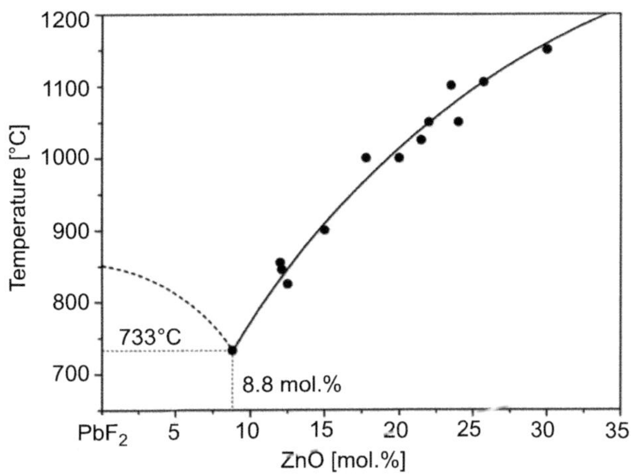

Figure 3.33 Solubility curve of ZnO in PbF_2. Reconstructed basing on data from Ref. [145], with some modifications and different visualization.

Nielsen and Dearborn [145] utilized resistance furnaces and 100 ml Pt crucibles covered with Pt lids. Starting materials were prepared from about 200 g of PbF_2 and 22 g of ZnO. The furnaces were heated up to 1150°C, held at that temperature for 2–4 h, and then slowly cooled down to about 800°C with the rate of 1–10 K/h. After reaching that temperature, the crucibles were removed from the furnaces and cooled down in air. During the growth experiments, the temperature was monitored with Pt-10%Rh thermocouples. Although the volatility of PbF_2 at 1150°C was relatively high, it enabled to perform growth experiments. As the growth atmosphere, both air and oxygen were used with no difference in growth habit. ZnO crystals were *a*-oriented plates when nucleated at 1150°C, which became more drum shaped when nucleated at lower temperature, 1050°C. The crystals were separated from the solidified flux mechanically. Their size in the largest direction (along *a*-direction) could approach even 5 cm.

Fischer and Sinn [197] performed similar experiments using PbF_2 as a flux and also PbO and PbO + PbF_2. The initial ZnO concentration was 20–25 mol.%. Cooling down was at the rate between 6 and 0.6 K/h started from 60 K above the liquids' temperature for the given

ZnO concentration (i.e., at temperatures between 850 and 1100°C). The authors noticed that the crucial parameter in the crystals' growth habit was the cooling rate, not the flux itself (PbF_2, PbO or PbO/PbF_2). For the cooling rate ≥0.9 K/h, the obtained crystals were plates of about 3 × 5 mm^2, hexagonal prisms of about 1–1.5 mm in diameter or hollow bodies. However, at smaller cooling rate, 0.6 K/h, the crystals were equidimensional. Experimentally, it was found that the deposition rate followed the relation 0.0222 × cooling rate. The obtained ZnO crystals were of yellow–brown coloration, but they became substantially colorless after annealing in air for several hours at the temperature the same as the growth temperature. Further, the authors applied flux reaction technique to hydrolyze ZnF_2 in PbF_2 with small cooling rate of 0.6 K/h. The initial starting composition consisted of 0.3 ZnF_2 and 0.7 PbF_2 having eutectic temperature of 730°C and the growth temperature was between 765 and 820°C. Here, two important reactions take place with water vapor introduced to a growth furnace:

$$ZnF_2 + H_2O \rightleftarrows ZnO + 2HF \tag{3.21}$$

$$PbF_2 + H_2O \rightleftarrows PbO + 2HF \tag{3.22}$$

The crystals obtained by the flux reaction technique with small cooling rate were also equidimensional, proving cooling rate as a critical growth parameter. In this case, the crystals were bluish-grey. The crystal coloration, however, did not change after annealing at the growth temperature. In both techniques, a typical size of the crystals was about 5–6 mm.

Li et al. [198–201] applied a starting material consisting of 78 mol.% of PbF_2 and 22 mol.% of ZnO contained within an enclosed Pt crucible of 20 mm diameter and 200 mm height. The growth experiments were performed in a vertical Bridgman configuration with defined high temperature, temperature gradient, and low-temperature zones. The starting material was first heated up to 1100°C and soaked at high-temperature zone for several hours and then the crucible was translated downward at the rate of 0.3 mm/h from the high-temperature zone through temperature gradient to the low-temperature zone. The PbF_2–ZnO system has the tendency to attack platinum; therefore, temperatures higher than 1100°C are not recommended. In this configuration, yellow, small and wedge-shaped ZnO crystals of about 1 mm diameter and 3 mm length were found attached to the crucible wall, as a result of growth from the vapor phase. The solidified body of PbF_2 and ZnO revealed (after stripping off the crucible) a thin ZnO plate at the top of the body and many ZnO grains inside the body. The ZnO plate of dimensions $10 \times 10 \times 0.7$ mm^3 was yellow and transparent and had pure ZnO phase. However, the crystal grains within the body of dimensions $5 \times 2 \times 4$ mm^3 were dark, due to contamination with the PbF_2 flux. In a modified version, the bottom of the crucible of 28 mm diameter and 280 mm height was additionally cooled with O_2 blowing thereon. The gas flow was initiated during the downward translation of the crucible, after homogenization of the starting material at 1100°C for 4 h. The optimum O_2 flow in terms of crystal size was found to be 3 l/min, which induced a temperature drop by 50 K. High-temperature gradients across the crucible bottom created nucleation centers for ZnO; therefore, a crystalline layer 5–8 mm thick was found at the bottom of the solidified body. The coloration of the crystalline layer was from pale yellow to pale tawny. The single-crystal grains were of about 7 mm in average. However, the structural quality of the crystals was questionable as growth striations, large PbF_2 inclusions of 30–90 μm in size, gas bubbles of about 15 μm in size, and grain boundaries were found. Moreover, the obtained ZnO crystals were contaminated with the flux elements: 390 and 40 wt. ppm of Pb and F, respectively.

Oka et al. [202] studied mixtures of $V_2O_5 + B_2O_3$ and $V_2O_5 + MoO_3$ as fluxes and two flux growth techniques: top-seeded solution growth (TSSG) and traveling-solvent floating zone (TSFZ). The TSSG growth furnace utilized an inductive heating and a Pt crucible of 55 mm diameter and 40 mm height. Basing on the phase diagrams of $ZnO - V_2O_5 + B_2O_3$ and $ZnO - V_2O_5 + MoO_3$, two different starting materials were chosen: 76–80 mol.% of ZnO and 20–24 mol.% of $V_2O_5 + B_2O_3$, and 52 mol.% of ZnO and 48 mol.% of $V_2O_5 + MoO_3$. The starting material was heated up to 1150°C, soaked at that temperature for 0.5 h, and then a seed was brought into a contact with the solution. The pulling and rotation rates were 0.5–1 mm/h and 20 rpm, respectively. During growth, the solution was cooled down at the rate of 2–5 K/h. Once the crystal was separated from the solution, the growth furnace was cooled down to room temperature at the rate of 300 K/h. The TSFZ growth furnace was equipped with four halogen lamps and four ellipsoidal mirrors as a heat source.

The starting materials were the same as in the case of the TSSG method. The feed and seed rods of 7 mm in diameter were prepared by isostatic pressing of the starting composition at 2000 bar and sintering at 1400°C in air for 15 h. The growth was performed in two steps: (i) pre-melting of the starting material, and (ii) growing crystals with the use of the solidified material in step (i) as a feed rod. In the first step (i), the translation rate of the molten zone was very high, 20 mm/h, while the counter rotation of the feed and seed rods was 20 rpm. In the second step (ii), the translation rate of the molten zone was 0.5–1 mm/h. After disconnecting the feed rod from the grown crystal, the furnace was cooled down within 1 h. All the growth experiments were performed in air. The TSSG crystals obtained from the V_2O_5 + B_2O_3 flux were brown, while those from V_2O_5 + MoO_3 were pale yellow. The maximum size of the crystal grains within a crystal boule was 10 × 5 × 2 mm^3. The TSFZ crystals were pale yellow with the maximum crystal grain size of 2 × 2 × 1 mm^3. During the TSFZ growth, also transparent whiskers were found on the feed rod, which were grown from the vapor phase. The crystals obtained from the V_2O_5 + B_2O_3 and V_2O_5 + MoO_3 fluxes contained V and Mo impurities at concentrations below 1.7 and 0.7 at.%, respectively.

Yet another flux for growing ZnO crystals, hydrous KOH, was used by Kashyap [203]. The idea came from hydrothermal growth of ZnO in the KOH–H_2O–ZnO system. The experiments were carried out in a vertical resistance furnace with the use of a silver crucible containing a mixture of 21 g of KOH, 9 g of H_2O, and 5.8 g of ZnO. The crucible was covered by a lid having a conical tip arranged centrally and downwardly. The melting point of such composition was 383°C; the growth temperature was between 450 and 480°C, while the cooling duration was about 50 h. The obtained transparent crystals were needle shaped with a diameter of 0.2 mm and length up to 6 mm. They were shorter for shorter cooling durations. The crystals grew at the lid's tip and at the crucible wall and were removed from the crucible by dissolving KOH with CH_3OH.

A more detailed study on the growth of ZnO in hydrous KOH and also in NaOH or KOH/NaOH fluxes was performed by Ushio and Sumiyoshi [204]. The growth experiments were performed in a resistance furnace with the use of Ir crucibles of 50 (small) and 400 (large) ml in volume.

In such environment, the following reactions take place:

$$KOH\ (\text{or } NaOH) \rightarrow K^+\ (\text{or } Na^+) + OH^- \qquad (3.23)$$

$$ZnO + 2OH^- \rightarrow ZnO_2^{2-} + H_2O \qquad (3.24)$$

$$ZnO_2^{2-} \rightarrow ZnO + O^{2-} \qquad (3.25)$$

In small crucibles, typical KOH and ZnO contents were 5.8 g and 30 g, respectively. In large crucibles, the content of both constituents was one order of magnitude larger. KOH and NaOH contained 28.5 and 25.4 wt.% of H_2O, respectively. Growth temperature varied from 450 to 800°C for small crucibles and from 500 to 900°C for large crucibles. Typical cooling duration was 1–20 h for small crucibles and 30–120 for large crucibles. The crystals grew at the crucible bottom and wall and were removed by the dissolution of the flux in water. The crystals were needle-shaped and ended with hexagonal pyramids. The needle diameter was 0.5 mm and 1 mm, while the maximum length was 7.5 mm and 18 mm for KOH and KOH + NaOH fluxes, respectively. The crystals obtained in small crucibles were brown and white when grown at 500°C and 800°C for 20 h, respectively, while when grown in large crucibles at 900°C for 30 h, they were light brown. The crystals obtained at higher temperatures were of better structural quality as compared to those obtained at lower temperatures. Operating temperatures of 800–900°C led to the evaporation of the flux and to the supersaturation of ZnO, which enables larger crystals. Indeed, the needle size was much larger in this case as compared to the needle size of ZnO grown by Kashyap at 450–480°C. On the other hand, high flux evaporation did not allow for long growth duration. Addition of NaOH to KOH decreases both flux evaporation and basicity, resulting in the largest crystal size. Basic outcome in growth of bulk ZnO single crystals with the use of flux techniques is shown in Table 3.2.

3.4.3 Hydrothermal

The hydrothermal method is a solution technique, as the flux method, but it proceeds at lower temperatures and at high pressures, while the solution comprises an alkaline aqua solution.

Table 3.2 Summary of results for flux techniques. *T*, *L*, and *D* indicate temperature, length, and diameter, respectively

Solvent	*T* [°C]	Crystal shape	Maximum single crystal size [mm]	Crystal coloration	Impurities	Method	Reference
PbF_2, PbF_2 + PbO, PbO	850–1150	Plates, blocks, prisms	*L*=50, *D*=7, *D*=1.5	Yellow, yellow–brown, brown, bluish-grey	Pb=390 wt. ppm, F=40 wt. ppm	Flux, Flux-Bridgman	145, 197, 198, 199, 200, 201
KOH, NaOH, KOH+NaOH	450–900	Needles	*D*=1, *L*=18	Brown, light brown, white	—	Flux	203, 204
V_2O_5+B_2O_3, V_2O_5+MoO_3	1150	Blocks	2×5×10	Brown, pale yellow	V<1.7 at.%, Mo<0.7 at.%	TSSG, TSFZ	202

There is a number of important factors in the hydrothermal growth of ZnO crystals that relate to both the autoclave design and growth conditions/operating parameters.

3.4.3.1 Autoclaves

Autoclaves for growing bulk ZnO single crystals by the hydrothermal method are made of high-strength steel or Ni-based alloys. To protect an autoclave's inner wall from the alkali solution, either it is lined with a metal resistant to alkali solutions, or the autoclave contains an inner container made of such metal. The space between the autoclave's inner wall and the container is filled up with distilled water to balance the pressure. A few lining and container materials are used: Ag, Au, Pt, and also Ti-based alloy, but the most frequently used metal is Pt, as it has a high resistance to alkaline aqua solutions. Different inner dimensions of the autoclaves (with lining) or the inner container were reported: an inner diameter from 22 mm to 280 mm, and a length from 250 mm to 3000 mm, resulting in a volume up to a few hundreds of liters. The interior of the autoclave/container is divided into dissolution and growth zones by a baffle having an open area of 5–15% of its total surface area. The baffle is typically made of the same metal as the lining or inner container. In the growth zone, a seed holder is situated to hold a number of crystal seeds. It is composed of wires typically made of the same metal as the lining/container/baffle. The seed holder should have a surface area as small as possible to minimize its impact on the mass transfer. High-volume autoclaves may accommodate dozens of seeds, e.g., around 100. Heaters are yet another important component of the autoclaves, which ensure a predefined temperature and temperature distribution in the dissolution and growth zones, particularly a temperature difference in-between. The temperature difference drives a convective heat and mass transfer from the dissolution zone to the growth zone; therefore, the amount of Zn-containing species to be converted into ZnO crystalline material. The heaters are typically segmented with an independent control and externally situated along the autoclave. In large autoclaves, as described by Dem'yanets and Lyutin [214], inner heaters are additionally utilized in the space between the container around the dissolution zone and the inner wall of the autoclave. The inner heaters enable energy saving in long-lasting growth runs and a better temperature control in the dissolution zone.

surfaces, which are positively and negatively charged, respectively, the growth on both surfaces proceeds in a different way:

Polar (0001):

$$\{Zn^{2+}{}_{(surface)}\} + ZnO_2{}^{2-}{}_{(solution)} \rightarrow 2ZnO_{(crystal)} \quad (3.28)$$

Polar (000$\bar{1}$):

$$\{O^{2-}{}_{(surface)}\} + ZnO_2{}^{2-}{}_{(solution)} + H_2O_{(solution)} \rightarrow ZnO_{(crystal)} + 4OH^- \quad (3.29)$$

The polarity-dependent surface reactions lead to differences in the growth rate on both polar surfaces of ZnO.

Moreover, the presence of Li^+ ions in the solution leads to a partial replacement of Zn by Li and, therefore, to a decrease in the positive charge of the (0001) surface, and thus the incorporation of Zn-containing negatively charged species into the crystal:

Polar (0001):

$$\{Zn^{2+}{}_{(surface)}\} + Li^+{}_{(solution)} \rightarrow \{Li^+{}_{Zn}\} + Zn^{2+}{}_{(i)} \quad (3.30)$$

Polar (000$\bar{1}$):

$$\{O^{2-}{}_{(surface)}\} + Li^+{}_{(solution)} \rightarrow \{O^{2-}\} + Li^+{}_{(i)} \quad (3.31)$$

So the presence of Li in the solution decreases the growth rate of the polar surface (0001) and improves the stoichiometry of the obtained crystals. Due to differences in the capture of Li^+ ions at Zn^{2+} and O^{2-} terminated polar surfaces, the tint of both surfaces may also be different: light yellow for (0001) and greenish for (000$\bar{1}$). Laudise and Ballman [144] found that the presence of Li^+ ions in the solution prevents the formation of hillocks in ZnO crystals and, thus, improves the crystals' structural quality. Good sources of Li^+ ions are LiOH, $LiBO_2$, and LiF at concentrations of 0.1–2 mol.%, but in the newest developments, LiOH is used. However, higher Li^+ incorporation at (000$\bar{1}$) also leads to a higher dislocation density as compared to (0001), which was measured at the level of 10^2–10^3 cm^{-2} [216].

Therefore, the growth rate and coloration of the obtained ZnO crystals will depend on the constituents in the alkali solution and their concentrations.

Different compositions of alkali aqueous solutions were used. Laudise et al. [205] obtained the best results with an alkali solution consisting of 5.45 M KOH + 0.7 M LiOH; Croxall et al. [206] used a mixture of 1 M LiOH and 6 M KOH; Sakagami [217] used, in addition to 1.5 M LiOH + 3 M KOH, also a strong oxidizer in the form of H_2O_2 (0.1–0.3 M); Suscavage et al. [208] used as a mineralizer a mixture of Li_2CO_3, 4N KOH, and 4N NaOH; Demianetes and Kostamarov [218] used KOH (5–5.4 M/l) + LiOH (0.8–1.2 M/l). In a modified version, Dem'yanets and Lyutin [214] utilized a hydrothermal solution consisting of NH_4OH (0.5–1 M), KOH (3.5 M), and LiOH (0.5–1 M); Ohshima et al. [215] used LiOH (1 mol/l) + KOH (3 mol/l) as an aqueous solution; Maeda et al. [219] used a mixture of LiOH (1 mol/(1 l H_2O) and KOH (3 mol/1 l H_2O). Yin et al. [213] tested several solvents: KOH (5 M) + LiOH (1 M), NaOH (5 M) + LiOH (1 M), and KOH (5 M) and concluded that the best results were obtained for the mixture of KOH (3–9 M) + LiOH (0.2–3 M). Mass et al. [220, 221], used a mixture of 3N NaOH, 1N KOH, and 0.5N Li_2CO_3 (or LiOH). In growth runs by Zhang et al. [209], the aqueous solution was a mixture of KOH (3 mol/l) + LiOH (1 mol/l) + H_2O_2 (1 wt.%). Wang et al. [211, 222, 223] used the following mineralizers for intentionally doped ZnO crystals: KOH (3–5 N) for doping with Mg, 3N NaOH + 1N KOH + 1N $LiNO_3$ or 20 vol.% NH_4OH + 80 vol.% KOH for doping with N, 4N KOH + 1N LiOH or 3N KOH + 1N NaOH +1N LiOH for doping with In, while Zuo et al. [226] used the aqueous solution as a mixture of KOH (3 mol/l) + LiOH (1 mol/l) + H_2O_2 (1 wt.%) in the case of doping with Sc.

A filling factor of the autoclaves with the solution is typically 65–90%. Higher fillings generate a higher pressure but provides more dissolved ZnO species. The most frequently used filling factor is 70–80%.

(iii) Seeds

In a single autoclave, a number of crystal seeds can be accommodated even about 100 in large-scale designs. Initially the seeds from other growth techniques were used, e.g., from the flux method [144], or from the gas phase [206, 209]. Then the hydrothermally grown crystals are used as seeds. The surface size of the seeds ranges from about 5 mm to 3 inch. The thickness of the seeds is from 0.5 mm to ca. 2 mm.

The surface of the as-cut seeds from a bulk crystal is damaged and negatively affects the resulted crystal quality. Normally, the as-cut seeds require chemical etching to remove the damaged layer. Laudise and Ballman [144] remarked that the thickness of the damaged layer to be removed is at least 0.05–0.07 mm. The etching anisotropy was observed with the etching rates in the following order: $\langle 10\bar{1}1\rangle > \langle 1000\rangle > \langle 000\bar{1}\rangle > \langle 0001\rangle$. Etching in NaOH at 90°C for 4 min was sufficient to remove the damaged layer from all of the faces except (0001) face (etching was too slow). For that face, another etching procedure was used: 100 cm^3 of saturated $K_2Cr_2O_7$ mixed with 2 cm^3 of saturated NaCl, then 10 cm^3 of 15% of H_2SO_4 was added to 1.7 cm^3 of saturated CrO_3 solution and 2.5 cm^3 of glacial acetic acid. This mixture was added to a mixture of 30 cm^3 of HCl and 20 cm^3 of HNO_3. The etching of the (0001) surface was conducted at 25°C for 90 min. To avoid too fast removal of the $(000\bar{1})$ surface, it was masked with a Teflon plate. The damaged layer of ZnO seeds can also be removed in autoclaves before the crystal growth run by applying a small temperature difference between the dissolution and growth zones, which leads to a slow seed dissolution. Dhanaraj et al. [224] stated that after the smooth dissolution of the seeds, no dislocations were found propagating from the seeds, mainly due to the so-called capping effect. After dissolution of the seed surface, a solvent layer is formed, which acts as a buffer layer preventing dislocation propagation from the seeds into the growing crystals. Another method of preparing high-quality seeds was described by Dem'yanets and Lyutin [214]. They prepared the seeds in a two-step manner ("turning seeds"): First the seeds were in the form of plates parallel to $\{10\bar{1}0\}$ or $\{11\bar{2}0\}$ planes on which the crystals were grown. From these crystals, seeds parallel to {0001} were prepared for next growth runs. Such procedure enabled to obtain crystals with extremely low dislocation density, just a few per cm^2.

(iv) Temperature and pressure

To achieve the supercritical state of water and dissolve solid ZnO (source material) in the alkali aqua solution at reasonable rate, a temperature of several hundred degrees Celsius and pressure over 10^7 Pa are required. A combination of both temperature and pressure varies depending on the chemical composition of the solution and filling factor. In practice, for undoped ZnO crystals, the temperature in the autoclaves is typically between 300 and 430°C, and the pressure between 0.7 and 2.55×10^8 Pa (Ehrentraut et al. [71]).

For example, Laudise and Ballman used growth temperature T_G = 350–430°C and pressure p = 1.4–2.6 × 10^8 Pa; Sakagami [217] used T_G = 370–400°C and p = 6.9–9.8 × 10^7 Pa; Dem'yanets and Lyutin [214] applied a dissolution temperature T_D = 330–360°C and p = 3–4 × 10^7 Pa; Ohshima et al. [215] and Maeda et al. [219] used T_G = 300–400°C and p = 0.8–1 × 10^8 Pa; Yin et al. [213] used T_G = 300 or 355°C and p = 1–1.5 × 10^8 Pa; Lin et al. [210] used T_G = 320°C and $p \leq 1 \times 10^8$ Pa. In the case of intentional doping, both growth parameters may be different from those used for undoped crystals due to different dopant concentration and chemistry of the dopant, particularly its solubility in the alkali solution and phase formation. Wang et al. [211, 222, 223] reported T_D = 650°C and p = 1.08–1.24 × 10^8 Pa for growing $Zn_{1-x}Mg_xO$ crystals ($x \leq 0.2$), and T_D = 365°C and $p = 2 \times 10^7$ Pa or T_D = 490°C and $p = 1.36 \times 10^8$ Pa for growing ZnO crystals doped with N, and T_D = 365°C and $p = 1.25 \times 10^8$ Pa for ZnO crystals doped with In (up to 0.55 mol.%), while Zuo et al. [226] used T_G = 360°C for doping ZnO crystals with Sc (up to 0.3 mol.% in the source material).

(v) Temperature difference between the dissolution and growth zones ($\Delta T_{D\text{-}G}$)

A temperature difference $\Delta T_{D\text{-}G}$ between the dissolution and growth zones is the driving force for mass transfer from the dissolution to the growth zone that transports Zn species from the dissolution zone toward the seeds in the growth zone. As a result, the growth rate will increase with $\Delta T_{D\text{-}G}$, as shown in Fig. 3.34. However, that parameter is a compromise between the growth rate and the structural quality of the obtained ZnO crystals. Too high $\Delta T_{D\text{-}G}$ values, roughly above 20 K, will lead to a spontaneous nucleation on the autoclave wall and lid and also to a deteriorated crystal quality. On the other hand, too low $\Delta T_{D\text{-}G}$ values, roughly below 10 K, will cause seeds to dissolve. Indeed, reports on the hydrothermal growth of ZnO most frequently point out to $\Delta T_{D\text{-}G}$ values between 10 and 20 K. For instance, Laudise et al. [205] stated that $\Delta T_{D\text{-}G}$ should rather not exceed 20 K; otherwise, there was a tendency toward spontaneous nucleation and flawed crystals; Croxall et al. [206] found $\Delta T_{D\text{-}G}$ = 10 K as optimum for growth runs; Dem'yanets and Lyutin [214] reported $\Delta T_{D\text{-}G}$ = 8–15 K; Yin et al. [213] and Zhang et al. [209] reported $\Delta T_{D\text{-}G}$ = 10–15 K; Mass et al. [220, 221] found $\Delta T_{D\text{-}G}$ = 10 K; and Lin et al. [210] found $\Delta T_{D\text{-}G}$ = 15–20 K. In the case of intentionally doped ZnO crystals, Wang et al. [211, 222, 223] reported $\Delta T_{D\text{-}G}$ = 50 K and 15 K for doping with Mg and N/In, respectively, while Zuo et al. [212] applied $\Delta T_{D\text{-}G}$ = 15–20 K for doping with Sc.

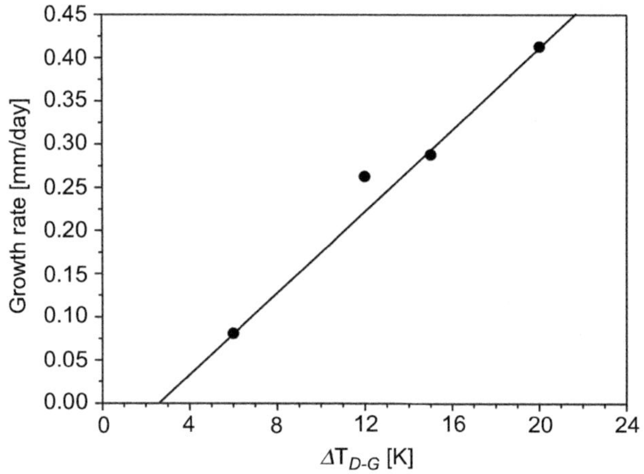

Figure 3.34 Growth rate versus temperature difference between the solution and growth zones, $\Delta T_{D\text{-}G}$, for the fastest growing (0001) face. Reconstructed basing on data from Ref. [205], with some modifications and different visualization.

3.4.3.3 Impurities and intentional doping

The main source of impurities in the hydrothermal growth is the applied alkali solution, i.e., Li^+ and K^+ ions. Other sources of impurities include ZnO source material, water, and lining/container material, which all have limited purity and may provide to the solution some of the impurities. Croxall et al. [206] reported that the green coloration of crystals may be partly associated with Fe, as its concentration increased from 0.15 through 4.5 to 30–60 wt. ppm for colorless, light green, and green sectors, respectively. All samples contained also other impurities at the level of about 1 wt. ppm or below, such as Cd, Ca, Al, K, and Na, wherein colorless samples contained lower level of impurities as compared to green samples. Reduction in coloration was achieved by using ultra-pure starting materials. Suscavage et al. [208] showed that the dark sectors of ZnO crystals contained 3–5 wt. ppm of Fe, about two times higher than the light sectors. On the other hand, the light sectors contained 5–6 wt. ppm of Al and Si, about two times higher than the dark sectors. The crystals reported by Ohshima et al. [215] had Li and K impurity concentration at the level of 0.5–12 and 0.04–0.2 wt. ppm, respectively, which originated from the mineralizer. Moreover, wafers cut from the $-c$ sector showed a few wt. ppm of Fe and Al impurities, while those form the $+c$ sector had much lower concentrations of these impurities. This points the Al and Fe impurities to the pale green coloration in the $-c$ sector. In the crystals reported by Maeda et al. [219], the Li and K concentrations in all sectors were of 0.9 wt. ppm and 0.3 wt. ppm, respectively, which originated from the solution; however, Fe and Al concentrations were higher in the $-c$ sectors as compared to the $+c$ sectors. Yin et al. [213] reported the following main impurity concentrations (in wt. ppm): K = 0.3–1.5, Na = 0.4–2.5, Li = 0.5–6.3, and Fe = 4.3–5.7 (4N purity ZnO powder)/32 (2N purity powder). In the measurements of impurities by Mass et al. [220, 221], the impurity concentrations in the $+c$ sectors (as revealed by SIMS) were as follows (in wt. ppm): Li = 1–10; K and Na ≤ 1–2; Al, Fe, Si, C = 1–10. The impurity concentration in the $-c$ sector was higher by a factor of 5 to 10, what makes this sector green in color. It is usually admitted that *m* sectors have the lowest density of defects and impurities. Lin et al. [210] revealed the following impurity concentrations in the crystals by GDMS (in wt. ppm): Li < 0.005, K = 0.16, Mg < 0.005, Ca = 0.01, Ba < 0.001, B < 0.01, Al = 0.18, Ga = 0.02, Si = 0.37, Pb = 0.97, P = 0.03, Ag = 0.06, and Cd < 0.01.

The distribution of main impurities (by ICP-MS) as a function of a distance from the seed for both surface polarities are shown in Fig. 3.35 [71]. Li (≤12 wt. ppm), Fe (≤11 wt. ppm), and Al (≤ 8 wt. ppm) impurity concentrations decrease on the $(000\bar{1})$ face with a distance from the seed, wherein the concentration of the impurities is larger on the $(000\overline{1})$ face than on the (0001) face. K concentration (≤0.3 wt. ppm) is substantially the same on both faces and does not change with the distance from the seed.

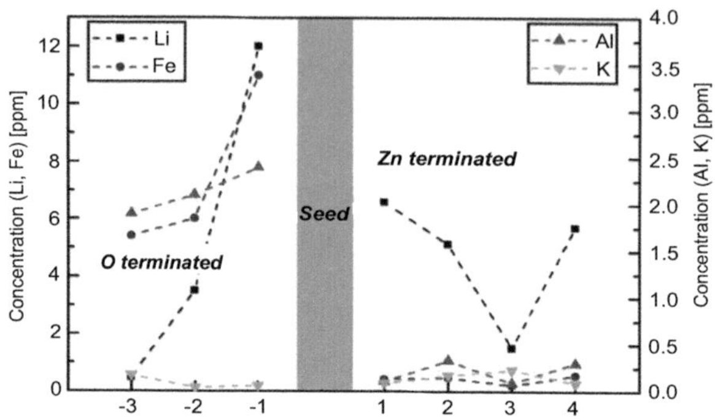

Figure 3.35 Li, Fe, K, and Al impurity distribution in a ZnO crystal obtained by the hydrothermal method with the use of a KOH + LiOH mineralizer. Wafer number indicates a position in the crystal from which it was prepared. Reprinted from Ref. [71], Copyright 2006, with permission from Elsevier.

Sekiguchi et al. [225] noticed that the crystal coloration varied between yellow, pale green, pale yellow, and colorless depending on temperature and pressure without an H_2O_2 oxidizer, while an excess of the oxidizer turned the crystals to yellow coloration. The most pronounced impurity was Li at a concentration between 7.8 and 10.5 wt. ppm. The Li concentration in the crystals grown with no oxidizer was much lower, between 1.7 and 4.6 wt. ppm. Other impurities, such as K, Na, Fe, Mn, and Pt, were at the level of about 0.5 wt. ppm in both groups of crystals.

In addition to unintentional impurities, also intentional doping of bulk ZnO crystals was practiced in the hydrothermal process. Wang et al. [211, 222, 223] doped ZnO crystals with Mg, N, and In, by adding to the alkali solution MgO, $LiNO_3$, and In_2O_3, respectively. The concentration of the incorporated N in ZnO crystals, which were slightly green, was about 10^{18} at. cm^{-3}, while In of 165–175 wt. ppm. Other impurities in In-doped ZnO crystals were (in wt. ppm) Fe = 35–37, Si = 1–1.3, Al = 22, and Li = 12–13. Both N- and In-doped ZnO crystals could be obtained at typical hydrothermal conditions as in the case of undoped crystals, however, not Mg-doped crystals. $Mg(OH)_2$ phase (brucite) can be formed in reaction with water in the presence of NaOH alkali solution at temperatures above 600°C and pressures above 1000 bar. Indeed, at typical temperatures of 350–400°C, no Mg was incorporated into ZnO crystals, but it did at 650 and 600°C in the dissolution and growth zones, respectively, at p = 1.08–1.24 × 10^8 Pa. The obtained crystals had Mg content, x = 0.03–0.055 (Mg concentration of 1.1 × 10^{21} at. cm^{-3} according to SIMS measurements). Li and K concentrations were around 10^{19} and 10^{16} at. cm^{-3}, respectively. The obtained $Zn_{1-x}Mg_xO$ crystals were green and had a thin layer of ZnO, which was formed during cooling. This is because ZnO can be easily formed from $Zn(OH)_2$–H_2O at low temperatures, even at 50°C, while MgO cannot be formed from $Mg(OH)_2$–H_2O at low temperatures.

Doping with Sc was performed by Zuo et al. [226], which was added to the alkali solution as Sc_2O_3. The crystals were grey-green in color. However, the concentration of Sc was rather low, 8.3 and 11.9 wt. pm for the $+c/-c$ sectors, respectively. The impurity concentration for the same sectors were as follows: Al = 0.82/0.94, Fe = 10.7/13.8, K = 0.21/0.4, Li = 1.1/2.1, Cu = 7.4/7.5, Si < 0.3, and Au = 0.85/0.17.

Dem'yanets et al. [227] performed doping ZnO crystals with elements having valency between +2 and +5: Cd^{2+}, Co^{2+}, Ni^{2+}, Sc^{3+}, In^{3+}, Ga^{3+}, Fe^{3+}, Te^{4+}, V^{5+}, and Mn^{3+}, which were added to the alkali solution as oxides. Temperatures in the dissolution and growth zones, as well as pressure, were in the range of those used for undoped crystals. The resulted concentrations of some of the dopants in the crystals were as follows (wt. %): Co^{2+} = 0.26, Ni^{2+}= 0.01, Sc^{3+} = 0.05, In^{3+} = 0.04, Fe^{3+} = 0.04, and Mn^{3+} = 0.76.

3.4.3.4 Growth morphology and growth rate

The growth habit and geometry of a growing ZnO crystal by the hydrothermal method are defined by different growth rates in different crystallographic directions, which, on the other hand, is the result of a growth environment governed by the chemical composition of the alkali solution, operating temperatures in the dissolution and growth zones, pressure, seed orientation, face polarity, and impurity/dopant concentration.

A typical growth morphology of ZnO crystals by the hydrothermal method with the use of (0001)-oriented seeds and the mineralizer consisting of LiOH + KOH and Li_2CO_3 + KOH + NaOH is shown in

Fig. 3.36a,b, respectively. There are five sectors named +*c* and −*c*, which correspond to (0001) and $(000\bar{1})$ monohedrons, respectively; +*p* and −*p* sectors, which correspond to {1011} hexagonal pyramids; and *m* sectors corresponding to (1010) hexagonal prisms. Typically, the crystals consist of 2 basal polar *c*-planes, which are Zn- and O-terminated (+*c* and −*c*, respectively), 6 prismatic *m*-planes, which are non-polar, and 12 polar prismatic *p*-faces, which are Zn- and O-terminated (−*p* and +*p*, respectively). These faces are called habit faces. The size of the faces declines in the following order: $(000\bar{1})$ > $\{10\bar{1}0\}$ > $\{10\bar{1}1\}$ > $\{10\bar{1}\bar{1}\}$ > (0001) [224].

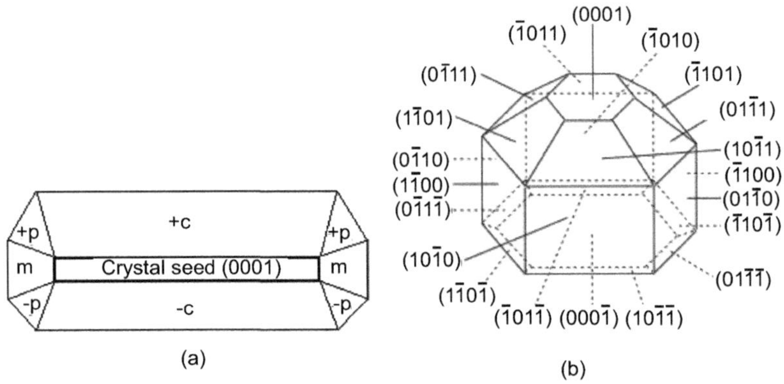

Figure 3.36 Growth morphology of ZnO crystals in the hydrothermal process from the alkaline solution: (a) LiOH + KOH, redrawn from Ref. [219] with some modifications; and (b) Li_2CO_3 + KOH + NaOH, reprinted from Ref. [224], Copyright 2006, with permission from Elsevier.

The $(00\bar{1}1)$ habit faces contained asymmetric hillocks [224] and $(000\bar{1})$ showed hillocks with ledges spreading across the habit face. The presence of these hillocks is pronounced at higher supersaturations.

The impurities and intentional dopants influence the growth rate, growth habit, and also crystal coloration. Sectors +*c*, *m*, and −*p* are typically highly transparent and typically light yellow, whereas sectors +*p* and −*c* are pale green. Different coloration of different sectors is the result of anisotropic impurity incorporation, in particular Li and Fe. Dem'yanets et al. [227] stated that doping with Fe promoted the growth of hexagonal pyramid faces, resulting in pyramidal shape of the crystals, while doping with Mn, Ni, and

In reduced the growth of monohedra, resulting in a tabular shape. Additionally, In promotes the growth of hexagonal prism faces (*m*-planes). Indeed, as shown by Wang et al. [211], in the presence of In^{3+} ions, the growth in the $[10\bar{1}0]$ direction is faster than in the [0001] and $[000\bar{1}]$ directions, i.e., there is a stagnant growth on both {0001} faces and fast growth on normal faces $\{10\bar{1}0\}$ (*m*- or *a*-planes) leading to a plate formation (Fig. 3.37). The surface morphology of the (0001) and $(000\bar{1})$ faces is not smooth but formed by hillocks due to In^{3+} ion incorporation, which blocks the growth on {0001} planes. This can be explained by the formation of growth units in the solution.

$$In_2O_3 + 2OH^- + 7H_2O \leftrightharpoons 2In(H_2O)_2(OH)_4^- \tag{3.32}$$

$$In(H_2O)_2(OH)_4^- \leftrightharpoons InO_2^- + 4H_2O \tag{3.33}$$

$$In(H_2O)_2(OH)_4^- \leftrightharpoons In^{3+} + H_2O + 4OH^- \tag{3.34}$$

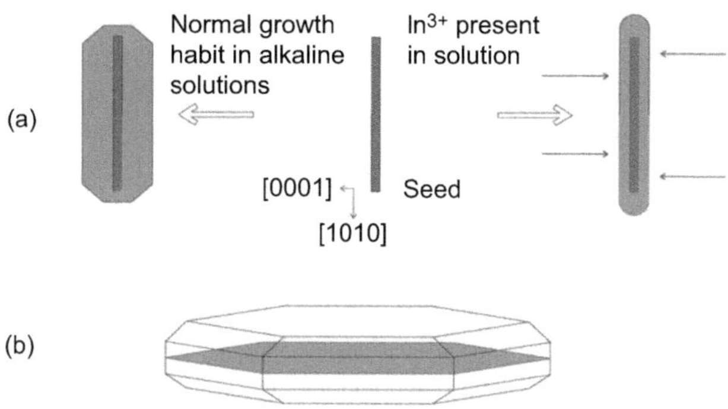

Figure 3.37 Growth morphology of hydrothermal ZnO crystals obtained from alkaline solutions containing Li and by using *c*-plane seeds: (a) a comparison of growth habit between undoped and In-doped crystals; (b) a perspective view of schematic In-doped ZnO crystal. Redrawn from Ref. [211] with some modifications.

InO_2^- tends to incorporate on the positively charged Zn-terminated surface (0001), while In^{3+} on the negatively charged O-terminated surface, which blocks the growth of ZnO (from growth unit $Zn(OH)_4^{2-}$) in these directions. However, ZnO still can growth in other directions, which are electrically neutral (nonpolar), if the impurity level is low (wt. ppm level). The same considerations relate to other dopants or impurities.

Doping with Sc (Zuo et al. [212]) changes the growth morphology in this way that pyramidal faces (*p*-planes) become smaller and prismatic faces (*m*-planes) become larger with increasing Sc_2O_3:ZnO ratio in the solution, as shown in Fig. 3.38. For the doping level of 0.3 mol.% Sc_2O_3, *p*-planes basically disappear. Here, in contrast to In-doping, the presence of Sc in the solution does not affect the growth rate along the [0001] direction and speeds up the growth along the $[000\bar{1}]$ direction, which is comparable to the growth rate along the [0001] direction.

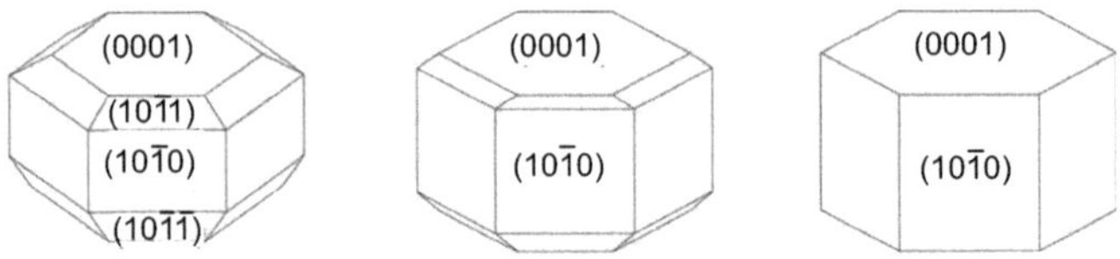

Figure 3.38 Schematic drawing of ZnO growth morphology when doped with Sc_2O_3 in the solution at the concentration of (a) 0.1 mol.%, (b) 0.2 mol.%, and (c) 0.3 mol.%. Redrawn from Ref. [212] with some modifications.

The growth rate of ZnO by the hydrothermal method spreads from a fraction of mm/day to about 10 mm/day for the fastest growing face (0001), depending on growth conditions. For example, Laudise and Ballman [144] reported the fastest growth rate of 10–15 mm/day along <0001>, slower rate of 5–10 mm/day along $<10\bar{1}1>$, and the slowest rate of 5 mm/day along $<10\bar{1}0>$. However, the fastest growing face (0001) was "*flawing*," while the slower growing faces were of good structural quality. According to Dem'yanets et al. [216] and Sakagami [217], the growth rate sequence is as follows: (0001) > $(000\bar{1})$ >> $\{10\bar{1}1\}$ > $\{10\bar{1}0\}$ ~ $\{11\bar{2}0\}$.

The experimental data revealed a difference in the growth rate along [0001] and $[000\bar{1}]$ directions, which is explained by the polar character of the *c*-axis, as expressed by Eqs. 3.28–3.31 and the presence of Li ions. Croxall et al. [206] reported the growth rate along [0001] and $[000\bar{1}]$ directions of 0.45 mm/day and 0.22 mm/day, while Zhang et al. [209] reported 0.17 and 0.09 mm/day, respectively. Demianetes and Kostamarov [218] noticed that the growth rate along the [0001] direction was larger by a factor of 1.5–2 than the growth along the $[000\bar{1}]$ direction; Dhanaraj et al. [224] found the growth rate ratio of 3 to 1, while Ehrentraut et al. [71]

about 2 to 1. Zhang et al. [209] reported the growth rates of 0.17 and 0.09 mm/day for [0001] and [$000\bar{1}$] directions, respectively. Dem'yanets and Lyutin [214] reported the growth rate of 0.15–0.2 mm/day when using seeds oriented in the [0001] direction. Along the $<10\bar{1}0>$ direction, Sakagami [217] measured the growth rate of 0.2 mm/day, and Lin et al. [210] reported 0.08 mm/day. Zuo et al. [226] reported that the growth rate of Sc-doped ZnO was comparable for both (0001) and ($000\bar{1}$) faces, 0.15–0.16 and 0.12–0.16 mm/day, respectively, pointing to much faster growth rate of the ($000\bar{1}$) faces as compared to undoped ZnO crystals.

3.4.3.5 Hydrothermally grown ZnO crystals and structural quality

The size and the structural quality of the obtained ZnO crystals depend on the growth environment that defines the growth rate and growth morphology, and on the growth time. High growth rates lead to poor structural quality of the crystals; therefore, low growth rates, typically below 1 mm/h, are used. To obtain ZnO crystals of sufficient thickness (around or above 1 cm), growth time must last for a number of days, usually between 15 and 30 days. The crystal diameter further depends on the seed size and the volume of the autoclave. A successful development of the hydrothermal method enabled to obtain 2- and 3-inch diameter and about 1 cm thick ZnO crystals, as shown in Fig. 3.39.

The structural quality of the obtained ZnO crystals by the hydrothermal method is relatively high. This can be concluded from the FWHM of the rocking curve and from the etch pit density (EPD) revealed by etching or by dislocation density visualized by X-ray topography. The FWHM of the rocking curve of the (0002) reflection of polished (0001) wafers was measured as low as 8–45 arcsec [71, 209, 210, 215, 219, 226]. Etch pits can be revealed by etching, e.g., in HCl, H_3PO_4, HNO_3, CH_3COOH aqueous solution, or NaOH. The ($000\bar{1}$) face is etched much faster as compared with the (0001) face and may become matt while the latter smooth and flat. As low values of the dislocation density as 10^2–10^3 cm^{-2} for as-grown (0001) faces were reported by Dem'yanets et al. [216] and EPD of 300 cm^{-2} was reported by Maeda et al. [219], which decreased to the value of 80 cm^{-2} after annealing. The lowest dislocation density of 1–5 cm^{-2} was reported by Dem'yanets and Lyutin [214] by using "turning seeds." The etch pits on the (0001) surface (Fig. 3.40) have a form of negative hexagonal prisms that keep crystallographic orientation of the crystal.

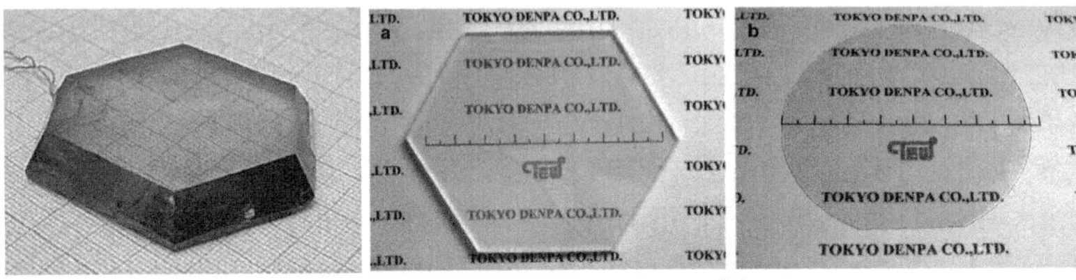

(a) (b)

Figure 3.39 ZnO crystals obtained by the hydrothermal method: (a) 2-inch-diameter and (b) 3-inch-diameter crystal and wafer. Figure (a) reprinted from Ref. [214], Copyright 2008, and (b) reprinted from Ref. [71], Copyright 2006, both with permission from Elsevier.

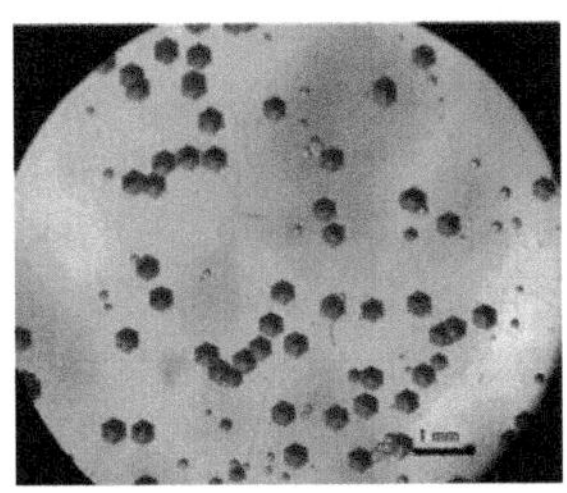

Figure 3.40 Etch pits on the (0001) surface of hydrothermally grown ZnO crystal. Reprinted from Ref. [214], Copyright 2008, with permission from Elsevier.

Due to the anisotropy of impurity incorporation, different sectors of the crystals reveal not only different coloration, but also different structural quality. The higher Li^+ incorporation at the $(000\bar{1})$ face leads to a higher dislocation density as compared to the (0001) face [216]. Due to the lowest impurity incorporation, *m*-planes are regarded as planes of high structural quality as a result of the lowest density of defects.

3.4.3.6 Effect of annealing on the surface quality

The (0001)-oriented wafers can be polished (usually by chemical–mechanical polishing, CMP) to a high flatness to values of about 0.155 nm of the root mean square (RMS) roughness [71]. Mechanical polishing only produced worst surface quality, as the FWHM of the rocking curve was about twice larger as compared to CMP [208]. In addition to polishing and etching, the wafer surface quality can be further modified by annealing at elevated or high temperatures in the presence of oxygen. After CMP, the (0001) and $(000\bar{1})$ surfaces do not show any step or terrace structure, but high flatness, with RMS roughness < 1 nm [219]. The terraces are formed by annealing in the presence of oxygen at 1050–1100°C for 2–4 h [219].

The step height depends on annealing temperature and a face orientation. After annealing at 1050°C and 1100°C for 2 h, the steps are monoatomic (about 0.5 nm) and diatomic (about 1 nm) on the *c*-planes, respectively, and multi-atomic (50 nm) on the *m*-plane, which requires lower thermal activation [71]. Annealing may also cause the formation of voids of 1–10 μm on the polished surface (0001).

Ko et al. [228] showed that annealing of the $(000\bar{1})$ surface in oxygen at 1000–1100°C for 1 h decreased both the FWHM of the rocking curve by a factor of almost 3 and the RMS roughness by factor of about 6 (Fig. 3.41) [228].

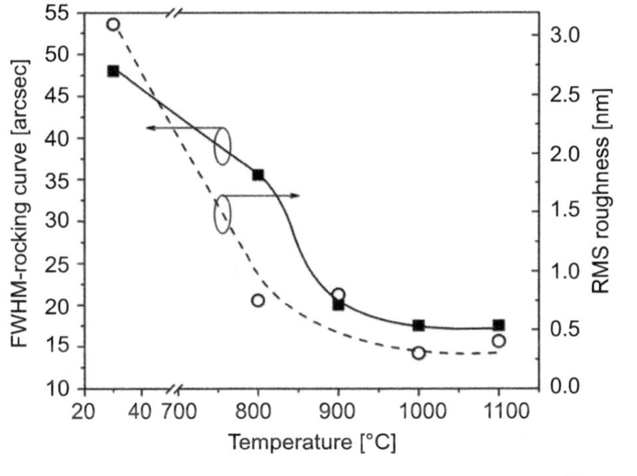

Figure 3.41 FWHM of the rocking curve and roughness of $(000\bar{1})$ surface as a function of annealing temperature in pure oxygen for 1 h. Reconstructed basing on data from Ref. [228], with some modifications and different visualization.

Børseth et al. [229] stated that the annealing of (0001)-oriented wafers in oxygen at 1 bar and 680–960°C improved the FWHM a bit but increased the roughness of both faces. On the other hand, annealing at a lower pressure, 0.5 bar, and 500–1050°C led to an increase in both FWHM and roughness. Annealing in air at the ambient pressure did not change the FWHM but improved the roughness. Further, annealing in ZnO powder substantially did not change the FWHM, but made the +*c*-plane flat with terraces. The surface improvement or degradation upon annealing depends on the equilibrium between Zn and O partial pressures in the crystals and partial pressures of Zn and O in the annealing environment.

3.4.3.7 Examples

This section shows detailed examples of growing bulk ZnO single crystals by the hydrothermal method with different autoclave volumes, growth conditions, and operating parameters that produced different outcome. Some of the examples are summarized in Table 3.3.

(i) For growing bulk ZnO crystals by the hydrothermal method, Laudise and Ballman [144] used a Pt-lined autoclave having an inner diameter of 25.4 mm and inner length of 305 mm. It could withstand pressures (p) up to 2.4×10^8 Pa (35,000 psi) at about 470°C. The autoclave was divided by a disc or baffle with an opening, into a dissolution zone, where the ZnO source material was located, and a growth zone, where ZnO seeds were situated. The seeds were prepared from ZnO crystals, which were spontaneously nucleated in previous hydrothermal growth runs, or which were obtained from PbF_2 flux by Nielsen and Dearborn [145]. In both cases, the seed orientation was <0001>. The most important growth parameters were identified as follows: a temperature of the growth zone (T_G), a temperature difference between the dissolution and growth zones ($\Delta T_{D\text{-}G}$), fill fraction (FF) of the autoclave, and a concentration of NaOH. The best results were obtained at the following conditions: $350 < T_G < 430$°C, $\Delta T_{D\text{-}G} \leq 10$–15 K, $FF \geq 70\%$, $p = 1.4$–2.6×10^8 Pa, and NaOH concentration of 1 M. Growth runs lasted for 2 to 40 days. The obtained crystals were relatively large, weighted several grams, and had an amber color. The growth habit during spontaneous nucleation or on (0001) seeds was hemimorphic, i.e., both ends of crystals developed in a different habit. They were likely heavily contaminated with iron. The fastest growth rate of 10–15 mm/day was along $<0001>$, slower of 5–10 mm/day along $<10\bar{1}1>$, and the slowest, 5 mm/day, along $<10\bar{1}0>$. The fastest growing face (0001) was "*flawing*," while slower growing faces were of good quality. In further experiments, Laudise et al. [205] used Morey-type autoclaves with Ag lining. Autoclaves of two different sizes were used: about 22 mm in diameter and about 70 mm long (short version), and 30 mm in diameter and 175 mm long (long version). A baffle separating the dissolution and growth zones had a 5% opening area, where half of this area was located centrally, while another half at the periphery. The seeds were mounted on a silver wire frame. The authors identified seven important factors in the growth of ZnO by the hydrothermal method:

a. **Base concentration:** Good solubility of ZnO was obtained at base concentrations from 4 to 6 mol and FF = 75–85%. The solubility of ZnO in KOH was similar to that of NaOH at the same concentration. The structural quality of the crystals was better when grown in KOH than in NaOH solution.
b. **Temperature difference:** $\Delta T_{D\text{-}G}$ should rather not exceed 20 K to achieve the supersaturation of about 0.1% in 6.47 M of KOH or NaOH at temperatures between 300 and 400°C under p = 550 bar. Otherwise, there was a tendency toward spontaneous nucleation and flawed crystals.
c. **Pre-saturation:** To avoid dissolution of the seeds during heating up, the solution was pre-saturated at 90°C. However, when using thicker seeds, the solution did not need to be pre-saturated as etching of the seeds during heating up had a positive effect on avoiding the spontaneous nucleation and flawed growth.
d. **Warm-up:** During warm-up, the solubility of ZnO is lower and temperature gradients are rather large; therefore, there was spontaneous nucleation and flawed growth during that period. To minimize these negative effects, suitable warm-up procedures were applied. The short autoclaves were warmed up within 2 h, and

Table 3.3 Selected hydrothermal growth environments for bulk ZnO single crystals. EPD, FWHM, T_D, T_G are etch pit density, full width at half maximum, temperature in the dissolution zone, and temperature in the growth zone, respectively

Property	Laudise et al. [205]	Dem'yanets and Lyutin [214]	Oshima et al. [215] Maeda et al. [219]	Zhang et al. [209]	Dhanaraj et al. [224]
Inner dimensions of the autoclave: diameter and length [mm×mm]	22×70 30×175	280×1000	50×1000 200×3000	55×1080	
Lining/container (liner)	Lining	Container	Container	Lining	Lining
Lining/container (liner) material	Ag	Ti-based alloy	Pt	Au	Pt
Open surface area of the baffle [%]	5			5–8	
Alkali solution	5.45 M KOH + 0.7 M LiOH	NH_4OH (0.5–1 M), KOH (3.5 M), and LiOH (0.5–1 M)	LiOH (1 mol/l) + KOH (3 mol/l)	KOH (3 mol/l) + LiOH (1 mol/l) + H_2O_2 (1 wt.%)	Li_2CO_3 + KOH + NaOH
ZnO source material	Hydrothermally grown ZnO	Sintered and pressed powder tablets + crystals (2–5 mm)	Polycrystals	Sintered powder	Sintered powder
Filling factor [%]	83			75	80
Temperature in the dissolution and/growth zone [C]	T_G=353	T_D=330–360	T_G=300–400	T_G=380	T_G=355

Property	Laudise et al. [205]	Dem'yanets and Lyutin [214]	Oshima et al. [215] Maeda et al. [219]	Zhang et al. [209]	Dhanaraj et al. [224]
Temperature difference between the dissolution and growth zones [K]	14	8–15		10–15	10
Pressure [Pa]	5.5×10^7	3–4×10^7	8–10×10^7	10×10^7	
Seed orientation	(0001)	(0001) "turning seeds"	(0001)	(0001)	(0001)
Growth time [days]	(a) 10 (b) 30			30	30
Maximum growth rate [mm/day]	(a) 0.35 (b) 0.3	0.15–0.2 in the [0001]		0.17/0.09 in the [0001]/[000$\bar{1}$]	10
Crystal size		3-inch diameter	(a) 25×15×12 (b) 50×50×15	30×38×8	
Crystal coloration	Amber		(a) pale yellow (b) pale green/ colorless	Yellow–green/brown	
Main impurities [wt. ppm]		Li, Na, K=100, K=0.04–2, Fe, Al, Ca, Si < 70, Ti < 5	Li=0.5–12, K=0.04–2, Fe, Al=a few ppm	Li=8–10, Au=10–12, Al=2–12, Fe=1– 30, Cu=1–6, Pb=2–12, Si=15–30	
Dislocation density or EPD [cm^{-2}]		1–5	80–300		100–1000
FWHM of the rocking curve [arcsec]		53	8	45	

when the temperature was above 250°C, the $\Delta T_{D\text{-}G}$ was not permitted to exceed 20 K. The long autoclaves were warmed up for 24 h, and the $\Delta T_{D\text{-}G}$ was not permitted to exceed 25 K above 150°C, and 20 K above 250°C.

e. **Addition of Li^+ ions:** It was found that the presence of Li^+ ions in the solution prevented the formation of hillocks in ZnO crystals and, therefore, improved the structural quality. Good sources of Li^+ ions were LiOH, $LiBO_2$, and LiF at concentrations of 0.1–2 M. Under the same growth conditions, the presence of Li^+ ions decreased the growth rate along the *c*-direction, and slightly increased the growth rate along the *a*-direction.

f. **Seed damage:** A correlation of the seed quality on the resulted structural quality of the obtained ZnO crystals was noticed as an important factor. The seed quality was achieved by chemical etching to remove at least 0.05–0.07 mm of a damaged surface formed after sawing the seeds.

g. **Nutrient size:** Very small particle size of the source material (nutrient) resulted in low growth rates and flawed crystals, as the small particles acted as nucleation centers. On the other hand, large bits of the source material had smaller surface area and, therefore, the process was growth-rate limited. Thus, the source material with intermediate size provided the best results.

The best results were obtained in long autoclaves, with an alkali solution consisting of 5.45 M KOH + 0.7 M LiOH, FF = 83%, $\Delta T_{D\text{-}G}$ = 14 K, T_G = 353°C, and the particle size of the source material slightly larger than 2 mm. Pre-saturation was applied to the seed prepared from the flux-grown crystals, but not for hydrothermally grown crystals. Under these conditions, the growth rate was about 0.35 mm/day in 10-day runs, and about 0.3 mm/day in 30-day runs. The growth rate of the fast-growing (0001) face linearly scaled with $\Delta T_{D\text{-}G}$. The obtained crystals were hemimorphic with smooth (0001) and $(000\bar{1})$ faces. The crystal coloration was amber. Li concentration in the crystals was about 50 wt. ppm.

When compared with the hydrothermal growth of SiO_2, a higher concentration of mineralizer $(OH)^-$ was required for

ZnO to achieve a similar solubility, as ZnO is less acidic than SiO_2. KOH produced less flawed crystals than NaOH because of reduced constitutional supercooling (K^+ has larger diffusion constant as compared to Na^+) that causes a tendency to a dendritic growth. The $\Delta T_{\text{D-G}} > 20$ K caused spontaneous nucleation and flawing. At these growth conditions, the (0001) face is not an equilibrium surface under hydrothermal conditions, and it grows by hillocks or faces with higher Miller indices having smaller surface energy. In such environment, the growth rate at the hillocks is higher than at the valleys due to higher supersaturation and low-temperature gradient at the growth interface leading to the constitutional supercooling. Such growth can be dendritic or crevice flawed with large voids or misoriented grains. The role of Li^+ ions is to increase the movement of steps in the *a*-direction relative to the nucleation of new steps in the *c*-direction. The presence of 0.1 mol Li^+ does not alter the solubility of ZnO but suppresses the dendritic growth.

(ii) Croxall et al. [206] used autoclaves made of steel with a 0.3 mm thick Pt liner. The inner diameter was 22 mm, while the inner length was 235 mm. The source material was ZnO powder, which was initially pressed into pellets, sintered at 900°C, washed in LiOH solution, and sintered again at 1200°C in air. As a solvent, a mixture of 1 M LiOH and 6 M KOH was used with the FF = 83%. Initially ZnO crystals grown from the vapor phase were used as seeds. Next the hydrothermally grown ZnO crystals with the basal plane (0001) were used as seeds. T_{D} = 365°C and $\Delta T_{\text{D-G}}$ = 10 K were found for optimum growth runs. If that temperature difference was too large, there was a spontaneous nucleation on the autoclave wall and lid, while in the case of too low temperature difference, the seeds were dissolved. The growth rate was 0.45 mm/day along the <0001> direction, and 0.22 mm/day in the opposite direction. The obtained crystals of up to 15 mm in cross section and 8 mm thick were facetted with green coloration found on the following faces: $(000\bar{1})$, $(\bar{1}011)$, and $(10\bar{1}1)$. The degree of green coloration was a bit reduced by annealing at temperatures up to 1000°C in air. Mass spectroscopic measurements revealed that the green coloration may be partly associated with iron, as its concentration increased from 0.15 through 4.5 to 30–60 wt. ppm for colorless, light green, and green sectors, respectively. All samples contained also other impurities at the level of about 1 ppm or below, such as Cd, Ca, Al, K, and Na, wherein colorless samples contained lower level of impurities as compared to green samples. Reduction in coloration was achieved by using ultra-pure source material. The observed defects were mainly dislocations, which were observed by X-ray topography and etching either in NaOH at 90°C or in 10% HNO_3 at RT. After etching, the $(000\bar{1})$ face (etched rapidly) became matt in appearance, while the (0001) face (etched slowly) remained smooth with no steps.

(iii) In addition to LiOH (1.5 M) and KOH (3 M) as a solvent for ZnO, Sakagami [217] used also a strong oxidizer in the form of H_2O_2 (0.1–0.3 M). Growth temperature was 370–400°C, and total pressure was 700–1000 atm. The growth duration was 15–20 days. With the growth rate along the $<10\bar{1}0>$ direction of 0.2 mm/day, the obtained ZnO crystals grown with the oxidizer were substantially colorless, while the crystals obtained at the same conditions, but with no oxidizer, were yellow. In another work, in which the crystals of size 10 mm grown on $5 \times 5 \times 0.5$ mm seeds for 2 weeks at similar conditions, Sekiguchi et al. [225] noticed that the crystal coloration varied between yellow, pale green, pale yellow, and colorless depending on temperature and pressure without an oxidizer, while an excess of the oxidizer turned the crystals to yellow coloration. The most pronounced impurity was Li at a concentration between 7.8 and 10.5 wt. ppm. The Li concentration in the crystals grown with no oxidizer was much lower, between 1.7 and 4.6 wt. ppm. Other impurities, such as K, Na, Fe, Mn, and Pt, were at the level of about 0.5 wt. ppm in both groups of crystals. Electrochemical coulometric examination revealed an excess of Zn with respect to stoichiometric composition. That Zn excess in the crystals grown with no oxidizer was between 13.5 and 21.6 wt. ppm, while in the crystals grown in the presence of the oxidizer, the Zn excess was only between 0.8 and 1.7 wt. ppm. The excess of Zn acting as a color centers was concluded as a source of yellow coloration. Further work with an oxidizer was performed by Sakagami et al. [207] at similar conditions. The source material was ZnO powder sintered at 1100°C for 24 h; the seeds were crystalline bars parallel to *c*-axis, while the growth runs lasted for 14 days. The obtained crystals of about 10–15 mm diameter and about 20 mm length were substantially colorless.

(iv) Suscavage et al. [208] used autoclaves made of high-strength steel with a sealed Pt liner. The mineralizer was a mixture of Li_2CO_3, 4N KOH, and 4N NaOH with FF = 80%. The source material was 99.99% ZnO powder with particle size below 3 μm, which was sintered at 1350°C in air for 4 h in a Pt crucible. The seeds were (0001) plates from previous hydrothermally grown crystals. The temperatures in the growth and dissolution zones were 355 and 365°C (the temperature difference was 10 K), respectively. The growth rate in the [0001] direction was 0.25 mm/day for the period of 30 days. The growth habit was anisotropic, with the ratio of the growth rates between +*c* and –*c* of 3 to 1, due to different atomic arrangement at their surfaces. The coloration of the +*c* sector was clear to light yellow–green, while that of the –*c* sector was typically dark green, due to different impurity incorporation in both sectors. The GDMS analysis showed that the dark sector contained 3–5 wt. ppm of Fe, about two times higher than the light sector. On the other hand, the light sector contained 5–6 wt. ppm of Al and Si, about two times higher than the dark sector. The FWHMs of the rocking curve of both –*c* and +*c* faces were 104 and 108 arcsec for mechanical polishing and 37 and 43 arcsec for CMP, respectively.

(v) Demianetes and Kostamarov [218] used autoclaves made of steel with an Ag lining. The solvent and the source material were KOH (5–5.4 M/l) + LiOH (0.8–1.2 M/l) and fine crystalline ZnO powder, respectively. T_D = 270–350°C in the dissolution zone and T_G = 250–300°C in the growth zone resulted in the temperature difference ΔT between both zones of 20–80 K. An open surface area of the baffle between both zones was 7–15%. The crystal seeds prepared from previously grown crystals were plates parallel to (0001), $(10\bar{1}0)$, and $(101\bar{1})$. The growth rate along the +*c*-direction was larger by a factor of 1.5–2 than the growth along the –*c*-direction, as the result of the polar character of the *c*-axis. The presence of Li in the solution decreases the growth rate of the polar surface (0001) and improves the stoichiometry of the obtained crystals. Due to differences in the capture of Li^+ ions at Zn^{2+}- and O^{2-}-terminated polar surfaces, the tint of both surfaces may also be different: light yellow for (0001) and greenish for $(000\bar{1})$. Higher Li^+ incorporation at $(000\bar{1})$ also leads to a higher dislocation density as compared to (0001), which was measured at the level of 10^2–10^3 cm^{-2} [216].

In a modified version, Dem'yanets and Lyutin [214] utilized autoclaves comprising a body and an inner container made of Ti-based alloy, which was divided by a baffle into dissolution and growth zones. The heaters were arranged outside the autoclave body and inside the body around the dissolution zone. Crystals of 2- and 3-inch diameter were grown in autoclaves having 3–5 l (80 mm diameter) and 500 l (280 mm diameter and 1000 mm length) capacity, respectively. Smaller autoclaves could accommodate 16–20, while larger ones 96–100 crystal seeds in the growth zone. The hydrothermal solution consisted of NH_4OH (0.5–1 M), KOH (3.5 M), and LiOH (0.5–1 M). The following operating parameters were used: $T_D = 330$–$360°C$, $\Delta T = 8$–15 K, and $p = 30$–40 MPa. The seeds oriented in the [0001] direction resulted in the growth rate of 0.15–0.2 mm/day. The source material (nutrient) in the dissolution zone consisted of cylindrical tablets made of pressed and sintered (at 800–1000°C) high-purity ZnO powder and of ZnO crystal bits (2–3 mm in size). The seeds along the +*c*-axis in the form of plates parallel to the (0001) face were prepared from previously grown crystals, which were lapped and etched prior use. To decrease the dislocation density, the seeds were also prepared in two steps: First the seeds were in the form of plates parallel to $\{10\bar{1}0\}$ or $\{11\bar{2}0\}$ on which the crystals were grown. From these crystals, seeds parallel to {0001} were prepared for the next growth runs. The best crystals had a very low dislocation density of 1–5 cm^{-2} only, particularly those prepared in the two-step manner. Other crystals had the dislocation density at the level of 10^3 cm^{-2}. Dem'yanets et al. [227] performed doping with elements having valency between +2 and +5: Cd^{2+}, Co^{2+}, Ni^{2+},

Sc^{3+}, In^{3+}, Ga^{3+}, Fe^{3+}, Te^{4+}, V^{5+}, and Mn^{3+}. The hydrothermal solution was KOH (5.15 M), LiOH (1.2 M), and H_2O. The doping ions were provided into the solution in the form of metal oxides in the following concentrations (wt. %): NiO = 1.5, CoO = 3.3, Sc_2O_3 = 2.9, In_2O_3 = 0.9, Fe_2O_3 = 1.2, Mn_2O_3 = 3. For growth runs T_D = 330–350°C, T_G = 230–300°C, and p = 50–250 MPa were applied. The seeds were prepared parallel to the (0001) face. The resulted concentrations of some of the dopants in the crystals were as follows (in wt. %): Co^{2+} = 0.26, Ni^{2+} = 0.01, Sc^{3+} = 0.05, In^{3+} = 0.04, Fe^{3+} = 0.04, and Mn^{3+} = 0.76. The dopants influenced the growth rate, growth habit, and also coloration. Doping with Fe promoted the growth of hexagonal pyramid faces, resulting in pyramidal shape of the crystals, while doping with Mn, Ni, and In reduced the growth of monohedra, resulting in a tabular shape. Additionally, In promoted the growth of hexagonal prism faces. Mn also changed the crystal coloration to orange or red.

(vi) Ohshima et al. [215] used ZnO polycrystals as the source material and LiOH (1 mol/l)/KOH (3 mol/l) as an aqueous solution (mineralizer). The autoclave contained a closed Pt container, which was divided by a Pt baffle into dissolution and growth zones, and after loading the mineralizer and seed therein, the container was sealed up. The space between the autoclave wall and the Pt container was filled with distilled water to balance the pressure. The temperature in the growth zone was 300–400°C, and the pressure in the autoclave was 80–100 MPa. The Pt container of 50 mm inner diameter and 1 m length produced transparent crystals with a pale yellow coloration and size of 25 × 15 × 12 mm^3, while the Pt inner container of 200 mm inner diameter and 3 m length produced transparent crystals with pale green coloration and size of 50 × 50 × 15 mm^3. The pale green coloration originated from the $-c$ sector, while the $+c$ sector was substantially colorless. The 2-inch-diameter crystals grown on c-oriented seeds had two large faces $+c$ and $-c$, $+/-p$ faces on the crystal periphery with m-faces between each of pair $+/-p$ faces. The FWHM of the rocking curve was measured as low as 8 arcsec for (002) reflections and 25 arcsec for (101) reflections. The wafers prepared perpendicularly to the c-axis revealed Li and K impurity concentration (by ICP-MS) at the level of 0.5–12 and 0.04–0.2 wt. ppm, respectively, which originated from the mineralizer. Moreover, wafers cut from the $-c$ sector showed a few wt. ppm of Fe and Al impurities, while those form the $+c$ sector had much lower concentrations of these impurities. This points the Al and Fe impurities to the pale green coloration in the $-c$ sector.

(vii) In growth runs of Maeda et al. [219], the solvent (mineralizer) was a mixture of LiOH (1 mol/1 l H_2O) and KOH (3 mol/1 l H_2O). Within the autoclave, a Pt container divided by a baffle into dissolution and growth zones was used. The crystals were grown from a supercritical water at temperature 300–400°C and pressure 80–100 MPa. The space between the Pt container and the autoclave was filled with distilled water to equalize the pressure between the Pt container and the autoclave. The obtained crystals were of 2 inch across the c-plane. The FWHM of the rocking curve of the (002) reflection was 18 arcsec. Li concentration of 0.9 wt. ppm and K concentration of 0.3 wt. ppm, which originate from the solution, were found in all sectors, wherein Fe and Al concentrations were higher in the $-c$ sector as compared to the $+c$ sector.

(viii) Yin et al. [213] tested several solvents: KOH (5 M) + LiOH (1 M), NaOH (5 M) + LiOH (1 M), and KOH (5 M). The ZnO source material was a powder of either 2N or 4N purity, which was pressed into pellets, sintered at 1350°C, and crushed into granules of 1–6 mm size. The autoclave was 50 mm in diameter and 600 mm long. The temperature in the growth zone was either 355 or 300°C; the temperature difference between the growth and dissolution zones was 10–15 K, and the pressure was 100–150 MPa. Both dissolution and growth zones were separated by a baffle with an open area of 5–6% with respect to the total surface area of the baffle. The seeds were prepared either from hydrothermally grown crystals or from the crystals obtained from the gas phase. The main impurities from the solution and starting material were at the following concentrations (in wt. ppm): K = 0.3–1.5, Na = 0.4–2.5, Li = 0.5–6.3, and Fe = 4.3–5.7 (32 for 2N purity powder). The crystals obtained from the higher purity starting material (4N) at 355°C were almost colorless, yellow, or pale yellow, and those from low purity were deep yellow. The crystals obtained at lower temperature, 300°C, were green. The authors concluded that the growth rate of ZnO was about 1.5 times higher from pure KOH solution than from pure NaOH solution, and pure LiOH was too weak for supersaturation and reasonable growth rate. The best results were obtained for the mixture of KOH (3 to 9 M) and LiOH (0.2–3 M).

(ix) Mass et al. [220, 221] used an autoclave made of stainless steel with a Pt liner. Alkali solvent was a mixture of 3N NaOH, 1N KOH, and 0.5N Li_2CO_3 (or LiOH). The filling was 80%. The temperatures in the growth and dissolution zones were 345 and 355°C, respectively. The seeds were (0001) of size 2 cm^2 × 0.5 mm. In addition to the basal planes $+c$ and $-c$, the obtained crystals had 6 prismatic faces (1010) (*m*-planes), which are nonpolar, and 12 prismatic faces (1011) (*p*-planes), which are either Zn-terminated ($-p$) or O-terminated ($+p$). The impurity concentrations in the $+c$ sector, as revealed by SIMS, were as follows (in wt. ppm): Li = 1–10; K and Na ≤ 1–2; Al, Fe, Si, C = 1–10. The impurity concentration in the $-c$ sector was higher by a factor of 5 to 10, which makes this sector green in color. It was admitted that the *m* sector has the lowest density of defects and impurities.

(x) Dhanaraj et al. [224] applied an autoclave made of steel with a Pt liner. The mineralizer solution was a mixture of Li_2CO_3, KOH, and NaOH, which filled 80% of the volume. The source material was 4N ZnO powder. The temperature of the growth zone was kept at 355°C. The growth proceeded on (0001) seed plates with the rate of 10 mm/day for 30 days. The growth was anisotropic with the ratio between fast $+c$– and slow $-c$-directions of 3 to 1. The crystals grew with (0001) and $(000\bar{1})$ monohedra, (1010) prism, and {1011} pyramid faces. The $(000\bar{1})$ sector of the crystals was dark green due to higher impurity incorporation, while (0001) sector was light green. X-ray topography revealed that the $(10\bar{1}0)$ slice (prism face) was strained and contained high density of dislocations propagating substantially normal to *c*-planes, which may reach the level of 10^6 cm^2, although in other crystals, it was at the level of 10^2–10^3 cm^2, depending on growth conditions. The dislocations propagated into the growing crystals from the seeds. Also newly created edge dislocations were noticed. After optimizing the growth conditions, by a smooth dissolution of the seeds, no dislocations were found propagating from the seeds, mainly due to the so-called capping effect. After dissolution of the seed surface, a solvent layer is formed, which acts as a buffer layer preventing dislocation propagation from the seeds into the growing crystals. In samples cut parallel to the basal *c*-plane, substantially no dislocations were observed in X-ray topographs. The dislocations, which are normal to the *c*-face, were then revealed by etching in dilute acetic acid (30%) and water (70%) at RT for 100–150 s. In some of the regions, a small density of hexagonal etch pits was present, but in others not, confirming very low dislocation density in the +*c* sectors. The etching was much faster on the –*c* surface than on the +*c* surface. Additionally, slip bands were found parallel to six prism faces, which could be created by anisotropy in thermal expansion and inclusions. This may indicate softness of the crystals. Indeed, the measured Vickers microhardness values were 340 and 270 kg/cm^2 on (0001) and $(10\bar{1}0)$ faces, respectively.

(xi) Ehrentraut et al. [71] worked with small autoclaves (16 mm inner diameter and 200 mm inner length) with a filling factor of 65–80%. The inner container was made of Pt. The space between the autoclave and the Pt inner container was filled up with a suitable amount of distilled water for balancing pressure. The baffle and the seed holder were made of Pt. This autoclave size could produce ZnO crystals with 1-inch diameter across the (0001) plane. Large crystals of 2- and 3-inch diameter across the (0001) plane were produced in containers of 200 mm and 300 mm diameter (3 m length), respectively. In one growth run, about 100 crystals of 2-inch diameter (across (0001)) and 1 cm thick could be obtained on (0001) and $(000\bar{1})$ seeds. Also 3-inch-diameter crystals were 1 cm thick. In all cases, the crystals were light yellow. The growth rate on the (0001) face was about two times greater than on the $(000\bar{1})$ face. The presence of Li^+ ions slows down the growth rate in the <0001> direction and slightly increases in the <1100> direction, because the positive charge on the surface decreases and, therefore, the incorporation of negatively charged Zn species decreases. The authors found that the molar ratio of KOH to LiOH at the level of 3 to 1 provided the best results. After CMP, the FWHM of the rocking curve using (0002) reflection of (0001) wafers was as low as 19–30 arcsec. The smallest RMS roughness of the wafers was measured by AFM at the level of 0.155 nm.

Etching revealed the EPD of 300 cm^{-2} and less than 80 cm^{-2} after annealing [219]. The etch pits on the (0001) face had six-fold symmetry with pyramidal facets, while on the $(000\bar{1})$ face were less faceted.

(xii) In growth runs by Zhang et al. [209], the autoclaves were made of high-strength steel with a gold liner of diameter 55 mm and length 1080 mm. An Au baffle was used with the opening area of 5–8%. The aqueous solution was a mixture of KOH (3 mol/l) + LiOH (1 mol/l) + H_2O_2 (1 wt.%). The filling factor was 75%. In the space between the liner and autoclave, distilled water was used. The temperature in the growth zone was 380°C; the temperature difference between the dissolution and growth zones was 10–15 K, and the pressure was about 100 MPa. The source material was a ZnO powder (4N), which was sintered at 1300°C in air for 24 h in a Pt crucible. The seed holder was also made of Au, and the seeds were prepared from CVT-grown crystals in the form of (0001) plates. The growth duration was 30 days, and the growth rates were 0.17 and 0.09 mm/day for [0001] and $[000\bar{1}]$ directions, respectively. The largest size of the obtained crystals was 30 × 38 × 8 mm^3. The FWHM of the rocking curve for the (0002) reflection was 45 arcsec. The impurity concentrations (in wt. ppm) for $+c/-c$ sectors were as follows: K < 1, Li = 8/10, Au = 10/12, Al = 2/12, Fe = 1/30, Cu = 1/6, Pb = 2/12, and Si = 15/30. The $+c$ sector was yellow-green and the –c sector was dark brown.

(xiii) Lin et al. [210] grew ZnO crystals from autoclaves (made of Ni–Cr alloy) with a liner having 66 mm inner diameter, 1000 mm height, and 3400 ml capacity. The baffle opening was 10%. The liner, the baffle, and the seed holder were made of low-cost material. The source material was 99.5% purity ZnO powder, which was pressed and sintered at 900°C for 4 h in a Pt crucible. The space between the liner and the autoclave was filled with distilled water. No information on the mineralizer was given, but the filling factor was 80% and the pressure was up to 100 MPa. Temperature in the growth zone was 320°C with the temperature difference between the dissolution and growth zones of 15–20 K. The heating time was 20 h. Over 100 plate seeds of 1 × 1 cm^2 were mounted onto a seed holder. The growth period was 30 days. Under new mineralizers (not disclosed), the fastest growth rate was 0.15 mm/day for $(10\bar{1}0)$, next 0.08 mm/day for (0001), and then 0.06 mm/day for $(000\bar{1})$. This is different as compared with conventional mineralizers, where the fastest growth rate was typically observed along <0001> direction, 0.25 mm/day. The obtained crystals were of 30 × 25 × 5 mm^3. The FWHM of the rocking curve of the (002) reflection was 36 arcsec. Impurity concentrations in the crystals by GDMS were as follows (in wt. ppm): Li < 0.005, K = 0.16, Mg < 0.005, Ca = 0.01, Ba < 0.001, B < 0.01, Al = 0.18, Ga = 0.02, Si = 0.37, Pb = 0.97, P = 0.03, Ag = 0.06, Cd < 0.01. The authors claim that they have a higher concentration of H in the solution and very low Li concentration in the crystals. The idea of a new mineralizer was based on the chemical potential, where impurities from the solution can hardly be incorporated into the crystal lattice if the chemical potential of an impurity in the crystal is higher than in solution. According to the authors, this can be done by modification of the pH of the solution (toward acidity) and adding a complexant. After the CMP, the Zn surface showed an RMS roughness of 0.916 nm, but no steps were observed. Annealing at 900°C in oxygen for 2 h revealed steps with the height of 2- or 3-unit cells extending along the $[10\bar{1}0]$ direction with the distance of 91–120 nm.

(xiv) Wang et al. [222] reported the growth of $Zn_{1-x}Mg_xO$ crystals. Nickel-based autoclaves were used to apply temperature greater than 600°C. Two autoclaves were used: one had a 140 cm^3 internal volume and 2.22 cm inner diameter, and the other one had a 260 cm^3 internal volume and 3.3 cm inner diameter. Both autoclaves could withstand 414 MPa at 600°C and 276 MPa at 700°C. A Pt liner was used to isolate the growth environment from the autoclaves' wall due to corrosive hydroxide used as a mineralizer. The Pt liner had an inner diameter of 14 mm, volume of 16–33 cm^3 with 45–60% fill. Between the Pt liner and the autoclave wall, the same filling factor of water was used to counterbalance the pressure. KOH (3–5 N) as the mineralizer, sintered ZnO powder as the source material, and pure ZnO hydrothermal crystals as seeds were used. Up to 20 wt.% of MgO (5N) was added into the solution. It was 9–20 g of ZnO and 0.5–5 g of MgO. In experiments made in alkaline solutions at 350–400°C, no Mg was incorporated into ZnO crystals. This is due to the fact that the MgO phase (periclase) can be formed only above 700°C at a pressure below 1 bar, while the $Mg(OH)_2$ phase (brucite) is stable below 700°C. This is why temperature for growing $Zn_{1-x}Mg_xO$ must be higher than that used for pure ZnO. Successful growth runs were performed at 650 and 600°C in the growth and dissolution zones, respectively, at the pressure of 103–124 MPa. If the temperature difference between both zones was below 10 K, ZnO seeds dissolved or no growth occurred. At the temperature range of 650–675°C, pressure of 169 MPa, and 40–60% fill, no $Zn_{1-x}Mg_xO$ crystals were formed, but they formed at lower pressure (124 MPa) at $x = 0.03$–0.05. The obtained crystals had Mg content, $x = 0.03$–0.055 (Mg concentration of 1.1×10^{21} at. cm^{-3} according to SIMS measurements). Li and K concentrations were around 10^{19} and 10^{16} at. cm^{-3}, respectively. The thermodynamic solubility limit of Mg in hexagonal MgZnO is about $x \approx 0.06$ at these conditions. The obtained ZnMgO crystals were green and had a thin layer of ZnO, which was formed during cooling. The FWHM of the rocking curve for the (0002) reflection was 100 arcsec.

(xv) Wang et al. [223] also grew ZnO crystals doped with N. The source material (nutrient) was 4N ZnO with particle size below 3 mm, which was sintered at 1150°C in air for 12 h in a Pt crucible. The seeds were (0001) plates prepared from previously grown ZnO crystals. Two different sets of conditions for an effective N incorporation were used: (i) 3N NaOH + 1N KOH + 1N $LiNO_3$ at temperatures of 350 and 365°C in the growth and dissolution zones, respectively, filling factor of 75%, and pressure of 20 MPa; and (ii) 20 vol.% NH_4OH + 80 vol.% KOH at temperatures of 475 and 490°C in the growth and dissolution zones, respectively, filling factor of 70%, and pressure of 136 MPa. The N-doped ZnO crystals were slightly greenish. In the ZnO crystals obtained from the solution containing NH_4OH, there was a dendritic growth due to the presence of NH_4^+. High-quality crystals were obtained from the solution containing $LiNO_3$. The concentration of incorporated N was about 10^{18} at. cm^{-3}, but *p*-type conductivity has not been achieved.

(xvi) Further, Wang et al. [211] grew ZnO crystals doped with In. High-strength steel autoclaves with a Pt liner were used. The source material was 4N ZnO powder, which was sintered at 1150°C in air for 12 h in a Pt crucible. The seeds were (0001) and $(11\bar{2}0)$ (*a*-plane) plates of undoped or In-doped ZnO crystals from previous hydrothermal growth. The solution was a mixture of either 4N KOH + 1N LiOH or 3N KOH + 1N NaOH + 1N LiOH with 0.01–0.04 M In_2O_3 (0.22–0.55 mol.% in the ZnO source material). The temperature was 350 and 365°C in the growth and dissolution zones, respectively; the filling factor of 80%, and the pressure of 124 MPa.

ZnO grown by the hydrothermal method typically has the following faces: (0001), $(000\bar{1})$, $(10\bar{1}1)$, $(10\bar{1}\bar{1})$, and $(10\bar{1}0)$ with the fastest growing face (0001) followed by the $(10\bar{1}1)$, $(10\bar{1}0)$, $(10\bar{1}\bar{1})$, and $(000\bar{1})$ faces. In the presence of In^{3+} ions, the growth in the $[10\bar{1}0]$ direction was faster than in the [0001] and $[000\bar{1}]$ directions, i.e., there was a stagnant growth on both {0001} faces and fast growth on normal faces $\{10\bar{1}0\}$ (*m*- or *a*-planes) leading to a plate formation. The surface morphology of the (0001) and $(000\bar{1})$ was not smooth, but contained hillocks due to blockage of the growth on the {0001} planes by In^{3+} ions.

On average, In-doped ZnO crystals contained (in wt. ppm) In = 165–175, Fe = 35–37, Si = 1–1.3, Al = 22, and Li = 12–13. It has been shown that the growth mechanism was strongly influenced by impurities.

(xvii) Zuo et al. [212] applied the hydrothermal method to grow ZnO crystals doped with Sc. Autoclaves were made of high-strength nickel alloy (inner diameter 22 mm and inner length 250 mm) with an Au liner and Au baffle having 5–10% opening. The source material was ZnO powder (4N) sintered at 1200°C in air for 48 h in a Pt crucible and Sc_2O_3 in the concentration of 0.1–0.3 mol.%. The seeds were (0001) plates prepared from previously hydrothermally grown ZnO crystals. The aqueous solution was a mixture of KOH (3 mol/l) + LiOH (1 mol/l) + H_2O_2 (1 wt.%). The filling factor was 75%. In the space between the liner and autoclave, distilled water was used. The temperature in the growth zone was 360°C, while the temperature difference between the dissolution and growth zones was 15–20°C. The growth duration was about 10 days. The growth rate of Sc-doped ZnO was 0.15–0.16 mm/day (similar to undoped crystals) and 0.12–0.16 mm/day (much faster as compared with undoped ZnO) in the case of (0001) and $(000\overline{1})$ faces, respectively. Doping with Sc changed the growth morphology as pyramidal faces (*p*-planes) become smaller with increasing the Sc_2O_3:ZnO ratio.

The newly grown sectors were grey–green in color. The size of the obtained crystals was 30 × 40 × 3.3 mm^3, and the FWHMs of the rocking curve of the (0002) reflection were 39 and 46 arcsec for $+c$ and $-c$ sectors, respectively. The impurity concentrations (in wt. ppm) for $+c/-c$ sectors were as follows: Al = 0.82/0.94, Fe = 10.7/13.8, K = 0.21/0.4, Li = 1.1/2.1, Cu = 7.4/7.5, Si < 0.3, Sc = 8.3/11.9, and Au = 0.85/0.17.

3.4.4 Cold Crucible

The cold crucible technique for growing bulk ZnO crystals from the melt was reported by Agarwal et al. [147], Nause [230], and Nause and Nemeth [231]. This technique was previously developed by Nause et al. [232, 233] at Cermet Inc., Atlanta, USA. Here, a 3-inch segmented, water-cooled copper crucible and surrounded RF coil were placed within a high-pressure growth chamber with a high-pressure window for temperature monitoring. The RF coil was energized with a 50 kW RF generator. Into the copper crucible, ZnO powder was loaded along with a 1.5-inch-diameter graphite disc acting as a susceptor. After evacuation for several minutes, the growth chamber was pressurized with oxygen at the pressure of 1 to 10 atm. An electromagnetic field from the RF coil induces an eddy current and produces joule heating within the susceptor, which heats

Experiments without a crystal seed resulted in ingots consisting of single-crystal grains due to spontaneous nucleation, while those with the seed resulted in single-crystal ingots. The ingots were removed from the crucible mechanically by core drilling. From ingots grown in 40 mm diameter crucibles, cylindrical single crystals of 33 mm diameter and 30–40 mm length could be prepared (Fig. 3.43a). The periphery of the obtained crystals, usually had small cracks arising from direct contact with the iridium crucible and different thermal expansion of both materials: ZnO and Ir.

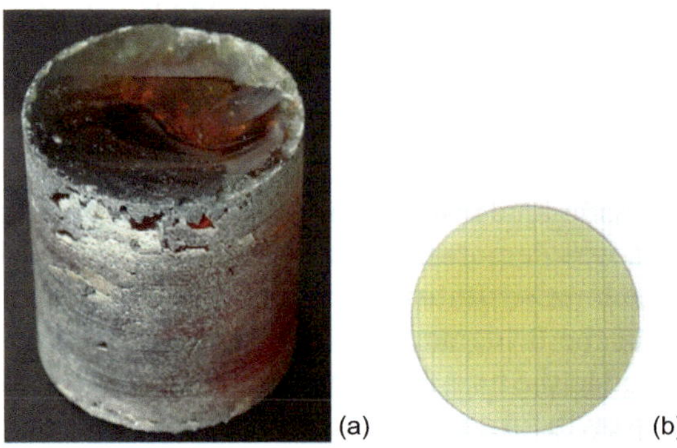

Figure 3.43 (a) Cylindrical part (diameter 33 mm) of an as-grown bulk ZnO crystal obtained by the Bridgman method; reprinted from Ref. [237], Copyright 2008, with permission from Elsevier; and (b) ZnO wafers fabricated from the bulk crystal and annealed at 900°C in the presence of oxygen for 24 h [241].

Despite the peripheral cracks, the rest of the crystal bodies were of high structural quality reflected in narrow rocking curves having FWHM of 22–47 arcsec. Further improvements to the growth furnace and cooling rates significantly reduced or eliminated the occurrence of the peripheral cracks. At the periphery of the conical part of the crystals of iridium, whiskers below 2 mm long were usually present, but not in other parts of the crystals. Occasionally, in some of the crystal samples, mosaic blocks with a disorientation of about 0.27° were observed, as well as local lattice strains likely arising from stacking faults [242].

When the crystals were grown at low $p(O_2)$ or were cooled down very fast (2 h), the resulted coloration was orange to brown, while keeping good transparency. The crystals obtained at lower temperature gradients and at higher $p(O_2)$ were typically red to orange. The dark coloration (brown and red) was attributed to particles with micrometer size (up to 2 μm) and nanometer size (about 3 nm), the amount of which (and thus darker coloration) increased with higher cooling rates. However, for the dark coloration, nanoparticles were responsible rather than microparticles. This is because the concentration of the microparticles, which would give rise to elastic Rayleigh scattering, was too small as compared to the concentration of the nanoparticles (4.6×10^{16} cm^{-3}) causing enhanced absorption [243]. The particles were likely Zn precipitates formed during ZnO decomposition: $ZnO \rightleftarrows Zn(g, l) + \frac{1}{2}O_2$. The dark coloration was minimized when using high $p(O_2)$ (19 bar of CO_2) and when the cooling rate did not exceed about 25 K/h at the high-temperature regime, i.e., between melting point and about 1000°C. Below about 1000°C, the cooling rate could be higher, about 100 K/h.

Annealing the crystals in air below 800°C did not affect the coloration. However, the crystal became even darker after annealing at 800°C. But annealing in air or oxygen at 900°C or higher for at least 1 h made the crystals yellowish (Fig. 3.43b) or almost colorless.

The undoped and as-grown ZnO single crystals obtained by the Bridgman method contained the following residual impurities as indicated by SIMS [244] (in at. cm^{-3}): Na ≈ 10^{17}–10^{18}; K, Mg, Ca ≈ 10^{16}–10^{18}; Cu ≈ 10^{18}; Al and F ≈ 10^{17}–10^{19}; Si ≈ 10^{18}–10^{19}; Cl ≈ 10^{19}–10^{20} at. cm^{-3}. EPR analysis also revealed traces of such impurities as Co, Mn, Li, and Na. ICP-OES results showed the following solid residual impurities [242] (in at. cm^{-3}): Cr, Co, Ni, Ga, Zr, Ag, Te, and Pb ≈ 10^{15}–10^{16} at. cm^{-3}, and Al ≈ 10^{17} at. cm^{-3} (in some of the measured samples). When using nuclear reaction analysis (NRA), also bound hydrogen at the concentration of about mid 10^{16} at. cm^{-3} was found [242]. Moreover, the same elements, such as Cu and Al, were indicated at different concentrations when using SIMS and ICP-OES methods. This is due to the different accuracies of both methods and different samples used for both analyses, as there are many sources of the residual impurities, namely, a powdered ZnO starting material (purity between 4N and 6N), Ir crucible (purity typically 3N), alumina and zirconia thermal insulation and growth atmosphere (usually 5N). As the materials used in crystal growth experiments vary between runs or between groups of runs, different crystals may have different levels of residual impurities.

Bridgman experiments were also performed by Schulz et al. [86] on the growth of $Zn_{1-x}Mg_xO$ crystals with x = 0.01, 0.02, and 0.03. The growth experiments of $Zn_{1-x}Mg_xO$ were carried out at the same growth furnace and conditions as pure ZnO. Also the size of the obtained crystals was similar. x = 0.01 and 0.02 resulted in single crystals, while x = 0.03 resulted in polycrystals. The profile of Mg content in the obtained crystals measured by ICP-OES is shown in Fig. 3.44. Fitting this experimental profile with the Scheil–Gulliver equation $C/C_0 = k_{\text{eff}}(1-g)^{(k_{\text{eff}}-1)}$, where C is the current Mg concentration in a crystal, C_0 is the initial Mg concentration, g is the crystallization fraction, and k_{eff} is the effective segregation coefficient, revealed k_{eff} between 1.32 and 1.4. The calculated phase diagram of ZnO–Mg [86] leads to a conclusion that the solubility limit of Mg in $Zn_{1-x}Mg_xO$ is about x = 0.07.

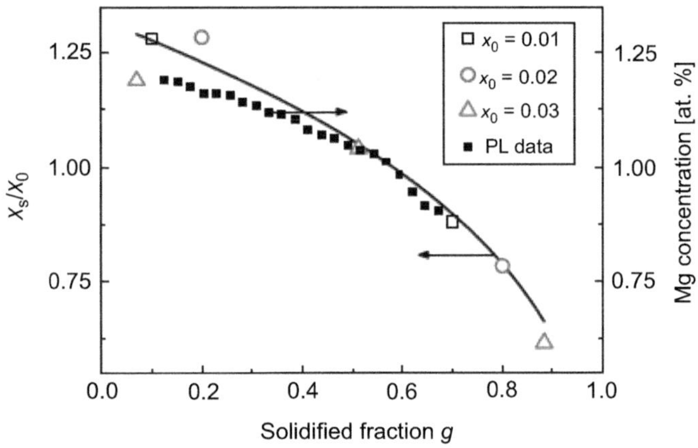

Figure 3.44 Normalized Mg concentration (open symbols, measured by ICP-OES) along the crystallized fraction g for starting concentrations x = 0.01 (squares), 0.02 (circles), and 0.03 (triangles). PL data (solid squares) are also shown for comparison. Reprinted from Ref. [86], Copyright 2011, with permission from Elsevier.

It is worth to mention the growth attempts of ZnO by the Czochralski method carried out by the same group [237]. Even when using CO_2 at 20 bar of overpressure, the decomposition rate of ZnO and thus evaporation rate were so high that they prevented the crystallization of a ZnO single crystal on an iridium seed. Instead, a dark-brown polycrystalline material was formed on the Ir seed.

Summary of basic growth conditions for growing bulk ZnO single crystals from the melt is included in Table 3.4.

Table 3.4 Basic growth conditions of bulk ZnO single crystals from the melt. D and L are diameter and length, respectively

Growth method	Metal crucible	Growth atmosphere	Pressure [bar]	Crystal size [mm]	Crystal coloration	Ref.
Cold crucible	None	Unknown	1–10	$D \approx 25$ $L = 15–25$	Colorless, yellow, brown	147, 230, 231
Bridgman	Ir	CO_2	20	$D \approx 33$ $L = 30–40$	Brown to red[a] Yellowish or almost colorless[b]	148, 236, 237, 238, 239, 240

[a]As-grown; [b]annealed in the presence of oxygen at ≥900°C for at least 1 h

3.5 Thermal and Mechanical Properties

Material preparation, device fabrication, and device functionality depend on a variety of material properties, including thermal and mechanical properties thereof. To derive such properties, a diversity of experimental techniques are employed, which are more recently aided by theoretical calculations. The availability of bulk ZnO single crystals enabled to measure their thermal and mechanical properties, which are quite consistent, wherein minor differences may arise from the crystal structural quality and purity. This section discusses fundamental thermal and mechanical properties of bulk ZnO single crystals with some inclusion of theoretical results.

3.5.1 Melting Point

Only the wurtzite phase of ZnO is stable up to melting point, MP = 1975°C = 2248 K [77], where it melts with a melting enthalpy ΔH_f = 54392 J/mol. For this solid phase and the liquid phase above

MP, thermal properties are compiled, e.g., in the FactSage database [245] and by Barin [246] as follows:

Solid: $\Delta H(298.15) = -350{,}460$ J/mol, $S(298.15) = 43.639$ J/molK

$$c_p = 48.24152 + 6.794816 \times 10^{-3}\,T - 1.03768355\times 10^{8}\,T^{-3} - 91.54741\,T^{-0.5}$$

Liquid: $\Delta H(298.15) = -296{,}068$ J/mol, $S(298.15) = 67.835$ J/molK

$$c_p = 60.668$$

Figure 3.45 compares the Gibbs free energy functions $G(T)$ for solid and liquid ZnO, which can be calculated from these experimental data with that of gaseous ZnO for total pressures p = 1 atm (solid) and p = 3 atm (dashed), respectively. Above MP = 1975°C, G for the liquid phase is lower than for the solid phase, almost independent on pressure. At 1 atm, however, $G(T)$ for the gas becomes even lower below MP. This is the reason why ZnO cannot be molten at ambient pressure. Already a slight overpressure (as shown for 3 atm) is sufficient to stabilize the condensed phases beyond MP.

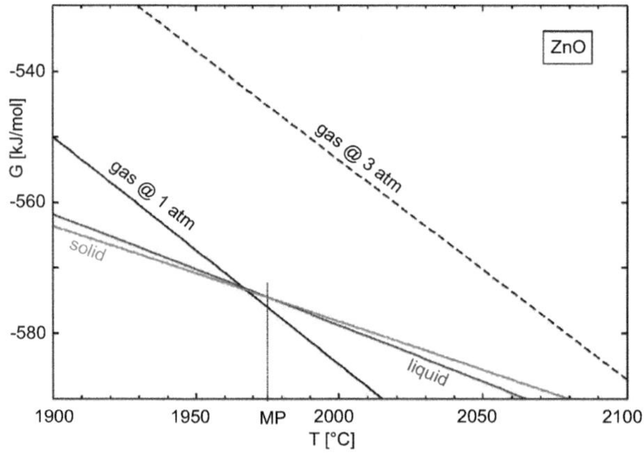

Figure 3.45 Gibbs free energy $G(T)$ for the three states of aggregation of ZnO. Only for the gas phase, G depends significantly also on the pressure p.

3.5.2 Thermal Conductivity and Specific Heat Capacity

Thermal conductivity of hexagonal ZnO is described by the second-order tensor with two independent thermal conductivity coefficients perpendicular $\lambda_{\perp c}$ and parallel $\lambda_{\parallel c}$ to the c-axis. Wolf and

Martin [247] reported thermal conductivities (with a steady-state longitudinal heat-flow apparatus) of bulk ZnO crystals (obtained by the hydrothermal and vapor phase methods) in a temperature range of 3.5–300 K with a maximum value of about 1000 W/mK below 50 K. At 300 K, the thermal conductivity is just below 100 W/mK. Also, the ratio of $\lambda_{\|c}/\lambda_{\perp c}$ was found to be 1.15, indicating a minor anisotropy of the thermal conductivity. The thermal conductivity is limited by phonon–phonon scattering at elevated and high temperatures.

Bulk ZnO single crystals obtained from the gas phase and measured by Florescu et al. [248] with scanning thermal microscopy showed at RT values of 102–116 ± 0.07(8) and 98–110 ± 0.08(9) W/mK for Zn-terminated and O-terminated faces of a *c*-oriented sample, respectively. For ZnO bulk crystals obtained by the cold crucible method, Özgür et al. [249] obtained thermal conductivities of 100 ± 0.08 and 95 ± 0.06 W/mK at RT for Zn-terminated and O-terminated faces, respectively. A thermal conductivity ratio for Zn-terminated and O-terminated faces is, therefore, between 1.04 and 1.05. Samples annealed in air (1050°C, 3 h) and N-plasma (750°C, 1 min) revealed the values of 135 ± 0.08 and 147 ± 0.08 W/mK, respectively. Higher values resulted likely from a change in surface electrical conductivity and surface morphology. Indeed, Gadzhiev [250] reported that thermal conductivities of ZnO ceramics with a porosity of 0–21% decrease from 35 to 14.3 W/mK at RT, which is less than half the value of bulk single crystals.

Specific heat capacity reported by Klimm et al. [77] increases from 41.05 at 298 K to 61.576 J/molK at the melting point of ZnO at 2248 K (see Section 3.4.1). Similar values were reported by Pässler [251] in this temperature range, with a fast drop thereof below RT, to a value of 12 J/molK at 80 K. Serrano et al. [252] concluded from experimental and theoretical studies that heat capacity depends on isotope masses of Zn and O atoms, which affect mainly acoustic phonons that are active at low temperatures, and optical phonons that are active at higher temperatures, respectively.

3.5.3 Thermal Expansion

The linear thermal expansion coefficients of bulk ZnO crystals were determined, e.g., by Iwanaga et al. [253], Albertsson et al. [254], and Khan [255].

Iwanaga et al. [253] found by X-ray powder diffractometry of a bulk ZnO crystal the following relation at a temperature range of 300–1373 K:

$$a = 3.2486 + 0.623 \times 10^{-5}T + 12.94 \times 10^{-9}T^2 \text{ along the } a\text{-axis} \quad (3.35)$$

$$c = 5.2042 + 0.522 \times 10^{-5}T + 12.13 \times 10^{-9}T^2 \text{ along the } c\text{-axis} \quad (3.36)$$

which yield at RT $\alpha_a^{\mathrm{RT}} = 4.31 \times 10^{-6}$ K^{-1} and $\alpha_c^{\mathrm{RT}} = 2.49 \times 10^{-6}$ K^{-1} with a ratio $\alpha_a^{\mathrm{RT}}/\alpha_c^{\mathrm{RT}} = 1.73$.

The c/a ratio decreases with temperature from 1.602 at 300 K to 1.596 at 1373 K. This means that the wurtzite-type structure of ZnO becomes more stable with an increase in temperature with respect to zincblende-type structure.

Along the a- and c-axes, the thermal expansion coefficient relations according to dilatometric measurements by Albertsson et al. [254] are as follows:

$$\alpha_a = \alpha_a^{\mathrm{RT}} \times (1 + 0.693 \times 10^{-3}\Delta T - 0.230 \times 10^{-6}\Delta T^2) \ [\mathrm{K}^{-1}]$$
$$\text{along the } a\text{-axis} \quad (3.37)$$

$$\alpha_c = \alpha_c^{\mathrm{RT}} \times (1 + 2.56 \times 10^{-3}\Delta T - 3.37 \times 10^{-6}\Delta T^2) \ [\mathrm{K}^{-1}]$$
$$\text{along the } c\text{-axis} \quad (3.38)$$

where $\Delta T = T - 298$ K, $\alpha_a^{\mathrm{RT}} = 6.511 \times 10^{-6}$ K^{-1} and $\alpha_c^{\mathrm{RT}} = 3.017 \times 10^{-6}$ K^{-1} for a- and c-axes, respectively. The ratio $\alpha_a^{\mathrm{RT}}/\alpha_c^{\mathrm{RT}} = 2.16$ is higher as compared with the above-mentioned study and may arise from the measurement technique.

For comparison, Khan [255] measured on powders in a temperature range of 27–619°C the following thermal expansion coefficients:

$$\alpha_a = 6.05 \times 10^{-6} + 2.20 \times 10^{-9}T + 2.29 \times 10^{-12}T^2 \ [°\mathrm{C}^{-1}]$$
$$\text{along the } a\text{-axis}$$

$$\alpha_c = 3.53 \times 10^{-6} + 2.38 \times 10^{-9}\Delta T + 9.24 \times 10^{-12}T^2 \ [°\mathrm{C}^{-1}]$$
$$\text{along the } c\text{-axis}$$

which result in $\alpha_a^{\mathrm{RT}} = 6.11 \times 10^{-6}$ K^{-1} and $\alpha_c^{\mathrm{RT}} = 3.59 \times 10^{-6}$ K^{-1} at RT for a- and c-axes, respectively.

3.5.4 Stiffness and Hardness

The elastic properties of ZnO are described by the fourth-rank tensors, with the stiffness coefficients c_{ij} forming a 6×6 matrix with five independent components: c_{11}, c_{12}, c_{13}, c_{33}, and c_{44}, wherein $c_{66} = (c_{11} - c_{12})/2$. Mechanical properties of ZnO were studied both experimentally and theoretically. Experimental results on elastic properties of bulk ZnO crystals can be found, e.g., in the works of Bateman [256], Solbrig [2], Kobiakov [257], and Azuhata et al. [258]. Theoretical study of mechanical properties was carried out, e.g., by Ahuja et al. [259], Catti et al. [260], Wu et al. [261], Tu and Hu [262], Gopal and Spaldin [263], Shein et al. [264], and Li et al. [265]. Selected experiments, determined mainly by ultrasonic techniques, and theoretical stiffness coefficients c_{ij} are listed in Table 3.5.

Using the Voigt approach and symmetry for the hexagonal system ($c_{11} = c_{22}$, $c_{13} = c_{23}$, $c_{44} = c_{55}$), the bulk modulus B and shear modulus G can be calculated from the following relations:

$$B = \frac{1}{9}[(2c_{11} + c_{33}) + 2(c_{12} + 2c_{13})] \tag{3.39}$$

$$G = \frac{1}{15}\left[(2c_{11} + c_{33}) - (c_{12} + 2c_{13}) + 3(2c_{44} + c_{66})\right] \tag{3.40}$$

and both Young's modulus and Poisson ratio can be calculated as follows:

$$E = \frac{9BG}{3B + G} \tag{3.41}$$

$$\nu = \frac{3B - 2G}{2(3B + G)} \tag{3.42}$$

The calculated moduli and Poisson ratio on the basis of the stiffness coefficients along with the anisotropy ratio [266] defined by $A = c_{44}/c_{66}$ are also included in Table 3.5.

The anisotropy ratio is unity or close to unity, which indicates an almost isotropic elastic behavior of ZnO. This means Young's modulus is nearly independent of the orientation. Bulk, shear, and Young's moduli derived from experimental stiffness coefficients lie in the range of 128–144 GPa, 43–46 GPa, and 115–126 GPa, while theoretical values are a bit overestimated in a majority of cases.

Table 3.5 Experimental and theoretical values of stiffness coefficients of bulk ZnO crystals. Exp.– experimental, Theor.– theoretical, A – anisotropy ratio, B – bulk modulus, G – shear modulus, E – Young's modulus, and ν – Possions ratio

c_{ij} [GPa]	Exp. Ref. [256]	Exp. Ref. [2]	Exp. Ref. [257]	Exp. Ref. [258]	Theor. Ref. [260]	Theor. Ref. [262]	Theor. Ref. [263]	Theor. Ref. [265]
c_{11}	209.7	194	207	190	246	218	217	191.5
c_{12}	121.1	102	117.7	110	127	137	117	108.7
c_{13}	105.1	94	106.1	90	105	121	121	95
c_{33}	210.9	217	209.5	196	246	229	225	206.7
c_{44}	42.47	—	44.8	39	56	38	50	41.4
c_{66}	44.29	46	44.6	40	59.5	40.5	50	38
A	0.96	—	1.00	0.98	0.94	0.94	1.00	1.09
B	144	132	143	128	157	158	153	132
G	46	—	46	43	61	42	50	44
E	124	—	126	115	162	117	135	118
ν	0.356	—	0.353	0.350	0.328	0.377	0.353	0.351

With respect to plastic deformation, ZnO is a comparably soft material. As a mineral, "zincite" has the hardness of 4 [267] on the Mohs scale, which means that it can be scratched by a knife. The Mohs hardness of bulk ZnO single crystals was measured at a bit higher value of 4.5 [80]. Microhardness of bulk ZnO crystals was studied by numerous researches. Yonenaga [268] found the microhardness of ZnO by the Vickers indentation method of 4.7 GPa, while Kucheyev et al. [269] by nanoindentation and atomic force microscopy of 5 ± 0.1 GPa at RT with Young's modulus of 111.26 ± 4.7 GPa. Multiple discontinuities (pop-ins) in force–displacement curves were observed as a result of a plastic slip on both the basal and pyramidal planes [269, 270]. Such pop-ins were also observed by Jian [271] with nanoindentation-induced deformation studies of bulk *c*-oriented ZnO crystals with microhardness and Young's modulus of 5.4 ± 0.7 GPa and 112.5 ± 8.4 GPa, respectively. The plastic slip occurs on the $(10\bar{1}1)$ plane [271, 272], which may lead to a non-pronounced cleaving parallel to $\{10\bar{1}0\}$. Coleman et al. [273] measured the microhardness and Young's modulus of bulk ZnO single crystals by nanoindentation of 2 ± 0.2 and 163 ± 6 (*a*-oriented) and 4.8 ± 0.2 and 143 ± 6 GPa (*c*-oriented), respectively.

Demyanets et al. [274] studied microhardness of hydrothermally grown bulk ZnO single crystals by the Vickers indentation method in a temperature range of −196 to 900°C on the planes of monohedra, hexagonal, prisms, and the pyramid. Indentation prints on the (0001) and $(000\bar{1})$ monohedra did not show any cracks, indicating a maximum hardness and plasticity over the temperature range. On the other hand, around the indentation prints on the $\{10\bar{1}0\}$ and $\{11\bar{2}0\}$ prisms and $\{10\bar{1}1\}$ pyramids, radial cracks were formed, which were aligned in the glide $<11\bar{2}0>$ direction. The microhardness was found to be relatively isotropic with a value of 1.7 GPa, while Young's modulus was 47.7 GPa.

The experimental density ρ_s of bulk ZnO crystals is 5.676 [256] or 5.642 g/cm^3 [275]. Having values of the solid density as well as bulk and shear moduli, one can estimate transverse and longitudinal sound velocity in ZnO crystals [264]:

$$v_T = \sqrt{\frac{G}{\rho_s}}, \quad v_L = \sqrt{\frac{4B + 3G}{4\rho_s}} \tag{3.43}$$

Taking experimental values of stiffness coefficients [256–258] and density [256], the semi-empirical sound velocities for bulk ZnO single crystals are v_T = 2742–2860 and v_L = 5317–5600 m/s. Bateman [256] measured sound velocities with a high-frequency ultrasonic buffer technique and found the values of 2735–2793 and 5939–6096 m/s for shear and longitudinal waves, respectively.

The experimental Debye temperature of ZnO mentioned by Abrahams and Bernstein [275] is Θ_D = 355–416 K. An example of the theoretical value Θ_D by Shein et al. [264] is 405 K.

More details on elastic properties of the wurtzite structure in general can be found in books of Adachi [276] and Gil [277].

3.5.5 Conclusion

Melting point of ZnO is high, 1975°C, but to obtain the liquid phase, an overpressure is required. Thermal conductivity of bulk ZnO single crystals is very high, about 100 W/mK with a minor anisotropy parallel and perpendicular to the *c*-axis of about 1.15, as well as for Zn- and O-terminated *c*-faces of 1.04–1.05. Also, the linear thermal expansion of ZnO single crystals shows anisotropy with values at RT of 4.31–6.511 × 10^{-6} K^{-1} along the *a*-axis and 2.49–3.017 × 10^{-6} K^{-1} along the *c*-axis. The ratio of the linear thermal conductivity along the *a*-axis to *c*-axis is in the range of 1.73–2.16. The experimental density is 5.642–5.676 g/cm^3. Basing on measured stiffness coefficients (five independent components) on bulk ZnO single crystals, bulk, shear, and Young's moduli are in the range of 128–144 GPa, 43–46 GPa, and 115 –126 GPa, respectively. Measured microhardness was found in the range of 4.7–5.4 GPa. Plastic slip occurs on both the basal and pyramidal planes. Semi-empirical transverse and longitudinal sound velocities in bulk ZnO single crystals are 2742–2860 and 5317–5600 m/s, respectively, while the Debye temperature is between 355 and 416 K.

Basic thermal and mechanical properties of bulk ZnO single crystals are collected in Table 3.6.

Table 3.6 Basic thermal and mechanical properties of bulk ZnO single crystals

Property	Unit	Value	Reference
Experimental solid density	g/cm^3	5.642–5.676	256, 275
Melting point	°C	1975	77
Thermal expansion coefficient of the lattice parameter at RT	K^{-1}	a: 4.31–6.511 × 10^{-6} c: 2.49–3.59 × 10^{-6}	253, 254, 255
Thermal conductivity at RT	W/mK	95–116	247, 248, 249
Molar heat capacity at RT	J/molK	41.05	77
Experimental stiffness coefficients	GPa	c_{11} = 210/194/207/190 c_{12} = 121/102/118/110 c_{13} = 105/94/106/90 c_{33} = 211/217/210/196 c_{44} = 42/ — /45/39	256/2/257/258
Bulk modulus[(a)]	GPa	144/132/143/128	256/2/257/258
Shear modulus[(a)]	GPa	46/ — /46/43	256/2/257/258
Young modulus[(a)]	GPa	124/ — /126/115	256/2/257/258
Poisson ratio[(a)]	—	0.356/ — /0.353/0.350	256/2/257/258
Microhardness	GPa	c-surface: 4.7–5.4 a-surface: 2	268, 269, 270, 271, 272, 273
Mohs hardness	—	4.5	80
Transverse sound velocity[(b)]	m/s	2742–2860	256, 257, 258
Longitudinal sound velocity[(b)]		5317–5600	
Debye temperature	K	355–416	275

[(a)]Semi-empirical values based on measured stiffness coefficients from corresponding references and Eqs. 3.39–3.42; [(b)]semi-empirical values based on measured stiffness coefficients and density from corresponding references, and Eq. 3.43

3.6 Basic Optical Properties

Optical properties of a material depend on its electronic structure, which in turn is affected by the chemistry and structural quality of the material. As bulk ZnO crystals are grown at different environments, one may expect different concentrations of residual impurities of "undoped" crystals, deviation from stoichiometry, and native point defect density, which are characteristic for a given growth method. The resulted optical properties of bulk ZnO crystals will, therefore, differ in-between.

3.6.1 Transmittance and Optical Energy Gap

Exemplary transmittance spectra in the visible spectrum of bulk ZnO crystals obtained from the gas phase (CVT, Mycielski et al. [183]), solution (hydrothermal method, Ehrentraut et al. [71]), and from the melt (Bridgman method, Irmscher et al. [278]), are shown in Fig. 3.46a, b, and c, respectively. For an as-grown crystal obtained from the gas phase (Fig. 3.46a), which is red in color, the absorption edge originates at about 450 nm. Annealing in O_2 made the crystal colorless and shifted the absorption edge to about 375 nm. Hydrothermally grown crystals from two different suppliers (Fig. 3.46b) show almost identical transmittance spectra with the absorption edge originated at about 390 nm. Those crystals had a yellow tint. The same relates to undoped and In-doped samples, although the In-doped sample has a lower total transmittance. The absorption edge of the as-grown ZnO sample obtained from the melt by the Bridgman method (Fig. 3.46c) originates at about 500 nm, similar to that obtained from the gas phase. In fact, Bridgman-grown crystals are also red. Annealing Bridgman-grown ZnO crystals in O_2 at 900°C or above has the same effect as in the case of CVT-grown crystals, as the absorption edge shifts to about 390 nm. The annealed ZnO crystals changed the coloration from red to yellowish, similar to the coloration of hydrothermally grown crystals. The transmittance spectra of Fig. 3.46a,b are shown for the visible spectrum; therefore, the free-carrier absorption, which appears in the near-infrared region (NIR), is not shown. However, Bridgman-grown crystals clearly reveal the free-carrier absorption in the NIR (Fig. 3.46c), whether the crystals are as-grown or annealed. Indeed, annealing did not remove the electrical conductivity of bulk ZnO crystals, only decreased the free electron concentration by a factor of 2–3.

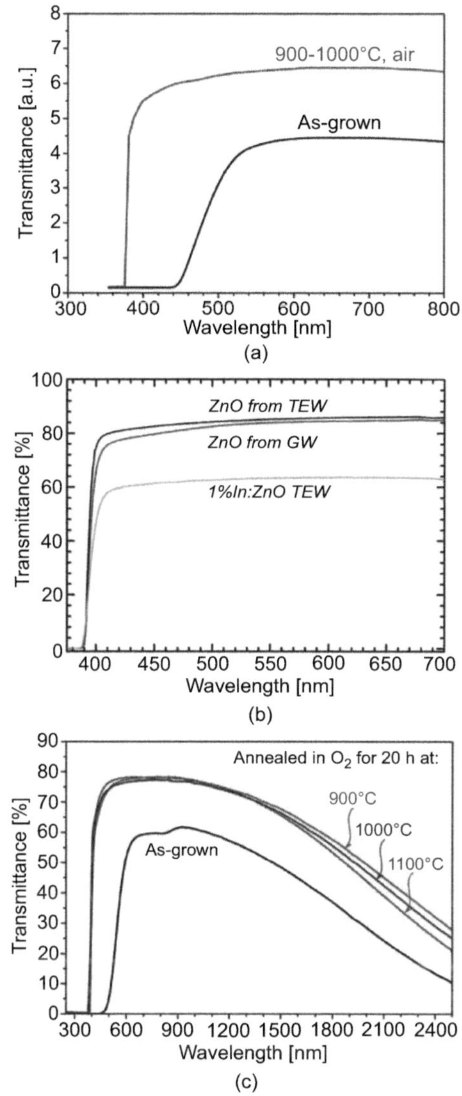

Figure 3.46 Transmittance spectra of undoped bulk ZnO crystals obtained from the gas phase (a), reconstructed basing on data from Ref. [183], with some modifications and different visualization; from the solution (b), TEW - Tokyo Denpa, GW - SPC Goodwill, reprinted from Ref. [71], Copyright 2006, with permission from Elsevier; and from the melt by the Bridgman method (c) [278].

The onset of the absorption of the red ZnO crystals in the 450–500 nm region was found to be due to metallic Zn nanoparticles (precipitates) present in melt-grown crystals (Irmscher et al. [243]). The size of the nanoparticles of about 3 nm and concentration of 4.6×10^{16} cm^{-3} well fits Mie's theory of light extinction on particle plasmons. The Zn nanoparticles are removed by annealing for at least several hours at 900°C or above in the presence of O_2. It is not surprising that some of the crystals obtained from the gas phase and by the Bridgman method contain metallic Zn precipitates, if we look at the growth environments. The exemplary CVT-grown crystal was obtained at Zn-rich growth conditions, while Bridgman-grown crystal at high temperature, in which decomposition of ZnO was significant due to insufficient oxygen partial pressure (Ir crucible). On the other hand, hydrothermally grown crystals were obtained at low temperature and high pressure, which ensure good crystal stoichiometry. Melt-grown crystals obtained by the cold crucible have the onset of the absorption similar to that of hydrothermally grown crystals. This is because although grown at high temperatures, the same as by Bridgman method, that process used high oxygen partial pressure (no metal crucible), which was sufficient to minimize the thermal decomposition and keep good crystal stoichiometry. Different dopants that were experimentally practiced will affect the transmittance depending on the dopant nature and concentration. For example, dopants such as Fe, Co, Mn will shift the absorption edge toward longer wavelengths, while other dopants, such as Mg, will shift the absorption edge toward shorter wavelengths. Some dopants will also affect the transmittance in the visible spectrum even if the absorption edge remains basically intact.

ZnO is a direct-gap oxide semiconductor with a dipole-allowed band-to-band transition [73]. Experimental bandgap values for bulk ZnO single crystals obtained with a use of a variety of techniques, such as transmittance/absorbance, reflection, ellipsometry, photoluminescence (PL), and Fourier transform infrared spectroscopy (FTIR) outputted different values, e.g., 3.18 [185], 3.21 [181], 3.3 [279, 280], and 3.38 [281]. Srikant and Clarke [280] stated that the true optical bandgap of ZnO is 3.3 eV, while lower measured values relate to a transition valence band–donor level (3.15 eV), which can dominate the absorption. Indeed, a number of studies deal with excitonic absorption through free and bound excitons, e.g., by Thomas [279], Liang and Yoffe [282], Meyer et al. [283], Klingshirn

[1, 73], and Klingshirn et al. [74]. Exciton states form hydrogen-like series with the main quantum number $n_B = 1, 2, 3, \ldots$ with excitonic binding energy of about 56–61 meV [74, 284]. Moreover, the exciton can be bound to different point defects, such as ionized donor (D^+), neutral donor (D^0), or neutral acceptor (A^0), forming bound excitons D^+X, D^0X, or A^0X [1, 73, 74].

An example of the absorption coefficient in power 2 for the direct transitions and the absorption coefficient are depicted in Fig. 3.47 for a melt-grown ZnO crystal (Bridgman method). Both the absorption coefficient (inset) and the relation $\alpha \propto (h\nu)^{1/2}$ reveal a very steep near-edge absorption with a very good linear fit at RT of the latter, which indicates, indeed, the direct bandgap. The estimated optical bandgap is 3.183 ± 0.005 eV, which can indicate a dominance of the above-mentioned excitonic absorption.

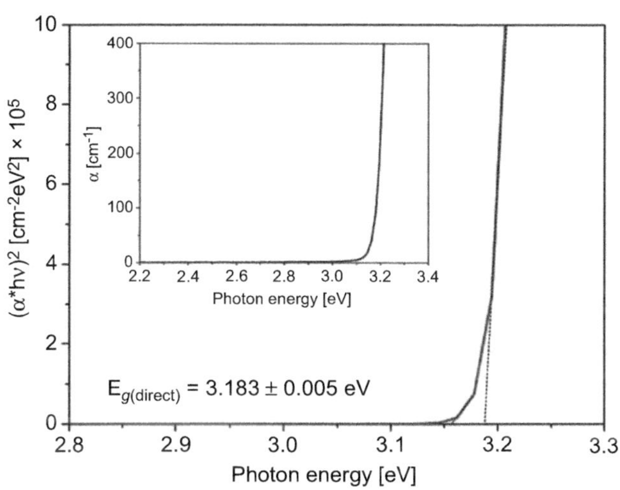

Figure 3.47 Absorption coefficient in power 2 versus photon energy at RT of a bulk ZnO crystal sample obtained from the melt by the Bridgman method after annealing at 1000°C for 24 h in air, which indicates direct transitions. Dot black line indicates extrapolation of nearly linear parts of the curve. The inset shows the absorption coefficient versus photon energy. Sample thickness was 150 μm [278].

Examples of the theoretical values of the direct bandgap for ZnO being close to experimental values are 3.21 [108], 3.27 [101], 3.4 [127], and 3.49 [125] eV.

It is less understood than the UV luminescence as likely many centers are involved in the green luminescence, including native and extrinsic point defects. The green luminescence position will, therefore, depend on the growth environment, such as O-rich or Zn-rich growth conditions, which will create a dominant luminescence involving either V_{Zn} or V_O point defects, respectively. Also, different impurities may play an important role. In contrast, the impurities affect the UV emissions to less extent. For example, in the case of ZnO crystals (melt-grown) doped with Ga, Er, Co, Ho, Fe, Mn, and co-doped with Mg and Li, the maxima of the UV emissions were observed at similar energy as undoped crystals, i.e., at 3.26–3.28 eV at RT (Feng et al. [235]).

In the case of the hydrothermal ZnO, the intensities and peak positions of the samples from the same crystal may slightly differ from each other depending on sectors and polarities for both UV and green emissions. As shown in Fig. 3.49a (Sugimura et al. [304]), the UV luminescence intensity of the +*c* sector (Zn-polar) is stronger than that from the −*c* sector (O-polar), while in the case of the green emission intensity, the situation is opposite (e.g., Oh et al. [305], Sasaki et al. [306]). At low temperature (4.2 K), the D^0X emission was observed from all faces, while the D^+X emission was observed from the O-terminated face only. The FX emission was observed from the +*c* sectors. Another report on hydrothermally grown ZnO (Sekiguchi et al. [225]) showed different relations of the emissions from different sectors, i.e., the emission intensity in the green region dominated over the UV emission on +*c* and −*p* sectors, while on the −*c*, +*p*, and *m* sectors, the UV dominated over the green emission (Fig. 3.49b). This is not surprising, as various sectors have different growth morphology and incorporate different concentrations of impurities depending on the solution composition. Annealing in the presence of oxygen at 700°C and 900°C decreased and increased the UV emission intensity (Dem'yanets and Lyutin [214]), respectively (Fig. 3.49c). The green emission intensities of annealed samples are in an opposite relation to UV emission intensities. Again, annealing redistributes point defects, especially those that have low formation energy and/or are mobile (low migration barrier). Santailler et al. [194] reported that the UV luminescence was much more intense in the hydrothermally grown ZnO than in the CVT-grown ZnO, while the green luminescence was in the opposite relation.

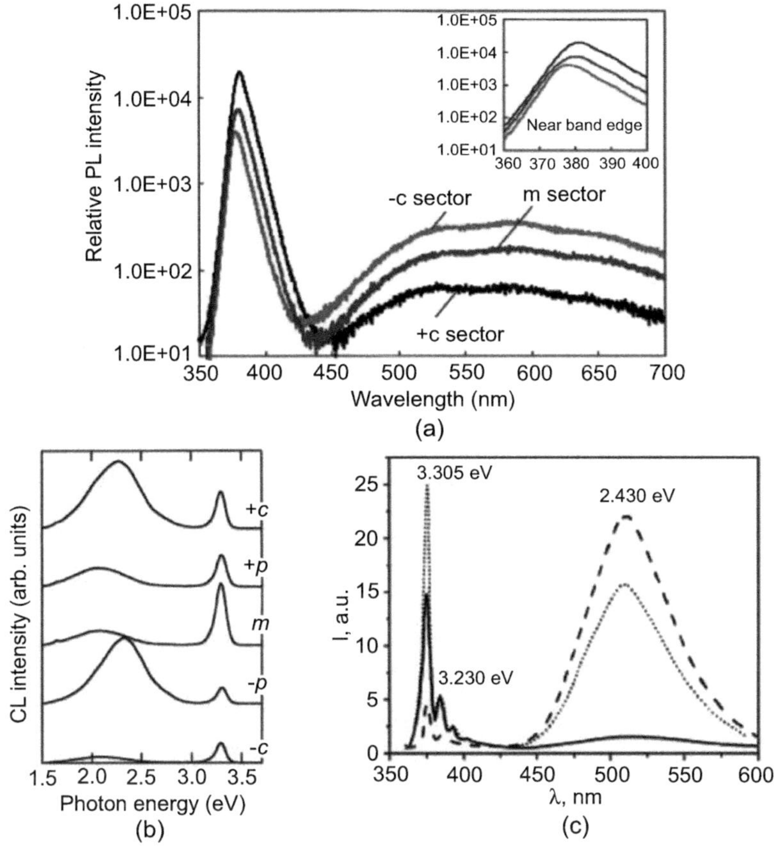

Figure 3.49 (a) PL spectra at RT of different hydrothermal ZnO crystal sectors; (b) CL spectra of different hydrothermal ZnO crystal sectors; and (c) CL spectra at RT of hydrothermal ZnO crystal samples (solid line: as-grown; dashed line: annealed in air at 700°C; dotted line: annealed in air at 900°C). Figure (a) reprinted from Ref. [304], Copyright 2011; (b) reprinted from Ref. [225], Copyright 2000; and (c) reprinted from Ref. [214], Copyright 2008, all with permission from Elsevier.

ZnO crystals obtained by the cold crucible method showed at RT the dominant near-band-edge emission on the O-polar surface as compared with the Zn-polar surface (Chevtchenko et al. [307]).

Although both UV and green emissions are typically present in all bulk crystals, various native and extrinsic point defects and their possible complexes may lead to different intensities and broadness of the emissions, as well as to a shift in peak positions depending on the growth environment, which defines the crystal stoichiometry and its purity. In other words, the emissions, especially their intensities, will strongly depend on the growth method and can be further modified by annealing, intentional doping, and ion implantation. In general, more intensive UV emission points to a higher density of radiative centers, while more intense green emission may indicate a higher density of deep defects in ZnO crystals.

3.6.4 Phonon Modes

Raman spectra have been frequently measured for bulk ZnO single crystals, e.g., by Damen et al. [308], Calleja and Cardona [309], Vinogradov et al. [310], Dem'yanets et al. [227], Feng et al. [235], and Serrano et al. [311], and calculated, e.g., by Calzolari and Nardelli [312], and Serrano et al. [311].

Phonon modes of ZnO are classified as follows [312]:

$$2A_1 + 2B_1 + 2E_1 + 2E_2 \tag{3.44}$$

Raman-active phonon modes are E_2 (low) and E_2 (high), which are IR inactive; A_1 longitudinal optical (LO), A_1 transverse optical (TO), E_1 (LO), and E_1 (TO), which are polar modes (atoms move parallel and perpendicular to the c-axis) [235, 312]. B_1 modes are IR and Raman inactive [312]. Experimental and theoretical phonon modes are listed in Table 3.7. Modes B_1 are Raman silent but could be revealed by inelastic neutron scattering. In general, both experimental and theoretical results are very close to each other.

3.6.5 Conclusion

The absorption edge of bulk ZnO single crystals originates at about 390 nm; however, it is shifted to 450–500 nm for red ZnO due to the presence of Zn nanoparticles, which can be removed by annealing in the presence of oxygen at ≥900°C for at least several hours. The bandgap is direct with a value of 3.3 eV; however, excitonic absorption may dominate and lower the optical bandgap. The refractive index shows uniaxial birefringence whether the polarization is parallel ($E \parallel c$) or perpendicular ($E \perp c$) to the c-axis. The birefringence is about 0.015 in the visible spectrum. The relative static dielectric constant also depends on polarization and shows anisotropy of about 14%. Undoped bulk ZnO single crystals exhibit two luminescence bands assigned to near-band-edge (3.27–3.37 eV) and green (2.3–2.5 eV) emissions, respectively. All phonon modes of the crystals have experimentally been determined. Basic optical properties of bulk ZnO single crystals are collected in Table 3.8.

Table 3.7 Exemplary experimental (bulk crystals) and theoretical phonon modes of ZnO. Ref. [308] – Raman spectroscopy; Ref. [311] – Raman spectroscopy (a), inelastic neutron scattering (b), and *ab initio* calculations (c); Ref. [312]

Phonon mode	Experimental Ref. [308] [cm^{-1}]	Experimental Ref. [311] (a) [cm^{-1}]	Experimental Ref. [311] (b) [cm^{-1}]	Theoretical Ref. [312] [cm^{-1}]	Theoretical Ref. [311] (c) [cm^{-1}]
E_2 (low)	101	100	98	106	91
B_1 (low)	—	—	259	283	261
A_1 (TO)	381	380	378	384	390
E_1(TO)	407	410	412	408	409
E_2 (high)	437	438	439	436	440
B_1 (high)	—	—	552	578	552
A_1(LO)	574	584	573	607	565
E_1 (LO)	583	595	593	615	557

Table 3.8 Basic empirical optical properties of bulk ZnO single crystals

Property	Unit	Value	Reference
Direct bandgap [eV]	eV	3.3	279, 280
Absorption edge	nm	≈390	71, 243
Emission band peak at RT	eV	near band edge: 3.27–3.37 green: 2.3–2.5	71, 182, 186, 214, 215, 219, 225, 235, 302, 303, 304, 305, 306, 307

(*Continued*)

Table 3.8 (*Continued*)

Property	Unit	Value	Reference
Static dielectric constant		$\parallel c$: 8.49/8.91 $\perp c$: 7.40/7.77	286/289
High-frequency dielectric constant		$\parallel c$: 3.72/3.78 $\perp c$: 3.68/3.70	286/289
Experimental refractive index at 400, 600, and 800 nm		Ordinary: 2.22, 2.00, 1.96 Extraordinary: 2.23, 2.01, 1.97	71
Phonon modes	cm^{-1}	E_2 (low): 101/100/98 B_1 (low): — / — /259 A_1 (TO): 381/380/378 E_1 (TO): 407/410/412 E_2 (high): 437/438/439 B_1 (high): — / — /552 A_1 (LO): 574/584/573 E_1 (LO): 583/595/593	308/311[a]/311[b]

[a]Raman spectroscopy; [b]inelastic neutron scattering

3.7 Basic Electrical Properties

Most important electrical properties include the free-carrier concentration, resistivity (or conductivity), carrier mobility, carrier activation energy, and carrier traps. The main techniques used to gather the electrical properties involve Hall effect, current–voltage characteristics, deep-level transient spectroscopy (DLTS), and others.

3.7.1 Origin of Electrical Conductivity

Electrical properties of the obtained crystals are the result of a growth environment (or growth method), which is defined by the phase (gas, flux, melt), growth temperature, and atmosphere, which differ between different growth methods, as well as by residual impurity and intentional dopant concentration. The obtained bulk ZnO single crystals are either *n*-type semiconductors or electrical insulators at RT, although *p*-type conductivity was reported as well, but for thin films [313–316]. A summary of *p*-type conductive ZnO films can be found in the review paper of Reynolds and Reynolds [79]. The electrical conductivity seems to have multiple origin, which resulted in a scientific debate. There is a number of suggestions, experimental proofs (whether direct or indirect), or theoretical predictions pointing to different sources of the underlying electrical conductivity, which may manifest differently in different growth environments. A number of factors have experimentally and theoretically been proposed to explain the electrical conductivity of ZnO crystals and its compensation. Here are a few examples.

Shallow donors, *n*-type conductivity:

- Trivalent solid impurities, such as Al, Ga, In, and Sc, which substitute Zn^{2+} ions in the crystal lattice (e.g., Helbig [146], Bogner and Mollwo [149], Abe et al. [193], Wang et al. [211], Weckstern et al. [317],)
- P substituting Zn, P_{Zn} (e.g., Weckstern et al. [317])
- Interstitial hydrogen, H_i^+ (e.g., Walle [318], Qiu et al. [319]), hydrogen at bond-centered site, H_{BC}, and hydrogen at oxygen vacancy, H_O (e.g., Lavrov et al. [320, 321], Janotti and Walle [322]). The presence of hydrogen has experimentally been confirmed, e.g., by Lavrov et al. [323, 324], Shi et al. [325], Jokela and McCluskey [326], Pearton et al. [70]
- Zinc interstitials, Zn_i (e.g., Look et al. [327], Sun and Wang [328])
- Cl^- ions that substitute O^{2-} ions in the crystal lattice (Ntep et al. [196])
- Zn_I–N_O complexes (Look et al. [329])
- N–H complexes (Jokela et al. [195])

Shallow acceptors, *p*-type conductivity:

- **General approaches:** doping a zinc site with group IA elements (Li, Na), doping an oxygen site with group VA elements (N, P, As, Sb), and co-doping of donors and acceptors, e.g., Al-2N, Ga-2N (Reynolds and Reynolds [79])

- N, As, P doping (e.g., Look et al. [330], Klingshirn [1], Pearton et al. [70])
- Complex zinc vacancy – nitrogen at oxygen site – interstitial hydrogen, V_{Zn}-N_O-H^+ (Reynolds et al. [331])
- Complex As or P occupying Zn site – 2 Zn-vacancies, As_{Zn}-$2V_{Zn}$ or P_{Zn}-$2V_{Zn}$ (Limpijumnong et al. [332], Lee et al. [333])

***n*-type and *p*-type conductivity compensators:**

- Zn-vacancies, V_{Zn} (e.g., Janotti and Van de Walle [133, 134])
- O-vacancies, V_O, and interstitials, O_i (e.g., Janotti and Van de Walle [133, 134])
- N in the O – sublattice, N_O (e.g., Lavrov et al. [331], Meyer et al. [334], Look [335])
- Hydrogen complex Zn-H + O-H, H_2^* (Karazhanov and Uyashin [336])
- Monovalent solid impurities, such as Li, Na, K, Cu that substitute Zn^{2+} ions in the crystal lattice or form interstitials (e.g., Helbig [146], Bogner and Mollwo [149], Look and Claflin [337], Grasza and Mycielski [171], Yan and Wei [338]).

Residual impurities cannot be avoided in any bulk crystal growth method, as they are incorporated from a thermal insulation, a starting material, carrier or reactive gases, a growth atmosphere, a solvent, seeds, and crucibles (if utilized). Depending on the growth method, different impurity (-ies) will dominate. Moreover, hydrogen is omnipresent and, as such, cannot be excluded from any of the growth method. Additionally, there are native point defects, which will form at different concentrations depending on oxidizing or reducing growth conditions and on the growth temperature. This can be referred to a deviation from stoichiometry. Although O-vacancies form deep donors, which cannot be directly responsible for the electrical conductivity, they may form complexes with other elements that act as shallow donors. The same relates to Zn-interstitials, when forming complexes with other elements.

This picture is really complex and should be considered in relation to a particular growth environment. Additionally, the balance of impurities and native point defects can be further affected by post-growth heat treatments with three degrees of freedom: atmosphere, temperature, and time.

In the case of gas phase, an impact on the electrical properties have the following factors: source material, transport method, oxidizing agent, transport agent (whether chemically active or not), reducing agent (if used), temperature, total pressure, type of growth furnace, furnace materials, seed type, and purity of chemicals/materials used in the growth process. All these conditions affect the ZnO stoichiometry, structural defect formation, and impurity incorporation, particularly those present in the surroundings of a growing crystal (from the transport gas, furnace materials, and seed, whether external or formed by an ampoule wall).

When considering solution methods (flux and hydrothermal), the main factors affecting the electrical properties include mineralizer or flux type, temperature, total pressure, and purity of chemicals/materials, including purity of the source material (nutrient). In an optimizing growth process, the dominating factors relate to the mineralizer/flux type and the purity of the starting material.

For melt growth methods, the main factors influencing the electrical properties are purity of the starting material, purity of crucible (Bridgman method), atmosphere type, oxygen partial pressure, total pressure, and cooling rate. The resulted electrical conductivity of the obtained crystals is a combination of native point defects and impurities/dopants.

In each of the growth methods, the electrical properties can be additionally tuned by intentional doping either by introducing shallow donors (increasing the electrical conductivity) or electrical conductivity compensators, as well as by post-growth heat treatment that changes the balance between point defects. Bulk ZnO crystals are either electrical insulators or semiconductors, and occasionally degenerate semiconductors. A consistent and reliable *p*-type conductivity of low resistivity is hard to achieve in bulk ZnO single crystals, due to the self-compensation effect. Therefore, below only *n*-type conductivity of the crystals is discussed.

3.7.2 Free Electron Concentration, Hall Mobility, and Resistivity/Conductivity at RT

The ZnO crystals obtained from the gas phase could be highly resistive (>10 Ωcm) or *n*-type semiconductors with the free electron concentration n_e up to about 3×10^{19} cm^{-3}, and the electron mobility μ approaching 220 cm^2/Vs. In addition to semi-insulating crystals (e.g., Ziegler et al. [162]), relatively high resistivity of the as-grown crystals at RT was reported by Schwarovsky [142] (about 200 Ωcm), Piekarczyk et al. [160] (50–300 Ωcm), Grasza et al. [171] and Nielson [166] (1–100 Ωcm), Helbig [146] (10–100 Ωcm), Matsushita et al. [177] (200–300 Ωcm), Ito et al. [178] (100–1000 Ωcm), and Dodson and Savage [167] (7–30 Ωcm). Mid or low resistivity was measured by Bogner and Mollwo [149] (0.1 and 10 Ωcm), Ziegler et al. [162] (0.4–3.7 Ωcm or 0.038–0.38 Ωcm if elemental Zn was added to the ZnO starting material), Tena-Zaera et al. [181] (0.4–5.8 Ωcm), Hong et al. [190] (0.28 Ωcm), and Mycielski et al. [183] (0.05 Ωcm). In terms of the free electron concentration, Ziegler et al. [162] reported the values of 1.6×10^{16} – 1.5×10^{18} cm^{-3}, Fischer [169] of about 10^{17} cm^{-3}, Hassani et al. [179] of 2×10^{18} cm^{-3}, Pfisterer et al. [182] of mid of 10^{16} cm^{-3}, Tena-Zaera et al. [181] of 9×10^{15} – 10^{18} cm^{-3}, Wei et al. [185] of 2×10^{17} – 2×10^{18} cm^{-3}, Hong et al. [190] of 2.8×10^{17} cm^{-3}, Santailler et al. [194] of 2×10^{17} cm^{-3}, and Ntep et al. [196] of 3×10^{19} cm^{-3}. Also the values of the electron mobility span in a wide range, e.g., of 70 to 200 cm^2/Vs (Ziegler et al. [162]), about 125 cm^2/Vs (Dodson and Savage [167]), 220 cm^2/Vs (Helbig [146]), 205 cm^2/Vs (Fischer [169]), 160 cm^2/Vs (Piekarczyk et al. [160]), 183 cm^2/Vs (Hassani et al. [179]), 14–120 cm^2/Vs (Tena-Zaera et al. [181]), 100–165 cm^2/Vs (Wei et al. [185]), 120 cm^2/Vs (Hong et al. [190]), 168 cm^2/Vs (Santailler et al. [194]), 100 cm^2/Vs (Jokela et al. [195]), and below 100 cm^2/Vs (Ntep et al. [196]). Fischer [169] reported that the compensation ratio of acceptors to donors is only 0.02, which is very low for a wide-gap material.

Flux-grown crystals are *n*-type semiconductors. For the as-grown ZnO crystals, Fisher and Sinn [197] reported the electrical resistivity of flux-grown ZnO crystals of 10 Ωcm at RT. The crystals obtained by Li et al. [199] showed high free electron concentration of 2.1 × 10^{18} cm^{-3} and a low resistivity of 0.024 Ωcm.

Hydrothermal ZnO crystals have usually high resistivity due to donor compensation by Li^{+} ions substituting Zn^{2+} ions. For instance, the resistivity of the as-grown crystals at RT reported by Laudise et al. [205] was 1 Ωcm, by Croxall et al. [206] about 10^3 Ωcm, by Sekiguchi et al. [225] 10^8 Ωcm when grown with the oxidizer, and between 1 and 100 Ωcm when grown with no oxidizer, by Meada et al. [219] 380 Ωcm, and by Zhang et al. [209] 80 Ωcm. A characteristic feature of hydrothermal crystals is a much higher conductivity of darker sectors (e.g., $-c$) as compared with light sectors (e.g., $+c$), due to higher impurity concentration. Suscavage et al. [208] noticed that the dark sectors of the crystals had a low resistivity with the electron concentration of 3×10^{18} cm^{-3}, while the light sectors had high resistivity with the electron concentration of only 2×10^{12} cm^{-3}. Also, Meada et al. [219] and Zhang et al. [209] reported low free electron concentration of light sectors of 10^{13} cm^{-3} and 10^{14} cm^{-3}, respectively. The electron mobility was between 100 and 200 cm^2/Vs (Zhang et al. [209], Suscavage et al. [208], Meada et al. [219]). Allen et al. [339] measured for hydrothermally grown crystals the free electron concentration of 3 × 10^{14} cm^{-3} to 4 × 10^{16} cm^{-3}, and Hall mobility of 190–230 cm^2/Vs at RT.

Melt-grown ZnO crystals have very similar properties to each other. At RT, the as-grown crystals are n-type semiconductors with the free electron concentration and electron mobility of mid 10^{17} cm^{-3} and 130–150 cm^2/Vs, respectively. For example, ZnO crystals obtained by the cold crucible method exhibit the free electron concentration, electron mobility, and resistivity of 5×10^{17} cm^{-3}, 131 cm^2/Vs, and 0.094 Ωcm, respectively (Nause [230]). In the case of Bridgman-grown crystals, these properties were 5×10^{17} cm^{-3}, 153 cm^2/Vs, and 0.08 Ωcm, respectively (Schulz et al. [236]).

A collection of basic electrical properties of bulk ZnO crystals is listed in Table 3.9.

Table 3.9 Selected examples of basic electrical properties at RT of as-grown and intentionally undoped bulk ZnO single crystals obtained from different phases

Growth method	Resistivity [Ωcm]	Free electron concentration [cm^{-3}]	Electron mobility [cm^2/Vs]	Growth environment	Reference
Gas phase	200	—	—	$Zn–N_2–H_2$	142
	0.4–3.7	$1.6–9.5×10^{16}$	70–200	HX and NH_4X	162
	0.038–0.38	$1.1×10^{17} – 1.5×10^{18}$	70–160	ZnO–Zn	162
	7–30	—	125	$ZnO–H_2$	167
	—	10^{17}	205	$ZnO–H_2$	169
	10–100	—	220	$ZnO–H_2–N_2$	146
	100–1000	—	—	ZnO–C	178
	0.4–5.8	$9×10^{15} – 1×10^{18}$	14–120	ZnO–C	181
	—	$2×10^{17} – 2×10^{18}$	100–165	ZnO–C	185
	0.21	$2.8×10^{17}$	120	ZnO–C	190
Flux	1	—	—	PbF_2	197
	0.024	$2.1×10^{18}$	—	PbF_2	199
Hydrothermal	1000/1[(a)]	— / —	— / 38	KOH + LiOH	206
	— / —	$2×10^{12}/3×10^{18}$ [(b)]	175	Li_2CO_3 + KOH + NaOH	208
	380	10^{13}	200	KOH + LiOH	219
	80	10^{14}	100	KOH + LiOH + H_2O_2	209
	10^8	—	—	KOH + LiOH + H_2O_2	225
	—	$3×10^{14} – 4×10^{16}$	190–230	—	339
Cold crucible	0.094	$5×10^{17}$	131	O_2, 1–10 atm	230
Bridgman	0.08	$5×10^{17}$	153	CO_2, 20 bar	236

[(a)]colorless/green sector; [(b)]light/dark sector

3.7.3 Temperature-Dependent Electrical Properties and Ionization Energy of Shallow Donors

An example of temperature-dependent free electron concentration, Hall mobility, and resistivity (from Hall effect measurements) of bulk ZnO crystals obtained by the CVT method is shown in Fig. 3.50 (Hong et al. [190]). The free electron concentration (Fig. 3.50a) increases with temperature due to the thermal excitation of electrons from shallow donors to the conduction band. At RT, it reaches a value of 2.8×10^{17} cm^{-3}. The character of the dependence remains unchanged after annealing ZnO crystals at 1000°C in O_2 for 5 h, but the free electron concentration drops by almost one order of magnitude. From the slope of the relation log(n) – T^{-1} in Fig. 3.50a, one can estimate the activation energy of isolated donors, which is E_{CD} = 42.8 meV at the free electron concentration of 2.8×10^{17} cm^{-3}. After annealing, the donor binding energy is almost the same. Ziegler et al. [162] gathered the activation energy between 29 and 49 meV and between 195 and 280 meV for moderately semiconducting and highly resistive samples, respectively. If the crystals were obtained at Zn-rich conditions (Ziegler et al. [340]), the donor activation energy was 15 meV, while at O-rich conditions, the activation energy was concluded at 40–50 or 200–280 meV. Ntep et al. [196] measured the donor activation energy of 23 meV at the free electron concentration of 3×10^{19} cm^{-3}, while Hassani et al. [179] found two donor levels at 2.4×10^{17} and 1.16×10^{18} cm^{-3} with the activation energies of 21 and 37 meV, respectively. Two donor levels with the activation energies of 32 and 65 meV were also noticed by Wei et al. [185]. The Hall electron mobility (Fig. 3.50b) is limited by the ionized impurity scattering ($T^{3/2}$) at low temperatures (30–70 K), and by lattice vibration scattering ($T^{-3/2}$) at higher temperatures (70 to 290 K). In general, the Hall mobility in bulk ZnO crystals is affected by acoustic phonons, optical phonons, ionized impurities, and piezoelectric effect, the latter being rather least influential as compared with other scattering mechanisms. The maximum of the Hall mobility is located at about 70 K, and its value is 358 and 765 cm^2/Vs for as-grown and annealed crystals, respectively. At RT, these values are in the range of 120–160 cm^2/Vs. All other reports also point similar temperatures at which the maximum of Hall mobility is reached. Fischer [169] measured the peak mobility of about 2000 cm^2/Vs at 50 K, Hong et al. [190] measured about 765 cm^2/Vs at 70 K, and Nause [230] (melt growth method) measured 298 cm^2/Vs at 78 K.

The resistivity versus inverse temperature (Fig. 3.50c) shows a typical behavior of semiconductors, which is governed by thermal excitation below RT and by phonon scattering around RT and above. At RT, the resistivities are 0.21 and 0.93 Ωcm for as-grown and annealed crystals, respectively.

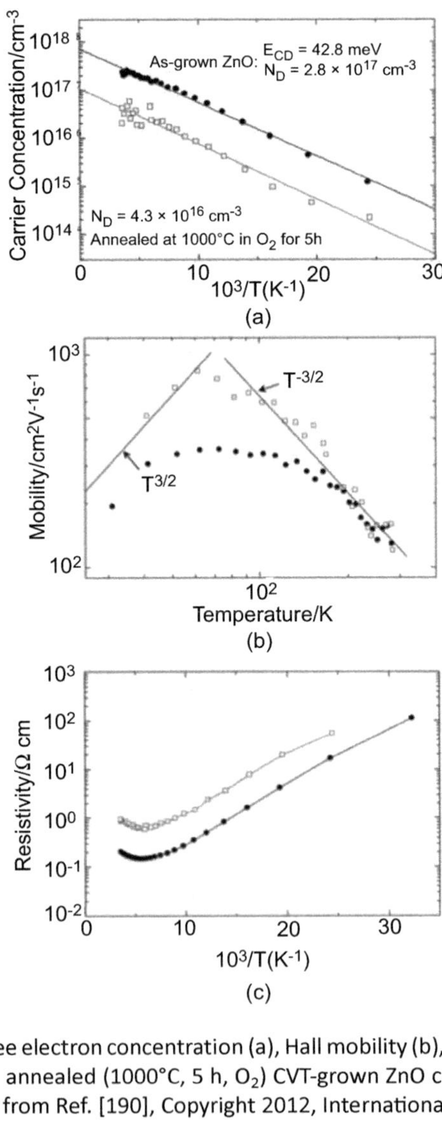

Figure 3.50 Free electron concentration (a), Hall mobility (b), and resistivity (c) of as-grown and annealed (1000°C, 5 h, O_2) CVT-grown ZnO crystals. Reprinted with permission from Ref. [190], Copyright 2012, International Organization for Ceramic Processing.

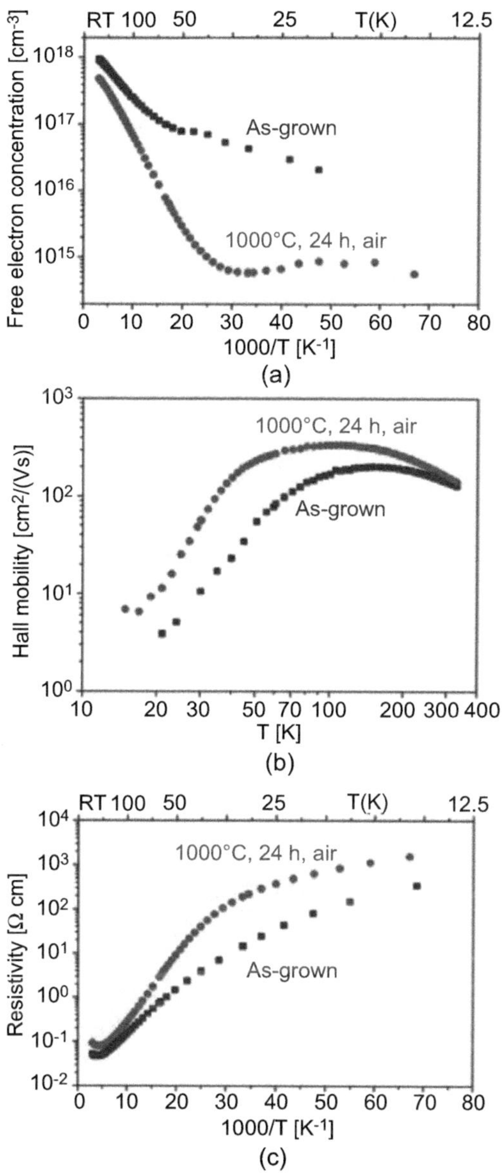

Figure 3.51 Free electron concentration (a), Hall mobility (b), and resistivity (c) of as-grown bulk ZnO crystals obtained by the Bridgman (melt) method [341].

Figure 3.51 [341] shows similar relations for melt-grown (Bridgman method) bulk ZnO single crystals, but with an extension to still lower temperatures. Here, the electrical properties of as-grown and air-annealed samples (1000°C, 24 h) at RT are very close to each other (e.g., the free electron concentration and resistivity differ by a factor of about two, only), but larger differences manifest at low temperatures. The maximum of the Hall mobility of 337 cm^2/Vs occurred at 100–120 K. A bit different behavior of melt-grown ZnO crystals (high growth temperature, metal crucible) is likely the result of different content of residual impurities and point defects as compared with ZnO crystals from the vapor phase (low temperature, no metal crucible).

3.7.4 Effective Electron and Polaron Masses

In reference to effective electron mass, it is described as bare (or quasi-free electrons of the conduction band) effective electron mass m_e^* or effective polaron (enhanced by the polaron effect) mass m_p^* as a result of coupling between electrons and longitudinal optical phonons. As an example, for hydrothermally grown bulk ZnO crystals, Baer [342] measured with Faraday rotation technique the bare effective electron mass $m_e^* = 0.23–0.25\ m_0$ for the free electron concentration in a range of $6.4 \times 10^{15}–1.5 \times 10^{19}\ cm^{-3}$, which yielded the effective polaron mass $m_p^* = 0.29\ m_0$. For effective electron mass determination of vapor-grown bulk ZnO crystals, Button et al. [343] utilized Zeeman splitting and cyclotron resonance. From Zeeman splitting, they obtained $m_p^* = 0.3$ and $0.34\ m_0$ for $H||c$ and $H\perp c$ (H is the magnetic field), respectively. In the case of the cyclotron resonance, $m_e^* = 0.24$ and $0.28\ m_0$ for $H\perp c$ and $H||c$, respectively. Cyclotron resonance measurements of CVT-grown ZnO by Oshikiri et al. [344] and Imanaka [345] yielded values of $m_p^* = 0.29$ and $m_e^* = 0.23\ m_0$, which are quite isotropic. Yet other vapor-grown ZnO crystals (doped with In) investigated by Weiher [346] with infrared reflectivity revealed the optical electron mass $m_e^{opt} = 0.25–0.26\ m_0$ for high free electron concentration of $1–5 \times 10^{19}\ cm^{-3}$.

For comparison, some thin films showed comparable values of effective bare electron and polaron masses, e.g., $m_p^* = 0.29$ and $m_e^* = 0.247\ m_0$ [347]. A relation between the effective electron mass and carrier concentration was studied for thin films, e.g., by Tang et al. [348]. For the free electron concentrations from $5.9 \times 10^{17}\ cm^{-3}$ to $4.0 \times 10^{19}\ cm^{-3}$, the effective electron mass (bare) increases from 0.23 m_0 to 0.26 m_0 as the result of the non-parabolicity of the conduction band. For yet higher free electron concentration (>10^{20} cm^{-3}), the effective electron mass increases to values of 0.29–0.32 m_0 [349].

Examples of theoretical values of conduction-band effective electron masses are $m_e^{\parallel c}$ = 0.23 m_0 and $m_e^{\perp c}$ = 0.21 m_0 by Lambrecht et al. [91], 0.24 m_0 by Oshikiri et al. [344], and 0.32 m_0 by Schleife et al. [108].

3.7.5 Effect of Annealing on Electrical Properties

Annealing, or post-growth heat treatment, can be conducted at different temperatures, atmospheres, and times. ZnO crystals obtained from the gas phase and annealed in the presence of oxygen at or above 600°C for several hours or more systematically increase their electrical resistivity and decrease the free electron concentration by about one order of magnitude. This is well shown in Fig. 3.50. Santailler et al. [194] stated that after annealing ZnO crystals at 1100°C in O_2 for 1 h, the free electron concentration dropped from 2 × 10^{17} cm^{-3} to 7 × 10^{15}–2 × 10^{16} cm^{-3}, wherein the electron mobility slightly increased from 168 to 182 cm^2/Vs. Also, Jokela et al. [195] reported a decrease in the free electron concentration from 5 × 10^{17} cm^{-3} to 3 × 10^{16} cm^{-3} after annealing at 745°C in air for 1 h. Here the electron mobility dropped from 100 to 60 cm^2/Vs. Mycielski et al. [183] showed that the resistivity increased from 0.05 Ωcm at RT to 5 Ωcm after annealing in an O_2-containing atmosphere at 900–1000°C. Wei et al. [185] stated that annealing ZnO crystals at 900°C in air for 20 h decreased the free electron concentration and electron mobility by 5–20%.

An opposite impact on the electrical properties was noticed for annealing at atmospheres containing elemental Zn (or Zn vapor). Schwarovsky [142] recorded a decrease in the resistivity from 200 Ωcm to 0.05 Ωcm after annealing in the Zn vapor at 800°C, while Pfisterer et al. [182] measured an increase in the free electron concentration from about 10^{16} to about 2 × 10^{16} cm^{-3}. Ziegler et al. [340] showed an impact of annealing of CVT-grown ZnO crystals (grown in O_2-rich conditions) in an atmosphere containing elemental Zn on electrical properties. As shown in Fig. 3.52, with increasing

When annealing Bridgman-grown bulk ZnO crystals in the presence of hydrogen (5% H_2 + Ar), whether as-grown or previously annealed in oxygen (1000°C, 20 h), the free electron concentration remains substantially unaffected, as shown in Fig. 3.53a [350]. After such annealing, the Hall mobility gradually decreases with annealing temperature but not too much, not more than 15% in a temperature range of 400–900°C (Fig. 3.53b).

3.7.6 Effect of Intentional Doping on Electrical Properties

Intentional doping with monovalent ions, such as Li^{1+}, Na^{1+}, and Cu^{1+}, increases the resistivity by one to three orders of magnitude or make the crystals semi-insulating (in the hydrothermal method due to Li^+ doping from the mineralizer). This is due to the substitution of Zn^{2+} ions by monovalent ions that act as compensating acceptors. Bogner and Mollwo [149] reported an increase in the resistivity from 0.1–10 Ωcm to 1000 Ωcm after doping with Cu. Similar resistivity increase, from 1–100 Ωcm to 10–1000 Ωcm, was also observed by Nielsen [166] after doping ZnO crystals with Cu. Savage and Dodson [168] showed that Cu doping concentration of 2.4×10^{18} –7×10^{19} at. cm^{-3} led to a resistivity increase from 7–30 Ωcm to 100–2000 Ωcm, and the electron mobility decrease from about 125 cm^2/Vs to below 20 cm^2/Vs, while Li doping concentration of 4.9×10^{18} at. cm^{-3} increased the resistivity to 10–50 Ωcm, and decreased the electron mobility below 10 cm^2/Vs. Helbig [146] measured the resistivity of undoped crystals between 10 and 100 Ωcm, and the electron mobility of 220 cm^2/Vs. Doping with Li (up to 200 wt. ppm) increased the resistivity to 10^4–10^{11} Ωcm, with Na (30–40 wt. ppm) to 10^2 and 10^4 Ωcm, and with Cu (4×10^{15}–8×10^{20} at. cm^{-3}) to 10^2–10^8 Ωcm. Grasza et al. [171] reported that doping with Cr, Cu, As increased the resistivity from 1–100 Ωcm to 10–1000 Ωcm.

Similar effect was obtained by doping hydrothermally grown ZnO crystals with N^{3-} (Wang et al. [223]), which substitute O^{2-} ions in the crystal lattice. The as-grown crystals had high resistivity at RT, at the level of 10^5 Ωcm, which increased by three orders of magnitude after annealing at 600°C for 2 h in air.

Doping with some of the trivalent ions, such as In^{3+}, Ga^{3+}, and Sc^{3+}, decreases the resistivity by one to three orders of magnitude. Here trivalent ions act as shallow donors when substituting Zn^{2+} ions.

Bogner and Mollwo [149] showed a significant decrease in the resistivity from 0.1–10 Ωcm to 0.05–0.003 Ωcm after doping ZnO crystals with In. However, above the In concentration at the level of 7×10^{19}, there was a saturation in electrical resistivity. Annealing In-doped ZnO crystals in air or vacuum for 30 min at 1200°C increased the resistivity by a factor of about 2.

Helbig [146] reported a decrease in the resistivity from 10–100 Ωcm to 10^2–10^{-3} Ωcm with Ga concentration of 1×10^{15}–1.3×10^{20} at. cm^{-3}, and to 3×10^{-2} Ωcm after doping with In. Doping with Mn substantially did not affect the resistivity. Abe et al. [193] reported that the free electron concentration could be increased to a value of 3×10^{19} cm^{-3} by doping with Ga_2O_3 at the level of 0.1–1 wt.%; however, the Hall mobility gradually decreases with Ga_2O_3 doping level to the value of about 30 cm^2/Vs at the free electron concentration of 3×10^{19} cm^{-3}. The free electron concentration saturates above the Ga_2O_3 content of 0.1 wt.% with a degradation of the structural quality of ZnO. Hydrothermally grown ZnO crystals doped with In (Wang et al. [211]) showed high free electron concentration of 1.1×10^{19} cm^{-3}, the same order of magnitude (although lower) as In concentration (1.3×10^{19} at. cm^{-3}). Sc-doped crystals showed a decrease in the resistivity from 80 to 0.056 Ωcm, and an increase in the electron concentration from 10^{14} to 10^{18} cm^{-3}. The electron mobility of Sc-doped ZnO was 121 cm^2/Vs (Zuo et al. [226]).

3.7.7 Deep Traps

Deep electron traps are typically gathered by the deep-level transient spectroscopy (DLTS). There are several reports on deep electron traps in bulk ZnO single crystals, which show different number, energetic distribution, and concentration of the traps, which are related to a growth environment. Following Scheffler et al. [351] (Table 3.10), as-grown crystals obtained by the CVT method revealed seven electron traps located at energy levels of 0.1–1.03 eV below the conduction band minimum (CBM). All traps, except E6, have the concentration at a level of 10^{13} cm^{-3}, while the latter one has about one order of magnitude larger. The hydrothermally and melt-grown crystals revealed only three electron traps, two of them located (E6 and E8) at the same level (0.6 and 1.03 eV below the CBM) as those in the crystal obtained from the gas phase, and another one (E3) at 0.28 eV below the CBM, which was not observed in the crystal obtained from

Table 3.10 Trap concentration in bulk ZnO crystals obtained from the gas phase, solution, and melt. All data are from Ref. [351]

Trap label **Activation energy** **[eV]**	**E1** **0.1** **[cm^{-3}]**	**E2** **0.18** **[cm^{-3}]**	**E3** **0.28** **[cm^{-3}]**	**E4** **0.4** **[cm^{-3}]**	**E5** **0.57** **[cm^{-3}]**	**E6** **0.6** **[cm^{-3}]**	**E7** **0.61** **[cm^{-3}]**	**E8** **1.03** **[cm^{-3}]**
Gas phase CVT	1.5×10^{13}	8.6×10^{13}	—	1.4×10^{13}	4.0×10^{13}	4.1×10^{13}	3.1×10^{14}	5.8×10^{13}
Solution[(a)] Hydrothermal	—	—	7.0×10^{14}	—	—	2.0×10^{14}	—	5.0×10^{14}
Melt Cold crucible	—	—	3.5×10^{15}	—	—	2.3×10^{14}	—	1.4×10^{15}

[(a)]Due to high resistivity, the sample was treated with hydrogen to increase the conductivity, enabling good Schottky contact formation for DLTS measurements.

Electrical properties		
Tunability of the free electron concentration at RT [cm^{-3}]	$\approx 10^{13}$ – 3×10^{19}	162, 169, 181, 185, 190, 193, 199, 208, 209, 211, 219, 230, 236, 339
Maximum experimental electron mobility at RT [cm^2/Vs]	200–230	146, 162, 169, 219, 339
Ionization energy of isolated shallow donors [meV] at $n_e \approx 10^{16}$–10^{19} cm^{-3}	15–50	162, 179, 190, 196, 340
Effective electron mass, m_e^*/m_0 at $n_e \approx 10^{16}$–10^{19} cm^{-3}	$\approx$0.23–0.25	342, 343, 344, 345, 346
Effective polaron mass, m_p^*/m_0	$\approx$0.29–0.34	342, 343, 344, 345
Concentration of deep-level traps [cm^{-3}]	10^{13} – 3×10^{15}	185, 351, 352

[(a)]A number of crystals in one growth run; [(b)]small anisotropy depending on the crystallographic direction ($\parallel c$ or $\perp c$) and surface polarity; [(c)]semi-empirical values based on measured stiffness coefficients from corresponding references and Eqs. 3.39–3.42

References

1. C. Klingshirn; *Phys. Status Solidi (b)* **244** (2007), 3027–3073.
2. C. Solbrig; *Z. Phys.* **184** (1965), 293–298.
3. I. B. Kobiakov; *Solid State Commun.* **35** (1980), 305–310.
4. Z. C. Tu, X. Hu; *Phys. Rev. B* **74** (2006), 035434.
5. G. Heiland, H. Ibach; *Solid State Commun.* **4** (1966), 353–354.
6. L. M. Trinca, A. C. Galca, A. G. Boni, M. Botea, L. Pintilie; *Appl. Surf. Sci. A* **427** (2018), 29–37.
7. Z.-H. Xiong, F.-Y. Jiang; *J. Phys. Chem. Solids* **68** (2007), 1500–1503.

8. A. Samariya, R. K. Singhal, S. Kumar, Y. T. Xing, M. Alzamora, S. N. Dolia, U. P. Deshpande, T. Shripathi, E. B. Saitovitch; *Mater. Chem. Phys.* **123** (2010), 678–684.
9. T. Chen, L. Cao, W. Zhang, W. Zhang, Y. Han, Z. Zheng, F. Xu, I. Kurash, H. Qian, J. Wang; *J. Appl. Phys.* **111** (2012), 123715.
10. M. Lorenz, R. Johne, T. Nobis, H. Hochmuth, J. Lenzner, M. Grundmann, H. P. D. Schenk, S. I. Borenstain, A. Schön, C. Bekeny, T. Voss, J. Gutowski; *Appl. Phys. Lett.* **89** (2006), 243510.
11. M. Kano, A. Wakamiya, K. Sakai, K. Yamanoi, M. Cadatal-Raduban, T. Nakazato, T. Shimizu, N. Sarukura, D. Ehrentraut, T. Fukuda; *J. Cryst. Growth* **318** (2011), 788–790.
12. M. Takehana, T. Nishino, K. Sugawara, T. Sugawara; *Mater. Sci. Eng. B* **41** (1996), 186–189.
13. J. Lozano, M. J. Fernández, J. L. Fontecha, M. Aleixandre, J. P. Santos, I. Sayago, T. Arroyo, J. M. Cabellos, F. J. Gutiérrez, M. C. Horrillo; *Sens. Actuator. B* **120** (2006), 166–171.
14. C.-C. Hsiao, K.-Y. Huang, Y.-C. Hu; *Sensors* **8** (2008), 185–192.
15. C. S. Wei, Y. Y. Lin, Y. C. Hu, C. W. Wu, C. K. Shih, C. T. Huang, S. H. Chang; *Sens. Actuator. A* **128** (2006), 18–24.
16. T. Nakazato, Y. Furukawa, M. Tanaka, T. Tatsumi, M. Nishikino, H. Yamatani, K. Nagashima, T. Kimura, H. Murakami, S. Saito, T. Shimizu, N. Sarukura, H. Nishimura, K. Mima, Y. Kagamitani, D. Ehrentraut, T. Fukuda; *J. Cryst. Growth* **311** (2009), 875–877.
17. A. V. Shumakov, A. S. Sviridov, S. V. Kolesnikov; *At. Energy* **110** (2011), 184–194.
18. T. Shimizu, K. Yamanoi, K. Sakai, M. Cadatal-Raduban, T. Nakazato, N. Sarukura, M. Kano, A. Wakamiya, D. Ehrentraut, T. Fukuda, M. Nagasono, T. Togashi, S. Matsubara, K. Tono, A. Higashiya, M. Yabashi, H. Kimura, H. Ohashi, T. Ishikawa; *Appl. Phys. Express* **4** (2011), 062701.
19. Y. R. Ryu, W. J. Kim, H. W. White; *J. Cryst. Growth* **219** (2000), 419–422.
20. D. C. Look, B. Claflin; *Phys. Status Solidi (b)* **241** (2004), 624–630.
21. L. J. Brillson, Y. Lu; *J. Appl. Phys.* **109** (2011), 121301.
22. M. W. Allen, S. M. Durbin, X. Weng, J. M. Redwing, K. Sarpatwari, S. E. Mohney, H. von Wenckstern, M. Grundmann; *IEEE Trans. Electron Devices* **56** (2009), 2160–2164.
23. S. Müller, H. von Wenckstern, M. Grundmann, R. Heinold, M. Allen; *J. Appl. Phys.* **116** (2014), 194506.
24. E. Fortunato, P. Barquinha, R. Martins; *Adv. Mater.* **24** (2012), 2945–2986.

25. N. Izyumskaya, V. Avrutin, Ü. Özgür, Y. I. Alivov, and H. Morkoç; *Phys. Status Solidi* (*b*) **244** (2007), 1439–1450.

26. M. Grundmann, H. Frenzel, A. Lajn, M. Lorenz, F. Schein, H. von Wenckstern; *Phys. Status Solidi* (**a**) 207 (2010), 1437–1449.

27. H. Frenzel, A. Lajn, H. von Wenckstern, M. Lorenz, F. Schein, Zh. Zhang, M. Grundmann; *Adv. Mater.* **22** (2010), 5332.

28. F. J. Klüpfel, F.L. Schein, M. Lorenz, H. Frenzel, H. von Wenckstern, M. Grundmann; *IEEE Trans. Electron Devices* **60** (2013), 1828–1833.

29. H. Frenzel, F. Schein, A. Lajn, H. von Wenckstern, M. Grundmann; *Appl. Phys. Lett.* **96** (2010), 113502.

30. S. Liang, H. Sheng, Y. Liu, Z. Huo, Y. Lu, H. Shen; *J. Cryst. Growth* **225** (2001), 110–113.

31. K. Liu, M. Sakurai, M. Aono; *Sensors* **10** (2010), 8604–8634.

32. Z. Zhang, H. von Wenckstern, M. Grundmann; *Appl. Phys. Lett.* **103** (2013), 171111.

33. D. C. Look, B. Claflin, Ya. I. Alivov, S. J. Park; *Phys. Status Solidi* (*a*) **201** (2004), 2203–2212.

34. A. Tsukazaki, A. Ohtomo, T. Onuma, M. Ohtani, T. Makino, M. Sumiya, K. Ohtani, S. F. Chichibu, S. Fuke, Y. Segawa, H. Ohno, H. Koinuma, M. Kawasaki; *Nat. Mater.* **4** (2005), 42–46.

35. A. Tsukazaki, M. Kubota, A. Ohtomo, T. Onuma, K. Ohtani, H. Ohno, S. F. Chichibu, M. Kawasaki; *Jpn. J. Appl. Phys.* **44** (2005), L643–L645.

36. S. Tüzemen, Emre Gür; *Opt. l Mater.* **30** (2007), 292–310.

37. C. Messaoudi, S Abd-lefdil, D. Sayah, M. Cadene; *Eur. Phys. J. AP* **1** (1998), 181–184.

38. R. Ghosh, B. Mallik, S. Fujihara, D. Basak; *Chem. Phys. Lett.* **403** (2005), 415–419.

39. H. Li, J. Wang, H. Liu, H. Zhang, X. Li; *J. Cryst. Growth* **275** (2005), e943–e946.

40. S. A. Kamaruddin, K.-Y. Chan, H.-K. Yow, M. Z. Sahdan, H. Saim, D. Knipp; *Appl. Phys. A* **104** (2011), 263–268.

41. L. Znaidi; *Mater. Sci. Eng. B* **174** (2010), 18–30.

42. P. Nunes, E. Fortunato, R. Martins; *Int. J. Inorg. Mater.* **3** (2001), 1125–1128.

43. A. Tecaru, A. - I. Danciu, V. Muşat, E. Fortunato, E. Elangovan; *J. Optoelectron. Adv. M.* **12** (2010), 1889–1893.

44. N. R. More, U. B Chanshetti; *Int. J. Chem. Phys. Sci.* **7** (2018), 51–54.

45. S. Zhu, C.-H. Su, S. L. Lehoczky, M. T. Harris, M. J. Callahan, P. McCarty, M. A. George; *J. Cryst. Growth* **219** (2000), 361–367.
46. A. Nahhas, H. Koo Kim J. Blachere; *Appl. Phys. Lett.* **78** (2001), 1511–1513.
47. J. C. Fan, Z. Xie, Q. Wan, Y. G. Wang; *J. Cryst. Growth* **304** (2007), 295–298.
48. O. Hamad, G. Braunstein, H. Patil, N. Dhere; *Thin Solid Films* **489** (2005), 303–309.
49. S.-H. Jeong, B.-S. Kim, B.-T. Lee; *Appl. Phys. Lett.* **82** (2003), 2625–2627.
50. S.-H. Jeong, I.-S. Kim, J.-K. Kim, B.-T. Lee; *J. Cryst. Growth* **264** (2004), 327–333.
51. Y. Zheng, J. C. Boulliard, D. Demaille, Y. Bernard, J. F. Pétroff; *J. Cryst. Growth* **274** (2005), 156–166.
52. H. von Wenckstern, H. Schmidt, C. Hanisch, M. Brandt, C. Czekalla, G. Benndorf, G. Biehne, A. Rahm, H. Hochmuth, M. Lorenz, M. Grundmann; *Phys. Status Solidi RRL* **1** (2007), 129–131.
53. H. von Wenckstern, M. Brandt, H. Schmidt, C. Hanisch, G. Benndorf, H. Hochmuth, M. Lorenz, M. Grundmann; *J. Korean Phys. Soc.* **53** (2008), 3064–3067.
54. D. J. Rogers, F. Hosseini Teherani, A. Largeteau, G. Demazeau, C. Moisson, D. Turover, J. Nausea, G. Garry, R. Kling, T. Gruber, A. Waag, F. Jomard, P. Galtier, A. Lusson, T. Monteiro, M. J. Soares, A. Neves, M. C. Carmo, M. Peres, G. Lerondel, C. Hubert; *Appl. Phys. A* **88** (2007), 49–56.
55. X. Q. Wei, Z. G. Zhang, M. Liu, C. S. Chen, G. Sun, C. S. Xue, H. Z. Zhuang, B. Y. Man; *Mater. Chem. Phys.* **101** (2007), 285–290.
56. M. Lorenz, G. Wagner, A. Rahm, H. Schmidt, H. Hochmuth, H. Schmid, W. Mader, M. Brandt, H. von Wenckstern, M. Grundmann; *Phys. Status Solidi* (*c*), **5** (2008), 3280.
57. K. Minegishi, Y. Koiwai, Y. Kikichi, K. Yano, M. Kasuga, A. Shimizu; *Jpn. J. Appl. Phys.* **36** (1997), L1453–L1455.
58. C. Neumann, S. Lautenschläger, S. Graubner, J. Sann, N. Volbers, B. K. Meyer, J. Bläsing, A. Krost, F. Bertram, J. Christen; *Phys. Status Solidi* (*b*) **244** (2007), 1451–1457.
59. S. Lautenschlaeger, S. Eisermann, B. K. Meyer, G. Callison, M. R. Wagner, A. Hoffmann; *Phys. Status Solidi RRL* **3** (2009), 16–18.
60. Th. Gruber, C. Kirchner, A. Waag; *Phys. Status Solidi* (*b*) **229** (2002), 841–844.

61. K. Maejima, S. Fujita; *J. Crystallogr. Growth* **293** (2006), 305–310.

62. Y. Kashiwaba, H. Kato, T. Kikuchi, I. Niikura, K. Matsushita, Y. Kashiwaba; *Appl. Surf. Sci.* **244** (2005), 373–376.

63. S. Heinze, A. Krtschil, J. Bläsing, T. Hempel, P. Veit, A. Dadgar, J. Christen, A. Krost; *J. Cryst. Growth* **308** (2007), 170–175.

64. H. Kato, M. Sano, K. Miyamoto, T. Yao; *Phys. Status Solidi (b)* **241** (2004), 612–615.

65. S. K. Hong, Y. Chen, H. J. Ko, H. Wenisch, T. Hanada, T. Yao; *J. Electron. Mater.* **30** (2001), 647–658.

66. H. Tampo, H. Shibata, P. Fons, A. Yamada, K. Matsubara, K. Iwata, K. Tamura, H. Takasu, S. Niki; *J. Cryst. Growth* **278** (2005), 268–272.

67. K. Ogata, K. Koike, T. Tanite, T. Komuro, F. Yan, S. Sasa, M. Inoue, M. Yano; *J. Cryst. Growth* **251** (2003), 623–627.

68. R. Triboulet; *Prog. Cryst. Growth Charact.* **60** (2014), 1–14.

69. R. Triboulet, J. Perriére; *Prog. Cryst. Growth Charact.* **47** (2003), 65–138.

70. S. J. Pearton, D. P. Norton, K. Ip, Y. W. Heo, T. Steiner; *Prog. Mater. Sci.* **50** (2005), 293–340.

71. D. Ehrentraut, H. Sato, Y. Kagamitani, H. Sato, A. Yoshikawa, T Fukuda; *Prog. Cryst. Growth Charact.* **52** (2006), 280–335.

72. Ü. Özgür, Ya. I. Alivov, C. Liu, A. Teke, M. A. Reshchikov, S. Doğan, V. Avrutin, S.-J. Cho, H. Morkoç; *J. Appl. Phys.* **98** (2005), 041301.

73. C. Klingshirn; *Chem. Phys. Chem.* **8** (2007), 782–803.

74. C. Klingshirn, J. Fallert, H. Zhou, J. Sartor, C. Thiele, F. Maier-Flaig, D. Schneider, H. Kalt; *Phys. Status Solidi (b)* **247** (2010), 1424–1447.

75. A. Janotti, C. G. Van de Walle; *Rep. Prog. Phys.* **72** (2009), 126501.

76. P. A. Rodnyi, I. V. Khodyuk; *Opt. Spectrosc.* **111** (2011), 776–785.

77. D. Klimm, D. Schulz, S. Ganschow; "Growth of bulk ZnO" in *Comprehensive Semiconductor Science and Technology*, edited by P. Bhattacharya, R. Fornari, and H. Kamimura, vol. 3 (2011), 302–338.

78. F. H. Teherani, C. W. Litton CW (editors); "Zinc Oxide Materials and Devices", *Proceedings of SPIE*, Bellingham, 2006.

79. J. G. Reynolds, C. L. Reynolds; *Adv. Cond. Matter Phys.* **2014** (2014), 457058.

80. Z. C. Feng (editor); *Handbook of Zinc Oxide and Related Materials: Volume Two, Devices and Nano-Engineering*, Taylor & Francis Group, 2013.

81. R. D. Shannon; *Acta Cryst. A.* **32** (1976), 751–767.

82. M. Schirra, R. Schneider, A. Reiser, G. M. Prinz, M. Feneberg, J. Biskupek, U. Kaiser, C. E. Krill, K. Thonke, R. Sauer; *Phys. Rev. B Condens. Mater. Phys.* **77** (2008), 1–10.

83. H. Schulz, K. H. Thiemann; *Solid State Commun.* **32** (1979), 783–785.

84. D. Klimm; "Growth of Bulk ZnO" in: S. Hashmi (Ed.), Ref. Modul. Mater. Sci. Mater. Eng., Elsevier, Oxford, 2016: pp. 1-18

85. F. Decremps, J. Zhang, R. C. Liebermann; *Europhys. Lett.* **51** (2000), 268.

86. D. Schulz, R. Bertram, D. Klimm, T. Schulz, E. Thiede; *J. Cryst. Growth.* **334** (2011), 118–121.

87. C.-Y. James Lu, Y.-T. Tu, T. Yan, A. Trampert, K. H. Ploog; *J. Chem. Phys.* **144** (2016), 214704.

88. P. Schröer, P. Krüger, J. Pollmann; *Phys. Rev. B* **47** (1993), 6971–6980.

89. Y. Mi, H. Odaka, S. Iwata; *Jpn. J. Appl. Phys.* **38** (1999), 3453–3458.

90. H. Q. Ni, Y. F. Lu, Z. M. Ren; *J. Appl. Phys.* **91** (2002), 1339–1343.

91. W. R. L. Lambrecht, A. V. Rodina, S. Limpijumnong, B. Segall, B. K. Meyer; *Phys. Rev. B* **65** (2002), 075207.

92. C.-H. Chien, S. H. Chiou, G. Y. Guo, Y.-D. Yao; *J. Magn. Magn. Mater.* **282** (2004), 275–278.

93. J. Hu, B. C. Pan; *J. Chem. Phys.* **129** (2008), 154706.

94. A. Janotti, C. G. Van de Walle; *Nat. Mater.* **6** (2007), 44–47.

95. J. L. Lyons, A. Janotti, C. G. Van de Walle; *Phys. Rev. B* **80** (2009), 205113.

96. W.-J. Lee, J. Kang, K. J. Chang; *Physica B* **376–377** (2006), 699–702.

97. E.-C. Lee, K. J. Chang; *Physica B* **376–377** (2006), 707–710.

98. E.-C. Lee, K. J. Chang; *Phys. Rev. B* **70** (2004), 115210.

99. P. Erhart, K. Albe, A. Klein; *Phys. Rev. B* **73** (2006), 205203.

100. X.-Y. Duan, Y.-J. Zhao, R.-H. Yao; *Solid State Commun.* **147** (2008), 194–197.

101. H. I. Berrezoug, A. E. Merad, M. Aillerie, A. Zerga; *Mater. Res. Express* **4** (2017), 035901.

102. T. Tamura, H. Ozaki; *J. Phys.: Condens. Matter.* **21** (2009), 026009.

103. Z. H. Xiong, S. Q. Shi, Q. X. Wan, F. Y. Jiang; *Phys. Scr.* **T129** (2007), 358–361.

104. L. Petit, T. C. Schulthess, A. Svane, Z. Szotek, W. M. Temmerman, A. Janotti; *Phys. Rev. B* **73** (2006), 045107.

105. S. Zh. Karazhanov, A. G. Ulyashin; *Phys. Rev. B* **78** (2008), 085213.

106. S. Guo, X. Zhang, Y. Huang, Y. Li, Z. Du; *Chem. Phys. Lett.* **459** (2008), 82–84.

107. H. Raebiger, S. Lany, A. Zunger; *Phys. Rev. B* **79** (2009), 165202.

108. A. Schleife, F. Fuchs, C. Rödl, J. Furthmüller, F. Bechstedt; *Phys. Status Solidi* (**b**) **246** (2009), 2150–2153.

109. H. Dixit, R. Saniz, D. Lamoen, B. Partoens; *J. Phys.: Condens. Matter.* **22** (2010), 125505.

110. R. Chowdhury, P. Rees, S. Adhikari, F. Scarpa, S. P. Wilks; *Physica B* **405** (2010), 1980–1985.

111. T. Han, F. Y. Meng, S. Zhang, X. M. Cheng, J. I. Oh; *J. Appl. Phys.* **110** (2011), 063724.

112. S. H. Deng, M. Y. Duan, M. Xu, L. He; *Physica B* **406** (2011), 2314–2318.

113. X. Qu, W. Wang, S. Lv, D. Jia; *Solid State Commun.* **151** (2011), 332–336.

114. X. Qu, S. Lu, L. Bai, Q. Meng, D. Jia; *Physica B* **407** (2012), 268–270.

115. Y. Liu, Q. Y. Hou, H. P. Xu, L. M. Li, Y. Zhang; *Physica B* **407** (2012), 2359–2364.

116. X. D. Zhang, M. L. Guo, Y. Y. Shen, C. L. Liu, Y. H. Xue, F. Zhu, L. H. Zhang; *Comput. Mater. Sci.* **54** (2012), 75–80.

117. G. Sharma, M. C. Mishra, M. S. Dhaka, R. K. Kothari, K. B. Joshi, B. K. Sharma; *J. Electron. Mater.* **42** (2013), 3429–3437.

118. M.-H. Lee, Y.-C. Peng, H.-C. Wu; *J. Alloy. Compd.* **616** (2014), 122–127.

119. X. J. Zhang, W. B. Mi, X. C. Wang, H. L. Bai; *J. Alloy. Compd.* **617** (2014), 828–833.

120. Y. Wang, W. Tang, J. Zhu, J. Liu; *Comput. Mater. Sci.* **99** (2015), 145–149.

121. Y.-C. Peng, C.-C. Chen, H.-C. Wu, J.-H. Lu; *Opt. Mater.* **39** (2015), 34–39.

122. M. Lorke, T. Frauenheim, A. L. da Rosa; *Phys. Rev. B* **93** (2016), 115132.

123. C.-C. Chen, H.-C. Wu; *Materials* **9** (2016), 164.

124. A. Mahmood, F. Tezcan, G. Kardas, F. Karadag; *J. Appl. Phys.* **122** (2017), 113102.

125. C. A. Ataide, R. R. Pelá, M. Marques, L. K. Teles, J. Furthmüller, F. Bechstedt; *Phys. Rev. B* **95** (2017), 045126.

126. M. Hammi, Y. Ziat, A. A. Raiss, M. Arejdal, Y. Sbai, O. El Rhazouani, A. Slassi; *Optik* **131** (2017), 399–405.

127. K. Bashyal, C. K. Pyles, S. Afroosheh, A. Lamichhane, A. T. Zayak; *J. Phys.: Condens. Matter.* **30** (2018), 065501.

128. K. Sawada, Y. Shirotori, K. Ozawa, K. Edamoto, M. Nakatake; *Appl. Surf. Sci.* **237** (2004), 343–347.

129. K. Ozawa, K. Sawada, Y. Shirotori, K. Edamoto; *J. Phys.: Condens. Matter.* **17** (2005), 1271–1278.

130. K. Ozawa, Y. Oba, K. Edamoto, M. Higashiguchi, Y. Miura, K. Tanaka, K. Shimada, H. Namatame, M. Taniguchi; *Phys. Rev. B* **79** (2009), 075314.

131. A. R. H. Preston, B. J. Ruck, L. F. J. Piper, A. DeMasi, K. E. Smith, A. Schleife, F. Fuchs, F. Bechstedt, J. Chai, S. M. Durbin; *Phys. Rev. B* **78** (2008), 155114.

132. P. D. C. King, T. D. Veal, A. Schleife, J. Zúñiga-Pérez, B. Martel, P. H. Jefferson, F. Fuchs, V. Muñoz-Sanjosé, F. Bechstedt, C. F. McConville; *Phys. Rev. B* **79** (2009), 205205.

133. A. Janotti, C. G. Van de Walle; *J. Cryst. Growth* **287** (2006), 58–65.

134. A. Janotti, C. G. Van de Walle; *Phys. Rev. B* **76** (2007), 165202.

135. Y. Sun, H. Wang; *Physica B* **325** (2003), 157–163.

136. C. G. Van de Walle; *Physica B* **376–377** (2006), 1–6.

137. C. G. Van de Walle; *Phys. Rev. Lett.* **85** (2000), 1012–1015.

138. M. G. Wardle, J. P. Goss, P. R. Briddon; *Phys. Rev. Lett.* **96** (2006), 205504.

139. M. G. Wardle, J. P. Goss, P. R. Briddon; *Phys. Rev. B* **72** (2005), 155108.

140. M.-H. Du, K. Biswas; *Phys. Rev. Lett.* **106** (2011), 115502.

141. O. Fritsch; *Ann. Phys.* **22** (1935), 375–401.

142. E. Scharovsky; *Z. Phys.* **135** (1953), 318–330.

143. A. C. Walker; *J. Amer. Ceram. Soc.* **36** (1953), 250–256.

144. R. A. Laudise, A. A. Ballman; *J. Phys. Chem.* **64** (1960), 688–691.

145. J. W. Nielsen, E. F. Dearborn; *J. Phys. Chem.* **64** (1960), 1762–1763.

146. R. Helbig; *J. Cryst. Growth* **15** (1972), 25–31.

147. G. Agarwal, J. E. Nause, D. N. Hill; *Mat. Res. Soc. Symp. Proc.* **512** (1988), 41–46.

148. D. Schulz, S. Ganschow, D. Klimm, M. Neubert, M. Rossberg, M. Schmidbauer, R. Fornari; *J. Cryst. Growth* **296** (2006), 27–30.

149. G. Bogner, E. Mollwo; *J. Phys. Chem. Solids* **6** (1958), 136–143.

150. I. Kubo; *Japan. J. Appl. Phys.* **4** (1965), 225–226.

151. N. Shibata, H. Iwanaga; *Jpn. J. Appl. Phys.* **11** (1972), 775–779.

152. E. A. Weaver; *J. Cryst. Growth* **1** (1967), 320–322.

153. T. Takahashi, A. Ebina, A. Kamiyama; *Jpn. J. Appl. Phys.* **5** (1966), 560–561.

154. M. Hirose, I. Kubo; *Jpn. J. Appl. Phys.* **8** (1969), 402.

155. M. Hirose, Y. Furuya; *Jpn. J. Appl. Phys.* **9** (1970), 423–424.

156. M. Hirose; *Jpn. J. Appl. Phys.* **10** (1971), 401–408.

157. M. Hirose, Y. Furuya; *Jpn. J. Appl. Phys.* **11** (1972), 423–424.

158. M. Hirose, Y. Furuya; *Jpn. J. Appl. Phys.* **9** (1970), 726–727.

159. Y. S. Park, D. C. Reynolds; *J. Appl. Phys.* **38** (1967), 756–760.

160. W. Piekarczyk, S. Gazda, T. Niemyski; *J. Cryst. Growth* **12** (1972), 272–276.

161. R. Widmer; *J. Cryst. Growth* **8** (1971), 216–218.

162. E. Ziegler, A. Heinrich, H. Oppermann, G. Stöver; *Phys. Status Solidi* **66** (1981), 635–648.

163. M. Shiloh, J. Gutman; *J. Cryst. Growth* (1971), 105–109.

164. K. Matsumoto, G. Shimaoka; *J. Cryst. Growth* **86** (1988), 410–414.

165. K. Matsumoto, K. Konemura, G. Shimaoka; *J. Cryst. Growth* **71** (1985), 99–103.

166. K. F. Nielsen; *J. Cryst. Growth* **3, 4** (1968), 141–145.

167. E. M. Dodson, J. A. Savage; *J. Mater. Sci.* **3** (1968), 19–25.

168. J. A. Savage, E. M. Dodson; *J. Mater. Sci.* **4** (1969), 809–813.

169. K. J. Fischer; *J. Cryst. Growth* **34** (1976), 139–144.

170. D. C. Look, D. C. Reynolds, J. R. Sizelove, R. L. Jones, C. W. Litton, G. Cantwell, W. C. Harsch; *Solid State Commun.* **105** (1998), 399–401.

171. K. Grasza, A. Mycielski; *Phys. Status Solidi* (*c*) **3** (2005), 1115–1118.

172. K. Grasza, A. Mycielski, J. Domagała, V. Domukhovski, W. Paszkowicz, H. Sakowska, W. Hofman; *Phys. Status Solidi* (*c*) **3** (2006), 793–796.

173. K. Grasza, E. Lusakowska, P. Skupinski, K. Kopalko, J. Bak-Misiuk, A. Mycielski; *Phys. Status Solidi* (*b*) **244** (2007), 1468–1472.

174. K. Grasza, P. Skupinski, A. Mycielski, E. Lusakowska, V. Domukhovski; *J. Cryst. Growth* **310** (2008), 1823–1826.

175. A. Shalimov, W. Paszkowicz, K. Grasza, P. Skupinski, A. Mycielski, J. Bak-Misiuk; *Phys. Status Solidi* (*b*) **244** (2007), 1573–1577.

176. P. Skupinski, K. Grasza, A. Mycielski, W. Paszkowicz, E. Luskowska, E. Tymicki, R. Jakiela, B. Witkowski; *Phys. Status. Solidi* (*b*) **247** (2010), 1457–1459.

177. T. Matsushita, K. Kodaira, J. Saito, R. Yoshida; *J. Cryst. Growth* **26** (1974), 147–148.

178. S. Ito, S. Shimada, A. Tsunashima, K. Kodaira, T. Matsushita; *Z. Anorg. Allg. Chem.* **439** (1978), 259–264.

179. S. Hassani, A. Tromson-Carli, A. Lusson, G. Didier, R. Triboulet; *Phys. Stat. Solidi* (*b*) (2002), 835–839.

180. R. Tena-Zaera, M. C. Martínez-Tomás, S. Hassani, R. Triboulet, V. Muňoz-Sajosé; *J. Cryst. Growth* **270** (2004), 711–721.

181. R. Tena-Zaera, M. C. Martínez-Tomás, C. J. Gómez-García,V. Muňoz-Sajosé; *Cryst. Res. Technol.* **41** (2006), 742–747.

182. D. Pfisterer, D. M. Hofmann, J. Sann, B. K. Meyer, R. Tena-Zaera, V. Munoz-Sanjose, T. frank, G. Pensl; *Physica B* **376–377** (2006), 767–770.

183. A. Mycielski, L. Kowalczyk, A, Szadkowski, B. Chwalisz, A. Wysmolek, R. Stepniewski, J. M. Baranowski, M. Potemski, A. Witowski, R. Jakiela, A. Barcz, B. Witkowska, W. Kaliszek, A. Jedrzejczak, A. Suchocki, E. Lusakowska, E. Kaminska; *J. Alloy. Compd.* **371** (2004), 150–152.

184. W. Polosz; *J. Cryst. Growth* **286** (2006), 42–49.

185. X. Wei, Y. Zhao, Z. Dong, J. Li; *J. Cryst. Growth* **310** (2008), 639–645.

186. M. Mikami, T. Eto, J. F. Wang, Y. Masa, M. Isshiki; *J. Cryst. Growth* **276** (2005), 389–392.

187. M. Mikami, T. Sato, J. F. Wang, Y. Masa, M. Isshiki; *J. Cryst. Growth* **286** (2006), 213–217.

188. M. Mikami, S. H. Hong, T. Sato, S. Abe, J. F. Wang, K. Masumoto, Y. Masa, M. Isshiki; *J. Cryst. Growth* **304** (2007), 37–41.

189. S.-H. Hong, M. Mikami, K. Mimura, M. Uchikoshi, A. Yasuo, S. Abe, K. Msumoto, M. Isshiki; *J. Cryst. Growth* **311** (2009), 3609–3612.

190. S. H. Hong, J. H. Mun, K. Mimura, M. Mikami, M. Uchikoshi, M. Isshiki; *J. Ceram. Process. Res.* **13** (2012), 105–109.

191. V. Munoz-Sanjosé, R. Tena-Zaera, C. Martínez-Tomás, J. Zuniga-Pérez, S. Hassani, R. Triboulet; *Phys. Status Solidi* (*c*) **2** (2005), 1106–1114.

192. K. Abe, Y. Banno, T. Sasayama, K. Koizumi; *Jpn. J. Appl. Phys.* **48** (2009), 021101.

193. K. Abe, Y. Banno, T. Sasayama, K. Koizumi; *J. Vac. Sci. Technol. B* **27** (2009), 1652–1654.

194. J. L. Santailler, C. Audoin, G. Chichignoud, R. Obrecht, B. Kaouache, P. Marotel, D. Pelenc, S. Brochen, J. Merlin, I. Bisotto, C. Granier, G. Feuillet, F. Levy; *J. Cryst. Growth* **312** (2010), 3417–3424.

195. S. J. Jokela, M. C. Tarun, M. D. McCluskey; *Physica B* **404** (2009), 4810–4812.

196. J. –M. Ntep, S. S. Hassani, A. Lusson, A. Tromson-Carli, D. Ballutaud, G. Didier, R. Triboulet; *J. Cryst. Growth* **207** (1999), 30–34.

197. K. Fischer, E. Sinn; *Cryst. Res. Technol.* **16** (1981), 689–894.

198. X. H. Li, J. Y. Xu, X. M. Li; *Appl. Phys. A* **82** (2006), 373–376.

199. X. H. Li, J. Y. Xu, M. Jin, H. Shen, X. M. Li; *Chin. Phys. Lett.* **23** (2006), 3356–3358.

200. X. Li, J. Xu, M. Jin, X. Wu, G. Qian, X. Li; *Cryst. Res. Technol.* **42** (2007), 221–226.

201. X. Li, J. Xu, H. Shen, A. Wu, X. Li; *Phys. Status Solidi (c)* **4** (2007), 1333–1336.

202. K. Oka, H. Shibata, S. Kashiwaya; *J. Cryst. Growth* **237–239** (2002), 509–513.

203. S. C. Kashyap; *J. Appl. Phys.* **44** (1973), 4381–4384.

204. M. Ushio, Y. Sumiyoshi; *J. Mater. Sci.* **28** (1993), 218–224.

205. R. A. Laudise, E. D. Kolb, A. J. Caporaso; *J. Am. Ceram. Soc.* **47** (1964), 9–12.

206. D. F. Croxall, R. C. C. Ward, C. A. Wallace, R. C. Kell; *J. Cryst. Growth* **22** (1974), 117–124.

207. N. Sakagami, M. Yamashita, T. Sekiguchi, S. Miyashita, K. Obara, T. Shishido; *J. Cryst. Growth* **229** (2001), 98–103.

208. M. Suscavage, M. Harris, D. Bliss, P. Yip, S.-Q. Wang, D. Scwall, L. Bouthillette, J. Bailey, M. Callahan, D. C. Look, D. C. Reynolds, R. L. Jones, C. W. Litton; *MRS Internet J. Nitride Semicond. Res.* **4S1**, G3.40 (1999).

209. C.-L. Zhang, W.-N. Zhou, Y. Hang, Z. Lü, H.-D. Huo, Y.-B. Zuo, S.-J. Qin, F.-H. Lu, S.-L. Gu; *J. Cryst. Growth* **310** (2008), 1819–1822.

210. W. Lin, D. Chen, J. Zhang, Z. Lin, J. Huang, W. Li, Y. Wang, F. Huang; *Cryst. Growth Des.* **9** (2009), 4378–4383.

211. B. Wang, M. J. Callahan, C. Xu, L. O. Bouthillette, N. C. Giles, D. F. Bliss; *J. Cryst. Growth* **304** (2007), 73–79.

212. Y.-B. Zuo, F.-H. Lu, C.-L. Zhang, Y. Hang, W.-N. Zhou, H.-D. Huo, S.-J. Qin, H.-T. Zhou, X.-L. He, D.-P. Li, H.-X. Zhang, Y.-X. Chen, S.-L. Gu; *J. Cryst. Growth* **318** (2011), 513–515.

213. L. Yin, L. Zhang, F. Li, M. Yu; *Mater. Res. Bull.* **40** (2005), 2219–2224.

214. L. N. Dem'yanets, V. I. Lyutin; *J. Cryst. Growth* **310** (2008), 993–999.

215. E. Ohshima, H. Ogino, I. Niikura, K. Maeda, M. Sato, M. Ito, T. Fukuda; *J. Cryst. Growth* **260** (2004), 166–170.

216. L. N. Dem'yanets, L. E. Li, T. G. Uvarova; *J. Mater. Sci.* **41** (2006), 1439–1444.

217. N. Sakagami; *J. Cryst. Growth* **99** (1990), 905–909.

218. L. N. Demianets, D. V. Kostamarov; *Ann. Chim. Sci. Mat.* **26** (2001), 193–198.

219. K. Maeda, M. Sato, I. Niikura, T. Fukuda; *Semicond. Sci. Technol.* **20** (2005), S49–S54.

220. J. Mass, M. Avella, J. Jiménez, M. Callahan, E. Grant, K. Rakes, D. Bliss, B. Wang; *Superlattice. Microst.* **38** (2005), 223–230.

221. J. Mass, M. Avella, J. Jiménez, M. Callahan, E. Grant, K. Rakes, D. Bliss, B. Wang; *Appl. Phys. A* **88** (2007), 95–98.

222. B. Wang, M. J. Callahan, L. O. Bouthillette; *Cryst. Growth Des.* **6** (2006), 1256–1260.

223. B. Wang, M. J. Callahan, L. O. Bouthillette, C. Xu, M. J. Suscavage; *J. Cryst. Growth* **287** (2006), 381–385.

224. D. Dhanaraj, M. Dudley, D. Bliss, M. Callahan, M. Harris; *J. Cryst. Growth* **297** (2006), 74–79.

225. T. Sekiguchi, S. Miyashita, K. Obara, T. Shishido, N. Sakagami; *J. Cryst. Growth* **214/215** (2000), 72–76.

226. Y.-B. Zuo, F.-H. Lu, C.-L. Zhang, Y. Hang, W.-N. Zhou, H.-D. Huo, S.-J. Qin, H.-T. Zhou, X.-L. He, D.-P. Li, H.-X. Zhang, Y.-X. Chen, S.-L. Gu; *J. Cryst. Growth* **318** (2011), 513–515.

227. L. N. Dem'yanets, R. M. Zakalyukin, B. N. Mavrin; *Inorg. Mater.* **47** (2011), 649–653.

228. H.-J. Ko, M.-S. Han, Y.-S. Park, Y.-S. Yu, B.-I. Kim, S.-S. Kim, J.-H. Kim; *J. Cryst. Growth* **269** (2004), 493–498.

229. T. M. Børseth, B. G. Svensson, A. Y. Kuznetsov; *Phys. Scr.* **T126** (2006), 10–14.

230. J. E. Nause; *III-Vs Review* **12** (1999), 28–31.

231. J. Nause, B. Nemeth; *Semicond. Sci. Technol.* **20** (2005), S45–S48.

232. J. E. Nause, D. Norman, D. N. Hill, S. G. Pope; US 5863326 (1999).

233. J. E. Nause, D. N. Hill, S. G. Pope; US 5900060 (1999).

234. R. Khanna, K. Ip, Y. W. Heo, D. P. Norton, S. J. Pearton, F. Ren; *Appl. Phys. Let.* **85** (2004), 3468–3470.

235. Z. C. Feng, J. W. Yu, J. B. Wang, R. Varatharajan, B. Nemeth, J. Nause, I. Ferguson, W. Lu, W. E. Collins; *Mater. Sci. Forum* **527–529** (2006), 1567–1570.

236. D. Schulz, S. Ganschow, D. Klimm, K. Struve; *J. Cryst. Growth* **310** (2008), 1832–1835.

237. D. Klimm, S. Ganschow, D. Schulz, R. Fornari; *J. Cryst. Growth* **310** (2008), 3009–3013.

238. D. Schulz, S. Ganschow, D. Klimm; *Meter. Res. Soc. Symp. Proc.* **1201** (2010), 120-H06-10.

239. K. Jacobs, D. Schulz, D. Klimm, S. Ganschow; *Solid State Sci.* **12** (2010), 307–310.

240. R. Fornari, S. Ganschow, D. Klimm, M. Neubert, D. Schulz; DE 10 2007 006 731 (2011).

241. D. Schulz; *Leibniz-Institut für Kristalzüchtung*, Berlin, Germany (2009), unpublished data.

242. W. Anwand, G. Brauer, R. I. Grynszpan, T. E. Cowan, D. Schulz, D. Klimm, J. Čižek, J. Kuriplach, I. Procházka, C. C. Ling, A. B. Djurišic, V. Klemm, G. Schreiber, D. Rafaja; *J. Appl. Phys.* **109** (2011), 063516.

243. K. Irmscher, M. Albrecht, B. Heimbrodt, M. Naumann, T. Remmele, D. Schulz, T. Schulz, R. Fornari; *Phys. Status Solidi* (*c*) **6** (2009), 2658–2660.

244. S. Eisermann, A. Laufer, S. Graubner, M. Pinnisch, J. Stehr, S. Lautenschläger, D. M. Hofmann, B. K. Meyer, D. Klimm, D. Schulz; *Phys. Status Solidi* (*a*) **208** (2011), 37–41.

245. www.factsage.com, FactSage 7.2, (2018).

246. I. Barin; *Thermodynamic Data of Pure Substances*, VCH, Weinheim 1993.

247. W. M. Wolf, J. J. Martin; *Phys. Status Solidi* (*a*) **17** (1973), 215–220.

248. D. I. Florescu, L. G. Mourokh, F. H. Pollak, D. C. Look, G. Cantwell, X. Li; *J. Appl. Phys.* **91** (2002), 890–892.

249. Ü. Özgür, X. Gu, S. Chevtchenko, J. Spradlin, S.-J. Cho, H. Morkoç, F. H. Pollak, H. O. Everitt, B. Nemeth, J. E. Nause; *J. Electron. Mater.* **35** (2006), 550–555.

250. G. G. Gadzhiev; *High Temp.* **41** (2003), 778–782.

251. R. Pässler; *Phys. Status Solidi* (*b*) **245** (2008), 1133–1146.

252. J. Serrano, R. K. Kremer, M. Cardona, G. Siegle, A. H. Romero, R. Lauck; *Phys. Rev. B* **73** (2006), 094303.

253. H. Iwanaga, A. Kunishige, S. Takeuchi; *J. Mater. Sci.* **35** (2000), 2451–2454.

254. J. Albertsson, S. C. Abrahams, Å. Kvick; *Acta Crystallogr. B* **45** (1989), 34–40.

255. A. A. Khan; *Acta Crystallogr. A* **24** (1968), 403.

256. T. B. Bateman; *J. Appl. Phys.* **33** (1962), 3309–3312.

257. I. B. Kobiakov; *Solid State Commun.* **35** (1980), 305–310.

258. T. Azuhata, M. Takesada, T. Yagi, A. Shikanai, S. F. Chichibu, K. Torii, A. Nakamura, T. Sota, G. Cantwell, D. B. Eason, C. W. Litton; *J. Appl. Phys.* **94** (2003), 968–972.

259. R. Ahuja, L. Fast, O. Eriksson, J. M. Wills, B. Johansson; *J. Appl. Phys.* **83** (1998), 8065–8067.

260. M. Catti, Y. Noel, R. Dovesi; *J. Phys. Chem. Solids* **64** (2003), 2183–2190.

261. X. Wu, D. Vanderbilt, D. R. Hamann; *Phys. Rev. B* **72** (2005), 035105.

262. Z. C. Tu, X. Hu; *Phys. Rev. B* **74** (2006), 035434.

263. P. Gopal, N. A. Spaldin; *J. Electron. Mater.* **35** (2006), 538–542.

264. I. R. Shein, V. S. Kiĭko, Yu. N. Makurin, M. A. Gorbunova, A. L. Ivanovskiĭ; *Phys. Solid State* **49** (2007), 1067–1073.

265. G. Li, U. Aydemir, S. I. Morozov, S. A. Miller, Q. An, W. A. Goddard III, P. Zhai, Q. Zhang, G. J. Snyder; *Acta Mater.* **149** (2018), 341–349.

266. Yu. I. Sirotin, M. P. Shaskolskaya; *Fundamentals of Crystal Physics*, Mir Publishers, Moscow, 1982.

267. E. H. Nickel, M. C. Nichols; *Mineral Reference Manual*, Van Nostrand Reinhold, New York, 1991.

268. I. Yonenaga; *Physica B* **308–310** (2001), 1150–1152.

269. S. O. Kucheyev, J. E. Bradby, J. S. Williams, C. Jagadish, M. V. Swain; *Appl. Phys. Lett.* **80** (2002), 956–958.

270. J. E. Bradby, S. O. Kucheyev, J. S. Williams, C. Jagadish, M. V. Swain, P. Munroe, M. R. Phillips; *Appl. Phys. Lett.* **80** (2002), 4537–4539.

271. S.-R. Jian; *J. Alloy. Compd.* **494** (2010), 214–218.

272. T. H. Sung, J. C. Huang, J. H. Hsu, S. R. Jian, T. G. Nieh; *Appl. Phys. Lett.* **100** (2012), 211903.

273. V. A. Coleman, J. E. Bradby, C. Jagadish, P. Munroe, Y. W. Heo, S. J. Pearton, D. P. Norton, M. Inoue, M. Yano; *Appl. Phys. Lett.* **86** (2005), 203105.

274. L. N. Demyanets, N. L. Sizova, L. E. Li; *Crystallogr. Rep.* **50** (2005), 646–653.

275. S. C. Abrahams, J. L. Bernstein; *Acta Crystallogr. B* **25** (1969), 1233–1236.

276. S. Adachi; "Properties of Group-IV, III-V and II-VI semiconductors," *Wiley Series in Materials for Electronic and Optoelectronic Applications*, edited by P. Capper, S. Kasap, and A. Willoughby, John Wiley & Sons, 2005.

277. B. Gil; *Physics of Wurtzite Nitrides and Oxides, Passport to Devices, Springer Series in Materials Science*, vol. 197, edited by R. Hull, C.

Jagadish, R. M. Osgood, J. Parisi, S. Uchida, and Z. M. Wang, Springer, 2014.

278. K. Irmscher, D. Schulz, Z. Galazka; *Leibniz-Institut für Kristalzüchtung*, Berlin, Germany (2009), unpublished data.

279. D. G. Thomas; *Phys. Chem. Solids* **15** (1960), 86–96.

280. V. Srikant, D. R. Clarke; *J. Appl. Phys.* **83** (1998), 5447–5451.

281. C. J. Youn, T. S. Jeong, M. S. Han, J. H. Kim; *J. Cryst. Growth* **261** (2004), 526–532.

282. W. Y. Liang, A. D. Yoffe; *Phys. Rev. Lett.* **20** (1968), 59–62.

283. B. K. Meyer, J. Sann, S. Lautenschläger, M. R. Wagner, A. Hoffmann; *Phys. Rev. B* **76** (2007), 184120.

284. D. C. Reynolds, C. W. Litton, D. C. Look, J. E. Hoelscher, B. Claflin, T. C. Collins, J. Nause B. Nemeth; *J. Appl. Phys.* **95** (2004), 4802–4805.

285. W. L. Bond; *J. Appl. Phys.* **36** (1965), 1674–1677.

286. H. Yoshikawa, S. Adachi; *Jpn. J. Appl. Phys.* **36** (1997), 6237–6243.

287. G. E. Jellison, Jr., L. A. Boatner; *Phys. Rev. B* **58** (1998), 3586–3589.

288. K. Postava, H. Sueki, M. Aoyama, T. Yamaguchi, Ch. Ino, Y. Igasaki, M. Horie; *J. Appl. Phys.* **87** (2000), 7820–7824.

289. N. Ashkenov, B. N. Mbenkum, C. Bundesmann, V. Riede, M. Lorenz, D. Spemann, E. M. Kaidashev, A. Kasic, M. Schubert, M. Grundmann, G. Wagner, H. Neumann, V. Darakchieva, H. Arwin, B. Monemar; *J. Appl. Phys.* **93** (2003), 126–133.

290. V. Val. Sobolev, V. V. Sobolev; *J. Appl. Spectrosc.* **70** (2003), 640–643.

291. A. B. Djurišić, Y. Chan, E. H. Li; *Appl. Phys. A* **76** (2003), 37–43.

292. C. Bundesmann, N. Ashkenov, M. Schubert, A. Rahm, H. von Wenckstern, E. M. Kaidashev, M. Lorenz, M. Grundmann; *Thin Solid Films* **455–456** (2004), 161–166.

293. Y. C. Liua, S. K. Tunga, J. H. Hsieh; *J. Cryst. Growth* **287** (2006), 105–111.

294. R. Schmidt-Grund, B. Rheinländer, E. M. Kaidashev, M. Lorenz, M. Grundmann, D. Fritsch, M. M. Schubert, H. Schmidt, C. M Herzinger; *J. Korean Phys. Soc.* **53** (2008), 88–93.

295. S. Shokhovets, L. Spieß, G. Gobsch; *J. Appl. Phys.* **107** (2010), 023509.

296. M. Cobet, C. Cobet, M. R. Wagner, N. Esser, C. Thomsen, A. Hoffmann; *Appl. Phys. Lett.* **96** (2010), 031904.

297. X. D. Li, T. P. Chen, P. Liu, Y. Liu, Z. Liu, K. C. Leong; *J. Appl. Phys.* **115** (2014), 103512.

298. E. Mammadov, N. Naghavi, Z. Jehl, G.Renou, T.Tiwald, N. Mamedov, D. Lincot J.-F.Guillemoles; *Thin Solid Films* **571** (2014), 593–596.

299. E. Agocs, B. Fodor, B. Pollakowski, B. Beckhoff, A. Nutsch, M. Jank, P. Petrik; *Thin Solid Films* **571** (2014), 684–688.

300. C. Singh, S. Nozaki, S. Rath; *J. Appl. Phys.* **118** (2015), 195305.

301. A. Calzolari, M. B. Nardelli; *Sci. Rep.* **3** (2013), 2999.

302. B. K. Meyer, H. Alves, D. M. Hofmann, W. Kriegseis, D. Forster, F. Bertram, J. Christen, A. Hoffmann, M. Straßburg, M. Dworzak, U. Haboeck, A. V. Rodina; *Phys. Status Solidi (b)* **241** (2004), 231–260.

303. P. A. Rodnyi, I. V. Khodyuk; *Opt. Spectrosc.* **111** (2011), 776–785.

304. S. Sugimura, H. Endo, Y. Kashiwaba, E. Sato; *Nucl. Instrum. Methods Phys. Res. A* **654** (2011), 314–317.

305. D. C. Oh, T. Kato, H. Goto, S. H. Park, T. Hanada, T. Yao, J. J. Kim; *Appl. Phys. Lett.* **93** (2008), 241907.

306. H. Sasaki, H. Kato, F. Izumida, H. Endo, K. Maeda, M. Ikeda, Y. Kashiwaba, I. Niikura, Y. Kashiwaba; *Phys. Status Solidi (c)* **3** (2006), 1034–1037.

307. S. A. Chevtchenko, J. C. Moore, Ü. Özgür, X. Gu, A. A. Baski, H. Morkoç, B. Nemeth, J. E. Nause; *Appl. Phys. Lett.* **89** (2006), 182111.

308. T. C. Damen, S. P. S. Porto, B. Tell; *Phys. Rev.* **142** (1966), 570–574.

309. J. M. Calleja, M. Cardona; *Phys. Rev. B* **16** (1977), 3753–3761.

310. E. A. Vinogradov, N. N. Mel'nik, A. E. Tsurkan, L. V. Kicherman; *Zh. Prikl. Spektrosk.* **26** (1977), 1059–1062.

311. J. Serrano, F. J. Manjón, A. H. Romero, A. Ivanov, M. Cardona, R. Lauck, A. Bosak, M. Krisch; *Phys. Rev. B* **81** (2010), 174304.

312. A. Calzolari, M. B. Nardelli; *Sci. Rep.* **3** (2013), 2999.

313. D. C. Look; *Semicond. Sci. Technol.* **20** (2005), S55–S61.

314. T. M. Barnes, K. Olson, C. A. Wolden; *Appl. Phys. Lett.* **86** (2005), 112112.

315. O. Maksimov; *Rev. Adv. Mater. Sci.* **24** (2010), 26–34.

316. F. Chen, Z. Ye, W. Xu, B. Zhao, L. Zhu, J. Lv; *J. Cryst. Growth* **281** (2005), 458–462.

317. H. von Wenckstern, H. Schmidt, M. Brandt, A. Lajn, R. Pickenhain, M. Lorenz, M. Grundmann, D. M. Hofmann, A. Polity, B. K. Meyer, H. Saal, M. Binnewies, A. Börger, K.-D. Becker, V. A. Tikhomirov, K. Jug; *Prog. Solid State Chem.* **37** (2009), 153–172.

318. C. G. Van de Walle; *Phys. Rev. Lett.* **85** (2000), 1012–1015.

319. H. Qiu, B. Meyer, Y. Wang, C. Wöll; *Phys. Rev. Lett.* **101** (2008), 236401.

320. E. V. Lavrov, F. Herklotz, J. Weber; *Phys. Rev. B* **79** (2009), 165210.

321. E. V. Lavrov, F. Herklotz, J. Weber; *Phys. Rev. Lett.* **102** (2009), 185502.

322. A. Janotti, C. G. Van de Walle; *Nat. Mater.* **6** (2007), 44–47.

323. E. V. Lavrov, J. Weber, F. Börrnert, Chris G. Van de Walle, R. Helbig; *Phys. Rev. B* **66** (2002), 165205.

324. E. V. Lavrov, F. Börrnert, J. Weber; *Phys. Rev. B* **72** (2005), 085212.

325. G. A. Shi, M. Stavola, S. J. Pearton, M. Thieme, E. V. Lavrov, J. Weber; *Phys. Rev. B* **72** (2005), 195211.

326. S. J. Jokela, M. D. McCluskey; *Phys. Rev. B* **72** (2005), 113201.

327. D. C. Look, J. W. Hemsky, J. R. Sizelove; *Phys. Rev. Lett.* **82** (1999), 2552-2555.

328. Y. Sun, H. Wang; *Physica B* **325** (2003), 157–163.

329. D. C. Look, G. C. Farlow, P. Reunchan, S. Limpijumnong, S. B. Zhang, K. Nordlund; *Phys. Rev. Lett.* **95** (2005), 225502.

330. D. C. Look, C. Coskun, B. Claflin, G. C. Farlow; *Physica B: Condens. Matter* **340–342** (2003), 32–38.

331. J. G. Reynolds, C. L. Reynolds, Jr., A. Mohanta, J. F. Muth, J. E. Rowe, H. O. Everitt, D. E. Aspnes; *Appl. Phys. Lett.* **102** (2013), 152114.

332. S. Limpijumnong, S. B. Zhang, S.-H. Wei, C. H. Park; *Phys. Rev. Lett.* **92** (2004), 155504.

333. W.-J. Lee, J. Kang, K. J. Chang; *Physica B* **376–377** (2006), 699–702.

334. B. K. Meyer, J. Sann, D. M. Hofmann, C. Neumann, A. Zeuner; *Semicond. Sci. Technol.* **20** (2005), S62–S66.

335. D. C. Look; *Semicond. Sci. Technol.* **20** (2005), S55–S61.

336. S. Zh. Karazhanov, A. G. Ulyashin; *Phys. Rev. B* **78** (2008), 085213.

337. D. C. Look, B. Claflin; *Phys. Status Solidi (b)* **241** (2004), 624–630.

338. Y. Yan, S.-H. Wei; *Phys. Status Solidi (b)* **4** (2008), 641–652.

339. M. W. Allen, P. Miller, R. J. Reeves, S. M. Durbin; *Appl. Phys. Lett.* **90** (2007), 062104.

340. E. Ziegler, A. Heinrich, H. Oppermann, G. Stöver; *Cryst. Res. Technol.* **18** (1983), 1165–1172.

341. K. Irmscher, D. Schulz, M. Pietsch; *Leibniz-Institut für Kristalzüchtung*, Berlin, Germany (2010), unpublished data.

342. W. S. Baer; *Phys. Rev.* **154** (1967), 785–789.

343. K. J. Button, D. R. Cohn, M. von Ortenverger, B. Lax, E. Mollwo, R. Helbig; *Phys. Rev. Lett.* **28** (1972), 1637–1639.

344. M. Oshikiri, Y. Imanaka, F. Aryasetiawan, G. Kido; *Physica B* **298** (2001), 472–476.

345. Y. Imanaka, O. Oshikiri, K. Takehana, T. Takamasu, G. Kido; *Physica B* **298** (2001), 211–215.

346. R. L. Wieher; *Phys. Rev.* **152** (1966), 736–739.

347. W. M. Kim, I. H. Kim, J. H. Ko, B. Cheong, T. S. Lee, K. S. Lee, D. Kim, T.-Y. Seong; *J. Phys. D: Appl. Phys.* **41** (2008), 195409.

348. J. Tang, L. Y. Deng, C. B. Tay, X. H. Zhang, J. W. Chai, H. Qin, H. W. Liu, T. Venkatesan, S. J. Chua; *J. Appl. Phys.* **115** (2014), 033111.

349. Y. Wang, J. Zhu, W. Tang; *Appl. Phys. Lett.* **104** (2014), 212103.

350. Z. Galazka, D. Schulz, K. Irmscher, M. Pietsch; *Leibniz-Institut für Kristalzüchtung*, Berlin, Germany (2015), unpublished data.

351. L. Scheffler, V. l. Kolkovsky, E. V. Lavrov, J Weber; *J. Phys.: Condens. Matter* **23** (2011), 334208.

352. H. von Wenckstern, R. Pickenhain, H. Schmidt, M. Brandt, G. Biehne, M. Lorenz, M. Grundmann; *Appl. Phys. Lett.* **89** (2006), 092122.

Chapter 4

Gallium Oxide (β-Ga_2O_3)

4.1 Introduction

β-Ga_2O_3 is an ultra-wide bandgap (UWBG) oxide semiconductor with a bandgap of about 4.85 eV that offers opportunities for developing novel devices, particularly in UV optoelectronics and high-power electronics. The availability of bulk crystals boosted an intensive worldwide activity on Ga_2O_3 in different areas of crystal growth technology (bulk crystals and epitaxial layers), material science (exploration of a material's properties), and device engineering. Such parallel research activity brings, year by year, β-Ga_2O_3 to a higher level that offers industry new opportunities. A fast development of β-Ga_2O_3 as such, its properties, and device demonstrators have been credited in a large number of research papers, several review articles (Stepanov et al. [1], Tsao et al. [2], Mastro et al. [3], Higashiwaki et al. [4], Baldini et al. [5], Pearton et al. [6], Razeghi et al. [7], Galazka [8]), and patents/patent applications. Development of bulk crystal growth has been intensively followed by the development of hetero- and homoepitaxial layers by all epitaxial techniques, particularly with the use of metal-organic chemical vapor deposition (MOCVD) or metal-organic vapor phase epitaxy (MOVPE), e.g., by Baldini et al. [9–12], Schewski et al. [13], Wagner et al. [14], Gogova et al. [15], Du et al. [16, 17], Tadjer et al. [18], Sbrockey et al. [19], Okur et al. [20],

Transparent Semiconducting Oxides: Bulk Crystal Growth and Fundamental Properties
Zbigniew Galazka

ISBN 978-981-4800-94-5 (Hardcover), 978-1-003-04520-5 (eBook)
www.jennystanford.com

Kim and Kim [21], Mi et al. [22–25], Liu et al. [26], Huang et al. [27], Chen et al. [28, 29], Hu et al. [30], and Feng et al. [31]; chemical vapor deposition (CVD), e.g., by Rafique et al. [32–36] and mist-CVD, e.g., by Lee et al. [37]; halide vapor phase epitaxy (HVPE), e.g., by Oshima et al. [38], Nomura et al. [39], Murakami et al. [40], Konishi et al. [41], and Thieu et al. [42]; molecular beam epitaxy (MBE), e.g., by Víllora et al. [43], Oshima et al. [44], Tsai et al. [45], Guo et al. [46], Vogt and Bierwagen [47–49], Ghose et al. [50, 51], Pratiyush et al. [52], Oshima et al. [53], Sasaki et al. [54–56], Okumura et al. [57], Kaun et al. [58], Oshima et al. [59, 60], Ahmadi et al. [61, 62], and Han et al. [63]; and pulsed laser deposition (PLD), e.g., by Orita et al. [64–66], Matsuzaki et al. [67, 68], Gollakota et al. [69], Goyal et al. [70], Müller et al. [71], Zhang et al. [72–74], Chen et al. [75–77], Wang et al. [78], Shiojiri et al. [79], Heinemann et al. [80], Wakabayashi et al. [81, 82], and Leedy et al. [83].

Extensive characterization of both bulk crystals and layers by a diversity of techniques and support from theoreticians not only revealed properties of obtained materials, but also indicated ways to control some of the properties and their limitations. Such cumulative knowledge and material availability enabled to demonstrate a diversity of devices for various applications. As the breakdown field of a material scales with the bandgap approximately parabolically, with its breakdown field of 8 MV/cm [84], β-Ga_2O_3 is a great candidate for high-power electronics, such as Schottky barrier diodes (SBDs), metal-oxide field-effect transistors (MOSFETs), and metal-semiconductor field-effect transistors (MESFETs). Indeed, such devices have been demonstrated in horizontal and vertical configurations, e.g., by Higashiwaki et al. [84–88], Sasaki et al. [89], Murakami et al. [90], Yang et al. [91], Oh et al. [92], Konishi et al. [93], Wong et al. [94–97], Green et al. [98, 99], Chabak et al. [100], Moser et al. [101, 102], Bae et al. [103], Hu et al. [104], and Hwang et al. [105]. Another group of extensively studied devices are solar- and visible-blind photodetectors, and flame detectors (e.g., Suzuki et al. [106, 107], Oshima et al. [108, 109], Guo et al. [110], Guo et al. [111], Guo et al. [112], Guo et al. [113], Liu et al. [114], Oh et al. [115], Nakagomi et al. [116, 117], Von Wenckstern et al. [118], Zhang et al. [119], Hu et al. [120], and Yu et al. [121]). Further, light-emitting diodes using β-Ga_2O_3 have been demonstrated by Schimamura and Villora

[122], and Li et al. [123]. In addition to electronic and optoelectronic devices, β-Ga_2O_3 has been demonstrated by Szalkai et al. [124], Yanagida et al. [125], Usui et al. [126, 127], and Drozdowski et al. [128] as a nuclear radiation (neutron, alpha, and gamma) detector. The sensitivity of β-Ga_2O_3 to reducing and oxidizing gaseous species led to a demonstration of gas-sensing devices, e.g., by Fleischer and Meixner [129–131], Fleischer et al. [132–134], Flingelli et al. [135], Frank et al. [136], Pohle et al. [137], Weh et al. [138], Schwebel et al. [139, 140], Stegmeier et al. [141], Ogita et al. [142–144], Baban et al. [145], Bartic et al. [146, 147], and Bartic [148], Nakagomi et al. [149], Trinchi et al. [150, 151], and Lee and Yan [152].

Unique properties and a broad spectrum of possible applications place β-Ga_2O_3 in the frontline of UWBG materials as a complementary material to (Al,Ga)N, AlN, and SiC.

This chapter discusses bulk β-Ga_2O_3 single crystals in terms of growth techniques, crystal size, and structural quality, as well as their thermal, mechanical, optical, and electrical properties. Some of the features of β-Ga_2O_3 are shown from the theoretical point view or compared therewith. The main objective is to show how to grow bulk β-Ga_2O_3 single crystals, which issues accompany crystal growth and which measures to apply to overcome the occurring issues, which methods enable to obtain large-size and high-quality crystals, and to reveal properties measured on bulk crystals obtained by different methods. In other words, this chapter shows the state of the art in the growth of bulk β-Ga_2O_3 single crystals and collection of their properties that may facilitate an assessment of this compound for further study, development, and applications.

4.2 Polymorphism of Ga_2O_3

Ga_2O_3 has several polymorphs that were investigated in detail both experimentally and theoretically by Roy et al. [153], Zinkevich and Aldinger [154], Penner et al. [155], Playford et al. [156], Yoshioka et al. [157], Delgado and Areán [158], Fornari et al. [159], and Cora et al. [160]. The studies revealed several polymorphs named α, β, γ, δ, ε, and κ. They have different crystallographic systems and formation conditions, as shown in Table 4.1 along with lattice parameters. However, the δ-phase could be a mixture of the β- and ε-phases (Playford et al. [156]), while the ε-phase is likely mimicking the κ-phase due to rotational grains formed on sapphire (Cora et al. [160]). Theoretical study of α, β, δ, and ε polymorphs of Ga_2O_3 by Furthmüller and Bechstedt [161] showed that electronic properties thereof are quite similar to each other, because the polymorphism is the result of different arrangements of Ga atoms that are coordinated tetrahedrally and octahedrally, whereas Ga–O bonding remains almost the same.

Table 4.1 Ga_2O_3 polymorphs

Polymorph	System	Space group	Lattice parameters [Å]	Reference
α	Hexagonal	$R\bar{3}c$	a = 4.9825 b = 13.433	162
β	Monoclinic	C2/m	a = 12.214 b = 3.0371 c = 5.7981 β = 103.83	163
γ	Cubic	$Fd\bar{3}m$	a = 8.22	164
κ	Orthorhombic	$Pna2_1$	a = 5.0463 b = 8.7020 c = 9.2833	160
δ	Cubic	Ia3	a = 9.491	157
ε	Hexagonal	$P6_3mc$	a = 2.9036 a = 9.2554	156

Among all the polymorphs, β-Ga_2O_3 is the most stable modification at high temperatures and can be grown directly from the melt. The melting point (MP) of β-Ga_2O_3 is around 1800°C. Other polymorphs can form at low temperature as metastable phases. The formation energy of the polymorphs has the following order β < ε < α < δ < γ [157]. Although other phases than β-phase cannot be grown as bulk crystals from the melt, flux, or gas phase, they could be obtained as thin films or thick layers. For instance, the α-Ga_2O_3 phase was obtained on sapphire substrates by mist chemical vapor deposition (mist-CVD) [165] (at 430–470°C), halide vapor phase epitaxy (HVPE) [166] (at 520–600°C), and laser molecular beam epitaxy (LMBE) [167]. Thin films of the γ-Ga_2O_3 phase were obtained by mist-CVD [168] and pulsed laser deposition (PLD) [169] with the use of sapphire or spinel substrates at temperatures up to 500 and 600°C, respectively. ε-Ga_2O_3 thin films were grown by the HVPE on nitride, β-Ga_2O_3, magnesia, and yttria-stabilized zirconia substrates with the growth temperature of 400–550°C [170, 171], by plasma-assisted molecular beam epitaxy (PAMBE) on sapphire at 700°C [172], and by metal-organic chemical vapor deposition (MOCVD) on sapphire [173].

4.3 Crystal Structure of β-Ga_2O_3

The metal gallium follows aluminum in the third main group of the periodic system, which implies some chemical and structural similarity of the element itself and of its compounds. Indeed, the oxides of both metals show a large variety of polytypes. The thermodynamically stable oxides of both metals have different crystal structures, and these are trigonal α-Al_2O_3 (mineral names corundum or less accurate sapphire, respectively) or monoclinic β-Ga_2O_3.

β-Ga_2O_3 crystallizes in the base-centered monoclinic system in the space group C2/m (Geller [174]). It is isomorphic to θ-Al_2O_3 [175]. The unit cell (Fig. 4.1) contains 20 atoms consisting of crystallographically inequivalent Ga^{3+} and O^{2-} atoms. In this low symmetry structure, Ga atoms are coordinated tetrahedrally and octahedrally (Ga(I) and Ga(II), respectively), while O atoms are coordinated three and fourfold (O(I)/O(II) and O(III), respectively). O(I) shares two bonds with Ga(II) and one bond with Ga(I); O(II) shares three bonds with Ga(II) and one bond with Ga(I), while O(III) shares two bonds with Ga(I) and one bond with Ga(II). In contrast to α-phases of Ga_2O_3 and Al_2O_3 with hexagonal dense oxygen packing, the oxygens obey in β-Ga_2O_3 and θ-Al_2O_3 a distorted face-centered cubic packing of oxygen atoms. The lattice parameters of β-Ga_2O_3 are listed in Table 4.1. There are two easy cleavage planes formed by O(III) atoms on the (100) plane and by O(I) atoms on the (001) plane.

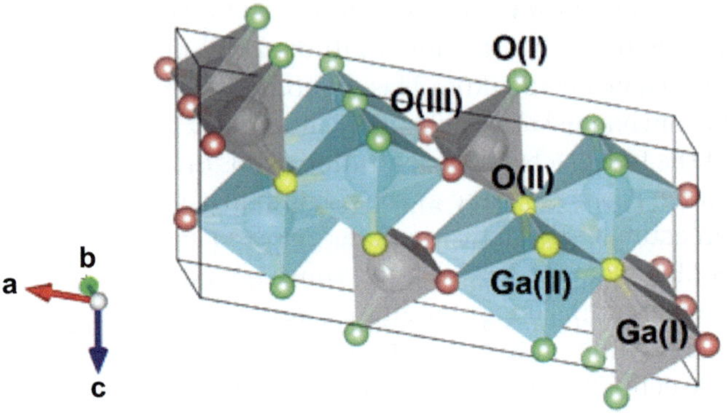

Figure 4.1 Unit cell of β-Ga_2O_3. Gallium atoms Ga(I) and Ga(II) are coordinated tetrahedrally and octahedrally, respectively. Inequivalent O atoms have threefold (O(I) and O(II)) and fourfold (O(III)) coordination.

4.4 Band Structure and Native Point Defects

4.4.1 Band Structure

The electronic band structure defines intrinsic electronic and optical properties of a material the knowledge of which enables to conclude a proper device design with desired functionality. Current computational power and advanced laboratory tools led to an explosion in studying material physics both theoretically and experimentally. Computational studies may provide hints to experimental research and explore structural, electronic, and optical properties, such as trapping effect, intrinsic defects, doping, conductivity type, energy gap and its nature, and optical transitions, as well as material limitations. On the other hand, experimental results in feedback may validate a mathematical model for electronic structure computation.

In the last decade, the electronic band structure of β-Ga_2O_3 was a subject of an intensive first-principles calculations using DFT (LDA, GGA) or hybrid-DFT, e.g. by Yamaguchi et al. [176] (LDA-PW), He et al. [177] (hybrid-B3LYP), Furthmüller et al. [161, 178] (hybrid-HSE+G0W0), Varley et al. [179–181] (hybrid-HSE06), Peelaers et al. [182] (hybrid-HSE06), Zacherle et al. [183] (GGA-PBE, hybrid-HSE06), Navarro-Quezada et al. [184] (GGA-HSE03+G0W0), Cocchi et al. [185] (LDA-PW), Deák et al. [186] (hybrid-HSE06), Dong et al. [187] (GGA-PBE), Mock et al. [188] (GGA-GauPBE), Wei et al. [189] (hybrid-HSE06), and measurements by Lovejoy et al. [190], Janowitz et al. [191], Mohamed et al. [192, 193], Zhang et al. [194], Michling and Schmeißer [195]. The band structure of β-Ga_2O_3 is shown in Fig. 4.2. The top of the valence band (VB) is composed mainly by O 2p states, which make it very flat. Very small dispersion of the VB maximum (VBM) results in a very high effective hole mass and consequently in a very low hole mobility making a possible *p*-type conductivity impractical. Additionally, theoretical results of Kyrtsos et al. [196] (GGA-PBE, hybrid-HSE) revealed that investigated mono- and divalent ions substituting Ga^{3+} in the crystal lattice introduce deep acceptor levels with ionization energies greater than 1eV, making such doping toward *p*-type conductivity not feasible.

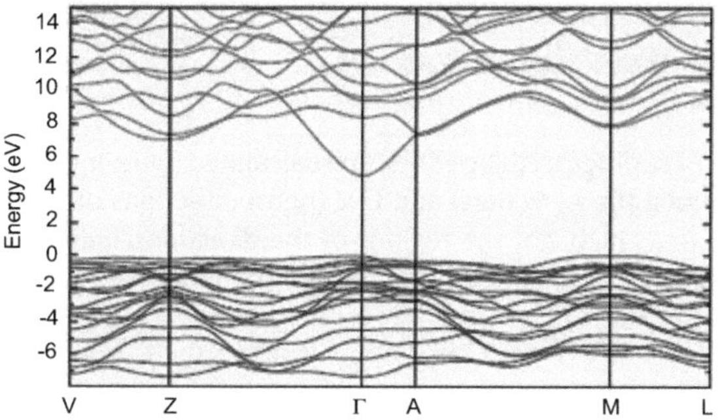

Figure 4.2 Band structure for β-Ga_2O_3 calculated using the primitive unit cell of base-centered monoclinic β-Ga_2O_3. Reprinted from Ref. [179], Copyright 2010, with the permission of AIP Publishing.

Local maxima in the VBM occur in the Γ and M points of the Brillouin zone (BZ). The M point is off the Γ point by only 30 meV; therefore, β-Ga_2O_3 can be regarded as a direct semiconductor (due to momentum conservation, the transition at Γ–Γ will be more probable than at M–Γ). The bottom part of the conduction band (CB) is dominated by hybridized Ga-s and O-p states with a well-distinguished local minimum at Γ point and high dispersion leading to a relatively low effective electron mass and high electron mobility.

Figure 4.3 shows the experimental VB band structure determined by angular-resolved photoemission spectroscopy (ARPES), which well fits the computed band structure. The experimental direct and indirect energy gaps are E_{gdir} = 4.9 ± 0.1 and E_{gindir} = 4.85 ± 0.1 eV, respectively [191] (difference = 50 meV).

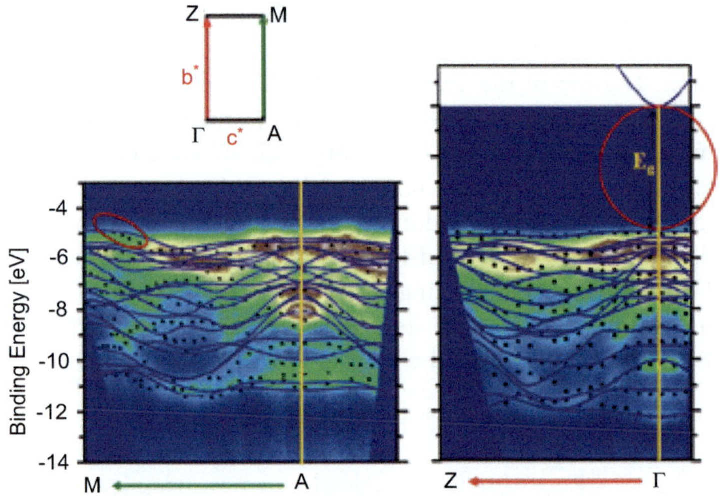

Figure 4.3 Experimental (black dots) and calculated (blue lines) band structure of β-Ga_2O_3 along the A–M (left) and Γ–Z (right) directions of the BZ (top). The red-marked areas highlight the regions of the direct and indirect energy gaps. Reprinted from Ref. [191] under Creative Commons license.

4.4.2 Native Point Defects

The native defects in β-Ga_2O_3 include Ga vacancies (V_{Ga}) at tetrahedral (Ga(I)) and octahedral (Ga(II)) sites as well as O vacancies at threefold coordinated sites (O(I) and O(II)) and fourfold coordinated sites (O(III)), as well as Ga (Ga_i) and O (O_i) interstitials. The oxygen vacancy (V_O) formation energies calculated by Varley et al. [179] are 3.31, 2.7, and 3.57 eV for O(I), O(II), and O(III), respectively. They are deep donors and do not directly contribute to the electrical conductivity. Oxygen vacancies occur as neutral (V_O) and double-ionized vacancies ($V_O^{\bullet\bullet}$) with V_O dominating at low oxygen partial pressure, $p(O_2)$, up to 0.01 bar, and their concentration decreases with $p(O_2)$ [183]. On the other hand, V_{Ga} form deep levels at −3 charge state that are located 1.62 eV (V_{Ga1}) and 1.83 eV (V_{Ga2}) below the CBM [180]. Moreover, the V_{Ga} concentration increases with $p(O_2)$ leading to a conductivity compensation. Theoretical results of Blanco et al. [197] showed that interstitial migration paths are "controlled" along the *b*-direction, while vacancy migration paths are "controlled" by atomic sheets perpendicular to *b*-direction in β-Ga_2O_3. It was also pointed out that dominating point defects in β-Ga_2O_3 are Frenkel pairs of V_O–O_i.

The calculations show that intrinsic defects alone rather cannot be responsible for the electrical conductivity of β-Ga_2O_3. However, the situation becomes different in the presence of intentional dopants or residual impurities. Varley et al. [180] revealed in their calculations that hydrogen forms shallow donor in isolated interstitial form (H_i^+) and in V_{Ga}–H complexes, which passivate compensating V_{Ga}. Such complexes have a low formation energy (lower than that of isolated V_{Ga}) and high binding energy, stable up to about 800°C. Wei et al. [189] concluded theoretically that V_{Ga} can bind up to four hydrogen atoms, and V_{Ga}-3H and V_{Ga}-4H complexes are very stable. Such complexes can be created at high concentration leading to a compensation of the underlying electrical conductivity by passivating V_{Ga}. Also, tetravalent ions Si^{4+}, Ge^{4+}, and Sn^{4+} substituting tetrahedrally and octahedrally coordinated Ga^{3+} ions in the crystals lattice, respectively, as well as monovalent F^- and Cl^- on threefold coordinated O^{2-} sites form shallow donors in β-Ga_2O_3 [178, 179, 198]. In terms of compensating acceptors, Lyons [199] concluded that Mg and Be are the most stable elements. Intentional doping with Si^{4+} and Sn^{4+} ions to enhance the electrical conductivity and with Mg^{2+} to compensate the electrical conductivity is commonly practiced for bulk β-Ga_2O_3 single crystals. VBM and CBM positions of Ga_2O_3 when doping with a number of nonmetal elements can be found in the work of Guo et al. [200].

4.5 Bulk Crystal Growth

The first available report on bulk β-Ga_2O_3 crystals was published in 1957 by Kohn et al. [201], where tiny crystals were obtained from the gas phase by the oxidation of molten elemental Ga. A few years later, in 1963, the first flux-grown crystals were obtained by Remeika [202], with a continuation by other groups in the next three decades (Katz et al. [203], Garton et al. [204], Fischer et al.

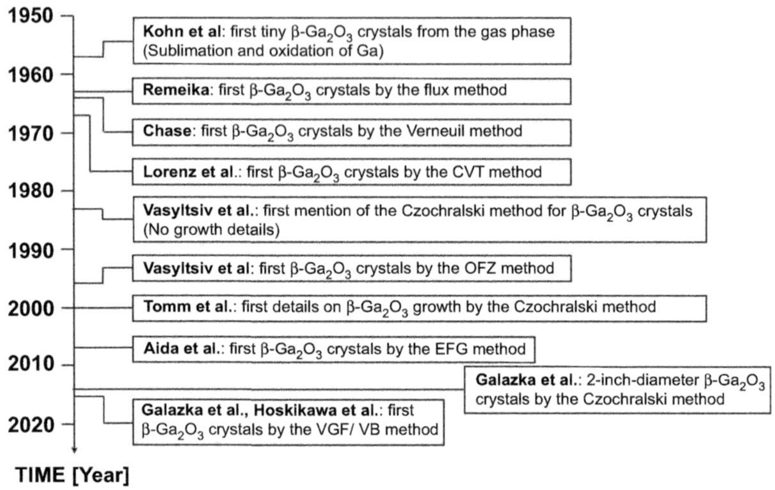

Figure 4.4 Main events in the growth of bulk β-Ga_2O_3 single crystals.

[205], Chani et al. [206]). In 1964, the first melt-grown crystals were demonstrated by Chase [207]. For that purpose, the Verneuil method was utilized, which was practiced from time to time for over one decade. Soon after, in 1967, first reports on the CVT growth of β-Ga_2O_3 were mentioned by Lorenz et al. [208]. The CVT method was continued in the next two decades by several research groups, including Matsumoto et al. [209, 210], Gerlach and Oppermann [211], Juskowiak and Pajączkowska [212, 213], and Pajączkowska and Juskowiak [214, 215]. Vasiltsiv et al. [216] mentioned in 1983 the use of the Czochralski method for growing β-Ga_2O_3 crystals; however, no details were given. In 1996, the growth of β-Ga_2O_3 by the optical floating zone (OFZ) method was reported by Vasyltsiv et al. [217], which was then successfully followed by numerous groups (Ueda et al. [218], Tomm et al. [219], Ko et al. [220], Villora et al. [221, 222], and Zhang et al. [223]). In 2000, some details of the growth of β-Ga_2O_3 crystals by the Czochralski was provided by Tomm et al. [224], which turned into an intensive development a decade later by Galazka et al. [225–229]. In 2008, the first detailed report on the EFG method was published by Aida et al. [230], who mentioned, however, that first attempts were performed by Shimamura et al. [231]. An intensive development of the EFG method has been then carried out by Kuramata et al. [232]. Successful application of melt growth

techniques (the Czochralski and EFG methods) opened the doors for other melt growth techniques. In 2015, the VGF and Bridgman methods were demonstrated by Galazka et al. [227], and in parallel (2016) by Hoshikawa et al. [233]. The summary of main events in the growth of β-Ga_2O_3 single crystals is given in Fig. 4.4.

4.5.1 Gas Phase

Bulk β-Ga_2O_3 crystals were grown from the gas phase in two ways: (i) by evaporation and oxidation of elemental Ga, which can be defined as the physical vapor transport (PVT) method, and (ii) by using a chemically active (reactive) transport gas carrying Ga from a source to a growth zone, which is called the chemical vapor transport (CVT) method.

At elevated temperatures, the partial pressure of Ga-containing species is not high enough for an efficient transport to the growth zone, where it is oxidized back to β-Ga_2O_3 crystals. Therefore, for the PVT method, rather sublimation of elemental Ga was used instead of Ga_2O_3. Much more efficient transport was achieved by the CVT method that was demonstrated as a dominating growth method of β-Ga_2O_3 crystals from the gas phase.

4.5.1.1 Physical vapor transport

In the PVT method, the most volatile species is Ga_2O, which is formed either by oxidation of elemental Ga or by thermal decomposition of Ga_2O_3 at high temperatures:

$$2Ga(s) + \tfrac{1}{2}O_2(g) \rightarrow Ga_2O(g) \tag{4.1}$$

$$Ga_2O_3(s) \rightarrow Ga_2O(g) + O_2(g) \tag{4.2}$$

Next, gallium suboxide $Ga_2O(g)$ is re-oxidized at lower temperatures and form β-Ga_2O_3 single crystals:

$$Ga_2O(g) + O_2(g) \rightarrow \beta\text{-}Ga_2O_3(s) \tag{4.3}$$

Using Ga_2O_3 as the source (starting) material for the PVT growth would require very high temperatures (close to the MP of Ga_2O_3) to obtain a partial pressure of Ga_2O, $p(Ga_2O)$ at a level of about 10^{-3} atm. Using elemental Ga as the starting material enables much higher partial pressure of Ga, p(Ga). For example, at 1600°C, $p(Ga_2O)$

≈ 1.1×10^{-4} atm, while $p(\text{Ga}) \approx 1.7 \times 10^{-2}$ atm when using Ga_2O_3 and elemental Ga as the starting material, respectively, and Ar as a growth atmosphere. Therefore, at lower temperatures (1000–1200°C), it is more sufficient to use elemental Ga for evaporation and transportation to the growth zone. However, such physical transport will not be very effective and only very small crystals are expected. This is why, PVT has not been developed for bulk β-Ga_2O_3 crystals.

As an example, Kohn et al. [201] obtained very small crystals of β-Ga_2O_3 by the PVT in three different ways. β-Ga_2O_3 crystals were formed during the growth of GaP crystals with the excess of Ga in a silica tube at 1100°C for 3 h, and then cooled down at the rate of 40 K/h. The crystals grew in a cold region on a tube wall above the GaP melt, likely by the reaction of Ga with residual oxygen. The crystals were light blue and had plate shape. In the second method, β-Ga_2O_3 crystals were obtained by heating elemental Ga to 1100°C in a stream of N_2. Residual O_2 in N_2 was sufficient to oxidize Ga and produce light-grey fibers of β-Ga_2O_3. In the third preparation method, elemental Ga was heated up in a quartz crucible to 700–800°C in air. Also in this case, tiny, light-grey crystals over molten Ga were formed.

4.5.1.2 Chemical vapor transport

In the CVT method used for growing bulk β-Ga_2O_3 crystals, different systems were used in terms of the starting material and chemically active (reactive) transport agent (CATA), which are discussed further below. Ga is efficiently transported from a source to a growth zone by a halogen (Cl, I), although additional gaseous elements, such as H and S, were used to affect the chemistry and transport of gaseous species in the system. The source of Ga can be either elemental Ga or Ga_2O_3.

Lorenz et al. [208] mentioned the use of I_2 as the chemically active transport agent. The obtained crystals were very small, needle shaped with light-blue coloration.

Matsumoto et al. [209, 210] grew the crystals in the GaCl–O_2 system and Ar as a carrier gas. Growth experiments were performed in a setup (Fig. 4.5) consisting of an outer tube, an inner quartz tube within the outer tube, and an external container with HCl, which is connected with the inner tube. The inner tube has a pocket accommodating elemental Ga and an enlarged end next to the pocket with Ga, where the growth occurs. Through the external container with HCl and then through the inner tube, the Ar carrier gas flows. Oxygen is introduced directly into the outer tube via a separate inlet.

Possible reactions producing Ga_2O_3 crystals are as follows:

$$HCl + Ga \leftrightharpoons GaCl + ½H_2 \tag{4.4}$$

$$2GaCl + 1½O_2 \leftrightharpoons Ga_2O_3 + Cl_2 \tag{4.5}$$

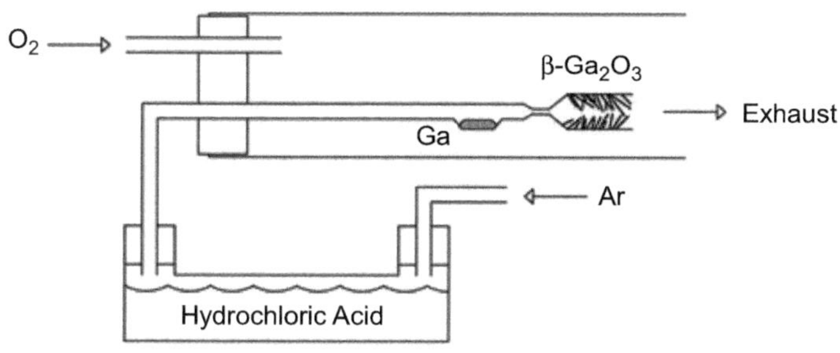

Figure 4.5 Crystal growth apparatus for the CVT method. Redrawn from Ref. [210] with some modifications.

The growth temperature T_G was 1100–1150°C, while Ar and O_2 flows were 20–30 ml/min and 2–4 ml/min, respectively. The obtained crystals were colorless and transparent. They were either (100)-oriented platelets with the size of about 3 × 3 × 0.2 mm^3 or needles of diameter about 0.2 mm and length up to 8 mm.

For growth experiments from the gas phase, Gerlach and Oppermann [211, 234] used highly purified Ga_2O_3 (obtained by vacuum evaporation at 1000°C) as the Ga source and $TeCl_4$ as a chemically active transport agent. Theoretically, the species dominating in Ga transport is $GaCl_3$, which has basically the highest partial pressure over other gaseous species (except $TeCl_2$) up to 1100°C except a temperature region of about 600–750°C. For Ga re-oxidizing, a sufficient $p(O_2)$ is required. This is fulfilled at temperatures above 750°C, in which other volatile species (except $TeCl_2$) have lower partial pressures than that of $GaCl_3$. The growth experiments were carried out in quartz ampoules of 12 mm in diameter and 15 cm in length, which were heated up in vacuum at 300°C after loading the starting materials. The ampoules were arranged within a two-zone furnace for the growth that lasted for several days. All the crystals were of β-phase. Most of them were colorless and transparent with the size of 2 × 2 × 2.5 mm^3 (Fig. 4.6). At high growth rate, the crystal size decreased and structural quality deteriorated.

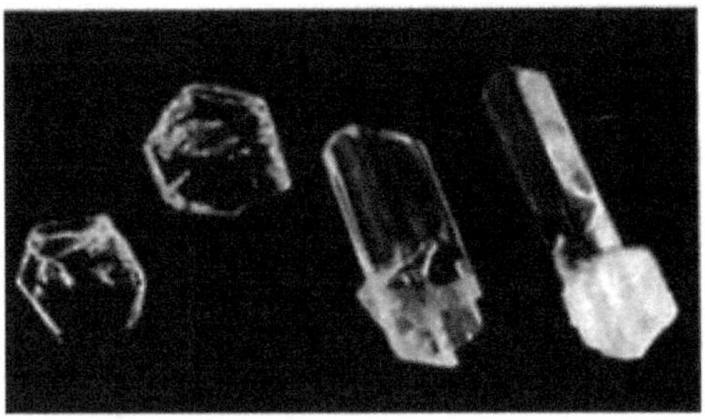

Figure 4.6 β-Ga_2O_3 single crystals obtained in the Ga_2O_3–$TeCl_4$ system. Reprinted with permission from Ref. [211], Copyright 1977, John Wiley and Sons.

At a pressure of 3 atm and a source temperature T_S = 750–1100°C, the transport was always from the hot to the cold zone for a wide spectrum of a temperature difference between the hot and cold zone (ΔT), ranging from 50 to 250 K. The growth rate increased with both T_S and ΔT.

Juskowiak and Pajączkowska used different chemically active transport agents in the CVT method: Cl_2 [212], H–Cl [214], N–H–Cl [215], and S–Cl [213]. The experiments were carried out in a closed setup within quartz ampoules having an inner diameter of 18 mm, length of 120 mm, and volume of about 30 cm^3. Prior growth experiments, the ampoules were annealed in air at 1130°C, while Ga_2O_3 starting material (powder) was annealed in air at 1030°C for 6 h. The ampoules loaded with Ga_2O_3 were annealed in vacuum at 180°C to remove traces of water.

(i) Ga_2O_3–Cl_2 system

The ampoules were sealed after introducing therein liquid chlorine in the amount of 0.5–4 mg/cm^3 and kept at reverse temperature gradient for 1 day. The growth experiments lasted for 3 to 6 days at growth temperatures T_G from 230 to 1030°C.

The main species responsible for Ga transport from a hot to a cold zone is $GaCl_3$ having the highest partial pressure. The chemical reactions for the mass transport and crystallization can be described as follows (neglecting the intermediate reactions):

$$Ga_2O_3 + 3Cl_2 \leftrightharpoons 2GaCl_3 + 1\tfrac{1}{2}O_2 \text{ (hot zone)} \quad (4.6)$$

$$2GaCl_3 + 1\tfrac{1}{2}O_2 \leftrightharpoons Ga_2O_3 + 3Cl_2 \text{ (cold zone)} \quad (4.7)$$

Experimental results revealed that below 630°C, only traces of Ga_2O_3 crystallized, but above that temperature, the growth rate increased significantly. This is likely due to the not-high-enough partial pressure of gaseous $GaCl_3$ below 630°C. Between 630 and 1030°C, the partial pressure of $GaCl_3$ increases significantly, enhancing the mass transport to the cold zone. The growth temperature T_G (cold zone) was lower than that of the source temperature T_S (hot zone) by 20 to 150 K. The most optimum growth conditions were found at a temperature range of T_G = 877–977°C and temperature difference between the hot (T_S) and cold (T_G) zones $\Delta T = T_S - T_G$ = 30–70 K. In this temperature range, both $p(GaCl_3)$ and $p(O_2)$ dominate over $p(Cl_2)$. Above 1030°C, the growth was limited by an intensive reaction of the dissociation products with the quartz ampoule. The obtained crystals were of pure β-Ga_2O_3 phase. They were colorless platelets of about 3 mm in size.

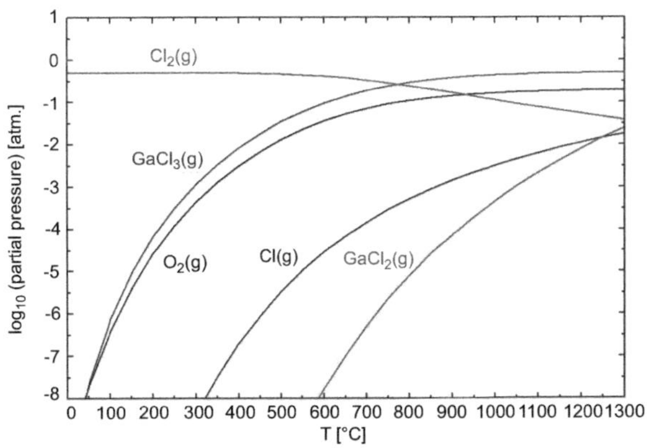

Figure 4.7 Partial pressures of most volatile species in the Ga_2O_3–Cl_2 system (for calculation, 1 mol of Ga_2O_3 and 1 mol of Cl_2 under total pressure of 1 atm were used).

Thermodynamic calculations (Fig. 4.7) show that the most volatile species in Ga_2O_3–Cl_2 system containing Ga is $GaCl_3$, which is responsible for the mass transport in the growth system. In a temperature range of T_G = 930–1230°C $p(GaCl_3)$ and $p(O_2)$ are higher than $p(Cl)$, $p(Cl_2)$, and $p(GaCl_2)$. This relation ensures an effective mass transport and re-oxidation of Ga.

(ii) Ga_2O_3–H/Cl system

Here the growth temperature T_G was between 530 and 1030°C, which corresponds to a pressure of 2–4 atm. The chemically active transport agents were Cl_2 and H_2O at different amounts to provide different ratios of H/Cl. To the Ga_2O_3 starting material elemental Ga was also added. Experimental results at T_G = 730°C showed that at H/Cl = 1, no transport of Ga-containing species was observed. Above that temperature, the transport in both directions from hot to cold and from cold to hot zones occurred. Small colorless crystals over the whole inner ampoule's wall were grown. For H/Cl ≠ 1 (9, 3, 1.2, and 0.33), there was mass transport always from the hot to cold zone due to a higher partial pressure of Ga-containing species. The obtained lump-like crystals of several millimeters (up to about 10 mm) in size and having β-phase were blue for H/Cl > 1.

(iii) Ga_2O_3–N-H-Cl system

NH_4Cl was sublimated into the ampoule after removing air therefrom. The source T_S and growth T_G temperatures were in the range of 550–1050°C and 700–980°C, respectively, corresponding to a total pressure of 2–8 atm. In the growth experiments, $\Delta T = T_S - T_G$ was between 50 and 150 K. Below 730°C, the mass transport was observed from the cold to the hot zone, as the result of the exothermic process by the creation of $GaCl_3$–·NH_3 complexes. Above 730°C, the mass transport proceeded from the hot to the cold zone. The obtained crystals were of single β-phase and different coloration, depending on temperature. The crystals were yellow, green–blue, and blue when obtained at T_G < 830°C, 830–930°C, and >930°C, respectively. Different coloration was explained in terms of doping with nitrogen, as its partial pressure was much higher than the oxygen partial pressure, and due to similar covalent radii of O (0.73 Å) and N (0.75 Å). The highest temperatures produced the largest crystal lumps, up to 10 mm in size; however, they were highly defected due to the presence of a block structure. The crystals did not react with the quartz ampoule, except those grown under exothermic reactions below 730°C.

(iv) Ga_2O_3–S/Cl system

The main possible reaction responsible for the transport is as follows:

$$Ga_2O_3(s) + 3/2Cl_2(g) + S_2(g) \rightarrow GaCl_3 + GaS + SO_2 + \tfrac{1}{2}O_2 \quad (4.8)$$

The experiments were done in 2 or 4 atm of a total pressure in closed ampoules with T_S = 550, 850, and 950°C, temperature difference between the source and growth zones of 50, 100, 150, and 220 K, and S/Cl ratios of 0.11 and 0.43.

Small and transparent crystals were obtained only in the case of small content of S (S/Cl = 0.11) for low and high T_S (550 and 950°C), and for both small and large temperature differences (ΔT = 50 and 220 K), wherein the transport proceeded from a cold to a hot zone at low T_S = 550°C and from the hot to the cold zone at high T_S = 950°C. That transition in the transport direction takes place at about 730°C. In the case of high S content, blue crystals were obtained only in the case of high T_S = 950°C and ΔT = 100 K. In all other cases, there was either no transport or a yellow substance was formed (sulfur or sulfides).

The main drivers in the system involving two transport agents S and Cl are temperature and the ratio of S/Cl. The experimental results can be explained from the partial pressures of the most volatile species shown in Fig. 4.8, containing a small and large amount of S. In both cases, Ga is transported mainly by $GaCl_3$. At low S content (Fig. 4.8a), $p(GaCl_3) > p(SO_2)$ and also $p(O_2)$ is relatively high, especially at T >750°C. Such conditions enable an effective Ga transport and its re-oxidation. However, at higher content of S (Fig. 4.8b), $p(GaCl_3) < p(SO_2)$, and more importantly, $p(O_2)$ is at much lower level, by about nine orders of magnitude. So we have a situation where Ga transport is less efficient, and additionally there is no enough O_2 for Ga re-oxidation. Ga re-oxidation may happen only at high temperatures, when $p(O_2)$ is higher. This explains obtaining sulfur or its compounds instead of Ga_2O_3 crystals, and the blue coloration of the crystals obtained at high source temperature due to the deficiency of O_2.

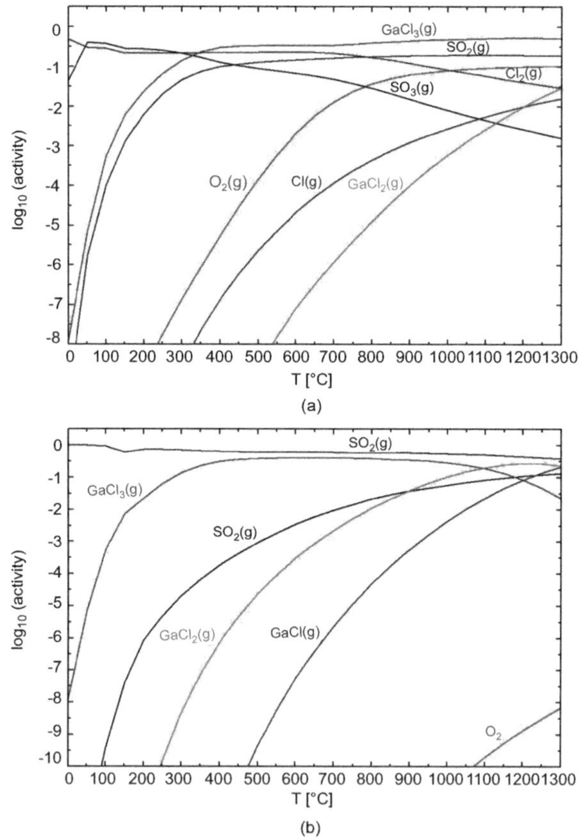

Figure 4.8 The most volatile species in the G_2O_3–Cl + S system: (a) S = 0.1 and Cl = 0.9 mol; (b) S = Cl = 0.5 mol. In both cases, Ga_2O_3 = 1 mol, and the pressure of 1 atm.

As summarized in Table 4.2, different systems in the CVT method require different growth temperatures to promote transport of Ga-containing species from the source zone to the growth zone. The crystal appearance (shape, size, and coloration) is the result of the growth environment, particularly of the chemically active transport agent, T_S, and T_G. Typically, the CVT-grown crystals are needles/fibers/bars or small plates, which are colorless, bluish, and occasionally greyish, yellowish, or greenish.

Table 4.2 Summary of results for gas phase techniques. T_S and T_G are source and growth temperatures, respectively, while $\Delta T = T_S - T_G$. D and L are diameter and length, respectively

System	T_S, T_G, ΔT [°C]	Crystal shape	Maximum single crystal size [mm]	Crystal coloration	Reference
Ga – N_2 Ga – air	T_S = 1100 T_S = 700–800	Fibers		Light grey	201
Ga – HCl – O_2	T_G = 1100–1150	Platelets Fibers	3×3×0.2 D = 0.2, L = 8	Colorless	209, 210
Ga_2O_3 – I_2	—	Needles	—	Light blue	208
Ga_2O_3 – Cl_2	T_G = 880–980 ΔT = 30–70	Platelets	3	Colorless	212
Ga_2O_3 – Cl_2 –H_2O	T_G = 530–1030 ΔT = 50	Lumps	D < 10	Colorless, blue	214
Ga_2O_3 – NH_4Cl	T_G = 700–980 ΔT = 50–150	Lumps	D < 10	Yellow, green–blue, blue	215
Ga_2O_3 – (Cl_2 + S)	T_S = 550–950 ΔT = 50–220	Bars	D = 1–2, $L \leq 5$	Colorless, blue	213
Ga_2O_3 – TCl_4	T_S = 750–1100 ΔT = 50–250	Bars	2×2×2.5	Colorless	211, 234

4.5.2 Flux

4.5.2.1 Crystal growth

Solutions at high temperatures and atmospheric pressure (fluxes) were used to grow bulk β-Ga_2O_3 crystals at temperatures far below the MP of β-Ga_2O_3, where it is thermally stable. Different fluxes were applied for growing β-Ga_2O_3, such as PbO + B_2O_3 (Remeika [202]), PbO + PbF_2 and MoO_3 (Schieber [235]), PbF_2 + $PbSO_4$ (Katz and Roy [236]), PbO + V_2O_5 and Bi_2O_3 + V_2O_5 (Garton et al. [237]), and B_2O_3 + Bi_2O_3 (Chani et al. [238]). Among these fluxes, PbO, PbF_2, and Bi_2O_3 are very good solvents, but they have a tendency to evaporate and to reduce to pure metals, forming alloys with Pt crucibles. On the other hand, V_2O_5 is not volatile at typical operating temperatures and reduces to lower oxidation state, but not to a pure metal. Therefore, fluxes containing V_2O_5 can be used even at higher temperatures (>1350°C) for a prolonged time. To the fluxes, Ga_2O_3 powder is added at certain proportions (typically 5–20 wt.% in relation to the flux), but Schieber [235] used $Sm_3Ga_5O_{12}$ garnet instead, which decomposes at elevated temperatures into Sm_2O_3 and β-Ga_2O_3. Growth experiments were usually performed in muffle furnaces with the use of Pt crucibles covered with Pt lids, and maximum temperature of 1300°C. In the growth of bulk β-Ga_2O_3 crystals with the use of fluxes, the Pt crucible was heated up to a maximum temperature T_{MAX}, kept at that temperature for at least several hours and then slowly cooled down at the rate of 2–4 K/h to a final growth temperature T_F of about 600–900°C. Next, the content of the crucible was either poured, quenched, or cooled down to RT. The crystals were typically released from the flux matrix in dilute acids. The obtained crystals had different coloration, shapes, size, and structural quality depending on the flux composition, growth temperature, and thermal history. The edge dimension of the plate crystals could reach even 16 mm, the thickness of which increases with the cooling rate (but the structural quality deteriorates). Typically, flux impurities were found in the crystals (even up to 0.2 wt.%), in particular B, Bi, and Pb.

4.5.2.2 Examples

(i) For growing bulk β-Ga_2O_3 crystals, Remeika [202] used a flux containing various mixtures of PbO and B_2O_3 in the weight ratio of 50:3. The starting material containing the flux (50 g PbO + 3 g B_2O_3) and Ga_2O_3 (10 g) was loaded into a Pt crucible covered with a lid, which then was placed within a muffle furnace. The furnace was heated to T_{MAX} = 1300°C and kept at this temperature for 8 h. A higher temperature led to an excessive evaporation of B_2O_3. Too high B_2O_3 content increases the flux viscosity that creates many nucleation centers and a large number of small crystals. The cooling rate was typically 2 K/h until reaching a temperature $T_F \approx 915$°C. Next the liquid was poured off, and the crucible with the crystals (presumably attached to the crucible wall/bottom) was immersed in a dilute solution of nitric acid and water to remove the residual flux and release the crystals. The obtained crystals were plates of edge size about 5 mm. They were colorless and transparent. The thickness of the crystal plates was a function of the cooling rate; the slower the cooling rate, the thinner the crystal plates. However, higher cooling rates led to entrapment of fluxes in the crystals.

(ii) Katz and Roy [203] used a powdered flux (100 g) consisted of 98 wt.% of PbF_2 and 2 wt.% of $PbSO_4$ that was mixed with powdered Ga_2O_3 (13.49 g) in the ratio of flux/solute = 85/15 mol.% and loaded into a Pt crucible of 80 ml in volume, which was covered with a lid to minimize volatilization. A vertical Kanthal furnace was equilibrated for 0.5–2 days at temperature T_{MAX} = 1225°C, which was then cooled down to T_F = 872°C at a rate of 4 K/h and next air-quenched. The crystals were released from the flux matrix in dilute HNO_3. The Pt crucible was not attacked by the flux during growth runs. β-Ga_2O_3 crystals, 3 to 10 mm long, with well-developed faces were typically contained in several intergrown clusters. Separate crystals were as long as 13–16 mm (0.182–0.233 g), but without developed faces. The crystals were prismatic in shape (Fig. 4.9A), which showed a perfect cleaving into a tubular shape (Fig. 4.9B).

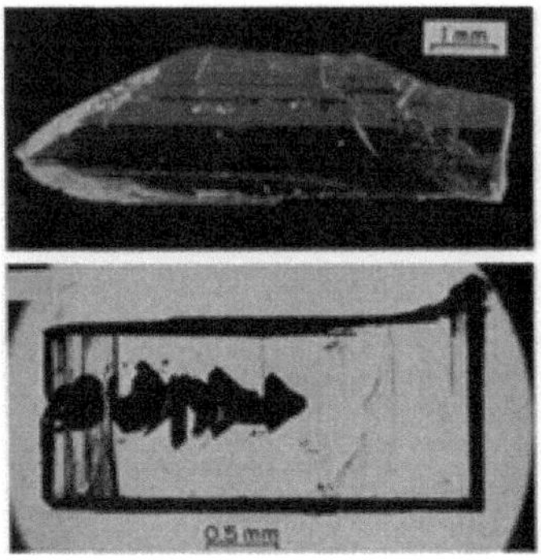

(a)

(b)

Figure 4.9 One of the more perfect prismatic crystals ($8 \times 3 \times 1$ mm^3) (a), and a tabular crystal with opaque regions representing flaws or inclusions (b). Reprinted with permission from Ref. [203], Copyright 2006, John Wiley and Sons.

The obtained crystals had twins on the (001) face, inclusions, and striations on the {011} faces. The inclusions could be originated by trapping Pb from the flux rather than substitutional incorporation into the crystal lattice, due to a large difference in ionic radii between Pb^{2+} (1.32 Å) and Ga^{3+} (0.62 Å). Pb impurity concentration in the crystals was between 0.07 and 0.2 wt.%. Pt was not detected, but it could be present at small concentration at the level of below a few wt. ppm.

(iii) Garton et al. [204] applied $4Bi_2O_3$–V_2O_5 fluxes for growing β-Ga_2O_3 crystals. The starting material consisting of the flux (84 g Bi_2O_3 + 8.5 g V_2O_3) and Ga_2O_3 (7 g) was loaded into a 50 ml Pt crucible with a tight lid, wherein the melt volume was not more than half of the crucible volume. Each crucible with the starting material was embedded in an alumina block, and any free spaces were filled up with powdered alumina. Such arrangement was heated up in a muffle furnace to T_{MAX} = 1300°C at which it was kept for 14 h. Then the furnace was cooled down to T_F = 900°C. The obtained crystals were colorless and transparent plates with an edge length up to 6 mm.

(iv) Chani et al. [238] doped β-Ga_2O_3 crystals with Cr and Er that were grown from the B_2O_3–Bi_2O_3 flux. The starting material consisted of 74.7 mol.% of B_2O_3, 7.78 mol.% of Bi_2O_3 (flux),

12.99 mol.% of Ga_2O_3, 4.38 mol.% of Er_2O_3, and 0.15 mol.% of Cr_2O_3 (the latter two were dopants). The starting material was loaded into a 50 ml Pt crucible, which was then heated to T_{MAX} = 1300°C for 12 h and kept at this temperature for 5 h. In the next step, the furnace temperature was decreased to T_{F1} = 1120°C for 0.5 h. That temperature was the starting point for crystal growth, during which the temperature was slowly decreased to T_{F2} = 600°C for 204 h. In the last step, the crucible was cooled down to RT for 10 h. The crystals were separated from the solidified flux with hot HNO_3 and the aid of ultrasonic bath. The obtained crystals were mostly bright green needles or platelets. The segregation coefficient of Cr^{3+} in the crystals was found very large, about 10, due to an ionic radius similar to Ga^{3+}, while those of Er^{3+} as well as flux impurities B^{3+} and Bi^{3+} were below unity as the result of large deviation of those ionic radii from that of Ga^{3+}.

(v) Schieber et al. [239] used PbO–PbF_2 or MoO_3 fluxes and $Sm_3Ga_5O_{12}$ garnet phase as a Ga source. At 900–1300°C and above 1050°C in the case of PbO–PbF_2 and MoO_3 fluxes, respectively, the garnet phase decomposes into oxides Sm_2O_3 and β-Ga_2O_3.

$$2Sm_3Ga_5O_{12} \rightarrow 3Sm_2O_3 + 5\beta\text{-}Ga_2O_3 \quad (4.9)$$

yielding bulk single crystals of β-Ga_2O_3.

The crystals were grown in Pt crucibles with lids to minimize flux vaporization.

The main growth conditions and obtained results are summarized in Table 4.3.

4.5.3 Verneuil

The Verneuil method was practiced for growing bulk β-Ga_2O_3 crystals only in the 1960s. Details of a growth furnace was described by Chase [207]. It consisted of a vertical burner (an oxyhydrogen torch), a powder hopper to supply a Ga_2O_3 powder to the burner, and a pedestal for mounting a rod onto which β-Ga_2O_3 crystal was grown. The hopper was equipped with a variable speed mechanism for controlling feed rates of powder. The burner defining a growth furnace was formed by a central tube carrying feed powder and

Table 4.3 Summary of growth conditions and results obtained by the flux techniques

Solvent	Growth temperature $T_{MAX} \rightarrow T_F$ [°C]	Cooling rate [K/h]	Crystal shape	Maximum single crystal size [mm]	Crystal coloration	Impurities	Reference
$PbO + B_2O_3$	1300 → 915	2	Plates	5 at edge	Colorless	—	202
$Bi_2O_3 + B_2O_3$	1300 → 1120 → 600	2.5	Needles Plates	—	Green	Bi, B Cr[(a)], Er[(a)]	238
$PbF_2 + PbSO_4$	1225 → 872	4	Prisms	16 at edge	Colorless	Pb = 0.07–0.2 wt.%	203
$V_2O_5 + Bi_2O_3$	1300 → 900	5	Plates	Up to 6×1	Colorless	—	204

[(a)]Intentional dopant

oxygen and by 20 tubes carrying hydrogen surrounding the central tube. Oxygen was also supplied into the space between the hydrogen tubes. Such arrangement improved gas mixing and more uniform radial temperature distribution across the flame. The growth furnace was 1.5 inch in diameter and 9 inch in length. It was found that using the ε-phase Ga_2O_3 powder as the starting material had a tendency to poor flow and strong vaporization in the flame. The obtained crystals were filled with bubbles. In next experiments, the ε-phase Ga_2O_3 powder was preliminary converted into the β-phase by sintering the powder at 1250°C for 18 to 24 h. The sintered β-phase Ga_2O_3 flowed better and was more stable in the flames. The crystals were grown on pedestals formed by polycrystalline alumina rods. First, a sintered cone of Ga_2O_3 was formed on the alumina rod, the tip of which was then melted by an increased gas flow. During the seeding process, several crystal grains were formed, but quickly a grain with a favorable orientation outgrew the others. Next the crystal diameter was enlarged by a further increase in the gas flow and powder supply. The most optimum flame composition was slightly oxidizing, obtained from inner and outer O_2 flow of 1 and 7.5 l/min, respectively, and H_2 flow of 14 l/min. The obtained crystals were about 1 cm in diameter and about 2.5 cm in length. The growth axis was typically parallel or close to the *b*-axis. The crystals had exhibited fine laminar twinning in the (100) plane.

Doping Verneuil-grown β-Ga_2O_3 crystals with Cr and Fe was mentioned by Tippins [240]. The Cr-doped β-Ga_2O_3 crystals weighted several grams; they were green and had a strong tendency for twinning. Lorenz et al. [208] concluded that crystals grown under oxidizing conditions were electrically insulating, while under reducing conditions were semiconducting with a light-blue color, which could be converted into insulating state after annealing in an oxidizing atmosphere at high temperatures for several hours.

4.5.4 Optical Floating Zone (OFZ)

4.5.4.1 Crystal growth

The OFZ method has been often used to grow bulk β-Ga_2O_3 crystals for basic research purposes. Although the OFZ method provides small-diameter crystals of not good structural quality, it enables fast and relatively cheap experiments particularly for checking the impact of growth atmosphere, doping, and annealing on electrical and optical properties of the obtained crystals.

The OFZ technique utilizes a feed rod and a seed rod aligned underneath with the feed rod. The feed rods are prepared from a high purity (4N–6N) Ga_2O_3 powder by cold pressing at 70–450 MPa and sintered at 1300–1600°C in air or O_2 for 10–24 h [218, 219, 221, 241, 243, 249, 254–263]. If doping is intended, an oxide of a required dopant is mixed prior pressing with the Ga_2O_3 powder at a desired concentration. As the seed rod, a single crystal is typically used. The heat source is light generated from two or four halogen lamps (e.g., of 1.5 kW each), which is focused onto the Ga_2O_3 feed rod tip by ellipsoidal mirrors. The lack of a metal crucible enables the use of any oxygen concentration in a growth atmosphere, which is typically a mixture of N_2 + O_2 enabling a control of oxygen concentration in the range 0–100 vol.%, or dry air. A total pressure in the growth chamber is 1 atm, but also 2 atm was reported.

The seeding is initiated by melting (with focused light) the tip of the rotating feed rod and bringing the seed rod rotating in an opposite direction into a contact with the molten tip of the feed rod. Next, both feed and seed rods are translated downward at fixed position of the lamps (alternatively the lamps are translated at fixed position of the rods). Once the growth is completed, the growth furnace is slowly cooled down to RT. The rotation and translation rates are typically 15–20 rpm and 1–15 mm/h [218, 219, 221, 241, 243, 249, 254–263], respectively. In such configuration, the crystal is grown on top of a seed rod, enabling the use of different seed orientations, also those perpendicular to the cleavage planes. Indeed, growth directions along the main crystallographic directions <100>, <010>, and <001> were successfully reported [221, 241, 252]. The OFZ-grown β-Ga_2O_3 crystals are cylindrical with the diameter and length of 5–8 mm and 30–70 mm, respectively. The crystal coloration depends mainly on oxygen partial pressure in the growth atmosphere and impurity/dopant concentration. Examples of β-Ga_2O_3 crystals obtained by the OFZ method are shown in Fig. 4.10. Although the crystal diameter is typically small, a larger diameter (25 mm) was also demonstrated by Villora et al. [221]. The crystals grown by the OFZ technique are subjected to relatively high temperature gradients arising from a small melt volume that is solidified, and from a furnace design. As a result, the expected structural quality of the crystals is not high, which was, indeed, not studied in detail. Occasionally reported values of the full width at half maximum (FWHM) of the rocking curve were, e.g., about 320 [221], 100 [242], and 22–36 [243] arcsec.

The purity of the obtained crystals is relatively high (no contamination from a metal crucible), but still it depends on the purity of the Ga_2O_3 starting material and the growth atmosphere, as well as on the thermal insulation type. Intentional doping of β-Ga_2O_3 is widely practiced with the use of the OFZ method to tune or induce electrical, optical, or magnetic properties of the obtained crystals. The OFZ-grown β-Ga_2O_3 crystals were doped with Sn [218, 244, 258, 264], Si [222, 245–247], Ge [219], Cr [248–250], Ce [251], Ti [219, 250], Mn [249, 250], In [252], Mg [253], Co, Cu, Ni, Fe [250].

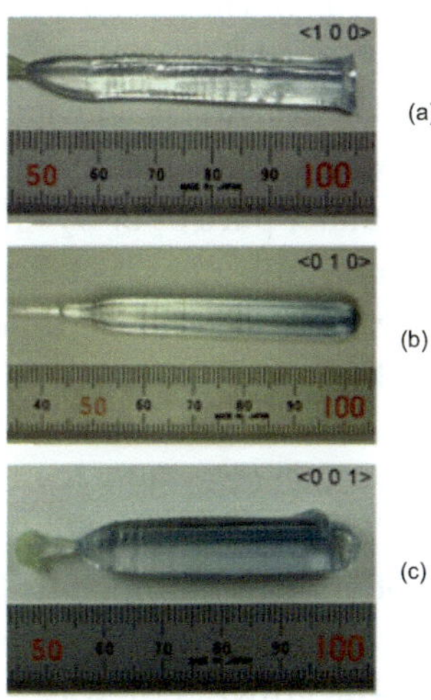

Figure 4.10 β-Ga_2O_3 crystals grown along the crystallographic axis <100> (a), <010> (b), and <001> (c). Reprinted from Ref. [221], Copyright 2004, with permission from Elsevier.

4.5.4.2 Examples

(i) Vasyltsiv et al. [217] grew β-$Ga_{2-x}In_xO_3$ (x = 0–0.2) crystals of 6–8 mm diameter and 30 mm length. The color of the crystals varied from light blue for β-Ga_2O_3 to yellow for β-$Ga_{2-x}In_xO_3$. The crystals were grown in the [010] or [100] crystallographic direction. With x = 0.2, the crystals still maintained the monoclinic system. Indium atoms substitute gallium atoms in octahedral sites.

(ii) In the growth experiments by Ueda et al. [218], the feed rods were prepared from 4N Ga_2O_3 powder by pressing in a cold isostatic press at a pressure of 280 MPa and sintering at 1300°C for 16 h. In the case of Sn-doping, 4N SnO_2 powder was added to Ga_2O_3 powder prior pressing and sintering. The resulted rods were about 5 mm in diameter and 60–70 mm in length. The light source consisted of 3.5 kW halogen lamps. The rotation rate of the feed rods was 17 rpm, while the translation speed of growing crystals was 15 mm/h. As a growth atmosphere, a mixture of N_2 and O_2 was used with the O_2/N_2 ratio of 0.25–0.75. The gas flow rate was 0.2 m^3/h, or in the case of no gas flow, the total pressure was 1 atm. For $O_2/N_2 < 0.25$, the crystal growth became unstable.

The undoped crystals were colorless and transparent or pale blue when grown under pure O_2 or $O_2 + N_2$, respectively. When doping with 3 mol.% SnO_2, the crystals were blue even when grown under pure O_2. The Sn concentration in a crystal sample was only 0.05 mol.%, which is likely due to the evaporation of Sn-containing species at high temperatures. Indeed, p(SnO) at MP of Ga_2O_3 is very high even at high $p(O_2)$.

(iii) Tomm et al. [219] prepared the feed rods by mixing Ga_2O_3 and GeO_2 or TiO_2 powders acting as dopants (all of 4N purity), and next by pressing and sintering the mixed powders in air at 1500°C for 10 h. The growth rate was 5 mm/h with counter-rotation of the feed and seed rods. As the growth atmosphere, dry air, pure O_2, or a mixture of 90% N_2 + 10%O_2 for undoped and dry air for doped crystals was used. The obtained crystals were transparent and colorless or pale purple when undoped and doped with Ge or with Ti, respectively.

(iv) Villora et al. [221, 254–257], Yamaga et al. [258], Shimamura et al. [259], and Lovejoy et al. [249] used an OFZ growth station equipped with four halogen lamps and ellipsoidal mirrors. They prepared feed rods from 4N Ga_2O_3 powder by cold pressing and sintering at 1450–1500°C for 10 h in air. The undoped β-Ga_2O_3 crystals were grown with the rate of 1–5 mm/h, under the atmosphere consisting of a mixture of $N_2 + O_2$ at different ratio and constant gas flow of 500 ml/min. The total pressure was either 1 or 2 atm. The authors mentioned that even a small concentration of O_2 in the growth atmosphere or overpressure significantly reduced the evaporation of gallium suboxides. At 1 atm pressure, the crystals could not be obtained under pure N_2, but they could under 2 atm. As seeds, ceramic rods and previously grown crystals with different crystallographic orientations were used. It was concluded that ceramic seeds produced crystals with cracks and twins, and the growth orientation was close to the *b*-axis. The crystals grown with the use of crystal seeds in the orientation parallel to one of the cleavage planes (100) and (001) were straight cylinders with two flat faces on the crystal sides, which were associated with the cleavage planes, and with striations on other round sides. The crystal seeds with the orientation perpendicular to one of the cleavage planes resulted in acentric and unstable growth, due to the weak bonding character of the cleavage planes {100} and {001}. The possibility of obtaining crystals was concluded for main crystallographic directions <100>, <010>, and <001>. Crystals grown along <010> and <001> directions had on the side parallel faces corresponding to the (100) cleavage plane, while that grown along the <100> direction had faces associated with the (001) cleavage plane.

Although the typical crystal diameter and length were 6–8 mm and 40–50 mm, respectively (Fig. 4.11), a 1-inch-diameter and 2-inch-long crystal grown along the <100> crystallographic direction was also obtained. The FWHM of the rocking curve was quite broad, about 0.09° (about 324 arcsec). A broad and multiple peak structure suggests the presence of twins or grains.

The obtained crystals were light blue, but those obtained at 1 atm contained bubbles and had coloration gradient along the growth axis with less coloration at the beginning of the crystal and more coloration at the end part thereof, which was assigned to a longer annealing time of the less colored part. Indeed, annealing experiments at 900°C in O_2 for 6 days at 1 atm removed the blue coloration. On the other hand, growth at 2 atm and O_2 concentration over 5 mol.% led to a homogeneous coloration of the crystals. The crystals were also doped with Sn by adding to the feed rods SnO_2 powder at the concentration of 3 mol.%.

In other growth experiments, high-purity 6N Ga_2O_3 powder was used as a starting material for the feed rods, which at similar growth conditions produced colorless crystals. Indeed, 6N purity powder contained only 0.05–1.8 wt. ppm of Si versus 10–20 wt. ppm for 4N purity powder. That residual Si in the starting material, especially in the 4N purity powder, was sufficient to induce electrical conductivity, which caused blue coloration of the crystals due to free carrier absorption in the red part of the visible spectrum.

Further experiments were performed with Cr and Mn-doping, by mixing Cr_2O_3 and MnO_2 powders with the Ga_2O_3 powder for a feed rod preparation. The undoped, Cr-doped, and Mn-doped crystals were colorless, greenish, and pink, respectively. The Mn-doped crystals showed a color gradient from the clearest at the beginning to most pink at the end of the crystal (purifying effect of the float zone method—the undissolved atoms are swept out of the crystal).

(v) A growth station used by Ohira et al. [241, 260] and Suzuki et al. [243] utilized halogen lamps of 1.5 kW and a double ellipsoidal infrared furnace comprising a quartz tube. Feed rods of 8–10 mm diameter and 40–50 mm length were prepared from 4N Ga_2O_3 powder, which was isostatically or hydrostatically pressed at 450 MPa or 80 MPa, respectively, and then sintered at 1600°C or 1500°C in air for 20 h or 10 h, respectively. For doping with Sn, SnO_2 was added to the Ga_2O_3 starting material at the concentration of 2–10 mol.%. The growth rate was 5 and 10–15 mm/h for the undoped and Sn-doped crystals (to minimize the evaporation of SnO), respectively, while the rotation of the seed and feed rods was 20 rpm in opposite directions. The growth atmosphere was dry air with the flow of 50 ml/min. The growth direction was either <100> or <001>. The obtained crystals of about 8 mm diameter and 30 mm length were colorless and transparent or bluish whether undoped or doped with Sn. Despite high Sn concentration in the starting material, the resulted Sn concentration in the crystals was 20–70 wt. ppm only. The surface roughness of CMP polished wafers was 0.33 nm rms, while the FWHM of the rocking curve was between 43–162 and 22–36 arcsec for Sn-doped and undoped β-Ga_2O_3 crystals.

(vi) A growth station used by Zhang et al. [261–263] was equipped with four halogen lamps, which were translated with respect to feed and seed rods. The feed rods of 6–7 mm diameter and 70–100 mm length were prepared from 5N purity Ga_2O_3 powder by cold pressing at 70 MPa, and sintering at 1400–1500°C in air for 10 h. The growth atmosphere was dry air flowing at the rate of 0.2 m^3/h. In the case of doping with Sn^{4+} or Cr^{3+}, 5N SnO_2 or 4N Cr_2O_3 powder was added to the Ga_2O_3 powder prior pressing at the concentration of 5 and 10 mol.%, or 0.1 at.%, respectively. The growth and rotation rates were 5–10 mm/h and 15 rpm, respectively. The seeds were prepared from crystals along the <100> direction.

The undoped crystals of diameter 6 mm and length 20 mm were colorless; Sn-doped crystals were pale blue (5 mol.%) or deep blue (10 mol.%), while Cr-doped were pale green. The Sn concentration in the obtained crystals was much smaller than the initial concentration in the feed rods: 0.089 and 0.092 mol.% for 5 and 10 mol.% doping, respectively. After annealing Sn-doped samples at 1000°C in air for 36 h, the blue coloration faded. The authors claim that after such annealing, the samples turned into electrical insulators. The annealed Cr-doped crystals at 1000°C in air for 72 h showed a bit higher absorption in the short wavelengths as compared with the as-grown crystals. The FWHM of the rocking curve of the undoped crystal sample was about 100 arcsec.

Examples of growth conditions, operating parameters, and the resulted crystal size and coloration are collected in Table 4.4.

Table 4.4 Selected growth conditions for growing pure β-Ga_2O_3 single crystals by the OFZ method, as well as crystal size and coloration

O_2/N_2 flows [l/h]/[l/h]	Feed rod preparation: pressure/ temperature/ time [MPa]/[°C]/[h]	Rotation rate/ growth rate [rpm]/ [mm/h]	Crystal diameter/ length [mm]/ [mm]	Crystal coloration	Ref.
40–86/ 160–114	280/1300/16	17/15	5/60–70	Pale blue	218, 264
200/0				Colorless	
15/15	500/1450/10	20/1–5	6–8/40–50	Pale blue	221, 265
0.63/2.34	450/1600/20	20/5	8/30	Colorless	241
42/158	70/1500/10	15/5–10	6/20	Colorless	261

4.5.5 Edge-Defined Film-Fed Growth (EFG)

In the EFG method, a β-Ga_2O_3 crystal is pulled up from the melt contained on a die top surface, the cross section of which defines a cross section of the growing crystal. The melt is transferred from a crucible to the die top surface through capillary forces. Both the crucible and the die are made of Ir, which are heated inductively by an RF coil. The cross section of the die top surface is rectangular; therefore, no crystal rotation is applied, while the growth rate is relatively high, up to about 10 mm/h. The growth direction was along the <010> crystallographic axis, parallel to both cleavage planes {100} and {001}, while the growth atmosphere consisted of 98% N_2 + 2% O_2 [232], or 50 vol. % Ar + 50 vol.% CO_2 [266]. The resulted crystals were blue or colorless. As the starting material, typically 5N Ga_2O_3 powder was used, which combines a relatively low price and high purity. To increase the electrical conductivity of β-Ga_2O_3 crystals, intentional doping with Sn or Si doping was used, by mixing SnO_2 or SiO_2 powders with the Ga_2O_3 starting material.

During the growth of β-Ga_2O_3 crystals by the EFG method, Aida et al. [230] concluded the importance of neck formation after seeding and before widening the crystal to a final width, which was carried out by a high seeding temperature and high growth rate, that was decreased to a fixed value for the rest of the growth. The widening of the seed (shoulder part) and the width of the crystals was controlled via crucible temperature. The relation between the neck cross section and seeding temperature is shown in Fig. 4.11a. High enough seeding temperature and small cross section of the neck always led to the formation of a single crystal (Fig. 4.11b). Otherwise, the obtained material was in the form of polycrystals. When considering the cooling rate during the widening of the growing crystal, it could be small to very high, between 18 and 120 K/h, if both conditions, i.e., the seeding temperature and a small cross section of the neck, are met for singe crystalline material. If not, a polycrystal was obtained even at low cooling rate.

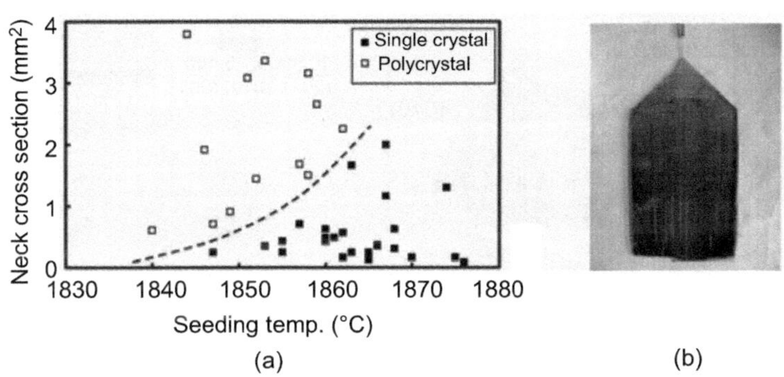

Figure 4.11 (a) Neck cross sections of the grown crystals plotted as a function of seeding temperature indicating a border (dashed line) separating conditions at which single crystals (black squares) or polycrystals (open squares) were obtained; (b) as-grown β-Ga_2O_3 single crystal slab. Reprinted with permission from Ref. [230], Copyright 2008, The Japan Society of Applied Physics.

Etching (100)-oriented crystal samples (in 10 wt.% KOH solution for 90 min at 90°C) of the early crystals revealed that the etch pit density (EPD) in the crystal obtained with much thinner neck (0.16 mm^2) was lower as compared to the crystal with a thicker neck (0.63 mm^2), 5×10^4 versus 4×10^5 cm^{-2}, respectively. The same relates to the FWHM of the rocking curve, which was 75 and 160 arcsec for the crystals with thinner and thicker neck, respectively. The crystals reported by Mu et al. [266] had the FWHM of the rocking curve of about 43–55 arcsec.

Different cross sections of the die top surface were used, such as 4 × 25 [266], 3 × 50 [230], 2.4 × 60, 18 × 60, and 6 × 110 mm × mm [2, 232] providing crystals with such cross sections and length of about 2–4 inch. The advantage of the EFG method is a large side surface enabling large wafers with the orientation defined by that surface, which is obtained by the rotation around the [010] direction, typically ($\bar{2}01$) or (001). Other wafer orientations (010) and (100) can be prepared from the cross section and smaller side surface of the crystal slab. A crystal slab with dimensions of 18 (thickness) × 60 (width) × 55 (length) mm^3 with indicated surface orientations and wafers prepared from the crystal slabs by CMP after annealing in N_2 at 1450°C for 6 h are shown in Fig. 4.12a,b, respectively.

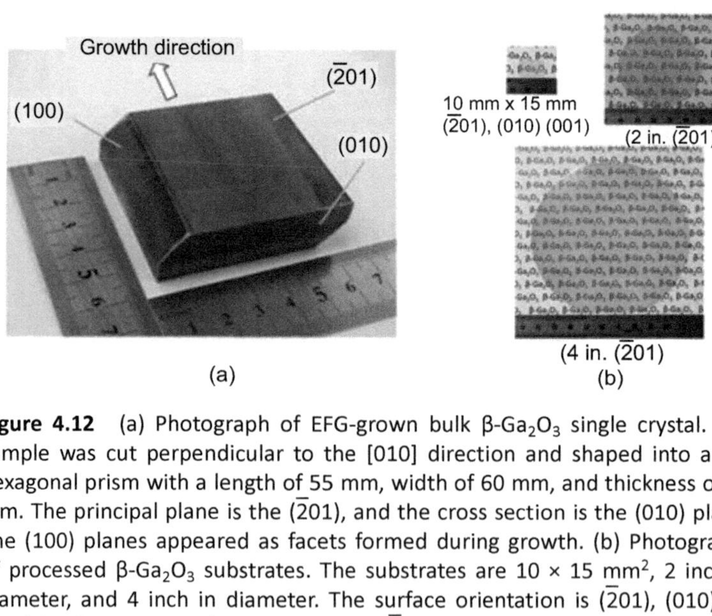

Figure 4.12 (a) Photograph of EFG-grown bulk β-Ga_2O_3 single crystal. The sample was cut perpendicular to the [010] direction and shaped into a flat hexagonal prism with a length of 55 mm, width of 60 mm, and thickness of 18 mm. The principal plane is the ($\bar{2}01$), and the cross section is the (010) plane. The (100) planes appeared as facets formed during growth. (b) Photographs of processed β-Ga_2O_3 substrates. The substrates are 10 × 15 mm^2, 2 inch in diameter, and 4 inch in diameter. The surface orientation is ($\bar{2}01$), (010), or (001) for the 10 × 15 substrates but only ($\bar{2}01$) for the 2- and 4-inch substrates. Reprinted with permission from Ref. [232], Copyright 2016, The Japan Society of Applied Physics.

Different defect types were identified in the EFG-grown crystals. One type is twin boundaries, which can be easily formed during the shouldering step (Fig. 4.13a). Once they are formed, they continuously propagate along the crystal. The shouldering step is, therefore, of high importance in the EFG method. The twin boundary is parallel to the (100) plane, which is the mirror plane for the ($\bar{2}$01) and ($\bar{1}$0$\bar{1}$) planes. They are easily visible under cross polarizers in the direction parallel to the [010] direction, as shown in Fig. 4.13b.

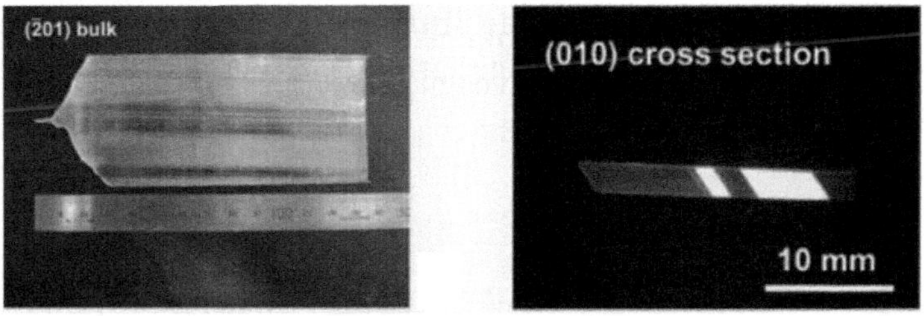

(a) (b)

Figure 4.13 (a) Photograph of EFG-grown crystal with twin boundaries and (b) polarization microscopy image of (010) cross section of another sample, including several twin boundaries. Reprinted with permission from Ref. [232], Copyright 2016, The Japan Society of Applied Physics.

Other types of structural defects are differently shaped pits revealed by etching CPM-polished crystal samples in H_3PO_4 at 130–140°C for up to 420 min [232, 267–269]. On the ($\bar{2}$01) plane, the pits typically extend in lines along the [010] direction with the density of about 10^4 cm^{-2} and the distance between the pit line of 20–100 μm (Fig. 4.14a). The pits relate to multiple edge dislocations, which are not exactly normal to the ($\bar{2}$01) plane, but slightly tilted toward the [102] direction. On the other hand, the etch pits on the (010) plane extend along the (100) plane (Fig. 4.14b).

More detailed inspection of the pits on the (010) plane revealed different types of the pits shown in Fig. 4.15. They are elongated grooves (type G), which convert with etching time into other types of the pits: A, B, C, and D. The length of the grooves ranges between 50 and 1200 nm in the [001] direction; the width is about 40 nm in the [100] direction, while the depth is about 30–100 nm in the [010] direction. The sidewalls and end walls of the grooves are (100) and likely (001) planes, which are thermodynamically stable. The grooves seen on the surface are hollow nanopipes in the crystal volume. Three possible reasons for the formation of these hollow defects (void-related) were proposed [269]: (i) stacking faults on the (100) plane due to multiple oxygen vacancies, (ii) penetration of tiny bubbles at the liquid–solid interface during crystal growth, and (iii) local melt-back at grown-in defects (e.g., screw dislocations). The E- and F-type etch pits do not change their shape nor position with etching time; they just increase in size. Both types of pits have a dislocation as its core in the [010] direction. F-type pits are strongly related to the leakage current in SBDs [268]. The grooves may also lead to an increase in the leakage current and thus a decrease of the breakdown voltage [267].

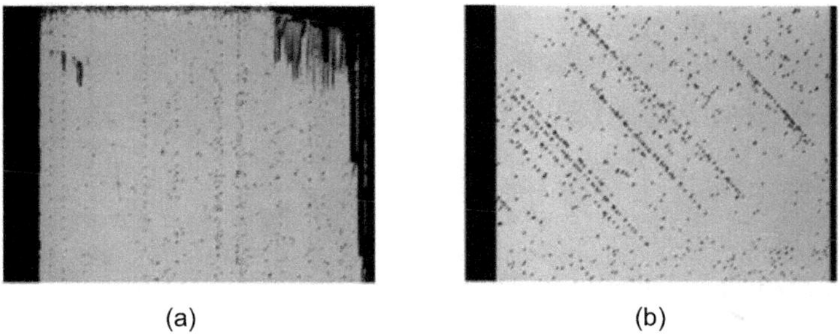

Figure 4.14 Optical micrographs of etch pits on a β-Ga_2O_3 crystal surface after etching by H_3PO_4 at 130°C for 420 min; (a) on the $(\bar{2}01)$ plane, and (b) on the (010) plane. Reprinted with permission from Ref. [269], Copyright 2016, The Japan Society of Applied Physics.

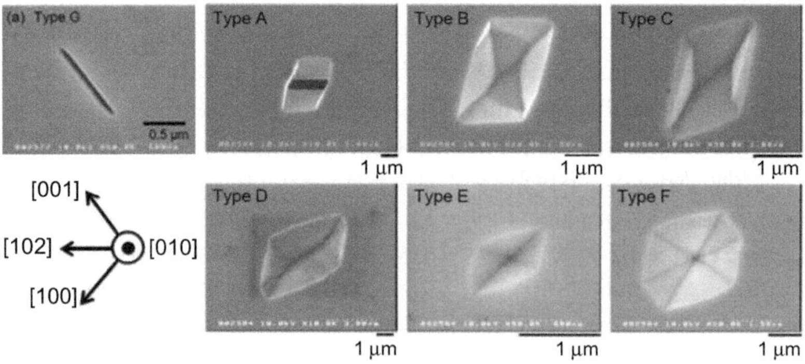

Figure 4.15 SEM images of seven etch pits of different shapes on β-Ga_2O_3 (010) single crystal surface. The etch pits were sorted into seven groups (types A, B, C, D, E, F, and G) according to shapes of the etch pits. Reprinted with permission from Ref. [268], Copyright 2016, The Japan Society of Applied Physics.

The revealed density of the dislocation-related pits was 10^3–10^4, 10^3, and 10^3 cm^{-2} for $(\bar{2}01)$, (010), and (001) planes, respectively [232]. At $(\bar{2}01)$ and (010) planes also line-shaped pits were observed with a density of one order of magnitude lower than the density of dislocation-related pits. The lowest value of the FWHM of the rocking curve was 17 arcsec on the $(\bar{2}01)$ plane. Crystals of poorer structural quality revealed the EPD of 10^6–10^7 cm^2 [270].

The undoped EFG-grown β-Ga_2O_3 crystals [232] had a residual Si concentration between 1.8 (central part) and 2.3 wt. ppm (near the crystal surface) as measured by the glow discharge mass spectrometry (GDMS), which corresponds to a value of 10^{17} cm^{-3} according to SIMS measurements. In Si-doped crystals, the upper limit for Si concentration is about 10^{19} cm^{-3}. In Sn-doped β-Ga_2O_3 (for the SnO_2 concentration in the starting material between 0.006 and 0.07 mol.%), the Sn concentration in the crystals was 36 wt. ppm. Too high Si or Sn concentration in the crystals deteriorates their structural quality by the formation of structural defects. Also, Ir at a concentration of 4.9 wt. ppm was found near the crystal surface, due to the use of oxygen in the growth atmosphere causing Ir oxidation.

4.5.6 Czochralski

Although the growth of bulk β-Ga_2O_3 crystals obtained by the Czochralski method was mentioned or demonstrated earlier by Vasiltsiv et al. [216] and Tomm et al. [224], an extensive and comprehensive development of that material was done by Galazka et al. [8, 225–229, 271, 324, 349].

4.5.6.1 Crystal growth

Czochralski growth stations used for growing bulk β-Ga_2O_3 single crystals consist of a stainless-steel chamber, a growth furnace inside the chamber, a puller with translation and rotation mechanisms, a balance coupled with a pulling rod passing into the growth furnace, an RF generator operating at 10–30 kHz for powering an RF coil around the growth furnace, and a control unit coupled with the pulling and rotation mechanism, the balance, and the generator. The control unit utilizes a PID (proportional, integral, differential) controller enabling an automatic diameter control of the growing crystal. The growth furnace consists of an Ir crucible and an active Ir afterheater ensuring low temperature gradients, both inductively heated by the RF coil, and a thermal insulation surrounding the crucible and afterheater from all sides. An interior of the afterheater defines a free space for the growing β-Ga_2O_3 crystal. The thermal insulation is a combination of zirconia, alumina, and quartz.

The crucibles are made of high-purity iridium (3N) of either 40 or 100 mm in diameter for growing 20 mm and 2-inch-diameter crystals, respectively, defining the crystal-to-crucible ratio of about 0.5. Large crucibles had the aspect ratio less than 1 (shallow crucibles). The growth rate was typically between 1 and 2 mm/h, while the crystal rotation of 8–15 and 4–8 rpm for 20 mm and 2-inch-diameter crystals, respectively. The length of the crystals was up to 70 and 100 mm for both crystal diameters, with the crystallization ratio below 50%.

The 20 mm diameter β-Ga_2O_3 crystals were typically grown at relatively low O_2 concentration, with the use of the growth atmosphere $(1 - x)$Ar + $x$$CO_2$ vol.% with $x = 0.1$–1.0, or $(1 - y)$Ar + $y$$O_2$ vol.% with $y = 0.02$–0.05 under atmospheric pressure or overpressure of 7 atm. Two-inch-diameter crystals were grown, however, under much higher O_2 concentration supplied by the growth atmosphere $(1 - y)$Ar + $y$$O_2$ vol.% with $y = 0.08$–0.35. The growth of bulk β-Ga_2O_3 single crystals by the Czochralski method is discussed in detail further below in connection with thermodynamics, which is crucial when growing this compound from the melt.

The growth takes place on a crystal seed that is dipped into the melt and after equilibration it is slowly pulled up while rotating. The seed diameter is enlarged to a predefined cylinder diameter, that after reaching a predefined length is separated from the melt and slowly cooled down to room temperature. In the Czochralski method the growth proceeds from the exposed melt surface.

4.5.6.2 Pulling direction

The presence of two easy cleavage planes {100} and {001} of β-Ga_2O_3 limits the pulling direction along the <010> crystallographic direction (i.e., along the *b*-axis), which is parallel to both cleavage planes. In the Czochralski method, a crystal is pulled upward and if any of the cleavage planes was substantially perpendicular to the pulling direction, the seed could easily crack parallel to the cleavage plane under the crystal's weight and thermal stress induced in the seed due to the existing temperature gradients. In such a situation, a crystal would simply fall down into the melt. This is crucial particularly during the growth of crystals with diameters exceeding 1 inch. For the same reasons also, EFG utilizes the pulling direction along the <010> direction. Moreover, numerical simulations of stress distribution in a real Czochralski growth furnace and growth conditions revealed a twice smaller maximum stress values for the <010> direction as compared with the <001> direction (Miller et al. [272]). The situation is different in the case of the OFZ and Bridgman techniques, where a crystal is grown on top of the seed; therefore, there is no risk of disconnection of the growing crystal from the seed. As shown in Section 4.5.4, growth directions parallel to all main crystallographic directions were successfully demonstrated.

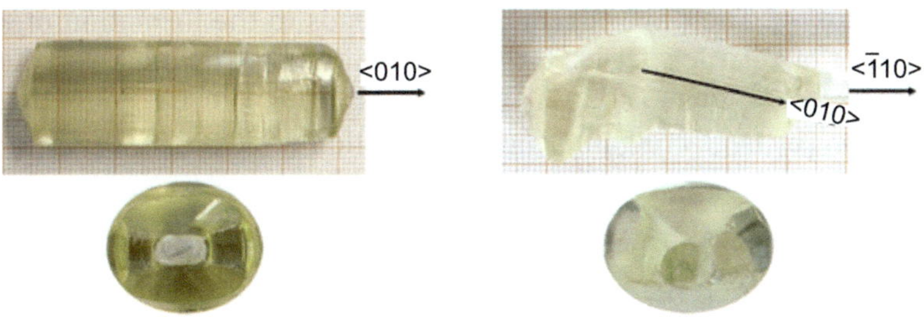

(a) (b)

Figure 4.16 Side and top views of bulk β-Ga_2O_3 single crystals grown by the Czochralski method along the <010> (a) and <$\bar{1}$10> (b) crystallographic directions.

Other than the <010> pulling directions, where the cleavage planes are inclined to the pulling direction, induced growth instabilities resulted from the growth kinetic. During growth along the <010> direction, the surface of a cylindrical crystal is stabilized by the cleavage planes that have lower formation energy as compared to the {010} plane. In other words, a deviation of the pulling direction from the <010> direction will lead to a crystal shape deviated from the straight cylinder. Figure 4.16 shows bulk β-Ga_2O_3 single crystals (side and top views) grown along the <010> (Fig. 4.16a) and <$\bar{1}$10> (Fig. 4.16b) crystallographic directions, respectively. The crystal grown along the <010> direction has a straight cylinder, which is kinetically stabilized by the cleavage planes {100} and {001}. As the result, its cross section is not circular, but a bit elliptical. However, the crystal pulled along the <$\bar{1}$10> direction had a tendency to grow along the <010> direction anyway. As the growth proceeded, the crystal was gradually approaching the crucible wall, which is the hottest part of growth furnaces. At a certain point, the crystal could not continue its journey to the crucible wall and returned toward the center of the crucible. The cross section of such a crystal is more elliptic and asymmetric. In conclusion, the most preferable growth direction of bulk β-Ga_2O_3 crystals for the Czochralski method is that parallel to the <010> crystallographic direction.

4.5.6.3 Impact of the oxygen partial pressure

Due to the high MP of Ga_2O_3, the growth of β-Ga_2O_3 crystals by the Czochralski method utilizes Ir crucibles. On the other hand, the thermal instability of Ga_2O_3 requires sufficient $p(O_2)$ in the growth atmosphere to decrease the Ga_2O_3 decomposition and resulted evaporation. The Ga_2O_3 decomposition leads to the formation of a number of volatile species: O_2, Ga_2O, GaO, and Ga. The most volatile species containing Ga is Ga_2O, which is formed in the following reaction (by neglecting intermediate reactions):

$$Ga_2O_3 \leftrightharpoons Ga_2O + O_2 \tag{4.10}$$

High $p(O_2)$ necessary to stabilize Ga_2O_3 contradicts the requirement for Ir that easily oxidizes, especially at low and moderate temperatures (below 1200°C). With Ir, only small O_2 concentration (up to about 2 vol.%) has been practiced. One of the solutions is the use of a growth atmosphere containing CO_2, which decomposes and provides an increasing O_2 concentration with temperature, according to the reaction:

$$CO_2 \leftrightharpoons CO + \tfrac{1}{2}O_2 \tag{4.11}$$

The growth atmosphere containing 10 vol.% of CO_2 and 90 vol.% Ar was used by Tomm et al. [224]. Such atmosphere provides at MP of Ga_2O_3 of about $p(O_2) = 0.002$ atm, which corresponds to about 0.2 vol.% of O_2. Indeed, such growth atmosphere enabled to obtain small single crystals (Fig. 4.17) of 1 cm in diameter and blue coloration; however, they had a tendency to grow in irregular shape.

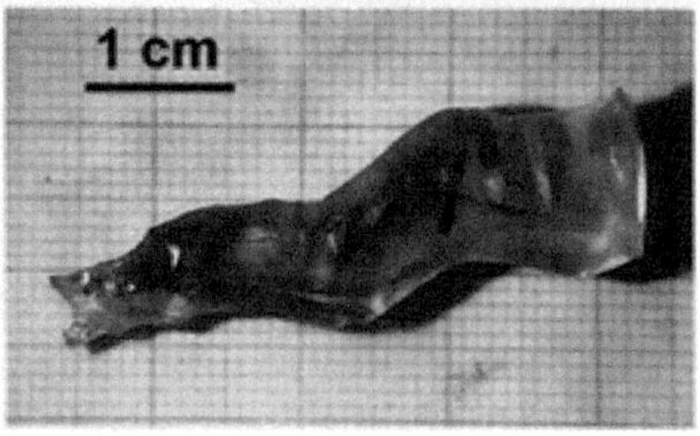

Figure 4.17 β-Ga_2O_3 crystal grown by the Czochralski method. Diameter = 1 cm, length = 3.5 cm.

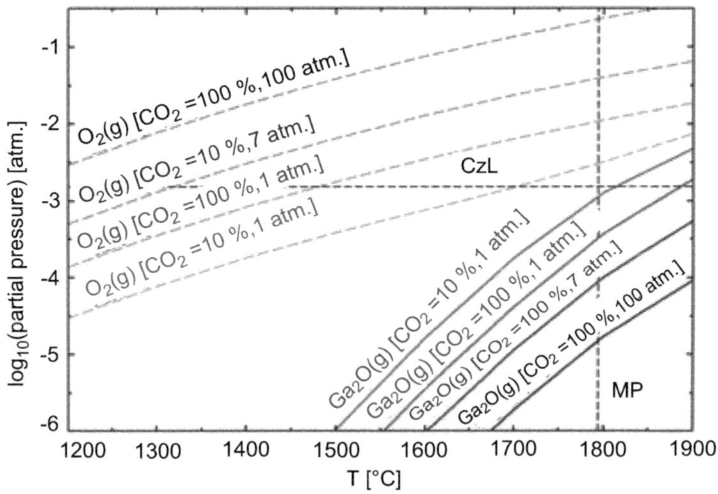

Figure 4.18 Partial pressure of the most volatile species Ga_2O (solid lines) during decomposition of Ga_2O_3 versus temperature at corresponding oxygen partial pressures (dashed lines). MP indicates the melting point of Ga_2O_3, while CzL is the limit of a partial pressure above which it is rather difficult to apply the Czochralski method.

Detailed study and further development of β-Ga_2O_3 growth by the Czochralski method with the use of the CO_2-containing atmosphere was conducted by Galazka et al. [225, 226]. The growth atmosphere was a mixture of $(1 - x)$Ar + $x CO_2$, where $x = 0.3–1$, and a total pressure of 1 or 7 atm. Additionally, a growth atmosphere containing Ar + 2% vol. O_2 was used for comparison as well. $p(Ga_2O)$ at different $p(O_2)$ obtained by applying CO_2-containing growth atmospheres at different pressures is shown in Fig. 4.18. From experiments, it was concluded that $p(Ga_2O)$ at the level of about 2 × 10^{-3} atm at the MP of Ga_2O_3 seems to be a limit of the Czochralski method (CzL), above which it is rather difficult to obtain a single crystal of an acceptable structural quality. Therefore, $p(Ga_2O)$ should be kept below that limit. Thermodynamic calculations clearly show that when using a neutral atmosphere (e.g., Ar or N_2) containing a few vol. ppm of residual O_2 (here 5 ppm), $p(Ga_2O)$ is above the limit line (2.8 × 10^{-3} atm) at the MP of Ga_2O_3, which is released from the Ga_2O_3 starting material during the decomposition. With the growth atmosphere of 90% Ar + 10% CO_2, $p(Ga_2O)$ is just about that limit, which enables to obtain only small crystals. Other CO_2-containing growth atmospheres (CO_2 = 30–100%) bring the $p(Ga_2O)$ below that limit line by almost one order of magnitude (100% CO_2, 1 atm). The atmosphere containing pure CO_2 provides at the MP of Ga_2O_3 a similar $p(O_2)$ as the growth atmosphere consisting of 98 vol.% Ar and 2 vol.% of O_2. Yet better results were obtained with the use of pure CO_2 under an overpressure of 7 bar. Further increase in the total pressure (e.g., 100 atm.) of CO_2 lowers $p(Ga_2O)$ to a level where the decomposition and evaporation of Ga_2O_3 is minor. Applying high pressure in the Czochralski method for high MP oxides is technically difficult and increases production costs of the crystals. Also, high pressure of CO_2 will result in carbon contamination, which affects electrical and optical properties of obtained crystals.

Figure 4.19 β-Ga_2O_3 single crystals obtained at the following growth atmospheres (from left-hand side): 70% Ar + 30% CO_2, 1 atm; 50% Ar + 50% CO_2, 1 atm; 0% Ar + 100% CO_2, 1 atm; 0% Ar + 100% CO_2, 7 atm.

The crystals obtained at different CO_2-containing growth atmospheres are shown in Fig. 4.19. The rough surface of the crystals is due to both decomposition of the crystals themselves and the growth of needle-shaped β-Ga_2O_3 crystals from the gas phase (by reaction of Ga_2O evaporated from the melt with O_2: $Ga_2O + ½O_2 \rightarrow$ β-Ga_2O_3) on a cooler β-Ga_2O_3 crystal surface, and by micro-cleaving of the surface due to thermal stress. Higher CO_2 concentration supplies higher O_2 concentration, which reduces $p(Ga_2O)$ and thus the growth from the gas phase. The crystal surface becomes smoother with a higher CO_2 concentration, while at 7 atm of CO_2 the crystal surface is entirely clear.

The advantage of using the CO_2-containing growth atmosphere is a reduction in Ir oxidation at low and moderate temperatures, which is in the form of solid IrO_2 that remains on the crucible. Alternatively, O_2 can be directly mixed with a neutral gas (e.g., Ar, N_2), but in practice the maximum of O_2 concentration in the combination with Ir is of about 2 vol.%, which is sufficient for growing β-Ga_2O_3 by the Czochralski method, but only when the evaporation is in quest.

Theoretically, also other growth atmospheres, such as NO or NO_2, could be considered to provide even higher O_2 concentration at high temperatures as compared to CO_2, but no crystal growth stations are adapted for using such toxic and aggressive gases, which are environmentally unfriendly.

So the evaporation of Ga_2O_3 can be easy minimized to the level that enabled the growth of β-Ga_2O_3 single crystals by the Czochralski method, simply by using oxygen concentration of about 0.2–2 vol.%. However, the formation of volatile species during Ga_2O_3 decomposition is not the only obstacle in growing bulk single crystals. During the decomposition of Ga_2O_3, metallic Ga is also formed, which is partly in the gas phase and evaporates, but partly remains in the Ga_2O_3 melt in the liquid phase, according to the reaction (by skipping intermediate reactions):

$$Ga_2O_3 \leftrightharpoons 2Ga(l, g) + 1\tfrac{1}{2}O_2 \qquad (4.12)$$

The presence of metallic Ga in the melt has many adverse effects on a growth run, such as eutectic formation between liquid Ga(l) and solid Ir(s) leading to a crucible damage, problems with seeding, and poor crystal quality (if obtained at all) due to the presence of Ga particles floating in the melt. The amount of Ga(l) in the melt depends on the melt volume; therefore, it is negligibly small during the growth of small-diameter crystals (below 1 inch), but it becomes crucial during the growth of larger crystals requiring larger melt volumes. Although a relatively small O_2 concentration (e.g., 2 vol.%) in the growth atmosphere is sufficient to significantly reduce the Ga_2O_3 decomposition to an acceptable level from the point of view of evaporation (although larger melt volumes in larger crucibles have a larger surface that promotes a higher evaporation rate), much higher O_2 concentration in the growth atmosphere is required to minimize the formation of metallic Ga in the melt with an increase in its volume. This is well illustrated in Fig. 4.20 showing the amount of Ga(l) in the Ga_2O_3 melt versus O_2 concentration during growth of 1-, 2-, 3-, and 4-inch-diameter crystals, which require 2, 18, 62, and 150 mol of Ga_2O_3, respectively, for the crucible aspect ratio that equals 1, and the crystal diameter, which is half of the crucible diameter.

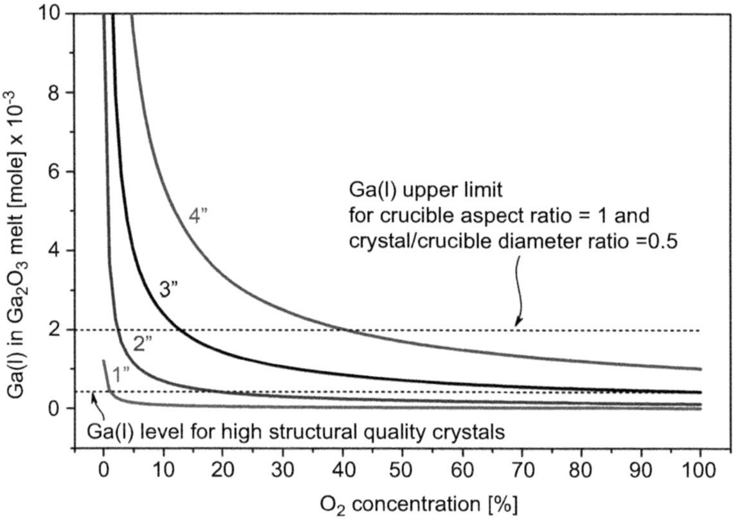

Figure 4.20 The amount of liquid Ga in the Ga_2O_3 melt versus O_2 concentration during the growth of 1-, 2-, 3-, and 4-inch-diameter crystals by the Czochralski method. Reprinted from Ref. [5], Copyright 2018, with permission from Elsevier.

It is clearly seen that in the case of 1-inch-diameter crystals, the amount of metallic Ga(l) in the melt is quite low and it is easy to obtain high structural quality crystals, as the amount of metallic Ga(l) is below the line indicating high-quality crystals even at small O_2 concentration. The situation drastically changes with increasing the crystal diameter (and thus the melt volume). When scaling up to 2-inch-diameter crystals, such small O_2 concentration as 2%, which is sufficient for 1-inch-diameter crystals, is rather the minimum at which a crystal can be grown, but the structural quality deteriorates due to high density of twin boundaries or the formation of a polycrystalline phase. Additionally, a large amount of metallic Ga(l) will form eutectic with Ir and may damage the crucible even after one single growth run. To decrease the Ga(l) amount to such low and save level as it is in the case of 1-inch-diameter crystals (about 1–2 vol.% of O_2), 2-inch-diameter crystals would require about 20 vol.% of O_2, 3-inch-diameter crystals pure O_2, while in the case of 4-inch-diameter crystals, such low level of Ga(l) will never be achieved. However, 4-inch-diameter crystals can still be grown, but at a higher level of Ga(l) in the melt, which must be kept below the upper limit. In large systems, the amount of Ga(l) in the melt can be reduced by using crucible with the aspect ratio of 0.5–0.8 (so-called shallow crucibles).

As mentioned earlier, only low O_2 concentrations are practiced in the combination with Ir, so the question arises how to combine the contradicting requirements to grow large-diameter β-Ga_2O_3 crystals: a high O_2 concentration to minimize the formation of Ga(l) in the melt, and a low O_2 concentration to minimize Ir oxidation. First, we need to look at the thermodynamics of the Ir–O_2 system. It is important to stress that at low and moderate temperatures, IrO_2 is formed:

$$Ir(s) + O_2(g) \rightarrow IrO_2(s) \quad (4.13)$$

which is in the solid phase and remains on the crucible surface. However, at high temperatures, IrO_3 is formed in favor of IrO_2:

$$Ir(s) + 1\tfrac{1}{2}O_2 \rightarrow IrO_3(g) \quad (4.14)$$

which is in the gas phase and evaporates.

The activities of both $IrO_2(s)$ and $IrO_3(g)$ (which is fugacity or partial pressure in the case of the gas phase) at different O_2 concentrations are shown in Fig. 4.21.

The activity of IrO_2 equals unity (meaning the solid phase is formed) at temperatures up to 1000–1200°C depending on the O_2 concentration. However, above those temperatures, the activity of IrO_2 drastically decreases, but instead, the fugacity of $IrO_3(g)$ increases. Even when using pure O_2, $p(IrO_3)$ is comparable or even lower than $p(Ga_2O)$ during the growth of small-diameter crystals (about 1 inch) at low O_2 concentration (1–2 vol.%). This means that the $p(IrO_3)$ will not be an obstacle in growing large Ga_2O_3 crystals at high O_2 concentrations.

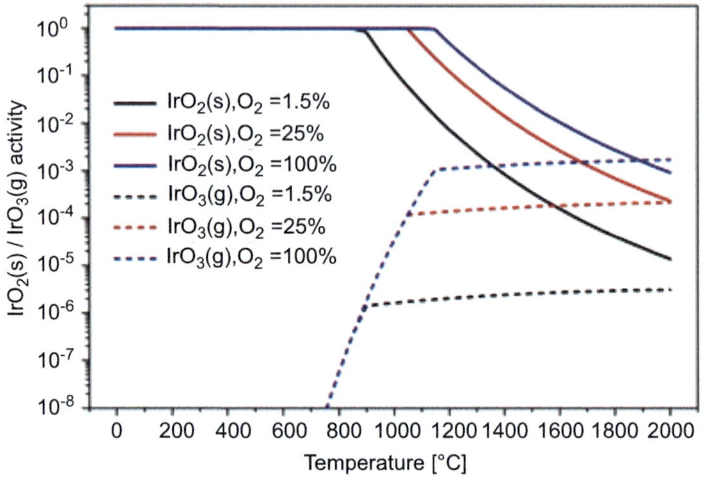

Figure 4.21 $IrO_2(s)$ and $IrO_3(g)$ activities versus temperature at different oxygen concentrations. Total pressure = 1 atm. Adapted from Ref. [228].

By combining the thermodynamics of Ir–O_2 and Ga_2O_3–O_2 systems, it is possible to design a scheme of oxygen supply to the growth atmosphere as shown in Fig. 4.22, which allows to use very high O_2 concentrations in the combination with Ir. This is done by using a low O_2 concentration in the growth atmosphere at low and moderate temperatures (between RT and transition point, TP), where $IrO_2(s)$ is formed, and high O_2 concentration at high temperatures (above TP), where $IrO_3(g)$ is formed.

During heating up of the growth furnace, at low and moderate temperatures (region (I) between RT and TP in Fig. 4.22), Ga_2O_3 is thermally stable, while Ir is sensitive to oxidation ($IrO_2(s)$); therefore, in that temperature range, O_2 can be used at small concentration, or even with no O_2 in the growth atmosphere at all. At higher temperatures (>TP, region (II)), Ga_2O_3 becomes thermally unstable (decomposition) and O_2 is required at concentration high enough to minimize the Ga_2O_3 decomposition. At the same time, $IrO_2(s)$ is not formed any longer, but instead $IrO_3(g)$ is formed, which simply evaporates. By a proper furnace design with low temperature gradients in a space of a growing crystal, $IrO_3(g)$ will be evacuated far from the melt and the growing crystal, which will next convert to IrO_2 and condensate at cooler areas of the furnace. So at higher temperatures, O_2 concentration increases with temperature until a predefined level at MP, depending on the crystal diameter to be grown (or melt volume). During crystal growth (region (III)), O_2 concentration can be maintained at the high predefined level or can further be increased/decreased to tune electrical properties of the growing crystal. During cooling down of the growth furnace, the O_2 concentration can be decreased in opposite manner as it was increased during heating up. Alternatively, during cooling down, O_2 can be maintained at high concentration, but this will lead to unnecessary Ir losses and possible contamination of the residual Ga_2O_3, what might be critical if the melt is planned to be reused for the next growth run. This technique allows to use high O_2 concentrations (including pure O_2) enabling large-diameter crystals of high structural quality with minimized crucible damage by the Ir–Ga eutectic formation and acceptable Ir losses. In the case of very large crystal diameter (4 inch), crystal growth may proceed at higher concentration of Ga(l) in the melt as compared with smaller-diameter crystals, but it still could be of good structural quality if Ga(l) is below the upper limit (Fig. 4.20) that can be achieved by both high O_2 concentration and crucible aspect ratio of 0.5–0.8.

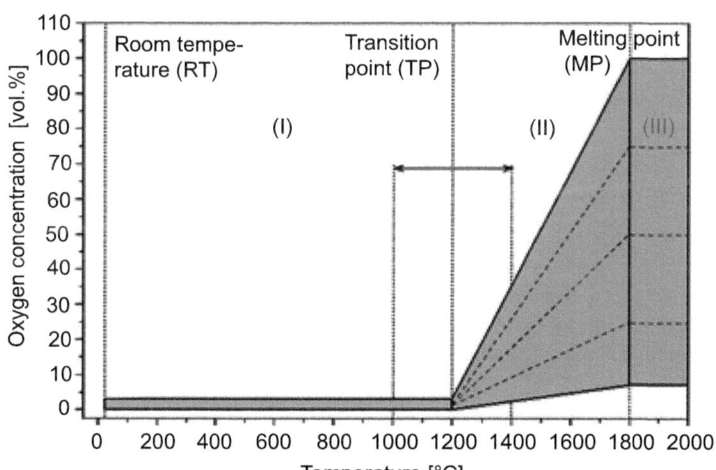

Figure 4.22 Scheme of oxygen supply to a growth atmosphere that enables to grow large-diameter β-Ga_2O_3 crystals by the Czochralski method. Regions (I), (II), and (III) indicate temperature ranges at which different oxygen concentrations are used. TP can indicate any temperature between 1000 and 1400°C as indicated by arrow, e.g., 1200°C. Adapted from Ref. [228].

O_2 can be supplied to the growth atmosphere in different ways. At low and moderate temperatures, the following gases can be used: Ar, N_2, CO_2, Ar or N_2 + CO_2, Ar/N_2 + O_2 (low), Ar/N_2 + O_2 (low) + CO_2. At higher temperatures, however, Ar/N_2 + O_2 must be used, as high O_2 concentration is required. The way how O_2 concentration increases in the growth atmosphere during heating up is not important, as this is an intermediate stage to achieve a high O_2 level around the MP of Ga_2O_3. O_2 concentration may, therefore, increase linearly, parabolically, or in a stepwise manner. An example of an O_2 concentration measured by a mass spectrometer in a growth chamber during heating up, growth, and cooling down of a 2-inch-diameter crystal is shown in Fig. 4.23.

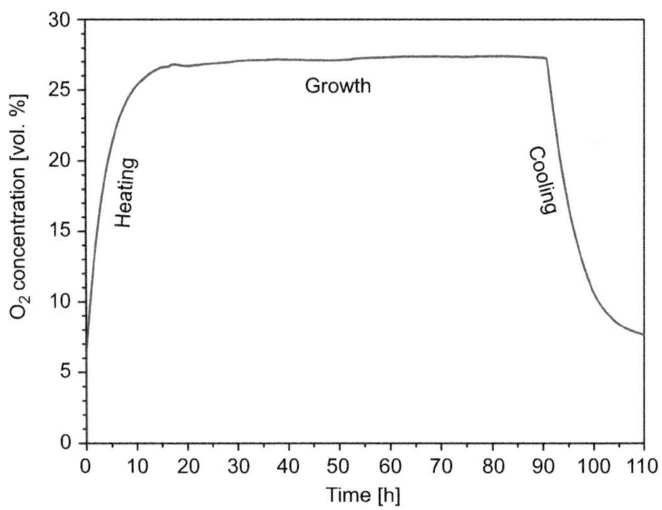

Figure 4.23 O_2 concentration profile measured by a mass spectrometer in a growth chamber during the growth of 2-inch-diameter β-Ga_2O_3 single crystal by the Czochralski method, which includes heating-up, growth, and cooling-down steps.

Total Ir losses per growth run depend on the duration of the growth run, including heating up, growth, and cooling down, and on O_2 concentration in the growth atmosphere. For the growth run lasted for about 80–100 h and O_2 concentration of 10–30%, the total Ir losses of unprotected crucibles are about 4–6 wt.% per growth run. In the case of pure O_2, Ir losses are higher and may reach the value of 10 wt.% in long runs. Therefore, the very high O_2 concentration is rather recommended for shorter growth runs, particularly for the Bridgman/VGF method and for protected crucibles.

The above-described considerations are fundamental and have an application to any melt growth techniques involving a noble metal crucible, particularly Ir or Ir alloys. These techniques include, in addition to the Czochralski method, the Kyropolous, all shaped techniques (EFG, Stephanov), Bridgman (vertical, horizontal), and VGF.

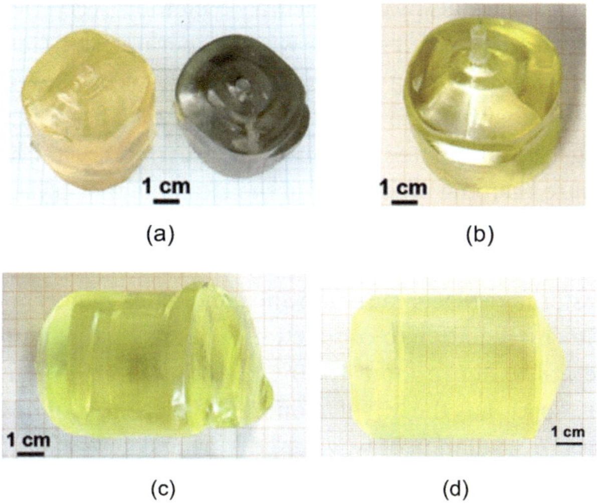

Figure 4.24 Two-inch-diameter β-Ga_2O_3 crystals obtained by the Czochralski method at high O_2 concentrations in the growth atmosphere: (a) >20 vol.% (left) and 8–10 vol.% (right); (b) >20 vol.%, doped with Al; (c) and (d) > 20 vol.% doped with Mg. Figure (b) reprinted from Ref. [229] and (c) reprinted from Ref. [5], Copyright 2018, with permission from Elsevier.

With low O_2 concentrations in the growth atmosphere (<2 vol.%), 2-inch-diameter β-Ga_2O_3 crystals of an acceptable quality cannot be easily obtained, as typically the obtained material is heavily twinned or simply it is a polycrystal. The reason is a high concentration of metallic Ga floating in the melt and then incorporated into the growing crystal. High O_2 concentration, above 5% (the best between 15 and 30%), enables obtaining 2-inch-diameter single crystals of high structural quality, as shown in Fig. 4.24. O_2 concentration in the growth atmosphere above 20 vol.% produces electrically insulating (or semi-insulating) crystals with a yellow tint, while lower O_2

concentration (8–10 vol.%) produces semiconducting crystals with a green tint (Fig. 4.24a). The crystals grown under O_2 concentration above 20 vol.% remain yellowish and electrically insulating whether doped with Al or Mg, as shown in Fig. 4.24a–d.

4.5.6.4 Free carriers and heat transport

Having thermodynamic issues understood and technically resolved, we can address other problems associated with the growth of β-Ga_2O_3 single crystals by the Czochralski method. A crystal is pulled out directly from the melt surface with temperature gradients above the melt surface enabling the liquid–solid phase transition. During that phase transition, the latent heat of crystallization is released, which must be dissipated. Because the growing crystal has lower temperature than the melt, most of the latent heat of crystallization will be transferred through the crystal and in small amount through the meniscus at the triple melt–solid–gas point. In solids, two types of heat transfer take place: radiation and conduction, wherein the ratio of heat flux by radiation to conduction depends on the temperature. At temperatures above about 1500°C, the dominating heat transfer is radiation. Ga_2O_3 has the MP of about 1800°C, so most of heat will be transferred by radiation at wavelengths around 1.38 μm (maximum of black body radiation) and longer, as shown in Fig. 4.25. In the case of semiconducting crystals, free carriers absorb radiation in the near-infrared (NIR) region. This is called free carrier absorption, which is reflected in the absorption spectra shown in Fig. 4.25. At a low free carrier concentration ($<10^{17}$ cm^{-3}), the absorption is minor and has a low impact on the radiative heat transport through the crystal. However, the absorption increases with the free carrier concentration, particularly at the level of 10^{18} cm^{-3} and above, and such high absorption has a high impact on the heat transport. The absorption spectra in Fig. 4.25 were taken at RT, but the absorption increases with temperature. For example, at 1400°C (dot line in Fig. 4.25), which is still far below the MP, the absorption in the NIR region increases by a factor of about 2. Moreover, even electrically insulating crystals at RT might be semiconducting at high temperatures during growth due to thermal excitation, where the free carriers will also induce some absorption, although smaller as compared with semiconducting crystals at RT. Indeed, high-temperature measurements of the absorption in crystals, which were electrically insulating at RT, showed the absorption typical for the free carrier absorption at temperatures greater than 1200°C.

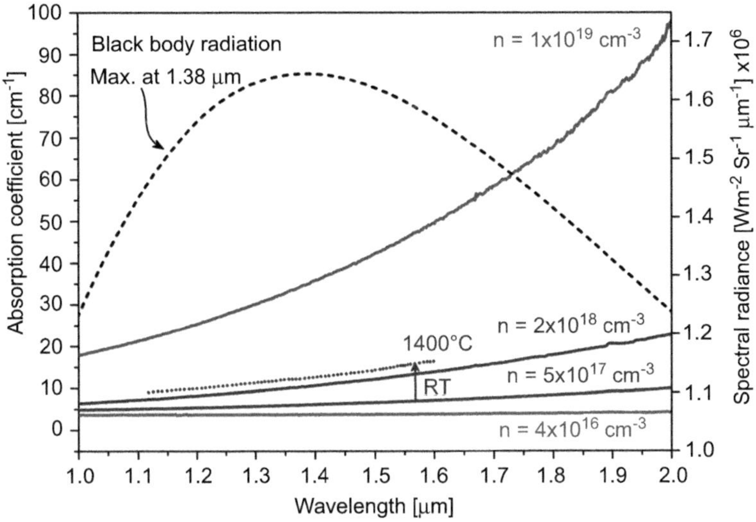

Figure 4.25 Absorption spectra of β-Ga_2O_3 crystals at different free carrier concentrations combined with the black body radiation spectrum at MP of Ga_2O_3. Reprinted from Ref. [226], Copyright 2014, with permission from Elsevier.

Now, if we combine the heat transfer via radiation with the free carrier absorption, we can easily conclude that heat will be absorbed in the growing crystal by the free carriers depending on their concentration and the crystal length. It means the heat will be less effectively dissipated from the growth interface as the growth proceeds and it will be accumulated in the crystal volume. This heat accumulation will flatten the interface (which is convex toward the melt at the beginning of the growth) and consequently lead to an inversion toward a concave shape. In this situation, the most effective region for the heat removal is the triple point at the crystal periphery contacting the melt through the meniscus. In the case of the automatic diameter control, the controller recognizes the interface inversion as a mass loss (melting back the crystal) and decreases the melt temperature. A combination of a more effective heat loss through the triple point and decrease in the melt temperature will force the crystal to growth laterally over the melt surface toward the crucible wall. Such deviation from the cylindrical shape is called a "foot." However, the crucible is the hottest part of the growth furnace and at certain point, the lateral growth will be stopped and returned toward its center. This effect, when combined with pulling the crystal upward and its rotation, will form a corkscrew or a spiral extending beyond the cylinder diameter. Such geometry effectively dissipates the latent heat of crystallization, because its large surface is exposed toward higher temperature gradients (it sees an upper part of the cooler insulation), and it has small enough thickness to accumulate heat. So the presence of the free carriers has a direct impact on the growth stability and the cylinder length. Such growth instability (spiral formation) is not observed in low-melting-point materials (e.g., some oxides, fluorides, halides, classical semiconductors, and compound semiconductors), in which conduction dominates over radiation. The spiral is a single crystal of high structural quality, often even better than that of the cylinder, but the usable volume is smaller, especially when large wafers are to be prepared.

The cylinder length of β-Ga_2O_3 crystals grown by the Czochralski method will, therefore, depend on the free carrier concentration and the crystal diameter. An impact of the free carrier concentration (at RT) on the cylinder length of 2 cm diameter crystals is demonstrated in Fig. 4.26. At low free carrier concentration ($\leq 5 \times 10^{17}$ cm^{-3}) or when the crystals are electrically insulating, the absorption is minor, and the latent heat of crystallization has no obstacles to be effectively dissipated through the growing crystal. As a result, the growth interface is convex toward the melt over the whole growth run and long cylinders can be grown (up to about 7 cm). With increasing the free carrier concentration ($\sim 10^{18}$ cm^{-3}), the absorption in the NIR region increases and the growth interface becomes concave toward the melt at smaller crystal length, about 5 cm. For very high electron concentrations ($\sim 10^{19}$ cm^{-3}), the spiral formation starts earlier, after about 1–3 cm of the cylinder. Long cylinders of highly conductive crystals cannot be easily grown by the Czochralski method, although it is possible when co-doped with Al (see Section 4.5.6.5).

In the case of large crystal diameters (≥ 2 inch), the optical thickness of the growing crystal is larger, and the spiral may start sooner as compared with the smaller diameter for the same free carrier concentration. Long cylinders can be obtained for the electron concentration $\leq 5 \times 10^{17}$ cm^{-3}.

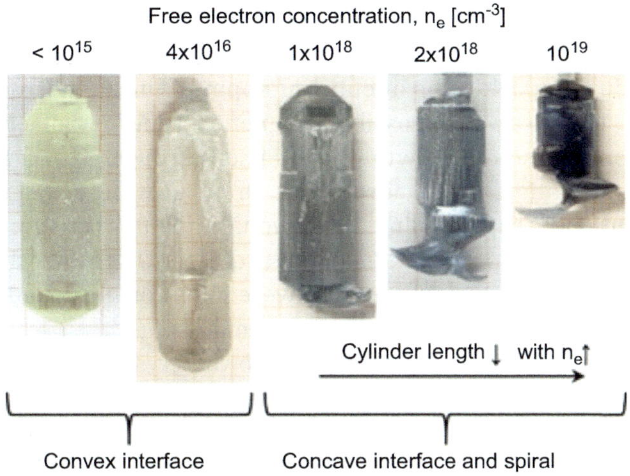

Figure 4.26 Impact of the free carrier concentration on the interface shape (convex or conceive) and cylinder length of β-Ga_2O_3 crystals grown by the Czochralski method.

Heat transfer through the growing crystal can be increased by higher temperature gradients above the melt surface in the space of the growing crystals. However, with high temperature gradients, the structural quality of β-Ga_2O_3 crystals significantly deteriorates due to the formation of high twin density and possible grains. Moreover, such conditions require higher input power to the crucible and thus temperature of the melt to compensate heat losses due to high temperature gradients. This will lead to a higher decomposition rate of Ga_2O_3, including higher concentration of Ga(I) in the melt. Furthermore, high temperature gradients lead to higher thermal stresses in the crystal, which is a typical source of cracking along the cleavage plane {100}, especially at large crystal diameters. To obtain high quality of β-Ga_2O_3 crystals, low temperature gradients are highly recommended.

In addition to the free carrier absorption, also other factors may contribute to the growth instability, such as impurities, metallic Ga in the crystals when grown at too low O_2 concentration, high structural defect density, growth kinetic due to asymmetric faceting and anisotropic heat conductivity, and seed misorientation.

4.5.6.5 Intentional doping and dopant segregation

An important aspect in crystal growth technology is the control of residual impurities or intentional doping, which affect not only optical, electrical, and possible magnetic properties of obtained crystals, but also the growth process itself through growth kinetic, defect formation, heat absorption in a growing crystal, and constitutional supercooling. Intentional doping constitutes an effective tool to expand the functionality of β-Ga_2O_3. The Czochralski-grown bulk β-Ga_2O_3 single crystals were intentionally doped with a number of mono-, di-, tri-, and tetravalent ions.

An incorporation of the dopant into the crystal lattice is quite complex, and it depends on (i) dopant properties (ionic radii, valency, coordination, and partial pressure), (ii) crystal properties (structure and ionic radius of an ion to be substituted by the dopant), (iii) dopant solubility, (iv) melt properties (kinematic viscosity, diffusion coefficient), (v) growth conditions (temperature gradients across the interface and oxygen partial pressure), (vi) growth kinetic (faceting), and (vii) operating parameters (growth and rotation rates).

For given growth conditions and operating parameters during the growth of bulk β-Ga_2O_3 single crystals by the Czochralski method, the amount of incorporated dopant into the crystal structure is mainly governed by two factors: (a) ionic radius of the dopant that supposes to substitute Ga^{3+} in either at tetrahedral (IV) or octahedral (VI) site; and (b) partial pressure of its volatile species, if any.

(a) Ionic radii of the dopants

Coordination- and valency-dependent ionic radii of various elements that were used for intentional doping bulk β-Ga_2O_3 single crystals grown from the melt by the Czochralski method are shown in Table 4.5. The list reveals that all the ions of interest, except Ce, have not too large difference in ionic radii with respect to Ga^{3+} ions, so we can expect their incorporation into β-Ga_2O_3 crystal lattice with certain segregation coefficient. As will be shown further below, Ce, indeed, is very poorly build in the Ga_2O_3 crystals, while others more easily. From the point of view of ionic radii, the most efficient dopant incorporation into the crystal lattice is expected for Mg^{2+}, Co^{2+}, Ni^{2+}, Cr^{3+}, Al^{3+}, Ge^{4+}, and Sn^{4+} ions. However, this picture is disturbed by the thermal stability of the dopants.

Table 4.5 Shannon ionic radii of elements used for doping β-Ga_2O_3 crystals at tetrahedral (IV) and octahedral (VI) sites

Element	Ionic radii	
	IV	VI
Ga^{3+}	0.47	0.62
Li^{1+}	0.59	0.76
Mg^{2+}	0.57	0.66
Cu^{1+}	0.6	0.77
Cu^{2+}	0.57	0.73
Co^{2+}	0.58	0.65/0.745
Ni^{2+}	0.55/0.49	0.69
Cr^{2+}	0.73	0.8
Cr^{3+}	—	0.615
Cr^{4+}	0.41	0.55
Al^{3+}	0.39	0.535
Ce^{3+}	—	1.01
Ce^{4+}	—	0.87
Si^{4+}	0.26	0.4
Ge^{2+}	—	0.73
Ge^{4+}	0.39	0.53
Sn^{4+}	0.55	0.69

(b) Thermal stability of the dopants

Dopants are added to the Ga_2O_3 starting material in the form of metal oxides or carbonates. However, not all the dopant oxides are thermally stable at high temperatures and will decompose during the growth of crystals in a similar manner to decomposition of Ga_2O_3. As a result, the dopant in quest will be partially or almost entirely lost from the melt due to evaporation in the form of an oxide, suboxide, or elemental metal, depending on metal oxide or carbonate. This issue is visualized in Fig. 4.27. The most stable dopants are SiO_2, CeO_2, Al_2O_3, and MgO (region (I) in Fig. 4.27), and their most volatile species (SiO_2, CeO, AlO, Mg) have low or very low partial pressures, much lower than those of Ga-containing species (Ga_2O, GaO). Losses of these dopants during growth will be neglectable. Another group of dopants consists of Cr_2O_3, CoO, and NiO (region (II) in Fig. 4.27), the volatile species of which (CrO_2, Co, Ni) have similar partial pressures to those of Ga_2O_3. They will partially evaporate during growth, but it is not critical. Volatile species of dopants CuO and Li_2CO_3 (region (III) in Fig. 4.27) have much higher partial pressures as compared to Ga-containing species. Their losses (as Cu and Li, respectively) from the melt during growth will be significant; therefore, it should be compensated by higher initial concentration, higher $p(O_2)$, and possible higher total pressure. And finally, volatile species of SnO_2 and GeO_2 (region (III) in Fig. 4.27) at the MP of Ga_2O_3 are very or extremely high. Even at external pressures of 150 and 1000 atm, the partial pressures of their volatile species (SnO and GeO, respectively) are still very high, which may lead to their high or almost entire loss during growth. These dopants are not well suited for the long growth runs and large growth systems. In addition to losses, their incorporation into the crystal lattice will be highly non-uniform with crystal length.

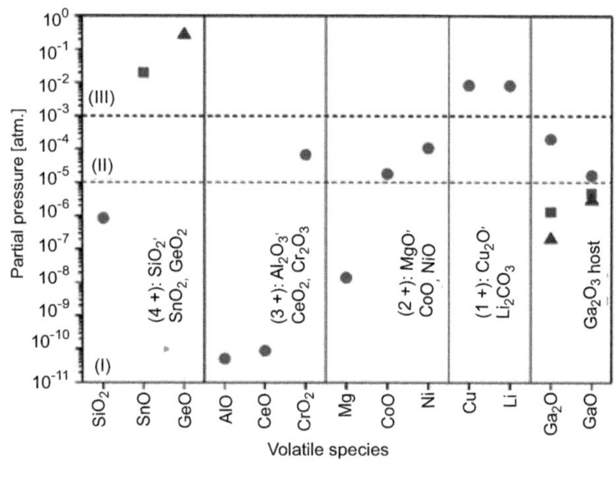

Figure 4.27 Partial pressures of volatile species of different metal oxides used as dopants for β-Ga_2O_3 single crystals grown by the Czochralski method. In all cases, $p(O_2) = 0.02$ atm. The circles, squares, and triangles refer to a total pressure of 1, 150, and 1000 atm. The dashed line at 10^{-3} atm indicates the limit above which there is intensive evaporation. Adapted from Ref. [324].

When summarizing possible dopants from the point of view of their thermal stability, the most suited dopants of different valency are Si^{4+}, Al^{3+}, Ce^{3+}, Mg^{2+}, which are added to the Ga_2O_3 starting material as SiO_2, Al_2O_3, CeO_2, and MgO. This is shown as region (I). Other dopants shown in region (II), i.e., Cr^{3+}, Co^{2+}, and Ni^{2+}, which are used in the form of Cr_2O_3, CoO, and NiO, can effectively dope β-Ga_2O_3 crystals, but dopant losses through evaporation and dopant concentration gradient along a crystal are expected. Dopants requiring technical attention (e.g., high initial concentration and short growth time) are Sn^{4+} (SnO_2), Cu^{1+} (Cu_2O), and Li^{1+} (Li_2CO_3), which are located in region (III). Moreover, Ge^{4+} (GeO_2) can be used only in very short growth runs; otherwise, it will entirely evaporate. In such cases, a high dopant concentration gradient will be produced in growing β-Ga_2O_3 crystals.

All the elements listed in Table 4.5 and Fig. 4.27 were used for doping bulk β-Ga_2O_3 single crystals grown by the Czochralski method, which are shown in Fig. 4.28. As the growth conditions were very similar to each other, particularly $p(O_2)$ in a growth atmosphere, we can clearly see changes in crystal appearance in terms of coloration, surface roughness, and shape as the result of doping. Undoped crystal is bluish, has a straight cylindrical shape, and convex interface. The free carrier concentration is not high enough for the interface inversion and spiral formation. Tetravalent ions Si, Sn, and Ge all were demonstrated to enhance the free carrier concentration, which has an impact on growth stability through free carrier absorption in the NIR spectral region (see Section 4.5.6.3). Indeed, this is valid for doping with Si and Sn, which make the crystals blue with a tendency for spiral formation after reaching a certain length. As a dopant, SiO_2 is thermally stable at operating temperatures during the growth of β-Ga_2O_3 by the Czochralski method, so its incorporation into the growing crystal is decent. SnO_2 is highly volatile, and less Sn^{4+} ions are incorporated into the crystal despite higher initial concentration, which is indicated by light blue coloration that almost disappears at the bottom part of the crystal as most of the SnO_2 dopant in the melt has evaporated. As shown in Fig. 4.29a, low SnO_2 doping level (0.2 mol.%) led to a colorless crystal, indicating very low Sn concentration in the obtained crystal, while higher SnO_2 doping concentration (1 mol.%) led to blue crystals due to relatively high free carrier concentration donated by Sn. Different results were obtained by doping with GeO_2. The crystal is substantially colorless, and the interface is nicely convex even at high doping level of GeO_2

(2 mol.%). This is due to the already discussed very high partial pressure of GeO and corresponding entire loss from the melt (no Ge was detected in the obtained crystals). Tri-, di-, and monovalent dopants do not enhance the free carrier concentration, but rather compensate it (at least some of them). Due to minimized free carrier absorption, the crystals have convex interface and can be grown as straight cylinders with a reasonable length. Doping with Ce, Mg, and Co resulted in yellow crystals, with Ni resulted in brownish, with Cr resulted in green, and with Al and Li resulted in colorless crystals. In the case of highly volatile Li dopant, the crystal appearance is very similar whether the Li_2O doping level was either 1 or 2 mol.% (Fig. 4.29b). A blue coloration of Cu-doped β-Ga_2O_3 crystal is associated with high free carrier concentration (lower oxygen concentration in the growth atmosphere) rather than by Cu itself (no Cu was found in the crystals).

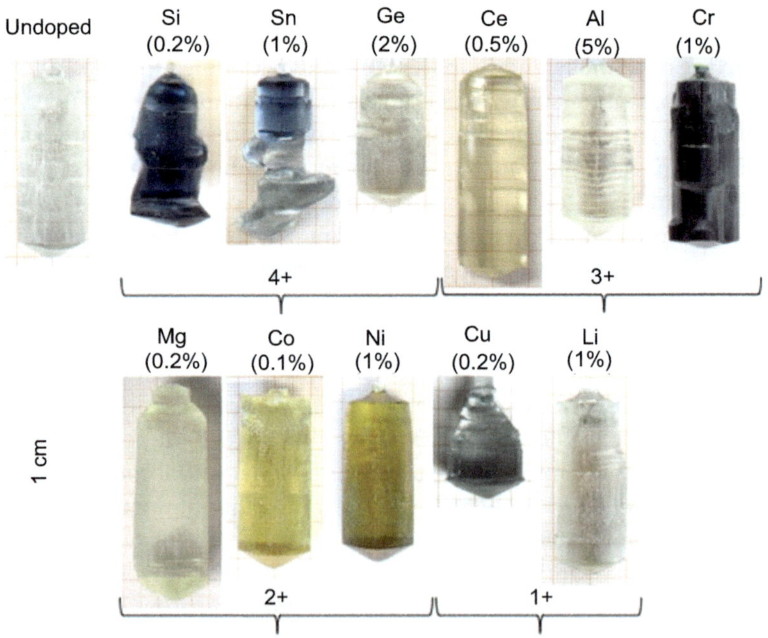

Figure 4.28 Czochralski-grown β-Ga_2O_3 crystals, undoped and doped with Si, Sn, Ge, Ce, Al, Cr, Mg, Co, Ni, Cu, and Li. Indicated dopant concentrations are in mol.% and relate to initial concentration in the Ga_2O_3 starting material. In all cases, oxygen concentration in a growth atmosphere was between 2 and 4 vol.%.

Figure 4.29 Czochralski-grown β-Ga_2O_3 crystals doped with Sn^{4+} and Li^{1+} at different concentrations (in mol.% in the melt as SnO_2 and Li_2O). Both dopant oxides are thermally unstable at high temperatures (see Fig. 4.27). Reprinted from Ref. [349], Copyright 2020, with permission from Elsevier.

At given growth conditions, all the crystals are clear and transparent, with enhanced absorption in Si- and Cr-doped crystals due to high free carrier absorption (red spectral region) and blue Cr absorption, respectively. The crystals have smooth or a bit rough (undoped, doped with Co, Ni, and Li) surface.

Tetravalent ions Si^{4+}, Sn^{4+}, and Ge^{4+} aim to increase the free electron concentration in β-Ga_2O_3 single crystals. Tested SiO_2 doping levels in the Ga_2O_3 melt for the Czochralski method were between 0.1 and 1 mol.%. Due to good thermal stability of SiO_2 at the MP of Ga_2O_3, low doping level of 0.1–0.2 mol.% is sufficient to achieve the maximum level of the free electron concentration, which is about 10^{19} cm^{-3}. SnO_2 concentrations in the Ga_2O_3 melt were between 0.1 and 2 mol.%, wherein high free electron concentrations ($>10^{18}$ cm^{-3}) could be obtained only for high SnO_2 doping level $\geq$1 mol.%. As expected from the thermal instability of GeO_2 at high temperatures, no Ge was found in the crystals even at high initial doping level of the starting material. Therefore, Ge did not only enhance the free electron concentration for GeO_2 doping levels between 0.1 and 2 mol.% but decreased as compared with undoped crystal obtained under substantially the same growth conditions. The reason is decomposition of GeO_2, the product of which is GeO and oxygen that is left inside the melt ($GeO_2 \rightarrow GeO(g) + \frac{1}{2}O_2(g)$). In other words, GeO_2 can be used as an effective oxygen source without any doping effect of Ge, which is entirely lost from the melt. The advantage is the oxygen production inside the melt.

Although there is an effective interaction of oxygen from the growth atmosphere and the melt surface, its diffusion to the melt volume is rather slow and limited. GeO_2 doping as an oxygen source can be particularly important during the growth of large-diameter β-Ga_2O_3 single crystals ($\geq$2 inch), which will stabilize the growth through the entire decomposition process of the dopant oxide. Similar effect has also CeO_2, which releases an oxygen in the melt, but through another process, i.e., valency change (see further below).

Trivalent dopants Ce^{3+}, Al^{3+}, Cr^{3+} rather do not have a direct impact on electrical properties of bulk Ga_2O_3 single crystals obtained by the Czochralski method but may have an indirect impact. Such ions influence optical properties of bulk Ga_2O_3 crystals. In particular, Ce^{3+} modifies scintillation properties under gamma quanta, Al^{3+} enlarges the bandgap, while Cr^{3+} induces electroluminescence [273] and could potentially constitute an active element in the Ga_2O_3 host for solid-state lasers. CeO_2 doping level in Ga_2O_3 melt was practiced between 0.1 and 2.5 mol.%. Due to very low segregation coefficient of Ce arising from a very large ionic radius (see further below), rather high CeO_2 concentration (>0.5 mol.%) is required to obtain a reasonable Ce concentration in the crystals (>20 wt. ppm). A concentration of Al_2O_3 in the melt was practiced at high levels, between 2 and 5 mol.%. Doping Ga_2O_3 crystals with 5 mol.% Al_2O_3 increases the bandgap by about 0.2 eV. Higher doping with Al_2O_3 is still possible (up to about 10 mol.%); however, it will be occupied by deteriorated structural quality, mainly due to constitutional supercooling. In the case of Cr_2O_3, its initial concentration in the melt was practiced in the range of 0.04–1 mol.%. Cr incorporates into the β-Ga_2O_3 crystal lattice easily (high segregation coefficient—see further below) and even the lowest initial Cr_2O_3 concentration makes the crystals greenish, i.e., it is sufficient to introduce absorption bands in the blue spectral region.

Divalent ions Mg^{2+}, Co^{2+}, and Ni^{2+} aim to compensate the electrical conductivity of bulk β-Ga_2O_3 crystals and make them electrically insulating or semi-insulating, what is often required for epitaxial purposes (as substrates), e.g., to fabricate horizontal electronic devices thereon. Indeed, Mg^{2+} was found to be a very efficient compensator of shallow donors. MgO concentration in the Ga_2O_3 melt of 0.1–0.2 mol.% was already sufficient to turn the crystals into electrically insulating (or semi-insulating) state. CoO and NiO at concentrations of 0.1–0.25 mol.% and 1 mol.% in the melt, respectively, also compensated the electrical conductivity in bulk β-Ga_2O_3 single crystals obtained by the Czochralski method, as the obtained crystals were electrical insulator or semi-insulators. Therefore, all the Mg^{2+}, Co^{2+}, Ni^{2+} dopants are practical dopants for bulk β-Ga_2O_3 crystals that can effectively be utilized as electrical conductivity compensators.

And finally, both Cu^{1+} and Li^{1+} are supposed to be the most effective shallow donor compensators in bulk β-Ga_2O_3. Due to the very high partial pressure of Li and Cu and their associated high losses, the initial Li_2O concentration in the melt (added to the Ga_2O_3 starting material as Li_2CO_3) was also high, 1 and 2 mol.%. Li_2CO_3 added to the starting material releases upon heating CO_2, which is an additional oxygen source. Here, indeed, the obtained crystals were electrically insulating. However, Cu_2O concentration of 0.2 mol.% in the melt did not compensate the shallow donors due to too high Cu losses by evaporation and possible marginal (below detection limit), if any, amount incorporated into the crystals, which was not sufficient to compensate the electrical conductivity of β-Ga_2O_3.

A concentration of the dopant/impurity (whether intentional or not) in the growing crystal is described by the Scheil–Gulliver equation for non-equilibrium solidification with the assumption of a complete mixing of the dopant in the melt and no diffusion of the dopant in the crystal (which are typically met in the growth of oxide crystals by the Czochralski method):

$$C_S = C_0 k_{\text{eff}} (1-g)^{k_{\text{eff}}-1} \tag{4.15}$$

where C_S is the dopant concentration in the crystal, C_0 is the initial dopant concentration in the melt, k_{eff} is the effective segregation coefficient, and g is the solidified fraction of the melt.

The effective segregation coefficient k_{eff} in the steady-state case is described by the Burton, Prim, and Slichter (BPS) model [274], while its more realistic description was given by Ostrogorsky and Müller [275], who have taken into account a convective dopant transport next to the solid–liquid growth interface.

Knowing the segregation coefficient, it is easy to select a proper initial dopant concentration in the melt to obtain a desired dopant concentration in the crystal at given growth conditions and operating parameters. This can be carried out by measurements of the dopant concentration at different crystal locations corresponding to the crystallization fraction and fitting the measured values with the Scheil–Gulliver equation. This works well for many oxides, such as $Y_3Al_5O_{12}$ (YAG) or sapphire. In the case of β-Ga_2O_3, which decomposes and evaporates, the fitting is less accurate, as the initial dopant concentration dynamically changes with growth time. Also, some of the dopant elements, their oxides or suboxides have a high vapor pressure and evaporate as well (see Fig. 4.26 and corresponding discussion). Moreover, some dopants may change their oxidation state depending on $p(O_2)$ and form different oxides having different partial pressures, which, due to different ionic radii, may change the effective segregation coefficient. In such cases, the segregation coefficient is not constant but may dynamically change with the crystallization fraction, causing a deviation from the Scheil–Gulliver equation.

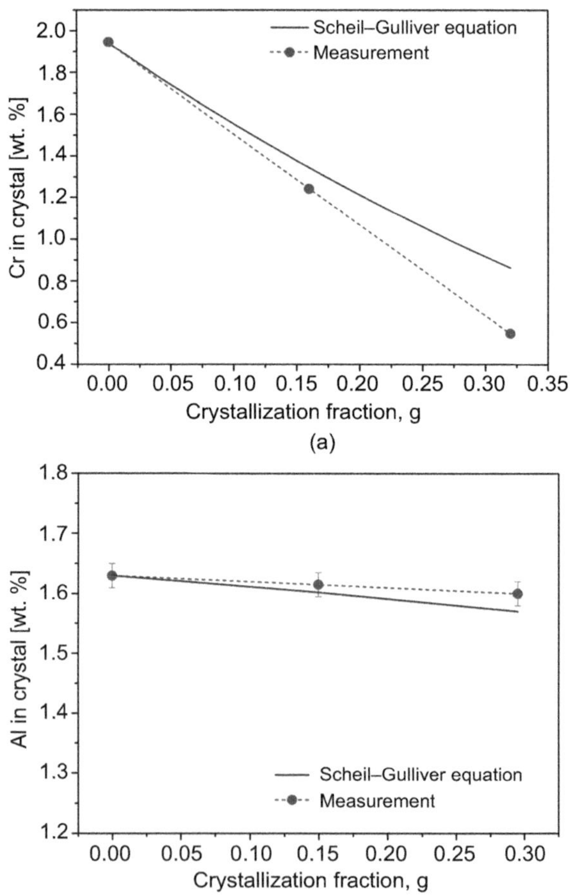

(b)

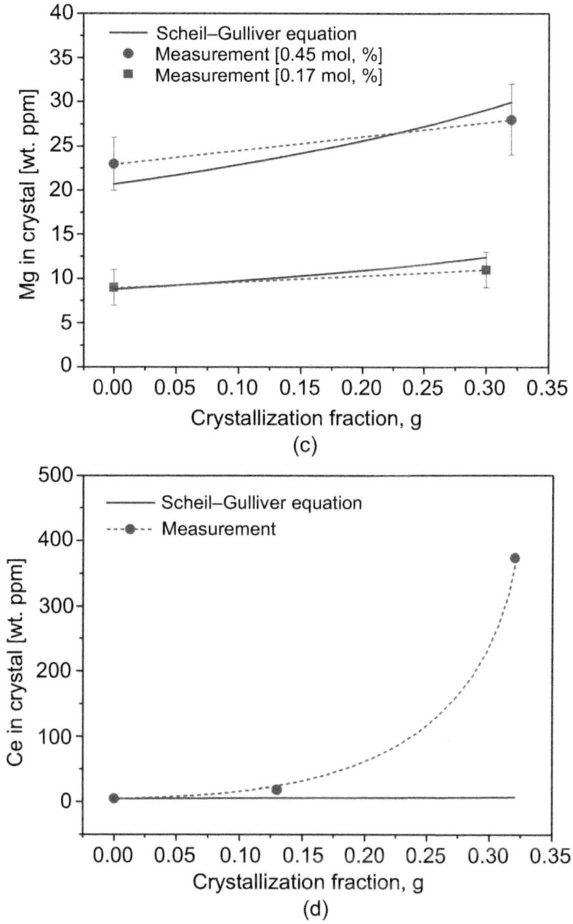

Figure 4.30 Dopant concentrations of Cr (a), Al (b), Mg (c), and Ce (d) of doped β-Ga_2O_3 crystals obtained by the Czochralski method. The dots are measured values, while solid lines represent fittings with the Scheil–Gulliver equation. Figures (a), (b), and (d) reprinted from Ref. [229], Copyright 2018, with permission from Elsevier.

If the segregation coefficient is greater than unity (e.g., Cr), the dopant concentration will decrease from the top to the bottom part over the crystal length. For segregation coefficients below unity (e.g., Mg, Sn, Si, Ce), the situation is reversed, while for those close to unity (e.g., Al) the dopant will be distributed in the crystal relatively uniformly. The segregation phenomenon with the segregation coefficients deviating from unity will cause a non-uniform dopant distribution along the crystal length. In addition to the longitudinal segregation, there is also a lateral dopant segregation (across the crystal) due to the shape of the growth interface that is different from a planar one and usually changes as the growth proceeds.

Examples of measured dopant concentrations in Czochralski-grown β-Ga_2O_3 crystals as a function of the crystallization fraction, compared with the fitting of the Scheil–Gulliver equation, are shown in Fig. 4.30 for Cr (a), Al (b), Mg (c), and Ce (d), which have the segregation coefficient greater than unity, close to unity, smaller than unity, and much smaller than unity, respectively.

The deviation of the Cr concentration in β-Ga_2O_3 crystals from the Scheil–Gulliver equation (Fig. 4.30a) arises from the change in the oxidation state of Cr that forms different oxides: CrO_2 and CrO_3, which have different partial pressure and evaporate at different rates. In addition, there is also evaporation of the main constituent forming the melt, Ga_2O. The formation and partial pressures of high-valency Cr oxides and Ga_2O depend on $p(O_2)$ in such a way that both $p(CrO_2)$ and $p(CrO_3)$ increase, while the $p(Ga_2O)$ decreases with $p(O_2)$. At the MP of Ga_2O_3, the $p(CrO_2)$ and $p(CrO_3)$ are higher than $p(Ga_2O_3)$ at high O_2 concentrations (e.g., 20 vol.%), but at lower O_2 concentrations (e.g., 7 vol.%), the situation is reversed, leading to different losses of Cr- and Ga-containing species. The formation of CrO_2 and CrO_3 in the Ga_2O_3 melt leads to a different incorporation rate into the β-Ga_2O_3 crystal lattice due to different ionic radii, as shown in Table 4.5. The mixture of Cr_2O_3, CrO_2, and CrO_3 in the Ga_2O_3 melt will, therefore, affect the effective segregation coefficient as less and less Cr^{3+} ions, which have the closest ionic radii to Ga^{3+}, are available as the growth proceeds. As a result, the effective segregation coefficient will change with the O_2 concentration in a growth atmosphere, as shown in Fig. 4.31a.

The segregation coefficient of Al^{3+} in β-Ga_2O_3 is close to unity and fits the Scheil–Gulliver equation well. This is the result of the thermal and oxidation state stability of Al, and low losses of Ga at relatively high O_2 concentration in the growth atmosphere. Moreover, Al^{3+} has the ionic radius very close to that of Ga^{3+}; therefore, the expected segregation coefficient is close to unity. As the result, Al distribution in β-Ga_2O_3 crystals is relatively uniform.

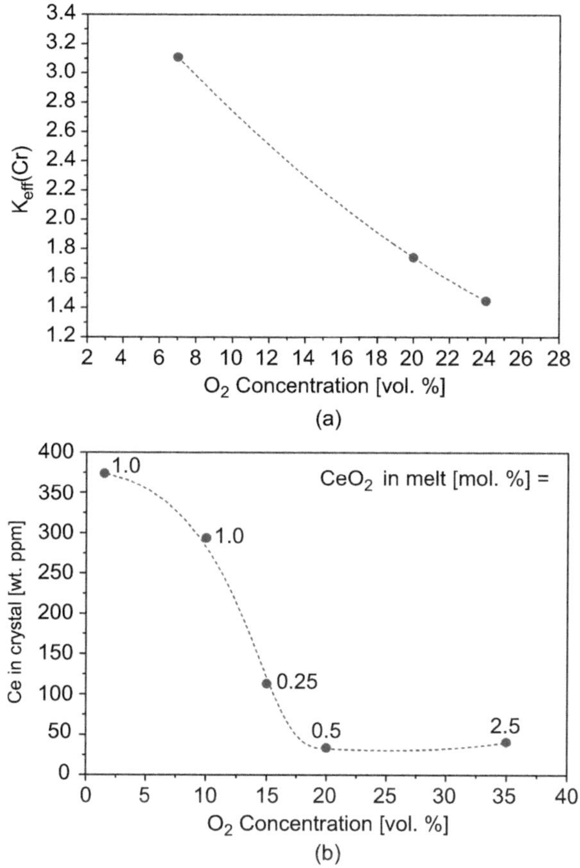

Figure 4.31 Segregation coefficient of Cr (a) and Ce concentration (b) in Czochralski-grown β-Ga_2O_3 crystals as a function of O_2 concentration in the growth atmosphere. Adapted from Ref. [229].

Mg^{2+} is stable in the same way as Al^{3+}, so we can expect a small deviation from the Scheil–Gulliver equation, as shown in Fig. 4.30c. A larger ionic radius of Mg^{2+} as compared with Ga^{3+} results in $k_{eff} < 1$, which is about 0.4.

An example of a large deviation of measured dopant concentration in the crystals from the Scheil–Gulliver equation is Ce (Fig. 4.30d). Generally, Ce^{3+} has a very large ionic radius as compared with the Ga^{3+} ionic radius, so the expected segregation coefficient is very low. Indeed, the measured Ce concentration in the crystals is very low as compared with the initial dopant concentration in the melt, but additionally we can observe an enormous deviation from the Scheil–Gulliver equation. The main reason is the tendency of Ce incorporation into structural defects (dislocations, twin boundaries) instead of substituting Ga^{3+} ions in the crystal lattice. Additionally, at high temperatures, Ce is stable in both Ce^{3+} and Ce^{4+} oxidation states depending on the O_2 concentration, which results in different ionic radii and a dynamic k_{eff} depending on the O_2 concentration in the growth atmosphere, as illustrated in Fig. 4.31b. Segregation coefficients of elements used as intentional dopants for β-Ga_2O_3 crystals grown by the Czochralski method, determined mainly by ICP-OES, but also by XRF and SIMS, are shown in Table 4.6.

Some of the thermally stable intentional dopants may have a positive impact on the growth stability. For example, it was found that the presence of Al_2O_3 and CeO_2 in the melt decreases the decomposition rate of Ga_2O_3 as compared with undoped melts even by a factor of 3 and makes the growth process more stable (Galazka et al. [229]). This is the entropy-stabilizing effect, on one hand, and improved heat transfer through the growing crystal, on the other hand, which keeps the growth interface convex toward the melt. Further, CeO_2 may provide extra oxygen by changing its oxidation state from 4+ to 3+ ($2CeO_2 \rightarrow Ce_2O_3 + \frac{1}{2}O_2$), while Al expands the bandgap of β-Ga_2O_3, which may lead to make the donor level deeper and consequently lower free carrier concentration at high temperatures (lower free carrier absorption). These dopants can be used as growth stabilizers for conducting crystals by doping with Si^{4+} or Sn^{4+} ions. Indeed, co-doping β-Ga_2O_3 single crystals with Al^{3+} and Si^{4+}, or with Al^{3+}, Ce^{3+}, and Si^{4+}, keeps the growth much more stable as compared to doping with Si^{4+} only, while still keeping similar electrical properties despite larger bandgap. Such triple-doped bulk β-Ga_2O_3 single crystals obtained by the Czochralski method are shown in Fig. 4.32. Stabilising effect on the growth have also divalent (Mg, Co, Ni) dopants by a decrease of heat absorption by free carriers during growth (see Section 4.5.6.4). In the same way as some of other highly unstable dopant oxides (SnO_2, GeO_2, Cu_2O), Li_2CO_3 also provides at high temperatures an extra oxygen inside the melt through its thermal decomposition: $Li_2CO_3 \rightarrow 2Li(g) + CO(g) + O_2(g)$.

Table 4.6 Segregation coefficients used as intentional doping in Czochralski-grown bulk single crystals of β-Ga_2O_3. Some of the data are from Ref. [324]. Note that for dopants with k_{eff} = 0 the Scheil–Gulliver equation does not apply, however, they are included here for a general overview

Dopant	Dopant concentration in the melt [mol.%]	O_2 concentration [vol.%]	Effective segregation coefficient, k_{eff}
Si^{4+}	0.1–1.0	1–2	0.07–0.12
Sn^{4+}	0.1–2.0	2–17	0.002–0.06[a]
Ge^{4+}	0.1	2–4	0[a]
Ce^{3+}	0.25–2.5	1.5–35	2×10^{-4}–10^{-2}
Al^{3+}	2.0–5.0	1.5–21	1.11
Cr^{3+}	0.04–1.0	7–24	1.45–3.11
Mg^{2+}	0.05–0.45	1–35	0.4
Co^{2+}	0.1–0.25	4	0.05–0.06[b]
Ni^{2+}	1	4	0.03[b]
Cu^{1+}	0.2	1	0[a]
Li^{1+}	1–2	2	0[a]

[a]Very high partial pressure at the melting point of Ga_2O_3 (k_{eff} governed by dopant evaporation), [b]high partial pressure at the melting point of Ga_2O_3 (k_{eff} governed by dopant evaporation to a large extent)

Concentrations of residual impurities in undoped crystals are also governed by the segregation phenomenon in the same way as intentional dopants. Residual impurity concentrations may differ between crystals depending on the purity of the starting material and Ir crucible, as well as on an insulation type, its composition, and reuse. The starting materials and Ir crucibles, although the same purity of 5N and 3N, respectively, may have different levels of residual impurities even from the same supplier. As a result, the obtained crystals may also have slightly different levels of residual impurities. Residual impurities may also come from insulation surrounding the crucible and a crystal pulling zone above the crucible. Typically used insulation are zirconia, alumina, and quartz (the most outer part of the furnace), which may contaminate the melt with trace concentrations, usually at the level of a few wt. ppm or below. The following residual impurities were identified in different undoped crystals (in wt. ppm): Si ≤ 2, Al ≤ 20, Zr ≤ 6, Ti ≤ 3, Fe ≤ 2, Mo ≤ 2, and occasionally traces of other elements, such as Co, Cr, Mg, and Ca. With larger crystal diameter, some of the impurities appear sometimes at larger concentrations, especially Fe and Al.

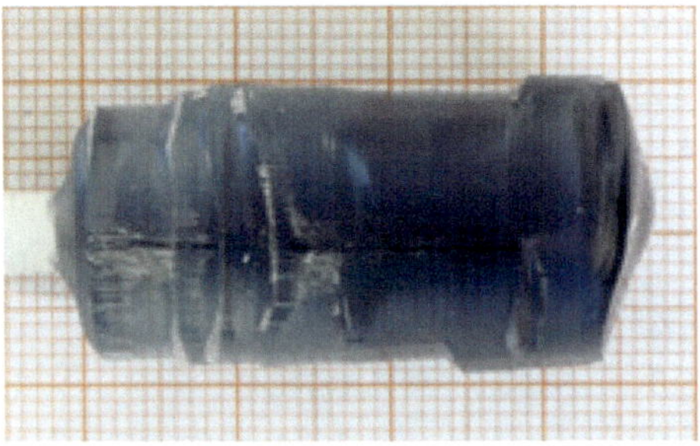

Figure 4.32 Triple-doped (Al^{3+}, Ce^{3+}, and Si^{4+}) bulk β-Ga_2O_3 single crystal obtained by the Czochralski method. Al_2O_3 = 5 mol.%, CeO_2 = 0.1%, and SiO_2 = 0.2 mol.% in the melt.

4.5.6.6 Examples

(i) Galazka et al. [225, 226, 324] used for undoped 20 mm diameter β-Ga_2O_3 single crystals 40 mm diameter and 40 mm high Ir crucibles, and an active afterheater above the crucible, which were surrounded by zirconia and alumina thermal insulation. Additionally, Ir lids were used inside the growth furnace to further decrease temperature gradients in a pulling zone. The growth atmosphere comprised $(1 - x)$Ar + CO_2 with $x = 0.3–1$. The growth and rotation rates were 1–2 mm/h and 8–15 rpm, respectively. The starting material was 5N purity Ga_2O_3 powder, which was dried, pressed, and sintered in air. The heating and cooling-down rates were 180–225 and 120–150 k/h, respectively. The obtained crystals were typically light blue with the length 50–70 mm.

(ii) To grow Mg, Si, Cr, Al, and Ce-doped 20 mm diameter β-Ga_2O_3 single crystals, Galazka et al. [226, 228, 229, 324] used similar furnace design and operating parameters as in the case of undoped crystals. Here, the growth atmosphere consisted of $(1 - y)Ar + yO_2$ with $y = 0.01–0.35$, or CO_2 in some of the growth experiments with Mg doping. The crystal coloration and length strongly depended on the dopant type and O_2 concentration in the growth atmosphere, as discussed in Sections 4.5.6.4 and 4.5.6.5.

(iii) Two-inch-diameter β-Ga_2O_3 single crystals obtained by Galazka et al. [228, 229] were grown from 100 mm diameter crucibles with the height smaller than 100 mm. The growth furnace also included an active Ir afterheater and Ir lids. The thermal insulation was composed of different types of the thermal insulation, such as zirconia, alumina, and quartz. The growth atmosphere was $(1 - y)Ar + yO_2$ with $y = 0.08–0.35$, preferably $y = 0.2–0.35$. The starting material was either pure or Mg-doped Ga_2O_3 5N purity powder, which was dried, pressed, and sintered prior loading into the crucible. The growth and rotation rates were 1–2 mm/h and 4–8 rpm, respectively. The heating and cooling rates were usually 120–150 K/h and 70–90 K/h. Undoped crystals grown at high O_2 concentration ($y > 0.15$) were yellowish, while at lower O_2 concentration, they were greenish. Mg-doped (0.1–0.2 mol.% in the melt) crystals were yellowish for the whole O_2 concentration range. Al-doped (2 mol.% in the melt) crystals at high O_2 concentration were yellowish as well. The length of the crystals varied from 50 to about 100 mm, depending on the O_2 concentration, dopant, and furnace design.

4.5.7 Vertical Gradient Freeze (VGF) and Vertical Bridgman (VB) Methods

VGF and VB methods were utilized by Galazka et al. [227, 228] to grow highly conducting β-Ga_2O_3 crystals as the Czochralski method shows some limitation in crystal volume due to limited heat transfer through a growing crystal by free carrier absorption, which leads to growth instability. In the VGF and VB methods, a crystal is not pulled out of the melt surface, but the melt just crystallizes step by step entirely inside a crucible during cooling down as the result of imposed vertical temperature gradients.

A growth furnace utilizes a 2-inch-diameter Ir crucible having a cylindrical portion, a seed portion in the form of a small-diameter tube at the bottom part to accommodate a crystal seed, and a shoulder portion connecting both the cylindrical and seed portions. The crucible was covered with an Ir lid with a small opening to minimize temperature gradients within the crucible, avoid contamination of the melt from insulation materials, and to create a path for a growth atmosphere communicating with the melt. The crucible with the lid was surrounded from all sides by thermal insulation comprising alumina and zirconia. The crucible was inductively heated by an RF coil disposed around the crucible and the thermal insulation. As the seed and shoulder portions have a smaller diameter than the cylindrical portion, they are inductively heated much less than the cylindrical portion. Additionally, locating the bottom part of the crucible at the bottom part of the induction coil decreases the probability of melting the seed sitting in the small-diameter tube of the crucible. The melting of the Ga_2O_3 starting material within the crucible can be detected by a thermocouple contacting the bottom or top part of the crucible or by a pyrometer monitoring a temperature of the lid. The melting is detected by the fast increase in the temperature typically by several degrees. After holding the melt for a predefined time for homogenization and stabilization in terms of melt flow (for about 1–2 h), the crucible with the melt is cooled down to RT. The imposed vertical temperature gradients will force solidification from the seed in the upwards direction. The initial cooling rate until full solidification should not be too high (below 100 K/h) in order to minimize the formation of high defect density, but on the other hand, very slow cooling rate (a few K/h) may enhance the formation of metallic Ga in large melt volumes, as discussed above. In the case of smaller-diameter crucibles (i.e., smaller melt volumes), the cooling rate can be very small as the formation of metallic Ga is minor. Once the melt is solidified, the cooling rate can be higher. In the discussed example, the initial cooling rate until melt solidification was about 75 K/h, while after solidification about 125 K/h, giving an average cooling rate of 100 K/h.

The seed was oriented parallel to the <010> crystallographic direction, and the growth atmosphere consisted of 17 vol.% O_2 +

83 vol.% Ar. High electrical conductivity was achieved by adding 0.1 mol.% of SnO_2 to the Ga_2O_3 starting material. The solidified material was a single crystal in the central part and polycrystal along the rim adjacent to the crucible wall. The obtained crystals of 55 mm diameter and 25 mm length (cylinder) were dark blue with visible twins, which could be formed due to high cooling rate and differences in thermal expansion of Ga_2O_3 and Ir crucible.

The VB and VGF methods to grow bulk β-Ga_2O_3 crystals were applied by Hoshikawa et al. [233] and Ohba et al. [276]. In the growth furnace, a crucible and heater surrounding the crucible were both made of Pt70%–Rh30% alloy. The heater with the crucible was surrounded from all sides by a zirconia and alumina thermal insulation, around which an RF coil was disposed operating at 15 kHz. The growth furnace was connected with rotation and translation mechanism to rotate and translate the crucible with respect to the coil. Temperature at the bottom of the crucible was monitored by a B-type thermocouple (PtRh30%–PtRh6%). Two types of crucible were used, cylindrical and conical, at the bottom both being 1 inch in diameter and 2 inch in length (cylindrical part). The advantage of the Pt–Rh alloy is the possibility of using high O_2 concentration without the risk of oxidation of the crucible and heater material. The MP of the Pt70%–Rh30% alloy is about 1900°C, i.e., 100 K higher than the MP of Ga_2O_3. This is sufficient for 1-inch- and perhaps for 2-inch-diameter crucibles, but it can be problematic for larger crucibles due to temperature gradients between the heater and crucible, as well as within the crucible. Additionally, Rh from the Pt–Rh alloy may contaminate the melt and the obtained crystals.

The rotation and translation rates were 3 rpm and 0.5 mm/h, respectively, while the growth atmosphere was air. The obtained crystals were removed from crucibles by destructive peeling off the crucible material from the crystal instead of core drilling as in the case of β-Ga_2O_3 grown by the VGF/VB methods from Ir crucibles. This was possible because of no adhesion between the β-Ga_2O_3 crystals and the Pt–Rh crucibles. However, the crucibles are of single use only.

At early attempts, no crystal seed was used, so the nucleation was spontaneous. As a result, some crystals were entirely polycrystals, some partly poly and partly single crystals, while others were entirely single crystals. The obtained crystals of size 2.5 × 2.5 cm (cylindrical part) had yellowish coloration with some color gradient depending on location within the crucible (Fig. 4.33).

Figure 4.33 β-Ga_2O_3 crystals obtained by the VB method from a cylindrical (left) and conical-bottom (right) crucibles. Reprinted from Ref. [233], Copyright 2016, with permission from Elsevier.

In the VB configuration, due to larger vertical temperature gradients as compared to horizontal gradients, and due to structural and thermal anisotropy of β-Ga_2O_3, the growth starts at the bottom with higher growth rate in the horizontal direction as compared with the vertical direction. With no seed, this will initiate a horizontal growth along the [010] direction and a vertical growth in a direction perpendicular to the (100) plane. Due to spontaneous nucleation, differently oriented grains may form, which explain the formation of polycrystalline material at early stage of crystal growth in some of the growth experiments. As the growth proceeds and temperature gradients change, one of the crystal grains may preferentially expand more in favor of others and lead to the formation of a single crystal. If growth conditions are optimized, this polycrystal–single crystal conversion may initiate at the very beginning of the growth. The formation of (001) facets on the crystal periphery, which form an angle of about 104° with the (100) or ($\bar{1}00$) planes, leads to a zig-zag structure (Fig. 4.34).

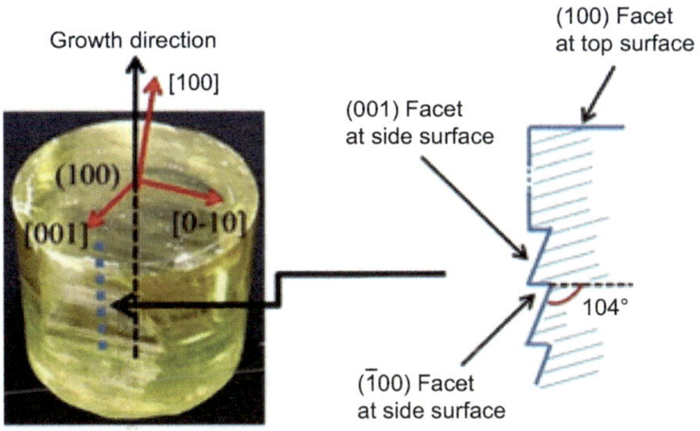

Figure 4.34 Facet formation at the crystal periphery during spontaneous growth of β-Ga_2O_3 by the VB method. Reprinted from Ref. [233], Copyright 2016, with permission from Elsevier.

The main impurities found in the crystals were (in wt. ppm) Rh = 3–30, Pt = 0.5–5, both coming from Pt–Rh crucibles, and Si = 1–10 likely from the Ga_2O_3 starting material.

In a similar configuration, the crystals were also grown by the VGF method, i.e., instead of the crucible translation, it was located at a fixed position and slowly cooled down to RT. The vertical temperature profile along the centerline was parabolic (lower temperature at the bottom and higher at the top) with the temperature difference between the top and bottom of about 50 K. The Ga_2O_3 starting material was hold for 1 h after melting, i.e., after reaching the MP of Ga_2O_3 at the bottom of the crucible. In growth experiments, the crucible was rotated at the rate of 5 rpm. Two types of crucibles were used: cylindrical and conical with the diameter and length of 1 and 2 inch, respectively. After melting, the cooling rate was between 0.75 and 1.5 K/h, while after solidification of Ga_2O_3, starting from 1650°C, the cooling rate was 45 K/h. The estimated growth rates arising from temperature gradients and cooling rates were between 0.5 and 4 mm/h. (100)-oriented samples prepared from different regions of the crystals grown at different rates, 3.3 and 1.1 mm/h, respectively, at corresponding temperature gradients of 4.5 and 7 K/cm, respectively, revealed different defect densities after etching in 30% KOH at 60°C for 2 h. Two types of defects were observed: dislocation-associated etch pits and line-shaped defects. Additionally, stressed areas at the periphery of the crystals were seen by X-ray topography. The samples grown at a higher rate, 3.3 mm/h, and lower temperature gradients (4.5 K/cm) had a mean etch pit density (EPD) of 2.3×10^3 cm^{-2} and a mean line-shaped defect density of 4.6×10^2 cm^{-2}. On the other hand, the sample grown at a lower rate, 1.1 mm/h and higher temperature gradients (7 K/cm), had a higher mean EPD, 2.9×10^4 cm^{-2}, and a lower mean line-shaped defect density, 0.5×10^2 cm^{-2}. This suggests that dislocation-related pits are preferably formed at higher temperature gradients, while line-shaped defects at a higher growth rate. The line-shaped defects had the size of 20–150 μm extending along the [010] direction, 0.05–1.2 μm along the [001] direction, and <0.04 μm along the [100] direction. The cross section of the line-shaped defects in [100] and [001] directions is a rhomboid, while lengths of all of the defect edges are inversely related with the lengths of edges of the β-Ga_2O_3 unit cell. This type of defects are likely microvoids [276], the same as in the case of the EFG method, resulted from a strong growth anisotropy of β-Ga_2O_3.

4.5.8 Conclusion

The growth of bulk β-Ga_2O_3 crystals from the gas phase and solution (flux) was carried out at an early stage of research, and nowadays it is not practiced, as large crystals can be grown directly from the melt. Although the Verneuil method was demonstrated for β-Ga_2O_3 decades ago, it is not practical due to small crystal size and poor structural quality. The OFZ technique (crucible free), due to its simplicity and low operational costs, was extensively utilized for growing bulk β-Ga_2O_3 crystals; however, a typical crystal diameter of about 5–6 mm is not sufficient for applications in a larger scale. However, it is enough for studying the impact of growth conditions (atmosphere, doping, orientation) on the crystal properties. Technologically important growth methods are those involving noble metal crucibles, including the Czochralski, EFG, and possible Bridgman/VGF methods. The advantage of the EFG method is a possibility of obtaining slabs with a large surface area (up to 4

Table 4.7 Summary of growth methods used to obtain bulk β-Ga_2O_3 crystals. *D* and *L* are diameter and length, respectively

Method	Noble metal crucible	Crystal shape	Maximum crystal size [mm]	Intentional dopant	Industrial importance	Reference
			Gas phase			
PVT	No	Fibers, plates	—	None	No	201
CVT	No	Fibers, bars, platelets, lumps	3×3×0.2, 2×2×2.5, D=2, L≤5	None	No	208, 209, 210, 211, 212, 213, 214, 215, 234
			Solution			
Flux	Yes	Needles, plates, prisms	16 at edge	Cr, Er	No	202, 203, 204, 235, 236, 237, 238, 239
			Melt			
Verneuil	No	Cylinders	D=10, L=25	Cr, Fe	No	207, 208, 240
OFZ	No	Cylinders	D=5–8, L≤70 also D=25, L=25	Sn, Si, Ge, In, Ce, Cr, Ti, Mn, Mg, Co, Cu, Ni, Fe	No	217, 218, 219, 220, 221, 222, 223, 241, 242, 243, 244, 245, 246, 247, 253, 248 249, 254, 255, 256, 257, 258, 259, 260, 261, 262, 263, 264, 265
EFG	Yes	Slabs	18×60, 6×110	Sn, Si, Fe	Yes	230, 232, 266
Czochralski	Yes	Cylinders	D=52, L=100	Si, Sn, Ge, Ce, Al, Cr, Mg, Co, Ni, Cu, Li	Yes	216, 224, 225, 226, 227, 228, 229, 8, 271, 324, 349
VB / VGF	Yes	Cylinders	D=55, L=25; D=25, L=25	Sn	Yes	227, 228, 233, 276

inch); however, a demanded (010) orientation is limited to a cross section of the slab. The advantages of the Czochralski method is large crystal volume and cylindrical shape enabling preparation of differently oriented wafers, including 2-inch-diameter (010) wafers. Both methods provide high structural quality crystals and allow for doping with a number of elements. More importantly, they enable scaling-up capabilities what is crucial for industrial-scale crystal production. Possibly the Bridgman/VGF method may become an important technique for growing bulk β-Ga_2O_3 crystals, but it requires more development. The basic summary of the growth techniques utilized to grow bulk β-Ga_2O_3 crystals is included in Table 4.7.

4.6 Surfaces

Surface properties arising from the physical and electronic structure, as well as surface preparation of wafers fabricated from bulk β-Ga_2O_3 crystals, are crucial for device operation and epitaxial growth. Surfaces may have different orientations with different terminations, may differently interact with an environment in which they operate, and may have different electronic structure. For epitaxial growth, the surfaces should be free from a damaged layer that is typically formed during wafer polishing.

4.6.1 Surface Properties

Theoretical study of the physical and electronic structures of the (100), (010), (001), and $(10\bar{1})$ faces of β-Ga_2O_3 was first performed by Bermudez [277]. The lowest energy plane is the easiest cleavage plane (100), which is non-polar and composed of two terminations named A and B (Fig. 4.35). The A-surface terminates in rows of O(II) with back-bonded two Ga(I) atoms, wherein between each O(II) row, there is a row of Ga(II) atoms. O(II) and Ga(II) atoms are singly unsaturated (missing a single nearest neighbor), while O(I), O(III), and Ga(I) in the surface plane are fully coordinated. The B-surface is terminated in rows of Ga(II) and O(III) atoms, each singly unsaturated. Ga(I) and O(I) atoms at the surface are fully coordinated, and there is no O(II) atoms at the surface plane. The (010) surface has only one termination containing all five atom types. All surface atoms are singly unsaturated except Ga(II), which is doubly unsaturated. The (001) surface is also non-polar and has two terminations as in the case of the (100) surface. Here, the A-surface terminates in rows of singly unsaturated O(I) atoms and back-bonded Ga(II) atoms. The O(I) rows are separated by rows of singly unsaturated Ga(I) atoms. Ga(II) and O(II) atoms in the surface plane are fully coordinated, and there are no O(II) atoms in the plane. The B-surface terminates in rows of doubly unsaturated O(III) atoms back-bonded to Ga(I) and Ga(II) atoms, and separated by rows of doubly unsaturated Ga(II) atoms. Ga(I) and O(II) atoms in the surface plane are fully coordinated, and there are no O(I) atoms in the surface plane.

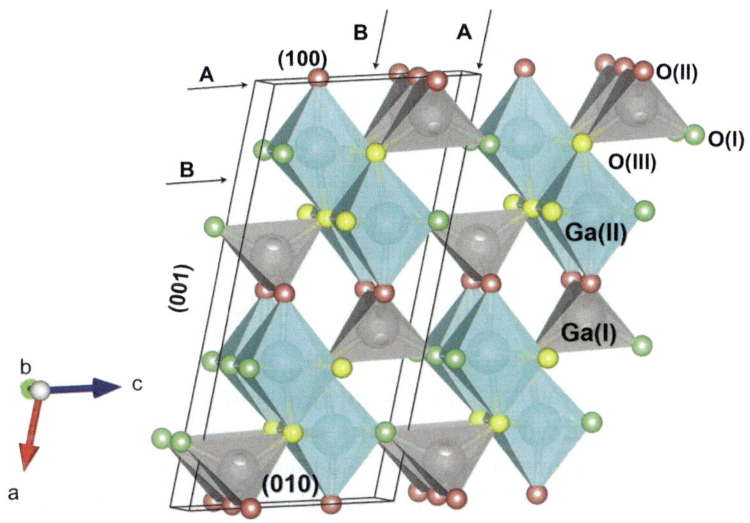

Figure 4.35 β-Ga_2O_3 structure viewed along the (010) direction. The labels "A" and "B" indicate the different (100) and (001) surface terminations. O(I), O(II), and O(III) are inequivalent oxygen sites, while Ga(I) and Ga(II) are inequivalent Ga sites. The box indicates the unit cell.

The presence of the two terminations A and B of the (100) surface that does not exhibit any reconstruction was experimentally confirmed by Lovejoy et al. [190] with the use of scanning tunneling microscopy (STM) and low-energy electron diffraction (LEED), as well as by Busch et al. [278] with the fast atom diffraction (H, H_2, He). The main termination of the (100) face is B-termination, while the contribution of the A-termination is below 5%. The dominating terraces arising from the termination have a height of 5.94 Å, and others of 4.4 and 1.5 Å. The terraces with the dominating height that corresponds to a half unit cell $(a/2)\sin(\beta)$ were also observed on epi-ready (100)-oriented wafers by Wagner et al. [279].

Another important surface property is the occurrence of upward/downward band bending, which may hinder ohmic/Schottky contact formation. Lovejoy et al. [190] experimentally demonstrated (hard X-ray photoemission spectroscopy, HXPS) the presence of upward band bending (electron depletion layer) by about 0.5 eV of the air-cleaved or vacuum-annealed (100) surface. The upward band bending could be a result of accumulation of negatively charged defects at the surface, either Ga vacancies or interstitial, or adsorbed oxygen [280]. The upward band bending has been confirmed by Navarro-Quezada et al. [184, 281] after applying X-ray photoemission spectroscopy (XPS) to as-cleaved and next annealed in UHV (100)-oriented samples. The values of upward band bending of 0.28 and 0.5 eV were found for as-cleaved and UHV-annealed samples, respectively. The upward band bending makes the surface insulating, while the bulk remains semiconducting. This has been experimentally proved by annealing bulk crystals by Galazka et al. [226] and Oshima et al. [282].

Furthermore, Navarro-Quezada et al. [281] showed that (100)-oriented wafers, after chemo-mechanical polishing (CMP), are contaminated with carbon, which can be removed by annealing in UHV ($<10^{-10}$ mbar) at 800°C for 30 min. The presence of carbon on the (100) surface may affect growth kinetics of β-Ga_2O_3 epi-layers and affect their electrical properties. Indeed, Wagner et al. [279] showed the negative impact of carbon (as CO_2) from Ga precursor tri-methyl-gallium (TMG) on the growth kinetics of homoepitaxial layers.

4.6.2 Surface Preparation

After cutting a bulk crystal into wafers, they are polished typically in the CMP process. For instance, Huang et al. [283] studied (100) surfaces of EFG-grown crystals after mechanical polishing (MP) and

CMP and concluded that very smooth surfaces can be obtained after the CMP process utilizing an alkaline slurry, where an important contribution was OH^- ions that react with the β-Ga_2O_3 (100) surface. The surface of the polished wafers has a damaged layer, which should be removed prior epitaxial growth. This can be done by wet or dry etching and/or by annealing. The wafer surface after such post-polishing treatment is epi-ready.

Ohira and Arai [284] studied a number of agents and temperatures for wet etching of (100) and (001)-oriented and CMP polished wafers fabricated from OFZ-grown bulk crystals. Among several solutions, only 47% HF at RT and 60.5% HNO_3 at 120°C were found to be effective with the etching rate of 31.3–58.7 and 86.6 nm/h, respectively. β-Ga_2O_3 seems to be resistant to other acid and basic solutions at RT. With the use of HF, the dissolution of Ga from Ga_2O_3 linearly increases with etching time and with HF concentration in the solution. The etching rate of the (100) was found to be almost twice higher as compared to the (001) plane. In the case of Sn-doped samples, the etching rate decreases with Sn-doping concentration.

Dry etching of (100), (010), and ($\bar{2}01$)-oriented β-Ga_2O_3 samples with the use of reactive ion etching (RIE) and inductively coupled plasma (ICP) was investigated by Hogan et al. [285]. Among different etching agents (BCl_3, CF_4/O_2, BCl_3/SF_6, and BCl_3/O_2), the highest etching rate was achieved by BCl_3. Both (010) and ($\bar{2}01$) planes revealed similar etching rates, while (100) showed lower etching rate. The highest etching rate was 19 and 33 nm/min for the RIE and ICP techniques, wherein the latter produced a lower surface roughness. However, Yang et al. [286] concluded that dry etching of ($\bar{2}01$)-oriented β-Ga_2O_3 in BCl_3/Ar by ICP may affect properties of Schottky diodes with Ni/Au contacts.

Thermal stability of the surface morphology of Sn-doped, (010)-oriented samples obtained from OFZ-grown bulk crystals was studied in detail by Togashi et al. [287]. Prior annealing, the samples were polished (CMP), etched in H_2SO_4 + H_2O_2 + deionized water (1:4:1) at RT for 5 min, rinsed in deionized water, and dried in flowing N_2. Annealing was performed in an RF furnace within a horizontal quartz tube with the samples placed on a pyrolytic barium nitride. The annealing experiments were carried out at temperatures 250 to 1450°C with flowing N_2 during heating up to a predefined temperature, and next the samples were treated at the predefined temperature for 60 min in a flowing gas mixture of H_2 + N_2. After the treatment at the predefined temperature, the gas was switched to N_2 during cooling down to RT. The total gas flow was 10000 sccm.

When the samples were annealed in N_2, there was no change in surface morphology after annealing at temperatures below 1150°C (Fig. 4.36a). Annealing at 1150°C or above revealed stripes parallel to the [001] direction, the number of which increased with temperature (Fig. 4.36b–e). These stripes are related to the decomposition of β-Ga_2O_3.

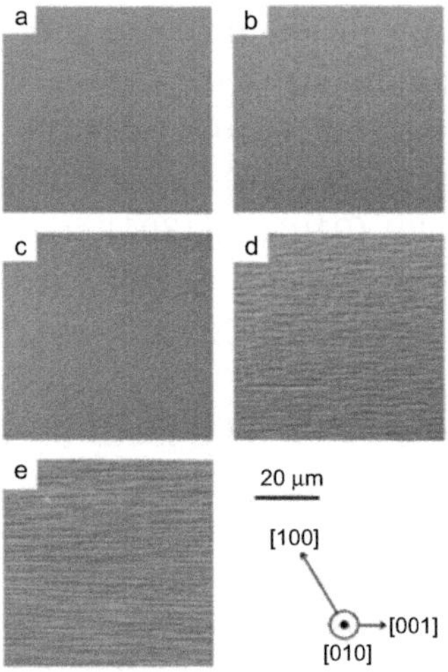

Figure 4.36 NDIC (Nomarski differential interference contrast) images of β-Ga_2O_3(010) surfaces after 60 min of heat treatment in N_2 flow at temperatures of (a) 1050, (b) 1150, (c) 1250, (d) 1350, and (e) 1450°C. Reprinted with permission from Ref. [287], Copyright 2015, The Japan Society of Applied Physics.

Annealing in the atmosphere containing H_2 ($H_2/(H_2 + N_2) = 0.75$ or $H_2/N_2 = 3:1$) at different temperatures showed no changes in the surface morphology after annealing at 250°C (Fig. 4.37a). However, a sample annealed at 350°C showed first signs of decomposition in the form of grains (Fig. 4.37b), which increased in size with annealing temperature (Fig. 4.37c–e). The samples that were annealed at 450°C and above revealed the formation of metallic Ga in the form of droplets on the surface.

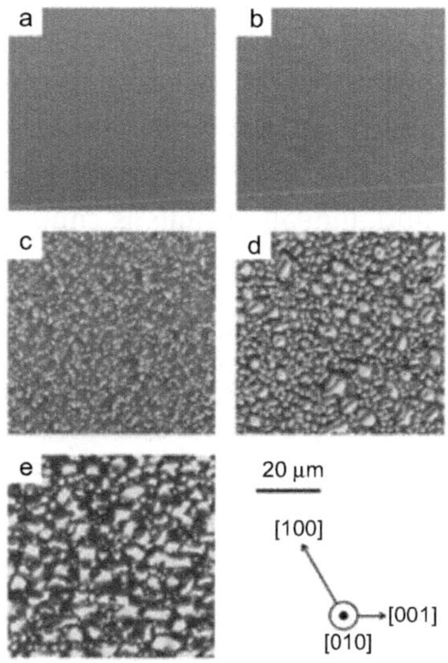

Figure 4.37 NDIC images of β-Ga_2O_3(010) surfaces after 60 min of heat treatment in a mixed flow of H_2 and N_2 (H_2/N_2 = 3:1) at temperatures of (a) 250, (b) 350, (c) 450, (d) 550, and (e) 650°C. Reprinted with permission from Ref. [287], Copyright 2015, The Japan Society of Applied Physics.

The decomposition rate increases with increasing the H_2 content in a similar way as with temperature at a constant $H_2/(H_2 + N_2)$ ratio.

The Czochralski-grown crystals showed thermal stability (10 h annealing time) up to about 1300°C in oxidizing atmosphere (O_2, air), up to about 1200°C in a neutral atmosphere (N_2, Ar), and up to about 600°C in a reducing atmosphere 5%H_2 + 95%Ar (H_2/Ar = 1:19) [8]. Here, thermal stability was indicated as a measurable mass loss during annealing and surface deterioration due to decomposition.

4.6.3 Conclusion

Surfaces (orientation, termination, band bending, cleanliness, stability) play a crucial role in epitaxy and device fabrication. Their preparation is, therefore, an important step in a technological chain. (100) and (001) surfaces of β-Ga_2O_3 are a mixture of so-called A and

B terminations with a domination of the latter (>95%). For polishing, the CMP process was found effective. Different wet and dry etching agents can be used for epi-ready surface preparation of β-Ga_2O_3, e.g., solutions of HF or HNO_3, and BCl_3. Thermal surface stability depends on atmosphere type, annealing temperature, and annealing time. In an oxidizing atmosphere, the crystals are substantially stable up to about 1300°C, in a neutral atmosphere up to 1150–1200°C, while in a reducing atmosphere up to 600°C depending on hydrogen content in the atmosphere and annealing time as conditions governing the surface decomposition. These basic surface properties are listed in Table 4.8.

Table 4.8 Basic surface features of bulk β-Ga_2O_3 single crystals

Feature	Description or value	Reference
Surface termination	A (<5%) + B (>95%) for (100) and (001) surfaces	277, 190, 278
Surface band bending	Upward for (100)	190, 280, 184, 281
Effective surface etchants	47% HF at RT, 60.5% HNO_3 at 120°C, BCl_3 (dry)	284, 285
Surface thermal stability in an oxidizing atmosphere (O_2) [°C]	1300[(a)]	8
Surface thermal stability in a neutral atmosphere (N_2) [°C]	1150[(b)]/1200[(a)]	287/8
Surface thermal stability in a reducing atmosphere (H_2) [°C]	350[(c)]/600[(d)]	287/8

Annealing time = [(a)]10 h; [(b)]1 h; [(c)]1 h (H_2/N_2 = 3:1); [(d)]10 h (H_2/Ar = 1:19)

4.7 Thermal and Mechanical Properties

Thermal and mechanical properties of β-Ga_2O_3 are important during bulk crystal growth (heat transport, growth orientation, stress, cleaving), epitaxial growth of layers (substrate orientation, heat transport, growth kinetic), as well as fabrication (cleaving) and operation (heat dissipation) of devices. Monoclinic system of

β-Ga_2O_3 determines the anisotropy of physical properties, which take a form of tensors instead of single values. Thermal properties (such as thermal conductivity and thermal expansion) can be described by a second rank tensor that requires four independent components. Elasticity must be described by a higher rank (4) tensor with considerably more components.

4.7.1 Melting Point and Density

The MP of β-Ga_2O_3 was typically measured with a pyrometer and resulted values span in a wide range, between 1725°C and 1850°C [153, 207, 225, 288, 289]. More accurate measurements of the MP were carried out by Hoshikawa et al. [233] with the use of type B thermocouple in the Bridgman configuration. The complete melting of β-Ga_2O_3 was recorded at 1793°C.

Liquid density of Ga_2O_3 was measured by Dingwell [290] using an Ir double-bob Archimedean method. The experimental liquid density is described by $\rho_l = 4.8374(84) - 0.00065(12)(T - 1950°C)$ [g/cm^3]. At 1800°C, which is around the MP of Ga_2O_3, the liquid density is 4.8736 g/cm^3, which is about 82% of the measured [163] and calculated β-Ga_2O_3 solid density of $\rho_s = 5.96$ g/cm^3.

4.7.2 Thermal Conductivity, Diffusivity, and Specific Heat

Thermal conductivity, diffusivity, and specific heat of bulk β-Ga_2O_3 grown from the melt were measured in a wide range from a low (25 K) to a high temperature of 1473 K by different techniques, such as laser flash method (Galazka et al. [226], Amgalan et al. [291], Villora et al. [292], Mu et al. [266]), 2ω and 3ω methods (Handwerg et al. [293, 294], Slomski et al. [295]), and time domain thermoreflectance TDTR (Guo et al. [296], Jiang et al. [297]). The measured crystals were either electrically insulating (Mg- and Fe-doping), or semiconducting (undoped and Sn-doped). It was found, however, that the intentional doping (whether toward low (Fe) or high (Sn) electrical conductivity) has a minor impact on the host thermal conductivity [295]. Measured and calculated [298] values of the thermal properties are listed in Table 4.9. Measured thermal conductivity is highly anisotropic with values at RT of 11–13, 13, 21–29, and 15–21 W/mK along the [100], [$\bar{2}$01], [010], and [001] directions, respectively. For the same orientation order, undoped Czochralski-grown bulk β-Ga_2O_3 single crystals showed at RT the following values: 12.2, 15, 24.6, and 17.6 W/mK, respectively (Fig. 4.38a). Although there are some differences in values gathered by different methods, the anisotropy trend is the same. The thermal conductivity decreases with temperature and at 1200°C approaches the value of about 40% of that at RT. The specific heat capacity has an inverse relation, and its value at RT equals 0.56 J/gK. The thermal diffusivity follows the anisotropy of the thermal conductivity with the values at RT of 3.7, 9.6, and 7.1 mm^2/s along the [100], [010], and [001] directions, respectively. The anisotropy factor for both thermal diffusivity $D[010]/D[001]$ and thermal conductivity $\lambda[010]/\lambda[001]$, which equals 1.4, was found to be temperature independent below RT [294].

Table 4.9 Orientation-dependent thermal conductivity of bulk β-Ga_2O_3 crystals at RT

Axis	Thermal conductivity [W/mK]	Method	Ref.
[100]	11 ± 1(a)	2ω	294
	12.2	Laser flash	291
	13	Laser flash	292
	10.9 ± 1	TDTR	296
	16	First principles	298
[010]	29 ± 2(a)	2ω	294
	29.21(b), 29.13(c), 29.06(d)	3ω	295
	24.6	Laser flash	291
	21(a)	Laser flash	226
	27 ± 2	TDTR	296
	22	First principles	298
[001]	21 ± 2(a)	2ω	294
	17.6	Laser flash	291
	15	TDTR	296
	21	First principles	298
[$\bar{2}$01]	15	Laser flash	291
	13	TDTR	296
	13	First principles	298

(a)Mg doped (inslating), (b)undoped ($n_e = 3 \times 10^{17}$ cm^{-3}), (c)Sn doped ($n_e = 3.5 \times 10^{18}$ cm^{-3}), (d)Fe doped (semi-insulating)

Low temperature measurements of Czochralski-grown crystals by Handwerg et al. [293] revealed that the thermal conductivity between 150 and 300 K follows phonon–phonon Umklapp scattering ($\propto T^{-1}$), while at lower temperatures, there is an increasing deviation from the phonon–phonon scattering, as shown in Fig. 4.38b. This deviation is assigned to point defect scattering, which might play an important role in thermal conductivity. On the other hand, Jiang et al. [297] claimed that the thermal conductivity of EFG-grown crystals between 80 and 400 K follows the $\propto T^{-1.3}$ relation.

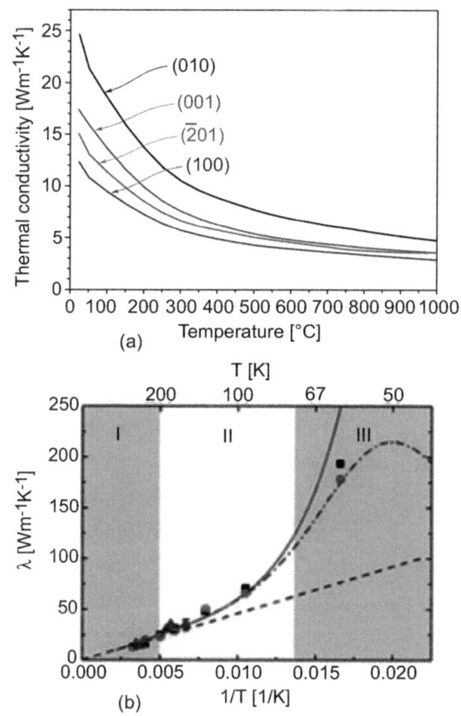

Figure 4.38 (a) High-temperature-dependent thermal conductivity (by laser flash method) of undoped, Czochralski-grown bulk β-Ga_2O_3 single crystals along directions perpendicular to the (100), (010), (001), and ($\bar{2}01$) surfaces, respectively [291]; (b) low-temperature-dependent thermal conductivity of Mg-doped (squares, circles) and undoped (triangles) bulk β-Ga_2O_3 single crystals along the [100] direction. Dashed line: phonon–phonon scattering for $T \gg \theta_D$, where θ_D is the Debye temperature. Solid line: phonon–phonon–Umklapp scattering, calculated for $T < \theta_D$. Dot-dashed line: additional scattering process using Matthiessens rule for a temperature-independent conductivity. The regions (I), (II), and (III) correspond to high-, intermediate-, and low-temperature regions, respectively. Reprinted from Ref. [293], Copyright 2015, with permission from IOP Publishing.

Thermal conductivity of β-Ga_2O_3 has four independent components, which are described in the standard orientation (twofold axis parallel [010]) by a tensor (basing on Refs. [294], [295], and [291]):

$$\lambda = \begin{pmatrix} \lambda_{[100]} & 0 & \lambda_{[\bar{2}01]} \\ 0 & \lambda_{[010]} & 0 \\ \lambda_{[\bar{2}01]} & 0 & \lambda_{[001]} \end{pmatrix} = \begin{pmatrix} 11 & 0 & 13 \\ 0 & 29 & 0 \\ 13 & 0 & 21 \end{pmatrix} [294, 295]$$

$$\text{or} \begin{pmatrix} 12.2 & 0 & 15 \\ 0 & 24.6 & 0 \\ 15 & 0 & 17.6 \end{pmatrix} [291] \frac{\text{W}}{\text{mK}} \text{ at RT}$$

4.7.3 Thermal Expansion

The thermal expansion of β-Ga_2O_3 is also anisotropic and can be described as a tensor with four independent components. The measurements were performed at low and elevated temperatures (between 5 and 700 K) using bulk crystals and powders. Several independent studies [272, 292, 299, 300] on the thermal expansion along main directions and the angle between *a*- and *c*-axes showed, however, discrepancy, as listed in Table 4.10. Lattice expansion measurements were carried out by powder X-ray diffraction [292, 299], high-resolution dilatometry [272], and synchrotron-based high-resolution X-ray diffraction [300]. At low-temperature region, the thermal expansion coefficients (with respect to those at RT) of 1.8×10^{-6} K^{-1} along the [100], and 4.2×10^{-6} K^{-1} along the [010] and [001] directions were measured, meaning a substantially homogenous expansion in the (100) plane (Villora et al. [292]). At an elevated temperature range, the following thermal expansion coefficients were measured (with respect to those at RT): 1.54×10^{-6} K^{-1}, 3.37×10^{-6} K^{-1}, and 3.15×10^{-6} K^{-1} along the [100], [010], and [001] directions, respectively, while the expansion of the angle β between the *a*- and *c*-axes is 2.23×10^{-5} K^{-1} (Orlandi et al. [299]). Detailed measurements of high-quality single crystals by Cheng et al. [300] revealed much smaller values of the thermal expansion coefficient as compared with other studies and also a nonlinear increase in the expansion coefficients with temperature (in a range of 298–1200 K).

Table 4.10 Thermal expansion coefficient of β-Ga_2O_3

Direction	Thermal expansion coefficient $[K^{-1}] \times 10^{-6}$			
	Ref. [292] below RT	Ref. [299] above RT	Ref. [272] at RT/at 650 K	Ref. [300] at RT/at 600 K/at 1200 K
[100]	1.8	1.54	2.92/4.85	0.1/1.24/2.78
[010]	4.2	3.37	6.27/8.33	1.68/4.51/5.84
[001]	4.2	3.15	5.89/8.47	1.74/4.79/6.27
β	—	22.3	0.865/0.0116	—

4.7.4 Stiffness and Hardness

Elastic stiffness coefficients for Czochralski-grown β-Ga_2O_3 crystals were measured by plate-resonance technique and resonant ultrasonic spectroscopy, as reported by Miller et al. [272]. The tensor of the coefficients c_{ij} around RT is as follows:

$$c_{ij} = \begin{pmatrix} 23.8 & 13.0 & 15.2 & 0 & -0.4 & 0 \\ & 35.9 & 7.8 & 0 & 0.2 & 0 \\ & & 34.6 & 0 & 1.9 & 0 \\ & & & 4.9 & 0 & 0.6 \\ & \text{Symmetry} & & & 9.1 & 0 \\ & & & & & 10.7 \end{pmatrix} \times 10^{10}\ [\text{Pa}]$$

Theoretical values of elastic stiffness coefficients reported by Furthmüller and Bechstedt [161] are quite close to those obtained experimentally by Miller et al. [272].

Basing on the experimental elastic tensor, the bulk modulus B_0, shear modulus G, Young's modulus E, and Poisson's ratio v can be calculated from the following Voigt relations:

$$B_0 = \frac{1}{9}[c_{11} + c_{22} + c_{33} + 2(c_{12} + c_{13} + c_{23})] \tag{4.16}$$

$$G = \frac{1}{15}[c_{11} + c_{22} + c_{33} + 3(c_{44} + c_{55} + c_{66}) - (c_{12} + c_{13} + c_{23})] \quad (4.17)$$

$$E = \frac{9B_0G}{(3B_0 + G)} \quad (4.18)$$

$$\nu = \frac{3B_0 - 2G}{(6B_0 + 2G)} \quad (4.19)$$

The obtained values are as follows: B_0 = 185 GPa, G = 88 GPa, E = 228 GPa, and ν = 0.295.

The above-described thermal properties along with the elastic stiffness coefficients enabled to calculate thermal stress distribution in 2-inch-diameter crystals grown by the Czochralski method (by using real thermal field in a growth furnace) at two different orientations along *b*- and *c*-axes [272]. The numerical results revealed that the maximum of von Mieses stress generated in a crystal grown parallel to the [010] direction is about 50% lower than that in a crystal grown parallel to the [001] direction.

Another mechanical property of bulk β-Ga_2O_3 crystals includes hardness. Microhardness of β-Ga_2O_3 measured by Guzilowa et al. [301] revealed an average value of 8.91 GPa on an (001)-oriented crystal sample. Microhardness and Mohs hardness reported by Mu et al. [266] have values of 8.32, 6.44, and 10.1 GPa, and of 6.39, 5.78, and 6.82 for (100), (010), and (001)-oriented samples, respectively.

4.7.5 Cleavage Planes

β-Ga_2O_3 suffers from two easy cleavage planes {100} and {001}, which are formed by O(III) and O(I) atoms, respectively, as shown in Fig. 4.39. The {100} plane is the lowest energy plane and constitutes the easiest cleavage plane [277].

On one hand, the cleavage planes enable easy preparation of crystal samples for fundamental study, as they are neither contaminated with polishing agents nor have a damaged layer. On the other hand, the cleavage planes make the bulk crystal growth and wafer fabrication more difficult. In both the Czochralski and EFG methods, the growth direction is limited to the <010> direction, which is parallel to both cleavage planes. Otherwise, the weight of a growing crystal combined with the thermal stress would easily break the seed along one of the cleavage planes. Slicing bulk crystals into thin wafers perpendicularly to the cleavage planes is quite difficult, especially from large-diameter crystals (>1 inch). Polishing such wafers is a challenging task as well. Finally, cutting such wafers with devices fabricated thereon into individual components would require much technical attention.

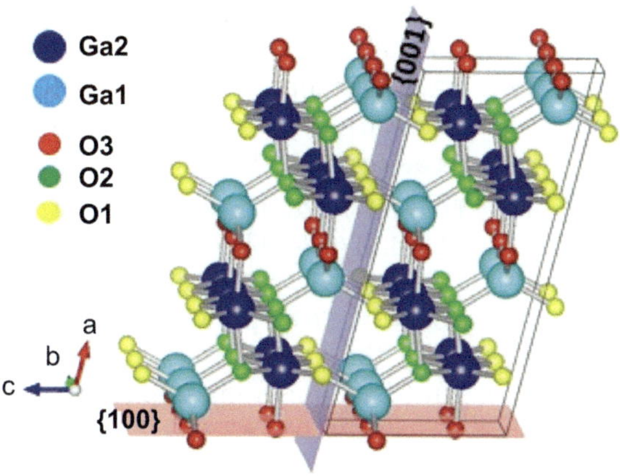

Figure 4.39 Unit cell of monoclinic β-Ga_2O_3 with indicated cleavage planes {100} and {001}. Reprinted from Ref. [8], Copyright 2018, with permission from IOP Publishing.

4.7.6 Conclusion

Thermal and mechanical properties (Table 4.11) of solid β-Ga_2O_3, i.e., thermal conductivity, diffusivity, specific heat, thermal expansion, and stiffness, are anisotropic and described by tensors. The thermal conductivity increases by a factor of about 2.5 when changing the direction from the [100] through [$\bar{2}$01] and [001] to the [010] direction. This shows that from the point of view of power device applications, the wafer orientation becomes important for heat dissipation. A complete set of thermal properties enables to calculate thermal stresses in growing crystals, which is important for large crystal diameters. The presence of two easy cleavage planes makes the crystal growth and wafer fabrication more challenging.

Table 4.11 Thermal and mechanical properties of β-Ga_2O_3

Property	Value	Reference
Melting point [°C]	1793	233
Density [g/cm^3]	Liquid = 4.87 Solid = 5.96	290 163
Thermal conductivity [W/mK]	10.9/11/12.2/13 along [100] 21/24.6/27/29 along [010] 15/17.6/21 along [001] 13/15 along [$\bar{2}$01]	296/294/291/292 226/291/296/294 296/291/294 296/291
Diffusivity [mm^2/s]	3.7 along [100] 9.6 along [010] 7.1 along [001]	294
Thermal expansion [K^{-1}] ×10^{-6} at 300/600/1200 K	1.04/1.24/2.78 along [100] 1.68/4.51/5.84 along [010] 1.74/4.79/6.27 along [001] 0.865/0.0116 for β	300 300 300 272
Bulk modulus [GPa]	185	272[a]
Shear modulus [GPa]	88	272[a]
Young's modulus [GPa]	228	272[a]
Poisson's ratio	0.295	272[a]
Microhardness [GPa]	8.32 for (100) 6.44 for (010) 10.1 for (001)	266
Mohs hardness	6.39 for (100) 5.78 for (010) 6.82 for (001)	266
Cleavage planes	{100}, {001}	

[a]Calculated on the basis of experimental data in the corresponding reference.

4.8 Basic Optical Properties

Basic optical properties of β-Ga_2O_3, such as transmittance, absorption, reflectance, refractive index, optical bandgap, vibrational modes, and the like, are determined by its band and crystal structures. Their exploration, understanding, and modification (e.g., through intentional doping) enable to design optoelectronic devices with desired functionality that may expand the area of applications. Optical properties provide an insight into the intrinsic properties of the material arising from the band (optical transitions) and crystal structure (e.g., phonon modes), and to some kind of defects (e.g., self-trapped excitons).

4.8.1 Transmittance, Absorption, Bandgap

The availability of bulk β-Ga_2O_3 single crystals of high perfection enables to prepare high-quality samples of a suitable size and different orientations for optical characterization. Such materials reflect well their intrinsic optical properties with a rather minor impact of the structural defects. Moreover, bulk crystals are also suitable for various dopants to tune the optical properties. As shown in Fig. 4.40a, transmittance spectra of bulk crystals reveal a steep absorption edge originating at about 260 nm and good transparency in the visible and near-infrared (NIR) regions at low free electron concentration or in the case of electrically insulating samples. However, with the increase in the free electron concentration, the transmittance deteriorates first in the NIR region and also in the visible part of the spectrum at very high free electron concentration, as a result of the free carrier absorption. The absorption in the red part of the visible spectrum makes the crystals blue.

Annealing of undoped Czochralski-grown crystals in non-reducing conditions at temperatures up to about 1400°C for at least several hours improves the transmittance in the infrared and red regions of the visible spectrum (Fig. 4.40b) by reducing the free electron concentration. The transmittance improves with the annealing temperature up to 1200°C, while at higher temperatures (1400°C), it slightly decreases due to surface deterioration caused by thermal decomposition. On the other hand, annealing in the reducing conditions (5%H_2 + Ar, 500–800°C, 10 h) deteriorates the transmittance in the spectral region 300–800 nm.

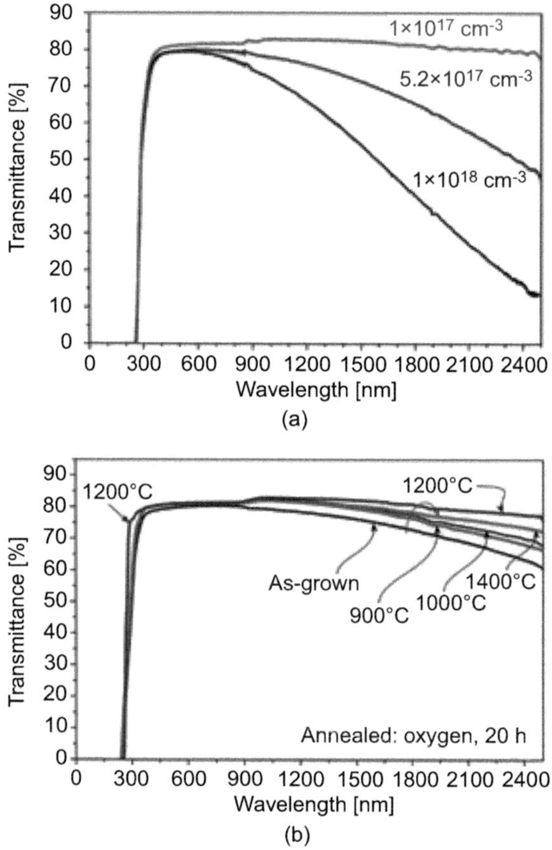

Figure 4.40 Transmittance spectra of as-grown (a) and annealed (b) bulk β-Ga_2O_3 crystal samples obtained by the Czochralski method and having different free electron concentrations. Sample thickness and orientation were 0.5 mm and (100), respectively. Figure (b) reprinted from Ref. [226], Copyright 2014, with permission from Elsevier.

β-Ga_2O_3 shows a polarization-dependent near-edge absorption, as presented in Fig. 4.41 [8]. By comparing different relations between measured near-edge absorption coefficient and the photon energy, it could be concluded that the optical bandgap of β-Ga_2O_3 can be regarded as a quasi-direct gap with the values of 4.56, 4.59, and 4.85 eV for the polarization parallel to *c*, *a*, and *b*, respectively. This is consistent with theoretical and experimental electronic structure [179, 191] revealing a very small energy difference of

40–50 meV between the direct and indirect transitions at Γ point of the BZ at CBM, and also a weakness of the indirect transition. Different energetic values of the absorption edge depending on the polarization arise from direct optical transitions at Γ point of the BZ between different VB levels and the conduction band that are dipole allowed [302] for E||*a*, E||*b*, and E||*c*.

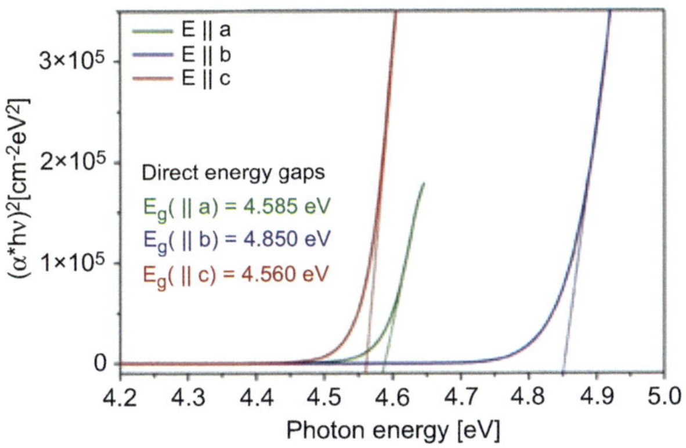

Figure 4.41 Polarization-dependent absorption edge of bulk β-Ga_2O_3 crystal samples obtained by the Czochralski method. Sample thickness = 120–124 μm. Reprinted from Ref. [8], Copyright 2018, with permission from IOP Publishing.

The transmittance or absorption spectra of bulk β-Ga_2O_3 crystals may also be affected by intentional dopants or residual impurities and their doping levels. The dopants/impurities may affect the transmittance (or absorption) through an increase in the free electron concentration (Si, Sn, Nb) or its decrease (Mg, Co, Ni, Li, Fe), as illustrated in Fig. 4.42 for some of the dopants. In such cases, the doping level is small, usually below 0.25 mol.% in the melt (for more details, see Section 4.5.6.5); therefore, they rather do not affect the fundamental absorption edge, although Shimamura et al. [259] reported a bandgap shift toward longer wavelengths by Si doping (~6×10^{18} cm^{-3}). To shift the absorption edge toward shorter wavelengths, a relatively high doping level (>1 mol.%) should be practiced. For example, doping of Czochralski-grown β-Ga_2O_3 with 5 mol.% Al_2O_3 resulted in the shift of the absorption edge by 10 nm [229]. Transition metals, such as Cr [229], Ti, V, Cu, Co, Ni, and Mn [250] may introduce an additional absorption in the visible spectrum and affect the fundamental absorption edge of bulk β-Ga_2O_3 crystals, if *d-d* transitions occur below the bandgap of β-Ga_2O_3. Indeed, as shown in Fig. 4.42, Co (also Ni, not shown) induces an additional absorption in the visible spectrum, in particular in its short-range wavelengths, while Cr induces absorption bands. Other dopants, such as Ce, do not have any impact on the transmittance/absorption in the UV, visible, and NIR regions of the spectrum when in concentrations below 300 wt. ppm in the crystals.

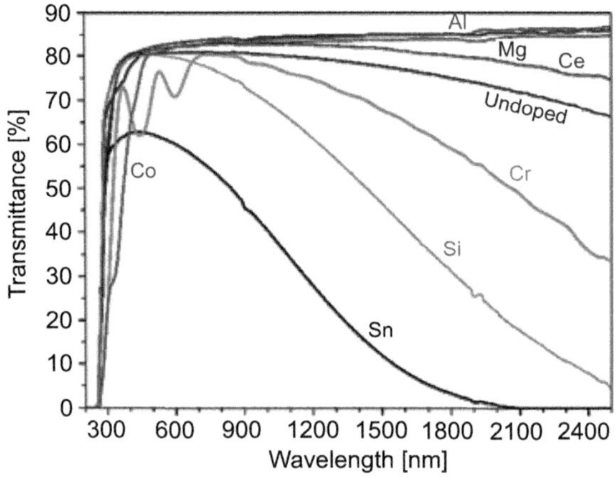

Figure 4.42 Transmittance spectra of Czochralski-grown bulk β-Ga_2O_3 crystal samples doped with Sn, Si, Cr, Al, Ce, Mg, and Co, in comparison with an undoped sample. Sample thickness and orientation were 0.5 mm and (100), respectively.

4.8.2 Refractive Index and Dielectric Constant

Details of the dielectric function and corresponding optical properties of bulk β-Ga_2O_3 single crystals, which are essential in operation with the electromagnetic radiation, can be found in a number of scientific reports, e.g., by Schubert et al. [309, 303], Mock et al. [188, 304], Knight et al. [336], Sturm et al. [305, 306], and Onuma et al. [307]. In a monoclinic system, it is described by four independent dielectric tensor elements [188]:

$$\varepsilon = \begin{pmatrix} \varepsilon_{xx} & \varepsilon_{xy} & 0 \\ \varepsilon_{xy} & \varepsilon_{yy} & 0 \\ 0 & 0 & \varepsilon_{zz} \end{pmatrix} \tag{4.20}$$

where *a* is parallel to *x*-axis, c^* is parallel to *y*-axis, while *b* is parallel to *z*-axis of the Cartesian coordinates.

Here we make a note on very basic parameters only, i.e., the refractive index and dielectric constants.

Figure 4.43 shows an example of the refractive index versus photon energy along different axes of β-Ga_2O_3, as obtained by Onuma et al. [307]. At low-energy region (<E_g of β-Ga_2O_3), the anisotropy of the refractive index is very small, and it decreases from about 2.1 at 310 nm to about 1.95 at 800 nm. However, the anisotropy of the refractive index becomes significant near the absorption edge with the highest value along the [010] direction.

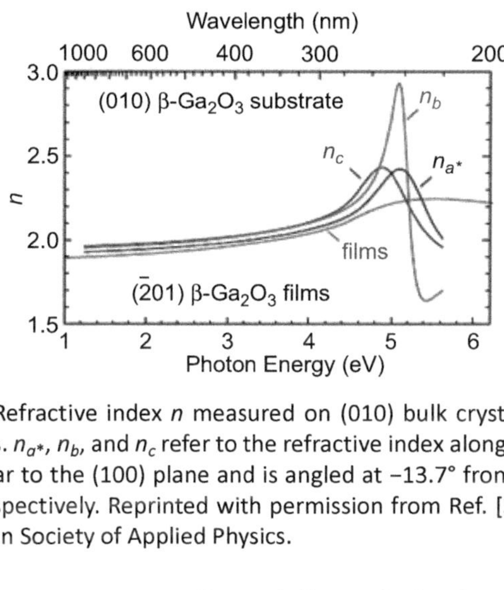

Figure 4.43 Refractive index *n* measured on (010) bulk crystals (substrates) and ($\bar{2}$01) films. n_{a*}, n_b, and n_c refer to the refractive index along a^* (the a^*-axis is perpendicular to the (100) plane and is angled at −13.7° from the *a*-axis), *b*- and *c*-axes, respectively. Reprinted with permission from Ref. [307], Copyright 2016, The Japan Society of Applied Physics.

In the measurement of partially polarized reflectance of differently oriented β-Ga_2O_3 samples, Onuma et al. [328] could match quite well features of the reflectance spectra with the onset of the absorption edge and also with other transitions from the VB to the CB.

The experimental relative static dielectric constant ε_r of Czochralski and EFG-grown semi-insulating β-Ga_2O_3 crystal samples as a function of temperature (25–500 K) was studied in detail by Fiedler et al. [308]. The outcome for directions perpendicular (⊥) to (100), (010), and (001) is illustrated in Fig. 4.44. ε_r shows a strong anisotropy with the highest values for the direction ⊥(001);

however, its increase over a temperature range 25–450 K is similar and equals approximately 0.5. Moreover, the capacitance and hence ε_r was found constant in a frequency range of 20 kHz to 1 MHz. At RT, $\varepsilon_r = 10.2 \pm 0.2$, 10.87 ± 0.08, and 12.4 ± 0.4 for directions ⊥(100), ⊥(010), and ⊥(001), respectively.

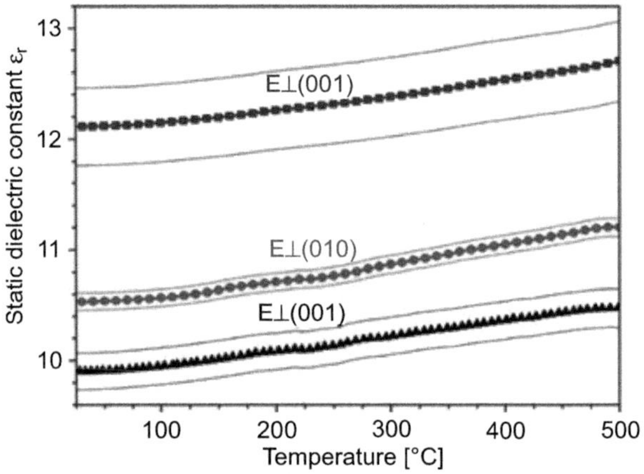

Figure 4.44 Static dielectric constant of Czochralski and EFG-grown semi-insulating β-Ga_2O_3 crystal samples as a function of temperature for directions perpendicular to (100), (010), and (001). Scatters are measured values, while thin solid lines confine error bars. Reprinted (with minor modifications) from Ref. [308] under Creative Commons license.

Other experimental values of ε_r measured on bulk crystals or epitaxial layers, and obtained theoretically, also showed anisotropy with similar values; however, they are assigned to the Cartesian coordinates *a* (ε_{xx}), *c** (ε_{yy}), and *b* (ε_{zz}). For example, ellipsometric measurements by Schubert et al. [309], Passlack et al. [310], Shan et al. [311], Schmitz et al. [312] revealed ε_r (ε_{xx}, ε_{yy}, ε_{zz}) ≈ 10–12, 9.93–10.2, 9–13, and 9.57, respectively. For comparison, theoretical ε_r values span between 10.8–13.9 [313] and 9.2–12.6 [314]. Relative high-frequency dielectric constant ε_∞ measured for bulk crystals and thin films is 3.2–3.7 [309], 3.53 [310], and 3.6 [312]. Exemplary theoretical values of the ε_∞ are 3.81–4.08 [313] and 2.78–2.86 [177].

4.8.3 Luminescence

Photoluminescence (PL) and cathodoluminescence (CL) measurements of bulk β-Ga_2O_3 crystals showed three emission bands at UV (3.2–3.6 eV), blue (2.8–3.0 eV), and green (2.4 eV) spectral regions, but no near band-edge emission was observed [254, 256]. These emissions involve self-trapped excitons (STEs), the stability of which has been confirmed theoretically [315] and experimentally [316]. According to Yamaga et al. [258], the UV and blue emissions occur through a recombination of a self-trapped hole and an electron trapped at a single oxygen vacancy, while the green emission occurs through a recombination of a self-trapped hole and an electron trapped at a cluster of oxygen vacancies. These emission bands were observed in bulk crystals obtained by different growth methods (different growth conditions) both as-grown and annealed, in differently oriented samples, and in samples doped with various elements. However, the intensities of particular bands may change depending on the growth method and a doping type. For instance, according to Shimamura et al. [259] and Nakazawa et al. [250], Si and some of the transition metals (Cr, Fe, Co, Ni, Mn, and Ti) decrease the intensity of the blue emission of PL spectra. Annealing Sn-doped β-Ga_2O_3 crystal samples in O_2 at 1100°C for 3 h led to an increase in the blue emission intensity [260].

More detailed study of temperature, orientation, dopant, and growth method-dependent CL spectra by Onuma et al. [253] led to the following conclusions (Fig. 4.45):

- In undoped samples, UV emission bands at 3.2, 3.4, and 3.6 eV dominate at low temperatures, whose intensities decrease with temperature.
- At RT, blue (2.8 and 3 eV) and green (2.4 eV) bands dominate.
- The peak positions are basically temperature and dopant independent.
- In Si-doped samples, blue and green bands are almost suppressed.
- In Mg-doped samples, there is an enhancement of the blue and green bands with rising temperature.
- At RT, the blue emission intensity is the lowest for Si doping (highly conducting) and the highest for Mg doping (electrically insulating) with undoped (likely moderately conducting) crystals in between.

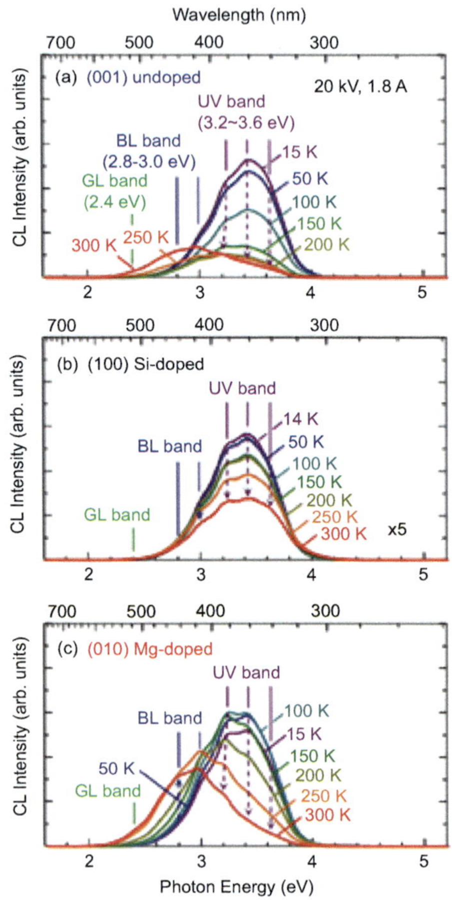

Figure 4.45 Temperature-dependent CL spectra of (a) (001) undoped, (b) (100) Si-doped (enlarged by factor of 5), and (c) (010) Mg-doped β-Ga_2O_3 crystal samples. Reprinted from Ref. [253], Copyright 2013, with permission from AIP Publishing.

The involvement of self-trapped holes (STHs) and conduction band electrons in the luminescence of bulk β-Ga_2O_3 crystals was experimentally (PL) and theoretically (*first-principles calculations*) confirmed by Wang et al. [317]. In Mg- and Ca-doped crystals, the highest UV emission was observed under 266 nm excitation that leads to a recombination of an electron and STHs formed by O_I and O_{II} sites in the β-Ga_2O_3 crystal lattice.

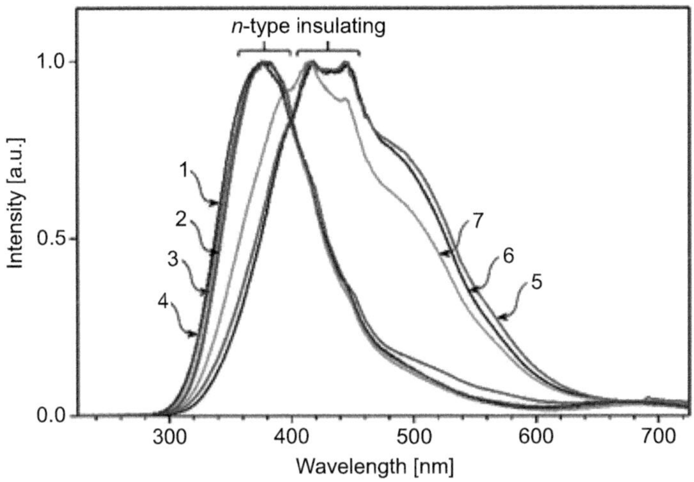

Figure 4.46 CL spectra of Czochralski-grown, (100)-oriented β-Ga_2O_3 crystal samples at RT. Semiconducting samples: (1) As-grown; (2) annealed in H_2 + Ar, 600°C, 10 h; (3) annealed in O_2, 800°C, 20 h and then in H_2 + Ar, 600°C, 10 h; (4) as-grown, doped with Sn. Electrically insulating samples: (5) annealed in O_2, 1200°C, 20 h; (6) as-grown, doped with Mg; (7) doped with Mg and annealed in H_2 + Ar, 600°C, 10 h. Reprinted (with minor modifications) from Ref. [8], Copyright 2018, with permission from IOP Publishing.

As the oxygen vacancies are likely involved in the emissions, the formation energy of which depends on Fermi level shift, the intensities of the luminescence will depend on the conductivity state of β-Ga_2O_3 crystals. This can be well recognized when plotting all together normalized CL spectra of different Czochralski-grown samples: as-grown, doped with Sn or Mg, annealed in the presence of O_2 or H_2, and double annealed in the presence of O_2 and then in H_2 (Fig. 4.46) [8]. We can observe, indeed, broad UV (peaking at ~380 and ~395 nm), blue (peaking at ~415 and ~445 nm), and green (peaking at ~495 nm) emissions, which are fundamentally the same as those observed in crystals obtained by the OFZ and EFG methods. The peak positions of the CL spectra at RT are located at the same positions, independent of growth conditions, doping or annealing conditions; however, their intensities change. The picture of all the CL spectra clearly shows two distinguished groups of samples: semiconducting and electrically insulating. The spectra in each of the group well overlap with each other independent of how the semiconducting or insulating states were achieved (through doping and/or annealing). In the case of electrically insulating samples, the blue emissions are enhanced in favor of the UV bands, in a similar way as in the case of OFZ- and EFG-grown crystals.

4.8.4 Phonon Modes

In β-Ga_2O_3 (monoclinic system, space group C2/m), there are 15 Raman ($10A_g + 5B_g$), 12 infrared ($4A_u + 8B_u$), and 3 acoustic ($2B_u + A_u$) active phonon modes, which are expressed by (Dohy et al. [318]):

$$10A_g + 5B_g + 10B_u + 5A_u \tag{4.21}$$

Table 4.12 Phonon modes in β-Ga_2O_3 measured at RT

Mode at RT	Phonon modes, Ref. [318], [cm^{-1}]	Phonon modes, Ref. [319], [cm^{-1}]	Phonon modes, Ref. [320], [cm^{-1}]
$A_g(1)$	111	112	111
$B_g(1)$	114	115	114.8
$B_g(2)$	147	149	144.8
$A_g(2)$	169	173	169.9
$A_g(3)$	199	205	200.2
$A_g(4)$	318	322	320
$A_g(5)$	346	350	346.6
$B_g(3)$	353	355	353.2
$A_g(6)$	415	421	416.2
$A_g(7)$	475	479	474.9
$B_g(4)$	475	480	474.9
$A_g(8)$	628	635	630
$B_g(5)$	651	659	652.3
$A_g(9)$	657	663	658.3
$A_g(10)$	763	772	766.7

Three phonon regions were identified in the Raman spectra that involve [318, 319]: (i) translation and liberation of tetrahedral chains Ga_IO_4 as a whole (low frequencies, 100–200 cm^{-1}); (ii) deformation of the tetrahedral and octahedral groups Ga_IO_4 and $Ga_{II}O_6$ (middle frequency, 300–500 cm^{-1}); and (iii) bending and stretching of tetrahedral groups Ga_IO_4 (high frequency, 600–800 cm^{-1}). Values of the phonon modes measured at RT by Dohy et al. [318], Onuma et al. [319], and Kranert et al. [320] are collected in Table 4.12.

The intensities and broadening of the Raman bands depend on lattice disorder caused by defects (e.g., at interstitial sites) and possible by off-stoichiometry [318], dopant concentration and its type [319], birefringence [320], and growth conditions (O_2 concentration in a growth atmosphere) [255]. Raman spectroscopy was found as an effective tool in assessing the structural quality of the homoepitaxial layers by comparing Raman spectra of both β-Ga_2O_3 substrates and obtained layers [321].

4.8.5 Conclusion

Basic optical properties of bulk β-Ga_2O_3 crystals are summarized in Table 4.13. β-Ga_2O_3 is transparent in the UV, visible, and NIR spectral regions with no absorption bands. The transmittance decreases in the NIR and also in the visible spectra with an increase in the free carrier concentration, whether intentionally doped or not. Other dopants, especially transition metals, may affect the transmittance or absorption depending on the dopant nature. The near-edge absorption is steep and orientation dependent with the values of about 4.56, 4.59, and 4.85 eV for E II [001], E II [100], and E II [010], respectively. The anisotropy is also reflected in the refractive index and reflectance, in particular around the near-edge absorption region.

The luminescence of bulk β-Ga_2O_3 under UV, electron, or X-ray excitations occurs in UV, blue, and green regions of the spectrum by involving STEs (recombination of an electron trapped at oxygen vacancy with a self-trapped hole). Semiconducting and electrically insulating crystal samples are easily distinguishable by the CL spectra. All phonon modes of β-Ga_2O_3 have been measured, and their values are substantially consistent between bulk crystals obtained by different growth methods.

Table 4.13 Basic optical properties of β-Ga_2O_3

Property	Unit	Value	Reference
Optical bandgap [eV]	eV	4.59 II [100] 4.85 II [010] 4.56 II [001]	8
Bandgap type		Quasi-direct	179, 191
Absorption edge	nm	≈260	226
Emission bands at RT	eV	3.2–3.6 2.8–3.0 2.4	8, 250, 253, 258, 259, 260, 316, 317
Relative static dielectric constant at RT		10.2 ± 0.2 ⊥(100) 10.87 ± 0.08 ⊥((010) 12.4 ± 0.4 ⊥((001)	308
		12.7 II *a* 11.2 II *b* 10.9 II *c**	309
Relative high-frequency dielectric constant at RT		3.7 II *a* 3.7 II *b* 3.2 II *c**	309
Refractive index at 310–800 nm		≈2.1–1.95	307
Phonon modes	cm^{-1}	$A_g(1) = 111$, $B_g(1) = 114$, $B_g(2) = 147$, $A_g(2) = 169$, $A_g(3) = 199$, $A_g(4) = 318$, $A_g(5) = 346$, $B_g(3) = 353$, $A_g(6) = 415$, $A_g(7) = 475$, $B_g(4) = 475$, $A_g(8) = 628$, $B_g(5) = 651$, $A_g(9) = 657$, $A_g(10) = 763$	318

4.9 Scintillation Properties

In addition to outstanding properties for electronic and optoelectronic applications, β-Ga_2O_3 has shown a scintillation response to X-ray and nuclear radiation, namely γ, α, and neutron radiations.

Scintillation properties of undoped, Sn-doped (1–10 mol.%), and Ce-doped (0.04–1.9 mol.%) β-Ga_2O_3 obtained by the EFG and OFZ methods were initially investigated by Yanagida et al. [125] and Usui et al. [126, 127]. Radioluminescence spectra showed peaks at around 380 nm (both EFG- and OFZ-grown crystals) [125, 126] at around 380–410 nm [126], and 400–420 nm (assigned to Ce^{3+} 5d-4f transitions) [127] for undoped, Sn-doped, and Ce-doped crystals, respectively. Scintillation time profiles under X-ray excitation were two-component for undoped EFG and OFZ-grown crystals, and three-component for Sn- and Ce-doped crystals (OFZ method). A fast decay time constant was measured around or below 10 ns for undoped and Ce-doped β-Ga_2O_3, and 15–19 ns for Sn-doped samples. The middle decay time constants were 156 ns, 48–60 ns, and 60–110 ns for undoped (OFZ), Ce-doped, and Sn-doped samples, respectively, while a long decay time constant was 977 ns, 221–994 ns, and 1.7–8.5 μs, for undoped (EFG), Ce-doped, and Sn-doped samples, respectively. The light output under γ-irradiation (^{137}Cs source, 662 keV) was measured at the level of 15,000 photons/MeV for undoped (EFG) crystals [125] (this value has, however, been indicated as overestimated [128]). Under α-irradiation (^{241}Am source), the light output was only 1500 ph/5.5 MeV for Sn-doped crystals [126]. Thermoluminescence spectra showed a peak at 400°C [125] and 180°C [127] for undoped crystals grown by the EFG and OFZ methods, respectively, and at 140°C for Ce-doped crystals (OFZ) [127]. The afterglow measured after 20 ms irradiation of β-Ga_2O_3 with 2 ms X-ray pulses was 57 ppm and 103–936 ppm for undoped and Sn-doped samples, respectively, which is quite high when compared with other scintillators, which could arise from not-so-good structural quality of Ce-doped crystals [127]. He et al. [322] studied OFZ-grown undoped β-Ga_2O_3 crystals under γ-irradiation from ^{60}Co source (1117 and 1330 keV) and showed a three-exponential decay times with a dominating very short time constant below 1 ns with a contribution of 92.6%; however, it is a device response rather than the β-Ga_2O_3 response, therefore incorrect.

Investigation of high structural quality undoped, Mg-, Si-, Sn-, and Ce-doped bulk β-Ga_2O_3 single crystals obtained by the Czochralski method and reported by Galazka et al. [229, 349], Drozdowski et al. [128], and Makowski at al. [350], did not confirm the results obtained for the EFG-grown (undoped) and OFZ-grown crystals (undoped, Sn-doped, Ce-doped). The maximum absolute light yield of about 7040 ph/MeV was measured for semiconducting β-Ga_2O_3 with the free electron concentration between 10^{16} and 10^{17} cm^{-3} for both undoped and Ce-doped crystals under ^{137}Cs (662 keV) γ-excitation [349, 350], as shown in Fig. 4.47a. For electrically insulating (Mg-doped) and highly conducting (>10^{18} cm^{-3}, doped with Si, Sn, Ce + Si, and Ce + Al + Si) crystals, light yield was either not measurable or very low, typically between 1000–2000 ph/MeV. It was also shown that light yield strongly depends on the sample thickness [128]. A relation between the light yield and the free electron concentration of Czochralski-grown crystals, whether undoped, or doped with Ce, (Ce + Al), (Ce + Si), and (Ce + Al + Si), is summarized in Fig. 4.47b. It shows a bell-like structure with a maximum of the light yield at low free electron concentrtions, just above 10^{16} cm^{-3}. For lower and higher free electron concentration the light yield decreases. The radioluminescence (RL) spectra with double-peak structure (peaking at 370 and 400 nm at 10 K) for undoped, Mg-, Si-, and Ce-doped crystals are very similar to each other (Fig. 4.47c), the same as in the case of CL spectra that are assigned to self-trapped excitons (see Section 4.8.3). Ce^{3+} 5d-4f transitions that have similar double-peak structure at similar energies to CL spectra were not observed in the RL spectra. Time profiles for undoped and doped (Ce, Ce + Si) crystals are typically two-, three- or four-exponential decays [128, 350], as exemplary shown in Fig. 4.47d. In the case of four-exponential decays, as discussed in detail by Makowski et al. [350], time constants are 3–6 ns, 13–36 ns, 38–182 ns, and 460–1200 ns. The fastest scintillation with a high contribution (> 50%) of the two shortest decay components were measured measured for Ce-doped, and in particular for (Ce + Si)-doped crystals. For the light yield above 6500 ph/MeV the energy resolution is good, between 9.2–10.5% [350]. Moreover, independent on doping (Ce, Al, Si, Sn) the Czochralski-grown crystals remain substantially transparent in the light emission under gamma excitation (340–410 nm) [349]. Also, there is basically no or very low afterglow. A great advantage of using semiconducting β-Ga_2O_3 as a scintillator is a possibility of formation thereon avalanche photodiodes for light collection, making the nuclear radiation detection modules efficient, compact, and cost-effective [349].

Szalkai et al. [124] demonstrated a detection capability of fast neutrons (14 MeV) by β-Ga_2O_3. Investigation of semiconducting and electrically insulating (doped with Mg) bulk β-Ga_2O_3 crystals obtained by the Czochralski method showed detection capability of insulating, Mg-doped β-Ga_2O_3, but not semiconducting. To collect electron–hole pairs, on the front and back sides of the bulk crystal (0.5 mm thick), Ti/Au ohmic contacts were deposited. The output signal was collected via a charge-sensitive preamplifier with a bias voltage. A peak-like structure of the interaction between the neutrons and β-Ga_2O_3 nuclei proceeds through several reactions: $^{16}O(n, n)^{16}C$, $^{16}O(n, \alpha)^{13}C$, $^{16}O(n, n+\alpha)^{12}C$, and $^{16}O(n, p)^{16}N$, where n, α, and p indicate a neutron, an alpha particle, and a proton, respectively. Reactions between neutrons and Ga isotopes could also be observed.

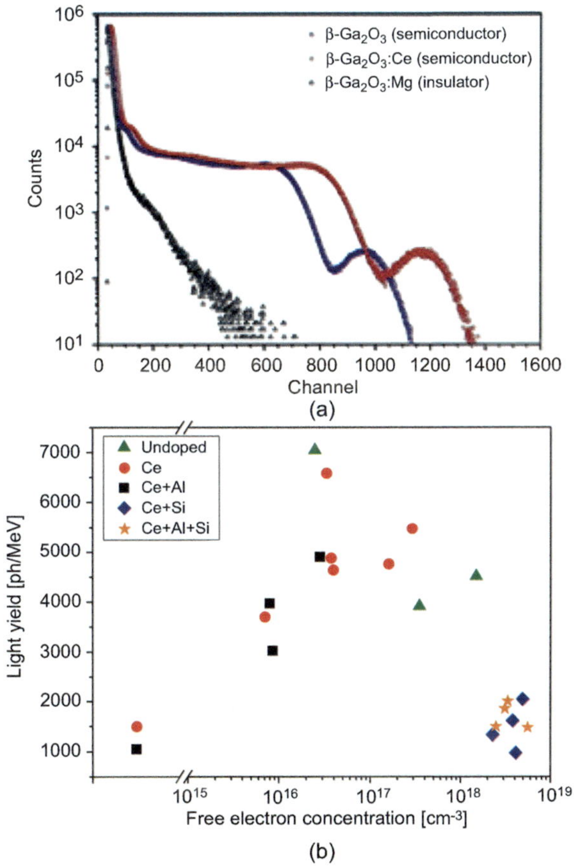

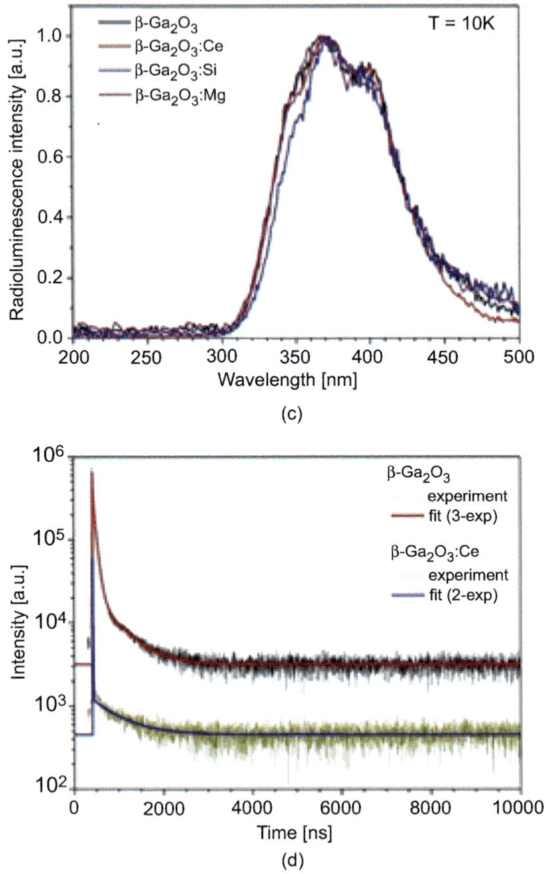

Figure 4.47 Scintillation light yield (a), relation between light yield and free electron concentration (b), radioluminescence spectra (c), and decay time profiles (d) of bulk β-Ga_2O_3 single crystals obtained by the Czochralski method. In (b), the free electron concentration of 10^0 cm^{-3} is only illustrative to indicate electrically insulating or semi-insulating state (in practice, the Hall effect was not measurable). Figures (a), (c), and (d) reprinted from Ref. [128], and (b) reprinted from Ref. [349], Copyright 2019, all with permission from Elsevier.

4.10 Basic Electrical Properties

Electrical properties are crucial for the operation of electronic and sensing devices. They include such properties as free carrier concentration, effective carrier mass and carrier mobility, resistivity (or conductivity), ionization energy of shallow donors, and deep traps. Temperature dependence as well as an impact of annealing on the electrical properties also brings additional information of physical phenomena involved in electrical conductivity in a semiconducting material in quest.

4.10.1 Origin of Electrical Conductivity

Bulk β-Ga_2O_3 single crystals are either electrical insulators (or semi-insulators), normal *n*-type semiconductors, or degenerate *n*-type semiconductors. *p*-type conductivity in bulk crystals has not been demonstrted so far. Growth experiments (OFZ method) showed a decrease in β-Ga_2O_3 conductivity with an increase in O_2 concentration in a growth atmosphere, as illustrated in Fig. 4.48. Such a direct relation led to a conclusion that oxygen vacancies (V_O) are responsible for the electrical conductivity. However, more recent numerical and experimental studies showed V_O as deep donors [179], which cannot directly contribute to the electrical conductivity.

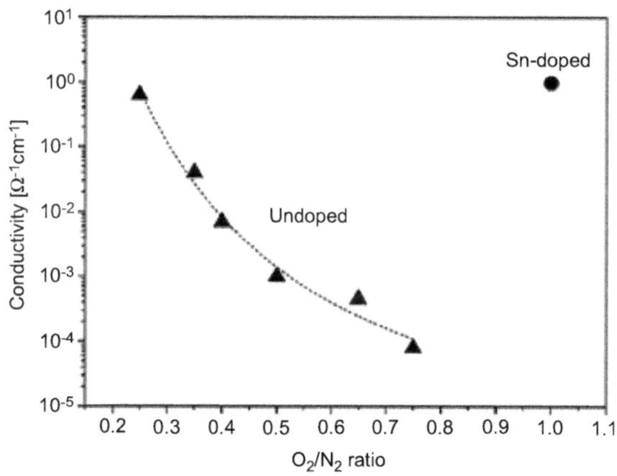

Figure 4.48 Electrical conductivity of the β-Ga_2O_3 single crystals along the *b*-axis as a function of the O_2/N_2 flow rate. The total gas flow rate (N_2 + O_2) was 0.2 m^3/h. Reconstructed basing on data from Ref. [264], with some modifications and different visualization.

(i) Experimental crystal growth studies clearly point to tetravalent Si^{4+} and Sn^{4+} ions that induce the electrical conductivity in β-Ga_2O_3. As demonstrated in Fig. 4.49 [257], the free electron concentration linearly increases with Si^{4+} doping concentration, and a concentration of residual Si impurity of a few wt. ppm in the Ga_2O_3 starting material may already produce the free electron concentration at the level of about 10^{17} cm^{-3}. Another source of Si might be quartz frequently used as an outer part of a growth furnace used in most of the melt grow methods. Intentional doping with Si^{4+} or Sn^{4+} ions to increase the free electron concentration in bulk β-Ga_2O_3 crystals was proved for the OFZ [241, 243, 257], EFG [232], VGF/VB [228], and Czochralski [226, 229, 324, 349] methods. The formation of shallow donors by Si^{4+} and Sn^{4+} ions by substituting Ga^{3+} in the β-Ga_2O_3 crystal lattice was confirmed by theoretical studies [178, 179, 198]. As shown in Section 4.5.6.5, Ge^{4+} ions are not present in Czochralski-grown crystals due to very high *p*(GeO) at the MP of Ga_2O_3, which leads to entire loss of Ge from the melt. Therefore, it is difficult to conclude whether Ge^{4+} enhances electrical conductivity in bulk β-Ga_2O_3 crystals grown from the melt in a similar manner as Si^{4+} and Sn^{4+}. However, Han et al. [323] showed that Ge^{4+} induces electrical conductivity of MBE-grown thin films of β-Ga_2O_3.

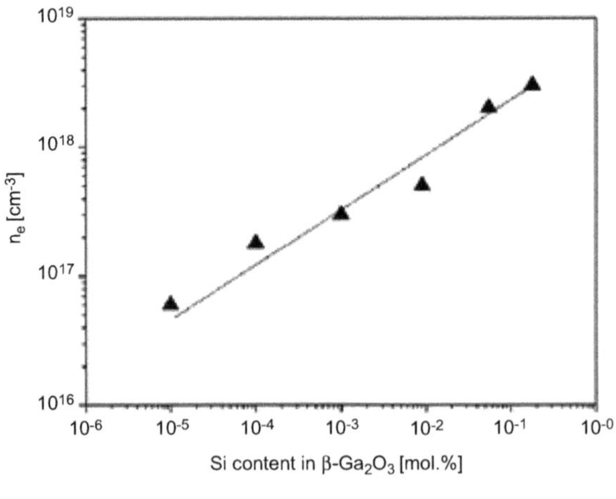

Figure 4.49 Free electron concentration as a function of a nominal Si concentration. Reconstructed basing on data from Ref. [257], with some modifications and different visualization.

Additionally, the shallow donors are also F^- and Cl^- on O^{2-} sites [179].

(ii) Another source of electrical conductivity in β-Ga_2O_3 is hydrogen, either as interstitial (H_i^+) or as complexes with gallium vacancies ($H_i^+ - V_{Ga}$) [180]. Hydrogen is omnipresent, and it is practically impossible to eliminate its presence during bulk crystal growth. For example, during the growth of β-Ga_2O_3 crystals by the Czochralski method, hydrogen was always measured in the growth atmosphere by a mass spectrometer connected to the growth station, at a concentration of a few to tens of wt. ppm. There are also direct evidences of hydrogen present in the crystals grown by the Czochralski method, e.g., in SIMS measurements and in Fourier-transform infrared spectroscopy (FTIR) as O–H line at 3437.6 cm^{-1}, as shown in Fig. 4.50 [324]. The absorbance of the O–H line decreases after annealing in O_2 at 800°C for 4 h, and it disappears after 28 h annealing at the same temperature. A subsequent annealing the sample in the presence of H_2 at 1200°C for 30 min recovers the absorption of the O–H line. Indeed, the free electron concentration decreased by about one order of magnitude after annealing in O_2, which was recovered by annealing in the presence of hydrogen. Weiser et al. [325] and Qin et al. [326] also showed by infrared spectroscopy the presence of O–H line (3437 cm^{-1} with E ||[102] on a ($\bar{2}01$) face) in EFG-grown β-Ga_2O_3 crystals that were assigned to V_{Ga}–2H complexes. They are likely shallow donors and can passivate deep compensating defects. The evidence of H-induced electrical conductivity was reported by King et al. [327] via muon-spin rotation and relaxation spectroscopy (μSR) after irradiation of β-Ga_2O_3 samples with 100% spin-polarized muon beam (hydrogen analog).

On the other hand, the electrically insulating state of bulk β-Ga_2O_3 crystals can be achieved by compensation of the electrical conductivity in two ways: (i) by creating V_{Ga} (high O_2 concentration in the growth atmosphere [264, 228]), and (ii) by introducing compensating acceptors, e.g., Mg^{2+} ions [226, 228], and other divalent (Co^{2+}, Ni^{2+}) ions discussed in Section 4.5.6.5.

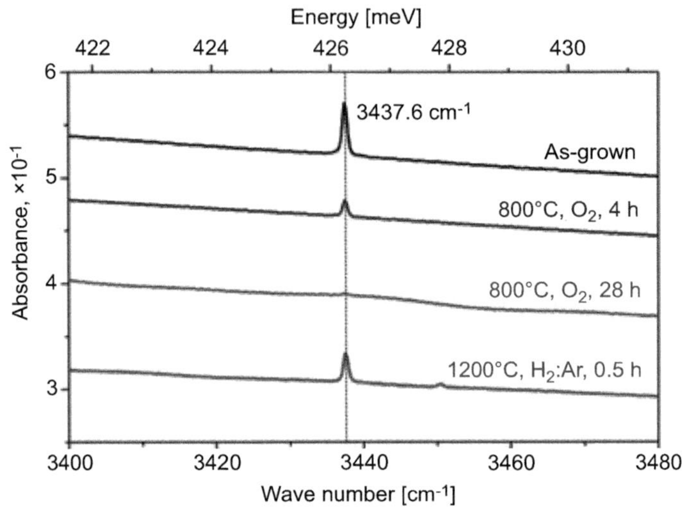

Figure 4.50 FTIR spectrum a β-Ga_2O_3 single crystal showing O–H line at 3437.6 cm^{-1} in the as-grown crystal obtained by the Czochralski method, and after subsequent annealing in O_2 (800°C, 4 and 28 h), and in the presence of hydrogen (1200°C, 0.5 h). Reprinted from Ref. [324], Copyright 2020, with permission from Elsevier.

It is, therefore, obvious that the electrical conductivity of bulk β-Ga_2O_3 crystals grown from the melt will depend on many factors: (i) purities of the Ga_2O_3 starting material and crucibles (if used); (ii) type, purity, and stability of thermal insulation; (iii) composition of the growth atmosphere and O_2 concentration; and (iv) intentional doping. Furthermore, the electrical conductivity can also be tuned by post-growth heat treatments as a function of atmosphere, temperature, and annealing time.

4.10.2 Free Electron Concentration, Hall Mobility, and Resistivity/Conductivity at RT

Keeping in mind the underlying electrical conductivity of β-Ga_2O_3, it is not surprising that basic electrical properties at RT will differ between crystals obtained not only by different growth techniques, but also between crystals obtained by the same method at different growth conditions.

Electrical properties of bulk crystals obtained from the gas phase or flux methods were not reported. The crystals obtained by the Verneuil method [208] were electrically insulating when grown under oxidizing conditions and semiconducting under reducing conditions with $n_e \approx 10^{18}$ cm^{-3} and $\mu \approx 80$ cm^2/Vs at RT. Annealing the crystals at high temperatures for several hours at oxidizing conditions turned the semiconducting crystals into the electrical insulators.

A similar conclusion was derived for the OFZ-grown crystals that were electrically insulating when grown at pure O_2 and semiconducting at $O_2 + N_2$ mixture forming a growth atmosphere (Fig. 4.48) [264]. The conductivity continuously increased with N_2 concentration reaching values of $n_e = 5 \times 10^{18}$ cm^{-3} and $\sigma = 38\ \Omega^{-1}$cm^{-1} at the O_2/N_2 ratio of 0.25. The Sn-doped sample showed the conductivity of 0.96 Ω^{-1}cm^{-1} even under pure O_2. The crystals became electrically insulating after annealing at 900°C for 136 h. Similar output was achieved in other OFZ growth experiments [255], in which the crystals obtained at O_2 = 0–20% and 1–2 atm were semiconductors with $n_e \approx 1.4 \times 10^{17}$–$1.3 \times 10^{18}$ cm^{-3}, μ = 60–100 cm^2/Vs, and $\rho = 6$–15×10^{-2} Ωcm, respectively, while O_2-annealed crystals were electrical insulators ($\rho > 10^6$ Ωcm). Controllable doping with Si (Fig. 4.49) enabled to control the free electron concentration within two orders of magnitude, between mid-10^{16} and mid-10^{18} cm^{-3} with $\mu \approx 100$ cm^2/Vs. The effectiveness of electrical activity of Si acting as shallow donors was concluded to be about 25–50%. Similar results were achieved by doping with Sn [241]. The crystals obtained by the OFZ method in dry air had the following electrical properties: $n_e = 1 \times 10^{18}$ cm^{-3}, $\mu = 80.4$ cm^2/Vs, and $\rho = 7.9 \times 10^{-2}$ Ωcm.

To obtain electrically insulating crystals, alternatively, or in addition to a high O_2 concentration in the growth atmosphere, which is readily obtainable in the OFZ method, Mg doping is used. Electrically insulating crystals ($\rho \approx 6 \times 10^{11}$ Ωcm) were obtained at Mg doping level of 4×10^{18}–2×10^{19} cm^{-3} [328].

In the case of EFG-grown β-Ga_2O_3 crystals (2 vol.% $O_2 + N_2$) [232], the free electron concentrations of 2×10^{17}, 2×10^{19}, and 1×10^{19} cm^{-3} were measured for undoped, Sn-, and Si-doped crystals, respectively. Here 2×10^{19} cm^{-3} is an upper limit of the free electron concentration for Sn doping. Si enables a higher doping limit, but the crystal quality significantly deteriorates when exceeding the doping level of 1×10^{19} cm^{-3}. The maximum free electron concentration in EFG-grown crystals by Si or Sn doping is higher than that in OFZ-grown crystals for the same dopants, perhaps due to lower O_2 concentration in the growth atmosphere. Similar conclusions were drawn for the VGF/Bridgman-grown crystals, in which the maximum free electron concentration of 1.1×10^{19} cm^{-3} was measured, independent of a wide doping level with Sn, between 0.1 and 2 mol.% in the melt [228, 324]. To obtain electrically insulating crystals by the EFG method, they are doped with Fe [329].

β-Ga_2O_3 crystals obtained by the Czochralski method showed a capability of tuning the free electron concentration within about two orders of magnitude, between 10^{16} and 10^{18} cm^{-3} as a result of O_2 concentration in the growth atmosphere, which can be extended by a further order of magnitude to 10^{19} cm^{-3} by doping with Si or Sn. Undoped crystals of 2 cm in diameter obtained by the Czochralski method at O_2 concentration less than 4.4 vol.% were always *n*-type semiconductors with $n_e = 4 \times 10^{16}$–2×10^{18} cm^3, μ = 80–152 cm^2/Vs, and ρ = 0.04–1.15 Ωcm [225, 226, 324]. Si-doped crystals showed n_e = 2.8–9 × 10^{18} cm^3, μ = 51–84 cm^2/Vs, and ρ = 0.013–0.027 Ωcm [324, 330]. Similar values were also measured for doubly (Ce + Si) and triply doped (Ce + Al + Si) β-Ga_2O_3 crystals obtained by the Czochralski method [349]. Sn-doped Czochralski-grown crystals showed in average similar electrical properties as compared with Si doping and the exemplary electrical properties thereof are n_e = 3 × 10^{18}–10^{19} cm^3, μ = 49–54 cm^2/Vs, and ρ = 0.012–0.040 Ωcm [324, 330] with a decrease in the electron concentration along the crystal length. Czochralski-grown 2-inch-diameter undoped crystals obtained at high O_2 concentration of 8–25 vol.% were also *n*-type semiconductors with $n_e = 5 \times 10^{16}$–6×10^{17} cm^{-3}, μ = 80–120 cm^2/Vs, and ρ = 0.16–0.77 Ωcm [228]. Higher O_2 concentration in the growth atmosphere usually led to electrically insulating crystals. Electrically insulating state can be obtained by Mg doping even at small O_2 concentration in the growth atmosphere (<2 vol.%). Other dopants used for Czochralski-grown β-Ga_2O_3 crystals that compensate the electrical conductivity even at low $p(O_2)$ in the growth atmosphere are Co and Ni. Al tends to lower the electrical conductivity, while Ce and Cr have rather a minor impact on the free electron concentration. For details on dopants and their concentrations used for the bulk β-Ga_2O_3 single crystals obtained by the Czochralski method, see Section 4.5.6.5.

The summary of basic electrical properties of bulk β-Ga_2O_3 crystals obtained from the melt is provided in Table 4.14.

Table 4.14 Basic electrical properties of melt-grown bulk β-Ga_2O_3 single crystals at RT

Growth method	Electron concentration [cm^{-3}]	Hall mobility [cm^2/Vs]	Resistivity [Ωcm]	Dopant	Reference
Verneuil	10^{18}	80	—	Undoped	208
OFZ	5×10^{18} (a)	—	10^4–10^0	Undoped	264
OFZ	10^{17}–10^{18}	60–100	6–0.15	Undoped	255
OFZ	~5×10^{16}–5×10^{18}	~100	—	Si	257
OFZ	5×10^{17}	88	0.143	Undoped	243
	2–7×10^{18}	49–65	1.8–4.3×10^{-2}	Sn	
OFZ	1×10^{18}	80	8×10^{-2}	Sn	241
OFZ	7.7×10^{16}	172	0.47	Undoped	328
	—	—	6×10^{11}	Mg	
EFG	2×10^{17}	—	—	Undoped	232
	2×10^{19} (a)	—	—	Sn	
	1×10^{19} (a)	—	—	Si	
EFG	4.9×10^{18}	93	0.014	Si	253
EFG	3.9×10^{16}	107	1.49	Undoped	266
VGF	4×10^{18}–1.1×10^{19}	28–62	1.9–2.3×10^{-2}	Sn	228
Czochralski	4×10^{16}–2×10^{18} (b)	80–152	0.04–1.15	Undoped	225, 226,
	5×10^{16}–6×10^{17} (c)	80–120	0.16–0.77	Undoped	228, 330
	3×10^{18}–1×10^{19}	49–54	0.012–0.040	Sn	324, 349
	2.8–9×10^{18}	51–84	0.013–0.027	Si	
	—	—	—	Mg, Co, Ni,	

(a)Maximum value, (b)low O_2 in the growth atmosphere, (c)high O_2 in the growth atmosphere

Despite anisotropy in structural, mechanical, thermal, and optical properties, bulk β-Ga_2O_3 crystals revealed substantially uniform electrical properties along the main crystallographic directions, as concluded by Villora et al. [221]. More detailed study by Golz et al. [331], who described the electrical conductivity of β-Ga_2O_3 by a second rank tensor and performed Hall effect measurements in the van der Pauw configuration on different Czochralski- and EFG-grown crystal samples, revealed that the principal conductivities do not differ in between by more than 5% in a temperature range of 150–375 K. Moreover, much higher anisotropies were measured in samples with extended structural defects (low-angle grain boundaries) for which energy barriers greater than 10 meV were determined. These studies show that, indeed, the electrical properties of bulk β-Ga_2O_3 crystals of high perfection are relatively uniform.

4.10.3 Impact of Annealing on Electrical Properties

As discussed earlier, the free electron concentration can be tuned by O_2 concentration in the growth atmosphere and intentional doping. Yet another tool for modifying the electrical properties of bulk β-Ga_2O_3 crystals includes post-growth heat treatment, or simply annealing, which offers three degrees of freedom: annealing atmosphere (oxidizing or reducing), temperature, and time.

An impact of an annealing atmosphere and temperature on the free electron concentration and Hall mobility of undoped Czochralski-grown β-Ga_2O_3 crystals is shown in Fig. 4.51 (Galazka et al. [226] and Galazka [8]). Annealing in an oxidizing atmosphere (O_2) for 20 h gradually decreases the free electron concentration in the bulk with annealing temperature, with a faster drop (by about one order of magnitude with respect to the as-grown sample) after annealing at 1200°C, or above. At the same time, a thin, electrically insulating layer at the crystal surface was formed, in the same way as in the case of OFZ-grown crystals upon annealing, which revealed an increase in the insulation layer thickness with annealing temperature and time [246]. If the initial electron concentration is smaller ($1–5 \times 10^{17}$ cm^{-3}), the decrease in the electron concentration in the bulk after annealing above 1000°C will be small or very small, typically ending at the level of about 10^{17} cm^{-3}. After annealing in the presence of O_2, the Hall mobility slightly increased. On the other hand, annealing of the as-grown crystals in the presence of H_2 (5% H_2 + 95% Ar) for 10 h did not affect the free electron concentration after annealing at temperatures up to 900°C (Fig. 4.51); however, the Hall mobility significantly decreased, leading to an increase in resistivity. After annealing in the presence of H_2 those samples that were previously annealed in O_2 at 1000°C for 20 h, the electron concentration increased to the same values as that of the as-grown crystals after annealing at 300–700°C, with an increase in the Hall mobility. The annealing in the presence of O_2 and H_2 is reversible in terms of the free electron concentration.

OFZ-grown Si-doped crystals also showed a decrease in the free electron concentration to the level of 10^{17} cm^{-3} with an insulating surface layer formed thereon after annealed in O_2 at 1150°C for 0.5–3 h, while annealing in N_2 at lower temperatures (800–1000°C) did not affect the free electron concentration [232].

The annealing experiments of undoped crystals suggest at least two sources for the electrical conductivity in β-Ga_2O_3 crystals: (i) residual solid impurities, likely Si from a Ga_2O_3 starting material, which provides the electron concentration at the level of about 10^{17} cm^{-3}, and (ii) hydrogen, which may provide the electron concentration at the level of about 10^{18} cm^{-3}. This is summarized in Fig. 4.51c that shows a relation between the free electron concentration and Si concentration (measured by SIMS) in undated and Si-doped crystals when as-grown and annealed in the presence of oxygen [324]. At $n_e < 2 \times 10^{18}$ cm^{-3}, the free electron concentration is higher than the Si concentration and they both are getting at a similar level after annealing. Such behavior points to both H and Si as sources for the electrical conductivity. At $n_e > 2 \times 10^{18}$ cm^{-3} the free electron concentration equals the Si concentration, as H cannot reach such high values. Hydrogen can be incorporated into the β-Ga_2O_3 crystal lattice during growth as complexes with V_{Ga} [180], and removed at high temperatures during annealing in the presence of O_2 (>1000°C). Annealing such crystals in the presence of H_2 may lead to incorporation of hydrogen as interstitials (H_i), which proceeds at lower temperatures (≥300°C) due to lower formation energy, and it can be removed at temperatures ≥700°C.

Interesting is the effect of annealing of Sn-doped bulk β-Ga_2O_3 crystals obtained by the VGF and Czochralski methods under high and low oxygen concentrations in the growth atmosphere, 17 vol.% and 2–4 vol.%, respectively. As shown in Fig. 4.52, if the Sn-doped crystals were grown at high oxygen concentration, they became entirely electrically insulating (if thin, <0.5 mm) or have low free electron concentration ($\approx 10^{16}$ cm^{-3}) with an insulating layer (when thick) after annealing in the presence of O_2 at 1000°C or above for 20 h. A decrease in the free electron concentration starts already after annealing at 800°C. On the other hand, the Sn-doped crystals obtained at low oxygen concentration remain entirely semiconducting (including the surface) with a lower free electron concentration by a factor of about 5. Such response of Sn-doped crystals to annealing is similar to that of undoped crystals (see Fig. 4.51).

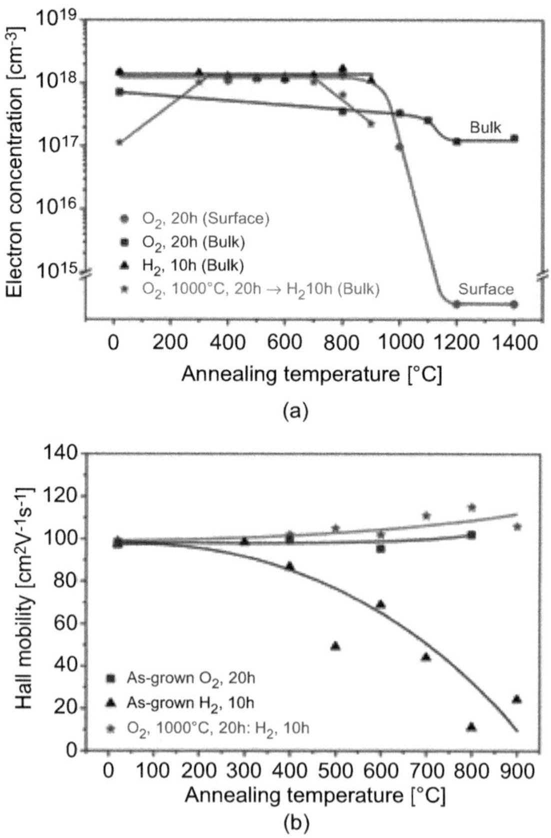

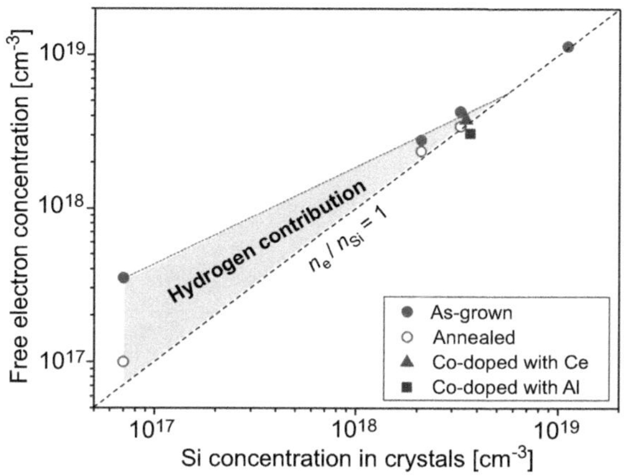

Figure 4.51 Free electron concentration (a) and Hall mobility (b) by Hall effect measurements of Czochralski-grown β-Ga_2O_3 crystals as a function of annealing temperature and atmosphere. For each single annealing experiment, another crystal sample was used, but from the same bulk crystal. (c) Free electron concentration as a function of Si concentration in different Czochralski-grown bulk β-Ga2O3 crystals. Figures (a) and (b) reprinted from Ref. [8], Copyright 2018, with permission from IOP Publishing and (c) reprinted from Ref. [324], Copyright 2020, with permission from Elsevier.

This phenomenon cannot be explained by evaporation of Sn during annealing, because it has too low partial pressure at 1000–1200°C ($<10^{-7}$ atm) and its concentrations were measured at the same level before and after annealing [8]. Moreover, annealing at the same conditions, Si-doped, Czochralski-grown crystals obtained at low oxygen concentration behave similarly to Sn-doped Ga_2O_3.

A full conversion into electrical insulators or a drastic decrease of the free electron concentration (by two orders of magnitude) in Sn-doped β-Ga_2O_3 crystals obtained at high oxygen concentration can be explained by the presence of Ga vacancies (V_{Ga}), hydrogen, and Sn dopant. During growth, V_{Ga} (acting as deep acceptor) are passivated by hydrogen and Sn forms shallow donors that create the electrical conductivity. During annealing in the presence of O_2 at high temperatures, hydrogen is removed and V_{Ga} passivate Sn, which leads to a drastic decrease in the free electron concentration or to a conversion into electrically insulating state. This would require high concentration of V_{Ga}, which is met by high O_2 concentration during growth, as V_{Ga} concentration increases with oxygen partial pressure [180]. The resulted free electron concentration is, therefore, a result of a difference in V_{Ga} and Sn concentrations. Sn prefers the octahedral site, the same as the gallium vacancy V_{Ga}, while Si prefers the tetrahedral site in the β-Ga_2O_3 crystal structure.

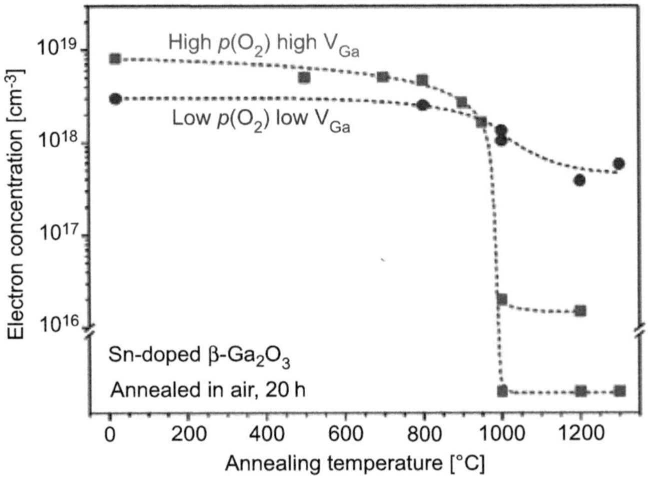

Figure 4.52 Free electron concentration measured at RT versus annealing temperature as a result of annealing of Sn-doped, Czochralski-grown bulk β-Ga_2O_3 crystals in the presence of O_2 for 20 h.

4.10.4 Temperature-Dependent Electrical Properties

The above-discussed electrical properties were measured at RT. Temperature-dependent resistivity, Hall electron mobility, and Hall electron concentration for bulk β-Ga_2O_3 crystals obtained by the Czochralski method and having different free electron concentrations at RT were reported by Irmscher et al. [332], and exemplary characteristics are presented in Fig. 4.53 [333]. The temperature-dependent resistivity (Fig. 4.53a) is typical for moderately doped semiconductors. It is governed by a hopping mechanism within the impurity band at low temperatures and no free electrons available (Fig. 4.53b), by a thermal excitation of the free electrons to the conduction band with rising temperature

(Fig. 4.53b), and by a decrease in the electron mobility around RT and above (Fig. 4.53c) with a limited increase in the free electron concentration (Fig. 4.53b). The maximum value of the Hall mobility (of about 550 cm^2/Vs) occurs at about 100 K for the samples with lowest free electron concentration at RT (6×10^{16} cm^{-3}). Above that temperature, scattering at phonons dominates, while below that temperature, scattering at ionized impurities, followed by hopping conductivity in the impurity band at very low temperatures.

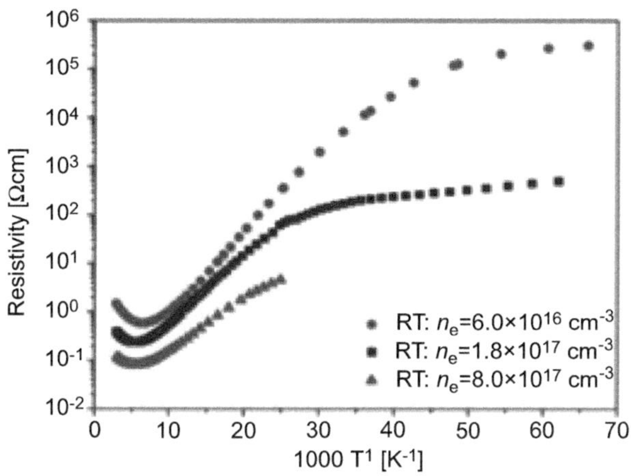

(a)

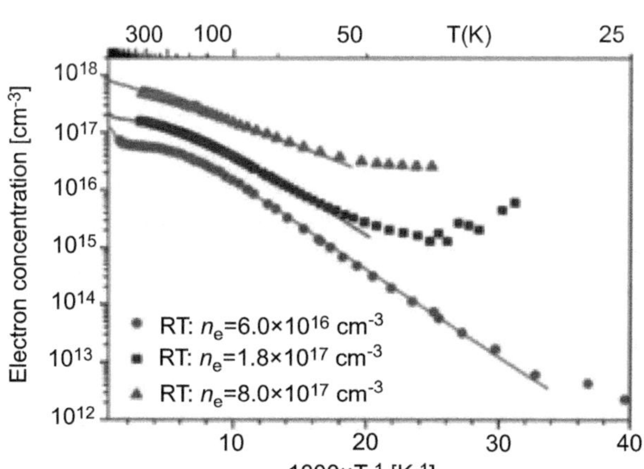

(b)

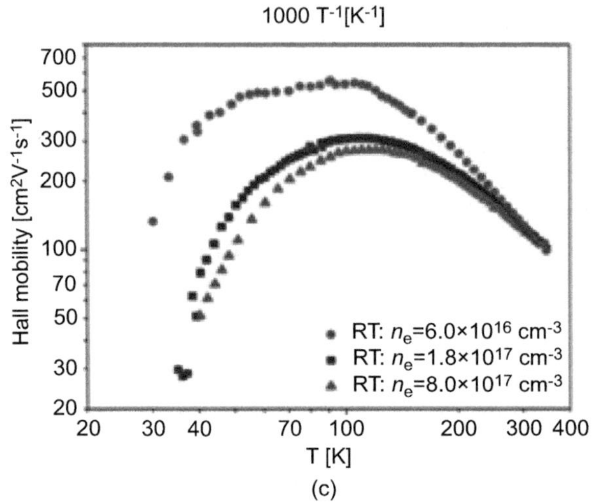

Figure 4.53 Temperature-dependent of (a) resistivity, (b) electron Hall concentration, and (c) electron Hall mobility of Czochralski-grown β-Ga_2O_3 crystals having different free electron concentrations at RT [333].

Similar temperature-dependent behavior of the electrical properties (at least one) was also measured for bulk β-Ga_2O_3 crystals obtained by the Verneuil (Lorenz et al. [208], Harwig et al. [334]) and OFZ (Ueda et al. [335], Villora et al. [257]) methods.

4.10.5 Effective Electron Mass and Electron Mobility

At the Γ point of the BZ or at its vicinity, the theoretical values of the effective electron mass (m_e*) described as m_e*/m_0 (where m_0 is the electron mass) span between 0.22 and 0.3 [176–179, 182, 188] with a possible anisotropy [176, 182, 188]. Experimental m_e*/m_0 values of bulk crystals are 0.28 [191, 192], as well as 0.283–0.288 for the (010) surface and 0.276–0.311 for the ($\bar{2}$01) surface [336], with the anisotropy of $m_e*(a)/m_e*(b) = 1.02$ and $m_e*(b)/m_e*(c) = 0.99$ for the (010) surface, and $m_e*(a)/m_e*(b) = 1.07$ and $m_e*(b)/m_e*(c) = 0.89$ for the ($\bar{2}$01) surface.

Low effective electron mass results in a relatively high electron mobility ($\mu \propto 1/m_e*$). Indeed, the maximum values of the Hall electron mobility of 152 cm^2/Vs (Czochralski method) [226] and 172 cm^2/Vs (OFZ method) [328] for $n_e < 10^{17}$ cm^{-3} were measured at RT. As shown in Fig. 4.54a, the electron mobility of Czochralski-grown crystals gradually decreases with the free electron concentration in a range of 10^{16}–10^{19} cm^{-3}. Similar correlation was observed by Sasaki et al. [337] for OFZ-grown crystals and homoepitaxial MBE layers. This correlation is very similar for both bulk crystals and layers obtained by different growth techniques. The electron mobility gradually decreases with the electron concentration from about 150 cm^2/Vs at $n_e = 10^{16}$–10^{17} cm^{-3} to about 40–50 cm^2/Vs at $n_e = 10^{19}$ cm^{-3} [226].

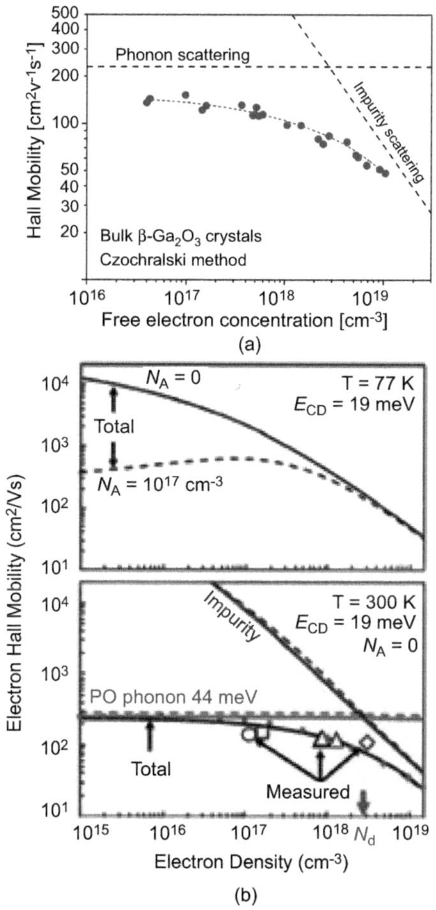

Figure 4.54 Electron mobility versus free electron concentration for (a) Czochralski-grown (100)-oriented single crystals, and (b) limiting factors, combined with measured values. Figure (b) reprinted from Ref. [338], Copyright 2016, with permission from AIP Publishing.

According to the studies of Ma et al. [338], Ghosh and Singisetti [339], the Hall electron mobility is limited by polar optical phonon scattering at low and moderate free electron concentrations ($<2 \times 10^{18}$ cm^{-3}), and by ionized impurity scattering at high electron concentrations (as shown in Fig. 4.54b), consistent with experimental values shown in Fig. 4.54a, and those reported by Sasaki et al. [337], and Parasini and Fornari [340]. A model for phonon-scattering carrier mobility in β-Ga_2O_3 crystals that includes polar and non-polar phonon–electron interactions was described by Parisini et al. [341]. The model showed a very good agreement with experimental results. The theoretical limit for the electron mobility in β-Ga_2O_3 at RT is about 220–230 cm^2/Vs [338].

4.10.6 Ionization Energy of Shallow Donors

The ionization energy (E_{CD}) of shallow donors, which can be obtained from electronic transport measurements, depends on the donor concentration (N_D) as illustrated in Fig. 4.55. It decreases from 25 to 16 meV with the measured free electron concentration (n_e) raising from 1.25×10^{17} to 9×10^{17} cm^{-3} in Czochralski-grown crystals (Fig. 4.55, Galazka [8], Irmscher et al. [332]). The extrapolation of this dependence to zero concentration N_D yielded for the ionization energy E_{CD} of isolated donors a value of about 36.3 meV [332]. At a lower level of the free electron concentration, $n_e = 1.4 \times 10^{17}$ cm^{-3}, the ionization energy of 31 meV was measured in EFG-grown crystals (Oishi et al. [342]), while at a higher level, $n_e = 10^{18}$ cm^{-3}, the ionization of 7 meV was reported for OFZ-grown crystals (Villora et al. [257]).

4.10.7 Deep Traps

Deep traps in bulk single crystals of β-Ga_2O_3 obtained by the Czochralski method were studied by Irmscher et al. [332] (undoped) with deep-level transient spectroscopy (DLTS), and later by Polyakov et al. [343] (Mg-doped) with thermally stimulated current (TSC), photoinduced current transient spectroscopy (PICTS), and micro-cathodoluminescence (MCL), while in crystals obtained by the EFG method, by Zhang et al. [344] with DLTS and deep-level optical spectroscopy (DLOS), and by Ingebrigtsen et al. [348] with the use of DLTS and theoretical calculations (DFT-HSE06).

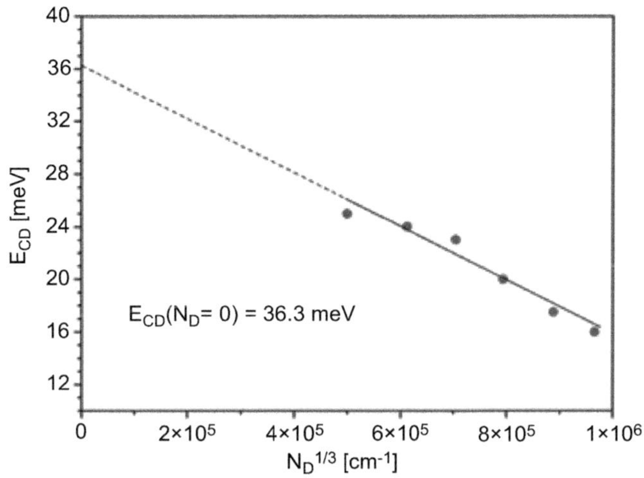

Figure 4.55 Measured donor ionization energies (E_{CD}) versus the cubic root of the donor concentrations (N_D). The solid line refers to a linear least-squares fit to the experimental points (solid circles) yielding E_{CD} = 36.3 MeV at N_D = 0 (dashed line). Reprinted from Ref. [8], Copyright 2018, with permission from IOP Publishing.

Deep trap distributions along with their concentrations for Czochralski-grown crystals are shown in Fig. 4.56 [333] and summarized in Table 4.15, including EFG-grown crystals. The trap distribution and concentrations are quite similar for crystals obtained by both melt growth methods. The deep trap E2 was present in all measured samples, and in a vast majority of cases, it was a dominating trap with the concentration of $1–7 \times 10^{16}$ cm^{-3}. Other less frequent traps are E3 with lower or even comparable concentration to the trap E2, and E1 having the concentration of one order of magnitude lower than that of E2 and E3 ($3 \times 10^{14}–6 \times 10^{15}$ cm^{-3}). Traps E4 and E5 (revealed for the EFG-grown crystals only) have a similar concentration to that of E1, while trap E6 has a comparable concentration to that of E2. Energetically, the traps are distributed across the whole energy gap from the vicinity of the CBM to the vicinity of the VBM. The deep electron traps possibly act as compensating centers, which are, indeed, at similar concentration as the concentration of compensating acceptors concluded from the Hall effect measurements [332]. At least some of the traps were suggested by Irmscher et al. [332] to have an extrinsic origin arising from transition metals (Fe^{3+} and Co^{2+}) that were observed by electron paramagnetic resonance (EPR). Indeed, E2 trap was found to be associated with Fe^{3+} impurities substituting Ga^{3+} in the crystal lattice as confirmed by EPR measurements (Son et al. [345]) and Fe concentration (Ingebrigtsen et al. [348]) in the investigated samples. The presence of Fe in the crystals obtained by the Czochralski and EFG methods is not surprising as both techniques utilize Ir crucibles, which always contain Fe residual impurities (and other transition metals, such as Co and Ti), in addition that to a possible presence in the Ga_2O_3 starting material. Additionally, as shown by Lenyk et al. [346] basing on EPR measurements, Ir^{4+} ions also introduce deep donor levels. E2 trap was found in Czochralski- and EFG-grown crystals in similar concentrations; however, epitaxial layers (MBE, HVPE) revealed about two orders of magnitude lower E2 concentration [347] (no metal crucible). In Mg-doped β-Ga_2O_3 crystals (electrically insulating), Polyakov et al. [343] identified traps at levels of E_C – 0.8 eV, E_C – 1.35 eV, E_C – 2.3 eV, and E_V + 1.06 eV, with the first and the last one assigned to Fe and Mg impurities, respectively.

Also another trap E2* close to E2 was identified, but its origin is not clear so far.

Table 4.15 Deep-level defect distribution in bulk β-Ga_2O_3 crystals

Deep trap	Bulk growth method	Ionization energy with respect to CBM, E_C – [eV]	Concentration [cm^{-3}]	Reference
E1	Czochralski	0.55	3×10^{14}–6×10^{15}	332
	EFG	0.56–0.62	3–4.7×10^{14}	344, 348
E2	Czochralski	0.74	2–5×10^{16}	332
	EFG	0.78–0.82	1–7×10^{16}	344, 348
E2*	EFG	0.72	—	348
E3	Czochralski	1.04	4×10^{16}	332
	EFG	1.00–1.01	4×10^{14}–3.7×10^{15}	344, 348
E4	EFG	1.48	5×10^{14}	348
E5	EFG	2.16	1×10^{15}	344
E6	EFG	4.40	1.5×10^{16}	344

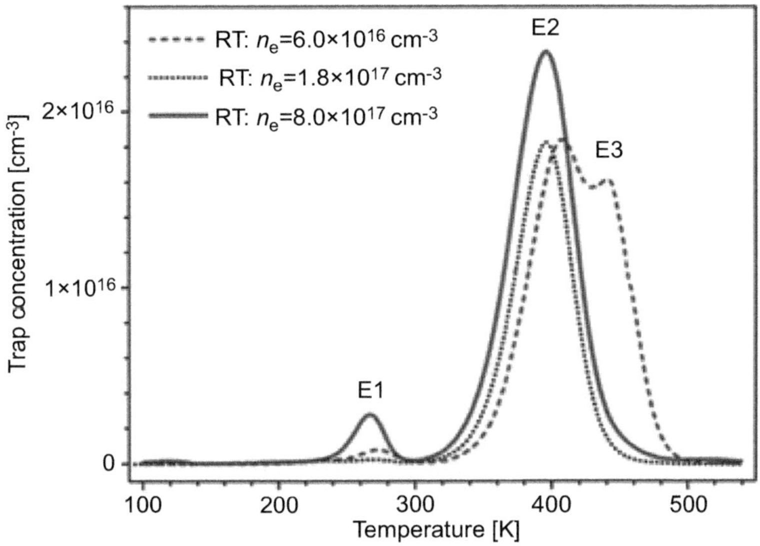

Figure 4.56 Typical DLTS spectra of β-Ga_2O_3 crystal samples obtained by the Czochralski method and having different free electron concentrations at RT. All the crystals were obtained under the same growth atmosphere consisting of 50 vol.%CO_2 + 50 vol.% Ar [333].

4.10.8 Conclusion

Bulk β-Ga_2O_3 single crystals can be obtained as electrical insulators, normal *n*-type semiconductors, or degenerate semiconductors. In crystals obtained from the melt, the free electron concentration (n_e) can be controlled within about three orders of magnitude (in the range of 10^{16}–10^{19} cm^{-3}) with the Hall mobility (μ) of up to about 172 cm^2/Vs. The free electron concentration can be tuned by growth conditions (particularly by oxygen concentration in the growth atmosphere), post-growth heat treatment (atmosphere, temperature, time), and/or intentional doping (Si^{4+}, Sn^{4+}, Mg^{2+}). Temperature dependence of the electrical properties showed a behavior typical for classical semiconductors.

The electron mobility in β-Ga_2O_3 is limited to a value of about 220–230 cm^2/Vs due to optical phonon scattering at $n_e \leq 2 \times 10^{18}$ cm^{-3}, and by scattering at ionized impurities at higher values of the free electron concentration. The effective electron mass at low and moderate free electron concentrations is about 0.28 of the free electron mass. The ionization energy of isolated donors is about 36.3 meV, which decreases with the electron concentration (e.g., it is 16 meV at $n_e \approx 10^{18}$ cm^{-3}).

Several deep defect traps have been identified in bulk β-Ga_2O_3 with a maximum concentration below 10^{17} cm^{-3}. The energetic trap distributions along with their concentrations are similar in bulk crystals obtained by the Czochralski and EFG methods that suggests a similar origin. The dominating trap E2 is assigned to Fe^{3+} impurities.

Basic electrical properties of bulk β-Ga_2O_3 single crystals are summarized in Table 4.16.

Table 4.16 Basic electrical properties of bulk β-Ga_2O_3 single crystals

Property	**Value**	**Reference**
Range of free electron concentration [cm^{-3}]	10^{16}–2×10^{19}	208, 225, 226, 228, 232, 241, 243, 253, 255, 257, 264, 266, 324, 328, 349
Experimental maximum electron mobility [cm^2/Vs]	152–172	226, 328
Theoretical maximum electron mobility [cm^2/Vs]	220–230	338
Experimental effective electron mass, m_e^*/m_0	$\approx$0.28	191, 192, 336
Ionization energy of isolated shallow donors [eV]	36.3	332
Deep-level traps: maximum concentration [cm^{-3}]	1–7×10^{16}	332, 344, 348
Deep-level traps: energy levels below CBM [eV]	0.55–0.62 (E1) 0.72 (E2*) 0.74–0.82 (E2) 1.00–1.04 (E3) 1.48 (E4) 2.16 (E5) 4.40 (E6)	332, 344, 348

4.11 Key Points

- Ga_2O_3 has several polymorphs with the monoclinic β-phase being the most stable modification at high temperatures.
- Large bulk single crystals of β-Ga_2O_3 are grown directly from the melt by the Czochralski and EFG methods. Doping with a number of mono-, di-, tri-, and tetravalent ions and its impact on growth and material's properties have been demonstrated.
- β-Ga_2O_3 is an ultra-wide bandgap *n*-type oxide semiconductor.
- β-Ga_2O_3 has a quasi-direct energy gap of about 4.85 eV and high transparency from deep UV to near-infrared spectral region that is only affected by the free carrier absorption and intentional dopants.
- Electrical properties of bulk β-Ga_2O_3 crystals can be tuned in a wide range by growth conditions, intentional doping, and post-growth heat treatment.
- β-Ga_2O_3 revealed anisotropic structural, thermal, optical properties; however, the electrical conductivity seems to be substantially isotropic.
- A unique combination of electrical and optical properties places β-Ga_2O_3 in the frontline of ultra-wide bandgap materials for future applications, particularly for high-power electronics and UV optoelectronics.

4.12 Summary

Table 4.17 Main properties of bulk β-Ga_2O_3 single crystals and growth techniques

	β-Ga_2O_3	
Property	**Value or description**	**Reference**
	Crystal system and structure	
Crystal system	Monoclinic	163
Lattice constants [Å], [°]	a=12.214 b=3.0371 c=5.7981 β=103.83	163
Space group	C2/m	163, 174
Cleavage planes	{100}, {001}	

Bulk crystal preparation		
Maximum size: diameter × length [mm×mm], or width × length × thickness [mm×mm×mm]	Czochralski (cylinder): 50×100	228
	EFG (slab): 110×110×6	232
	OFZ (cylinder): 25×50	221
	VGF (cylinder): 50×30	228
	Bridgman (cylinder): 25×25	233
Thermal and physical properties		
Solid density [g/cm^3]	5.96	
Liquid density [g/cm^3]	4.87	290
Melting point [°C]	1793	233
Thermal conductivity at RT [W/mK]	Along [100] = 10.9–13 Along [010] = 21–29 Along [001] = 15–21 Along [$\bar{2}$01] = 13–15	226, 291, 292, 294, 296
Thermal diffusivity at RT [mm^2/s]	Along [100] = 3.7 Along [010] = 9.6 Along [001] = 7.1	294
Thermal expansion [K^{-1}] At 300/600/1200 K	Along [100] = 1.04/1.24/2.78 Along [010] = 1.68/4.51/5.84 Along [100] = 1.74/4.79/6.27 0.865/0.0116 for β	300, 272
Mechanical properties		
Bulk modulus [GPa]	185	272
Shear modulus [GPa]	88	
Young's modulus [GPa]	228	
Poisson's ratio	0.295	
Microhardness [GPa]	8.32 for (100) 6.44 for (010) 10.1 for (001)	266
Mohs hardness [GPa]	6.39 for (100) 5.78 for (010) 6.82 for (001)	266

Optical properties		
Optical bandgap [eV]	Parallel to [100] = 4.59 Parallel to [010] = 4.85 Parallel to [001] = 4.56	8
Luminescence bands at RT [eV]	3.2–3.6 2.8–3.0 2.4	8, 254, 256
Static dielectric constant at RT	10.2 ± 0.2 for ⊥(100) 10.87 ± 0.08 for ⊥(010) 12.4 ± 0.4 for ⊥(001)	308
Refractive index at 2–4 eV	≈1.9–2.1	307
Scintillation properties		
Nuclear radiation detection	γ, α, neutron	124, 125, 126, 127, 128, 349, 350
Light output under 662 keV excitation [ph/MeV]	> 7000	128, 349, 350
Decay time constant (fastest) under 662 keV excitation [ns]	5–10 (40%)	128
Electrical properties		
Tunability of the free electron concentration at RT [cm^{-3}]	$\approx 10^{16}$–10^{19}	208, 225, 226, 228, 230, 232, 241, 243, 253, 255, 257, 264, 266, 328
Maximum experimental electron mobility at RT [cm^2/Vs]	172	328
Maximum theoretical electron mobility at RT [cm^2/Vs]	220–230	338
Donor activation energy [MeV]	36.3 for isolated donors 16–30 for $n_e \approx 10^{17}$–10^{18} cm^{-3}	332

Deep electron traps [eV]	E1=0.55–0.62 E2=0.74–0.82 E2*=0.72 E3=1.00–1.05 E4=1.20–1.48 E5=2.16–2.30 E6=4.40	332, 344, 348
Effective electron mass for a moderate free electron concentration (10^{17} cm^{-3}), m_e^*/m_0	0.28	191, 192
Breakdown field [MV/cm]	8	84

References

1. S. I. Stepanov, V. I. Nikolaev, V. E. Bougrov, A. E. Romanov; *Rev. Adv. Mater. Sci.* **44** (2016), 63–86.
2. J. Y. Tsao, S. Chowdhury, M. A. Hollis, D. Jena, N. M. Johnson, K. A. Jones, R. J. Kaplar, S. Rajan, C. G. Van de Walle, E. Bellotti, C. L. Chua, R. Collazo, M. E. Coltrin, J. A. Cooper, K. R. Evans, S. Graham, T. A. Grotjohn, E. R. Heller, M. Higashiwaki, M. S. Islam, P. W. Juodawlkis, M. A. Khan, A. D. Koehler, J. H. Leach, U. K. Mishra, R. J. Nemanich, R. C. N. Pilawa-Podgurski, J. B. Shealy, Z. Sitar, M. J. Tadjer, A. F. Witulski, M. Wraback, J. A. Simmons; *Adv. Electron. Mater.* **4** (2018), 1600501.
3. M. A. Mastro, A. Kuramata, J. Calkins, J. Kim, F. Ren, S. J. Peartong; *ECS J. Solid State Sci. Technol.* **6** (2017), P356–P359.
4. M. Higashiwaki, H. Murakami, Y. Kumagai, A. Kuramata; *Jpn. J. Appl. Phys.* **55** (2016), 1202A1.
5. M. Baldini, Z. Galazka, G. Wagner; *Mat. Sci. Semicon. Proc.* **78** (2018) 132–146.
6. S. J. Pearton, J. Yang, P. H. Cary IV, F. Ren, J. Kim, M. J. Tadjer, M. A. Mastro; *Appl. Phys. Rev.* **5** (2018), 011301.
7. M. Razeghi, J. H. Park, R. McClintock, D. Pavlidis, F. H. Teherani, D. J. Rogers, B. A. Magill, G. A. Khodaparast, Y. Xu, J. Wu, Vinayak, P. Dravid; *Proc. of SPIE* **10533** (2018), 105330R1–105330R23.
8. Z. Galazka; *Semicond. Sci. Technol.* **33** (2018), 113001.
9. M. Baldini, D. Gogova, K. Irmscher, M. Schmidbauer, G. Wagner, R. Fornari; *Cryst. Res. Technol.* **49** (2014), 552–557.

10. M. Baldini, M. Albrecht, D. Gogova, R. Schewski, G. Wagner; *Semicond. Sci. Technol.* **30** (2015), 024013.
11. M. Baldini, M. Albrecht, A. Fiedler, K. Irmscher, D. Klimm, R. Schewski, G. Wagner; *J. Mater. Sci.* **51** (2016), 3650–3656.
12. M. Baldini, M. Albrecht, A. Fiedler, K. Irmscher, R. Schewski, G. Wagner; *ECS J. Solid State Sci. Technol.* **6** (2017), Q3040–Q3044.
13. R. Schewski, G. Wagner, M. Baldini, D. Gogova, Z. Galazka, T. Schulz, T. Remmele, T. Markurt, H. von Wenckstern, M. Grundmann, O. Bierwagen, P. Vogt, M. Albrecht; *Appl. Phys. Express* **8** (2015), 011101.
14. G. Wagner, M. Baldini, D. Gogova, M. Schmidbauer, R. Schewski, M. Albrecht, Z. Galazka, D. Klimm, R. Fornari; *Phys. Status Solidi (a)* **211** (2014), 27–33.
15. D. Gogova, G. Wagner, M. Baldini, M. Schmidbauer, K. Irmscher, R. Schewski, Z. Galazka, M. Albrecht, R. Fornari; *J. Cryst. Growth* **401** (2014), 665–669.
16. X. Du, W. Mi, C. Luan, Z. Li, C. Xia, J. Ma; *J. Cryst. Growth* **404** (2014), 75–79.
17. X. Du, Z. Li, C. Luan, W. Wang, M. Wang, X. Feng, H. Xiao, J. Ma; *J. Mater. Sci.* **50** (2015), 3252–3257.
18. M. J. Tadjer, M. A. Mastro, N. A. Mahadik, M. Currie, V. D. Wheeler, J. A. Freitas Jr., J. D. Greenlee, J. K. Hite, K. D. Hobart, C. R. Eddy Jr., F. J. Kub; *J. Electron. Mater.* **45** (2016), 2031–2037.
19. N. M. Sbrockey, T. Salagaj, E. Coleman, G. S. Tompa, Y. Moon, M. S. Kim; *J. Electron. Mater.* **44** (2015), 1357–1360.
20. N. M. Sbrockey, T. Salagaj, E. Coleman, G. S. Tompa, Y. Moon, M. S. Kim; *J. Electron. Mater.* **44** (2015), 1357–1360.
21. H. W. Kim, N. H. Kim; *Mater. Sci. Eng. B* **110** (2004), 34–37.
22. W. Mi, J. Ma, C. Luan, H. Xiao; *J. Lumin.* **146** (2014), 1–5.
23. W. Mi, Z. Li, C. Luan, H. Xiao, C. Zhao, J. Ma; *Ceram. Int.* **41** (2015), 2572–2575.
24. W. Mi, X. Du, C. Luan, H. Xiao, J. Ma; *RSC Adv.***4** (2014), 30579–30583.
25. W. Mi, J. Ma, Z. Li, C. Luan, H. Xiao; *J. Mater. Sci.: Mater. Electron.* **26** (2015), 7889–7894.
26. Y. Liu, X. Xia, H. Liang, H. Zhang, J. Bian, Y. Liu, R. Shen, Y. Luo, G. Du; *J. Mater. Sci: Mater. Electron.* **23** (2012), 542–545.
27. C. Y. Huang, R. H. Horng, D. S. Wuu, L. W. Tu, H. S. Kao; *Appl. Phys. Lett.* **102** (2013), 011119.

28. Y. Chen, H. Liang, X. Xia, R. Shen, Y. Liu, Y. Luo, G. Du; *Appl. Surf. Sci.* **325** (2015), 258–261.
29. Y. Chen, H. Liang, X. Xia, P. Tao, R. Shen, Y. Liu, Y. Feng, Y. Zheng, X. Li, G. Du; *J. Mater. Sci.: Mater. Electron.* **26** (2015), 3231–3235.
30. D. Hu, S. Zhuang, Z. Ma, X. Dong, G. Du, B. Zhang, Y. Zhang, J. Yin; *J. Mater. Sci.* **28** (2017), 10997–11001.
31. X. Feng, Z. Li, W. Mi, Y. Luo, J. Ma; *Mat. Sci. Semicon. Proc.* **34** (2015), 52–57.
32. S. Rafique, L. Han, A. T. Neal, S. Mou, M. J. Tadjer, R. H. French, H. Zhao; *Appl. Phys. Lett.* **109** (2016), 132103.
33. S. Rafique, L. Han, H. Zhao; *Phys. Status Solidi (a)* **213** (2016), 1002–1009.
34. S. Rafique, L. Han, S. Mou, H. Zhao; *Opt. Mater. Express* **7** (2017), 3561–3570.
35. S. Rafique, L. Han, M. J. Tadjer, J. A. Freitas, Jr., N. A. Mahadik, H. Zhao; *Appl. Phys. Lett.* **108** (2016), 182105.
36. S. Rafique, M. R. Karim, J. M. Johnson, J. Hwang, H. Zhao; *Appl. Phys. Lett.* **112** (2018), 052104.
37. S. D. Lee, K. Kaneko, S. Fujita; *Jpn. J. Appl. Phys.* **55** (2016), 1202B8.
38. Y. Oshima, E. G. Víllora, K. Shimamura; *J. Cryst. Growth* **410** (2015), 53–58.
39. K. Nomura, K. Goto, R. Togashi, H. Murakami, Y. Kumagai, A. Kuramata, S. Yamakoshi, A. Koukitu; *J. Cryst. Growth* **405** (2014), 19–22.
40. H. Murakami, K. Nomura, K. Goto, K. Sasaki, K. Kawara, Q. T. Thieu, R. Togashi, Y. Kumagai, M. Higashiwaki, A. Kuramata; *Appl. Phys. Express* **8** (2015), 015503.
41. K. Konishi, K. Goto, R. Togashi, H. Murakami, M. Higashiwaki, A. Kuramata, S. Yamakoshi, B. Monemar, Y. Kumagai; *J. Cryst. Growth* **492** (2018), 39–44.
42. Q. T. Thieu, D. Wakimoto, Y. Koishikawa, K. Sasaki, K. Goto, K. Konishi, H. Murakami, A. Kuramata, Y. Kumagai, S. Yamakoshi; *Jpn. J. Appl. Phys.* **56** (2017), 110310.
43. E. G. Villora, K. Shimamura, K. Kitamura, K. Aoki; *Appl. Phys. Lett.* **88** (2006), 031105.
44. T. Oshima, T. Okuno, S. Fujita; *Jpn. J. Appl. Phys.* **46** (2007), 7217–7220.
45. M. Y. Tsai, O. Bierwagen, M. E. White, J. S. Speck; *J. Vac. Sci. Technol. A* **28** (2010), 354–359.

46. D. Guo, Z. Wu, P. Li, Y. An, H. Liu, X. Guo, H. Yan, G. Wang, C. Sun, L. Li, W. Tang; *Opt. Mater. Express* **4** (2014), 1067–1076.

47. P. Vogt, O. Bierwagen; *Appl. Phys. Lett.* **106** (2015), 081910.

48. P. Vogt, O. Bierwagen; *Appl. Phys. Lett.* **108** (2016), 072101.

49. P. Vogt, O. Bierwagen; *Appl. Phys. Lett.* **109** (2016), 062103.

50. S. Ghose, S. Rahman, J. S. Rojas-Ramirez, M. Caro, R. Droopad; *J. Vac. Sci. Technol. B* **34** (2016), 02L109.

51. S. Ghose, S. Rahman, L. Hong, J. S. Rojas-Ramirez, H. Jin, K. Park, R. Klie, R. Droopad; *J. Appl. Phys.* **122** (2017), 095302.

52. A. S. Pratiyush, S. Krishnamoorthy, S. V. Solanke, Z. Xia, R. Muralidharan, S. Rajan, D. N. Nath; *Appl. Phys. Lett.* **110** (2017), 221107.

53. T. Oshima, N. Arai, N. Suzuki, S. Ohira, S. Fujita; *Thin Solid Films* **516** (2008), 5768–5771.

54. K. Sasaki, A. Kuramata, T. Masui, E. G. Villora, K. Shimamura, S. Yamakoshi; *Appl. Phys. Express* **5** (2012), 035502.

55. K. Sasaki, M. Higashiwaki, A. Kuramata, T. Masui, S. Yamakoshi; *J. Cryst. Growth* **378** (2013), 591–595.

56. K. Sasaki, M. Higashiwaki, A. Kuramata, T. Masui, S. Yamakoshi; *J. Cryst. Growth* **392** (2014), 30–33.

57. H. Okumura, M. Kita, K. Sasaki, A. Kuramata, M. Higashiwaki, J. S. Speck; *Appl. Phys. Express* **7** (2014), 095501.

58. S. W. Kaun, F. Wu, J. S. Speck; *J. Vac. Sci. Technol. A* **33** (2015), 041508.

59. Y. Oshima, E. Ahmadi, S. C. Badescu, F. Wu, J. S. Speck; *Appl. Phys. Express* **9** (2016), 061102.

60. Y. Oshima, E. Ahmadi, S. Kaun, F. Wu, J. S. Speck; *Semicond. Sci. Technol.* **33** (2018), 015013.

61. E. Ahmadi, O. S. Koksaldi, S. W. Kaun, Y. Oshima, D. B. Short, U. K. Mishra, J. S. Speck; *Appl. Phys. Express* **10** (2017), 041102.

62. E. Ahmadi, O. S. Koksaldi, X. Zheng, T. Mates, Y. Oshima, U. K. Mishra, J. S. Speck; *Appl. Phys. Express* **10** (2017), 071101.

63. S.-H. Han, A. Mauze, E. Ahmadi, T. Mates, Y. Oshima, J. S. Speck; *Semicond. Sci. Technol.* **33** (2018), 045001.

64. M. Orita, H. Ohta, M. Hirano, H. Hosono; *Appl. Phys. Lett.* **77** (2000), 4166–4168.

65. M. Orita, H. Ohta, H. Hiramatsu, M. Hirano, S. Den, M. Sasaki, T. Katagiri, H. Mimura, H. Hosono; *Rev. Sci. Instrum.* **72** (2001), 3340–3343.

66. M. Orita, H. Hiramatsu, H. Ohta, M. Hirano, H. Hosono; *Thin Solid Films* **411** (2002), 134–139.

67. K. Matsuzaki, H. Yanagi, T. Kamiya, H. Hiramatsu, K. Nomura, M. Hirano, H. Hosono; *Appl. Phys. Lett.* **88** (2006), 092106.

68. K. Matsuzaki, H. Hiramatsu, K. Nomura, H. Yanagi, T. Kamiya, M. Hirano, H. Hosono; *Thin Solid Films* **496** (2006), 37–41.

69. P. Gollakota, A. Dhawan, P. Wellenius, L. M. Lunardi, J. F. Muth, Y. N. Saripalli, H. Y. Peng, H. O. Everitt; *Appl. Phys. Lett.* **88** (2006), 221906.

70. A. Goyal, B. S. Yadav, O. P. Thakur, A. K. Kapoor, R. Muralidharan; *J. Alloy. Compd.* **583** (2014), 214–219.

71. S. Müller, H. von Wenckstern, D. Splith, F. Schmidt, M. Grundmann; *Phys. Status Solidi (a)* **211** (2014), 34–39.

72. F. B. Zhang, K. Saito, T. Tanaka, M. Nishio, Q. X. Guo; *J. Cryst. Growth* **387** (2014), 96–100.

73. F. Zhang, K. Saito, T. Tanaka, M. Nishio, Q. X. Guo; *J. Mater. Sci.* **26** (2015), 9624–9629.

74. F. Zhang, M. Arita, X. Wang, Z. Chen, K. Saito, T. Tanaka, M. Nishio, T. Motooka, Q. Guo; *Appl. Phys. Lett.* **109** (2016), 102105.

75. Z. Chen, K. Saito, T. Tanaka, M. Nishio, M. Arita, Q. Guo; *J. Cryst. Growth* **430** (2015), 28–33.

76. Z. Chen, X. Wang, F. Zhang, S. Noda, K. Saito, T. Tanaka, M. Nishio, Q. Guo; *J. Lumin.* **177** (2016), 48–53.

77. Z. Chen, X. Wang, S. Noda, K. Saito, T. Tanaka, M. Nishio, M. Arita, Q. Guo; *Superlattice. Microst.* **90** (2016), 207–214.

78. X. H. Wang, F. B. Zhang, K. Saito, T. Tanaka, M. Nishio, Q. X. Guo; *J. Phys. Chem. Solids* **75** (2014), 1201–1204.

79. D. Shiojiri, R. Yamauchi, D. Fukuda, N. Tsuchimine, S. Kaneko, A. Matsuda, M. Yoshimoto; *J. Cryst. Growth* **424** (2015), 38–41.

80. M. D. Heinemann, J. Berry, G. Teeter, T. Unold, D. Ginley; *Appl. Phys. Lett.* **108** (2016), 022107.

81. R. Wakabayashi, K. Yoshimatsu, M. Hattori, A. Ohtomo; *Appl. Phys. Lett.* **111** (2017), 162101.

82. R. Wakabayashi, T. Oshima, M. Hattori, K. Sasaki, T. Masui, A. Kuramata, S. Yamakoshi, K. Yoshimatsu, A. Ohtomo; *J. Cryst. Growth* **424** (2015), 77–79.

83. K. D. Leedy, K. D. Chabak, V. Vasilyev, D. C. Look, J. J. Boeckl, J. L. Brown, S. E. Tetlak, A. J. Green, N. A. Moser, A. Crespo, D. B. Thomson, R. C. Fitch, J. P. McCandless, G. H. Jessen; *Appl. Phys. Lett.* **111** (2017), 012103.

84. M. Higashiwaki, K. Sasaki, A. Kuramata, T. Masui, S. Yamakoshi; *Appl. Phys. Lett.* **100** (2012), 013504.

85. M. Higashiwaki, K. Sasaki, T. Kamimura, M. H. Wong, D. Krishnamurthy, A. Kuramata, T. Masui, S. Yamakoshi; *Appl. Phys. Lett.* **103** (2013), 123511.

86. M. Higashiwaki, K. Sasaki, A. Kuramata, T. Masui, S. Yamakoshi; *Phys. Status Solidi (a)* **211** (2014), 21–26.

87. M. Higashiwaki, H. Murakami, Y. Kumagai, A. Kuramata; *Jpn. J. Appl. Phys.* **55** (2016), 1202A1.

88. M. Higashiwaki, G. H. Jessen; *Appl. Phys. Lett.* **112** (2018), 060401.

89. K. Sasaki, M. Higashiwaki, A. Kuramata, T. Masui, S. Yamakoshi; *IEEE Electr. Device Lett.* **34** (2013), 493–495.

90. H. Murakami, K. Nomura, K. Goto, K. Sasaki, K. Kawara, Q. T. Thieu, R. Togashi, Y. Kumagai, M. Higashiwaki, A. Kuramata; *Appl. Phys. Express* **8** (2015), 015503.

91. J. Yang, S. Ahn, F. Ren, S. J. Pearton, S. Jang, J. Kim, A. Kuramata; *Appl. Phys. Lett.* **110** (2017), 192101.

92. S. Oh, G. Yang, J. Kim; *ECS J. Solid State Sci. Technol.* **6** (2017), Q3022–Q3025.

93. K. Konishi, K. Goto, H. Murakami, Y. Kumagai, A. Kuramata, S. Yamakoshi, M. Higashiwaki; *Appl. Phys. Lett.* **110** (2017), 103506.

94. M. H. Wong, K. Sasaki, A. Kuramata, S. Yamakoshi, M. Higashiwaki; *Jpn. J. Appl. Phys.* **55** (2016), 1202B9.

95. M. H. Wong, K. Sasaki, A. Kuramata, S. Yamakoshi, M. Higashiwaki; *Appl. Phys. Lett.* **106** (2015), 032105.

96. M. H. Wong, K. Sasaki, A. Kuramata, S. Yamakoshi, M. Higashiwaki; *IEEE Electr. Device Lett.* **37** (2016), 212–215.

97. M. H. Wong, Y. Morikawa, K. Sasaki, A. Kuramata, S. Yamakoshi, M. Higashiwaki; *Appl. Phys. Lett.* **109** (2016), 193503.

98. A. J. Green, K. D. Chabak, E. R. Heller, R. C. Fitch, Jr., M. Baldini, A. Fiedler, K. Irmscher, G. Wagner, Z. Galazka, S. E. Tetlak, A. Crespo, K. Leedy, G. H. Jessen; *IEEE Electron Device Lett.* **37** (2016), 902–905.

99. A. J. Green, K. D. Chabak, M. Baldini, N. Moser, R. Gilbert, R. C. Fitch, Jr., G. Wagner, Z. Galazka, J. McCandless, A. Crespo, K. Leedy, G. H. Jessen; *IEEE Electron Device Lett.* **38** (2017), 790–793.

100. K. D. Chabak, N. Moser, A. J. Green, D. E. Walker, Jr., S. E. Tetlak, E. Heller, A. Crespo, R. Fitch, J. P. McCandless, K. Leedy, M. Baldini, G. Wagner, Z. Galazka, X. Li, G. Jessen; *Appl. Phys. Lett.* **109** (2016), 213501.

101. N. A. Moser, J. P. McCandless, A. Crespo, K. D. Leedy, A. J. Green, E. R. Heller, K. D. Chabak, N. Peixoto, G. H. Jessen; *Appl. Phys. Lett.* **110** (2017), 143505.

102. N. Moser, J. McCandless, A. Crespo, K. Leedy, A. Green, A. Neal, S. Mou, E. Ahmadi, J. Speck, K. Chabak, N. Peixoto, G. Jessen; *IEEE Electron Device Lett.* **38** (2017), 775–778.

103. J. Bae, H. W. Kim, I. H. Kang, G. Yang, J. Kim; *Appl. Phys. Lett.* **112** (2018), 122102.

104. Z. Hu, K. Nomoto, W. Li, N. Tanen, K. Sasaki, A. Kuramata, T. Nakamura, D. Jena, H. G. Xing; *IEEE Electron Device Lett.* **39** (2018), 869–872.

105. W. S. Hwang, A. Verma, H. Peelaers, V. Protasenko, S. Rouvimov, H. (Grace) Xing, A. Seabaugh, W. Haensch, C. Van de Walle, Z. Galazka, M. Albrecht, R. Fornari, D. Jena; *Appl. Phys. Lett.* **104** (2014), 203111.

106. R. Suzuki, S. Nakagomi, Y. Kokubun, N. Arai, S. Ohira; *Appl. Phys. Lett.* **94** (2009), 222102.

107. R. Suzuki, S. Nakagomi, Y. Kokubun; *Appl. Phys. Lett.* **98** (2011), 131114.

108. T. Oshima, T. Okuno, S. Fujita; *Jpn. J. Appl. Phys.* **46** (2007), 7217–7220.

109. T. Oshima, T. Okuno, N. Arai, N. Suzuki, H. Hino, S. Fujita; *Jpn. J. Appl. Phys.* **48** (2009), 011605.

110. P. Guo, J. Xiong, X. Zhao, T. Sheng, C. Yue, B. Tao, X. Liu; *J. Mater. Sci. Mater. El.* **25** (2014), 3629–3632.

111. D. Y. Guo, Z. P. Wu, Y. H. An, X. C. Guo, X. L. Chu, C. L. Sun, L. H. Li, P. G. Li, W. H. Tang; *Appl. Phys. Lett.* **105** (2014), 023507.

112. X. C. Guo, N. H. Hao, D. Y. Guo, Z. P. Wu, Y. H. An, X. L. Chu, L. H. Li, P. G. Li, M. Lei, W. H. Tang; *J. Alloy. Compd.* **660** (2016), 136–140.

113. D. Guo, Z. Wu, P. Li, Y. An, H. Liu, X. Guo, H. Yan, G. Wang, C. Sun, L. Li, W. Tang; *Opt. Mater. Express* **4** (2014), 1067–1076.

114. X. Z. Liu, P. Guo, T. Sheng, L. X. Qian, W. L. Zhang, Y. R. Li; *Opt. Mater.* **51** (2016), 203–207.

115. S. Oh, Y. Jung, M. A. Mastro, J. K. Hite, C. R. Eddy Jr., J. Kim; *Opt. Express* **23** (2015), 249734.

116. S. Nakagomi, T. Momo, S. Takahashi, Y. Kokubun; *Appl. Phys. Lett.* **103** (2013), 072105.

117. S. Nakagomi, T. Sato, Y. Takahashi, Y. Kokubun; *Sensor. Actuator. A* **232** (2015), 208–213.

118. H. von Wenckstern, D. Splith, M. Purfürst, Z. Zhang, Ch. Kranert, S. Müller, M. Lorenz, M. Grundmann; *Semicond. Sci. Technol.* **30** (2015), 024005.

119. Z. Zhang, H. von Wenckstern, J. Lenzner, M. Lorenz, M. Grundmann; *Appl. Phys. Lett.* **108** (2016), 123503.

120. G. C. Hu, C. X. Shan, N. Zhang, M. M. Jiang, S. P. Wang, D. Z. Shen; *Opt. Express* **23** (2015), 13554.

121. F.-P. Yu, S.-L. Ou, D.-S. Wuu; *Opt. Mater. Express* **5** (2015), 1240–1249.

122. K. Shimamura, E. G. Víllora; *J. Fluorine Chem.* **132** (2011), 1040–1046.

123. D. Li, V. Hoffmann, E. Richter, T. Tessaro, Z. Galazka, M. Weyers, G. Tränkle; *J. Cryst. Gowth* **478** (2017), 212–215.

124. D. Szalkai, Z. Galazka, K. Irmscher, P. Tüttő, A. Klix, D. Gehre; *IEEE Trans. Nucl. Sci.* **64** (2017), 1574–1579.

125. T. Yanagida, G. Okada, T. Kato, D. Nakauchi, S. Yanagida; *Appl. Phys. Express* **9** (2016), 042601.

126. Y. Usui, D. Nakauchi, N. Kawano, G. Okada, N. Kawaguchi, T. Yanagida; *J. Phys. Chem. Solids* **117** (2018), 36–41.

127. Y. Usui, T. Oya, G. Okada, N. Kawaguchi, T. Yanagida; *Optik* **143** (2017), 150–157.

128. W. Drozdowski, M. Makowski, M. E. Witkowski, A. J. Wojtowicz, Z. Galazka, K. Irmscher, R. Schewski; *Radiat. Meas.* **121** (2019), 49–53.

129. M. Fleischer, H. Meixner; *Sensor. Actuat. B* **4** (1991), 437–441.

130. M. Fleischer, H. Meixner; *Sensor. Actuat. B* **13** (1993), 259–263.

131. M. Fleischer, H. Meixner; *Sensor. Actuat. B* **25** (1995), 544–547.

132. M. Fleischer, M. Seth, C. D. Kohl, H. Meixner; *Sensor. Actuat. B* **36** (1996), 290–296.

133. M. Fleischer, M. Seth, C. D. Kohl, H. Meixner; *Sensor. Actuat. B* **36** (1996), 297–302.

134. M. Fleischer, V. Wagner, G. Hacker, H. Meixner; *Sensor. Actuat. B* **26** (1995), 85–88.

135. G. K. Flingelli, M. Fleischer, H. Meixner; *Sensor. Actuat. B* **48** (1998), 258–262.

136. J. Frank, M. Fleischer, H. Meixner, A. Feltz; *Sensor. Actuat. B* **49** (1998), 110–114.

137. R. Pohle, M. Fleischer, H. Meixner; *Sensor. Actuat. B* **68** (2000), 151–156.

138. T. Weh, J. Frank, M. Fleischer, H. Meixner; *Sensor. Actuat. B* **78** (2001), 202–207.

139. T. Schwebel, M. Fleischer, H. Meixner, C. D. Kohl; *Sensor. Actuat. B* **49** (1998), 46–51.

140. T. Schwebel, M. Fleischer, H. Meixner; *Sensor. Actuat. B* **65** (2000), 176–180.

141. S. Stegmeier, M. Fleischer, P. Hauptmann; *Sensor. Actuat. B* **148** (2010), 439–449.

142. M. Ogita, N. Saika, Y. Nakanishi, Y. Hatanaka; *Appl. Surf. Sci.* **142** (1999), 188–191.

143. M. Ogita, K. Higo, Y. Nakanishi, Y. Hatanaka; *Appl. Surf. Sci.* **175–176** (2001), 721–725.

144. M. Ogita, S. Yuasa, K. Kobayashi, Y. Yamada, Y. Nakanishi, Y. Hatanaka; *Appl. Surf. Sci.* **212–213** (2003), 397–401.

145. C. I. Baban, Y. Toyoda, M. Ogita; *Jpn. J. Appl. Phys.* **43** (2004), 7213–7216.

146. M. Bartic, Y. Toyoda, C. I. Baban, M. Ogita; *Jpn. J. Appl. Phys.* **45** (2006), 5186–5188.

147. M. Bartic, C. I. Baban, H. Suzuki, M. Ogita, M. Isai; *J. Am. Ceram. Soc.* **90** (2007), 2879–2884.

148. M. Bartic; *Phys. Status Solidi* **213** (2016), 457–462.

149. S. Nakagomi, T. Sai, Y. Kokubun; *Sensor. Actuat. B* **187** (2013), 413–419.

150. A. Trinchi, K. Galatsis, W. Wlodarski, Y. X. Li; *IEEE Sens. J.* **3** (2003), 548–553.

151. A. Trinchi, S. Kaciulis, L. Pandolfi, M. K. Ghantasala, Y. X. Li, W. Wlodarski, S. Viticoli, E. Comini, G. Sberveglieri; *Sensor. Actuat. B* **103** (2004), 129–135.

152. C. T. Lee, J. T. Yan; *Sensor. Actuat. B* **147** (2010), 723–729.

153. R. Roy, V. G. Hill, E. F. Osborn; *J. Am. Ceram. Soc.* **74** (1952), 719–722.

154. M. Zinkevich, F. Aldinger; *J. Am. Ceram. Soc.* **87** (2004), 683–691.

155. S. Penner, C. Zhuo, R. Thalinger, M. Grünbacher, C. Hejny, S. Vanicek, M. Noisternig; *Monatsh. Chem.* **147** (2016), 289–300.

156. H. Y. Playford, A. C. Hannon, E. R. Barney, R. I. Walton; *Chem. Eur. J.* **19** (2013), 2803–2813.

157. S. Yoshioka, H. Hayashi, A. Kuwabara, F. Oba, K. Matsunaga, I. Tanaka; *J. Phys.: Condens. Matter.* **19** (2007), 346211.

158. M. R. Delgado, C. O. Areán; *Z. Anorg. Allg. Chem.* **631** (2005), 2115-2120.

159. R. Fornari, M. Pavesi, V. Montedoro, D. Klimm, F. Mezzadri, I. Cora, B. Pécz, F. Boschi, A. Parisini, A. Baraldi, C. Ferrari, E. Gombia, M. Bosi; *Acta Mater.* **140** (2017), 411–416.

160. I. Cora, F. Mezzadri, F. Boschi, M. Bosi, M. Čaplovičová, G. Calestani, I. Dódony, B. Pécza, R. Fornari; *CrystEngComm.* **19** (2017), 1509–1516.

161. J. Furthmüller, F. Bechstedt; *Phys. Rev. B* **93** (2016), 115204.

162. M. Marezio, J. P. Remeika; *J. Chem. Phys.* **46** (1967), 1862–1865.

163. J. Åhman, G. Svensson, J. Albertsson; *Acta Cryst. C* **52** (1996), 1336–1338.

164. K. Pohl; *Naturwiss* **55** (1968), 82.

165. D. Shinohara, S. Fujita; *Jpn. J. Appl. Phys.* **47** (2008), 7311–7313.

166. Y. Oshima, E. G. Víllora, K. Shimamura; *Appl. Phys. Express* **8** (2015), 055501.

167. D. Y. Guo, X. L. Zhao, Y. S. Zhi, W. Cui, Y. Q. Huang, Y. H. An, P. G. Li, Z. P. Wu, W. H. Tang; *Mater. Lett.* **164** (2016), 364–367.

168. T. Oshima, T. Nakazono, A. Mukai, A. Ohtomo; *J. Cryst. Growth* **359** (2012), 60–63.

169. T. Oshima, K. Matsuyama, K. Yoshimatsu, A. Ohtomo; *J. Cryst. Growth* **421** (2015), 23–26.

170. Y. Oshima, E. G. Vıllora, Y. Matsushita, S. Yamamoto, K. Shimamura; *J. Appl. Phys.* **118** (2015), 085301.

171. H. Nishinaka, D. Tahara, M. Yoshimoto; *Jpn. J. Appl. Phys.* **55** (2016), 1202BC.

172. M. Kracht, A. Karg, J. Schörmann, M. Weinhold, D. Zink, F. Michel, M. Rohnke, M. Schowalter, B. Gerken, A. Rosenauer, P. J. Klar, J. Janek, M. Eickhoff; *Phys. Rev. Appl.* **8** (2017), 054002.

173. F. Mezzadri, G. Calestani, F. Boschi, D. Delmonte, M. Bosi, R. Fornari; *Inorg. Chem.* **55** *(*2016), 12079–12084.

174. S. Geller; *J. Chem. Phys.* **33** (1966), 676–684.

175. Z. Łodziana, K. Parliński; *Phys. Rev. B.* **67** (2003), 174106.

176. K. Yamaguchi; *Solid State Commun.* **131** (2004), 739–774.

177. H. He, R. Orlando, M. A. Blanco, R. Pandey, E. Amzallag, I. Baraille, M. Rérat; *Phys. Rev. B* **74** (2006), 195123.

178. J. Furthmüller, F. Bechstedt; *Phys. Rev. B* **93** (2016), 115204.

179. J. B. Varley, J. R. Weber, A. Janotti, C. G. Van de Walle; *Appl. Phys. Lett.* **97** (2010), 142106.

180. J. B. Varley, H. Peelaers, A. Janotti, C. G. Van de Walle; *J. Phys.: Condens. Matter* **23** (2011), 334212.

181. J. B Varley, A. Schleife; *Semicond. Sci. Technol.* **30** (2015), 024010.

182. H. Peelaers, C. G. Van de Walle; *Phys. Status Solidi (b)* **252** (2015), 828–832.

183. T. Zacherle, P. C. Schmidt, M. Martin; *Phys. Rev. B* **87** (2013), 235206.

184. A. Navarro-Quezada, S. Alamé, N. Esser, J. Furthmüller, F. Bechstedt, Z. Galazka, D. Skuridina, P. Vogt; *Phys. Rev. B* **92** (2015), 195306.

185. C. Cocchi, H. Zschiesche, D. Nabok, A. Mogilatenko, M Albrecht, Z. Galazka, H. Kirmse, C. Draxl, C T. Koch; *Phys. Rev. B* **94** (2016), 075147.

186. P. Deák, Q. D. Ho, F. Seemann, B. Aradi, M. Lorke, T. Frauenheim; *Phys. Rev. B* **95** (2017), 075208.

187. L. Dong, R. Jia, C. Li, B. Xin, Y. Zhang; *J. Alloy. Compd.* **712** (2017), 379–385.

188. A. Mock, R. Korlacki, C. Briley, V. Darakchieva, B. Monemar, Y. Kumagai, K. Goto, M. Higashiwaki, M. Schubert; *Phys. Rev. B* **96** (2017), 245205.

189. Y. Wei, X. Li, J. Yang, C. Liu, J. Zhao, Y. Liu, S. Dong; *Sci. Rep.* **8** (2018), 10142.

190. T. C. Lovejoy, E. N. Yitamben, N. Shamir, J. Morales, E. G. Villora, K. Shimamura, S. Zheng, F. S. Ohuchi, M. A. Olmstead; *Appl. Phys. Lett.* **94** (2009), 081906.

191. C. Janowitz, V. Scherer, M. Mohamed, A. Krapf, H. Dwelk, R. Manzke, Z. Galazka, R. Uecker, K. Irmscher, R. Fornari, M. Michling, D. Schmeißer, J. R. Weber, J. B. Varley, C. G Van de Walle; *New J. Phys.* **13** (2011), 085014.

192. M. Mohamed, C. Janowitz, I. Unger, R. Manzke, Z. Galazka, R. Uecker, R. Fornari, J. R. Weber, J. B. Varley, C. G. Van de Walle; *Appl. Phys. Lett.* **97** (2010), 211903.

193. M. Mohamed, I. Unger, C. Janowitz, R. Manzke, Z. Galazka, R. Uecker, R. Fornari; *J. Phys.: Conference Series* **286** (2011), 012027.

194. G.-L. Li, F. Zhang, Y.-T. Cui, H. Oji, J.-Y. Son, Q. Guo; *Appl. Phys. Lett.* **107** (2015), 022109.

195. M. Michling, D. Schmeißer; *IOP Conf. Series: Mater. Sci. Eng.* **34** (2012), 012002.

196. A. Kyrtsos, M. Matsubara, E. Bellotti; *Appl. Phys. Lett.* **112** (2018), 032108.

197. M. A. Blanco, M. B. Sahariah, H. Jiang, A. Costales, R. Pandey; *Phys. Rev. B* **72** (2005), 184103.

198. S. Lany; *APL Mater.* **6** (2018), 046103.

199. J. L. Lyons; *Semicond. Sci. Technol.* **33** (2018), 05LT02.

200. W. Guo, Y. Guo, H. Dong, X. Zhou; *Phys. Chem. Chem. Phys.* **17** (2015), 5817–5825.

201. J. A. Kohn, G. Katz, J. D. Broder; *Am. Mineral.* **42** (1957), 398–407.

202. J. P. Remeika, US 3075831 (1963).

203. G. Katz, R. Roy; *J. Am. Ceram. Soc.* **49** (1966), 168–169.

204. G. Garton, S. H. Smith, B. M. Wanklyn; *J. Cryst. Growth* **13/14** (1972), 588–592.

205. K. Fischer, D. Linzen, E. Sinn, H. Stieff, F. Voigt; *Kristall und Technik* **14** (1979), 1521–1524.

206. V. I. Chani, K. Inoue, K. Shimamura, K. Sugiyama, T. Fukuda; *J. Cryst. Growth* **132** (1993), 335–336.

207. A. B. Chase; *J. Am. Ceram. Soc.* **47** (1964), 470.

208. M. R. Lorenz, J. F. Woods, R. J. Gambino; *J. Phys. Chem. Solids* **28** (1967), 403–404.

209. T. Matsumoto, M. Aoki, A. Kinoshita, T. Aono; *Jpn. J. Appl. Phys.* **13** (1974), 737–738.

210. T. Matsumoto, M. Aoki, A. Kinoshita, T. Aono; *Jpn. J. Appl. Phys.* **13** (1974), 1578–1582.

211. Von U. Gerlach, H. Oppermann; *Z. Anorg. Allg. Chem.* **432** (1977), 17–25.

212. H. Juskowiak, A. Pajączkowska; *J. Mater. Sci.* **21** (1986), 3430–3434.

213. H. Juskowiak, A. Pajączkowska; *Mater. Res. Bull.* **23** (1988), 1071–1077.

214. A. Pajączkowska, H. Juskowiak; *J. Mater. Sci.* **21** (1986), 3435–3439.

215. A. Pajączkowska, H. Juskowiak; *J. Cryst. Growth* **79** (1986), 421–426.

216. V. I. Vasiltsiv, Y. Zakarko; *Zhurnal Prikladnoi Spektroskopii* **39** (1983), 423–428.

217. V. I. Vasyltsiv, Y. I. Rym, Y. Zakharko; *Phys. Stat. Solidi (b)* **195** (1996), 653–658.

218. N. Ueda, H. Hosono, R. Waseda, H. Kawazoe; *Appl. Phys. Lett.* **71** (1997), 933–935.

219. Y. Tomm, J. M. Ko, A. Yoshikawa, T. Fukuda; *Sol. Energy Mater. Sol. Cells* **66** (2001), 369–374.

220. J. M. Ko, A. Yoshikawa, T. Fukuda; *Sol. Energy Mater. Sol. Cells* **66** (2001), 369–374.

221. E. G. Villora, K. Shimamura, Y. Yoshikawa, K. Aoki, N. Ichinose; *J. Cryst. Growth* **270** (2004), 420–426.

222. E. G. Víllora, K. Shimamura, Y. Yoshikawa, T. Ujiie, K. Aoki; *Appl. Phys. Lett.* **92** (2008), 202120.

223. J. Zhang, B. Li, C. Xia, G. Pei, Q. Deng, Z. Yang, W. Xu, H. Shi, F. Wu, Y. Wu, J. Xu; *J. Phys. Chem. Solids* **67** (2006), 2448–2451.

224. Y. Tomm, P. Reiche, D. Klimm, T. Fukuda; *J. Cryst. Growth* **220** (2000), 510–514.

225. Z. Galazka, R. Uecker, K. Irmscher, M. Albrecht, D. Klimm, M. Pietsch, M. Brutzam, R. Bertram, S. Ganschow, R. Fornari; *Cryst. Res. Technol.* **45** (2010), 1229–1236.

226. Z. Galazka, K. Irmscher, R. Uecker, R. Bertram, M. Pietsch, A. Kwasniewski, M. Naumann, T. Schulz, R. Schewski, D. Klimm, M. Bickermann; *J. Cryst. Growth* **404** (2014), 184–191.

227. Z. Galazka, R. Uecker, D. Klimm, M. Bickermann; PCT / EP2015 / 079938 (WO/2016/110385) (2015).

228. Z. Galazka, R. Uecker, D. Klimm, K. Irmscher, M. Naumann, M. Pietsch, A. Kwasniewski, R. Bertram, S. Ganschow, M. Bickermann; *ECS J. Solid State Sci. Technol.* **6** (2017), Q3007–Q3011.

229. Z. Galazka, S. Ganschow, A. Fiedler, R. Bertram, D. Klimm, K. Irmscher, R. Schewski, M. Pietsch, M. Albrecht, M. Bickermann; *J. Cryst. Growth* **486** (2018), 82–90.

230. H. Aida, K. Nishiguchi, H. Takeda, N. Aota, K. Sunakawa, Y. Yaguchi; *Jpn. J. Appl. Phys.* **47** (2008), 8506–8509.

231. K. Shimamura, E. G. Villora, K. Matsumura, K. Aoki, M. Nakamura, S. Takekawa, N. Ichinose, and K. Kitamura; *Nihon Kessho Seicho Gakkaishi* **33** (2006), 147 [in Japanese].

232. A. Kuramata, K. Koshi, S. Watanabe, Y. Yamaoka, T. Masui, S. Yamakoshi; *Jpn. J. Appl. Phys.* **55** (2016), 1202A2.

233. K. Hoshikawa, E. Ohba, T. Kobayashi, J. Yanagisawa, C. Miyagawa, Y. Nakamura; *J. Cryst. Growth* **447** (2016), 36–41.

234. Von U. Gerlach, H. Oppermann; *Z. Anorg. Allg. Chem.* **429** (1977), 25–33.

235. M. Schieber; *J. Am. Ceram. Soc.* **47** (1964), 537–538.

236. G. Katz, R. Roy; *J. Am. Ceram. Soc.* **49** (1966), 168–169.

237. G. Garton, S. H. Smith, B. M. Wanklyn; *J. Cryst. Growth* **13/14** (1972), 588–592.

238. V. I. Chani, K. Inoue, K. Shimamura, K. Sugiyama, T. Fukuda; *J. Cryst. Growth* **132** (1993), 335–336.

239. M. Schieber; *J. Am. Ceram. Soc.* **47** (1964), 537–538.

240. H. H. Tippins; *Phys. Rev.* **137** (1965), A865–A871.

241. S. Ohira, M. Yoshioka, T. Sugawara, K. Nakajima, T. Shishido; *Thin Solid Films* **496** (2006), 53–57.

242. J. Zhang, C. Xia, Q. Deng, W. Xu, H. Shi, F. Wu, Y. Wu; *J. Phys. Chem. Solids* **67** (2006), 1656–1659.

243. N. Suzuki, S. Ohira, M. Tanaka, T. Sugawara, K. Nakajima, T. Shishido; *Phys. Status Solidi* (*c*) **4** (2007), 2310–2313.

244. S. Ohira, N. Arai; *Phys. Status Solidi* (*c*) **9** (2008), 3116–3118.

245. K. Shimamura, E. G. Víllora, T. Ujiie, K. Aoki; *Appl. Phys. Lett.* **92** (2008), 201914.

246. T. Oshima, K. Kaminaga, A. Mukai, K. Sasaki, T. Masui, A. Kuramata, S. Yamakoshi, S. Fujita, A. Ohtomo; *Jpn. J. Appl. Phys.* **52** (2013), 05101.

247. A. Kuramata, K. Koshi, S. Watanabe, Y. Yamaoka, T. Masui, S. Yamakoshi; *Jpn. J. Appl. Phys.* **55** (2016), 1202A2.

248. J. Zhang, B. Li, C. Xia, J. Xu, Q. Deng, X. Xu, F. Wu, W. Xu, H. Shi, G. Pei, Y. Wu; *Sci. China Ser. E-Tech. Sci.* **50** (2007), 51–56.

249. T. C. Lovejoy, R. Chen, E. N. Yitamben, V. Shutthanadan, S. M. Heald, E. G. Villora, K. Shimamura, S. Zheng, S. T. Dunham, F. S. Ohuchi, M. A. Olmstead; *J. Appl. Phys.* **111** (2012), 123716.

250. A. Nakazawa, D. Yasukawa, H. Wakai, H. Oda, A. Yamanaka; *Phys. Staus Solidi* (*c*) **11** (2013), 1584–1587.

251. Y. Usui, T. Oya, G. Okada, N. Kawaguchi, T. Yanagida; *Optik* **143** (2017), 150–157.

252. V. I. Vasyltsiv, Y. I. Rym, Y. Zakharko; *Phys. Stat. Solidi* (*b*) **195** (1996), 653–658.

253. T. Onuma, S. Fujioka, T. Yamaguchi, M. Higashiwaki, K. Sasaki, T. Masui, T. Honda; *Appl. Phys. Lett.* **103** (2013), 041910.

254. E. G. Villora, M. Yamaga, T. Inoue, S. Yabasi, Y. Masui, T. Sugawara, T. Fukuda; *Jpn. J. Appl. Phys.* **41** (2002), L622–L625.

255. E. G. Villora, Y. Morioka, T. Atou, T. Sugawara, M. Kikuchi, T. Fukuda; *Phys. Status Solidi* (*a*) **193** (2002), 187–195.

256. E. G. Villora, K. Hatanaka, H. Odaka, T. Sugawara, T. Miura, H. Fukumura, T. Fukuda; *Solid State Commun.* **127** (2003), 385–388.

257. E. G. Villora, K. Shimamura, Y. Yoshikawa, T. Ujiie, K. Aoki; *Appl. Phys. Lett.* **92** (2008), 202120.

258. M. Yamaga, E. G. Villora, K. Shimamura, N. Ichinose, M. Honda; *Phys. Rev. B* **68** (2003), 155207.

259. K. Shimamura, E. G. Víllora, T. Ujiie, K. Aoki; *Appl. Phys. Lett.* **92** (2008), 201914.

260. S. Ohira, N. Suzuki, N. Arai, M. Tanaka, T. Sugawara, K. Nakajima, T. Shishido; *Thin Solid Films* **516** (2008), 5763–5767.

261. J. Zhang, B. Li, C. Xia, G. Pei, Q. Deng, Z. Yang, W. Xu, H. Shi, F. Wu, Y. Wu, J. Xu; *J. Phys. Chem. Solids* **67** (2006), 2448–2451.

262. J. Zhang, C. Xia, Q. Deng, W. Xu, H. Shi, F. Wu, Y. Wu; *J. Phys. Chem. Solids* **67** (2006), 1656–1659.

263. J. Zhang, B. Li, C. Xia, J. Xu, Q. Deng, X. Xu, F. Wu, W. Xu, H. Shi, G. Pei, Y. Wu; *Sci. China Ser. E-Tech. Sci.* **50** (2007), 51–56.

264. N. Ueda, H. Hosono, R. Waseda, H. Kawazoe; *Appl. Phys. Lett.* **70** (1997), 3561–3563.

265. K. Shimamura, E. G. Víllora; *J. Fluorine Chem.* **132** (2011), 1040–1046.

266. W. Mu, Z. Jia, Y. Yin, Q. Hu, Y. Li, B. Wu, J. Zhang, X. Tao; *J. Alloy. Compd.* **714** (2017), 453–458.

267. K. Hanada, T. Moribayashi, T. Uematsu, S. Masuya, K. Koshi, K. Sasaki, A. Kuramata, O. Ueda, M. Kasu; *Jpn. J. Appl. Phys.* **55** (2016), 030303.

268. K. Hanada, T. Moribayashi, K. Koshi, K. Sasaki, A. Kuramata, O. Ueda, M. Kasu; *Jpn. J. Appl. Phys.* **55** (2016), 1202BG.

269. O. Ueda, N. Ikenaga, K. Koshi, K. Iizuka, A. Kuramata, K. Hanada, T. Moribayashi, S. Yamakoshi, M. Kasu; *Jpn. J. Appl. Phys.* **55** (2016), 1202BD.

270. K. Nakai, T. Nagai, K. Noami, T Futagi; *Jpn. J. Appl. Phys.* **54** (2015), 051103.

271. Z. Galazka, I. M. Hanke; *Leibniz-Institut für Kristallzüchtung*, Berlin, Germany (2018–2019), unpublished data.

272. W. Miller, K. Böttcher, Z. Galazka, J. Schreuer; *Crystals* **7** (2017), 26.

273. A. Fiedler, Z. Galazka, K. Irmscher; *J. Appl. Phys.* **126** (2019), 213104.

274. A. Burton, R. C. Prim, W. P. Slichter; *J. Chem. Phys.* **21** (1953), 1987–1991.

275. A. G. Ostrogorsky and G. Müller; *J. Cryst. Growth* **121** (1992), 587–598.

276. E. Ohba, T. Kobayashi, M. Kado, K. Hoshikawa; *Jpn. J. Appl. Phys.* **55** (2016), 1202BF.

277. V. M. Bermudez; *Chem. Phys.* **323** (2006), 193–203.

278. M. Busch, E. Meyer, K. Irmscher, Z. Galazka, K. Gärtner, H. Winter; *Appl. Phys. Lett.* **105** (2014), 051603.

279. G. Wagner, M. Baldini, D. Gogova, M. Schmidbauer, R. Schewski, M. Albrecht, Z. Galazka, D. Klimm, R. Fornari; *Phys. Status Solidi (a)* **211** (2014), 27–33.

280. T. C. Lovejoy, R. Chen, X. Zheng, E. G. Villora, K. Shimamura, H. Yoshikawa, Y. Yamashita, S. Ueda, K. Kobayashi, S. T. Dunham, F. S. Ohuchi, M. A. Olmstead; *Appl. Phys. Lett.* **100** (2012), 181602.

281. A. Navarro-Quezada, Z. Galazka, S. Alamé, D. Skuridina, P. Vogt, N. Esser; *Appl. Surf. Sci.* **349** (2015), 368–373.

282. T. Oshima, K. Kaminaga, A. Mukai, K. Sasaki, T. Masui, A. Kuramata, S. Yamakoshi, S. Fujita, A. Ohtomo; *Jpn. J. Appl. Phys.* **52** (2013), 05101.

283. C. Huang, W. Mu, H. Zhou, Y. Zhu, X. Xu, Z. Jia, L. Zheng, X. Tao; *RSC Adv.* **8** (2018), 6544–6550.

284. S. Ohira, N. Arai; *Phys. Status Solidi (c)* **9** (2008), 3116–3118.

285. J. E. Hogan, S. W. Kaun, E. Ahmadi, Y. Oshima, J. S. Speck; *Semicond. Sci. Technol.* **31** (2016), 065006.

286. J. Yang, S. Ahn, F. Ren, R. Khanna, K. Bevlin, D. Geerpuram, S. J. Pearton, A. Kuramata; *Appl. Phys. Lett.* **110** (2017), 142101.

287. R. Togashi, K. Nomura, C. Eguchi, T. Fukizawa, K. Goto, Q. T. Thieu, H. Murakami, Y. Kumagai, A. Kuramata, S. Yamakoshi, B. Monemar, A. Koukitu; *Jpn. J. Appl. Phys.* **54** (2015), 041102.

288. S. J. Schneider, J. L. Waring; *J. Res. Natl. Bur. Stand.* **67A** (1963), 19–25.

289. M. Zinkevich, F. Aldinger; *J. Am. Ceram. Soc.* **87** (2004), 683–691.

290. D. B. Dingwell; *J. Am. Ceram. Soc.* **75** (1992), 1656-1657.

291 B. Amgalan, D. Klimm, Z. Galazka; *Leibniz-Institut für Kristallzüchtung*, Berlin, Germany (2019), unpublished data.

292. E. G. Villora, K. Shimamura, T. Ujiie, K. Aoki; *Appl. Phys. Lett.* **92** (2008), 202118.

293. M. Handwerg, R. Mitdank, Z. Galazka, S. F. Fischer; *Semicond. Sci. Technol.* **30** (2015), 024006.

294. M. Handwerg, R. Mitdank, Z. Galazka, S. F. Fischer; *Semicond. Sci. Technol.* **31** (2016), 125006.

295. M. Slomski, N. Blumenschein, P. P. Paskov, J. F. Muth, T. Paskova; *J Appl. Phys.* **121** (2017), 235104.

296. Z. Guo, A. Verma, X. Wu, F. Sun, A. Hickman, T. Masui, A. Kuramata, M. Higashiwaki, D. Jena, T. Luo; *Appl. Phys. Lett.* **106** (2015), 111909.

297. P. Jiang, X. Qian, X. Li, R. Yang; *Appl. Phys. Lett.* **113** (2018), 232105.

298. M. D. Santia, N. Tandon, J. D. Albrecht; *Appl. Phys. Lett.* **107** (2015), 041907.

299. F. Orlandi, F. Mezzadri, G. Calestani, F. Boschi, R. Fornari; *Appl. Phys. Express* **8** (2015), 111101.

300. Z. Cheng, M. Hanke, Z. Galazka, A. Trampert; *Appl. Phys Lett.* **113** (2018), 182102.

301. L. I. Guzilova, V. N. Maslov, K. E. Aifantis, A. E. Romanov, V. I. Nikolaev; *Sci. Tech. J. Inf. Technol. Mech. Opt.* **15** (2015), 546–549.

302. K. Yamaguchi; *Solid State Commun.* **131** (2004), 739–774.

303. M. Schubert, A. Mock, R. Korlacki, S. Knight, Z. Galazka, G. Wagner, V. Wheeler, M. Tadjer, K. Goto, V. Darakchieva; *Appl. Phys. Lett.* **114** (2019), 102102.

304. A. Mock, J. VanDerslice, R. Korlacki, J. A. Woollam, M. Schubert; *Appl. Phys. Lett.* **112** (2018), 041905.

305. C. Sturm, J. Furthmüller, F. Bechstedt, R. Schmidt-Grund, M. Grundmann; *APL Materials* **3** (2015), 106106.

306. C. Sturm, M. Grundmann; *Phys. Rev. A* **93** (2016), 053839.

307. T. Onuma, S. Saito, K. Sasaki, T. Masui, T. Yamaguchi, T. Honda, A. Kuramata, M. Higashiwaki; *Jpn. J. Appl. Phys.* **55** (2016), 1202B2.

308. A. Fiedler, R. Schewski, Z. Galazka, K. Irmscher; *ECS J. Solid State Sci. Technol.* **8** (2019), Q3083–Q3085.

309. M. Schubert, R. Korlacki, S. Knight, T. Hofmann, S. Schöche, V. Darakchieva, E. Janzén, B. Monemar, D. Gogova, Q.-T. Thieu, R. Togashi, H. Murakami, Y. Kumagai, K. Goto, A. Kuramata, S. Yamakoshi, M. Higashiwaki; *Phys. Rev. B* **93** (2016), 125209.

310. M. Passlack, N. E. J. Hunt, E. F. Schubert, G. J. Zydzik, M. Hong, J. P. Mannaerts, R. L. Opila, R. J. Fischer; *Appl. Phys. Lett.* **64** (1994), 2715-2717.

311. F. K. Shan, G. X. Liu, W. J. Lee, G. H. Lee, I. S. Kim, B. C. Shin; *J. Appl. Phys.* **98** (2005), 023504.

312. G. Schmitz, P. Gassmann, R. Franchy; *J. Appl. Phys.* **83** (1998), 2533–2538.

313. B. Liu, M. Gu, X. Liu; *Appl. Phys. Lett.* **91** (2007), 172102.

314. Y. Kang, K. Krishnaswamy, H. Peelaers, C.G. Van De Walle; *J. Phys. Condens. Matter.* **29** (2017), 234001.

315. J. B. Varley, A. Janotti, C. Franchini, C. G. Van de Walle; *Phys. Rev. B* **85** (2012), 081109.

316. S. Yamaoka, M. Nakayama; *Phys. Status Solidi* (*c*) **13** (2016), 93–96.

317. Y. Wang, P. T. Dickens, J. B. Varley, X. Ni, E. Lotubai, S. Sprawls, F. Liu, V. Lordi, S. Krishnamoorthy, S. Blair, K. G. Lynn, M. Scarpulla, B. Sensale-Rodriguez; *Sci. Rep.* **8** (2018), 18075.

318. D. Dohy, G. Lucazeau, A. Revcolevschi; *J. Solid State Chem.* **45** (1982), 180–192.

319. T. Onuma, S. Fujioka, T. Yamaguchi, Y. Itoh, M. Higashiwaki, K. Sasaki, T. Masui, and T. Honda; *J. Cryst. Growth* **401** (2014), 330–333.

320. C. Kranert, C. Sturm, R. Schmidt-Grund, Marius Grundmann; *Sci. Rep.* **6** (2016), 35964.

321. S. Rafique, L. Han, M. J. Tadjer, J. A. Freitas, Jr., N. A. Mahadik, H. Zhao; *Appl. Phys. Lett.* **108** (2016), 182105.

322. N. He, H. Tang, B. Liu, Z. Zhu, Q. Li, C. Guo, M. Gu, J. Xu, J. Liu, M. Xu, L. Chen, X. Ouyang; *Nucl. Instrum. Methods Phys. Res. A* **888** (2018), 9–12.

323. S.-H. Han, A. Mauze, E. Ahmadi, T. Mates, Y. Oshima, J. S. Speck; *Semicond. Sci. Technol.* **33** (2018), 045001.

324. Z. Galazka, K. Irmscher, R. Schewski, I. M. Hanke, M. Pietsch, S. Ganschow, D. Klimm, A. Dittmar, A. Fiedler, T. Schroeder, M. Bickermann; *J. Cryst. Growth* **529** (2020), 125297.

325. P. Weiser, M. Stavola, W. B. Fowler, Y. Qin, S. Pearton; *Appl. Phys. Lett.* **112** (2018), 232104.

326. Y. Qin, M. Stavola, W. B. Fowler, P. Weiser, S. J. Pearton; *ECS J. Solid State Sci. Technol.* **8** (2019), Q3103–Q3110.

327. P. D. C. King, I. McKenzie, T. D. Veal; *Appl. Phys. Lett.* **96** (2010), 062110.

328. T. Onuma, S. Saito, K. Sasaki, T. Masui, T. Yamaguchi, T. Honda, M Higashiwaki; *Jpn. J. Appl. Phys.* **54** (2015), 112601.

329. M. H. Wong, K. Sasaki, A. Kuramata, S. Yamakoshi, M. Higashiwaki; *Appl. Phys. Lett.* **106** (2015), 032105.

330. K. Irmscher, Z. Galazka, M. Pietsch; *Leibniz-Institut für Kristallzüchtung*, Berlin, Germany (2018), unpublished data.

331. C. Golz, Z. Galazka, J. Lähnemann, V. Hortelano, F. Hatami, W. T. Masselink, O. Bierwagen; *Phys. Rev. Mater.* **3** (2019), 124604.

332. K. Irmscher, Z. Galazka, M. Pietsch, R. Uecker, R. Fornari; *J. Appl. Phys.* **110** (2011), 063720.

333. K. Irmscher, Z. Galazka, M. Pietsch; *Leibniz-Institut für Kristallzüchtung*, Berlin, Germany (2011), unpublished data.

334. T. Harwig, G. J. Wubs, G. J. Dirksen; *Solid State Commun.* **18** (1976), 1223–1225.

335. N. Ueda, H. Hosono, R. Waseda, H. Kawazoe; *Appl. Phys. Lett.* **71** (1997), 933–935.

336. S. Knight, A. Mock, R. Korlacki, V. Darakchieva, B. Monemar, Y. Kumagai, K. Goto, M. Higashiwaki, M. Schubert; *Appl. Phys. Lett.* **112** (2018), 012103.

337. K. Sasaki, A. Kuramata, T. Masui, E. G. Villora, K. Shimamura, S. Yamakoshi; *Appl. Phys. Express* **5** (2012), 035502.

338. N. Ma, N. Tanen, A. Verma, Z. Guo, T. Luo, H. (Grace) Xing, D. Jena; *Appl. Phys. Lett.* **109** (2016), 212101.

339. K. Ghosh, U. Singisetti; *Appl. Phys. Lett.* **109** (2016), 072102.

340. A. Parisini, R. Fornari; *Semicond. Sci. Technol.* **31** (2016), 035023.

341. A. Parisini, K. Ghosh, U. Singisetti, R. Fornari; *Semicond. Sci. Technol.* **33** (2018), 105008.

342. T. Oishi, Y. Koga, K. Harada, M. Kasu; *Appl. Phys. Express* **8** (2015), 031101.

343. A. Y. Polyakov, N. B. Smirnov, I. V. Shchemerov, E. B. Yakimov, S. J. Pearton, Fan Ren, A. V. Chernykh, D. Gogova, A. I. Kochkova; *ECS J. Solid State Sci. Technol.* **8** (2019), Q3019–Q3023.

344. Z. Zhang, E. Farzana, A. R. Arehart, S. A. Ringel; *Appl. Phys. Lett.* **108** (2016), 052105.

345. N. T. Son, K. Goto, K. Nomura, Q. T. Thieu, R. Togashi, H. Murakami, Y. Kumagai, A. Kuramata, M. Higashiwaki, A. Koukitu, S. Yamakoshi, B. Monemar, E. Janzén; *J. Appl. Phys.* **120** (2016), 235703.

346. C. A. Lenyk, N. C. Giles, E. M. Scherrer, B. E. Kananen, L. E. Halliburton, K. T. Stevens, G. K. Foundos, J. D. Blevins, D. L. Dorsey, S. Mou; *J. Appl. Phys.* **125** (2019), 045703.

347. A. Y. Polyakov, N. B. Smirnov, I. V. Shchemerov, E. B. Yakimov, J. Yang, F. Ren, G. Yang, J. Kim, A. Kuramata, S. J. Pearton; *Appl. Phys. Lett.* **112** (2018), 032107.

348. M. E. Ingebrigtsen, J. B. Varley, A. Yu. Kuznetsov, B. G. Svensson, G. Alfieri, A. Mihaila, U. Badstübner, L. Vines; *Appl. Phys. Lett.* **112** (2018), 042104.

349. Z. Galazka, R. Schewski, K. Irmscher, W. Drozdowski, M. E. Witkowski, M. Makowski, A. J. Wojtowicz, I. M. Hanke, M. Pietsch, T. Schulz, D. Klimm, S. Ganschow, A. Dittmar, A. Fiedler, T. Schroeder, M. Bickermann; *J. Alloy. Compd.* **818** (2020), 152842.

350. M. Makowski, W. Drozdowski, M. E. Witkowski, A. J. Wojtowicz, K. Irmscher, R. Schewski, Z. Galazka; *Opt. Mater. Express* **9** (2019), 3738–3743.

Chapter 5

Indium Oxide (In_2O_3)

5.1 Introduction

Research on In_2O_3 started in the 1950s, when first thin films and tiny bulk crystals were obtained. In_2O_3 has interesting electrical and optical properties that, in combination with the possibility of preparing coatings, layers, or thin films in an easy and cheap manner, make this binary oxide compound technologically important material in many applications. Particularly Sn-doped In_2O_3, called indium tin oxide (ITO), is a well-suited material for energy-efficient windows, as heat reflective coatings [1–4] or electrochromic coatings [5]. Other applications involve photovoltaics (solar cells) where In_2O_3 is used in energy conversion heterojunctions, metal–insulator–semiconductor (MIS), or semiconductor–insulator–semiconductor (SIS) structures [6–9] as transparent electrodes in flat panel displays (Sn- or H-doped) [10], or as conductive layers in touchscreens of smartphones, tablets, and other consumer electronics. Such applications are the result of a high electrical conductivity, optical transparency, and easy mass production as thin films of ITO. Additionally, the use of In_2O_3 was demonstrated for transparent electronics as *p-n* junctions [11] or thin-film transistors (channel material) [12]. In light-emitting diodes, In_2O_3 doped with Tb was demonstrated as a transparent light converter [13] and

Transparent Semiconducting Oxides: Bulk Crystal Growth and Fundamental Properties
Zbigniew Galazka

ISBN 978-981-4800-94-5 (Hardcover), 978-1-003-04520-5 (eBook)
www.jennystanford.com

doped with Sn as a cladding layer [14]. Moreover, In_2O_3 surfaces show interesting sensing properties of oxidizing and reducing species, as well as different organic substances, so it can operate as a gas sensor [15–29]. Further, In_2O_3 doped with transition metals (Ni, Fe, Cr, Mn, Cu) shows ferromagnetic/antiferromagnetic properties [30–37] (although it may be related to a secondary phase formation [37]). As such, transparent magnetic semiconductor In_2O_3 may find applications in spintronics [38].

Thin films of In_2O_3 both undoped and doped with various elements have frequently been grown by a diversity of techniques, such as sol-gel [39–41], spray pyrolysis [42–50], evaporation (reactive, electron beam) [51–59], sputtering [60–69], metal-organic chemical vapor deposition (MOCVD) [70, 71], halide vapor phase epitaxy (HVPE) [72], pulsed laser deposition (PLD) [73–80], and molecular beam epitaxy (MBE) [81–92]. A general review on In_2O_3 can be found in the work of Bierwagen [93].

Demonstration by Galazka et al. [94–96] of relatively large bulk single crystals of high-structural-quality In_2O_3 obtained directly from the melt for the first time extends application of this compound to substrates for homo- and heteroepitaxy [97, 98]. Moreover, bulk In_2O_3 single crystals are well suited for exploring different properties that might be difficult to achieve with thin films. Therefore, bulk In_2O_3 single crystals may serve as reference samples for thin films.

This chapter focuses on bulk crystal growth of In_2O_3, particularly on growth from the melt, and on basic properties measured on bulk crystals. The properties are also referenced in some cases to thin films and theory to obtain a complete picture of that compound.

5.2 Crystal Structure

In_2O_3 crystallizes in a body-centered cubic structure (space group $Ia\overline{3}$), which is named after the mineral bixbyite $(Fe,Mn)_2O_3$. The structure can be derived from the fluorite ($CaF_2 = \frac{1}{2}Ca_2F_4$) structure by removing one-quarter of the anions and a subsequent shift in the remaining ions [99]. As a consequence of this shift, the cations are surrounded by distorted octahedra. Three-fourths of the cations are on the Wyckoff position *d* (site symmetry 2) and one-fourth on the Wyckoff position *b* (site symmetry $\overline{3}$); these positions are marked in Fig. 5.1 by blue or red circles, respectively. It results in two inequivalent In sites In1 (¼, ¼, ¼), and In2 (*u*, 0, ¼) and one O site. Both In1 and In2 are six-fold coordinated, surrounded by O atoms, while all O atoms are four-fold coordinated to In atoms. The local structure of In1 is highly symmetric, while the local structure of In2 is less symmetric. The local structure of O forms a distorted tetrahedron with all four In–O bonds inequivalent [100]. Indeed, the refined crystal structure given by Marezio [101] revealed equal In1–O bond lengths of 2.18 Å, and different In2–O bond lengths of 2.13, 2.19, and 2.23 Å. The measured lattice parameter of a single crystal is 10.117 ± 0.001 Å. The In_2O_3 structure contains 80 atoms in a unit cell.

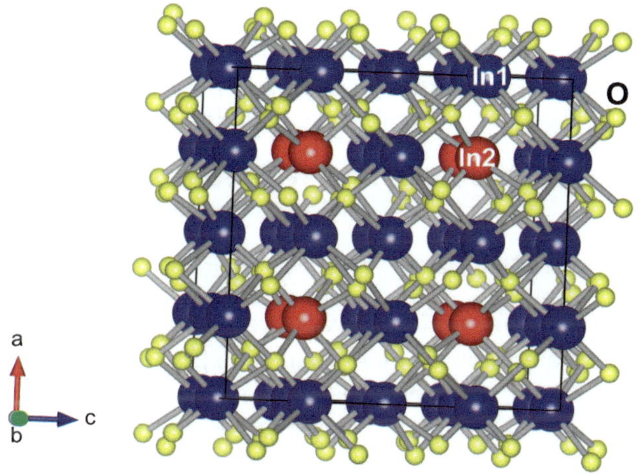

Figure 5.1 Cubic bixbyite crystal structure of In_2O_3 where In^{3+} ions (In1 and In2) occupy two nonequivalent positions.

Bixbyite is, hence, isomorphous with many other trivalent metal oxides, and it is especially identical to the C-structure of many rare-earth oxides [102, 103]. Correspondingly, the solubility of some other metal 3+ ions in the In_2O_3 structure is remarkable: 4–10% for Ga [104] and ca. 28% for Eu [105]. Cava et al. [106] reported with Ga_2O_3 the intermediate phase $GaInO_3$, which crystallizes also in the β-Ga_2O_3 structure. Correspondingly, the different Ga sites in this structure are assumed to be occupied alternately by Ga^{3+} and In^{3+}.

For technical applications also, tin-doped indium oxide (ITO) is relevant. The isomorphous solubility of SnO_2 in In_2O_3 varies with temperature between 1.1 mol.% and 13.1 mol.%; besides some intermediate phases were found, which are stable only at elevated temperature [107]. Recently a simplified thermodynamic assessment of the SnO_2–In_2O_3 system was published [108].

Bixbyite is the stable structure of In_2O_3 under ambient pressure. However, under high pressures, In_2O_3 transforms to other metastable polymorphs. According to an accurate theoretical model of Walsh et al. [109], which is based on interatomic potentials with appropriate parametrization, several high-pressure phases of In_2O_3 may exist up to 70 GPa: (i) corundum (*R-3c*), (ii) α-Gd_2S_3 (*Pnam*), (iii) Rh_2O_3 (*Pbna*), (iv) β-Ga_2O_3 (*C-2m*), and La_2O_3 (*P-3m1*). Some of these polymorphs were found experimentally under high pressure [110, 111]. At zero pressure, the energetic ordering was found as follows: bixbyite < corundum < Rh_2O_3 < α-Gd_2S_3 < β-Ga_2O_3 < La_2O_3.

5.3 Band Structure and Native Point Defects

5.3.1 Band Structure

The band structure of In_2O_3 was investigated both theoretically, e.g., by Kompany et al. [112], Aliabad et al. [113], Medvedeva [114], Erhart et al. [115], Walsh et al. [116], Reunchan et al. [117], Sabino et al. [118], and experimentally, e.g., by Walsh et al. [116], Körber et al. [119], Janowitz et al. [120], and Scherer et al. [121, 122]. According to Walsh et al. [116] (Fig. 5.2), the valence band maximum (VBM) state at Γ point of the Brillouin zone has O 2*p* and In 4*d* character, while the conduction band minimum (CBM) state is a mixture of In 5*s* and O 2*s* orbitals. The VBM exhibits a very small dispersion, about 19 meV along the Γ-N line, in contrast to CBM, which shows a large dispersion. Due to symmetry requirements in the bixbyite structure (inversion center), a direct VBM–CBM absorption is parity forbidden, but it may weakly contribute to the photon absorption under lattice vibrations. Strong optical absorptions can occur for Γ_8–Γ_1 transition, which is 0.81 eV below the VBM (i.e., at 2.89 eV). The onset of the optical absorption can, therefore, be observed at 3.70 eV.

The above conditions explain the differences between the optical and fundamental band gaps in In_2O_3 [118]. As a result, In_2O_3 can be regarded as a "forbidden" gap material with a fundamental band gap 0.81 eV lower in energy than the onset of strong optical absorption [116].

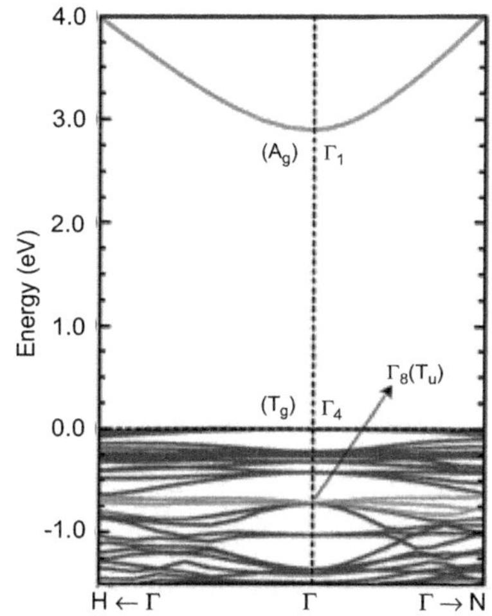

Figure 5.2 Band structure of In_2O_3 (along the H-Γ-N lines). Reprinted with permission from Ref. [116], Copyright 2019, American Physical Society.

Experimental electronic structure revealed by Janowitz et al. [120] with the use of the angle-resolved photoemission spectroscopy (ARPES) and X-ray photoelectron spectroscopy (XPS) confirmed the theoretical conclusions on the VB orbital character (O 2*p* and In 4*d*) with an In 5*d* character at higher conduction bands. The derived energy gaps are 2.63 (or 2.7 ± 0.1 by Scherer et al. [121]) and 3.71 eV [120], which fit well the theoretical predictions.

5.3.2 Native Point Defects

Native (intrinsic) point defects in In_2O_3 were derived theoretically by Reunchan et al. [117], Ágoston et al. [123, 124], Limpijumnong et al. [125], Palandage and Fernando [126], and Varley et al. [127]. According to Reunchan et al. [117], most probable intrinsic point defects in In_2O_3 are oxygen and indium vacancies, which are double donors (V_O^{2+}) and triple acceptors (V_{In}^{3-}) [117]. V_O^{2+} counteract the *p*-type doping, while V_{In}^{3-} compensate the *n*-type doping. V_O^{2+} are formed at about 200 meV below the CBM, while V_{In}^{3-} are located 0.79 eV above the VBM. Limpijumnong et al. [125] claimed that both V_O^+ and V_O^{2+} are stable and can be formed in In_2O_3 as deep donors. Varley et al. [127] stated that V_{In}^{3-} is stable in both inequivalent In sites, but due to high formation energy, it will rather not be formed at high concentrations. Ágoston et al. [123, 124] performed an extended study of all possible intrinsic donor (In_i, In_O, V_O) and acceptor (O_i, V_i) defects at different charge states in In_2O_3. They concluded that V_O has a low formation energy and forms a shallow donor, while In_i and In_O have higher formation energies. O_i is thermodynamically stable at oxidizing conditions, but it has a large formation energy. V_i has yet higher formation energy and, rather, plays a minor role as a compensating acceptor. In-related defects have generally higher migration energies as compared with O-related defects. The overall defect analysis pointed to V_O as the main contributor to *n*-type conductivity in In_2O_3 (intrinsic character of the conductivity). However, Palandage and Fernando [126] concluded from their theoretical work that V_O alone is not a strong native donor, but it is when combined with In_i or a substitutional transition metal.

Another, extrinsic point defect type that is likely always present in In_2O_3 relates to hydrogen. Hydrogen is omnipresent and has an impact on conductivity in most, if not all, of the TSOs, including In_2O_3, as the charge neutrality level (CNL) and the H(+/−) level lie within the conduction band (King et al. [128]). Indeed, all studies point hydrogen to be a shallow donor, whether as interstitial (H_i) or substituting oxygen (H_O). H_i^+ is stable within the band gap [125], has a low migration barrier (0.76 eV), is mobile, and can be present at high concentrations [127]. H_i can also form stable complexes with indium vacancies $(H_i\text{–}V_{In})^{2-}$ as their formation energy is relatively low and binding energy large (3–3.1 eV) [127]. H_O is stable in 1+ charge state (H_O^+) and has only slightly higher formation energy as compared with that of H_i [125]. According to the above theoretical studies, hydrogen seems to be another contributor to *n*-type conductivity in In_2O_3.

Positron doppler spectroscopy performed by Korhonen et al. [129] on thin films and bulk In_2O_3 crystals revealed the presence of V_{In} and V_O in Mg- and Sn-doped thin films, but not in bulk crystals obtained from the melt. The density of these compensating defects depends on the doping type and annealing conditions.

An experimental work of Braun et al. [130] on bulk crystals with scanning tunneling spectroscopy (STS) revealed in-gap state defects, which were identified, after combining with ARPES and DFT calculation results, mainly as V_O, and possible as In_i and O_i.

5.4 Bulk Crystal Growth

5.4.1 Stability Fields of In–O System

In_2O_3 has a high melting point (MP) and is highly unstable at high temperatures. A combination of these two characteristics restricts the conversion of In_2O_3 to the liquid phase. Flux and vapor growth methods, however, offer much lower operating temperatures although the size of such crystals is very small. The stability diagram of the In–O system shown in Fig. 5.3 was obtained with the FactSage software combined with experimental data of Galazka et al. [95, 96]. At very low oxygen partial pressure $p(O_2)$, roughly below 10^{-9} atm and temperatures up to about 2080°C, only elemental In in the liquid phase In(liq) is stable. The stability of the solid phase of In_2O_3(s) requires increasing $p(O_2)$ with rising temperature; otherwise, a gas phase (ideal gas) will form. At high temperatures, roughly above 2080°C, only gas is present in the system for $p(O_2)$ values up to 1 atm. The liquid phase of In_2O_3(liq) is confined only to very small window limited by a temperature range of about 1950–2080°C and $p(O_2)$ of 0.03–1 atm. This estimated Ellingham diagram points to obtaining In_2O_3 crystals from the melt as really challenging, although possible, as demonstrated further below. Growth from the gas phase can be practiced at low or high temperatures in a wide range of $p(O_2)$ in the system, while flux growth would require temperatures high enough to melt the flux and a higher $p(O_2)$ at similar operating temperatures as compared with the gas phase growth methods.

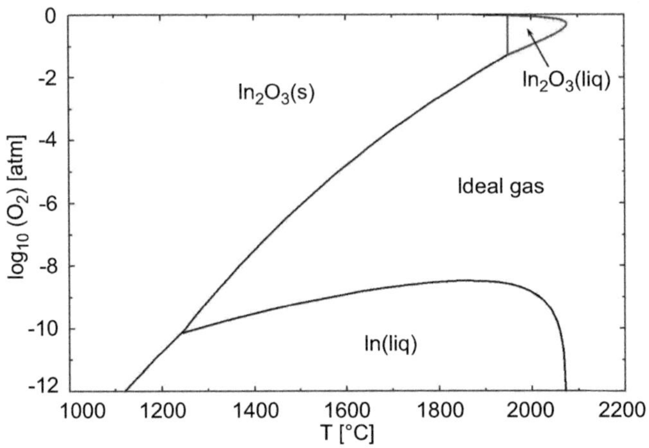

Figure 5.3 Diagram of stability fields of In_2O_3 (Ellingham-type diagram).

5.4.2 Historical Events

First bulk In_2O_3 single crystals were reported in 1954 by Roy and Shafer [131] with the use of the hydrothermal method. Gas phase (PVT) grown bulk crystals were first obtained by Weiher [132] in 1962, while the flux method was first applied by Remeika and Spencer [133] in 1964. A typical chemical vapor transport (CVT) method for growing bulk In_2O_3 crystals was likely first utilized by De Wit [134] in 1972. Until 2013, In_2O_3 single crystals have not been grown directly from the melt, when Galazka et al. [94] defined and developed a novel crystal growth technique under the name Levitation-Assisted Self-Seeding Crystal Growth Method (LASSCGM), which enabled to obtain large In_2O_3 single crystals. That short history of bulk crystal growth of In_2O_3 is depicted in Fig. 5.4.

5.4.3 Vapor Phase

The growth of In_2O_3 from the gas phase can be divided into three groups depending on the creation and transportation of In-containing species from a source zone to a growth zone, where the In-containing species is oxidized to solid In_2O_3 in the form of crystals:

(i) A direct thermal decomposition of In_2O_3 starting material that involves creation of volatile In_2O, its transportation, and re-oxidation to In_2O_3 solid (crystal). This technique is purely physical vapor transport (PVT) method.

(ii) Use of a reducing agent (graphite) and elemental In or the reducing agent in a combination with In_2O_3 or In_2O_3 + In starting material that facilitates the formation of volatile In_2O at lower temperatures. It is also called a vapor reaction method.

(iii) Use a chemically active transport agent, such as S + I_2 or Cl_2, for transporting In-containing species, and its oxidation to In_2O_3 solid (crystal). This technique defines the CVT method.

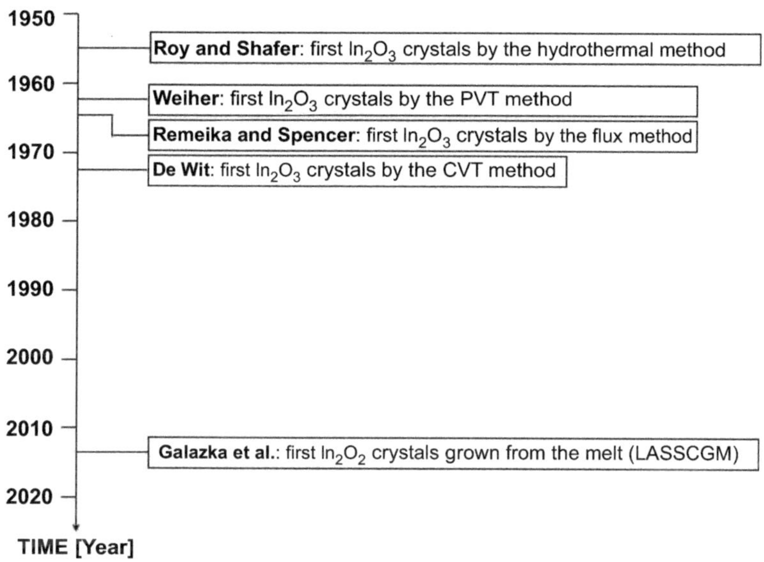

Figure 5.4 Main events in bulk crystal growth of In_2O_3.

5.4.3.1 Thermal decomposition of In_2O_3 (PVT)

The PVT growth experiments of In_2O_3 crystals by thermal evaporation (sublimation) were conducted by Galazka et al. [95]. Solid In_2O_3(s) decomposes at high temperatures (source temperature, T_S) and forms mainly gaseous In_2O(g) of high partial pressure (plus other species, but at much lower partial pressure):

$$In_2O_3(s) \xrightarrow{T_S} In_2O(g) + O_2(g) \qquad (5.1)$$

which is transported to a cold region ($T_S - \Delta T$) at the substrate (seed) where it oxides into solid $In_2O_3(s)$:

$$In_2O(g) + O_2(g) \xrightarrow{T_S-\Delta T} In_2O_3(s) \qquad (5.2)$$

The growth experiments were performed in a growth furnace consisting of a 40 mm diameter Ir crucible, a sapphire substrate on the top of the crucible, and a thermal insulation surrounding the crucible with the substrate from all sides. The crucible was inductively heated by an RF coil. The starting material was dried, pressed, and sintered In_2O_3 powder of 4N purity. The growth experiments were performed with the use of CO_2 growth atmosphere and about 1600°C growth temperature (at sapphire substrate). The temperature of the crucible was about 1700–1800°C. Such high temperature was sufficient to produce high partial pressure of $In_2O(g)$ resulting from the $In_2O_3(s)$ decomposition, which was transported to sapphire and reacted with oxygen from CO_2 and In_2O_3 decompositions. The distance between the starting material and the substrate was about 2 cm.

The obtained In_2O_3 crystals of almost 40 mm in diameter and 3 mm thickness were green and transparent in visible light (Fig. 5.5). They consisted of small crystal grains of 0.1–0.2 mm having two different orientations of <111> and <100>. Interestingly, the crystal coloration and grain structure were not affected by any annealing in an oxidizing atmosphere at 400–1400°C for 20 h.

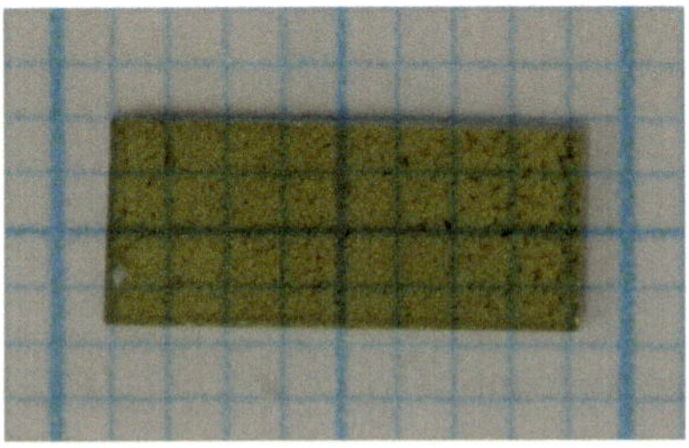

Figure 5.5 In_2O_3 crystal obtained from the gas phase by the PVT method.

5.4.3.2 Use of a reducing agent

This technique has a few variations depending on the source material, which can be solid In(s), solid $In_2O_3(s)$, or a mixture of In(s) + In_2O_3(s). With the use of a reducing agent and the presence of residual oxygen, volatile In_2O(g) is created at relatively high temperatures, but much lower as compared with thermal decomposition of solid In_2O_3. Volatile In_2O(g) is transported to a cold region, where it is oxidized to solid In_2O_3 in the form of crystals.

Weiher [132] and Weiher and Ley [135] used as a starting material a mixture of In metal and carbon. The growth experiments were performed in a covered porcelain crucible that was heated in ambient oxygen at 1000°C for 24 h. Needle-shaped, pale-yellow crystals approximately of 0.5 × 0.5 × 5 mm^3 or plates with an area of approximately 5 mm^2 and thickness of 30–100 μm in size grew on the crucible wall and the cover. The cross section of the needle-shaped crystals was either square or hexagonal, corresponding to [100] and [111] growth directions, respectively. The main identified impurities were Si and Mo at concentration of about 1 wt. ppm.

Shimada et al. [136] used different systems for growing In_2O_3 crystals from the gas phase: graphite–In_2O_3, active carbon–In_2O_3, graphite–In, graphite–In_2O_3–In, and In_2O_3–In.

In the C–In_2O_3 systems, solid In_2O_3 is first reduced to gaseous In_2O at high temperatures T_S, which is transported to a lower temperature zone $T_S - \Delta T$, where it is re-oxidized to solid In_2O_3. Thermodynamically favorable reactions are as follows:

$$In_2O_3(s) + 2C(s) \rightarrow In_2O(g) + 2CO(g) \tag{5.3}$$

$$In_2O_3(s) + 3C(s) \rightarrow 2In(g) + 3CO(g) \tag{5.4}$$

$$In_2O(g) + CO(g) \rightarrow 2In(g) + CO_2(g) \tag{5.5}$$

wherein $p(In_2O) > p(In)$.

In the systems with metallic In at high temperatures, liquid In with oxygen and solid In_2O_3 form gaseous In_2O:

$$4In(l) + O_2(g) \rightarrow 2In_2O(g) \tag{5.6}$$

$$In_2O_3(s) + 4In(l) \rightarrow 3In_2O(g) \tag{5.7}$$

which is next re-oxidized to form In_2O_3 crystals according to Eq. 5.2.

Powders of these systems of 1.3–4 g were mixed in an appropriate molar ratio, placed in 5 ml porcelain crucibles and heated in air at 800–1300°C for 12 h. Tiny crystals deposited along the junction between the crucible rim and a cover as well as on inner walls of the crucible. That junction is the passage to the interior of the crucible for oxygen, which reacts with $In_2O(g)$ to form In_2O_3 crystals. As at the operating temperature $p(O_2)$ is more than 10 orders of magnitude smaller than $p(In_2O)$, additional oxygen in the growth furnace is likely needed for the formation of In_2O_3 crystals.

The graphite–In_2O_3 system produced at temperatures 960 and 1000°C transparent pale-yellow needle-shaped crystals of size 0.1 × 0.1 × 3, 0.3 × 0.3 × 3, and 0.5 × 0.5 × 7/8 mm^3, depending on the temperature and molar ratio of graphite/In_2O_3, which was between 20/1 and 40/1. Above 1040°C, fine crystals were formed between the cover and the crucible rim. In the system, active carbon–In_2O_3 fine yellow crystals were obtained only at 900°C. It was found that the favorable amount of the starting material for the 5 ml crucible was 2.4–2.9 g; otherwise, the crystals became much smaller. When using the graphite–In system, transparent crystals of size 0.3 × 0.3 × 10 mm^3 were obtained at 1000°C. They were yellow or blue/black if metallic In was used in the form of powder or granules, respectively. The In_2O_3–In system produced fine yellow crystals at 1100°C and opaque yellow at 1200°C, but no crystals grew below 1100°C or above 1200°C. When using the graphite–In_2O_3–In system as the starting material, the obtained crystals were fine and yellow for the temperature range 1000–1080°C. Above 1080°C, no crystals were obtained. Investigation of the largest crystals obtained from the graphite–In_2O_3 system showed the growth direction along the *a*-axis.

In other growth experiments from the gas phase, Shimada et al. [137] used the In_2O_3–SnO_2–graphite system. The starting material was In_2O_3 and SnO_2 in the same molar content, and graphite in the amount of 5.6, 10, and 15 wt.%. The powders were mixed in a mortar and placed in a silica boat of size 40 × 10 × 7 mm^3, which was covered by a similar silica boat. The boats with the starting material were placed in a tube furnace of 25 mm diameter, through which N_2 was flowing at the rate of 100 ml/min. The furnace was heated up to 860–1100°C at the rate of about 8 K/min and held at these temperatures for 3–6 h.

Crystal growth did not occur if a flowing air or uncovered boat was used, or at temperatures below 960°C. At temperatures above

960°C, needle-shaped or dendrite crystals were obtained on the walls and the bottom of the boat. Typically, white, cotton-like SnO_2 crystals were grown, which were overlapped by In_2O_3 crystals. The size and coloration of In_2O_3 crystals were strongly correlated with the growth temperature: at temperatures up to 1000°C, the In_2O_3 crystals were yellowish-green of size $1 \times 0.05 \times 0.05$ mm^3, which became green and of larger size, $10 \times 0.1 \times 0.1$ mm^3, when grown at 1045°C. The needle-shaped In_2O_3 crystals were square in a cross section, which contained, the same as dendrites, a varying concentration of Sn, between about 7 and 30%. The most contaminated In_2O_3 crystals with Sn were located at the bottom of the silica boat in the proximity of cotton-like SnO_2 crystals.

The growth of In_2O_3 in the In_2O_3–graphite system (i.e., without SnO_2) occurred only at 1045°C, with the resulted very small crystals. This indicates an important role of SnO_2 in the transport mechanism. Here, solid SnO_2 is reduced by graphite to gaseous SnO in the same way as In_2O_3:

$$SnO_2(s) + C(s) \rightarrow SnO(g) + CO(g) \tag{5.8}$$

Both $In_2O(g)$ and $SnO(g)$ react with each other to form oxides in the solid phase:

$$In_2O(g) + 2SnO(g) \rightarrow In_2O_3(s) + 2Sn(l) \tag{5.9}$$

$$In_2O(g) + SnO(g) \rightarrow SnO_2(s) + 2In(l) \tag{5.10}$$

Analysis of the products after growth experiments revealed, indeed, Sn metal and Sn–In alloy, indicating the formation of both Sn and In metals.

5.4.3.3 Use of chemically active transport agent (CVT)

In this type of vapor growth, the transport agent aims to bind In and transport it to a growth zone, where it is oxidized to solid In_2O_3. De Wit [134] used HCl as the chemically active transport agent. One gram of solid In_2O_3 was placed in one end of a quartz ampoule of 1 cm inner diameter and 20 cm length. The ampoule was evacuated and then filled with HCl vapor at a pressure of 2000 Pa at room temperature (RT). The ampoule was next sealed off and horizontally placed within a furnace in such a way that its end with In_2O_3 was in the hot (reaction) zone at 950°C, while the other end was in a cold (growth) zone at 680–720°C. The transport rate was 4 mg/h. The growth took place at the growth zone for several days. The obtained small crystals of 1 mm^3 in volume were green and octahedrally shaped. At lower HCl pressure, the growth habit was different, and the crystals were in the shape of cubes.

The transport reaction at high temperatures in the In_2O_3–HCl system and a reverse reaction at lower temperatures can be written as follows:

$$In_2O_3(s) + 6HCl(g) \leftrightharpoons 2InCl_3(g) + 3H_2O(g) \tag{5.11}$$

Cl_2 as the chemically active transport agent was used by Jozefowicz and Piekarczyk [138]. The experiments were made in a closed quartz ampoule with an inner diameter of 19 mm and length of 110 mm. Three grams of In_2O_3 was loaded into the ampoule, which was then evacuated to about 10^{-3} Pa. Next the ampoule was filled with about 0.15 g of chlorine. The ampoule was sealed off and placed in a two-zone furnace. For 24 h, the ampoule was kept in a reverse temperature gradient (850°C in the reaction zone and 950°C in the growth zone) to clean the ampoule in the growth zone. After that the temperature was changed to 950°C in the reaction zone and 900°C in the growth zone. The growth runs lasted for about 100 h. The transport did not start if the amount of Cl_2 was smaller than 0.12 g and the temperature in the reaction zone was smaller than 950°C. The temperature difference between the reaction and growth zones, ΔT, should be larger than 30 K; however, the larger the ΔT, the smaller the crystals obtained. Temperature in the growth zone larger than 900°C resulted in a strong crystal adherence to the ampoule walls.

The obtained crystals were cubes in average size of 3 × 3 × 3 mm^3. They were yellow and transparent. The transport rate was about 9 mg/h. The obtained crystals were much larger than those grown with HCl or $I_2 + Cl_2$.

Although different species of indium chlorides are formed, the dominating reaction is:

$$In_2O_3(s) + 3Cl_2(g) \rightarrow 2InCl_3(g) + 1\tfrac{1}{2}O_2(g) \tag{5.12}$$

At the operating temperatures (<1000°C), the highest partial pressure in the In_2O_3–Cl_2 system has $InCl_3$, O_2, and Cl_2, as shown in Fig. 5.6. The thermodynamic calculations clearly show that In is transported by $InCl_3(g)$, which has the highest partial pressure in the system above 650°C. Comparable $p(O_2)$ and $p(InCl_3)$ enable re-oxidation of In to form In_2O_3 single crystals. The transport proceeds from the hot (reaction) zone to the cold (growth) zone. In the case of using HCl instead of Cl_2, $p(O_2)$ is smaller by about three orders of magnitude than $p(InCl_3)$ at 850°C at the same precursor amounts and pressure. This would explain larger crystals in the case of In_2O_3–Cl_2 as compared with the In_2O_3–HCl system. When considering the In_2O_3–Cl_2–I_2 system, the ratio of $p(InCl_3)/p(O_2)$ is higher than that in the In_2O_3–Cl_2 system. From the thermodynamic point of view, the In_2O_3–Cl_2 system seems to be very effective in growing In_2O_3 single crystals from the gas phase.

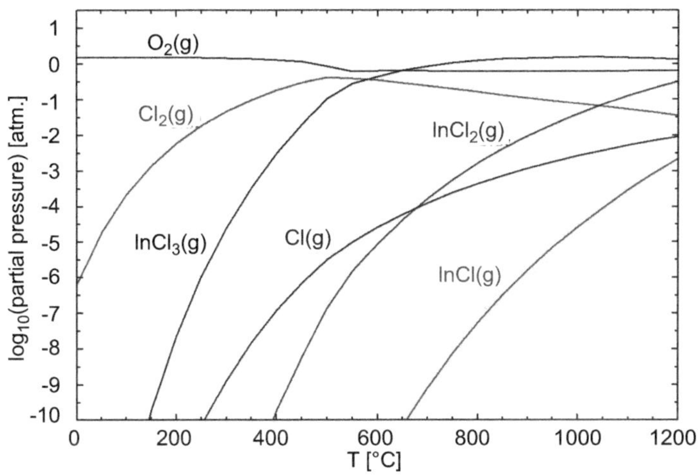

Figure 5.6 Calculated partial pressures of the most volatile species in the In_2O_3–Cl_2 system with In_2O_3 = 3 g and Cl_2 = 0.15 g, the same as in Ref. [138], and p = 5 atm.

Werner et al. [139] and Behr et al. [140] used 4N purity of SnO_2 and In_2O_3 powders as a source material to grow Sn-doped In_2O_3 and In-doped SnO_2 crystals and S + I_2 or Cl_2 as chemically active transport agent. The source material in the amount of about 1.5 g was placed, together with the transport agent, in closed silica ampoules of 1.5 cm diameter and 15 cm length. The ampoules with the S + I_2 transport agent were evacuated at low temperature to about 10^{-3} bar and then sealed off. The ampoules with gaseous Cl_2 were filled up to 450–550 mbar at RT. The total concentration of the transport medium was 20 μmol/ml, i.e., 10 S + 10 I_2, and 20 Cl_2 μmol/ml. The experiments were performed in a horizontal configuration for about 10 days, where the source material was kept at a temperature of 827–1027°C, while the growth zone at a lower temperature by 50, 100, and 150 K.

For the S + I_2 transport agent, which has a reducing character, I_2 is the transport medium for metals, while S is the transport medium for oxygen:

$$2In_2O_3 + 3S + 3I_2 \rightarrow 3InI + InI_3 + 3SO_2 \quad (5.13)$$

$$SnO_2 + S + I_2 \rightarrow SnI_2 + SO_2 \quad (5.14)$$

No sulfides were grown, unless the ratio of S/I_2 was 2, when very small platelets of SnS_2 were found. The residual impurity level of the transport agent in the crystals was less than 10 wt. ppm.

For the Cl_2 transport medium, Cl is the transport agent for metals with the release of free oxygen:

$$2In_2O_3 + 3Cl_2 \rightarrow InCl_2 + InCl_3 + In_2Cl_6 + 3O_2 \quad (5.15)$$

$$SnO_2 + Cl_2 \rightarrow SnCl_4 + O_2 \quad (5.16)$$

$p(SnCl_4)$ is about 3 orders of magnitude smaller than $p(InCl_3)$ when Cl_2 is used as the chemically active transport agent, so Sn is almost not transported and the obtained crystals are In_2O_3 doped with Sn. This is visible in Fig. 5.7, showing $p(InCl_3)$, $p(SnCl_4)$, and $p(O_2)$, as the $p(InCl_3)/p(SnCl_4)$ ratio is almost 10^3 above 600°C. In_2O_3 crystals doped with Sn (4 mol.%) were up to 5 mm in size.

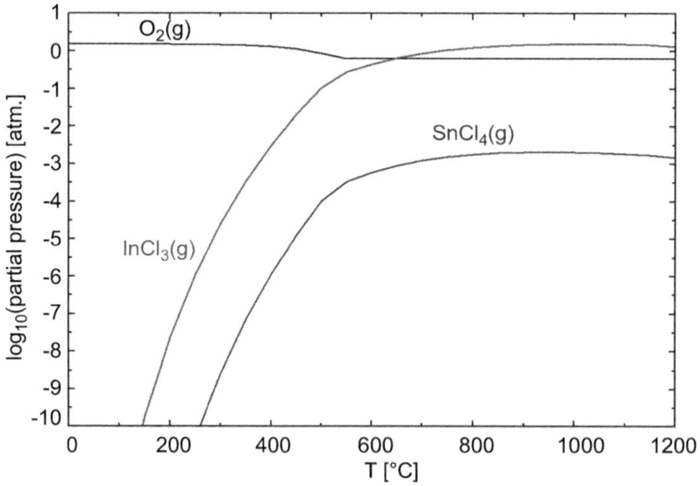

Figure 5.7 Calculated partial pressures of $InCl_3$, $SnCl_4$, and O_2 in the In_2O_3–SnO_2–Cl_2 system with 1 mol of each of In_2O_3, SnO_2, and Cl_2, and p = 5 atm.

In the case of the $S + I_2$ chemically active transport agent, In is transported by InI_3 and InI, while Sn by SnI_2. The ratio of $p(InI, InI_3)/p(SnI_2)$ is about 160 and 43 at 850°C and 1000°C, respectively, which enables a higher doping of In_2O_3 crystals with Sn at higher growth temperatures as compared with the Cl_2. Crystal stoichiometry in terms of oxygen vacancies seems to be better for the Cl_2 as compared with the $S + I_2$. This is due to the much higher $p(O_2)$ in the case of using Cl_2, by several orders of magnitude, because oxygen is consumed by sulfur in the case of $S + I_2$.

In_2O_3 and SnO_2 solid solutions always lead to an In_2O_3-rich phase. The SnO_2 phase can be obtained only with the use of the $S + I_2$ transport agent.

The phase boundaries have been determined to be 8.2 ± 0.2 mol.% SnO_2 in In_2O_3 and 2.4 ± 0.1 mol.% In_2O_3 in SnO_2. In the case of liquid phase epitaxy on sapphire grown from Sn–In melts, the solubility of Sn in In_2O_3 and solubility of In in SnO_2 of 7 ± 2 at.% and 4 ± 2 at.% at 600–800°C was determined [141].

Figure 5.8 In_2O_3 single crystals obtained by the CVT method. Reprinted from Ref. [121], Copyright 2012, with permission from AIP Publishing.

The I_2 + S transport medium for growing In_2O_3 crystals from the gas phase was also used by Scherer et al. [142]. As the source material, 5N purity In_2O_3 powder was used. A four-zone horizontal furnace was utilized, which was tilted to the horizontal axis by 7° to facilitate vapor transport via gas convection. Two outer zones were buffer zones to minimize the influence of the surrounding on the temperature distribution in the reaction and growth zones.

The obtained crystals, as shown in Fig. 5.8, were transparent and green plates or prisms with the maximum size of $3 \times 3 \times 1$ cm^3.

5.4.4 Solution Methods

The solution techniques utilized to grow bulk In_2O_3 crystals include hydrothermal, flux, and salt electrolysis. Although the hydrothermal conditions were studied for In_2O_3 with the use of water or alkaline solutions, the growth of bulk In_2O_3 crystals by the hydrothermal method has not been developed. Salt electrolysis was discussed rarely. The most frequently used solution technique for bulk In_2O_3 crystals is the flux method.

5.4.4.1 Hydrothermal

A few studies on the growth of bulk In_2O_3 crystals by the hydrothermal method were reported. Roy and Shafer [143] investigated the In_2O_3–H_2O system in hydrothermal conditions, at a temperature range of 25–800°C and pressure of 3.45×10^6 –1.39×10^8 Pa. It was concluded that at least three phases exist in the In_2O_3–H_2O system in a pressure range of 1.38×10^7 Pa to 9.65×10^7 Pa: (a) $In(OH)_3$, which is stable up to about 260°C; (b) InOOH, which seems to be substantially stable between about 260 and 440°C; and (c) In_2O_3, which is stable above 440°C. The reactions and, therefore, the formation of In_2O_3 proceed as follows:

$$In(OH)_3 \leftrightharpoons InOOH + H_2O \tag{5.17}$$

$$2InOOH \leftrightharpoons In_2O_3 + H_2O \tag{5.18}$$

Best results for In_2O_3 crystals were obtained at a temperature range of 432–449°C, pressure 6.89×10^7 Pa, and growth time of 48–96 h. At a higher pressure, 8.27×10^7 Pa, the growth time could be shorter (14 h). However, no information on crystal size and quality was given.

Christensen et al. [144] continued research on the growth of In_2O_3 of Roy and Shafer by the hydrothermal method in the In_2O_3–H_2O–Na_2O system. Cubic In_2O_3 was obtained at temperatures between 307 and 500°C, pressure of 9.5×10^6–8.1×10^7 Pa atm, Na_2O concentration of 0–3.3 M, and growth time of 17–173 h. At certain growth conditions, also rhombohedral In_2O_3 was obtained, usually in combination with InOOH. Another investigated system was In_2O_3–D_2O–Na_2O. Here, both cubic and/or rhombohedral In_2O_3

was obtained as well. Information on crystal size, perfection, and growth habit was, however, not provided.

5.4.4.2 Flux

For growing In_2O_3 crystals from high-temperature solutions, only $PbO + B_2O_3$ flaxes were used (Remeika and Spencer [133], Chase and Wilcox [145], Chase and Tippins [146], Chase [147], Chase and Teviotdale [148], Wen et al. [149], and Hagleitner et al. [150]). The starting material consisting of In_2O_3, the $PbO + B_2O_3$ flux, and possibly a dopant was loaded after mechanical mixing into a Pt crucible with a Pt lid, which was next placed into a furnace heated to the maximum temperature T_{MAX} = 1200–1320°C and held at such maximum temperature for several hours. Next, the furnace was slowly cooled down to a final growth temperature T_F = 500–1100°C at the rate of 1–10 K/h, and during that cooling, In_2O_3 crystals were formed. At the final growth temperature T_F, the crucible was removed from the furnace and it naturally cooled down to RT, or the flux was poured after removal. The crystals were released from the flux by immersing the solidified flux in a hot solution of HNO_3 + H_2O, which dissolves the flux but does not attack the In_2O_3 crystals. The obtained undoped crystals were typically yellowish or greenish plates with the edge size approaching even 10 mm [133], wherein the thickness was typically 1 mm, with a maximum volume of about 0.1 cm^3. The reported dopants were Mg [146, 150] and Sn [149].

5.4.4.2.1 *Examples*

(i) Remeika and Spencer [133] used as a starting material powders of In_2O_3 (4 g), B_2O_3 (4 g), and 50 g of PbO (50 g), which were loaded into a 100 cm^3 Pt crucible covered with a Pt lid. The crucible with the powders was placed in a horizontal, resistive furnace, which was heated up to T_{MAX} = 1200°C and held at that temperature for 4 h. Next, the furnace was cooled down at the rate of 10 K/h to T_F = 500°C at which the crucible was removed from the furnace and cooled down to RT. The crystals were removed from the flux with a hot solution of $HNO_3 + H_2O$. The obtained crystals were slightly yellow–green and transparent prismatic plates of $10 \times 10 \times 1$ mm^3 in size.

(ii) Chase and Wilcox [145] grew In_2O_3 crystals from a 200 g starting material containing 5 mol.% In_2O_3, 20 mol.% B_2O_3, and 75 mol.% PbO. The starting material was loaded in a 100 ml Pt crucible with a Pt lid having an opening for a thermocouple. The crucible was held at T_{MAX} = 1250°C for 4 h in a large muffle furnace with very low vertical temperature gradients, cooled down to T_F = 1000°C at the rate of 4–6 K/h, and then the crucible was removed from the furnace. It was found that at constant furnace temperature (1250°C), there were temperature fluctuations of 0.3 K for about 20 min, while during cooling down, the fluctuations were periodic steps of about 0.14 K approximately every 5 min.

At T_{MAX} = 1250°C, crystals did not grow, but they did during cooling down to T_F. In the crystals, striations were observed, the period of which corresponded to the period of temperature fluctuations.

Using the same starting composition and maximum temperature, Chase [147] and Chase and Teviotdale [148] additionally used smaller, 50 ml crucibles, held time at T_{MAX} of 4 to 10 h, T_F = 900, 1050, and 1100°C to which the crucible was cooled down at the rate of 1 to 10 K/h, and at which the crucibles were removed from the furnace. In some cases, the flux was poured from the crucibles after removal from the furnace, while in other cases, the crystals were separated from solidified fluxes by soaking in hot 20% HNO_3 for 4 to 24 h.

(iii) Using a similar flux composition, crucible volume, maximum temperature, and holding time at the maximum temperature, Chase and Tippins [146] performed growth experiments of Mg-doped In_2O_3 crystals. MgO concentration in the starting composition was 0–2 mol.% in favor of the PbO content. The furnace was cooled down to T_F = 900°C with the rate of 2.3 to 4 K/h. At 900°C, the crucibles were removed from the furnace and naturally cooled down to RT. The crystals were grown at the crucible bottom and wall as well as on the flux surface in the form of plates. The In_2O_3 crystals obtained from an MgO-free flux were black and had a size of 5 × 5×3 mm^3, while those from the flux containing MgO were green to yellow with a size of 2 × 2 × 1 mm^3.

(iv) Mg-doped In_2O_3 crystals were also grown from the flux by Hagleitner et al. [150]. A Pt crucible with the starting material was heated up in a furnace to T_{MAX} = 1200°C and held at that temperature for 4–10 h. Next, the temperature of the furnace was slowly decreased to T_F = 500°C at the rate of 3 K/h, and then the furnace was turned off. The crystals were extracted from the solidified flux with a water solution of HNO_3 (HNO_3:H_2O = 1:4). The obtained crystals were yellow plates of edge size 1–2 mm, as shown in Fig. 5.9. The crystals had multiple domains and high level of stresses (stress birefringence shown in Fig. 5.9).

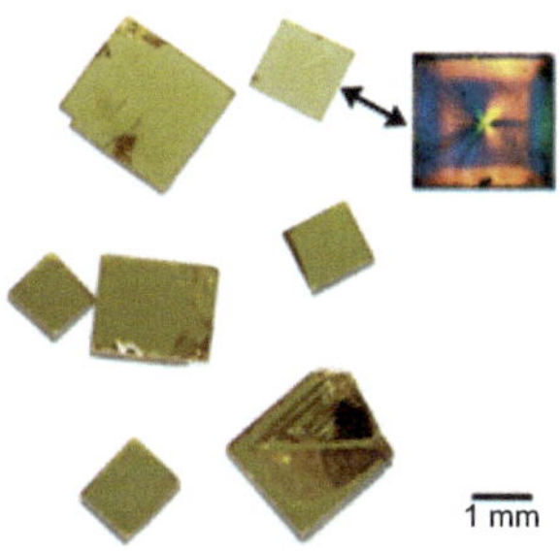

Figure 5.9 In_2O_3 crystals obtained from the PbO + B_2O_3 flux containing MgO. Reprinted with permission from Ref. [150], Copyright 2012, American Physical Society.

(v) Wen et al. [149] conducted the flux growth of pure and Sn-doped In_2O_3 crystals. The starting material containing powders of In_2O_3, SnO_2, as well as flux components B_2O_3 and PbO (total amount 43.5 g), was mixed and loaded into a Pt crucible covered with a lid to minimize evaporation of the flux. SnO_2/In_2O_3 molar fraction in the starting material ranged from 0 (pure In_2O_3) to 20% (Sn-doped In_2O_3). The crucible with the starting material was heated in a furnace up to T_{MAX} = 1320°C in ambient atmosphere at the rate of 60 K/h and held for 6 or 12 h. Next, the furnace temperature was lowered down to T_F = 850°C at the rate of 5 K/h and then the crucible was removed from the furnace. The crystals were removed from the solidified flux by soaking the Pt crucible in an aqueous HNO_3 solution. The obtained crystals were typically 2 × 2 × 1 mm^3 in size and contained Pb impurities. In the case of Sn doping, the ratio of Sn/In in In_2O_3 crystals never exceeded 1% pointing to the solubility limit of SnO_2 in In_2O_3.

5.4.4.2.2 *Growth kinetics and crystal quality*

Chase [147] noticed that In_2O_3 crystals grown from the flux have {100} habit at high supersaturation modified by small {211} faces at low supersaturations, because {100} faces are more stable at high supersaturations, while {211} at low supersaturations. The growth proceeds by two-dimensional corner and edge nucleation followed by growth at screw dislocations. Two-dimensional nucleation requires much higher supersaturation than the growth at a screw dislocation. This is manifested by hillocks, the apex of which revealed after etching at least one pit that is associated with the screw dislocation. Growth on such screw dislocations requires lower supersaturations as compared with the nucleation on edges or corners.

The major defects are trapped flux elements and striations shown in Fig. 5.10. The striations have a thickness from <10 μm to 1 mm. The striations occur radially and are related to growth mechanism and growth rate. The reason for the striations could be temperature oscillations: one of about 0.07 K magnitude with a period of 5 min, and the other of random oscillations with a magnitude of 0.5 K. During the initial nucleation from edges and corners under high supersaturation, rapping of flux elements occurs. As the growth proceeds, the supersaturation decreases and the growth at screw dislocations dominates. Due to low growth rate at low supersaturation, there are lower local changes in the supersaturation and less flux impurities are trapped with a more uniform manner. Such growth mechanisms promoting faster initial growth at the edges and corners followed by a slower growth at screw dislocations generate different structure of striations and different concentration of flux elements from the edges and corners toward the center of the face, which in fact is concave at the central part with formed steps.

By etching the crystals in hot (90°C) 50% HNO_3 or in 20% HCl/80% HNO_3 for 0.5 min to 12 h, Chase and Teviotdale [148] revealed etch pits that are parallel to the intersection of the {111} and {100} planes. Extended etching period in HNO_3 (12 h or more) produced etch tubes at the bottom of the surface pits, which intersect the {100} faces parallel to the {111} planes. These etch tubes are dislocation related, caused likely by screw dislocations. The etch pits are always elongated parallel to the <100> crystallographic direction. About 5% of the crystals showed twin boundaries, which were 90° rotated about the <100> axis. The twins were generated during nucleation or shortly thereafter. The distribution of the pits was found uniform, and its density was at the level of 10^3 cm^{-2}. An extended etching caused disappearing of most of the pits with remaining etch tubes. The source of the etch tubes extending from the etch pits is screw dislocations located either at the center of a dendrite (i.e., they were formed during crystal nucleation) or from inclusions trapped from the flux. Some of the etch tubes propagated not perpendicularly to a growing face, but also laterally and parallel to the growth direction, which can be associated with dislocation climb during crystal growth. There are also etch tubes, which changed propagation direction by about 90°. This may happen if a dislocation propagates parallel to the growth direction near the crystal edge, and it is captured by a slower growing face. Such dislocation trapping may cause changes in the relative growth rate of adjacent faces. The dislocations are growth centers and can be trapped by adjacent faces, which explain the observation that in solution growth, the majority of dislocations are located on the slowest growing faces. If the flux was poured after removal from the furnaces, extra etch pits were observed, which were formed likely by defects due to rapid growth since they were present mainly at the surface.

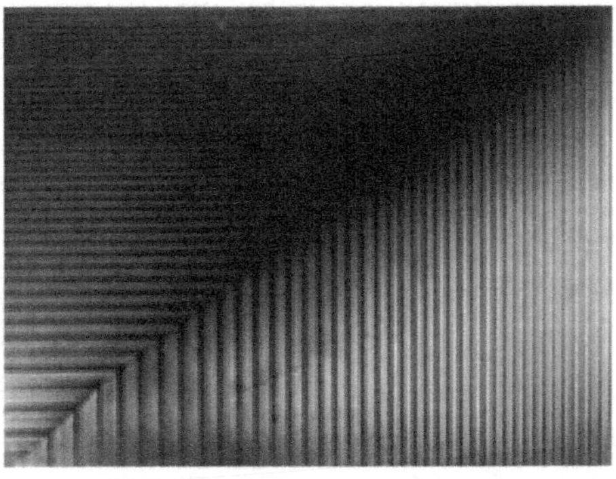

Figure 5.10 Striations in flux-grown In_2O_3 crystals. Reprinted with permission from Ref. [147], Copyright 2006, John Wiley and Sons.

According to growth experiments with Mg-doped In_2O_3 crystals by Chase and Tippins [146], the crystals were grown at the crucible bottom and wall as well as on the flux surface. Those grown on the flux surface were plates with all {100} faces present, but in the case of the crystals at the crucible wall or bottom, one {100} face was missing. In_2O_3 crystals grown from an MgO-free flux had striations and flux elements trapped therein due to supersaturation. The crystals grown from the flux containing MgO had no striations and trapped flux elements. These differences in growth habit can be explained by MgO acting as nucleation centers, therefore avoiding high supersaturation necessary to grow nominally pure In_2O_3 crystals. Lower supersaturation in the case of the MgO-containing flux allows for a slow and more uniform growth, minimizing at the same time the incorporation of flux elements and creation of striations, which are related to temperature fluctuations and associated incorporation of impurities (flux elements) following those temperature fluctuations. The concentration of Mg in the obtained crystals was 0.16 and 0.23 wt.% for Mg concentration of 1.75 and 3.5 wt.% in the flux, respectively. This leads to the segregation coefficient below 0.1.

The undoped In_2O_3 crystals showed in absorption spectra free carrier absorption indicating semiconducting behavior, while those doped with Mg showed no free carrier absorption. This means that Mg acts as a compensator for electrical conductivity.

Hagleitner et al. [150] noticed that the flux-grown In_2O_3 crystals had multiple domains and high level of stresses. The crystals also had a very high concentration of impurities (in wt. ppm): Pb = 4307 ± 74, Mg = 1388 ± 42, Pt = 155 ± 28, and other impurities (Zr, Sn, Sb, Nd, and Bi) < 50.

5.4.4.3 Salt electrolysis

Application of molten salt electrolysis to grow In_2O_3 crystals was demonstrated by Teweldemedhin et al. [151]. The salt contained fine powders of In_2O_3, Li_2MoO_4, and MoO_3 in the molar ratio of 0.4:1:1 and total weight of about 39 g. The oxides were loaded into a porcelain or alumina crucible with an immersed Pt cathode of 0.5 × 2 cm size and Pt anode of 1 × 1 cm size. Electrolysis was carried out at 840°C for 6 days by passing a constant current of 35 mA.

The crystals obtained on the cathode from alumina crucibles were green–yellow, while those obtained from porcelain crucibles were dark green. If the amount of In_2O_3 in the starting material was less than 0.4 mol, at a higher temperature (900°C) only very small crystals grew; however, at lower temperatures, MoO_2 crystals were formed. Generally, larger In_2O_3 crystals and larger quantity thereof were obtained with higher In_2O_3 molar ratio, and at lower operating temperature. Usually In_2O_3 crystals grew on the upper part of the cathode, which was not entirely immersed in the salt. The In_2O_3 crystals were of size ~1.6 × 1.1 × 1 mm^3. The crystals obtained in porcelain crucibles contained (in wt.%): Mo < 0.09, Li < 0.04, Al ≈ 0.13, Si ≈ 0.2, and Sn < 0.06. The crystals obtained in alumina crucibles had similar impurities, but at twice lower concentration, except Sn. The color of the crystals did not change after annealing in non-reducing conditions (air, O_2, inert gas) at 950°C for several hours.

5.4.5 Levitation-Assisted Self-Seeding Crystal Growth Method (LASSCGM)

This novel crystal growth method that made it possible to obtain bulk In_2O_3 single crystals from the melt for the first time was defined and developed by Galazka et al. [94–96, 152]. The method is applicable particularly for oxide materials having high electrical conductivity in the liquid phase.

5.4.5.1 Thermodynamics

In_2O_3 is highly unstable at high temperatures and proceeds a strong decomposition with the formation of O_2(g), In_2O(g), and In(g) as the most volatile species. Decomposition reactions can be written as follows (when neglecting intermediate reactions):

$$In_2O_3(s) \leftrightharpoons In_2O(g) + O_2(g) \tag{5.19}$$

$$In_2O_3(s) \leftrightharpoons In(g) + 1\tfrac{1}{2}O_2(g) \tag{5.20}$$

Partial pressures of volatile species depend on $p(O_2)$ in the system, but this dependence becomes weaker as temperature increases. Figure 5.11 shows the partial pressures of gaseous In_2O (solid lines) and In (dashed lines) at different oxygen partial pressures supplied by CO_2 at 1 and 20 bar, and by pure O_2 at 1 bar. Intensive decomposition of In_2O_3 starts already at about 1600°C, which is far below the MP (= 1950°C) of In_2O_3. When approaching the MP, the differences in partial pressures of the volatile species of the same kind are very similar to each other regardless of O_2 partial pressure, e.g., $p(In_2O) = 0.31$ and 0.11 atm, when using CO_2 ($p(O_2) = 0.014$ atm) and O_2 ($p(O_2) = 1$ atm), respectively. At such high partial pressure of decomposition products, melting of In_2O_3 is rather difficult. Additionally, due to the high MP of In_2O_3, only Ir crucible can be used, which limits the use of very high O_2 partial pressure, although it is possible as shown in the case of Ga_2O_3. Extremely high $p(In_2O)$ and high MP of In_2O_3 limited the growth of In_2O_3 crystals to low-temperature techniques: from the vapor phase and solution (flux, hydrothermal, electrolysis), which produced very small crystal size, sufficient for basic material characterization, but not sufficient to be used as substrates for epitaxy.

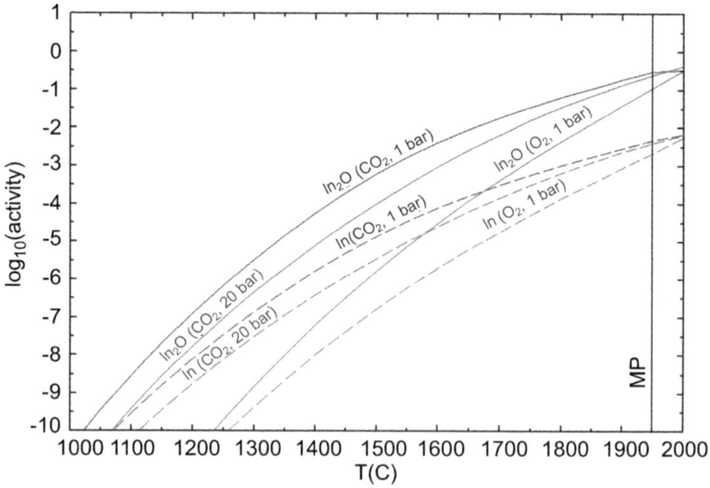

Figure 5.11 Partial pressure of volatile In_2O (solid lines) and In (dashed lines) during decomposition of In_2O_3 at different atmospheres: CO_2/1 bar, CO_2/20 bar, and O_2/1 bar. MP indicates the melting point of In_2O_3.

One of the important features of the In_2O_3 decomposition is the formation of metallic In, the concentration of which in the In_2O_3 melt is much higher as compared to the concentration of metallic species in the liquid phase formed during the decomposition of ZnO and Ga_2O_3, by about one and two orders of magnitude, respectively. This is an additional difficulty, since liquid In will immediately form eutectic with an Ir crucible. This basic thermodynamic consideration makes the growth of In_2O_3 crystals from the melt almost impossible when one depends on the current state of the art of crystal growth technologies. As a result, a novel melt crystal growth method has been defined and developed by Galazka et al. [94, 96], which enabled the growth of large In_2O_3 single crystals directly from the melt.

5.4.5.2 Concept of LASSCGM

The underlying concept for growing In_2O_3 crystals from the melt is based on a problem inversion, i.e., on using the decomposition of In_2O_3 as a driver for melting and crystallization. High concentration of metallic In in the In_2O_3 starting material at high temperatures makes the In_2O_3 starting material highly conducting. This enables creation of an additional heat source from direct coupling of an alternating magnetic field with a highly conducting In_2O_3 starting material through a crucible wall, which must be thinner than the penetration depth of the electromagnetic field. The electromagnetic coupling also enables separation of the In_2O_3 melt from the crucible wall by electromagnetic levitation. The electromagnetic levitation of In_2O_3 melt performs two functions: (i) prevents the formation of eutectic between metallic In and Ir crucible, and (ii) creates a seed by levitating a portion of the In_2O_3 melt that connects two liquid regions. In this growth process, two crucial phenomena are involved, electromagnetic levitation and self-creation of the seed, which both formulate the name of the method: Levitation-Assisted Self-Seeding Crystal Growth Method (LASSCGM).

The concept of LASSCGM is schematically shown in Fig. 5.12, which reveals for the sake of clarity only an Ir crucible with an Ir lid thereon, In_2O_3 starting material inside the crucible, and an RF induction coil. The thermal insulation in Fig. 5.12 is omitted.

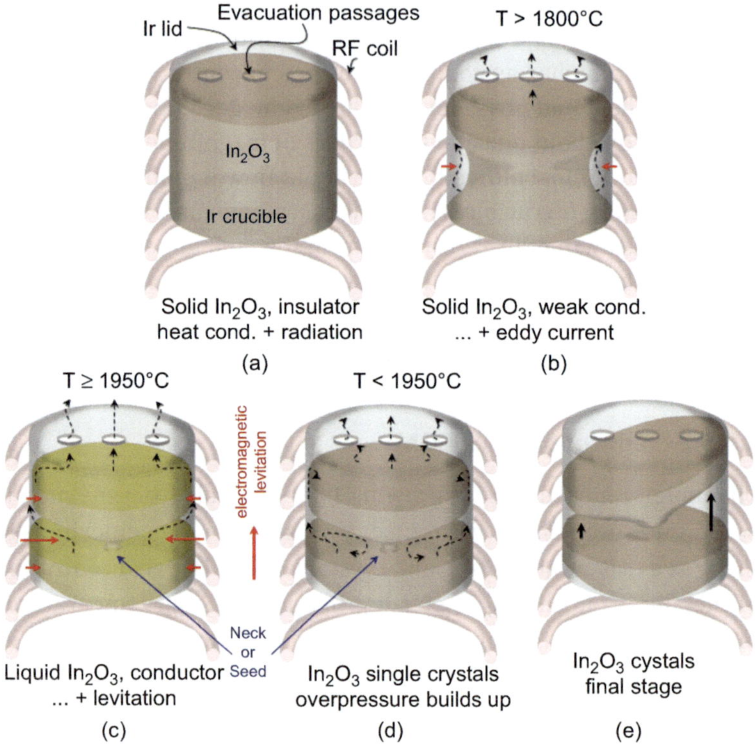

Figure 5.12 Successive stages of the LASSCGM:(a) In_2O_3 at initial stage, before heating up; (b) In_2O_3 before melting; (c) In_2O_3 in the liquid phase; (d) In_2O_3 single crystals after crystallization; and (e) In_2O_3 crystals in the final stage. Dashed arrows indicate decomposition products, horizontal solid arrows—an eddy current, while vertical solid arrows—overpressure. Reprinted from Ref. [96], Copyright 2014, with permission from Elsevier.

At the beginning (Fig. 5.12a), the In_2O_3 starting material is substantially an electrical insulator, so it is heated up by the crucible wall via radiation and conduction, which in turn is heated up inductively by the RF coil. As temperature increases in the course of heating up, the In_2O_3 starting material becomes more conductive due to its decomposition with releasing an increasing amount of metallic In. At high temperatures, roughly just above 1800°C, the concentration of metallic In in the starting material at the hottest part of the crucible is high enough to induce an electrical conductivity of the In_2O_3 starting material capable of being coupled with the electromagnetic field from the RF coil through the crucible wall (Fig. 5.12b). Here the crucible wall thickness must be smaller than the penetration depth of the electromagnetic field, or skin depth, which is defined by:

$$th \leq \delta = \sqrt{1/(\pi f \mu \sigma)} \quad (5.21)$$

where *th* is the crucible wall thickness, δ is the penetration depth of the electromagnetic field, f is the generator frequency, μ is the permeability, and σ is the electrical conductivity of the metal crucible. For typical generator frequencies of 10 and 30 kHz, the penetration depths are about 3.8 and 2.2 mm, respectively; therefore, the crucible thickness of 1–1.5 mm is sufficient to meet the condition defined by Eq. 5.21. The penetration depth is defined by a distance, over which the electromagnetic field decreases to the value of $1/e \approx 0.37$, i.e., to the level of about 37% of its initial value. Even if the crucible wall thickness is equal to the penetration depth, the electromagnetic field will still penetrate the starting material, but with less intensity. With the cylindrical crucible, the maximum of heat generation at the crucible wall by inductive heating takes place at the lower cylinder part, wherein the crucible bottom is inductively heated much less than the crucible wall. As a result, the In_2O_3 starting material, especially at its lower part, receives an additional heat source in the form of an eddy current, which accelerates melting of the starting material at the lower part of the crucible.

Further heating up increases the decomposition rate of the In_2O_3 starting material. The gaseous decomposition products are partly evacuated through the evacuation passages in the lid and partly deposit on the underside of the lid. Higher decomposition rate increases the electrical conductivity of the In_2O_3 starting material (due to higher concentration of metallic In) and facilitates coupling with the electromagnetic field that induces a higher eddy current, which additionally heats up the In_2O_3 starting material by Joule heat. This additional heat source (to radiation and conduction) causes instant melting of the In_2O_3 starting material. The eddy current flowing through liquid In_2O_3 in the vicinity of the crucible wall induces its own electromagnetic field that opposes the electromagnetic field being the source for the eddy current in the crucible. Counteracting electromagnetic fields from the RF coil and from the In_2O_3 starting material lead to the levitation of a portion of the liquid In_2O_3 starting material, because it is loose inside the crucible. The amount of liquid In_2O_3 being levitated is limited by the strength of the electromagnetic field that overcomes the gravity of the levitated material. Generally, it depends on the mutual relation and geometrical configuration of both the crucible and the coil as well as the amount of the In_2O_3 starting material and its electrical properties, which in turn depend on temperature and growth atmosphere defining a decomposition rate of the In_2O_3 starting material. As the heat source is located along the crucible wall (the crucible wall itself and the eddy current in the vicinity of the crucible wall), while the bottom of the crucible as well as the lid at the crucible top are heated to a less extent toward their centers, temperature along the centerline of the crucible is lower than that along the crucible wall. The distribution of temperature and electromagnetic field penetration inside the crucible make the levitation more effective at the crucible wall than along the centerline. The resultant melt structure will consist of two liquid regions connected by a liquid neck along the centerline, as shown in Fig. 5.12c, which acts as the seed during solidification. Obtaining and keeping the neck as thin as possible without breaking is a very challenging task as there is no visual access to the interior of the crucible. Cooling down the melt structure of Fig. 5.12c leads to the formation of one or two crystals on the opposite sides of the seed, as depicted in Fig. 5.12d.

5.4.5.3 Operation of LASSCGM

Although there is no visual access to the interior of the crucible, the fingerprints of the main phenomena can be detected by external devices: (i) a pyrometer or thermocouple that monitors the temperature of the lid or the crucible bottom; (ii) a mass spectrometer or oxygen analyzer connected to a growth chamber and monitoring oxygen changes in the growth atmosphere; and (iii) a scale weighting the furnace and indicating mass losses due to evaporation.

In the temperature profile (Fig. 5.13), recorded on the outer surface of the crucible lid, there is a sudden temperature jump by 20 to 80 K and its subsequent decrease. Visually it looks like a light flash. This is due to rapid phase transition from solid to liquid, separation of the melt from the crucible wall (at least partly), and change in view factors inside the crucible, i.e., the lid sees now a hotter melt than the cooler solid. If the temperature is being monitored in the vicinity or at the opening in the lid, high temperature fluctuations (±1–5 K) appear as the result of strong evaporation of the In_2O_3 starting material. The evaporation of the melt is much higher than the solid, which is detectable by a higher rate of mass loss of the starting material, if the crucible with the In_2O_3 starting material is weighted. Upon melting and when in the liquid phase, the rate of mass loss increases by a factor of about 2. Further increase in the temperature to a maximum temperature, called overheating point (OP > MP), aims to make the liquid neck thinner. However, this is not an obligatory step. The solidification point (SP) of In_2O_3 during cooling down of the growth furnace is indicated as a slower cooling rate due to the latent heat of crystallization, which is released during liquid–solid phase transition. At a certain point (EP in Fig. 5.13) during the cooling down step with the solidified In_2O_3 in the crucible, some temperature oscillations are typically observed. Such oscillations occur always after SP, and their position and amplitude vary between growth runs depending on the geometrical configuration of the crucible lid, thermal history of the growth run, and the free space between both top and bottom crystals. In the geometrical crystal configuration in Fig. 5.12d, there is a free space between the top and bottom crystals (structure similar to an hourglass) that communicates with the surrounding via a thin slit between the top crystal and the crucible wall. As the cooling proceeds, more and more In_2O_3 is grown from the gas phase on the cooler top crystal that eventually blocks the slit between the top crystal and the crucible wall. As a result, an overpressure builds up in the free space between the crystals, because the gaseous decomposition products of In_2O_3 cannot be evacuated any longer. Once the overpressure reaches a critical value, it is released by pushing up the top crystal, which enables escape of the entrapped gas through the evacuation passages in the lid. The internal overpressure of the entrapped gas tilts the top crystal to find a way for escape Fig. 5.12e. If the entrapped gas cannot be evacuated at a high rate, it will lead to an explosion, which was experimentally observed. This is why the point EP was named the explosion point. In other words, EP is an indication of the melt levitation. If there is no melt levitation, such temperature oscillations at EP would not be observed.

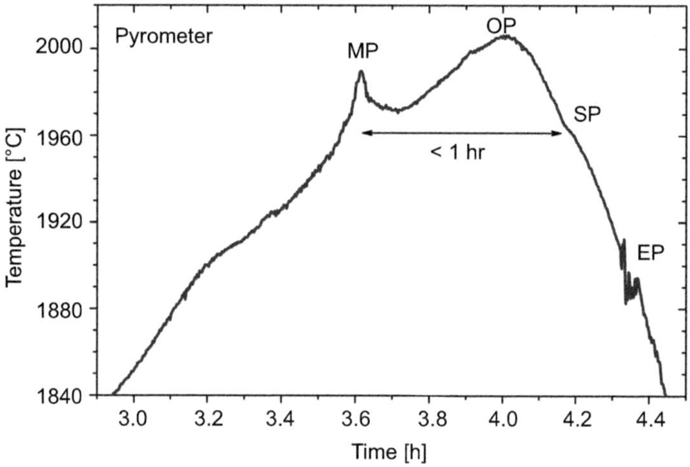

Figure 5.13 Temperature profile monitored on the crucible lid during the growth of In_2O_3 single crystals from the melt by LASSCGM. Here the MP, OP, SP, and EP indicate the melting, overheating, solidification, and explosion points, respectively. Reprinted from Ref. [96], Copyright 2014, with permission from Elsevier.

During the decomposition of In_2O_3, which is much more pronounced in the liquid phase than in the solid phase, oxygen is released (e.g., $In_2O_3 \rightarrow In_2O + \mathbf{O_2}$), which can be easily detected by the mass spectrometer or oxygen analyzer. When using CO_2 as the growth atmosphere, there is a gradual increase in oxygen with temperature due to the CO_2 decomposition ($CO_2 \rightarrow CO + ½O_2$), as in Fig. 5.14a. At high temperatures, the O_2 concentration from the In_2O_3 decomposition increases faster than O_2 from the CO_2 decomposition and this difference increases with temperature (dashed area between $O_2(CO_2 + In_2O_3)$ and $O_2(CO_2)$). Oxygen monitored by the mass spectrometer (Fig. 5.14b) revealed an increased O_2 concentration with time as the heating proceeded. A well-pronounced change in the rate of O_2 increase is an indication of melting of In_2O_3. Maintaining the In_2O_3 in the liquid phase and

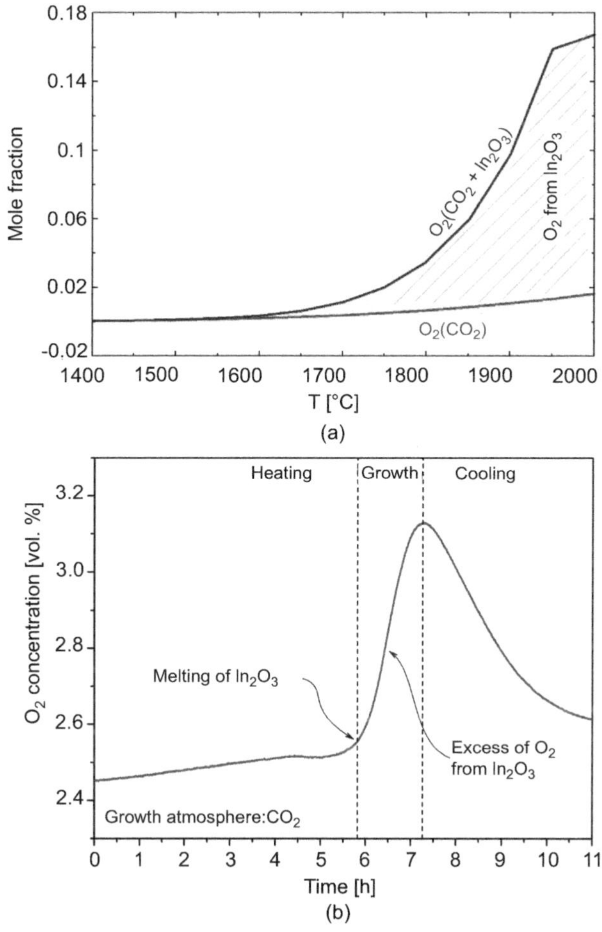

Figure 5.14 (a) Theoretical amount of O_2 produced only by the growth atmosphere CO_2 and by the system growth atmosphere—starting material (CO_2–In_2O_3) versus temperature; and (b) O_2 concentration measured in the growth chamber during heating up of the growth furnace, crystal growth, and cooling down of the growth furnace when using CO_2 growth atmosphere. Figure (b) reprinted (with minor modification) from Ref. [96], Copyright 2014, with permission from Elsevier.

its overheating increase the measured O_2 concentration. During cooling down, the oxygen concentration correspondingly decreases. Note the different units in both Fig. 5.14a and b, which are expressed in molar fraction versus temperature (Fig. 5.14a) as a result of calculations and in volume percentage versus time (Fig. 5.14b) as a result of measurement in a real growth run. Applying at least one of the tools monitoring different physical properties, i.e., temperature, mass, or oxygen concentration, provides a very useful information on melting of In_2O_3. MP in the space of temperature and time is a reference point for further actions in this growth method.

At low temperature gradients inside the crucible, the liquid neck connecting the two liquid regions may break very easily. To have better control on the formation and maintenance of the neck, which can be concluded on experimental basis, it is favorable to design higher temperature gradients inside the crucible (e.g., defined by openings in the lid on the top of the crucible), which should initially create a relatively thick neck, which can be thinned in a more controlled way by predefined overheating for a predefined time. The overheating is typically between 20 and 50 K above the MP of In_2O_3, while the overheating time (i.e., once reaching the overheating point) is between 5 and 10 min. Higher and longer overheating increases the risk of breaking the liquid neck prior crystallization. This overheating (OP) is shown in the temperature profile in Fig. 5.13.

The next step is cooling the liquid structure down at relatively high rate, typically between 300 and 600 K/h. This is to minimize losses of In_2O_3 by evaporation in both liquid (melt) and solid (crystal) phases. High cooling rate should be applied at least to the SP of In_2O_3, since In_2O_3 losses are very high in the liquid phase, about 20 wt.%/h. Solidification takes place approximately at the same temperature as melting, and it is indicated in the temperature profile as a tiny shoulder (SP). The total time between melting and solidification should preferably be not longer than 1 hr. At certain point after solidification, an overpressure releases from the crucible leading to relatively temperature oscillations (EP), as described above.

Once the system is cooled down to RT, the crystals are removed from the crucible with an aid of mechanical tools, but sometimes the crystals are loose within the crucible and can easily taken out therefrom.

To bring this new growth method into practice, special care must be taken in the growth furnace design (furnace components, configuration), growth conditions (temperature gradients, starting material, atmosphere, pressure), and operating parameters (heating and cooling rates, temperature, time).

5.4.5.4 Furnace design

Due to high MP of In_2O_3 (about 1950°C), the natural choice of the noble metal crucible is Ir. The crucible wall thickness should be as thin as possible to enable penetration of the electromagnetic field into the In_2O_3 starting material. Because Ir is brittle, too thin crucible wall may break during operation or handling. The most suitable crucible wall thickness is between 0.8 and 1.5 mm for generator frequencies between 10 and 30 kHz. A crucible shape can be different, such as cylindrical (Fig. 5.15a,b), barrel-like (Fig. 5.15c), trapezoidal (Fig. 5.15d), with a flat or convex bottom, and brings various functionality in terms of inductive heating, temperature gradients within the crucible, and removal of the crystals from the crucible. The cylindrical crucible wall is easier to fabricate and facilitate crystal removal; trapezoidal crucible wall facilitates crystal removal, while the barrel-like crucible wall decreases temperature gradients inside the crucible and also improves electromagnetic levitation.

An inductive coil may have a different configuration, corresponding to the crucible shape and/or to enhance the electromagnetic levitation. The inductive coil can be cylindrical (Fig. 5.15a,b), conical (Fig. 5.15d), or a combination of both, with possible additional coil turns at the bottom part (Fig. 5.15c), which will enhance the electromagnetic levitation. Additionally, the distance between the individual coil turns may decrease along the vertical axis toward the bottom part thereof. Instead of one, two or even more independent coils coupled with generators having different frequencies can be utilized, which will penetrate the melt at different depths.

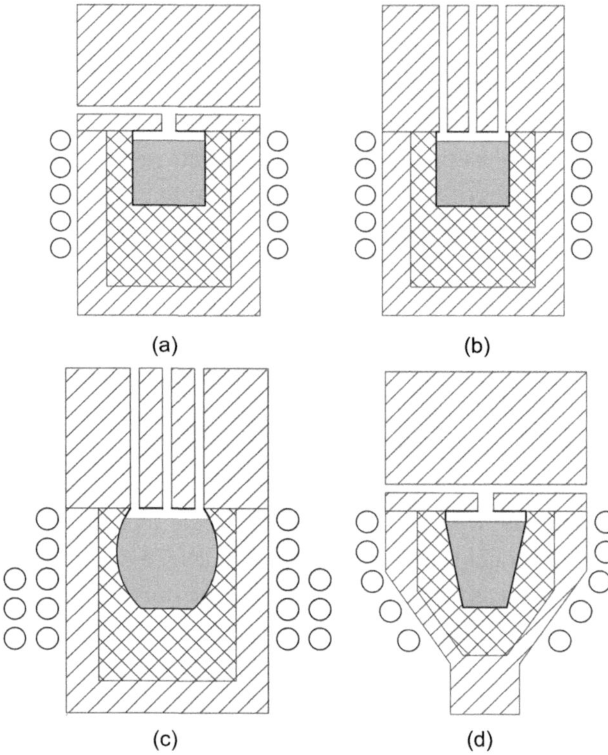

Figure 5.15 Examples of growth furnace design for LASSCGM showing different configurations of the thermal insulation, crucible, and RF coil. Reprinted from Ref. [96], Copyright 2014, with permission from Elsevier.

Another crucial furnace component is the crucible lid having well-defined openings functioning as (i) evacuation passages for gaseous products of the In_2O_3 decomposition, (ii) oxygen supply passages, and (iii) a controller of temperature gradients inside the crucible. If the total area of the openings is too large, there will be incomplete melting of the In_2O_3 starting material and too high decomposition rate, while if the total area is too small, the overpressure will built up during cooling down, which may lead to an explosion due to insufficient evacuation rate of the gaseous products of the In_2O_3 decomposition. It was experimentally found that the total surface area of the openings in the crucible lid should be between 0.25 and 10% of the crucible cross-sectional surface area. The distribution and shapes/configurations of the openings in the lid can be different. They can be circular, oval, rectangular, polygonal, slits, or other and can be located centrally and/or between the lid's center and periphery. The openings can also be formed between the underside of the lid and crucible edge. The crucible lid can be flat, convex, or concave toward the interior of the crucible, can be spherical, conical or wave like, which forms the openings between the underside of the lid and the crucible edge. The most suitable material for the lid is Ir, the same as crucible material. In this case, the thickness of the crucible lid can be similar to the thickness of the crucible wall or thicker, e.g., 2 mm. Instead of Ir, another refractory material can be used for the lid's material, such as sapphire; however, it will increase the temperature gradients inside the crucible, as it is transparent to infrared radiation, and it may easily crack as growth environment is quite aggressive, due to strong evaporation of In_2O_3 and fast changes in temperature gradients (high heating and cooling rates).

The remaining part of the growth furnace is a thermal insulation that surrounds from all sides the crucible with the In_2O_3 starting material therein and the lid thereon. To avoid excessive overheating of the In_2O_3 starting material, the thermal insulation should be well designed to decrease heat losses from the crucible. The thermal insulation around the crucible wall consists of zirconia (inner) and alumina (outer) insulation of minimum 3 cm thickness for a 40 mm crucible diameter. The thermal insulation on the top of the crucible should have channels in communication with the openings in the crucible's lid to effectively evacuate gaseous products from the In_2O_3 decomposition, particularly during cooling down when the overpressure builds up. The channels in the top thermal insulation can be vertical or horizontal. Examples of complete growth furnaces adapted for LASSCGM are illustrated in Fig. 5.15.

5.4.5.5 Growth conditions

Temperature gradients are the result of the furnace design that defines a crystal growth environment. Temperature gradient inside the crucible should be very low to avoid an excessive decomposition of In_2O_3. This is realized by a good thermal insulation around the crucible with the In_2O_3 starting material and the lid entirely covering the crucible. However, the crucible lid must have well-defined openings or passages for communication of the growth atmosphere with the In_2O_3 starting material inside the crucible and to evacuate gaseous decomposition products of In_2O_3. Such openings or passages have, at the same time, an impact on temperature gradients inside the crucible. They cannot be too large as the In_2O_3 starting material may not entirely melt due to too high temperature gradients. So the surface area of the openings/passages is a compromise between relatively low temperature gradients inside the crucible assuring a complete melting of the In_2O_3 starting material and effective evacuation of gaseous decomposition products accumulated between the top and bottom crystals after solidification. Additionally, too low temperature gradients inside the crucible make it difficult to create and maintain the liquid neck between two liquid In_2O_3 regions. Well-defined openings/passages in the lid enable achievement of optimal temperature gradients, which in turn allows for a better control of the liquid neck, e.g., by a small overheating.

The In_2O_3 starting material inside the crucible should be as pure as possible and as stoichiometric as possible. Some impurities at relatively high concentrations (>500 wt. ppm), such as Ca and Mg, may compensate the electrical conductivity of In_2O_3 at high temperatures and consequently suppress the electromagnetic levitation. Other dopants can be neutral or increase the electrical conductivity. Nonetheless, too high concentration of impurities or intentional dopants may suppress the formation of single crystals due to limited solubility. 4N purity of the In_2O_3 starting material in the form of powder is generally sufficient for the present method to function. Also the stoichiometry of the In_2O_3 starting material is important. If there is an excess of In (or deficiency of oxygen), In may react with Ir to form eutectic prior melting of the In_2O_3 starting material. This may happen, e.g., when using as the starting material In_2O_3 crystal/polycrystals from previous growth experiments, which typically contain an excess of In. Stoichiometric or near-stoichiometric In_2O_3 starting material is an electrical insulator or a weak semiconductor, and at the beginning, it is heated up normally by conduction and radiation from the crucible wall, which in turn is heated up inductively.

The growth atmosphere must contain some oxygen; otherwise, the In_2O_3 starting material may decompose to a large extent prior melting, which will decrease the volume of the grown crystals and damage the Ir crucible due to the formation of eutectic between the In released in large amount during heating and the Ir crucible, as shown in Fig. 5.16a. On the other hand, too high oxygen concentration may decrease the amount of metallic In formed during the In_2O_3 decomposition and suppress, to some extent, the electromagnetic levitation. Oxygen concentration or oxygen partial pressure controls, to some extent, the decomposition of In_2O_3 and resulted conductivity of the liquid phase. Experimentally, it was found that the minimum and maximum oxygen partial pressures in a growth furnace are about 0.005 and 0.2 bar (corresponding to about 0.5 and about 20 vol.%), respectively. The required oxygen partial pressure can be supplied to the growth furnace in different ways: as a mixture of oxygen with neutral gases, such as N_2 or Ar, from temperature-dependent decomposition of CO_2, or a combination thereof. The advantage of CO_2 is the minimization of Ir oxidation at low and moderate temperatures (low O_2 partial pressure) and supply of higher O_2 partial pressure (about 0.02 bar) at the MP of In_2O_3. O_2 partial pressure from the CO_2 decomposition is sufficient to keep the decomposition of In_2O_3 at high temperatures at the level enabling to obtain In_2O_3 single crystals by the electromagnetic levitation of the liquid phase.

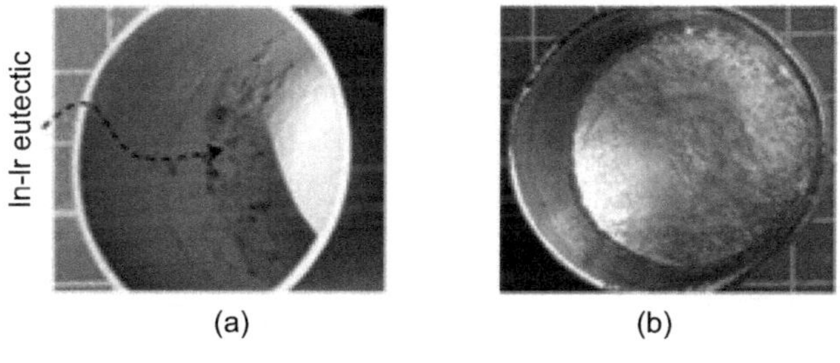

Figure 5.16 Ir crucible interior after growth of In_2O_3 single crystals with no levitation and/or excess of In in the In_2O_3 starting material (a), and with levitation and stoichiometric In_2O_3 starting material (b). Reprinted from Ref. [96], Copyright 2014, with permission from Elsevier.

An optimal pressure inside the growth furnace is atmospheric pressure. Underpressure often leads to an explosion due a larger pressure difference between the enclosed space formed after solidification between the top and bottom crystals and the crucible surrounding. On the other hand, an overpressure may have a negative impact on the electromagnetic levitation as the levitation forces must overcome not only the gravitational forces exerted on the material to be levitated, but also an external overpressure.

5.4.5.6 Operating parameters

Operating parameters are those quantities that are controlled during a growth run. Here we can distinguish heating and cooling rates, maximum temperature and time at maximum temperature, and time between melting and solidification. These quantities define a degree of the In_2O_3 decomposition and thus the operation of the method and volume of the obtained crystals. As discussed earlier, In_2O_3 is highly unstable at high temperatures and proceeds decomposition at high rates. It becomes clear that all growth run steps should be as short as possible, including the heating rate, maintaining the In_2O_3 starting material in the liquid phase, solidification, and cooling rate. Among all, the most critical is the liquid phase, which decomposes at the rate of about 5–20 wt.%/h depending on temperature gradients inside the crucible, oxygen partial pressure, and external pressure. An optimum heating time from RT to the MP of In_2O_3 (about 1950°C) is between 4 and 8 h, preferably 6 h, which corresponds to heating rates of 480 and 240 K/h, preferably 320 K/h. Overheating is typically between 1 and 4% with respect to the MP (20–80 K). Time between melting and solidification should not exceed about 1 h, as very high decomposition rate may significantly decrease the volume of the obtained crystals, or in extreme cases all In_2O_3 may evaporate. The initial cooling rate until reaching the SP should also be very high, between 200 and 500 K/h, to minimize high losses of In_2O_3 upon decomposition. As the decomposition rate of solid In_2O_3 is much smaller than that of liquid In_2O_3, further cooling can be lower in order to minimize the generation of an excessive stress and cracking of the obtained crystals, as well as formation of point and structural defects. In particular, low cooling rate of about 50–100 K/h can be applied below 1600°C.

5.4.5.7 Bulk single crystals

A single growth run produces one (typically top) or two (both top and bottom) single crystals on both sides of the seed, which correspond to liquid In_2O_3 regions connected by a liquid neck. Occasionally, three crystals are obtained due to the formation of three liquid regions connected by two liquid necks, but this is rather rare. A growth yield, defined as a usable crystal weight with respect to the weight of the starting material, is between 30 and 50%. Crystals obtained from 40 mm diameter crucibles are typically arc-shaped discs with a diameter of about 32–36 mm and thickness between 5 and 12 mm. They are brown, dark brown, or almost black depending on growth conditions and operating parameters, which is related to a concentration of point defect as a result of the In_2O_3 decomposition discussed further below. Examples of as-grown crystals and as-cut wafers are shown in Fig. 5.17.

Figure 5.17 Bulk In_2O_3 single crystals (a) and as-cut wafers (b) obtained by LASSCGM. Reprinted (with minor modifications) from Ref. [96], Copyright 2014, with permission from Elsevier.

The dark coloration of the crystals is the result of visible light extinction due to the presence of In nanoparticles (Fig. 5.18a). With each of the In nanoparticles, a cluster of oxygen vacancies is associated, indicated as a void, the volume of which corresponds exactly to the shrinkage volume of a unit cell after removing all oxygen atoms, as schematically shown in Fig. 5.18b [153]. The decomposition of In_2O_3 at high temperatures during growth leads to oxygen loss, forming oxygen vacancies in unrelaxed state. During rapid cooling down, there is a supersaturation of oxygen vacancies leading to precipitation of In particles. Indium unit cell is tetragonal and has lower volume than the volume of cubic In_2O_3, which causes the formation of a free space (void). The theoretical ratio of the void volume to the total volume of the In particle and void is 0.19. The experimental ratio of 0.1875 has been found, which matches well the theoretical value. The clustering of oxygen vacancies takes place if the diffusion time of oxygen vacancies is shorter than the cooling time of the crystal to RT. The particle size and distribution are presented in Table 5.1 [153]. The higher the density of particle size below 100 nm, the darker the coloration of the crystal. The density of In particles is in the range of 10^{11}–10^{15}, about, and below

10^{11} cm^{-3} for the particle sizes of 2–20, 20–100, and above 100 nm, respectively. Annealing in the presence of oxygen at a temperature of at least 700°C for several hours causes oxygen diffusion into the crystal, oxidation of In particles, and refilling open volume. In this way, the In_2O_3 crystal lattice is reconstructed to a large extent and the concentration of In particles decreases to the level of about 10^{11} cm^{-3} with a particle size of 2–20 nm only. As a result of such annealing, the crystals become yellow and transparent in the visible spectrum as shown in Fig. 5.19.

Table 5.1 Indium particle size and concentration in In_2O_3 single crystals obtained by LASSCGM. Data from Ref. [153]

	Size [nm]		
Sample coloration	**2–20**	**20–100**	**>100**
Red	10^{11}–10^{15}	$\approx 10^{11}$	$<10^{11}$
Brown	$\approx 10^{11}$	10^{11}–10^{13}	$\approx 10^{11}$
Black	$\approx 10^{15}$	10^{11}–10^{13}	$\approx 10^{11}$
Yellow (annealed)	$\approx 10^{11}$	—	—

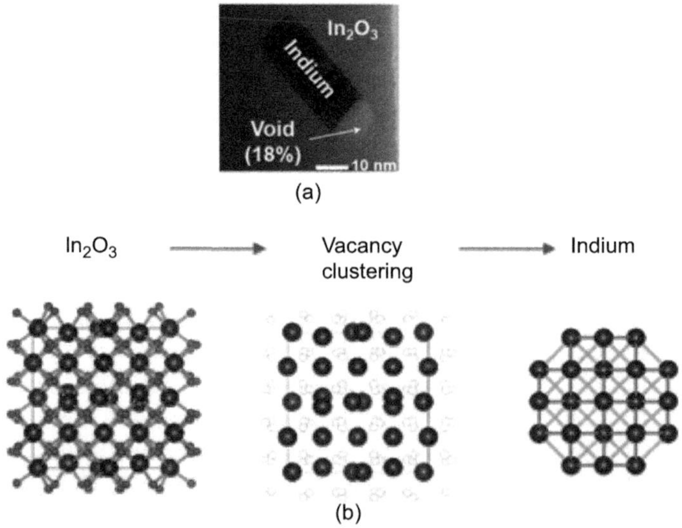

Figure 5.18 (a) Image of an In nanoparticle in the In_2O_3 host and (b) a schematic reconstruction of the formation of the In nanoparticle. Figure (a) reprinted from Ref. [96], Copyright 2014, with permission from Elsevier and (b) reprinted from Ref. [153], Copyright 2014, with permission from AIP Publishing.

The green deposition on the top crystals is In_2O_3 is from the gas phase, according to the reaction:

$$In_2O_3(s) \xrightarrow{T} In_2O(g) + O_2(g) \xrightarrow{T-\Delta T} In_2O_3(s) \qquad (5.22)$$

where ΔT is the temperature difference between the interior of the crucible and the top surface of the top crystal.

The top surface of the top crystal is cooler than the interior of the crucible; therefore, gaseous In_2O from the decomposition of the bottom part of the top crystal and the entire bottom crystal passes through the slit between the top crystal and the crucible wall, reaches a space above the top crystal, reacts with oxygen from the growth atmosphere, and deposits on the top surface of the top crystal, which acts as a seed for vapor growth. When the crucible becomes cooler and cooler, deposition also takes place in that slit, which is eventually blocked. This is the reason for the overpressure that builds up in the free space between the top and bottom crystals. In_2O_3 grown from the gas phase on the top crystals is formed by small, typically vertically elongated single crystals 1–2×3–5 mm^2. The coloration of such In_2O_3 is yellow or greenish, which is different from the coloration of the crystals obtained from the melt. Namely, the gas phase-grown In_2O_3 does not contain an excess of metallic In, which is present in melt-grown In_2O_3 crystals. The growth of In_2O_3 from the gas phase rather does not occur on the top surface of the bottom crystal.

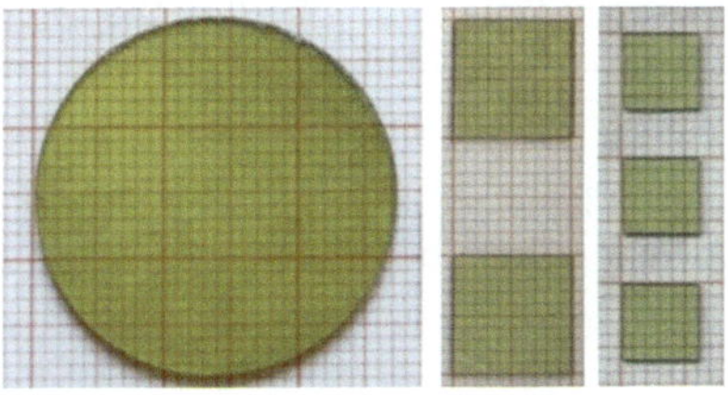

Figure 5.19 In_2O_3 wafers prepared from bulk crystals (LASSCGM) after annealing at 1000°C for 20 h in O_2. Reprinted from Ref. [96], Copyright 2014, with permission from Elsevier.

As in other oxide crystals grown from the melt, different residual impurities are found in In_2O_3 crystals, which come from the powdered starting material, Ir crucible, and alumina and/or zirconia insulation. The impurity type and concentration may differ between crystals depending on the material supplier and starting material preparation. For example, a crystal obtained from a new crucible may contain more transition metal impurities than the next crystals obtained from the same crucible. In an exemplary In_2O_3 crystal obtained by LASSCGM, the main impurities (by ICP-OES) were as follows (in wt. ppm): Ca = 56, Fe = 16, Al = 11, Ti = 8, Cr = 5, Sn = 5, Co = 3, Ni = 1.6, Cu < 2, Zr < 1.3. In terms of intentional doping, LASSCGM works the best with "pure" In_2O_3. It has experimentally been found that Ca and Mg, at a concentration above 500 wt. ppm, suppress the levitation. Nd, at a concentration of 0.1–0.4 wt.%, weakens the levitation and deteriorates the structural quality, but crystal grains of about 5–10 mm could be obtained.

The XRD pattern of powdered melt-grown In_2O_3 crystals, shown in Fig. 5.20 revealed a single phase of cubic In_2O_3, matching with the PDF 00-006-0416.

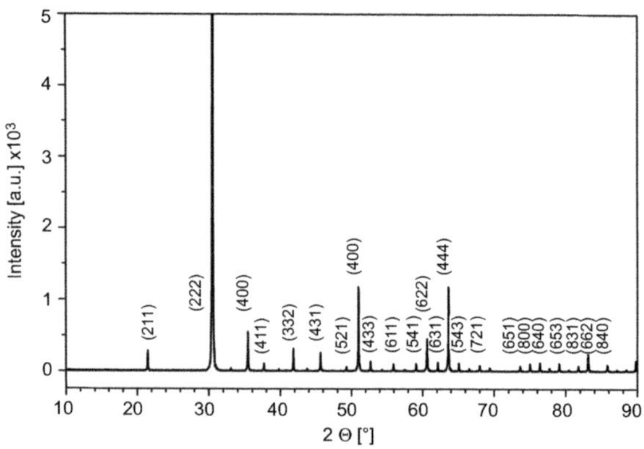

Figure 5.20 XRD powder diffraction of bulk In_2O_3 single crystals obtained by the LASSCGM.

Despite the extreme conditions at which In_2O_3 crystals are grown, their structural quality is surprisingly good. After annealing in the presence of oxygen, the FWHM of the rocking curve is typically below 50 arcsec, such as 28 arcsec shown in Fig. 5.21a with no grain boundaries, as indicated by an X-ray texture in Fig. 5.21b. The average etch pit density (EPD) after etching in HCl is between 5×10^4 and 3×10^5 cm^{-2} at both faces (001) and (111). The pits are pyramids with an elongated triangle base on (111)- and square base on (001)-oriented samples, as shown in Fig. 5.21c. A good structural quality of melt-grown bulk In_2O_3 crystals was also confirmed by a TEM image of the (111) face (Fig. 5.21d).

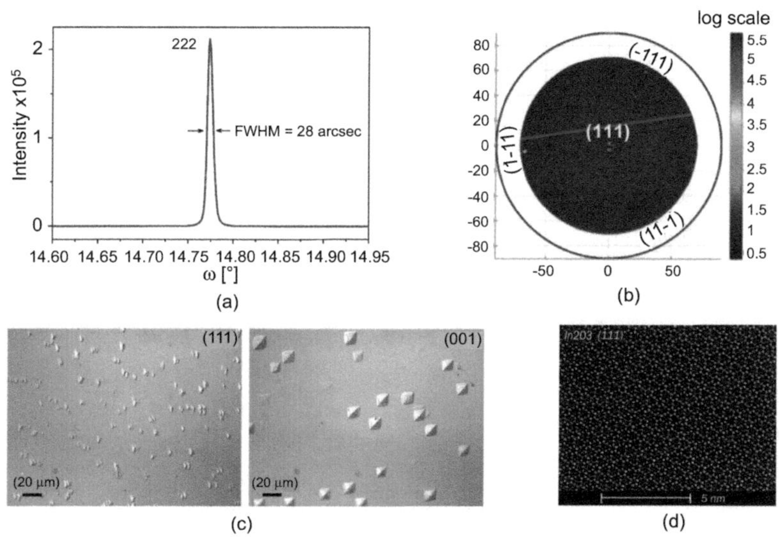

Figure 5.21 (a) Rocking curve of (111) oriented In_2O_3; (b) X-ray texture of (111)-oriented In_2O_3; (c) etch pits of (111)- and (001)-oriented In_2O_3 (LASSCGM); and (d) TEM image of (111)-oriented In_2O_3. Figure (d) reprinted from Ref. [95], Copyright 2013, with permission from Elsevier.

5.4.4.8 Thermal stability of melt-grown bulk single crystals

The instability of In_2O_3 at high temperatures required to grow crystals from the melt has been discussed in Section 5.4.3.1. Also important is the thermal stability of the obtained single crystals for high-temperature operation or for post-growth heat treatment to tune electrical and optical properties. To establish the stability range, the melt-grown crystals were annealed at different temperatures (up to 1400°C) and atmospheres (oxidizing, neutral, reducing, vacuum). The thermal instability can be roughly concluded from the mass loss in the course of annealing (Fig. 5.22), which is caused by the In_2O_3 decomposition and evaporation of volatile species. More precisely, the beginning of the decomposition can be noticed from a surface inspection, the deterioration of which clearly indicates the beginning of the decomposition (Fig. 5.23), even if the mass loss has not been detected. The surface inspection is crucial when preparing substrates for epitaxy. Nonetheless, mass loss is a good decomposition indicator.

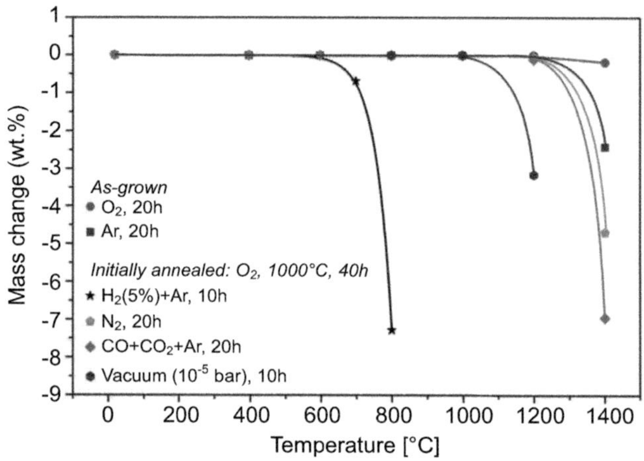

Figure 5.22 Mass loss of bulk In_2O_3 single crystals (LASSCGM) after annealing at different temperatures (400–1400°C), atmospheres (O_2, Ar, N_2, CO_2 + CO + Ar, vacuum, and H_2 + Ar), and times (10 or 20 h). Adapted (extra data included) from Ref. [152], Copyright 2013, with permission from The Royal Society of Chemistry.

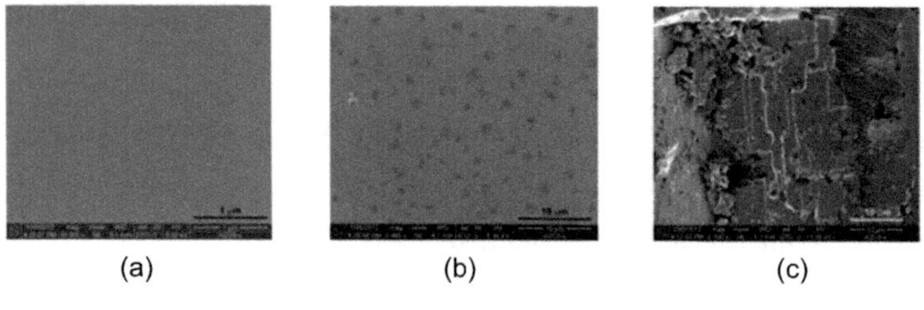

Figure 5.23 SEM images of annealed bulk In_2O_3 single crystals (LASSCGM): (a) 1000°C, O_2, 40 h; (b) 600°C, 5% H_2 + Ar, 10 h; (c) 800°C, 5% H_2 + Ar, 10 h. Reproduced from Ref. [152], Copyright 2013, with permission from The Royal Society of Chemistry.

For 20 h annealing time in non-reducing atmospheres (O_2, Ar, N_2, CO_2 + CO + Ar) and atmospheric pressure, the crystals did not show any mass loss up to 1200°C (within accuracy of about 0.1%), whether the crystals were as-grown or initially annealed in O_2. Annealing at temperatures above 1200°C leads to a mass loss, which indicates the decomposition of In_2O_3. After annealing at 1400°C, the mass loss reaches a value of 2 to 7 wt.% for O_2 and CO + CO_2 + Ar, respectively. In the case of vacuum, the decomposition starts at temperatures just above 1000°C for 10 h annealing time. In_2O_3 single crystals are the least stable in highly reducing atmosphere containing H_2 (here 5% H_2 + 95% Ar). For 10 h annealing time, the crystals are stable up to about 600°C, where the surface starts to decompose (Fig. 5.23b). Higher annealing temperatures, such as 800°C, lead to a surface damage (Fig. 5.23c) and high mass loss of about 7 wt.% (Fig. 5.22).

This relates to long annealing times, and if needed, the crystals can be subjected to temperatures and atmospheres above the stability range for a short time (e.g., several minutes) without a noticeable decomposition.

This basic stability diagram may be considered as a roadmap for thin-film growth, when using In_2O_3 as substrates. Decomposition of the substrate's surface will affect the quality of a film grown thereon.

5.4.6 Conclusion

In growth experiments of bulk In_2O_3 single crystals for over half a century, the most frequently used methods were those from the gas phase and flux. This was due to very high thermal instability of In_2O_3, which did not enable the growth of crystals at high temperatures directly from the melt. The gas and flux methods operate at moderate temperatures, below the significant decomposition of In_2O_3. The volume of the crystals did not exceed the value of about 0.1 cm^3. The invention and development of a novel crystal growth method by Galazka et al. [94–96, 152], which is based on melt levitation within an Ir crucible and self-created seed, made a technological revolution that enabled to obtain large single crystals directly from the melt, with the volume up to 10 cm^3, which is 100 times larger than that from the previous growth methods. The outputs of different methods utilized to grow In_2O_3 single crystals are shown in Table 5.2. Large In_2O_3 single crystals of high structural quality, which were obtained from the melt, allowed for an extensive characterization of that compound and prepare substrates for epitaxy, e.g., for InN [154, 155]. The melt-grown bulk In_2O_3 crystals are thermally stable up to about 1200 and 600°C in non-reducing (20 h annealing time) and reducing (10 h annealing time) conditions, respectively.

Table 5.2 Summary of bulk In_2O_3 crystals obtained by different growth methods. EPD and FWHM indicate etch pit density and full width at half maximum of the rocking curve, respectively

Phase	Method	Maximum crystal size [mm]	Basic structural quality	Feature	Reference
Gas	CVT	Cubes 3×3×3 Plates/prisms 3×3×1	—	Small size	134, 138, 139, 140, 141, 142
	Vapor reaction	Needles 0.5×0.5×8 Plates 10×10×0.1	—	Small size	132, 135, 136, 137
	PVT	Disks 40×3	Grains of two orientations	Large size, grains	95
Solution	Hydrothermal	—	—	Very small size	143, 144
	Flux	Plates 10×10×1 Prisms 5×5×3	EPD=10^3 cm^{-2}, striations, twins	Small size, impurities	133, 145, 146, 147, 148, 149, 150
	Salt electrolysis	Prisms 1.6×1.1×1	—	Very small size	151
Melt	LASSCGM	Disks 36×12	FWHM<50 arcsec EPD=5×10^4–3×10^5 cm^{-2}	Large size, high quality	95, 96, 94, 152

5.5 Thermal and Mechanical Properties

5.5.1 Thermal Properties

The MP of In_2O_3 was measured indirectly during crystal growth experiments from the melt by LASSCGM. Although the temperature was monitored on an Ir lid covering the crucible, a high distinguished temperature peak (Fig. 5.13) related to the instant melting and levitation of a portion of the melt can be taken as a MP of In_2O_3 with an accuracy of ±30 K. The base of the temperature peak that can be regarded as the MP of In_2O_3 is located at about 1950°C.

Thermal expansion of the lattice parameter a of both bulk In_2O_3 crystals and sintered powder measured by Weiher and Ley [156] in a temperature range of 0–700°C revealed the following relation:

$$\Delta a/a_0(0°\text{C}) = (6.15 \pm 0.11) \times 10^{-6}\,°\text{C}^{-1}\,T + (3.15 \pm 0.32) \times 10^{-9}\,°\text{C}^{-2}\,T^2.$$

Kundra and Ali [157] used powder for thermal expansion measurements of the lattice parameter a in a temperature range of 0–968°C, which showed the relation:

$$\Delta a/a_0(0°\text{C}) = (7.2 \pm 0.06) \times 10^{-6}\,°\text{C}^{-1}\,T + (1.15 \pm 0.08) \times 10^{-9}\,°\text{C}^{-2}\,T^2.$$

Thermal expansion of the lattice parameter a of melt-grown In_2O_3 crystals measured by Gesing [158] showed the following relation in a temperature range of 25–1000°C:

$$\Delta a/a_0(25°\text{C}) = 8.088 \times 10^{-6}\,°\text{C}^{-1}\,T + 6.301 \times 10^{-10}\,°\text{C}^{-2}\,T^2$$

As shown in Fig. 5.24, the lattice parameter versus temperature reveals almost a linear relation. At RT (25°C), the lattice parameter a = 10.12121 Å, which is a bit larger as compared with the value of 10.117 ± 0.001 Å measured by Marezio [101], but the same as calculated by Walsh et al. [109] (10.121 Å). In the temperature range of 25–1000°C, the expansion coefficient for the lattice parameter and unit cell volume with respect to RT (25°C) is 8.66×10^{-6} and 2.62×10^{-5} K^{-1}, respectively.

The solid density of In_2O_3 reported by Marezio [101] is $\rho_s = 7.12$ g/cm^3. This value is widely used in the literature.

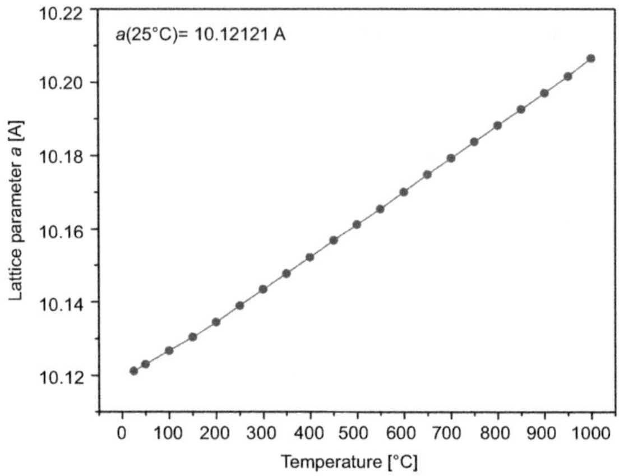

Figure 5.24 Expansion of the lattice parameter *a* of melt-grown In_2O_3 single crystal obtained by the LASSCGM [158].

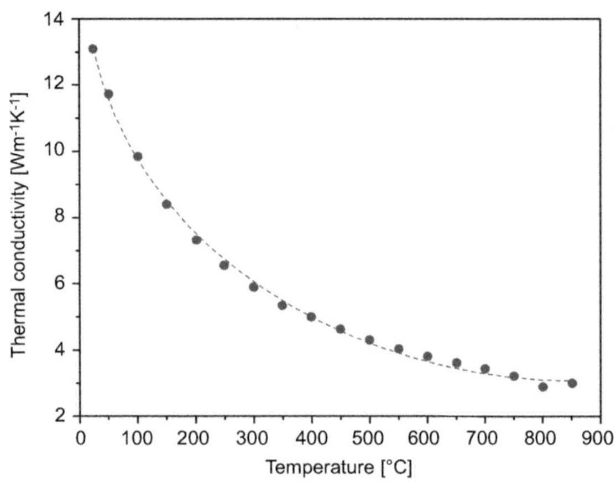

Figure 5.25 Temperature-dependent thermal conductivity of a melt-grown In_2O_3 crystal obtained by the LASSCGM [159].

Thermal conductivity of a melt-grown In_2O_3 crystal measured by Klimm and Galazka [159] is shown in Fig. 5.25. Its value is 13.09 W/mK at RT and decreases to 2.89 W/mK at 800°C. It is in the range of other oxides, such as $Y_3Al_5O_{12}$ (12 W/mK) [160], and is almost half as sapphire (30–32.5 W/mK) [161]. Molar heat capacity of In_2O_3 is 104.2 J/molK according to the database of FactSage software [162].

5.5.2 Mechanical Properties

In a cubic system, a complete tensor of elastic stiffness coefficients c_{ij} is reduced to three independent components c_{11}, c_{12}, and c_{44}. Theoretical elastic constants of In_2O_3 derived by Walsh et al. [109] are (in GPa) c_{11} = 297.75, c_{12} = 141.78, and c_{44} = 76.42. From these coefficients, one can estimate other mechanical properties, i.e., bulk modulus B_0, Young's modulus E, shear modulus G, and Poisson's ratio ν from the following Voigt relations:

$$B_0 = \frac{1}{3}(c_{11} + 2c_{12}) \tag{5.23}$$

$$E = \frac{(c_{11} - c_{12}) \times (c_{11} + 2c_{12})}{(c_{11} + c_{12})} \tag{5.24}$$

$$G = c_{44} \tag{5.25}$$

$$\nu = \frac{1}{2}\left(1 - \frac{c_{11} - c_{12}}{c_{11} + c_{12}}\right) \tag{5.26}$$

Basing on the c_{ij} coefficients given by Walsh et al. [109], the above-mentioned mechanical properties yield: B_0 = 193.8 GPa, E = 206.3 GPa, G = 76.4 GPa, and ν = 0.323 [163].

For comparison, other theoretical values of bulk modulus B_0 calculated by Ramzan et al. [164], Aliabad et al. [113], Erhart et al. [115] are 192.66, 186.6, and 180 GPa, respectively, while an experimental value measured by Liu et al. [110] is $B_0 = 194.24 \pm 3.26$ GPa.

From the elastic stiffness coefficient c_{11}, one can calculate the longitudinal sound velocity v_L in In_2O_3 from a relation of $v_L = \sqrt{c_{11}/\rho_s}$, which yields 6.47×10^3 m/s. The transverse sound velocity has two components: $v_{T1} = \sqrt{c_{44}/\rho_s}$ = 3.28×10^3 m/s, and $v_{T2} = \sqrt{(c_{11} - c_{12})/(2\rho_s)}$ = 3.31×10^3 m/s.

The Debye temperature concluded by Preissler et al. [165] is 700 K. The above-discussed physical properties of In_2O_3 are summarized in Table 5.3.

Table 5.3 Thermal and mechanical properties of bulk In_2O_3 crystals

Property	Unit	Value	Reference
Solid density	g/cm^3	7.12	101
Melting point	°C	1950±30	95
Thermal expansion coefficient of the lattice parameter (25–1000°C)	K^{-1}	8.66×10^{-6}	158
Thermal conductivity at RT	W/mK	13.09	159
Molar heat capacity at RT	J/molK	104.2 (a)	162
Elastic stiffness coefficients	GPa	c_{11}=297.8 (a) c_{12}=141.8 (a) c_{44}=76.4 (a)	109
Bulk modulus B_0	GPa	193.8 (b)/192.7	109/164
Young's modulus E	GPa	206.3 (b)	109
Shear modulus G	GPa	76.4 (b)	109
Poisson's ratio ν	—	0.323 (b)	109
Longitudinal sound velocity	m/s	6.47×10^3 (c)	101, 109
Transverse sound velocity		$3.28\times10^3/3.31\times10^3$ (c)	
Debye temperature	K	700	165

(a)theoretical value; (b)calculated from Eqs. (5.23–5.26) and theoretical data from Ref. [109]; (c)calculated from data in Refs. [101, 109]

5.6 Optical Properties and Effect of Annealing

The availability of large bulk In_2O_3 single crystals of high structural quality obtained directly from the melt made it possible to accurately reveal basic optical properties of that binary compound, including transmittance, optical bandgap, refractive index, and luminescence.

5.6.1 Transmittance, Absorption Edge, and Optical Bandgap

The presence of In particles in the as-grown crystals causes light extinction, which has a reflection in transmittance spectra, marked "as-grown" in Fig. 5.26a. The absorption edge originates at 440 nm, but there is an intensive absorption in whole visible spectrum, which increases with the density and size of In particles. A high absorption in the NIR spectrum is due to the free carrier absorption, as the free electron concentration in the as-grown crystals is high, at the level of 10^{18} cm^{-3}. Annealing in any oxidizing atmosphere (including N_2 and Ar, which typically contain a few ppm of O_2) at temperatures 700–1200°C for at least several hours significantly improves the transmittance of melt-grown In_2O_3 in the visible spectrum (by oxidizing In particles), making the absorption edge very steep. After such annealing, the crystals become yellowish and transparent, as shown in Fig. 5.19 and Fig. 5.26a. The improvement in transmittance in the NIR spectrum is due to a decrease in the free electron concentration. The best transmittance spectra were obtained after annealing at 800–1000°C for 20 h annealing time. Annealing at or above 1200°C for a long time leads to deterioration of the crystal's structural quality, particularly its surface as a result of some decomposition. Consequently, the transmittance also deteriorates a bit in the visible spectrum. However, lower transmittance in the visible and NIR spectra of samples annealed at 1200°C and 1400°C is partly due to higher initial electron concentration.

Annealing in the presence of H_2 increases, as discussed further below, the free electron concentration causing an enhanced absorption in the NIR and red part of the visible spectrum. The change in the transmittance with annealing temperature up to 600°C is associated with an increase in the free electron concentration (Fig. 5.26b). Higher annealing temperatures (up to 800°C) lead, in addition to the increase in the free electron concentration, to deterioration of the crystal's structural quality, since In_2O_3 crystals become thermally unstable and proceed decomposition at such conditions (see Fig. 5.23). This causes a decrease in the transmittance also in the visible spectral region.

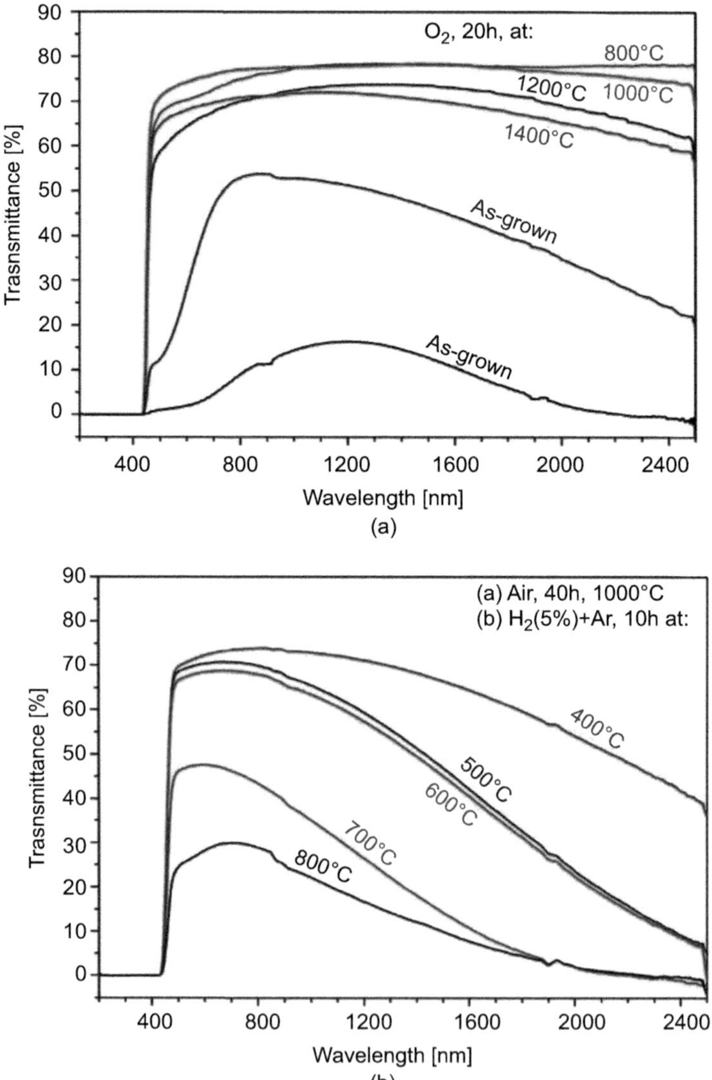

Figure 5.26 Transmittance spectra of (a) as-grown and annealed in O_2 bulk In_2O_3 single crystals obtained by the LASSCGM, and (b) double-annealed bulk In_2O_3 single crystals, first in air at 1000°C for 40 h, and next in H_2 (5%) + Ar for 10 h at different temperatures (400–800°C) [166].

An inside to the optical bandgap of bulk In_2O_3 single crystals by analysis of the absorption coefficient was given by Irmscher et al. [167]. A transmittance *TR* measured on a thin sample enables calculation of the absorption coefficient α with the use of the following formula:

$$\alpha = -\frac{1}{d}\ln\left[\frac{\sqrt{(1-R)^4 + 4(TR)^2R^2} - (1-R)^2}{2TR^2}\right] \tag{5.27}$$

where d is the sample thickness and R is the reflectance, which can be approximated by $R = (n - 1)^2/(n + 1)^2$ assuming a constant value of the refractive index $n = 2$ (at 800 nm).

From the absorption coefficient, we can easily conclude the nature of the optical energy gap, whether direct or indirect, allowed or forbidden. The best fitting of the absorption coefficient was obtained for the indirect-allowed transition:

$$\alpha \propto (h\nu \pm E_{\text{phonon}} - E_{\text{g,ind}})^2 \tag{5.28}$$

where $h\nu$ is the photon energy, E_{phonon} is the phonon conserving energy in indirect transitions, and $E_{\text{g,ind}}$ is the indirect energy gap. As shown in Fig. 5.27, room- and low-temperature measurements of the transmittance and hence absorption coefficient enabled extrapolation of the optical bandgap, which is 2.72 eV and 2.90 at 300 and 9 K, respectively.

Weiher and Ley [135] concluded from absorption coefficient measurements optical bandgaps of 2.62 and 3.75 eV, which were assigned to indirect forbidden and direct-allowed transitions, respectively.

ARPES measurements of CVT-grown In_2O_3 by Janowitz et al. [120] revealed the optical bandgaps of 2.63 and 3.71 eV. ARPES and STS measurements of CVT-grown In_2O_3 crystals by Scherer et al. [142] and Braun et al. [130] showed the fundamental bandgap of 2.7 $\pm$ 0.1 eV and 2.78 $\pm$ 0.04 eV, respectively.

Collection of the absorption spectra from 9 to 1300 K and determination of the bandgap in the same manner as above enables an evaluation of the bandgap versus temperature, which is depicted in Fig. 5.28 [167]. Above RT up to 1300 K, the bandgap decreases with temperature linearly with the rate as large as of 1.34 meV/K. In this temperature range, the bandgap decreases by about half. Weiher and Ley [135] concluded for the crystals grown from the gas phase that the rate of bandgap decrease with temperature is about 1 meV/K.

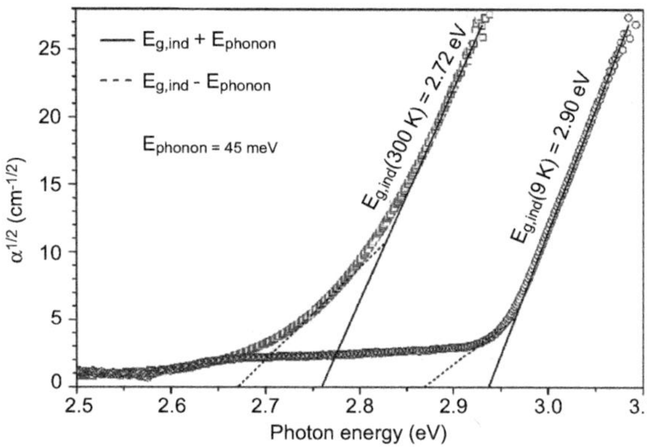

Figure 5.27 Absorption coefficient in power ½ versus photon energy at RT and 9 K that exhibits indirect-allowed transitions. Solid and dot black lines indicate extrapolations of nearly linear parts of the curves. The thickness of measured melt-grown In_2O_3 sample (LASSCGM) was 45 μm. Reprinted with permission from Ref. [167], Copyright 2014, John Wiley and Sons.

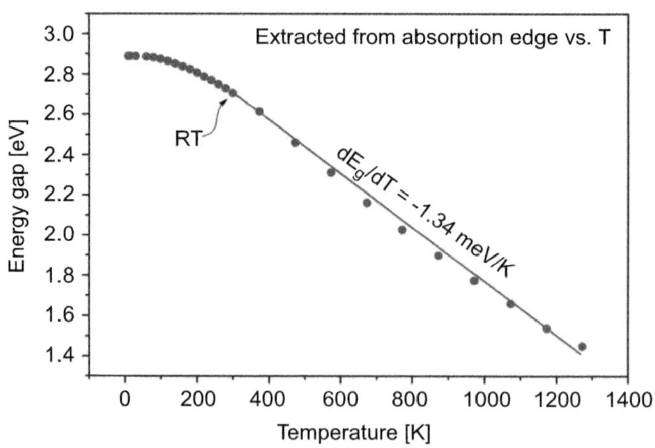

Figure 5.28 Temperature-dependent indirect energy gap of a bulk In_2O_3 single crystal (LASSCGM). Based on data from Ref. [167].

5.6.2 Dielectric Constant, Refractive Index, Reflectance

Dielectric function in its complex form $\varepsilon = \varepsilon_1 + i\varepsilon_2$ is frequency and temperature dependent. It was experimentally explored for In_2O_3 by a number of researches with the use of ellipsometry for thin films and bulk crystals, e.g., by Hamberg and Granqvist [168–170], Qiao and Mergel [171], Schmidt-Grund et al. [172], Feneberg et al. [173], and Schleife et al. [174] who found an excellent quantitative agreement between experiment and computation.

The values of relative static (ε_0) and high-frequency (ε_∞) dielectric constants of In_2O_3 defined by

$$\varepsilon_0 = \lim_{\hbar\omega\to\infty} \varepsilon_1 \text{, and } \varepsilon_\infty = \lim_{\hbar\omega\to 0} \varepsilon_1 \qquad (5.29)$$

and determined by Hamberg and Granqvist [168] as $\varepsilon_0 = 8.9$ and $\varepsilon_\infty =$ 3.95, respectively, the latter being measured by Feneberg et al. [173] at a bit higher value of $\varepsilon_\infty = 4.08 \pm 0.02$.

Aliabad et al. [113] calculated the refractive index of In_2O_3 as 1.78–1.67 for the wavelengths of 500–1000 nm. Weiher and Ley [135] concluded the refractive index of 2.08 at 589 nm from their measurement of optical properties. Schmidt-Grund et al. [172] derived the refractive index n from In_2O_3 thin films based on Couchy equation, which reads:

$$n(\lambda, T) = (1.9982 + 3 \times 10^{-6}\,\text{K}^{-1}\,T + 7 \times 10^{-8}\,\text{K}^{-2}\,T^2) + (0.02825\,\mu\text{m}^2 + 3 \times 10^{-6}\,\mu\text{m}^2/\text{K}\,T)\,/\lambda^2$$

The refractive index and reflectance derived from ellipsometric measurements of bulk In_2O_3 crystals obtained from the melt by LASSCGM are shown in Fig. 5.29a and Fig. 5.29b, respectively (Anooz and Galazka [175], and Irmscher and Galazka [176]). The refractive index n has a maximum of 2.58 at 330 nm (3.76 eV), which corresponds to a direct-allowed transition (strong absorption onset). With wavelength, the refractive index quickly decreases to 2.17 at 500 nm, 2.13 at 600 nm, and 2.1 at 800 nm. In the reflectance spectrum (Fig. 5.29b), we can clearly distinguish three features located at about 2.78, 3.8, and 4.75 eV. Weiher and Ley [135] also observed similar features at 2.75 and 3.75 eV, which were assigned to indirect-allowed and direct-allowed transitions, respectively.

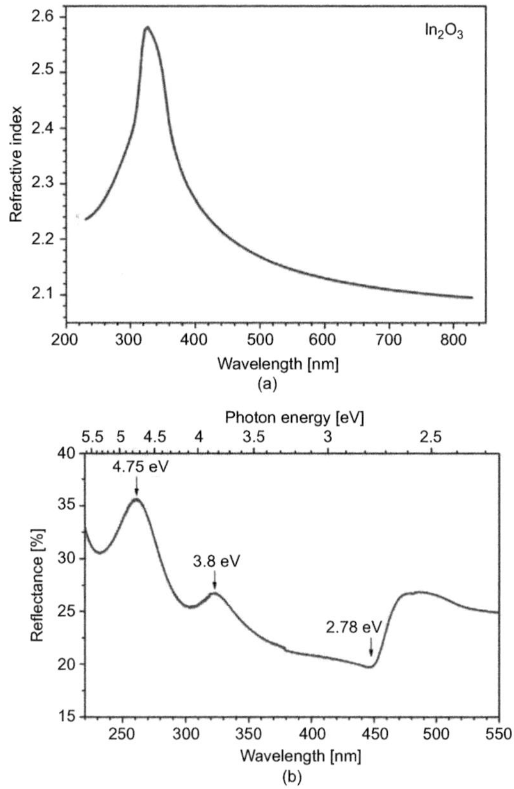

Figure 5.29 Refractive index (a) and reflectance (b) of bulk In_2O_3 single crystal obtained from the melt by the LASSCGM [175, 176].

5.6.3 Luminescence

As illustrated in Fig. 5.30, the cathodoluminescence (CL) spectra of melt-grown In_2O_3 single crystals obtained by LASSCGM show a broad emission within the energy gap, peaking at about 630 nm (1.97 eV) and 650 nm (1.91 eV) at RT and 7 K, respectively (orange emission) [152, 177]. At RT the intensity depends on annealing, atmosphere, and orientation. The most intense emission was observed in a (111)-oriented sample annealed at 1000°C in air for 40 h, and lower from a sample annealed at 400°C in the same atmosphere for 20 h.

(001) revealed lower emission as compared with the (111)-oriented sample after annealing at the same conditions. Additional annealing of a sample in a reducing atmosphere (H_2 + Ar) at 500°C, which was previously annealed in air at 1000°C, also decreases the CL intensity. Interestingly, at RT all the spectra, whether the crystals were as-grown or annealed in oxidizing/reducing conditions, perfectly overlap with each other when normalized [152]. At low temperature (7 K), the CL spectrum is narrower and shows the lowest intensity. Such broad luminescence is commonly assigned to the presence of self-trapped excitons, as a result of a recombination of an electron and a self-trapped hole (STH).

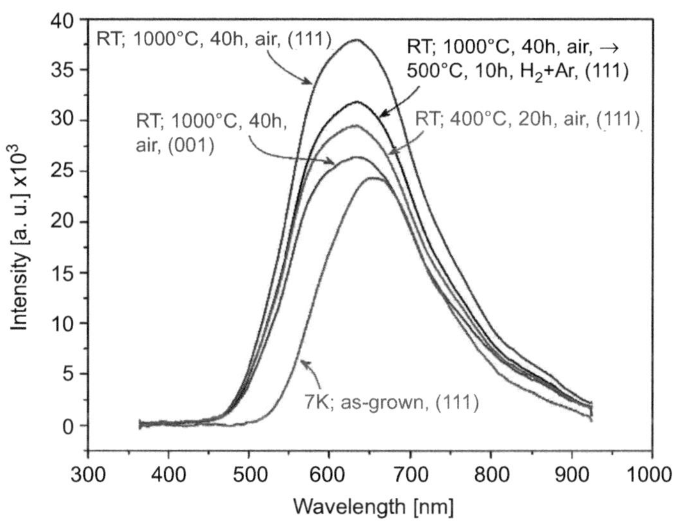

Figure 5.30 Cathodoluminescence (CL) spectra of as-grown and annealed In_2O_3 single crystals (by the LASSCGM) at room temperature (RT) and 7 K. [177].

Indeed, Varley et al. [178] concluded from their theoretical studies that the STH formation is a fundamental property of many oxide systems and creates an obstacle for *p*-type conductivity. The calculated emission wavelength for In_2O_3 is 821 nm (1.51 eV), which is longer than that obtained experimentally from the CL spectra.

Radioluminescence (X-ray-induced emission) spectra at 10 K and RT of melt-grown In_2O_3 crystal, measured by Drozdowski et al. [179], are shown in Fig. 5.31. At 10 K, the spectrum shows a maximum at about 620 nm, close to that seen in the CL spectra, while no emission was recorded at RT. Temperature-dependent radioluminescence showed the emission peak gradually decreasing with temperature and disappearing at RT.

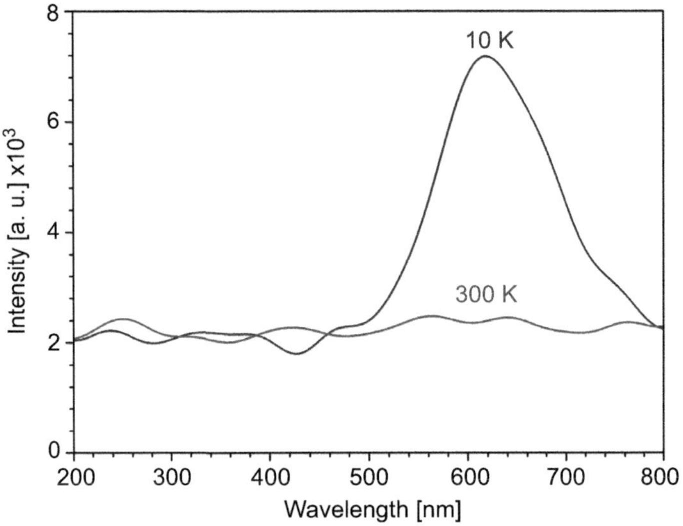

Figure 5.31 Uncorrected X-ray-induced emission spectra at 10 and 300 K of melt-grown In_2O_3 single crystals (LASSCGM) [179].

5.6.4 Phonon Modes

In_2O_3 has nine fundamental phonon modes that are described by:

$$5A_1 + 1A_2 + 1B_1 + 2B_2$$

where A_1 and B_2 correspond to stretching, ring deformation, and in-plane vibrations, while A_2 and B_1 relate to torsion and out-of-plane bending vibrations [180]. The calculated and observed phonon modes in thin films according to Panneerdoss et al. [180] are shown in Table 5.4.

Table 5.4 Theoretical and experimental (thin films) phonon modes of In_2O_3 based on Ref. [180]

Assignment	Theoretical [cm^{-1}]	Experimental [cm^{-1}]
B_2	606–610	600
B_2	585–593	580
A_1	496–498	490
A_1	395–402	390
A_1	312–314	300
A_1	258–271	260
A_1	110–225	210
A_2	78–90	200
B_1	50–78	110

Note: The ranges of the theoretical values arise from different calculation methods.

5.6.5 Conclusion

The availability of large bulk In_2O_3 single crystals of high structural quality and purity obtained from the melt enabled exploration of optical properties of that compound in more detail in addition to optical characterization made on crystals obtained from the gas phase or flux. The optical properties are sensitive to the structural quality and purity of the material; therefore, a selection of suitable crystal samples is crucial in revealing the intrinsic optical properties of In_2O_3.

From the absorption coefficient evaluation, the fundamental bandgap of 2.72 eV was found to be indirect, wherein a direct bandgap has a value of about 3.8 eV. The as-grown crystals (from the melt) show an extensive absorption in the visible spectral region due to the presence of In nanoparticles as the result of the In_2O_3 decomposition at high temperatures, and high absorption in the NIR region due to free carrier absorption. Annealing the crystals in an oxidizing atmosphere removes the absorption in the visible region and significantly reduces the absorption in the NIR region, making the crystals transparent in the visible region. The absorption edge that originates at about 440 nm becomes very steep. Annealing the transparent crystals in a reducing atmosphere increases the absorption in the NIR region and eventually in the red part of visible spectrum due to enhanced free carrier absorption. At RT the CL spectra show one broad emission peaking at about 630 nm, and at about 620 nm at 10 K under X-ray excitation. It is likely due to self-trapped excitons. Relative static and high-frequency dielectric constants are 8.9 and 4.08, respectively. The refractive index of In_2O_3 in the visible spectral region 500–800 nm is in the range of 2.122–2.043. Phonon modes of In_2O_3 have also been explored, but for thin films or nanostructures.

A collection of empirical optical properties of In_2O_3 single crystals is provided in Table 5.5.

Table 5.5 Basic empirical optical properties of In_2O_3 crystals

Property	Unit	Value	Reference
Fundamental bandgap	eV	2.63–2.78	120, 130, 167, 142
Fundamental bandgap type		Indirect	167
Direct bandgap	eV	3.71–3.8	120, 135, 176
Bandgap shrinkage at 300–1300°C	meV/K	−1.34	167
Absorption edge	nm	≈440	95, 96
Emission band peak at RT	eV	1.97	152, 177
Relative static dielectric constant		8.9	168
High-frequency dielectric constant		4.08±0.02	173
Refractive index at 500–800 nm		2.122–2.043	175
Phonon modes[a]	cm^{-1}	$B_1 = 110$, $A_2 = 200$, $A_1 = 210$, $A_1 = 260$, $A_1 = 300$, $A_1 = 390$, $A_1 = 490$, $B_2 = 580$, $B_2 = 600$	180

[a]measured on thin films

5.7 Electrical Properties

Bulk In_2O_3 crystals obtained by any of the growth techniques typically show *n*-type conductivity whether with or without intentional doping. The crystals are *n*-type semiconductors with high free electron concentrations. The underlying origin of the conductivity in In_2O_3 is not clear up to now; however, the following have been suggested, both theoretically and experimentally:

- Oxygen vacancies, V_O^{2+} (e.g., Ágoston et al. [123, 124], Braun et al. [130]);
- Oxygen vacancies + indium vacancies, $V_O + V_{In}$ (Palandage and Fernando [181]);
- Hydrogen, H_i, H_O, V_{In}-H_i (e.g., Limpijumnong et al. [125], Varley et al. [127], King et al. [128])
- Solid impurities, such as Sn (e.g., Behr et al. [140], Wen et al. [149, 182]) and possible Si.

As the native point defects (always formed) along with solid impurities (unavoidable), and likely hydrogen (omnipresent), are present in In_2O_3 crystals independent of the growth method, perhaps there is an interplay of all these defect types with different "output" depending on particular concentration of the defects, which in turn depends on particular growth environment. Pure In_2O_3 crystals, i.e., without any intentional doping and with low concentration of residual impurities always show electrical conductivity. High-resistivity bulk crystals could be obtained either by intentional doping with Mg, or with high concentration of other residual impurities. Moreover, as discussed further below, bulk In_2O_3 crystals are sensitive to annealing in terms of the temperature, atmosphere, and time. In particular, even long annealing in the presence of oxygen at a wide spectrum of temperatures did not turn the crystals into electrically insulating state; the crystals remained semiconductors, but with lower free electron concentrations. On the other hand, annealing in the presence of hydrogen significantly increases the free electron concentration.

5.7.1 Free Electron Concentration, Hall Mobility, and Resistivity/Conductivity at RT

Crystals grown from the gas phase by Weiher [132] with no intentional doping showed at RT the free electron concentration and resistivity of $n_e = 5 \times 10^{17}$–1.1×10^{18} cm^{-3} and $\rho = 0.2$–0.02 Ωcm, respectively. Scherer et al. [142] measured the values of $n_e = 1.3 \times 10^{19}$ cm^3 and $\rho = 8.5 \times 10^{-3}$ Ωcm, respectively, and the Hall mobility of $\mu = 66$ cm^2/Vs. Shimada et al. [136] and De Wit [134] reported at RT $\rho = 5.5 \times 10^{-2}$ and 4×10^4 Ωcm, respectively. Doping with Sn (Behr et al. [140]) resulted in $n_e = 5 \times 10^{19}$–10^{21} cm^{-3} and $\mu = 20$–100 cm^2/Vs, respectively. The free electron concentration scales almost 1 to 1 with the Sn concentration in the In_2O_3 crystals up to the solubility limit of about 8.2 mol.%.

In_2O_3 crystals obtained by Galazka et al. [95] from the gas phase by the PVT at high temperatures were highly conducting at RT with $n_e = 1.2$–1.5×10^{19} cm^{-3}, $\mu = 110$ cm^2/Vs, and $\rho = 4$–5×10^{-3} Ωcm.

Flux-grown undoped In_2O_3 crystals reported by Wen et al. [149, 182] were characterized by the following electrical properties at RT: $n_e = 2.5 \times 10^{18}$–1.8×10^{19} cm^{-3}, $\mu = 45$–85 cm^2/Vs, and $\rho = 2.9 \times 10^{-2}$–$7.7 \times 10^{-3}$ Ωcm. The conductivity of In_2O_3 crystals increased with Sn doping, and the corresponding electrical properties were $n_e = 3.8 \times 10^{19}$–8.8×10^{20} cm^{-3}, $\mu = 48$–100 cm^2/Vs, and $\rho = 3.4 \times 10^{-3}$–$3 \times 10^{-4}$ Ωcm. At the highest Sn concentrations, only a part of Sn atoms was electrically active. The authors also mentioned that Pb impurities are electrically inactive in In_2O_3. However, other impurities in flux-grown crystals could be electrically active. As an example, Hagleitner et al. [150] reported a very high concentration of impurities (in wt. ppm): Pb = 4307 ± 74 (flux constituent), Mg = 1388 ± 42 (intentional dopant), Pt = 155 ± 28 (crucible material), and other impurities (Zr, Sn, Sb, Nd, and Bi) < 50. The resistivity of such crystals at RT was very high, 2×10^5 Ωcm, although still showing some residual conductivity.

Chase and Tippins [146] concluded that undoped In_2O_3 crystals obtained from the flux were semiconductors, while those doped with Mg were rather electrical insulators. Mg acts as a conductivity compensator, which was also concluded for epitaxial films.

In_2O_3 crystals obtained by salt electrolysis (Teweldemedhin et al. [151]) showed resistivities of 10^{-3}–10^{-4} and 10^{-2} Ωcm at RT for dark-green and green–yellow crystals, respectively.

Bulk In_2O_3 crystals grown from the melt by Galazka et al. [94–96, 152] were always semiconducting, characterized by the free electron concentration, Hall mobility, and resistivity of $n_e = 1–4 \times 10^{18}$ cm^{-3}, μ = 130–170 cm^2/Vs, and $\rho = 1–3 \times 10^{-2}$ Ωcm, respectively.

Table 5.6 Basic electrical properties of as-grown bulk In_2O_3 crystals at RT. n_e, μ, and ρ are the free electron concentration, electron mobility, and resistivity, respectively

Method	n_e [cm^{-3}]	μ [cm^2/Vs]	ρ [Ωcm]	Reference
Gas phase (CVT)	5×10^{19} – 1.0×10^{21} (a)	20–100	—	140
	1.3×10^{19}	66	8.5×10^{-3}	142
Gas phase (Vapor reaction)	5×10^{17} – 1.1×10^{18}	—	2.0×10^{-1} – 2.0×10^{-2}	132
Gas phase (PVT)(b)	$1.2–1.5\times10^{19}$	110	$4–5\times10^{-3}$	95
Flux	2.5×10^{18} – 1.8×10^{19}	45–85	2.9×10^{-2} – 7.7×10^{-3}	149, 182
	3.8×10^{19} – 8.8×10^{20} (c)	48–100	3.4×10^{-3} – 2.8×10^{-4}	149, 182
Salt electrolysis	—	—	10^{-2}–10^{-4}	151
Melt (LASSCGM)	$1–4\times10^{18}$	130–170	$1–3\times10^{-2}$	94, 95, 96, 152

(a)Reducing agent, (b)In_2O_3 thermal decomposition, (c)Doped with Sn

From this overview of the basic electrical properties at RT that are summarized in Table 5.6, it can be concluded that undoped bulk In_2O_3 crystals have a high free electron concentration independent of the growth method, roughly between 10^{18} and 10^{19} cm^{-3}, which can further be increased to the level of 10^{21} cm^{-3} by doping with Sn. On the other hand, the electrical conductivity of flux-grown crystals can be significantly decreased by doping with Mg. The electron mobility is also high at such high free electron concentration, up to 170 cm^2/Vs. The highest electron mobility at similar free electron concentrations was measured for melt-grown In_2O_3 crystals, about twice higher as compared with other growth techniques. This may suggest high structural quality and purity of melt-grown crystals as compared with those obtained by other methods. High electron mobility (up to about 110 cm^2/Vs) was measured even at very high electron concentrations exceeding 10^{19} cm^{-3}, which is almost twice higher as compared with Ga_2O_3 at such free electron concentration. Moreover, very high free electron concentration $>10^{20}$ cm^{-3} in bulk In_2O_3 crystals is not achievable for bulk β-Ga_2O_3 crystals in which doping limit is about 10^{19} cm^{-3}.

5.7.2 Temperature-Dependent Electrical Properties

Bulk In_2O_3 samples prepared in a different way, i.e., from the gas phase, flux, and by pressing and sintering In_2O_3 powder, were studied by Weiher [132], Scherer et al. [142], Remeika and Spencer [133], and De Wit [183, 184]. However, the outcome was different, likely due to the insufficient quality of the samples (e.g., polycrystalline), deviation from stoichiometry, and/or different levels of impurities. For instance, Weiher [132] reported for In_2O_3 samples obtained from the gas phase a decrease in the conductivity between RT and about 800°C, and its rapid increase in a temperature range of 800–1100°C.

The resistivity, free electron concentration, and Hall mobility versus temperature of as-grown, oxygen-, and hydrogen-annealed melt-grown In_2O_3 crystals are shown Fig. 5.32a, Fig. 5.32b, and Fig. 5.32c, respectively [185]. The resistivity ρ and free electron concentration n_e behavior both indicate a typical semiconducting behavior; however, the temperature dependence of as-grown and hydrogen-annealed samples is rather weak and similar to each other, due to high $n_e = 1\text{–}3 \times 10^{18}$ cm^{-3}; therefore, they are close to the degenerate state. The dependence is much stronger in the case of oxygen-annealed sample, which resulted in a decrease in n_e to the level of about 10^{17} cm^{-3}, and an increase in ρ by almost two orders of magnitude.

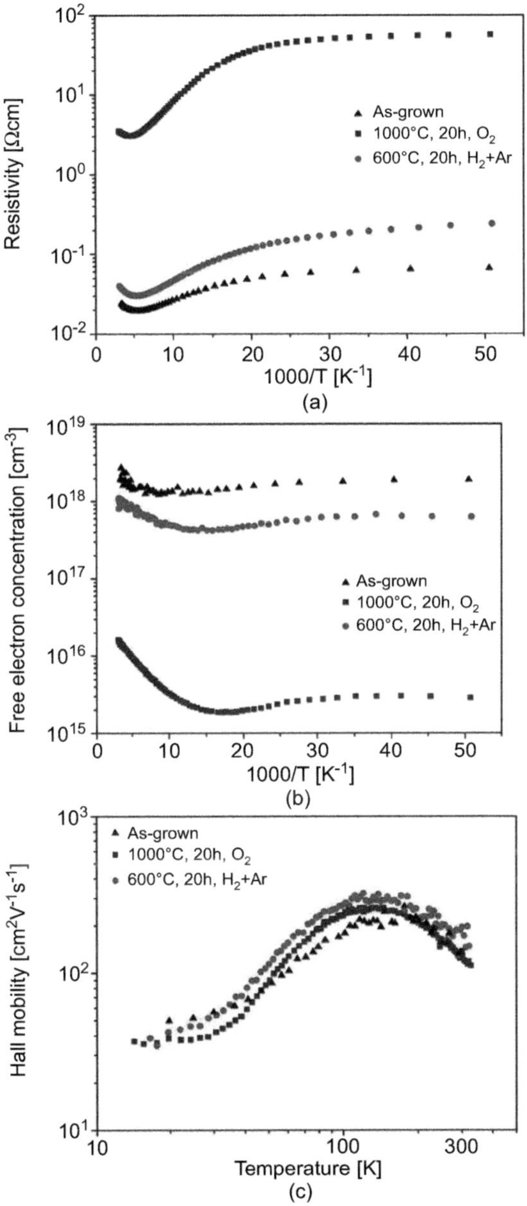

Figure 5.32 Resistivity (a), free electron concentration (b), and Hall mobility (c) of bulk In_2O_3 single crystals obtained by the LASSCGM. Triangles, squares, and circles relate to samples: as-grown and annealed in oxidizing (1200°C, 20 h, O_2) and reducing (600°C, 10h, H_2 + Ar) atmospheres, respectively [185].

The Hall mobility of the as-grown sample at RT (Fig. 5.32c) is about 130–140 cm^2/Vs, which increases to its maximum value of about 270 cm^2/Vs at about 175 K. Below that temperature, scattering at ionized impurities dominates, while above that temperature, scattering at phonons. An oxygen-annealed sample shows at RT a similar Hall mobility as the as-grown one, but hydrogen-annealed sample reveals a bit higher values at RT, about 200 cm^2/Vs. Moreover, the maximum Hall mobility of both annealed samples shifts to lower temperatures of 120–140 K, with a maximum of 320 cm^2/Vs for the hydrogen-annealed sample. For the samples obtained from the gas phase, the maximum of the Hall mobility was measured in the range of 100–150 K [132] and 180 K [142].

5.7.3 Impact of Annealing on Electrical Properties

The electrical properties of melt-grown bulk In_2O_3 crystals can be easily tuned by post-growth heat treatment (annealing) as a function of annealing atmosphere, temperature (Fig. 5.33), and time at oxidizing (Fig. 5.34) and reducing atmosphere (Fig. 5.35). After annealing in O_2, or in any other oxidizing atmosphere for 20 h, the free electron concentration gradually decreases with annealing temperature, in total by about one order of magnitude, from about 2×10^{18} to about 2×10^{17} cm^{-3} (Fig. 5.33a). The fastest decrease in the free electron concentration takes place in samples annealed in the temperature range of 600 to 1000°C, above which saturation is observed. The Hall mobility of free electrons gradually increases with annealing temperature and reaches its maximum after annealing at 800°C (Fig. 5.33b). The mobility increase is about 25–30% (from about 140 to about 180 cm^2/Vs), but occasionally a higher Hall mobility was measured, about 200 cm^2/Vs. After reaching its maximum, the Hall mobility decreases by about 50% after annealing at 1400°C, which can be linked to a deterioration of the crystal perfection by the thermal decomposition. The corresponding resistivity (Fig. 5.33c) changes in an inverse manner with respect to the electron concentration and increases by almost two orders of magnitude, from about 0.01 to about 0.8 Ωcm.

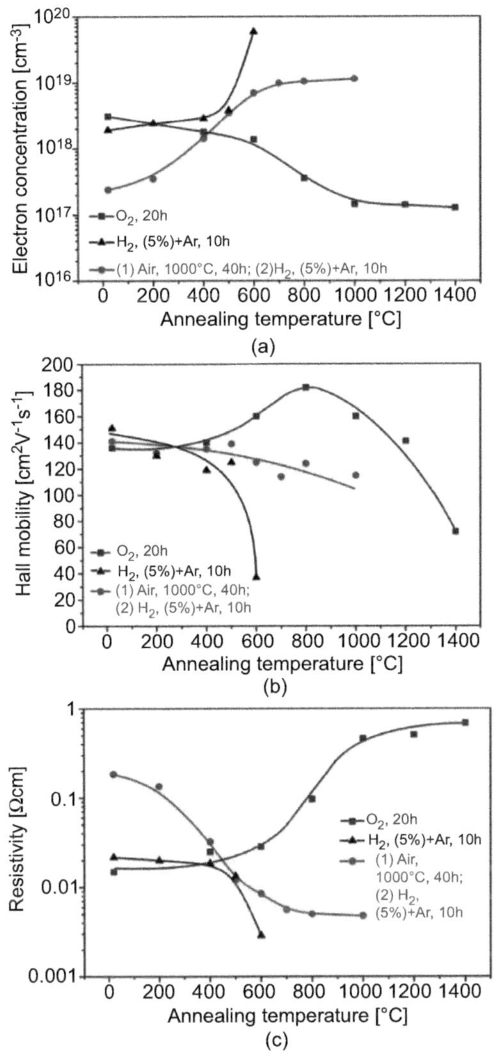

Figure 5.33 Free electron concentration (a), Hall mobility (b), and resistivity (c) of melt-grown In_2O_3 crystal samples (LASSCGM) as a function of annealing temperature at fixed annealing time (depending on annealing atmosphere). The annealing was done in oxidizing (O_2), reducing (H_2 + Ar), and first oxidizing (to oxidize In nanoparticles) and then in reducing (air → H_2 + Ar) conditions. Reproduced (with minor modifications) from Ref. [152], Copyright 2013, with permission from The Royal Society of Chemistry.

Annealing as-grown In_2O_3 crystals in the presence of H_2 (reducing agent) for 10 h produces opposite results to the annealing in the presence of O_2, i.e., the free electron concentration increases with annealing temperature, in total by more than one order of magnitude, from about 2×10^{18} to about 6×10^{19} cm^{-3} (Fig. 5.33a). A very fast increase in the electron concentration takes place after annealing in the temperature range between 500 and 600°C. The annealing temperature of 600°C in the atmosphere containing H_2 is the upper limit for as-grown crystals, due to strong decomposition of the crystals. The crystals remain, however, black and opaque in the visible spectrum. The Hall mobility significantly decreases after such annealing from about 140 cm^2/Vs in the case of as-grown crystals to about 40 cm^2/Vs after annealing at 600°C (Fig. 5.33b). The resistivity decreases inversely to the free electron concentration by one order of magnitude, from about 0.02 to about 0.002 Ωcm (Fig. 5.33c).

To obtain a high free electron concentration combined with good transparency, a combined annealing can be utilized: (i) in an oxidizing atmosphere (e.g., at 1000°C for 20–40 h) to minimize the concentration of metallic In particles (and, therefore, to improve the transmittance), and next (ii) in the presence of H_2 to increase the free electron concentration while keeping good transparency. Such combined annealing enables fine tuning of the free electron concentration in the range of 2×10^{17} and 1×10^{19} cm^{-3} (Fig. 5.33a). In this case, the thermal stability of the crystals also increases, but there is saturation in the free electron concentration after annealing in the presence of hydrogen at 700°C or above. The corresponding Hall mobility gradually decreases with the annealing temperature, but not as much as in the case of as-grown crystals annealed in the presence of H_2. It decreases by about 25–30% in the whole annealing range up to 1000°C (Fig. 5.33b). It is worth to note that annealing in the presence of H_2 and O_2 is fully revisable, i.e., it is possible to switch the free electron concentration between low (2×10^{17} cm^{-3}) and high (10^{18}–10^{19} cm^{-3}) levels by alternating annealing in the presence of hydrogen and oxygen. Therefore, In_2O_3 is a good material for high-temperature gas sensing.

An impact of annealing time on the electrical properties of melt-grown In_2O_3 crystals is shown in Fig. 5.34 and Fig. 5.35 for an oxidizing (O_2) and reducing (H_2 + Ar) atmosphere, respectively. In the case of the oxidizing atmosphere, annealing at 800°C leads to a gradual decrease in the free electron concentration with annealing time up to 12 h, above which saturation takes place (Fig. 5.34a). The Hall mobility increases almost linearly with annealing time reaching the maximum value of 180–190 cm^2/Vs (Fig. 5.34b). Occasionally, a yet higher Hall mobility of 200 cm^2/Vs was measured (Galazka et al. [185]). The resistivity increases inversely to the free electron concentration, in total by about one order of magnitude for the annealing time of 12–20 h (Fig. 5.34c). At fixed but higher temperature, the annealing time has similar impact as annealing temperature at fixed, but long annealing time.

After annealing in the presence of H_2 at 400°C of previously annealed samples in the presence of O_2, the free electron concentration increases with annealing time relatively fast, but 20 h annealing seems to be a turning point, above which the free electron concentration increases slowly up to 80 h annealing time (Fig. 5.35a). At this annealing temperature, the Hall mobility decreases by 20–25% after annealing time of 20–40 h, and then it seems to slightly increase with annealing time (Fig. 5.35b). The resistivity decreases fast by one order of magnitude for annealing time up to 20 h and next decreases slowly for longer annealing time (Fig. 5.35c).

By fixing annealing temperature that is lower than the temperature causing decomposition and using either oxidizing or reducing atmosphere, it is possible to controllably tune the electrical properties by adjusting annealing time. Annealing offers three degrees of freedom to modify the electrical properties of melt-grown In_2O_3 in a broad spectrum, which is atmosphere, temperature, and time.

In terms of In_2O_3 crystals obtained from the gas phase by CVT, the electrical properties were affected by annealing in oxygen at 1000°C, i.e., the resistivity was a function of annealing time (scale of days) [130]. However, the PVT-grown crystals were not affected by annealing in oxidizing conditions in the temperature range of 700–1400°C [95].

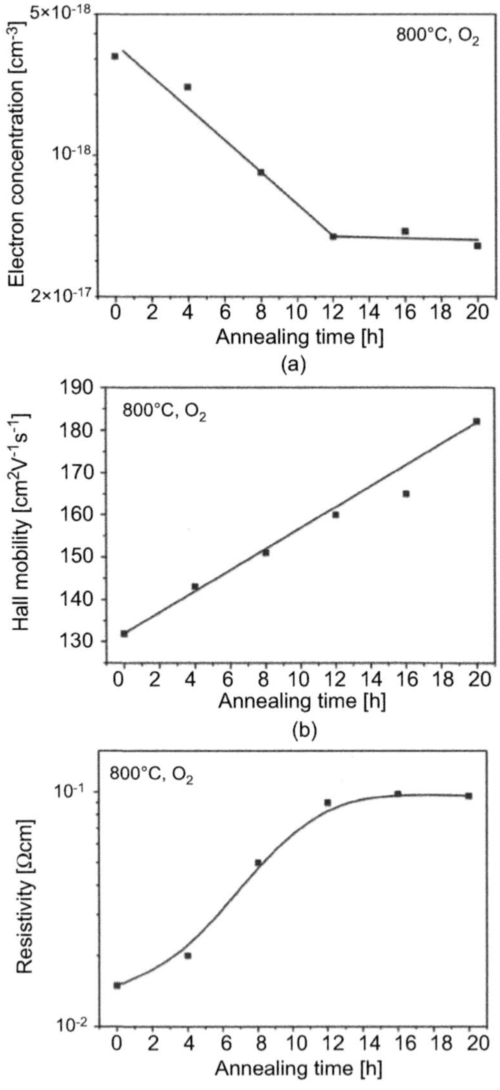

Figure 5.34 Free electron concentration (a), reproduced (with minor modifications) from Ref. [152], Copyright 2013, with permission from The Royal Society of Chemistry; Hall mobility (b) [185]; and resistivity (c) [185] of melt-grown In_2O_3 crystal samples (LASSCGM) as a function of annealing time at fixed temperature (800°C) in the oxidizing atmosphere.

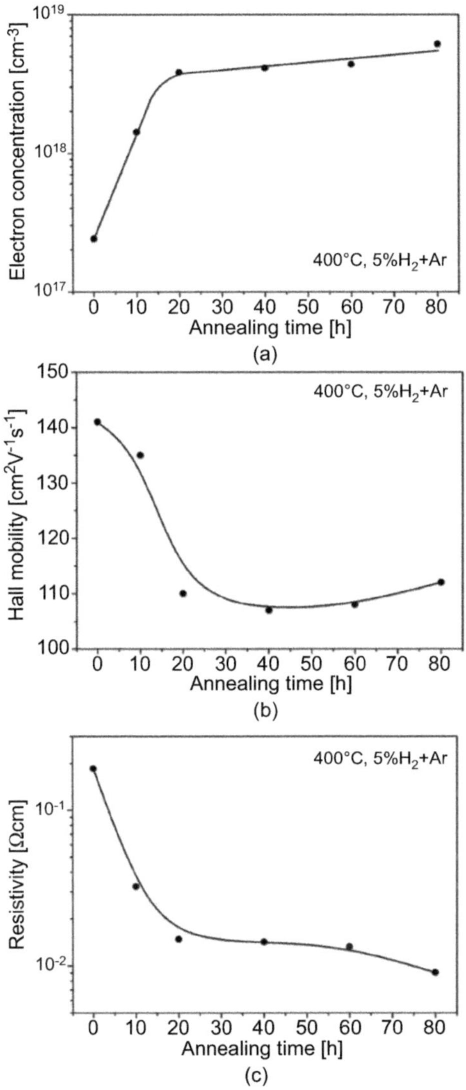

Figure 5.35 Free electron concentration (a), adapted from Ref. [152], Copyright 2013, with permission from The Royal Society of Chemistry; Hall mobility (b) [185]; and resistivity (c) [185] of melt-grown In_2O_3 crystal samples (LASSCGM) as a function of annealing time at fixed temperature (400°C) in the reducing atmosphere (H_2+Ar). Before annealing in the reducing atmosphere, the samples were first annealed in air at 1000°C for 40 h to oxidize In nanoparticles.

5.7.4 Surface Electron Accumulation Layer

In the literature, there was a debate whether In_2O_3 has the surface electron accumulation layer (SEAL) or not, as the conclusions were unclear. This could arise from not-too-high structural quality of the In_2O_3 films, which are frequently investigated, and different preparation conditions. The availability of melt-grown In_2O_3 crystals enabled resolution of the SEAL issue. The results by Nagata et al. [186], who used soft and hard X-ray photoelectron spectroscopy (SXPES and HXPES), clearly showed that chemo-mechanical polished (CMP), (001)-oriented wafers, whether as-grown or previously annealed in an oxidizing atmosphere, reveal SEAL. This is visible in Fig. 5.36, illustrating a downward band bending by about 1 eV, and the corresponding high electron concentration at the surface of about 10^{20} cm^{-3}, peaking at about 0.6 nm below the surface. The surface electron concentration is about one or two orders of magnitude higher than that in the bulk of as-grown and annealed crystals, respectively. The corresponding surface electron density is 2.6×10^{13} cm^2. Further, Nagata et al. [187] also showed a similar behavior of in situ cleaved (111) surface; however, it revealed a smaller band bending of about 0.76 eV with a smaller surface free electron concentration of 6×10^{19} cm^{-3} after cleaving in high vacuum of 5×10^{-7} mbar. Theoretical work of Walsh [188] points to the formation of delocalized oxygen vacancies on the (111) surface with low formation energy comparing to the bulk that can exceed the Mott criterion (metal–insulator transition). The downward band bending in In_2O_3 films was also observed by Körber et al. [119], and King et al. [189, 190] (range 0.35–0.46$\pm$0.16 eV for both (111) and (001) surfaces). Bierwagen et al. [191] and Berthold et al. [192] concluded that the SEAL observed in MBE-grown layers can be depleted with oxygen plasma treatment, and moreover, SEAL can be recovered by annealing in vacuum. Zhang et al. [193] claim on the basis of ARPES measurements that high conductivity of In_2O_3 (111) is the result of the 2D electron gas (2DEG) formed by double-ionized oxygen vacancies on the surface, which have much lower formation energy as compared to the bulk. The formation of 2DEG on the In_2O_3 (111) surface covered with a very thin In film thereon was also observed by Nazarzahdemoafi et al. [194]. In flux-grown crystals, Hagleitner et al. [150] concluded a downward band bending

(0.4 eV) and a small upward band bending for reduced and oxidized surfaces, respectively. These results are inconsistent with those for melt-grown In_2O_3 discussed above, where both reducing and oxidizing conditions resulted in downward band bending. It should be stressed that the flux-grown crystals contained a high density of impurities and were of not high structural quality.

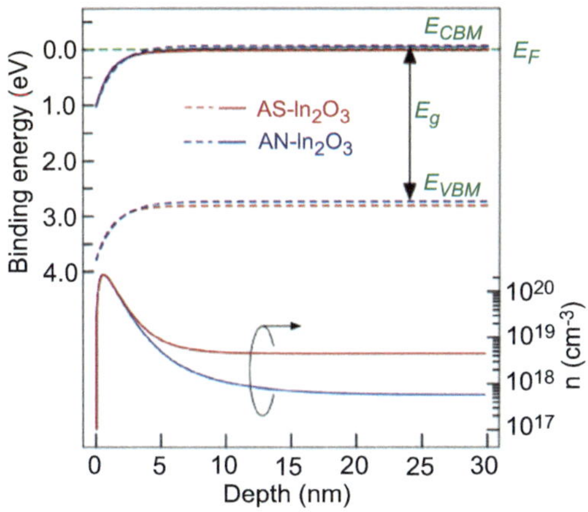

Figure 5.36 Binding energy (upper panel) and free electron concentration (lower panel) of air-cleaved melt-grown In_2O_3 single crystals (LASSCGM) as a function of the sample depth from its surface indicated by "0." AS and AN refer to as-grown and air-annealed samples, respectively. Reproduced with permission from Ref. [186], Copyright 2017, The Japan Society of Applied Physics.

Another study by Scherer et al. [122] proved that when cleaving melt-grown In_2O_3 crystals in ultrahigh vacuum (10^{-11} mbar), SEAL is not formed. The same relates to CVT-grown In_2O_3 crystals. The results obtained so far do not clarify the origin of SEAL, whether it is caused by oxygen vacancies on the surface or by adsorbents from the surrounding.

With SEAL, Schottky contacts are not possible; only ohmic contacts can be fabricated. Preparing Schottky contacts requires a freshly cleaved surface of In_2O_3 in ultrahigh vacuum. Indeed, ultrahigh vacuum was the key point to fabricate Schottky barrier between melt-grown In_2O_3 and Ag, as demonstrated by Nazarzadehmoafi et al. [195]. The Schottky barrier height of Ag with In_2O_3 is 0.22 ± 0.08 eV [195]. Schottky barriers could be fabricated on epitaxial In_2O_3 films (PA-MBE) with Au, Pd, and Pt [196].

5.7.5 Effective Electron Mass and Electron Mobility

In metrology of electrical properties and operation of electronic devices, an important property of a material is the effective electron mass m_e*. Ellipsometric measurements by Feneberg et al. [173] of melt-grown In_2O_3 and thin films in a broad range of free electron concentrations enabled determination of the effective mass as the function of the free electron concentration.

The experimental results (scatters) are shown in Fig. 5.37 together with the theoretical curve (dashed line) approximated with the following equation:

$$m_e^* = \sqrt{m_0^{*2} + 2C\hbar^2 m_0^*(3\pi^2 n_e)^{2/3}} \qquad (5.30)$$

where m_0* is the effective electron mass at the CBM (0.18 m_0), C is an empirical material system-dependent parameter (0.5 eV^{-1}), $\hbar$ is the Planck constant, and n_e is the free electron concentration.

The relation is non-parabolic, and the effective electron mass increases by a factor of over 2 for the electron concentration at the level of 10^{21} cm^{-3} as compared to the electron concentration of about 10^{18} cm^{-3}. Low effective electron mass enables high electron mobility.

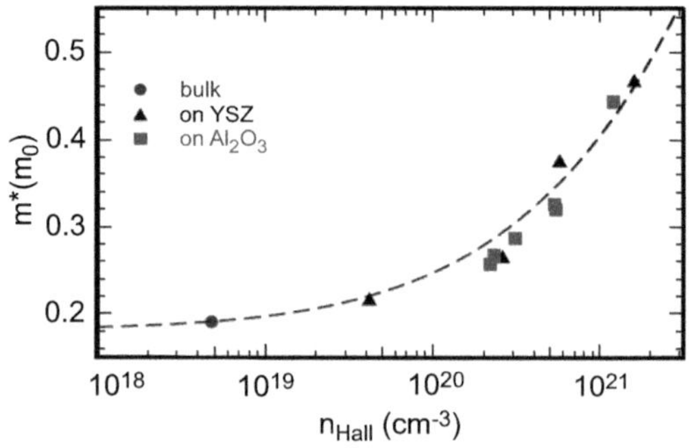

Figure 5.37 Effective electron mass of a bulk In_2O_3 crystal (LASSCGM) and epilayers as a function of the free electron concentration. The epilayers were grown on yttrium-stabilized zirconia (YSZ) and Al_2O_3 substrates. The free electron concentration was measured by the Hall effect, while the effective electron mass was gathered from ellipsometric measurements. Reprinted with permission from Ref. [173], Copyright 2016, American Physical Society.

According to theoretical and empirical data on epitaxial In_2O_3 by Preissler et al. [165] and Bierwagen and Speck [197], the drift and Hall mobilities at RT are limited to 190 cm^2/Vs and 270–274 cm^2/Vs, respectively, by polar optical phonon scattering. The maximum Hall mobility of melt-grown bulk In_2O_3 crystals of 200 and 320 cm^2/Vs was measured at RT and at about 130 K, respectively. The maximum Hall mobilities of high structural thin films of In_2O_3 grown by the PA-MBE were 226 and 1290 cm^2/Vs at RT and at about 100 K, respectively [197].

5.7.6 Ionization Energy of Shallow Donors

Weiher [132] concluded the donor ionization energy of 28–8.5 meV for an electron concentration range of 4.95×10^{17}–1.09×10^{18} cm^{-3} in bulk In_2O_3 crystals obtained from the gas phase. The approximation of the ionization energy to zero electron concentration yields 93 meV.

A temperature-dependent Hall effect analysis of a number of melt-grown bulk In_2O_3 crystals (as-grown and annealed at different atmospheres) by Galazka et al. [185] revealed the ionization energy of isolated donors of 38.6 meV, which is much smaller than that obtained by Weiher [132]. Moreover, the donor ionization energy values at the electron concentration at the level of 2×10^{17}, 1.1×10^{18}, and 2×10^{18} cm^{-3} are of 20, 12, and 8 meV, respectively.

5.7.7 Conclusion

The experimental data of electrical properties revealed *n*-type conductivity of bulk In_2O_3 crystals with a high free electron concentration at RT, typically in a range of 10^{18}–10^{19} cm^{-3} without any intentional doping or post-growth heat treatments. Intentional doping with Sn or annealing in the presence of H_2 increases the free electron concentration by one or even two orders of magnitude. On the other hand, annealing in the presence of oxygen or doping with Mg decreases the free electron concentration by about one order of magnitude or may even turn the crystals into a high resistivity (Mg). Melt-grown In_2O_3 crystals always show a semiconducting behavior, with high sensitivity to oxidizing and reducing atmospheres that induce capability of tuning the free electron concentration within almost three orders of magnitude (10^{17}–6×10^{19} cm^{-3}). It makes In_2O_3 a very good high-temperature gas sensor. Despite high free electron concentration, the Hall mobility is quite high, which may approach the value of 200 cm^2/Vs, with a limit of about 270 cm^2/Vs at RT. The high electron mobility is the result of a high dispersion of the CBM and low effective electron mass (0.18–0.2 at $n_e = 10^{17}$–10^{19} cm^{-3}). The ionization energy of isolated donors is about 38.6 meV, which decreases to about 8 meV at $n_e > 10^{18}$ cm^{-3}. Bulk In_2O_3 crystals show SEAL causing a downward band bending of 0.35–1 eV. The SEAL is not formed in ultrahigh vacuum and in oxygen plasma.

A summary of electrical properties of bulk In_2O_3 single crystals is shown in Table 5.7.

Table 5.7 Basic electrical properties of bulk In_2O_3 crystals

Property	Unit	Value	Reference
Free electron concentration n_e at RT	cm^{-3}	Gas phase: 5×10^{17} – 1.5×10^{19}, 5×10^{19} – 1.0×10^{21} (a)	132, 140, 142, 95
		Flux: 2.5×10^{18} – 1.8×10^{19}, 3.8×10^{19} – 8.8×10^{20} (a)	149, 182
		Melt: 1–4×10^{18}, 1–7×10^{17} (b), 3×10^{18}–6×10^{19} (c)	95, 96, 152
Maximum Hall mobility at RT	cm^2/Vs	Gas phase: 110	95
		Flux: 100	149, 182
		Melt: 200	152, 185
Hall mobility limit at RT	cm^2/Vs	270–274	165, 197
Effective electron mass at $n_e = 1\times10^{18}$–2×10^{21}		0.18–0.5	173
Ionization energy of isolated donors	meV	≈38.6	185

(Continued)

Table 5.7 (*Continued*)

Property	Unit	Value	Reference
Ionization energy of shallow donors at n_e = 2×10^{17}–2×10^{18}	meV	20–8	185
Surface electron accumulation layer (SEAL)		Present in ambient and high vacuum. Not present in ultrahigh vacuum, or in oxygen plasma.	186, 187, 191, 192
Downward band bending by SEAL	eV	0.35–1	186, 187, 189, 190, 150

[(a)]doped with Sn, [(b)]annealed in the presence of O_2, [(c)]annealed in the presence of H_2

5.8 Key Points

- Large bulk In_2O_3 single crystals with a volume up to 10 cm^3 were grown directly from the melt by using a novel growth technique named levitation-assisted self-seeding crystal growth method (LASSCGM).
- In_2O_3 is an indirect wide bandgap (2.72 eV) oxide semiconductor with the SEAL at ambient conditions.
- In_2O_3 has a steep absorption edge (at 440 nm) and is transparent in the visible and near-infrared spectral regions, the latter being affected by the free carrier absorption.
- The free electron concentration in bulk crystals at RT can be tuned in a range of about 10^{17}–10^{20} cm^{-3}. The Hall mobility is high, typically in a range of 130–200 cm^2/Vs.

5.9 Summary

Table 5.8 Main properties of bulk In_2O_3 single crystals and growth techniques

	In_2O_3	
Property	**Value or description**	**Reference**
	Crystal system and structure	
Lattice parameter [Å]	a = 10.117 (crystal – flux)	101
	a = 10.12121 (crystal – melt)	158
System, structure	Cubic, bixbyite	101
Space group	$Ia\bar{3}$	101
Cleavage plane	{111}	
	Bulk crystal preparation	
Maximum size: diameter × length [mm×mm], or width × length × thickness [mm×mm×mm]	LASSCGM: 36×12	96
	Flux: 10×10×1, 5×5×3	133, 146
	PVT: 35×3	95
	CVT: 3×3×3	138
	Thermal and mechanical properties	
Solid density [g/cm³]	7.12	101
Melting point [°C]	1950±30	95
Thermal conductivity at RT [W/mK]	13.09	159
Thermal diffusivity at RT [J/molK]	104.2	162
Debye temperature [K]	700	165
Thermal expansion coefficient for lattice parameter with respect to RT [K^{-1}]	8.66×10^{-6}	158
Elastic stiffness coefficients [GPa]	c_{11} = 297.8[(a)] c_{12} = 141.8[(a)] c_{44} = 76.4[(a)]	109
Bulk modulus B_0 [GPa]	193.8[(b)]/192.7	109/164
Young's modulus E [GPa]	206.3[(b)]	109
Shear modulus G [GPa]	76.4[(b)]	109
Poisson's ratio ν	0.323[(b)]	109

Optical properties		
Bandgap [eV]	2.63–2.78 (indirect, fundamental) 3.71–3.8 (direct)	120, 130, 167, 142 120, 135, 176
Bandgap shrinkage at 300–1300°C [meV/K]	−1.34	167
Luminescence band at RT [eV]	1.97	152, 177
Relative static dielectric constant	8.9	168
High-frequency dielectric constant	4.08±0.02	173
Refractive index at 1.55–2.48 eV (800–500 nm)	2.043–2.122	175
Electrical properties		
Tunability of the free electron concentration at RT [cm^{-3}]	10^{17}–10^{21}	95, 96, 152, 132, 140, 142, 149, 182
Maximum experimental electron mobility at RT [cm^2/Vs]	200	185
Maximum theoretical electron mobility at RT [cm^2/Vs]	≈270–274	165, 197
Donor activation energy [meV]	38.6 for isolated donors 20–8.5 for $n = 2\times10^{17}$–2×10^{18} cm^{-3}	185
Effective electron mass for a moderate free electron concentration (10^{17}–10^{18} cm^{-3})	0.18	173
Downward band bending [eV]	0.35–1	186, 187, 189, 190, 150

[a]theoretical value, [b]calculated from Eqs. (5.23–5.26), and theoretical data from Ref. [109]

References

1. I. Hamberg, A. Hjortsberg, C. G. Granqvist; *Appl. Phys. Lett.* **40** (1982), 362–364.
2. I. Hamberg, C. G. Granqvist; *Appl. Phys. Lett.* **44** (1984), 721–722.
3. I. Hamberg, C. G. Granqvist; *J. Appl. Phys.* **60** (1986), R123–R159.
4. S. Kulaszewicz, W. Jarmoc, I. Lasocka, Z. Lasocki, C. Michalski, K. Turowska; *Thin Solid Films* **418** (1987), L55–L57.
5. C. G. Granqvist, A. Hultåker; *Thin Solid Films* **411** (2002), 1–5.
6. T. Mizrah, D. Adler; *Appl. Phys. Lett.* **29** (1976), 682–684.
7. G. Cheek, N. Inoue, S. Goodnick, A. Genis, C. Wilmsen, J. B. DuBow; *Appl. Phys. Lett.* **33** (1978), 643–645.
8. K. Ito, T. Nakazawa; *Surf. Sci.* **86** (1979), 492–497.
9. S. P. Singh, A. K. Saxena, L. M. Tiwari, O. P. Agnihotri; *Thin Solid Films;* **127** (1985), 77–84.
10. T. Koida, H. Sai, M. Kondo; *Thin Solid Films* **518** (2010), 2930–2933.
11. B. V. Mistry, P. Bhatt, K. H. Bhavsar, S. J. Trivedi, U. N. Trivedi, U. S. Joshi; *Thin Solid Films* **519** (2011), 3840–3843.
12. Dhananjay, S.-S. Cheng, C.-Y. Yang, C.-W. Ou, Y.-C. Chuang, M. C. Wu, Chih-Wei Chu; *J. Phys. D: Appl. Phys.* **41** (2008), 092006.
13. L.-C. Chen, C.-H. Tien, W.-C. Liao; *J. Phys. D: Appl. Phys.* **44** (2011), 165101.
14. M. T. Hardy, C. O. Holder, D. F. Feezell, S. Nakamura, J. S. Speck, D. A. Cohen, S. P. DenBaars; *Appl. Phys. Lett.* **103** (2013), 081103.
15. S. N. Malchenko, Y. N. Lychkovsky, M. V. Baykov; *Sens. Actuators B* **13–14** (1993), 159–161.
16. A. Gurlo, N. Bârsan, M. Ivanovskaya, U. Weimar, W. Göpel; *Sens. Actuators B* **47** (1998), 92–99.
17. M. Z Atashbar, B. Gong, H. T Sun, W. Wlodarski, R. Lamb; *Thin Solid Films* **354** (1999), 222–226.
18. S.-R. Kim, H.-K. Hong, C. H. Kwon, D. H. Yun, K. Lee, Y. K. Sung; *Sens. Actuators B* **66** (2000), 59–62.
19. M. Ivanovskaya, A. Gurlo, P. Bogdanov; *Sens. Actuators B* **77** (2001), 264–267.
20. G. Korotcenkov, V. Brinzari, A. Cerneavschi, M. Ivanov, A. Cornet, J. Morante, A. Cabot, J. Arbiol; *Sens. Actuators B* **98** (2004), 122–129.

21. G. Korotcenkov, V. Brinzari, V. Golovanov, A. Cerneavschi, V. Matolin, A. Tadd; *Appl. Surf. Sci.* **227** (2004), 122–131.

22. G. Korotcenkov, M. Ivanov, I. Blinov, J. R. Stetter; *Thin Solid Films* **515** (2007), 3987–3996.

23. P. Guo, H. Pan; *Sens. Actuators B* **114** (2006), 762–767.

24. L. Francioso, M. Russo, A. M. Taurino, P. Siciliano; *Sens. Actuators B* **119** (2006), 159–166.

25. M. Ali, Ch. Y. Wang, C.-C. Röhlig, V. Cimalla, Th. Stauden, O. Ambacher; *Sens. Actuators B* **129** (2008), 467–472.

26. S. Kannan, H. Steinebach, L. Rieth, F. Solzbacher; *Sens. Actuators B* **148** (2010), 126–134.

27. S. Kannan, L. Rieth, F. Solzbacher; *Sens. Actuators B* **149** (2010), 8–19.

28. L. Filipovic, Siegfried Selberherr; *Sensors* **15** (2015), 7206–7227.

29. T. Berthold, S. Katzer, J. Rombach, S. Krischok, O. Bierwagen, M. Himmerlich; *Phys. Status Solidi (b)* **255** (2018), 1700324.

30. M. Sasaki, K. Yasui, S. Kohiki, H. Deguchi, S. Matsushima, M. Oku, T. Shishido; *J. Alloy. Compd.* **334** (2002), 205–210.

31. S. Kohiki, M. Sasaki, Y. Murakawa, K. Hori, K. Okada, H. Shimooka, T. Tajiri, H. Deguchi, S. Matsushima, M. Oku, T. Shishido, M. Arai, M. Mitome, Y. Bando; *Thin Solid Films* **505** (2006), 122–125.

32. G. Peleckis, X. L. Wang, S. X. Dou; *J. Magn. Magn. Mater.* **301** (2006), 308–311.

33. X. Li, C. Xia, G. Pei, X. He; *J. Phys. Chem. Solids* **68** (2007), 1836–1840.

34. H. Raebiger, S. Lany, A. Zunger; *Phys. Rev. Lett.* **101** (2008), 027203.

35. R. K. Singhal, A. Samariya, S. Kumar, S. C. Sharma, Y. T. Xing, U. P. Deshpande, T. Shripathi, E. Saitovitch; *Appl. Surf. Sci.* **257** (2010), 1053–1057.

36. R. K. Gupta, K. Ghosh, P. K. Kahol; *Mater. Lett.* **64** (2010), 2022–2024.

37. D. Bérardan, E. Guilmeau, D. Pelloquin; *J. Magn. Magn. Mater.* **320** (2008), 983–989.

38. S. H. Babu, S.Kaleemulla, N. M. Rao, C. Krishnamoorthi; *J. Magn. Magn. Mater.* **416** (2016), 66–74.

39. A. Gurlo, M. Ivanovskaya, A. Pfau, U. Weimar, W. Göpel; *Thin Solid Films* **307** (1997), 288–293.

40. M. A. Flores-Mendoza, R. Castanedo-Perez, G. Torres-Delgado, J. Márquez Marín, O. Zelaya-Angel; *Thin Solid Films* **517** (2008), 681–685.

41. E. Savarimuthu, K. C. Lalithambika, A. M. E. Raj, L. C. Nehru, S. Ramamurthy, A. Thayumanavan, C. Sanjeeviraja, M. Jayachandran; *J. Phys. Chem. Solids* **68** (2007), 1380–1389.

42. W. Siefert; *Thin Solid Films* **121**(1984), 215–282.

43. L. A. Ryabova, V. S. Salun, I. A. Serbinov; *Thin Solid Films* **92** (1982), 327–332.

44. S. P. Singh, A. Raza, A. K. Sharma, O. P. Agnihotriand, L. M. Tewari; *Thin Solid Films* **105** (1983), 131–138.

45. J. J. Prince, S. Ramamurthy, B. Subramanian, C. Sanjeeviraja, M. Jayachandran; *J. Cryst. Growth* **240** (2002), 142–151.

46. G. Korotcenkov, V. Brinzari, A. Cerneavschi, A. Cornet, J. Morante, A. Cabot, J. Arbiol; *Sens. Actuators B* **84** (2002), 37–42.

47. J.-H. Lee, B.-O. Park; *Surf. Coat. Tech.* **184** (2004), 102–107.

48. S. M. Rozati, T. Ganj; *Renew. Energ.* **29** (2004), 1665–1669.

49. P. Prathap, Y. P. V. Subbaiah, M. Devika, K. T. Ramakrishna Reddy; *Mater. Chem. Phys.* **100** (2006), 375–379.

50. S. Parthiban, E. Elangovan, K. Ramamurthi, R. Martins, E. Fortunato; *Sol. Energy Mater. Sol. Cells* **94** (2010), 406–412.

51. P. Nath, R. F. Bunshah; *Thin Solid Films* **69** (1980), 63–68.

52. P. Natah, R. F. Bunshah, B. M. Basol, O. M. Staffsud; *Thin Solid Films* **72** (1980), 463–468.

53. M. Mizuhashi; *Thin Solid Films* **76** (1981), 97–105.

54. E. B. Ali, H. El Maliki, J. C. Bernede, M. Sahnoun, A. Khelil, O. Saadane; *Mater. Chem. Phys.* **73** (2002), 78–85.

55. S. Kaleemulla, A. S. Reddy, S. Uthanna, P. S. Reddy; *J. Alloy. Compd.* **479** (2009), 589–593.

56. S. Kaleemulla, N. M. Rao, M. G. Joshi, A. S. Reddy, S. Uthanna, P. S. Reddy; *J. Alloy. Compd.* **504** (2010), 351–356.

57. H. Sieber, St. Senz, D. Hesse; *Thin Solid Films* **303** (1997), 216–221.

58. N. Mori, S. Ooki, N. Masubuchi, A. Tanaka, M. Kogoma, T. Ito; *Thin Solid Films* **411** (2002), 6–11.

59. A. S. A. C. Diniz; *Renew. Energ.* **36** (2011), 1153–1165.

60. M. Fujinaka, A. A. Berezin; *Thin Solid Films* **101** (1983), 7–10.

61. A. J. P. Theuwissen, G. J. Declerck; *Thin Solid Films* **121** (1984), 109–119.

62. D. Mergela, M. Schenkela, M. Ghebrea, M. Sulkowski; *Thin Solid Films* **392** (2001), 91–97.

63. Y. Gassenbauer, A. Klein; *Solid State Ion.* **173** (2004), 141–145.

64. Z. Qiao, R. Latz, D. Mergel; *Thin Solid Films* **466** (2004), 250–258.

65. L. Lin, F. Lai, Y. Qu, R. Gai, Z. Huang; *Mat. Sci. Eng. B* **138** (2007), 166–171.

66. Y. Sato, M. Taketomo, N. Ito, Y. Shigesato; *Thin Solid Films* **516** (2008), 5868–5871.

67. I. Hotovy, J. Pezoldt, M. Kadlecikova, T. Kups, L. Spiess, J. Breza, E. Sakalauskas, R. Goldhahn, V. Rehacek; *Thin Solid Films* **518** (2010), 4508–4511.

68. A. Rogozin, M. Vinnichenko, N. Shevchenko, U. Kreissig, A. Kolitsch, W. Möller; *Scr. Mater.* **60** (2009), 199–202.

69. A. Kono, Z. Feng, N. Nouchi, F. Shoji; *Vacuum* **83** (2009), 548–551.

70. F. Yang, J. Ma, X. Feng, L. Kong; *J. Cryst. Growth* **310** (2008), 4054–4057.

71. L. Kong, J. Ma, F. Yang, Z. Zhu, C. Luan, H. Xiao; *Appl. Surf. Sci.* **257** (2010), 518–522.

72. R. Togashi, S. Numata, M. Hayashida, T. Suga, K. Goto, A. Kuramata, S. Yamakoshi, P. Paskov, B. Monemar, Y. Kumagai; *Jpn. J. Appl. Phys.* **55** (2016), 1202B3.

73. N. B. Ukah, R. K. Gupta, P. K. Kahol, K. Ghosh; *Appl. Surf. Sci.* **255** (2009), 9420–9424.

74. R. K. Gupta, K. Ghosh, R. Patel, S. R. Mishra, P. K. Kahol; *Mater. Chem. Phys.* **112** (2008), 136–139.

75. R. K. Gupta, K. Ghosh, S. R. Mishra, P. K. Kahol; *Appl. Surf. Sci.* **254** (2008), 4018–4023.

76. R. K. Gupta, K. Ghosh, S. R. Mishra, P. K. Kahol; *Mater. Lett.* **62** (2008), 1033–1035.

77. P. F. Newhouse, C.-H. Park, and D. A. Keszler, J. Tate, P. S. Nyholm; *Appl. Phys. Lett.* **87** (2005), 112108.

78. S. H. Kim, N.-M. Park, T. Y. Kim, G. Y. Sung; *Thin Solid Films* **475** (2005), 262–266.

79. N. Tripathi, S. Rath, V. Ganesan, R. J. Choudhary; *Appl. Surf. Sci.* **256** (2010), 7091–7095.

80. F. Schmidt, D. Splith, S. Müller, H. von Wenckstern, M. Grundmann; *Phys. Status Solidi (b)* **252** (2015), 2304–2308.

81. A. Bourlange, D. J. Payne, R. G. Palgrave, J. S. Foord, R. G. Egdell, R. M. J. Jacobs, A. Schertel, J. L. Hutchison, P. J. Dobson; *Thin Solid Films* **517** (2009), 4286–4294.

82. E. H. Morales, U. Diebold; *Appl. Phys. Lett.* **95** (2009), 253105.
83. O. Bierwagen, M. E. White, M.-Y. Tsai, J. S. Speck; *Appl. Phys. Lett.* **95** (2009), 262105.
84. O. Bierwagen, J. S. Speck; *J. Appl. Phys.* **107** (2010), 113519.
85. O. Bierwagen, J. S. Speck; *Appl. Phys. Lett.* **97** (2010), 072103.
86. O. Bierwagen, J. S. Speck; *Phys. Status Solidi* (*a*) (**2013**), 1–6.
87. O. Bierwagen, J. S. Speck; *Appl. Phys. Lett.* **101** (2012), 102107.
88. O. Bierwagen, J. S. Speck, T. Nagata, T. Chikyow, Y. Yamashita, H. Yoshikawa, K. Kobayashi; *Appl. Phys. Lett.* **98** (2011), 172101.
89. P. Vogt, O. Bierwagen; *Appl. Phys. Lett.* **109** (2016), 062103.
90. P. Vogt, A. Trampert, M. Ramsteiner, O. Bierwagen; *Phys. Status Solidi* (*a*) **212** (2015), 1433–1439.
91. O. Bierwagen, J. Rombach, J. S. Speck; *J. Phys.: Condens. Matter.* **28** (2016), 224006.
92. A. Papadogianni, L. Kirste, O. Bierwagen; *Appl. Phys. Lett.* **111** (2017), 262103.
93. O. Bierwagen; *Semicond. Sci. Technol.* **30** (2015), 024001.
94. Z. Galazka, R. Uecker, R. Fornari; PCT/EP2012/057447 (WO 2013/159808) (2012).
95. Z. Galazka, R. Uecker, K. Irmscher, D. Schulz, D. Klimm, M. Albrecht, M. Pietsch, S. Ganschow, A. Kwasniewski, R. Fornari; *J. Cryst. Growth* **362** (2013), 349–352.
96. Z. Galazka, R. Uecker, R. Fornari; *J. Cryst. Growth* **388** (2014), 61–69.
97. S. Sadofev, Y. J. Cho, O. Brandt, M. Ramsteiner, R. Calarco, H. Riechert, S. C. Erwin, Z. Galazka, M. Korytov, M. Albrecht, R. Uecker, R. Fornari; *Appl. Phys. Lett.* **101** (2012), 172102.
98. Y. J. Cho, S. Sadofev, S. Fernández-Garrido, R. Calarco, H. Riechert, Z. Galazka, R. Uecker, O. Brandt; *Appl. Surf. Sci.* **369** (2016), 159–162.
99. D. S. Ginley (Ed.); *Handbook of Transparent Conductors*, Springer, 2011.
100. P. Reunchan, X. Zhou, S. Limpijumnong, A. Janotti, C. G. Van de Walle; *Curr. Appl. Phys.* **11** (2011), S296–S300.
101. M. Marezio; *Acta Crystallogr.* **20** (1966), 723–728.
102. L. Pauling, M. D. Shappell; *Z. Krist.* **75** (1930), 128–142.
103. T. Hirsch, R. Uecker, D. Klimm; *Cryst. Res. Technol.* **52** (2017), 1600237.
104. D. D. Edwards, P. E. Folkins, W. O. Mason; *J. Am. Ceram. Soc.* **80** (1997), 253–257.

105. American Ceramic Society, Phase Diagrams for Ceramists, entry 340.

106. R. J. Cava, J. M. Phillips, J. Kwo, G. A. Thomas, R. B. van Dover, S. A. Carter, J. J. Krajewski, W. F. Peck, Jr., J. H. Marshall, S. A. Carter; *Appl. Phys. Lett.* **64** (1994), 2071–2072.

107. W. J. Heward, D. J. Swenson; *J. Mater. Sci.* **42** (2007), 7135–7140.

108. I. Isomäki, M. Hämäläinen, W. Gierlotka, B. Onderka, K. Fitzner; *J. Alloy. Comp.* **422** (2006), 173–177.

109. A. Walsh, C. R. A. Catlow, A. A. Sokol, S. M. Woodley; *Chem. Mater.* **21** (2009), 4962–4969.

110. D. Liu, W. W. Lei, B. Zou, S. D. Yu, J. Hao, K. Wang, B. B. Liu, Q. L. Cui, G. T. Zou; *J. Appl. Phys.* **104** (2008), 083506.

111. M. F. Bekheet, M. R. Schwarz, S. Lauterbach, H.-J. Kleebe, P. Kroll, A. Stewart, U. Kolb, R. Riedel, A. Gurlo; *High Pressure Res.* **33** (2013), 697–711.

112. A. Kompany, H. A. Rahnamaye Aliabad, S. M. Hosseini, J. Baedi; *Phys. Status Solidi (b)* **244** (2007), 619–628.

113. H. A. R. Aliabad, S. M. Hosseini, A. Kompany, A. Youssefi, E. A. Kakhki; *Phys. Status Solidi (b)* **246** (2009), 1072–1081.

114. J. E. Medvedeva; *Phys. Rev. Lett.* **97** (2006), 086401.

115. P. Erhart, A. Klein, R. G. Egdell, K. Albel; *Phys. Rev. B* **75** (2007), 153205.

116. A. Walsh, J. L. F. Da Silva, S.-H. Wei, C. Körber, A. Klein, L. F. J. Piper, A. DeMasi, K. E. Smith, G. Panaccione, P. Torelli, D. J. Payne, A. Bourlange, R. G. Egdell; *Phys. Rev. Lett.* **100** (2008), 167402.

117. P. Reunchan, X. Zhou, S. Limpijumnong, A. Janotti, C. G. Van de Walle; *Curr. Appl. Phys.* **11** (2011), S296–S300.

118. F. P. Sabino, R. Besse, L. N. Oliveira, S.-H. Wei, J. L. F. Da Silva; *Phys. Rev. B* **92** (2015), 205308.

119. C. Körber, V. Krishnakumar, A. Klein, G. Panaccione, P. Torelli, A. Walsh, J. L. F. Da Silva, S.-H. Wie, R. G. Egdell, D. J. Payne; *Phys. Rev. B* **81** (2010), 165207.

120. C. Janowitz, V. Scherer, M. Mohamed, A. Krapf, H. Dwelk, R. Manzke, Z. Galazka, R. Uecker, K. Irmscher, R. Fornari, M. Michling, D. Schmeißer, J. R. Weber, J. B. Varley, C. G Van de Walle; *New J. Phys.* **13** (2011), 085014.

121. V. Scherer, C. Janowitz, A. Krapf, H. Dwelk, D. Braun, R. Manzke; *Appl. Phys. Lett.* **100** (2012), 212108.

122. V. Scherer, C. Janowitz, Z. Galazka, M. Nazarzadehmoafi, R. Manzke; *Europhys. Lett.* **113** (2016), 26003.

123. P. Ágoston, P. Erhart, A. Klein. K. Albe; *J. Phys.: Condens. Matter* **21** (2009), 455801.

124. P. Ágoston, K. Albe; *Phys. Rev. B* **81** (2010), 195205.

125. S. Limpijumnong, P. Reunchan, A. Janotti, C. G. Van de Walle; *Phys. Rev. B* **80** (2009), 193202.

126. K. Palandage, G. W. Fernando; *Phys. Lett. A* **374** (2010), 2879–2885.

127. J. B. Varley, H. Peelaers, A. Janotti, C. G. Van deWalle; *J. Phys.: Condens. Matter* **23** (2011), 334212.

128. P. D. C. King, R. L. Lichti, Y. G. Celebi, J. M. Gil, R. C. Vilão, H. V. Alberto, J. Piroto Duarte, D. J. Payne, R. G. Egdell, I. McKenzie, C. F. McConville, S. F. J. Cox, T. D. Veal; *Phys. Rev. B* **80** (2009), 081201.

129. E. Korhonen, F. Tuomisto, O. Bierwagen, J. S. Speck, Z. Galazka; *Phys. Rev. B* **90** (2014), 245307.

130. D. Braun, V. Scherer, C. Janowitz, Z. Galazka, R. Fornari, R. Manzke; *Phys. Status Solidi* (*a*) **211** (2014), 59–65.

131. R. Roy, M. W. Shafer; *J. Phys. Chem.* **58** (1954), 372–375.

132. R. L. Weiher; *J. Appl. Phys.* **33** (1962), 2834–2839.

133. J. P. Remeika, E. G. Spencer; *J. Appl. Phys.* **35** (1964), 2803.

134. J. H. W. De Wit; *J. Cryst. Growth* **12** (1972), 183–184.

135. R. L. Weiher, R. P. Ley; *J. Appl. Phys.* **37** (1966), 299–302.

136. S. Shimada, O. Sato, A. Tsunashima, K. Kodaira, T. Matsushita; *J. Cryst. Growth* **80** (1987), 366–370.

137. S. Shimada, K. J. D. MacKenzie; *J. Cryst. Growth* **55** (1981), 453–456.

138. M. Jozefowicz, W. Piekarczyk; *Mater. Res. Bull.* **22** (1987), 775–780.

139. J. Werner, G. Behr, W. Bieger, G. Krabbes; *J. Cryst. Growth* **165** (1996), 258–267.

140. G. Behr, J. Werner, S. Oswald, G. Krabbes, P. Dordor, D. Elefant, W. Pitschke; *Solid State Ion.* **101–103** (1997), 1183–1187.

141. G. Frank, L. Brock, H. D. Bausen; *J. Cryst. Growth* **36** (1976), 179–180.

142. V. Scherer, C. Janowitz, A. Krapf, H. Dwelk, D. Braun, R. Manzke; *Appl. Phys. Lett.* **100** (2012), 212108.

143. R. Roy, M. W. Shafer; *J. Phys. Chem.* **58** (1954), 372–375.

144. A. N. Christensen, N. C. Broch, O. von Heidenstam, Å. Nilsson; *Acta Chem. Scand.* **21** (1967), 1046–1056.

145. A. B. Chase, W. R. Wilcox; *J. Am. Ceram. Soc.* **50** (1967), 332–333.

146. A. B. Chase, H. H. Tippins; *J. Appl. Phys.* **38** (1967), 2469–2471.

147. A. B. Chase; *J. Am. Ceram. Soc.* **51** (1968), 507–510.

148. A. B. Chase, R. Teviotdale; *J. Appl. Phys.* **39** (1968), 3574–3576.

149. S. J. Wen, G. Couturier, J. P. Chaminade, E. Marquestaut, J. Claverie, P. Hagenmuller; *J. Solid State Chem.* **101** (1992), 203–210.

150. D. R. Hagleitner, M. Menhart, P. Jacobsen, S. Blomberg, K. Schulte, E. Lundgren, M. Kubicek, J. Fleig, F. Kubel, C. Puls, A. Limbeck, H. Hutter, L. A. Boatner, M. Schmid, U. Diebold; *Phys. Rev. B* **85** (2012), 115441.

151. Z. S. Teweldemedhin, K. V. Ramanujachary, M. Greenblatt; *J. Solid State Chem.* **86** (1990), 109–115.

152. Z. Galazka, K. Irmscher, M. Pietsch, T. Schulz, R. Uecker, D. Klimm, R. Fornari; *CrystEngComm* **15** (2013), 2220–2226.

153. M. Albrecht, R. Schewski, K. Irmscher, Z. Galazka, T. Markut, M. Naumann, T. Schulz, R. Uecker, R. Fornari, S. Meuret, M. Kociak; *J. Appl. Phys.* **115** (2014), 053504.

154. S. Sadofev, Y. Cho, O. Brandt, M. Ramsteiner, R. Calarco, H. Riechert, S. C. Erwin, Z. Galazka, M. Korytov, M. Albrecht, R. Uecker, R. Fornari; *Appl. Phys. Lett.* **101** (2012), 172102.

155. Y. J. Cho, S. Sadofev, S. Fernández-Garrido, R. Calarco, H. Riechert, Z. Galazka, R. Uecker, O. Brandt; *Appl. Surf. Sci.* **369** (2016), 159–162.

156. R. L. Weiher, R. P. Ley; *J. Appl. Phys.* **34** (1963), 1833–1834.

157. K. D. Kundra, S. Z. Ali; *J. Appl. Cryst.* **3** (1970), 543–545.

158. T. M. Gesing; *Universität Bremen*, Bremen, Germany (2011), unpublished data.

159. D. Klimm, Z. Galazka; *Leibniz-Institut für Kristallzüchtung*, Berlin, Germany (2018), unpublished data.

160. A. I. Zagumennyi, G. B. Lutts, P. A. Popov, N. N. Sirota, I. A. Shcherbakov; *Laser Phys.* **3** (1993), 1064–1065.

161. E. R. Dobrovinskaya, A. L. Lytvynov, V. Pishchnik; *Sapphire: Material, Manufacturing, Applications*, Springer, 2009.

162. www.factsage.com.

163. K. H. L. Zhang, V. K. Lazarov, T. D. Veal, F. E. Oropeza, C. F. McConville, R. G. Egdell, A. Walsh; *J. Phys.: Condens. Matter.* **23** (2011), 334211.

164. M. Ramzan, Y. Li, R. Ahuja; *Solid State Commun.* **172** (2013), 37–40.

165. N. Preissler, O. Bierwagen, A. T. Ramu, J. S. Speck; *Phys. Rev. B* **86** (2013), 085305.

166. K. Irmscher, Z. Galazka, M. Pietsch; *Leibniz-Institut für Kristallzüchtung*, Berlin, Germany (2012), unpublished data.

167. K. Irmscher, M. Naumann, M. Pietsch, Z. Galazka, R. Uecker, T. Schulz, R. Schewski, M. Albrecht, R. Fornari; *Phys. Status Solidi (a)* **211** (2014), 54–58.

168. I. Hamberg, C. G. Granqvist; *J. Appl. Phys.* **60** (1986), R123–R159.

169. I. Hamberg, C. G. Granqvist; *Sol. Energy Mater.* **14** (1986), 241–256.

170. I. Hamberg, C. G. Granqvist; *Thin Solid Films* **105** (1983), L83–L86.

171. Z. Qiao, D. Mergel; *Phys. Status Solidi (a)* **207** (2010), 1543–1548.

172. R. Schmidt-Grund, H. Krauβ, C. Kranert, M. Bonholzer, M. Grundmann; *Appl. Phys. Lett.* **105** (2014), 111906.

173. M. Feneberg, J. Nixdorf, C. Lidig, R. Goldhahn, Z. Galazka, O. Bierwagen, J. S. Speck; *Phys. Rev. B* **93** (2016), 045203.

174. A. Schleife, M. D. Neumann, N. Esser, Z. Galazka, A. Gottwald , J. Nixdorf, R. Goldhahn, M. Feneberg; *New J. Phys.* **20** (2018), 053016.

175. S. Bin Anooz, Z. Galazka; *Leibniz-Institut für Kristallzüchtung*, Berlin, Germany (2019), unpublished data.

176. K. Irmscher, Z. Galazka; *Leibniz-Institut für Kristallzüchtung*, Berlin, Germany (2014), unpublished data.

177. T. Schulz, K. Irmscher, Z. Galazka; *Leibniz-Institut für Kristallzüchtung*, Berlin, Germany (2012), unpublished data.

178. J. B. Varley, A. Janotti, C. Franchini, C. G. Van de Walle; *Phys. Rev. B* **85** (2012), 081109.

179. W. Drozdowski, W. Łachmański, K. Brylew, A. J. Wojtowicz; *Nicolaus Copernicus University*, Toruń, Poland (2015), unpublished data.

180. I. J. Panneerdoss, S. J. Jeyakumar, S. Ramalingam, M. Jothibas; *Spectrochim. Acta A Mol. Biomol. Spectrosc.* **147** (2015), 1–13.

181. K. Palandage, G. W. Fernando; *Phys. Lett. A* **374** (2010), 2879–2885.

182. S. J. Wen, G. Campet; *Active Passive Elec. Comp.* **16** (1993), 41–47.

183. J. H. W. De Wit; *J. Solid State Chem.* **8** (1973), 142–149.

184. J. H. W. De Wit; *J. Solid State Chem.* **13** (1975), 192–200.

185. Z. Galazka, K. Irmscher, M. Pietsch; *Leibniz-Institut für Kristallzüchtung*, Berlin, Germany (2013), unpublished data.

186. T. Nagata, O. Bierwagen, Z. Galazka, M. Imura, S. Ueda, H. Yoshikawa, Y. Yamashita, T Chikyow; *Appl. Phys. Express* **10** (2017), 011102.

187. T. Nagata, O. Bierwagen, Z. Galazka, M. Imura, S. Ueda, Y. Yamashita, T. Chikyow; *Jpn. J. Appl. Phys.* **58** (2019), SDDG06.

188. A. Walsh; *Appl. Phys. Lett.* **98** (2011), 261910.

189. P. D. C. King, T. D. Veal, D. J. Payne, A. Bourlange, R. G. Egdell, C. F. McConville; *Phys. Rev. Lett.* **101** (2008), 116808.

190. P. D. C. King, T. D. Veal, F. Fuchs, C. Y. Wang, D. J. Payne, A. Bourlange, H. Zhang, G. R. Bell, V. Cimalla, O. Ambacher, R. G. Edgell, F. Bechstedt, C. F. McConville; *Phys. Rev. B* **79** (2009), 205211.

191. O. Bierwagen, J. S. Speck, T. Nagata, T. Chikyow, Y. Yamashita, H. Yoshikawa, K. Kobayashi; *Appl. Phys. Lett.* **98** (2011), 172101.

192. T. Berthold, J. Rombach, T. Stauden, V. Polyakov, V. Cimalla, S. Krischok, O. Bierwagen, M. Himmerlich; *J. Appl. Phys.* **120** (2016), 245301.

193. K. H. L. Zhang, R. G. Egdell, F. Offi, S. Iacobucci, L. Petaccia, S. Gorovikov, P. D. C. King; *Phys. Rev. Lett.* **110** (2013), 056803.

194. M. Nazarzahdemoafi, F. Titze, S. Machulik, C. Janowitz, Z. Galazka, R. Manzke, M. Mulazzi; *Phys. Rev. B* **93** (2016), 081303.

195. M. Nazarzadehmoafi, S. Machulik, F. Neske, V. Scherer, C. Janowitz, Z. Galazka, M. Mulazzi, R. Manzke; *Appl. Phys. Lett.* **105** (2014), 162104.

196. H. von Wenckstern, D. Splith, F. Schmidt, M. Grundmann, O. Bierwagen, J. S. Speck; *APL Mater.* **2** (2014), 046104.

197. O. Bierwagen, J. S. Speck; *Appl. Phys. Lett.* **97** (2010), 072103.

Chapter 6

Tin Oxide (SnO_2)

6.1 Introduction

SnO_2 reveals typical features of wide bandgap transparent semiconducting oxides (TSOs), i.e., *n*-type semiconducting behavior with high electron mobility and high transparency in the visible spectrum. It is also chemically and thermally stable at room and elevated temperatures. Moreover, it can be easily deposited as coatings and amorphous, polycrystalline, or nanocrystalline layers with cost-effective methods, such as (not exhaustive): sol-gel [1–5], spray pyrolysis [6–9], evaporation [10–14], magnetron sputtering [15–19], as well as epitaxial layers with the use of pulsed laser deposition (PLD) [20–23], molecular-beam epitaxy (MBE) [24–27], chemical vapor deposition (CVD) [28–31], and metal organic chemical vapor deposition (MOCVD) [32–37].

The features of SnO_2 are suitable as transparent electrodes for silicon solar cells (SnO_2:F or SnO_2:Sb) [38, 39], hybrid organic–inorganic solar cells [40], and for perovskite solar cells as functional layers (electron transport layers) [41, 42]. Sb-doped SnO_2 was also shown as a suitable transparent electrode for GaN-based light-emitting diodes [43, 44]. SnO_2 was demonstrated in *p*–*n* junctions [45, 46], as a channel material (SnO_2:Sb) in transparent thin-film transistors [47–50], in different heterostructure architectures for

Transparent Semiconducting Oxides: Bulk Crystal Growth and Fundamental Properties
Zbigniew Galazka

ISBN 978-981-4800-94-5 (Hardcover), 978-1-003-04520-5 (eBook)
www.jennystanford.com

UV LEDs [51], memories [52], and photo-catalysis [53]. Magnetic properties of SnO_2 doped with a diversity of elements (Ni, Co, Cu, Cr, Mn, Fe) were intensively investigated [54–60] as well. Further, SnO_2 showed good sensing capabilities for different gases (CH_4, NH_3, H_2, H_2S, CO, O_2, O_3, humidity) [6, 61–68].

Other features of SnO_2 layers, thin films, or nanostructures, including the growth, surfaces, physical properties, and applications, can be found in review articles of Jarzębski [69], Chorpa et al. [70], Zorina et al. [71], Das and Jayaraman [72], Xiong et al. [41], Batzill and Diebold [73], and Dixon et al. [74].

Bulk single crystals of SnO_2 are not readily available as they cannot be grown directly from the melt due to very high melting point (>2100°C) and very high thermal instability at temperature above 1500°C, which leads to its strong thermal decomposition prior melting. Although other growth techniques (vapor phase, flux) were successfully utilized for growing bulk SnO_2 single crystals, the size of resulted crystals is rather small (typically a few millimeters), which is sufficient to reveal fundamental physical properties, but not sufficient for large-scale applications, e.g., as substrates for epitaxy, or as active components in electronic/optoelectronic devices. Nonetheless, bulk SnO_2 single crystals have often higher structural quality as compared to thin films (often amorphous or polycrystalline) and enable exploration of real intrinsic properties of that binary compound. Moreover, different attempts undertaken to grow bulk SnO_2 single crystals showed the possibility of obtaining relatively large-size crystals (1 inch in diameter) with further scale-up capabilities, which may bring further development for large-scale applications in the future. This chapter discusses in detail growth methods utilized to obtain bulk SnO_2 single crystals along with their thermal, mechanical, electrical, and optical properties. In other words, it reflects the results obtained in several decades of research on bulk SnO_2 single crystals and their properties (occasionally, properties of thin films are cited for comparison). The physical properties of bulk SnO_2 single crystals are described in values and graphs rather than in scientific discussion of underlying physics, which can be easily found in the appended references.

6.2 Crystal Structure

SnO_2 has the tetragonal rutile structure with space group $P4_2/mnm$. It is found for many other four-valent binary oxide systems, such as TiO_2, MnO_2, GeO_2, and PbO_2. Some fluorides such as MgF_2, MnF_2, FeF_2, and ZnF_2 are crystallizing in this structure as well. "Rutile" is the mineral name of the most common natural form of TiO_2. It is the stable form under ambient conditions not only for TiO_2 but also for SnO_2. Other metastable forms such as anatase are available easily for TiO_2 but can be obtained for tin(IV) oxide only as solid solution with TiO_2 [75].

In the structure (Fig. 6.1), the Sn^{4+} ions are coordinated by distorted oxygen octahedra. These octahedra are connected via two edges to their neighbors, thus forming chains along the *c*-axis of the crystal. By using neutron diffraction, Pannetier and Denes [76] obtained the lattice parameters of a = 4.8382 (8) Å and c = 3.8029 (5), while Bolzan et al. [77] derived lattice parameters of a = 4.7374 (1) and c = 3.1864 (1) Å. Two sets of O-Sn-O angles of the tetrahedral SnO_2 are 117.3° with the Sn-O distance of 2.224 (8) Å [76].

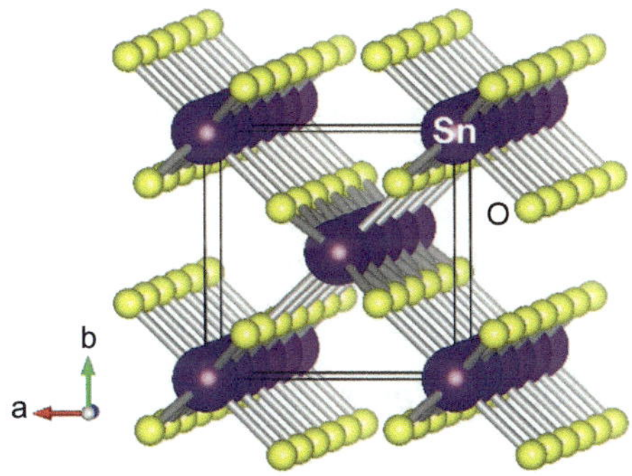

Figure 6.1 Rutile structure of SnO_2.

6.3 Band Structure and Native Point Defects

6.3.1 Band Structure

The electronic structure was theoretically studied, e.g., by Mäki-Jaskari and Rantala [78], Sensato et al. [79], Liu et al. [80], Borges et al. [81], Schleife et al. [82], Scanlon and Watson [83], and Dou and Persson [84]. All the calculations revealed a direct transition at Γ-point of the BZ. Although the fundamental bandgap values depend on the calculation method (bandgap underestimation), Schleife et al. [82] as well as Scanlon and Watson [83] calculated the value of 3.65 and 3.6 eV, respectively, which correspond to typical empirical values. According to the theoretical results of Schleife et al. [82], the direct optical gap between the valence band maximum (VBM) and the conduction band minimum (CBM) at 344 nm (3.6 eV) is dipole forbidden, and direct-allowed transitions only occur from lower valence bands (VBs) for wavelengths shorter than 288 nm (4.3 eV).

The bottom of the conduction band (CB) is formed by the Sn *s* states, while the top of the VB by Sn *4d* and O *2s* states.

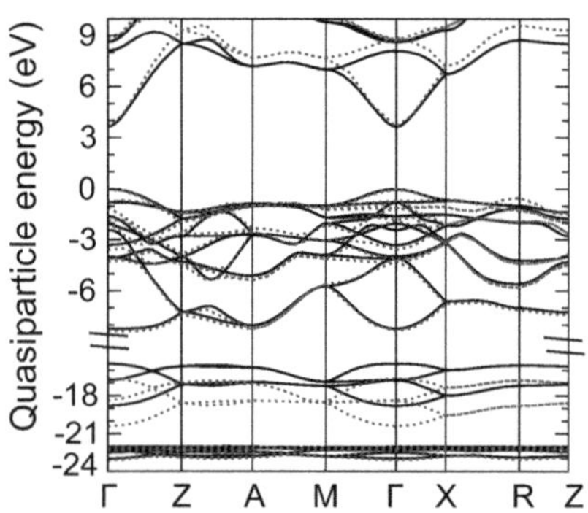

Figure 6.2 Band structure of SnO_2. Reprinted with permission from Ref. [82], Copyright 2011, American Physical Society.

6.3.2 Native Point Defects

Theoretically predicted native point defects [83, 85, 86] include oxygen vacancies (V_O), oxygen interstitials (O_i), tin vacancies (V_{Sn}), tin interstitials (Sn_i), and tin antisites (Sn_O). V_O is a deep donor with the ionization energy of about 1.8 eV. In Sn-rich conditions, V_O is stable in the +2 and neutral charge state for the Fermi level E_F near the VBM and CBM, respectively. A doubly ionized oxygen vacancy as a native defect in SnO_2 was experimentally concluded by Samson and Fonstad [87]. According to Singh et al. [85], Sn_i and Sn_O both act as a shallow donor, which is stable only in the +4 charge state; however, their formation energy is very high (about 12 eV); therefore, their concentration is expected to be very low under equilibrium conditions. Due to low migration barrier of 0.43 eV, Sn_i^{4+} is highly mobile and migrates through the interstitial channels along the [001] direction. V_{Sn} and O_i are native acceptors in SnO_2. The formation energy of V_{Sn} is very high (7.9 eV) [83], so its concentration at equilibrium concentration will be negligible. However, Korhonen et al. [88] concluded from positron annihilation spectroscopy of Sb-doped SnO_2 thin films the presence of V_{Sn}-related vacancies, likely V_{Sn}–V_O complexes. O_i has a lower formation energy (3.1 eV) [83], but its ionization levels are deep in the conduction band. Due to the formation of deep levels (V_O) or high formation energy (Sn_i, Sn_O), none of the native point defects can explain the underlying *n*-type conductivity of SnO_2.

On the other hand, hydrogen was theoretically found to be a source of *n*-type conductivity [83, 85, 86, 89]. Interstitial hydrogen (H_i) has a low formation energy, low migration energy barrier (0.57 eV), and forms a strong bond with oxygen (O–H). It acts as a shallow donor with a concentration not depending on the oxygen partial pressure $p(O_2)$ in a growth environment. Hydrogen-substituting oxygen (H_O) has a bit higher formation energy and also acts as a shallow donor. The migration barrier of H_O^+ is about 2.2 eV. As it substitutes oxygen in the crystal lattice, it depends on the $p(O_2)$ in the growth or annealing environment [87, 90]. H_O is more stable than H_i. The electrical conductivity of SnO_2 can be induced or increased by a partial substitution of oxygen ions (O^{2-}) with fluorine ions (F^{1-}) to achieve fluorine doped tin oxide (SnO_2:F) [91, 92] and of Sn ions (Sn^{4+}) with antimony (Sb^{5+}), phosphorus (P^{5+}), and arsenic (As^{5+}) [93]. Indeed, an increase in the electrical conductivity by doping with Sb was proved experimentally for both bulk crystals [94, 95] and thin films [96–98].

6.3.3 Toward *p*-Type Conductivity

Singh et al. [85] have theoretically found that different elements of group IIIA (In, Ga, Al) or group IIB (Zn) act as shallow acceptors in SnO_2. Under Sn-poor conditions, the formation energy of substitutional acceptors is low enough and, simultaneously, the formation energy of compensating donors (V_O, Sn_i, Sn_O) is high enough to achieve *p*-type doping. However, Scanlon et al. [83] showed that *p*-type doping of SnO_2 creates localized holes acting as deep acceptors, so an efficient *p*-type doping is not achievable. In, Ga, Al, and Li substituting Sn as well as N substituting O (N_O) in the SnO_2 lattice produce localized polarons centered on single oxygen sites (hole polarons). Creation of deep and localized states by doping Sn-site with acceptors was also concluded by Lany and Zunger [99]. Theoretical results by Varley et al. [100] confirmed the formation of a very deep acceptor by N_O, while other elements in group V, i.e., P, As, and Sb, substitute Sn in the SnO_2 crystal lattice and act as shallow donors. Deep and localized states can also trap multiple holes. Acceptor dopants will trap the Fermi level near the mid gap, meaning that a high concentration of acceptors can only serve to annihilate free electrons and produce a highly resistive SnO_2. This is the result of the VBM composition with O *2p* states that favor the formation of localized oxygen-hole polarons when acceptor doped. Localized oxygen holes, deep acceptor levels, and dopant self-compensation will lead to electrically insulating state instead of *p*-type conductivity, even in the case of heavy doping with acceptors. Although there are some experimental results claiming *p*-type conductivity, e.g., by doping SnO_2 layers with Li (*p*-type conductivity present at low temperatures only) [101], Zn [102], or N [103], there are no solid confirmations of such findings for *p*-type conductivity of SnO_2 at room temperature.

6.4 Bulk Crystal Growth

6.4.1 Stability Fields of the Sn–O System

To grow bulk SnO_2 single crystals, it is important to conclude how SnO_2 is thermally stable. The stability fields of the Sn–O system are shown in Fig. 6.3. The calculations show no liquid phase of SnO_2(l), which leads to the conclusion that growing SnO_2 crystals from the melt is technically difficult, if possible at all. Although there are reports indicating melting point of 1999°C [104], this value may not relate to stoichiometric or near stoichiometric SnO_2 but rather to Sn+SnO_2, where metallic Sn acts as a self-flux for SnO_2 and lowers its melting point. Indeed, more comprehensive experiments by Galazka et al. [90], performed at high temperatures (>2000°C) and high oxygen partial pressure, $p(O_2)$ (>100 atm.), did not show any melting signs of SnO_2. SnO_2 seems to be one of the most unstable binary compounds at high temperatures among other TSOs.

High thermal instability of SnO_2 promotes, therefore, the growth from the gas phase, as it can be easily converted into gas phase. Indeed, the stability diagram clearly shows that the gas phase (ideal gas field) can be easily reached just above 1500°C at low $p(O_2) = 10^{-5}$ bar. The gas phase field can be reached from both sides, when using SnO_2 or metallic Sn (e.g., as solid Sn(s), which is heated up to liquid Sn(l)). At very low $p(O_2) < 10^{-5}$ bar, metallic Sn is formed at temperatures even below 1550°C. The higher $p(O_2)$ used for growing SnO_2 crystals, the higher the operating temperature utilized.

These considerations are valid for the physical vapor transport (PVT) method, but not necessary for the chemical vapor transport (CVT) method, where Sn is transported via a chemically active transport agent at lower temperatures.

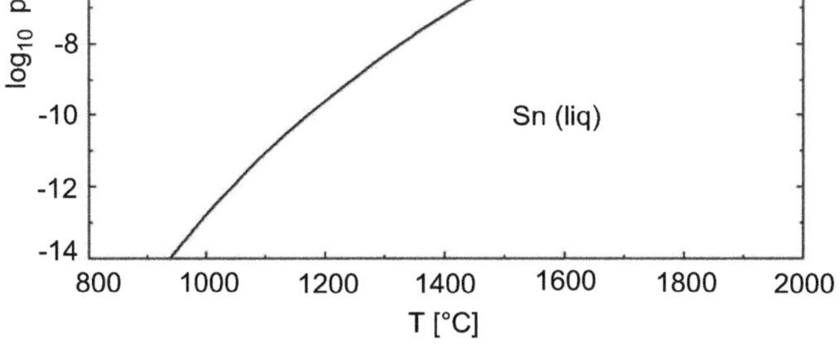

Figure 6.3 Stability diagram of SnO_2 versus temperature. The field "ideal gas" enables the growth of SnO_2 crystals from the gas phase.

6.4.2 Historical Events in Bulk SnO_2 Crystal Growth

The first report on growing bulk SnO_2 single crystals from the gas phase by the PVT method was published in 1961 by Marley and MacAvoy [105]. In 1965, the first SnO_2 single crystals were grown by the flux and hydrothermal methods by Kunkle et al. [106] and Harvill and Roy [107], respectively. In that year the first SnO_2 crystals obtained by the CVT method were reported by Nagasawa et al. [108] as well. In 2014, large, 1-inch diameter SnO_2 crystals obtained by the PVT method were reported by Galazka et al. [90]. The dominating growth methods for SnO_2 crystals are those involving gas phase, mainly PVT. Main historical events in growing bulk SnO_2 single crystals are collected in Fig. 6.4.

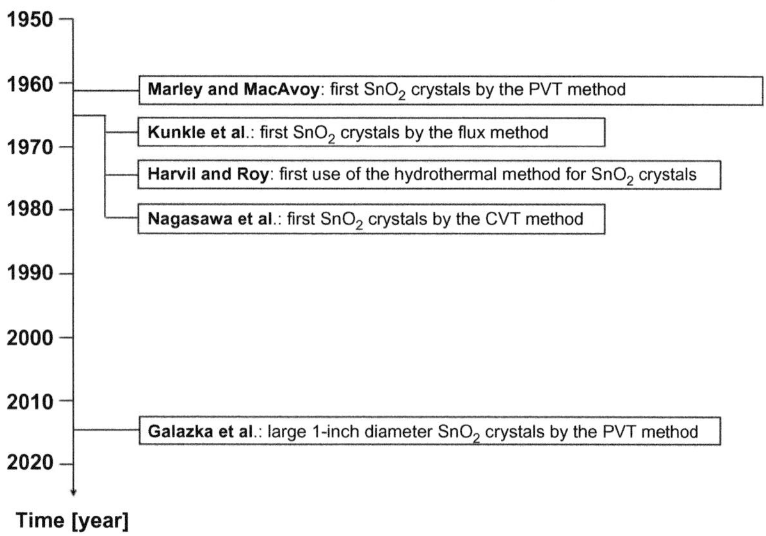

Figure 6.4 Main events in obtaining bulk SnO_2 single crystals.

6.4.3 Vapor Phase

The growth of SnO_2 single crystals from the vapor (gas) phase can be divided into two main groups:

(i) Physical vapor transport (PVT) or sublimation technique, where gaseous SnO(g) is created at high temperatures in an evaporation zone and transported to a cooler growth zone, where it is oxidized to solid SnO_2(s) in the form of single crystals. Here, we can distinguish two subgroups, where the starting (source) material is:

a. SnO_2(s), which thermally decomposes and creates SnO(g), or
b. metallic Sn(s), which is oxidized at high temperatures to SnO(g);

(ii) Chemical vapor transport (CVT), where Sn is transported via a chemically active transport agent, such as Cl or I.

6.4.3.1 Physical vapor transport (PVT)

In the PVT method, volatile SnO(g) of high partial pressure is created and transported to a cold (growth) zone, where it is oxidized to single crystals of SnO_2(s) (Eq. 6.3). The formation of SnO(g) can be achieved by either thermal decomposition of SnO_2(s) (Eq. 6.1) or oxidizing metallic Sn (Eq. 6.2) at high temperatures:

$$\mathrm{SnO_2(s)} \xrightarrow{T_{\mathrm{HIGH}}} \mathrm{SnO(g)} + \frac{1}{2}\mathrm{O_2} \tag{6.1}$$

$$\mathrm{Sn(s)} + \frac{1}{2}\mathrm{O_2} \xrightarrow{T_{\mathrm{HIGH}}} \mathrm{SnO(g)} \tag{6.2}$$

$$\mathrm{SnO(g)} + \frac{1}{2}\mathrm{O_2} \xrightarrow{T_{\mathrm{HIGH}} - \Delta T} \mathrm{SnO_2(s)} \tag{6.3}$$

where s and g stand for the solid and gas phases, respectively.

6.4.3.1.1 *Crystal growth*

Typically, SnO_2 crystals were grown by PVT in resistively heated horizontal furnaces consisting of ceramic thermal insulation. The furnace has at least two zones of high temperature (T_S) and lower temperature ($T_G = T_S - \Delta T$), named evaporation (or source) and growth zones, respectively. If the starting material is SnO_2, it is placed at the evaporation zone, which has typically a temperature of T_S = 1500–1650°C [90, 105, 109, 110], while the growth proceeds in the growth zone at temperature lower than the evaporation temperature by ΔT = 20–350 K [90, 105, 109]. Such temperature arrangement enables the reactions 6.1–6.3 to proceed. If the starting material is Sn, the evaporation zone temperature between 900 and 1450°C was reported [95, 111, 112].

In addition to the evaporation and growth temperatures, a key property defining the growth environment is the growth atmosphere, often called a carrier gas in the literature. The growth atmosphere, more precisely $p(O_2)$, defines the decomposition rate of SnO_2, or oxidation rate of Sn, and also re-oxidation rate during growth. Marley and MacAvoy [105] stated that the type of carrier gas had a high impact on the transport of SnO, as well as on the chemistry and quality of grown SnO_2 crystals. Neutral carrier gases such as N_2 and Ar led to a rapid growth, but the quality of the obtained crystals was poor, as they contained many voids and other imperfections. On the other hand, O_2 suppresses the SnO_2 decomposition and SnO evaporation, therefore making the crystal growth rather slow. As a result, the common practice was to introduce a neutral (N_2 or Ar) atmosphere or even with an admixture of H_2 (N_2 + H_2) to the evaporation zone, and separately O_2-containing atmosphere (e.g., O_2 or N_2 + O_2/Ar + O_2) to the growth zone [105, 109, 110]. Such atmospheres enable control of $p(O_2)$ in both evaporation and growth zones at predefined operating temperatures, thus the growth rate of SnO_2 crystals. Oxidizing and/or neutral atmosphere for the evaporation zone and oxidizing atmosphere for the growth zone were also utilized for Sn acting as the starting (source) material. To increase dissipation of the latent heat of crystallization, He and He + O_2 growth atmospheres were used as well [105]. A different concept of the growth atmosphere was used by Galazka et al. [90], who used CO_2 that produces, through decomposition, an increasing $p(O_2)$ with temperature.

When using SnO_2 as the starting material, a typical growth setup has an inner mullite or alumina tube extending through the furnace, which is surrounded by outer alumina insulation [105, 109, 110]. All the furnaces are electric furnaces with resistance heating, e.g., Kanthal heaters. The starting material is placed directly within an inner tube or within an alumina boat inside the inner tube in the evaporation zone, while the crystals are grown on an inner surface of the inner tube in the growth zone.

As metallic Sn starting material enables lower operating temperatures, the inner tube made of quartz can be utilized [111], although alumina-based furnaces were used as well [95]. Sn starting material was placed in quartz [113] or SiC or porcelain [112] boats/crucibles. Some of the growth furnaces had also a separate small diameter tube to introduce O_2-containing atmosphere to the growth zone.

A completely different furnace design was proposed by Galazka et al. [90, 114]. Here the starting material was loaded into an Ir crucible that was heated inductively. The Ir crucible covered from the top by a crystal substrate acted as a seed. The crucible with the substrate was surrounded from all sides by a thermal insulation consisting of zirconia and alumina. More details on the growth furnace design, growth conditions, and resulted crystals for the PVT method are given in Section 6.4.3.1.3.

The growth time is typically between 1 and 15 days; however, Galazka et al. [90] reported even shorter growth times of 10–30 h.

The obtained crystals had different shapes depending on growth environment, such as rods (e.g., $30 \times 4 \times 2$ mm^3, $20 \times 7 \times 7$ mm^3, $10 \times 1.5 \times 1.5$ mm^3) [95, 105, 110, 111, 113], plates (e.g., $15 \times 3 \times 0.1$ mm^3) [95, 105, 110, 111, 113], needles/fibres/whiskers [105, 112, 113], and large hemispheres (1 inch diameter) [90]. They were typically colorless, occasionally with a bluish, pink, or violet tint. Large SnO_2 crystals reported by Galazka et al. [90] were brown, dark brown, or even black depending on growth conditions.

Different residual impurity concentrations were reported for PVT-grown SnO_2 crystals. For example, Marley and MacAvoy [105] revealed total metal impurities (Fe, V, Mn, Cr, Cu, and Ti) at the level below 20 wt. ppm, while Reed et al. [111] stated that the crystals were highly contaminated with Si at a concentration of 0.05–0.2 wt.%, along with other impurities (Mg, Ca, Cr, Al, Cu). The crystals of De Murcia and Fillard [109] had high concentration of (in wt. ppm) Al = 209, Fe = 1022, and Cr = 30. Thiel and Helbig [110] found the following impurities in their SnO_2 crystals (in wt. ppm): Na = 150; Fe = 20–80; Si = 9–20; Ca = 10; Mg, Ni, Co = 1–4 each. Relatively high or very high residual impurity concentration in SnO_2 crystals is the result of a direct contact of the starting material and the crystals with non-metal boats or crucibles, and furnace insulation. The starting material in the design of Galazka et al. [90] had no direct contact with any insulation, while the crystals were grown on high-purity crystal seeds (sapphire); therefore, their purity was very high as compared with the purity of crystals obtained in standard electric furnaces. The following main residual impurities were detected by ICP-OES (in wt. ppm): Al = 10; Si, Fe = 2–5; Zr, Mn, Cu, Ca, Mg < 1 [90].

6.4.3.1.2 *Crystal growth habit, morphology, and structural quality*

When using SnO_2 as the starting material, Marley and MacAvoy [105] pointed out that the crystal growth habit is temperature dependent in the growth zone. With the vaporization temperature of 1650°C, the crystals grown at a temperature range of 1620–1570°C were rode shaped, at 1570–1460°C twinned plates, while at 1460–1300°C needles, as shown in Fig. 6.5. De Murcia and Fillard [109] obtained platelets and faceted bars at a growth temperature of 1520–1480 and 1400–1450°C, respectively. Facetted bars were also reported by Thiel and Helbig [110] at a growth temperature of 1500°C.

Figure 6.5 Different shapes of SnO_2 crystals grown by the PVT at different temperature ranges. Reprinted from Ref. [105], Copyright 1961, with permission from AIP Publishing.

Similar conclusions on the growth habit when using Sn as the starting material were drawn by Takizawa and Sakurai [113], who observed the growth of plates and rods at higher temperature (1270°C) and whiskers at lower growth temperature (1250°C), by Reed et al. [111], who obtained rods or plates at 1350°C, and by Morgan and Wright [95], who grew twinned plates and rectangular rods at 1450°C.

Usually the bar- and fiber-shaped SnO_2 crystals grow along the crystallographic axis *c* [109, 110], although growth directions along other main crystallographic axes *a* and *b* were also reported [112, 113]. Typically, the bars are fully facetted, as shown in Fig. 6.6. Often observed side faces are {110}, {010}, {011}, and {101} [95, 109, 110, 112, 113]. The cross section of such bars is either hexagonal (Fig. 6.6a) or rhombic (Fig. 6.6b) depending on the growth direction. In the case of the plate-like crystals, the flat surface was either (110) or (100) plane [113]. The plates could be rhombic [105] or elongated [113] in the top view.

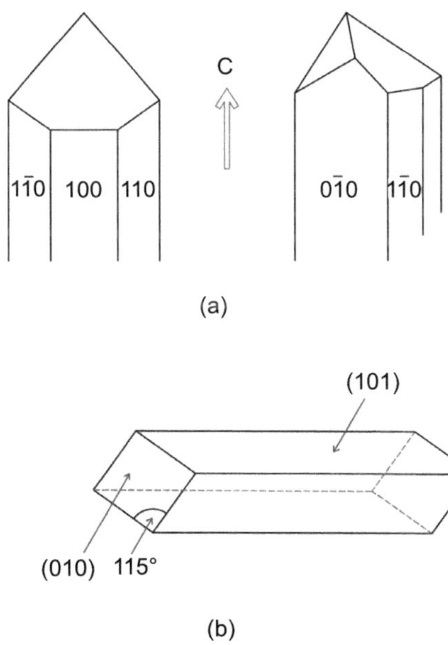

Figure 6.6 Bar-shaped SnO_2 crystals obtained by the PVT method along the *c*-axis (a) and *b*-axis (b). Redrawn with some modifications from Refs. [109] and [113], respectively.

On the other hand, SnO_2 crystals grown on *R*-plane sapphire substrates [90] were hemispheres (Fig. 6.7a) with no occurring facets. However, the growth is initiated by forming isolated SnO_2 prisms that are fully facetted (Fig. 6.7b left). The initial growth (seeding) starts at lower temperature than the final growth temperature with a relatively low growth rate. The prisms revealed a stepwise growth morphology (Fig. 6.7b right). Interestingly, when the growth rate was kept low, the prisms were grown separately leading to a polycrystalline phase. High growth rates enabled to switch the growth mode from island to "laminar flow" growth.

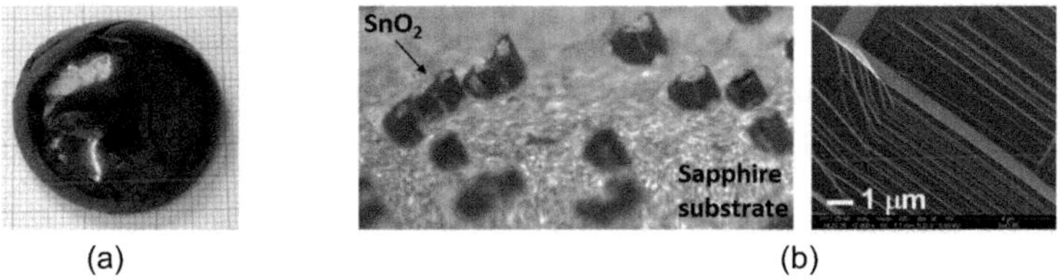

Figure 6.7 Bulk SnO_2 crystal (a) and initial growth of SnO_2 crystals (b) by the PVT method on an *R*-plane sapphire revealing isolated SnO_2 prisms (left) and step-like growth morphology (right).

The structural quality of SnO_2 crystals obtained by PVT in a resistive furnace was studied in detail by Koffyberg [115]. The investigated crystals were either cooled down to RT for 6–8 h in 80% He – 20% O_2 or quenched for 1–2 min (removed from the furnace and air cooled). The total impurity content (mainly Al and Fe) was below 50 wt. ppm.

A number of etching reagents were tested, and some of them, such as H_2SO_4, AgCl, or $SnCl_2$, did not reveal any pits, even after high-temperature etching. Elongated pits on the {110} face were formed parallel to the [001] direction after etching in H_3PO_4 (380°C, 20 min), $Na_2B_4O_7$ (730°C, 10 min), and KOH (350°C, 5 min), but the best etchant was boiling 47% HI (125°C, 5 min), which produced square or elongated pits. This etchant was also efficient for the {100}, {210}, {111}, and {101} faces. In general, the crystals are chemically stable and cannot be easily dissolved in acids or in alkaline solutions at RT.

The etch pit density (EPD) on the {100} faces was of the order of 10^8 cm^{-2} independent of whether the crystals were quenched or slowly cooled down. However, such highly dislocated layer was limited to a 3–5 μm in the case of quenched crystals and up to 30 μm in the case of slowly cooled down crystals. Below that layer, the EPD was at a much smaller level, 10^2–10^3 cm^{-3}. On the other hand, the EPD on the {110} faces, also confined to thin layer, was much smaller, 10^4–10^5 cm^{-2} for both quenched and slowly cooled down crystals. Inside the bulk after cleaving, the EPD was between 0 and 10^3 cm^{-3}, with the highest density at the base through which the crystals were anchored to the mullite substrate (likely due to stress induced by differences in lattice expansion). Sb-doped crystals showed the same concentration of dislocation at the surface and inside the bulk 0–100 cm^{-2}. The dislocation density is related to the stoichiometry gradient in the crystal. It can, therefore, be reduced by doping.

Boiling HI solution as the effective etchant for SnO_2 crystals was also confirmed by Takizawa and Sakurai [113], as well as by Thiel and Helbig [110]. The outlines of etch pits on the {110} and {100} faces, which are dislocation related, were rectangular and oval, respectively [115, 110].

The slip planes are {110} and {101} with the slip direction <001> and <10$\overline{1}$>, respectively [115]. The crystals can be cleaved parallel to the {110} plane [115].

The SnO_2 crystals obtained by Galazka et al. [90] on the sapphire substrates had usually low-angle grain boundaries that are formed likely by coalescence of individual SnO_2 prisms (islands) on sapphire substrate, the density of which decreases with distance from the substrate, or even vanish. High structural quality was confirmed by the rocking curve, the full width at half maximum (FWHM) of which can be as low as about 35 arcsec (Fig. 6.8a), and by X-ray texture (Fig. 6.8b). In both cases, low-angle grain boundary can be identified as a shoulder and double peak on the rocking curve and X-ray texture, respectively. A defect-free area of an SnO_2 crystal sample seen along the <001> direction is imaged in Fig. 6.8c. The bright spots are Sn columns.

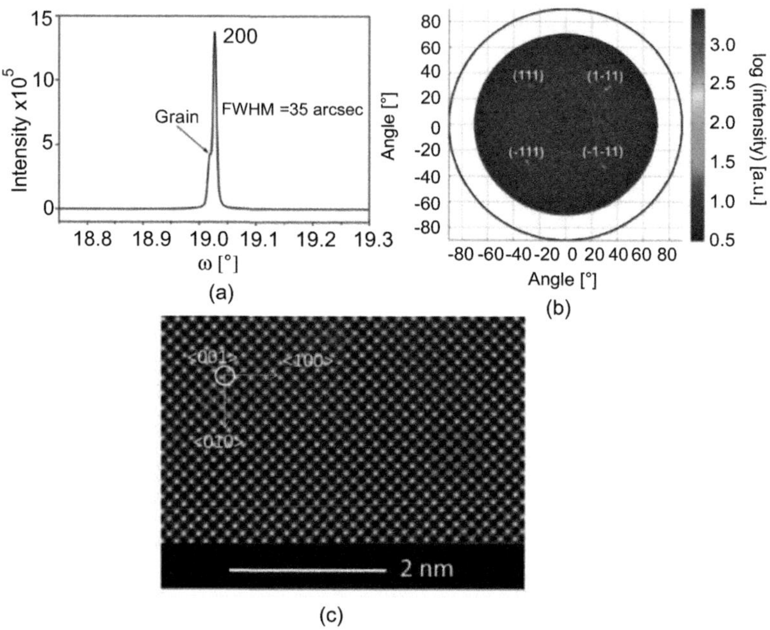

Figure 6.8 (a) Rocking curve; (b) X-ray texture (pole figure) of (100)-oriented SnO_2 wafer obtained by the PVT method on an *R*-plane sapphire substrate; and (c) HRTEM image of a (001)-oriented SnO_2 crystal sample obtained by the PVT method. Figures (a) and (c) reprinted with permission from Ref. [90], Copyright 2014, John Wiley and Sons.

6.4.3.1.3 *Examples*

(i) Marley and MacAvoy [105] used SnO_2 powder as the starting material for the PVT method. The SnO_2 starting material was prepared from zone-refined tin of 6N purity and high-quality HNO_3. The obtained stannic acid was then annealed in oxygen at 1450°C to obtain SnO_2 powder.

Growth experiments were performed in a horizontal two-zone furnace made of alumina muffle ceramics with a Pt-20%Rh resistance heating element. The inner diameter and length of the furnace were 13.8 and 91 cm, respectively. In the deposition area, however, a removable tube of high-temperature mullite was used, which acted as a substrate for growing crystals, since it becomes plastic to firmly anchor the growing crystals. The hot zone of length 7.6 cm had a constant temperature of 1650°C, while the transport and growth zones were 25-cm long and had a decreasing temperature gradient of about 8 K/cm.

The SnO_2 starting material (50 g) was placed in the hot zone at the temperature of 1650°C, at which it was subjected to evaporation. Over the SnO_2 starting material, a carrier gas was passed at the flow rate of 100 to 250 cm^3/min. The type of carrier gas was found to have a high impact on the transport of SnO, as well as on the chemistry and quality of grown SnO_2 crystals. For effective decomposition and evaporation when using O_2 as the carrier gas, a higher temperature would be required. One of the solutions to overcome the slow growth process was to carry out the evaporation of SnO_2 in a neutral gas (N_2 or Ar) and crystal growth in a mixture of N_2 or Ar + O_2, the latter being separately introduced into the transport zone in the growth furnace. Indeed, a rapid crystal growth occurred, as within 24 h relatively large crystals were obtained, 25 × 2 × 2 mm^3. However, they were full of large voids, likely due to a localized re-vaporization because of insufficient heat removal produced during gas–solid phase transition. Thermal conductivity of oxides, including that of SnO_2, is not high, and the problems with heat removal, especially from larger crystals, lead to heat accumulation and local vaporization, leaving behind voids. Substituting a neutral carrier gas with a gas of higher conductivity, He, significantly improved the quality of crystals, as the generated heat could be easily dissipated through the carrier gas of high thermal conductivity. Large single crystals of size 30 × 4 × 2 mm^3 were obtained after 28 h growth time. The crystals were colorless, and total metal impurities (Fe, V, Mn, Cr, Cu, and Ti) were at the level below 20 wt. ppm.

Other experiments under the He + O_2 atmosphere by Marley and Dockerty [94] revealed that SnO_2 crystals slowly cooled down were substantially colorless, while those quenched were brown, but transparent. The brown coloration could be removed by annealing at about 1000°C for several hours in argon or oxygen. The obtained crystals had low dislocation density in the range of 10–1000 cm^{-2} and low impurity level of Fe and Al (about 20 wt. ppm). The crystals were also doped with Sb at a level of about 300 wt. ppm, which had blue coloration.

(ii) De Murcia et al. [109, 116, 117] used SnO_2 powder as a starting material for the PVT growth of SnO_2 crystals. The growth furnace comprised an alumina tube of 1200 cm length and 4 cm inner diameter, which was heated in a three-zone Kanthal furnace (Fig. 6.9). Ar and O_2 were introduced into the alumina tube from the same end but separately in a way that Ar was introduced into an evaporation zone, where an alumina boat with 15 g of the SnO_2 starting material was placed, while oxygen was introduced into a growth zone beyond the boat with the SnO_2 starting material, which is defined by a growth tube inside the alumina tube. The source temperature in the evaporation zone was 1540°C, while the temperature gradient in the growth zone was 4.6 K/cm. The flows of Ar and O_2 were 50–80 and 20–30 cm^3/min, respectively.

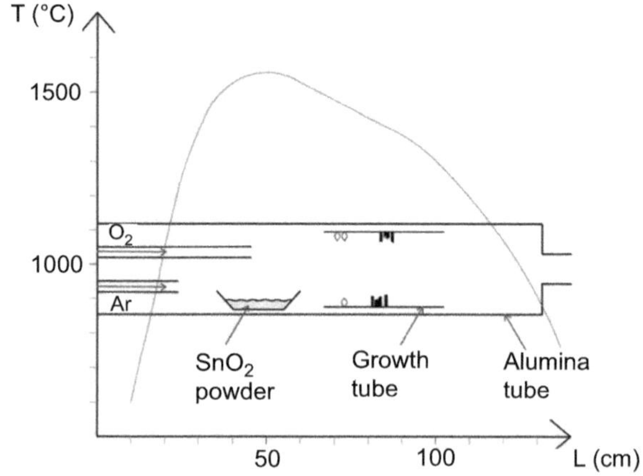

Figure 6.9 PVT growth furnace. Redrawn from Ref. [109] with some modifications.

In a temperature range in the growth zone of 1400–1450°C, the crystals were faceted bars of 2–4 mm length and about 1 mm diameter, while in a temperature range of 1520–1480°C, the crystals were platelets of about 2 × 2 – 4 × 3 mm^2 with a thickness of 0.5–1 mm. All the crystals were colorless and transparent. The crystals contained the following impurities (in wt. ppm): Al = 209, Fe = 1022, Cr = 30.

(iii) A PVT growth furnace utilized by Thiel and Helbig [110] (Fig. 6.10) was composed of a muffle thermal insulation, an alumina growth tube of 120 mm diameter and 220 mm length, which defines a growth zone inside the muffle thermal insulation, and of an alumina transport tube of 48 mm diameter and 1000 mm length, which extends toward the interior of the growth tube and defines a transport path. Inside the transport tube, pressed SnO_2 powder was loaded through which N_2 and H_2 streams were passed at rates of 100 and 1–1.4 l/h, respectively. The purpose of H_2 was to increase $p(SnO)$ at the source temperature of 1500°C. The weight of the SnO_2 starting material was 700 g, and after a growth run lasted for 70 h, the weight of the SnO_2 residual material was 100 g.

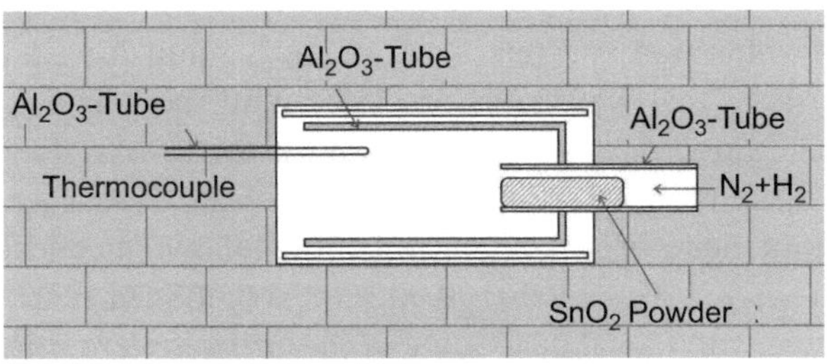

(a)

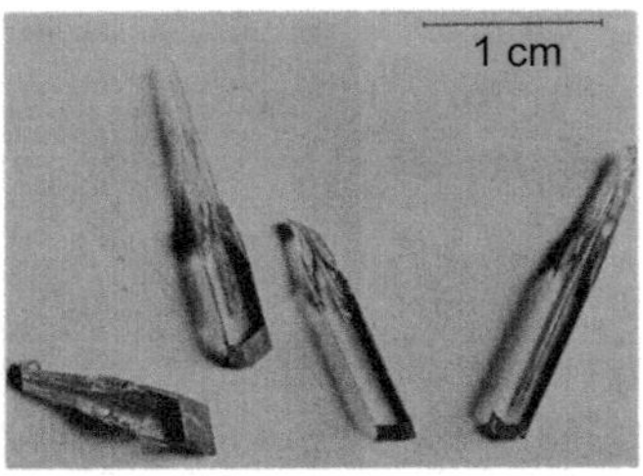

(b)

Figure 6.10 (a) PVT growth furnace, redrawn from Ref. [110] with some modifications, and (b) obtained SnO_2 crystals, reprinted from Ref. [110], Copyright 1976, with permission from Elsevier.

The obtained crystals (Fig. 6.10b) were bars with a square cross section and average dimensions of 3 × 3 × 15 mm^3, maximum 7 × 7 × 20 mm^3. The crystallographic *c*-axis was lying along the growth direction, while the side faces were {110}. Most of the crystals were clear and colorless, but some of them were slightly pink or violet. The obtained SnO_2 crystals contained the following impurities (in wt. ppm): Na = 150; Fe = 20–80; Si = 9–20; Ca = 10; Mg, Ni, Co = 1–4 each.

(iv) Reed et al. [111] used in their PVT growth experiments elemental tin as the starting material instead of SnO_2 powder. Vaporization of Sn (through the formation of volatile SnO) enables lower temperatures and, therefore, the use of other furnace materials, such as quartz. The growth experiments were performed in a furnace containing a quartz crucible of 5 cm diameter and 8 cm height. The crucible was connected with an oxygen source (air) acting as an oxidizing and carrier gas. Within the quartz crucible, a few grams of zone-refined Sn was placed, which was kept within the furnace at 1350°C for several days.

The obtained crystals were up to 30 mm long and 5 mm across. They were rods or plates with the *c*-axis parallel or sometimes perpendicular to the growth axis. The crystals were contaminated with Si at the level of 0.05–0.2 wt.%. Other impurities were also detected at the following concentrations (in wt.%): Mg = 0.001–0.003, Cr < 0.005, Al < 0.005, Cu < 0.001, and Ca < 0.001. Some of the crystals were intentionally doped with Cr, Nb, Mn, and Co by placing a corresponding metal oxide on top of Sn.

(v) In a furnace design of Takizawa and Sakurai [113], elemental Sn of 5N purity was placed inside a quartz crucible of 22 mm diameter with a neck at the top part in order to achieve a gradient of oxygen concentration. Such crucible shape was intended to obtain SnO(g) at low $p(O_2)$ by Sn oxidation, and SnO_2(s) at high $p(O_2)$. The most favorable growth conditions were obtained with a quartz crucible having a length of 10 cm with a 20 cm long neck, 1 g of Sn, a growth temperature of 1270°C, and a growth time of 7 days.

The obtained crystals were either plates up to 15 × 3 × 0.1 mm^3 or rods up to 20 (length) mm × 1 mm (cross section).

At lower growth temperature (1250°C), whiskers of 1–30 μm diameter and 10–15 mm length were grown in the crucible.

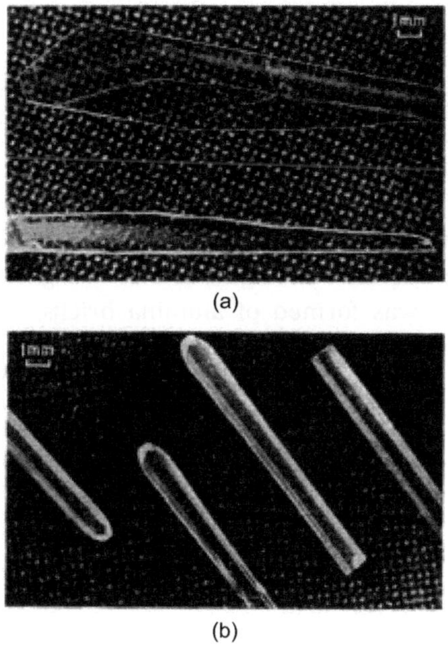

Figure 6.11 PVT-grown SnO_2 crystals in the form of plates (a) and rods (b). Reprinted with permission from Ref. [113], Copyright 1973, The Physical Society of Japan and The Japan Society of Applied Physics.

The plate-like crystals were colorless and transparent or milky due to the presence of micro-voids. The flat surface was either (110) or (100) plane. The rod-shaped crystals were colorless and transparent with no voids. The lengthwise direction was perpendicular to the (010) plane (end planes), while the wide side walls were formed by the {101} planes. The EPD was at the level of 10^3 cm^{-3}. The main impurity was Si at a concentration of about 0.001 wt.%.

(vi) Ghare [112] used a mixture of elemental Sn and Sb as the starting material, which was placed either into closed SiC or porcelain crucibles, which were heated up in a porcelain tunnel kiln. Heating from RT to 900°C was performed in an oxidizing atmosphere, in a temperature range of 900–1300°C in a reducing atmosphere, while cooling down from 1300°C to

RT was performed in an oxidizing atmosphere. The growth in the SiC crucibles was more pronounced than in the porcelain crucibles.

The obtained crystals were fibers of 0.1–0.5 mm in cross section and 10–50 mm in length. They were transparent with varying bluish tint, but they were almost colorless at top parts of the crucibles. The fiber crystals had rutile structure and grew along main crystallographic axes *c* and *a*.

(vii) Elemental Sn as the starting material and elemental Sb as a dopant were also used by Morgan and Wright [95]. The concept of a furnace design is shown in Fig. 6.12. The thermal insulation was formed of alumina bricks, through which a horizontal alumina tube was passing. Inside that alumina tube, a closed horizontal mullite tube was situated with separate alumina boats containing elemental Sn and Sb, respectively. The boat with Sn (5N purity) was situated close to the central part of the mullite tube that had a maximum temperature of about 1450°C, while the boat with Sb was situated at an inlet end of the mullite tube. Through the inlet end of the mullite tube, Ar and O_2 were introduced through separate mullite tubes at constant rates of 100–150 cm^3/min and 20–40 cm^3/min, respectively. The opposite end of the furnace had an outlet defining exhaust for the gases flowing through the furnace. In such configuration, Ar passed first through Sb and next through Sn, while the oxygen outlet was positioned behind the boats with Sn and Sb. A main growth zone was formed at the O_2 outlet of the oxygen mullite tube, where O_2 met Ar carrying evaporated Sn and Sb. The best crystals grew, however, in the boat with metallic Sn, due to an access to O_2 being carried back to the boat with Sn by convection, likely because of insufficient evacuation of gases from the furnace.

After growth experiments lasted for 10–15 days at 1450°C, the obtained crystals were of size up to $1.5 \times 1.5 \times 10$ mm^3. They were in the shape of twinned plates or rectangular rods. The rod-shaped crystals typically grew parallel to the *c*-axis. The end and side faces of such rods were {101} and {110}, respectively. Some of the rods grew with the *c*-axis perpendicular to their longest side with a hexagonal cross section and sides formed by the {011}, {010}, and {110} faces.

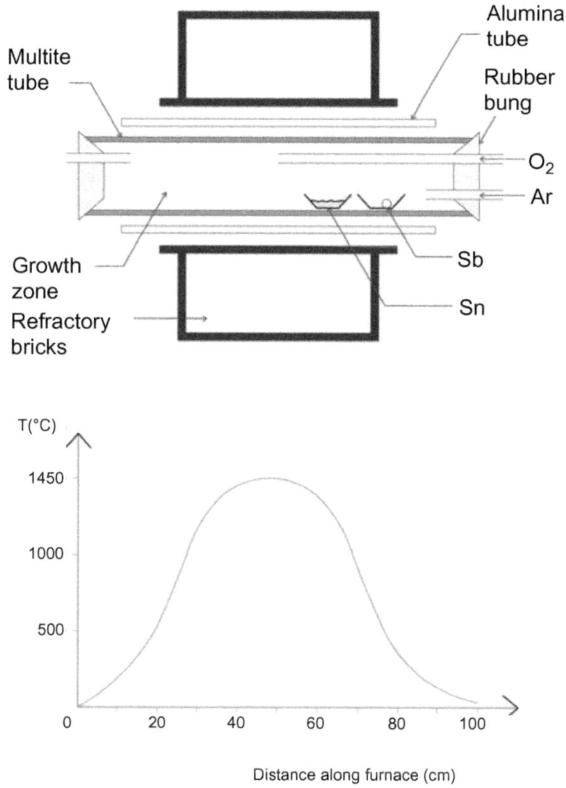

Figure 6.12 PVT growth furnace with a temperature profile for growing SnO_2 crystals. Redrawn from Ref. [95] with some modifications.

(viii) A different concept of a growth furnace and atmosphere for growing SnO_2 crystals by the PVT method was proposed by Galazka et al. [90, 114]. Crystal growth experiments were performed in a furnace shown in Fig. 6.13, which was assembled in a sealed stainless steel chamber. The growth furnace consists of an Ir crucible (40 mm diameter and 40 or 60 mm long) with an SnO_2 starting material (source), an Ir substrate holder on the top of the crucible with a crystal substrate on which the SnO_2 crystal is grown, a thermal insulation surrounding the crucible with the substrate from all sides, and an RF coil for inductive heating the Ir crucible connected to an RF generator powering the coil. The thermal insulation above the crucible determines the temperature difference (ΔT) between the SnO_2 source (T_S) and the substrate (T_G). ΔT can be adjusted by a suitable design of the top thermal insulation, e.g., by its thickness, thermal conductivity, or density gradient. In Fig. 6.13, the top insulation has a vertical channel (10 mm in diameter) that increases vertical temperature gradients at the central part of the substrate and thus adjusts ΔT.

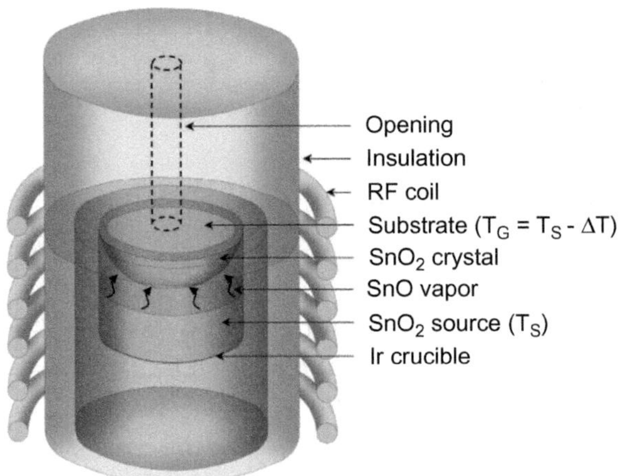

Figure 6.13 Furnace for growing SnO_2 crystals by the PVT method with use of an Ir crucible and inductive heating. ΔT = 100–150 K. Redrawn from Ref. [114] with some modifications.

The diameter and length of the thermal insulation were 90 and 150 mm, respectively. The Ir substrate holder on the crucible top was in the form of a ring with a recess of 30 mm in diameter to accommodate the substrate and a central opening of 25 mm facing the SnO_2 starting material. The substrate was either *R*-plane sapphire or rutile (TiO_2). Despite the fact that SnO_2 is isostructural to TiO_2, this type of substrates did not produce good results.

A special attention was paid to the growth atmosphere that defines the decomposition (or dissociation) rate of SnO_2(s) and re-oxidation of SnO(g) to SnO_2(s). Figure 6.14 shows p(SnO) or fugacity as a result of SnO_2(s) decomposition at different $p(O_2)$ in the growth atmosphere versus source temperature.

p(SnO) can be easily adjusted by temperature at fixed $p(O_2)$, or by $p(O_2)$ at fixed temperature. Experimentally, it was found that

$p(SnO)$ at the level of about 0.01–0.1 bar enabled an effective $SnO(g)$ transport to the growth zone defined by the substrate. For example, $p(SnO) = 0.03$ bar, as shown by the dashed line in Fig. 6.14, requires the growth temperature T_G of 1650, 1750, and about 1950°C when using as the growth atmosphere CO_2 at 1 bar, 100% O_2 at 1 bar, and 100% O_2 at 100 bar, respectively.

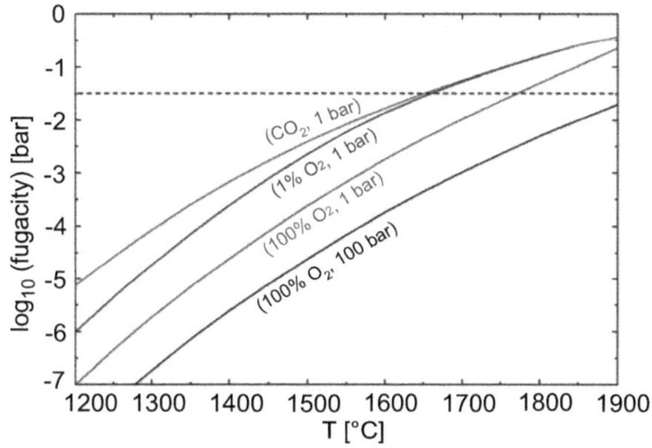

Figure 6.14 $p(SnO)$ during decomposition of SnO_2 at different $p(O_2)$ versus source temperature. The dashed line corresponds to $p(SnO) = 0.03$ bar.

There is, therefore, a number of degrees of freedom for using a set of $p(O_2)$, T_S, and ΔT for efficient crystal growth by the PVT method. To avoid very high temperatures, in which there is a faster deterioration of the Ir crucibles by oxidation and thermal insulation, optimal conditions would involve $T_S < 1700$°C and $p(O_2) < 0.1$ bar. At $p(O_2) \approx 0.01$ bar (1% O_2 or CO_2 at 1 bar), the optimal $T_S = 1600$–1700°C. Indeed, thermogravimetric analysis (TGA) of ceramic SnO_2 (pressed at 2000 bar) revealed at 1640°C in air (21% O_2) a high mass loss due to evaporation, about 0.25 wt.%/min. At lower $p(O_2)$, the mass loss will be higher.

The gaseous $SnO(g)$ must be next converted into a crystalline SnO_2. This is performed by transporting $SnO(g)$ toward a cooler region, where $SnO_2(s)$ is more stable than $SnO(g)$. This inverse reaction is defined by ΔT in addition to $p(O_2)$. For example, Fig. 6.15 shows the amount of $SnO(g)$ and $SnO_2(s)$ as a function of ΔT, where the source temperature T_S was set to 1600°C. If ΔT is small (high T_G), below 50 K, the growth rate of $SnO_2(s)$ will be very small due to the domination of $SnO(g)$. In such a situation, the growth will proceed at regions of lower temperatures, e.g., at leaking slits in the furnace. At very high values of $\Delta T > 200$ K, SnO_2 will grow on the substrate at very high rate, which leads to deterioration of the structural quality of the crystals. As a result, the optimal ΔT has been found in the range of 100–150 K.

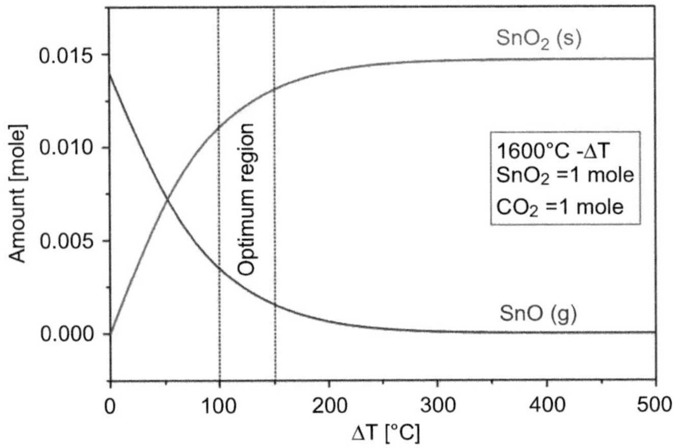

Figure 6.15 Amount of available SnO(g) and SnO_2(s) at a substrate versus temperature difference (ΔT) between T_S (1600°C) and T_G. In these calculations, the amounts of SnO_2 as the starting (source) material and CO_2 as the growth atmosphere are 1 mol each.

The oxygen source for controlling SnO_2(s) dissociation and re-oxidation of SnO(g) may have different forms, such as a mixture of O_2 and neutral gas (O_2 + Ar/N_2), CO_2, a mixture of CO_2 and a neutral gas (CO_2 + Ar/N_2), a mixture of O_2, CO_2, and a neutral gas (O_2 + CO_2 + Ar/N_2) at- or at higher/lower pressure than the atmospheric pressure. Additionally, water can be added to the growth atmosphere, e.g., by passing the gases through water.

An interesting oxygen source is CO_2, which provides through decomposition an increasing amount of oxygen with increasing temperature:

$$CO_2(g) \rightarrow CO(g) + ½O_2(g) \tag{6.4}$$

In addition to O_2, also CO is formed, which acts as a reducing agent for SnO_2:

$$SnO_2(s) + CO(g) \rightarrow SnO(g) + CO_2(g) \tag{6.5}$$

When water is present in the atmosphere, hydrogen (from H_2O dissociation) is also a strong reducing agent for SnO_2:

$$SnO_2(s) + H_2(g) \rightarrow SnO(g) + H_2O(g) \quad (6.6)$$

Both CO_2 and H_2O speed up dissociation of SnO_2; therefore, yet lower source temperatures can be applied. In addition to H_2O, H_2 is omnipresent, and the crystal growth may be strongly affected by their presence in the growth atmosphere composition. Also, other residual gases in the growth atmosphere may have an impact on the growth chemistry of SnO_2 crystals by the PVT method.

Good results were obtained by using CO_2 as the growth atmosphere, which increases $p(O_2)$ with temperature, as indicated by $p(O_2)$ in Fig. 6.16 along with $p(SnO)$, which also increases with temperature. Both $p(O_2)$ and $p(SnO)$ cross at about 1480°C and define O_2-rich or SnO-rich growth conditions. The crystals obtained at O_2-rich conditions were electrical insulators, while at SnO-rich conditions, they were semiconductors with a typical free electron concentration of 10^{17}–10^{18} cm^{-3}.

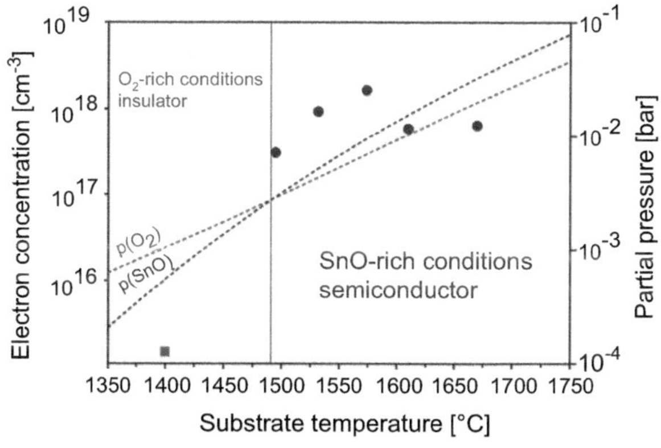

Figure 6.16 Calculated partial pressure of O_2 and SnO versus substrate temperature during SnO_2 crystal growth by the PVT method when using CO_2 as the growth atmosphere. Reprinted (with minor modifications) with permission from Ref. [90], Copyright 2014, John Wiley and Sons.

In operation, before each of the growth experiments, the growth chamber with an assembled growth furnace therein was evacuated to remove residual gases, and then CO_2 growth atmosphere was introduced in the growth furnace at a pressure of 0.5–2 atm, although the best results were obtained at 1 atm. Next, the growth furnace was heated up to a predefined temperature for 6 h, held at that predefined temperature for 10–30 h, and then cooled down for 6 h.

The obtained crystals of 25 mm in diameter and about 10 mm thick are shown in Fig. 6.17. The coloration of the crystals strongly depends on the relation between $p(O_2)$ and $p(SnO)$ near the substrate. At O_2-rich conditions ($p(O_2) > p(SnO)$), i.e., at temperature below 1500°C, the crystals were pink–orange (Fig. 6.17a) and of poor structural quality. The growth rate was small and after about 20 h of growth, only 1–2 mm thick crystals were obtained. Much better results were obtained at $p(O_2) \approx p(SnO)$. The crystals were brownish (Fig. 6.17b) and transparent, as indicated by the prepared wafers (Fig. 6.17f). In the case of SnO-rich conditions ($p(O_2) < p(SnO)$), the crystals were dark brown or black (Fig. 6.17c–e) and typically opaque in the visible spectrum (Fig. 6.17g), that became substantially colorless and transparent after annealing at high temperatures in the presence of oxygen (Fig. 6.17h). The crystals obtained at CO_2 as the growth atmosphere had relatively smooth and often shiny surface (Fig. 6.17b–d); however, addition of water to the growth atmosphere produced a rough surface (Fig. 6.17e).

The obtained SnO_2 crystals had a single rutile phase with no other phases (e.g., elemental Sn or SnO) as demonstrated in Fig. 6.18 by X-ray powder diffraction (accuracy of about 1%). The diffraction peaks are aligned with a powder diffraction file (PDF) nr. 00-041-1445 [118].

Figure 6.17 SnO_2 crystals obtained by the PVT method with the use of Ir crucibles and CO_2-containing growth atmospheres. The corresponding ratio of $p(O_2)/p(SnO)$ was as follows: (a) > 1; (b) ≈ 1; (c–e) < 1. The crystals are shown from the top (a–c) and (e) and from the side (d). Wafers (f) and (g–h) were prepared from the crystals obtained at $p(O_2)/p(SnO) \approx 1$ and $p(O_2)/p(SnO) < 1$, respectively, wherein wafer (h) was additionally annealed at 1200°C in air for 30 h.

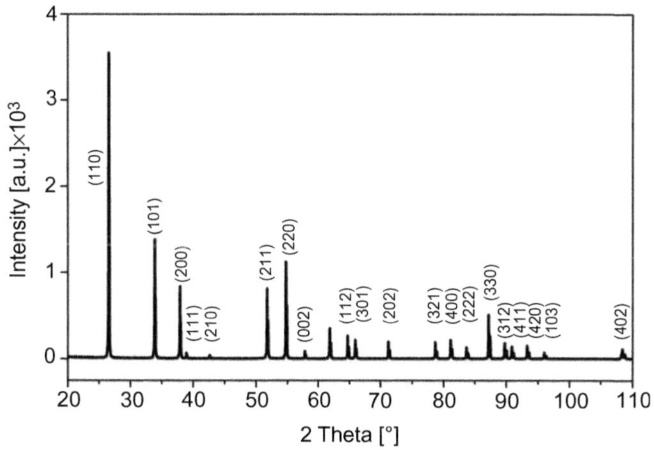

Figure 6.18 X-ray diffraction of powdered SnO_2 crystals obtained by the PVT method with the use of an Ir crucible and sapphire substrate. Reprinted with Reprinted with permission from Ref. [90], Copyright 2014, John Wiley and Sons.

The crystals had good structural quality. The FWHM of the rocking curve was typically below 100, often below 50 arcsec; however, the crystals contained low-angle grain boundaries. Indeed, there is a relatively high density of grain boundaries at the beginning of growth, which decreases with the crystal thickness. This is associated with the growth mechanism, when at the beginning individual SnO_2 islands are formed (Fig. 6.7b), which then expanded and coalesced in between to form a layer. Such SnO_2 islands are grown likely during heating up of the growth furnace at relatively low p(SnO). With increasing temperature, both p(SnO) and the growth rate increase, which speed up the expansion of the SnO_2 islands.

No intentional dopants were added to the SnO_2 starting material. However, some of the residual impurities were detected (in wt. ppm): Al = 10; Si, Fe = 2–5; Zr, Mn, Cu, Ca, Mg < 1. When comparing the purity of these SnO_2 crystals with the purity of those crystals obtained by other PVT approaches, it is easy to conclude a significant purity improvement. This is the result of the furnace design, where neither the SnO_2 starting material (inside a noble metal crucible), nor the growing crystals (on the sapphire crystal substrate) had any direct contact with the thermal insulation and the growth proceeded in the space surrounded by Ir and sapphire.

6.4.3.2 Chemical vapor transport (CVT)

In the CVT method, Sn is transported from a source zone to a growth zone not by SnO as in PVT, but by a halogen (Cl or I) and/or chalcogen (S) elements. The most frequently used chemically active transport agent is Cl_2, which forms with tin $SnCl_4$. $SnCl_4$ is transported to a high-temperature zone, where it reacts with oxygen (combustion, Eq. 6.7), water (hydrolysis, Eq. 6.8), or a mixture of oxygen and hydrogen (Eq. 6.9), and forms solid SnO_2(s) in the form of crystals:

$$SnCl_4(g)/SnI_4(g) + O_2(g) \xrightarrow{T_{HIGH}} SnO_2(s) + 2Cl_2(g)/4I(g) \quad (6.7)$$

$$SnCl_4(g)/SnI_4(g) + 2H_2O(g) \xrightarrow{T_{HIGH}} SnO_2(s) + 4HCl(g)/4HI(g) \quad (6.8)$$

$$SnCl_4(g)/SnI_4(g) + 2H_2(g) + O_2(g) \xrightarrow{T_{HIGH}} SnO_2(s) + 4HCl(g)/4HI(g) \quad (6.9)$$

where *s* and *g* indicate the solid and gas phases, respectively.

$SnCl_4$ (or SnI_4) can be used as a solution outside a growth furnace, and its vapor is injected into a growth furnace, or it can be formed inside the furnace by chlorinating SnO_2(s) (Eq. 6.10) or metallic Sn(s) (Eq. 6.11):

$$SnO_2(s) + 2Cl_2(g)/4I(g) \longrightarrow SnCl_4(g)/SnI_4(g) + O_2(g) \quad (6.10)$$

$$Sn(s) + 2Cl_2(g)/4HI(g) \longrightarrow SnCl_4(g)/SnI_4(g) \quad (6.11)$$

6.4.3.2.1 *Crystal growth*

SnO_2 crystals in the CVT configuration were grown in resistively heated horizontal furnaces of similar designs to those used for the PVT method. Due to environmentally unfriendly reaction products, the furnaces must be sealed off and the reaction products must be safely evacuated. Through the furnace horizontally extends an alumina [119–122], mullite [123], or quartz [124, 125] tube that defines a growth environment for SnO_2 crystals. Inside the horizontal tube, other furnace components are arranged if needed, e.g., a quartz tube [120, 123] inside the alumina or mullite tube, or an alumina liner [119] to form a “substrate” for growing crystals instead of alumina [123], minimize contamination from mullite tube [123], or to separate the Sn source from oxidation [119]. A carrier gas for vapor of the Sn source and a reaction (oxygen source) gas are injected into the furnace separately either by alumina [119] or quartz [120, 123–125] tubes. The Sn source in the form of $SnCl_4$ or SnI_4 vapor is introduced from outside of the furnace [119, 120, 123–125] by an inert gas (carrier gas) of Ar or N_2, or the vapor is formed inside the furnace from an $SnCl_4$ or SnI_4 powder at elevated temperatures [121, 122]. If a dopant is considered, it is also injected into the furnace as a chloride [123–125]. The reaction gas containing oxygen is O_2 [119, 121, 122], H_2O [121, 122, 124, 125], $O_2 + H_2O$ [120], or $O_2 + H_2$ [123]. A different design of a growth environment includes a sealed-off quartz ampoule with loaded therein an SnO_2 starting material, and $I_2 + S_2$ functioning as a chemically active transport gas [126]. In contrast to the aforementioned growth systems, this system is closed and there is no continuous supply of Sn and O_2.

In contrast to the PVT method, $SnCl_4$ and SnI_4 vapors are created at low temperatures (at about 100 and 200°C for $SnCl_4$ and SnI_4, respectively [121–125]), while the growth proceeds at high temperatures, but still lower than that in the PVT method. The reported growth temperatures (T_G) are in the range of 1000–1400°C [119–125]. Different temperature profiles must be applied in the case of closed ampoules and the SnO_2 starting material, where the formation of SnI_4 vapor proceeds at high temperatures of 700–1180°C, while the growth proceeds at lower temperatures by 100–300 K [126].

The obtained crystals were usually small, smaller than those obtained by the PVT method. Exemplary dimensions of the crystals are $2 \times 2 \times 12$ mm^3 [119], $2 \times 5 \times 15$ mm^3 [120], and $2.5 \times 5 \times 5$ mm^3 [123]. The reported growth time spans between 5 h and 10 days [120–125]. The CVT-grown SnO_2 crystals had different shapes: bars or rods, plates, prisms, granules, butterflies, leaves, and dendrites [119–123]. They are usually twinned [120–122]. Undoped crystals were typically colorless and transparent [119–123], while doping with Sb and Eu made the crystals bluish [123] and pink [124, 125], respectively. The undoped crystals were reported to be contaminated the most with Si, in the range of 10–50 wt. ppm. [119, 120, 123]. Other impurities include Fe, Ca, Cu, and possibly Al at concentrations of several wt. ppm.

6.4.3.2.2 *Crystal growth habit, morphology, and structural quality*

The shape and size of SnO_2 crystals grown by the CVT method strongly depend on particular growth conditions, i.e., on temperature, supply of Sn and O_2 (concentrations or flow rates), type of the reaction (combustion or hydrolysis), and geometrical furnace design that defines the location of crystal growth, which is usually defined by a space where streams supplying Sn and O_2 meet. This resulted in a diversity of growth morphology, as reported by Nagasawa et al. [120], Hirose and Furuya [121], and Furuya et al. [122].

Reported by Nagasawa et al. [120] rod-, plate-, butterfly-, and dendrite-type SnO_2 crystals are schematically shown in Fig. 6.19. The crystals were facetted with the most developed faces {110} and {111}, although in some of the crystals also the {100}, {133}, and {011} faces were identified as well. Pyramid sections of the crystals were formed by the {110} faces. All the crystals showed a twin structure with the cleavage plane {011}, which is indicated by a horizontal dot-line in Fig. 6.19.

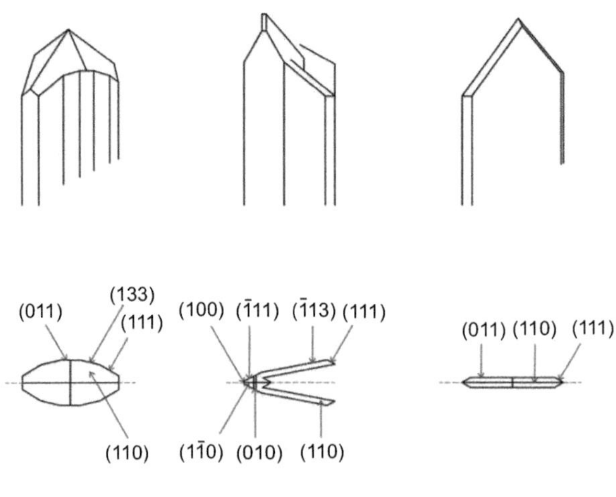

Figure 6.19 Schematic illustration of CVT-grown SnO_2 crystals in the form of a rod (left), butterfly (middle), and plate (right). Redrawn from Ref. [120] with some modifications.

Hirose and Furuya [121] and Furuya et al. [122] described the obtained crystals as leaves, dendrites, pyramids, plates, and hollow bodies. Leaf-type crystals had complex surfaces with many small {101} and {011} faces forming gables-like structure. The pyramid-type crystals had eight {101} and {011} faces forming an octahedron. In both crystal types, the lengthwise direction was [110] (growth direction). The plate-type crystals had well-developed {011} faces, while the lateral faces were formed either by the {100} and {110} or by the {100}, {110}, and {101} faces depending on the lengthwise growth directions of <011> or <111> (Fig. 6.20). Both plate types showed twins along the {011} plane. The twinning plane {011} was also observed in hollow crystals. The leaf-type crystals were grown by the formation of a primary needle (one-dimensional growth) along the [110] direction, and then by the formation of secondary needles in the perpendicular directions. Elongating and thickening the needles leads to a dendrite formation. The pyramid-type crystals were formed from rod-like structure having four side faces. The growth proceeded in the [110] direction by elongation and thickening in two dimensions on the {101} and {011} faces (2D layer growth), where the supersaturation was higher. The plate-type crystals also were grown through the 2D layer growth mechanism on the (011) surfaces.

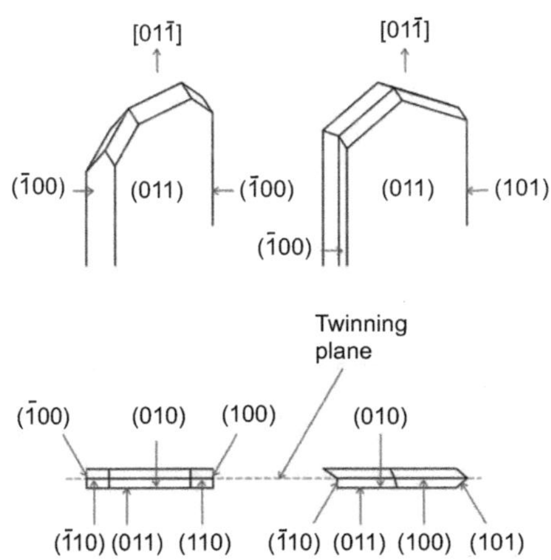

Figure 6.20 Schematic illustration of plate-type SnO_2 crystals obtained by the CVT method along the <011> and <111> crystallographic directions. Redrawn from Ref. [122] with some modifications.

The CVT-grown SnO_2 crystals had often more complex growth habit and smaller size as compared with PVT-grown crystals, likely due to a different chemistry and lower growth temperature. Although the faceting and some of the shapes (rods, plates) are similar in both growth methods from the gas phase, the growth environment containing halogen (CVT) seems to have a large impact on the growth morphology.

6.4.3.2.3 *Examples*

(i) For growing SnO_2 crystals by CVT, Crabtree et al. [119] used $SnCl_4$ and elemental Sn as the Sn source, Ar as the carrier gas, and O_2 as the oxygen source. A growth furnace for this purpose is shown in Fig. 6.21. Through the furnace an alumina tube extends horizontally. The advantage of the alumina tube as compared with the mullite one is minimized contamination with Si. Inside the alumina tube, there is an alumina liner with a boat containing Sn. Ar is introduced into the alumina tube,

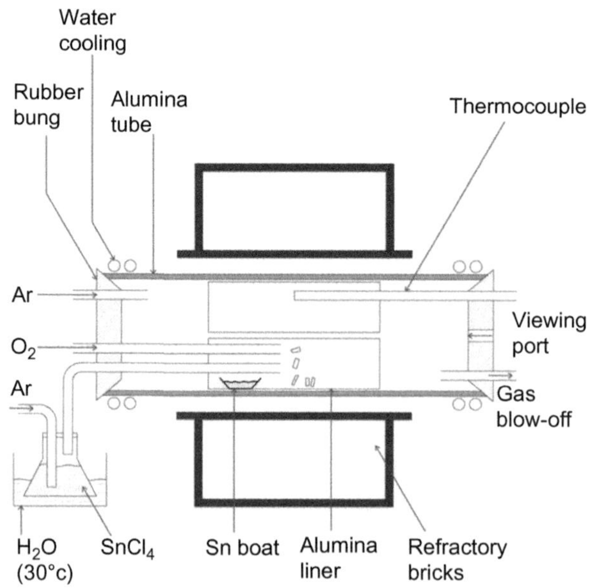

Figure 6.21 Schematic furnace for SnO_2 growth by the CVT method. Redrawn from Ref. [119] with some modifications.

while into the alumina liner both Ar carrying $SnCl_4$, and O_2, by means of separate transport alumina tubes. Ar carrying $SnCl_4$ flows directly over the boat with Sn and meets the O_2 stream beyond the boat with Sn in the alumina liner, which defines the growth zone. Flowing $SnCl_4$ over the boat with Sn prevents Sn oxidation. The gas outlet is situated on the opposite end of the furnace. The growth temperature was 1400°C.

The obtained crystals were clear and transparent, with dimensions up to $2 \times 2 \times 12$ mm^3. Contamination with impurities was significantly reduced, as Si and Na content was below 50 wt. ppm each, while the content of other impurities was below 10 wt. ppm.

(ii) Another growth furnace design applied by Crabtree [124, 125] is depicted in Fig. 6.22. Here all alumina tubes were replaced with quartz tubes. Into the horizontal quartz tube, two transport quartz tubes supply, separately, hot $SnCl_4$ and H_2O steam via Ar transport gas. The Sn and O_2 sources were $SnCl_4$ and H_2O, respectively. Hydrolysis of $SnCl_4$ produced SnO_2 single crystals on the wall of the horizontal quartz tube downstream of the outlets of the transport quartz tubes in the reaction described by Eq. 6.8, wherein the HCl vapor was evacuated on the opposite end of the horizontal quartz tube.

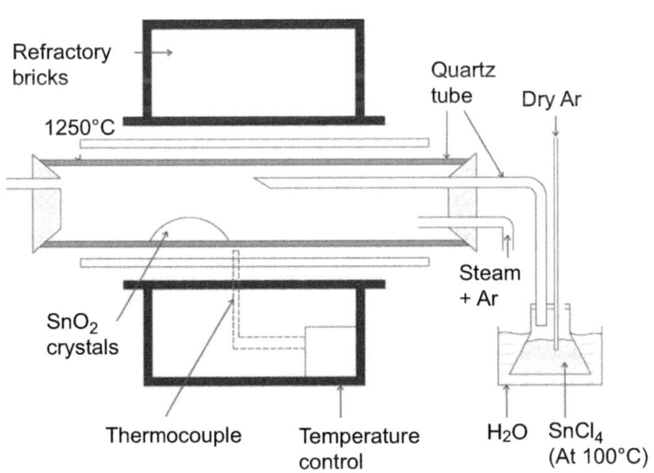

Figure 6.22 Schematic furnace for growing SnO_2 crystals by the CVT method. Redrawn from Ref. [124] with some modifications.

The growth runs were performed at 1250°C for 2–3 days. The crystals were doped with Eu or Tb by using $EuCl_3$ or $TbCl_3$ sources. Eu and Tb concentrations in the SnO_2 crystals were 100–1000 and 1000–10000 wt. ppm, respectively. The crystals with low Eu concentrations were colorless, while those with higher concentrations were pink. Eu and Tb are incorporated into the crystal lattice as Eu^{3+} and Tb^{3+}.

(iii) To grow SnO_2 crystals by the CVT method, Nagasawa et al. [120] used anhydrous $SnCl_4$ as the Sn source in a furnace shown in Fig. 6.23. The resistively heated horizontal furnace contains an alumina tube (10 cm inner diameter and 100 cm length) with two inlets for N_2 carrying $SnCl_4$ and wet O_2 on one end, and exhaust on the opposite end. The inlets are connected with quartz tubes (3 cm inner diameter and 30 cm length) extending inside the alumina tube, and with hoses outside the furnace that are connected with two glass vessels containing $SnCl_4$ and H_2O, respectively. Through the $SnCl_4$ and H_2O vessels, N_2 and O_2 pass with the function of a carrier and oxidizing gas, respectively. The temperature of the $SnCl_4$ and H_2O sources was either 20 or 50°C, while the flow rates of N_2 and O_2 were between 25 and 150 cc/min. The $H_2O/SnCl_4$ molar ratio was between 1 and 2. The temperature in a growth zone situated at the central part of the furnace was between 1100 and 1300°C. SnO_2 crystals grew through the hydrolysis reaction described by Eq. 6.8 at two locations: at the outlet of the quartz tube supplying $SnCl_4$, and on the inner surface of the alumina tube. SnO_2 crystals on the alumina tube were highly defected granules, while those grown at the outlet of the quartz tube were colorless and transparent with smooth faces. The crystals were rods, plates, butterflies, and dendrites. The rod-shaped crystals dominated at high temperatures and high flow rates of N_2 and O_2, while the butterflies and dendrites dominated at low flow rates of these gases. The largest crystals after 24 h growth time had size of $2 \times 5 \times 15$ mm^3. The main impurity was Si at the concentration of 50 wt. ppm, Fe and Ca at the concentration of 5 wt. ppm, and possibly Al.

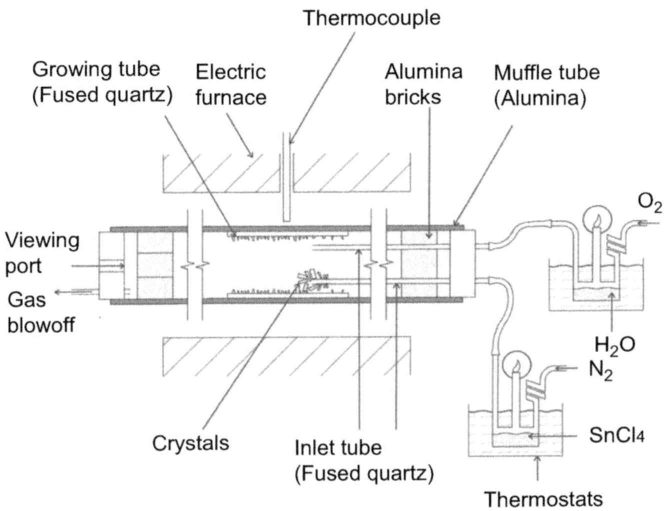

Figure 6.23 Schematic illustration of a CVT growth furnace. Redrawn form Ref. [120] with some modifications.

(iv) Fonstad et al. [123] used in SnO_2 growth experiments $SnCl_4$ and O_2 + H_2 as the Sn and oxygen source, respectively, which leads to the formation of SnO_2(s) and HCl (Eq. 6.9). $SnCl_4$ was produced outside the growth furnace by chlorination of elemental Sn (Eq. 6.11).

The growth of SnO_2 crystals was carried out in a resistively heated furnace as shown in Fig. 6.24. It consists of a 5 cm diameter and 90 cm long mullite tube with endcaps on both ends, and of a quartz tube within the mullite tube, which acts as a substrate for growing SnO_2 crystals with no contact with the mullite tube (minimization of contamination from the mullite tube). The furnace is connected with an $SnCl_4$ reactor, an $SnCl_4$ + dopant reactor, and with individual Cl_2, H_2, and O_2 flows controlled by valves and flowmeters. The $SnCl_4$ reactor is a funnel containing Sn, which is heated with a heater tape to 100°C, through which Cl_2 is passing. At 100°C and pressure of 1.33 kPa, there is a complete reaction between Sn and Cl_2. The second parallel reactor ($SnCl_4$ + dopant) is similar to the first reactor, but the starting material contains, in addition to Sn, a small proportion of an electrically active dopant, e.g., Sb. The total gas flow was typically 100 cm^3/min at 120 kPa and

300 K, while the ratio of individual Cl_2, H_2, and O_2 flows was in the proportions of 2:10:5, respectively. $SnCl_4$ and additional Cl_2 were injected into the furnace through a zirconia nozzle coaxial with the furnace tube and H_2 through a quartz tube concentric with the zirconia nozzle, while O_2 was injected near the periphery of the furnace tube. H_2 stream shields the zirconia nozzle with $SnCl_4$ + Cl_2 from O_2 and, therefore, prevents the growth of SnO_2 crystals at the injector. The individual Cl_2 added to the $SnCl_4$ flow slows the reaction of SnO_2 formation, i.e., it controls the growth rate. The crystal nucleation occurred at the region downstream from the furnace center, where the temperature was decreasing from 1250 to 1175°C at 10 K/cm.

Before the growth, the whole growth system with hot furnace was purged with He gas at the pressure of 1.33 kPa. Then the He flow was turned off, and the flow of other gases (H_2, O_2, and Cl_2) was started. The growth of crystals could be observed through a viewing window on the opposite end of the furnace. The additional Cl_2 flow (at the rate of 1 cm^3/min) acts as the growth regulator, which was reduced to zero when the crystals grew larger. If doping was desired, also a flow from the reactor $SnCl_4$ + dopant was initiated. The exhaust gases passed through liquid nitrogen and pumped into an exhaust. Once the growth was completed, the gas flows were stopped and the whole system was purged with He again, with an increased pressure of 1 MPa. Then the crystals were removed from the furnace. This growth system enabled (i) control of the growth rate (additional Cl_2 flow); (ii) reverse reaction (high additional Cl_2 flow) to remove unwanted nucleation or etch the seeds; (iii) high-purity crystals; (iv) controlled doping at different growth stages (a separate reactor for dopants, which can be switched on/off at any time during growth).

A typical size of the obtained crystals was $2.5 \times 5 \times 5$ mm^3 for 40 h growth runs. Undoped crystals were clear and colorless. Sb-doped crystals were slightly bluish. In the case of a small Sb concentration, the growth habit was the same as in the case of undoped crystals. A few impurities were found at small concentrations, namely Si = 10 wt. ppm and Ca, Cu, Fe below 1 wt. ppm each. Si concentration at the base of the crystals, which were anchored to the quartz tube, was high, 0.1 wt.%.

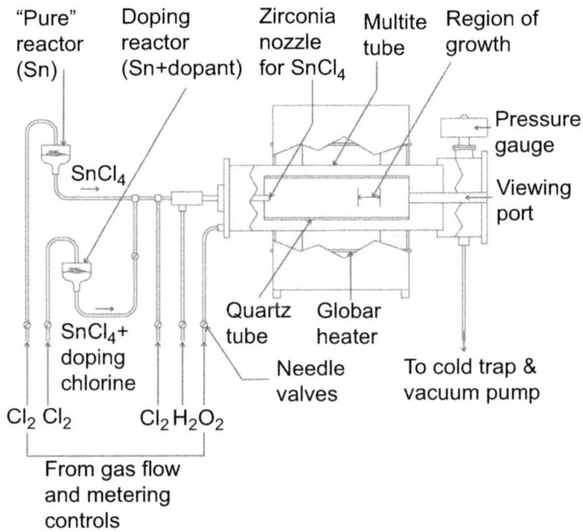

Figure 6.24 Visualization of a CVT growth furnace for SnO_2 crystals. Redrawn from Ref. [123] with some modifications.

(v) Hirose and Furuya [121] and Furuya et al. [122] used SnI_4 powder as the Sn source and O_2 or H_2O as the O_2 source (reaction gas). The growth furnace consists of two separate electrically heated zones, called furnace I and furnace II (Fig. 6.25). Through both furnaces extends a horizontal alumina tube of size 50 mm in diameter and 1000 mm in length. Inside the alumina tube, there is a combustion tube of 20 mm in diameter, which extends through furnace I and partly into furnace II. At the central part of furnace II, next to the end of the combustion tube, there is a quartz tube of 35 mm inner diameter and 60 mm length. At opposite ends of the alumina tube, there are an inlet glass tube and an exhaust glass tube. Inside the combustion tube in the central part of furnace I, there is a combustion boat with SnI_4 starting material of 3N purity (30 g). SnI_4 was heated to 190–200°C, and its vapor was transported via dried N_2 as a carrier gas to the growth zone at the central part of furnace II. From the opposite end, dried O_2 or H_2O vapor (carried by dried N_2 or wet N_2) was transported in the upstream direction to the growth zone having a temperature of 1000–1300°C. Wet N_2 was saturated with H_2O at high and low level, by passing dried N_2 through water at 100°C or at room temperature. The flow rates of dried N_2 that carried SnI_4 and O_2/wet N_2 were 230 and 150 cm^3/min, respectively. The growth time of SnO_2 crystals was several hours, typically 5 h, and the growth proceeded at the tip of the combustion tube.

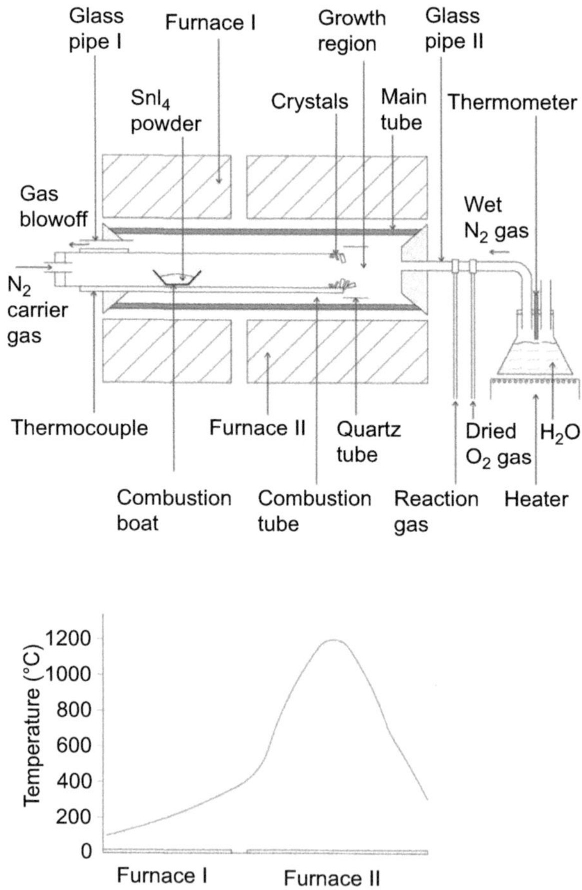

Figure 6.25 Furnace for growing SnO_2 crystals by the CVT method and the temperature profile. Redrawn from Ref. [121] with some modifications.

When dry O_2 was used as the reaction gas at the optimum growth temperature of 1200–1250°C, after 5 h growth only very small and dendritic SnO_2 crystals with SnO_2 powder were obtained. All SnO_2 was white. In this case, $SnO_2(s)$ was obtained by the oxidizing (combustion) reaction described by Eq. 6.8. Larger crystals up to 10 mm long were obtained with the use of wet N_2 at the growth temperature of 1100–1150°C in the growth zone. SnO_2 crystals were obtained by the hydrolysis reactions shown by Eq. 6.9.

The crystals grown by the hydrolysis reaction within 5 h were of different shapes: leaves (low and high saturation of N_2 with H_2O), dendrite (low saturation of N_2 with H_2O), pyramid (higher saturation of N_2 with H_2O), plate and hollow bodies (low and high saturation of N_2 with H_2O). They were all colorless and transparent or semi-transparent.

(vi) In growing SnO_2 crystals by the CVT method, Matsumoto et al. [126] used SnO_2 powder as an Sn source and I + S as a chemically active transport gas, as I itself is not an adequate transport agent for Sn. In such environment, the transport reaction proceeds as follows:

$$SnO_2(s) + I_2(g) + \tfrac{1}{2}S_2(g) \leftrightharpoons SnI_2(g) + SO_2(g) \qquad (6.12)$$

Standard free energy in this reaction changes from positive to negative values, so it is thermodynamically allowed. Further, the standard enthalpy change is positive, so the direction of the transport is expected to be from higher to lower temperatures.

For optimal transport, the equilibrium constant for the reaction 6.12 should be around unity.

Growth experiments were performed in a sealed ampoule located on a movable support inside an alumina tube of a resistively heated horizontal furnace with a predefined temperature profile, as shown in Fig. 6.26. The ampoule was 15 mm in diameter and 170 mm in length (volume 30 cm^3). The amount of the SnO_2 starting (source) material was 1, 3, or 5 g, the amount of iodine was 50, 150, or 450 mg, and the amount of sulfur was 3.5, 10, or 30 mg. After loading the source materials, the ampoules were evacuated to $1.3 \times 10^{-6} - 1.3 \times 10^{-7}$ bar and sealed off. The temperature at a hot zone was varied between 700 and 1180°C, while the temperature difference between the hot and cold (growth) zones was 100, 200, or 300 K. The growth time was typically

2–3 days (sometimes 10 days), and the cooling time to room temperature was half a day.

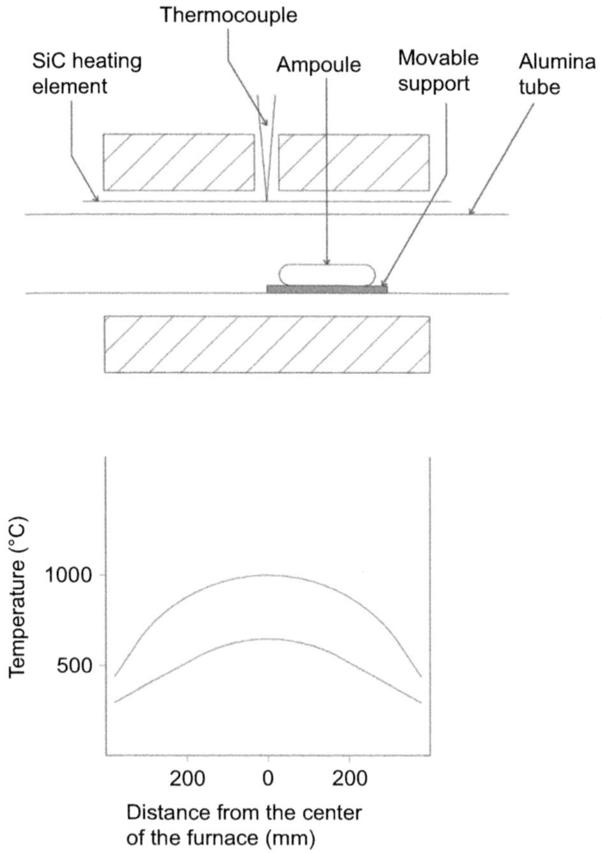

Figure 6.26 Growth furnace for SnO_2 and temperature distribution therein. Redrawn from Ref. [126] with some modifications.

In the case of using iodine alone as the chemically active transport gas, no reaction occurred and the SnO_2 starting material remained powdered. Using both I and S enabled both reaction and transport, which are necessary for growing crystals. The optimum molar ratio of S:I was found to be 1:4. If it was larger, golden films of, likely, SnS_2 were deposited in the ampoule, while when lower, the transport rate was very low. The transport rate could also be controlled either by temperature at the hot zone at fixed temperature difference between the hot and cold zones, or by higher temperature difference at fixed temperature at the hot zone. The growth rate increased with temperature at the hot zone or with larger temperature difference. The optimum operating temperatures were as follows: temperature at the hot zone of 1100°C and the temperature difference of 200 K.

The obtained crystals were small (about 1 mm size) and slightly colored. Some of these crystals were used as seeds for next growth runs, which lasted for 10 days. The obtained crystals were still small, up to 2 mm in size. Some of them were clear and others colored. The coloration could be partly removed by annealing at 1000°C in O_2 at 1 atm for 7 days.

6.4.4 Solution

Among solution techniques, mainly flux method was frequently used to grow bulk SnO_2 crystals, although hydrothermal method was also occasionally utilized. As SnO_2 is highly unstable at high temperatures, the flux method seems to be well suited for growing small single crystals for fundamental studies. In fact, this technique was frequently used for that purpose.

6.4.4.1 Hydrothermal

The hydrothermal method for growing SnO_2 was applied by Kuzmina and Litvin [127], Harvil and Roy [128], and Fujiki and Suzuki [129].

In growth experiments by Fujiki and Suzuki [129], a pressed (490 bar, RT) and sintered (1000°C, air, 20 h) SnO_2 source material (nutrient) was placed within a small gold capsule (inner diameter 4 mm, length 4 mm) with a small hole, which acted as a baffle. The small Au capsule was placed in a dissolution zone of a larger Au capsule (outer diameter 6 mm, length 70 mm) along with a solvent, which was next sealed by welding. The larger Au capsule was then loaded into a pressure vessel. A number of different solvents were 10% solutions of NaOH, KOH, Na_2CO_3, K_2CO_3, NH_4HCO_3, NaCl, LiCl, NH_4Cl, NH_4F, HF, KF, and 3.5% solution of NaF. The pressure was typically 1000 atm, but 500 atm was applied as well. Growth duration was between 65 and 190 h. The growth temperature at the crystallization zone was between 500 and 600°C, while the temperature difference between the crystallization and dissolution zones was between 60 and 70 K. In each case, the largest crystals of 2–3 mm size were grown at 600°C at 1000 atm for at least 100 h with the use of either Na_2CO_3 or NaOH, or NH_4 or HF. It was pointed out that the growth habit and crystal coloration strongly depend on the solvent. Crystals obtained from oxyhydride and carbonate solvents were acicular or polyhedral with developed {100} and {101} faces, while from chloride solvent they were prismatic with similar faces, often twinned on the $(01\bar{1})$ plane. They were colorless in both cases. Fluoride solutions produced polyhedral-shaped crystals with a blue to black coloration and developed {110}, {111}, and {322} faces.

6.4.4.2 Flux

6.4.4.2.1 *Crystal growth*

SnO_2 crystals were grown from the flux in standard resistively heated furnaces, by placing therein a Pt crucible with a starting material consisting of an Sn source, flux, and dopant (if used). The reported fluxes are Cu_2O [106, 130–134], B_2O_3–V_2O_5 [135, 136], and PbF_2, $Na_2B_4O_7$, $Na_2B_4O_7$–KF [129], while the Sn source is typically SnO_2 powder, although Zn_2SnO_4 [135] was tested as well. The maximum temperature T_{MAX} to which the flux was heated is 990–1200°C [129] for $Na_2B_4O_7$, $Na_2B_4O_7$–KF, PbF_2, 1250–1290°C [106, 130–132] for Cu_2O and 1300°C [136] for B_2O_3–V_2O_3 fluxes. In the case of Cu_2O flux, higher temperatures 1300°C lead to the formation of metallic Cu that forms with Pt eutectic cause crucible damage. The growth proceeded either by soaking the crucible with the starting material at the T_{MAX} (e.g., 1250°C) for a predefined time and then cooled down [106], or by slowly cooling down (e.g., at 2–4 or 5 K/h) from the maximum temperature (e.g., 990–1200 or 1300°C) to a lower final temperature T_F (e.g., 790–890 or 900–1000°C) [129, 136]. The growth duration was between several hours [130, 131] and several days [106, 136]. Cooling down was performed by a predefined decrease in the furnace temperature [130, 131], by natural cooling (furnace shut off) [106], or by pouring the liquid flux on a try [136]. The crystals were removed from the Cu_2O flux in a hot HCl aqua solution [106, 130, 131], and in aqua regia or aqua ammonia [133, 134]. In the case of the B_2O_3–V_2O_3 flux, SnO_2 crystals were removed in hot 50% nitric acid [136], while in the case of the PbF_2, $Na_2B_4O_7$, $Na_2B_4O_7$–KF fluxes in 20% nitric acid [129].

The obtained SnO_2 crystals were small, from micrometer to millimeter size, in the form of needles, rods, plates, and prisms. For instance, Kawamura et al. [130, 131] reported the crystal size of about 10–20 μm (prisms), Kunkle and Kohnke [106] reported needles, platelets, and rods of 5–10 mm length and 1 mm^2 cross section, Shimada et al. [135, 136] reported needles of 1 mm^2 cross section, while Hidenori and Kenji [133, 134] reported lumps of $3 \times 3 \times 3$ mm^3. Undoped (i.e., with no intentional doping), crystals were colorless when grown from the Cu_2O flux [106], and pale brown, brown, or pale red in the case of the B_2O_3–V_2O_3 flux [135, 136].

The flux-grown crystals usually contained flux and furnace impurities, e.g., Kunkle and Kohnke [106] reported the presence of Cu and Si in SnO_2 crystals obtained from the Cu_2O flux. In terms of intentional doping, Hidenori and Kenji [133, 134] used Sb^{5+} ions to increase the electrical conductivity of SnO_2 crystals, while Kawamura et al. [130–132] used a number of different elements ranging from mono- to hexavalent ions to study their impact on crystal growth habit and size.

6.4.4.2.2 *Growth morphology*

Kawamura et al. [130–132] performed an extensive study of the impact of different mono- to hexavalent ions on the growth morphology of SnO_2 crystals by the flux method. Undoped SnO_2 crystals (i.e., grown from pure Cu_2O flux) were square in a cross section with prismatic tips at both ends, similar to a theoretically predicted habit with the use of the periodic bond chain (PBC) model, as demonstrated in Fig. 6.27. The preferred growth direction (longwise) is parallel to the crystallographic axis *c*. The side walls of the theoretical crystals are formed by the {110}, while the pyramidal tips by the {101} faces. However, in growth experiments, the faces formed at the crystal tips were {111} instead of {101}, probably due to the impact of the flux. Similar crystal growth morphology was also observed in some of the PVT growth experiments.

Mono-, di-, tetra-, and hexavalent cations had no impact on the growth morphology; however, tri- and pentavalent cations with ionic radii similar to Sn^{4+} affected the growth morphology and crystal size. They accelerated (except Fe^{3+} and Al^{3+}) the growth along the *c*-axis and transformed {111} faces from flat and smooth to rounded and rough (from S face to K face). The development

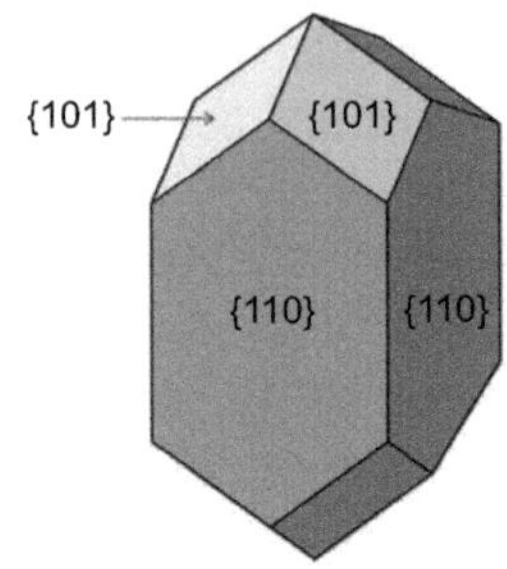

Figure 6.27 Structural form of SnO_2 concluded from the periodic bond chain model (PBC). Redrawn from Ref. [131] with some modifications.

of S and K faces was also observed by Fujiki and Suzuki [129] in hydrothermally grown SnO_2 crystals. Only {111} faces were affected by the impurities. The transformation of the {111} faces is due to surface reaction with impurities, and it is attributed to the braking of the PBCs by impurity cations through substitution of Sn^{4+} ions by trivalent cations, which results in the desorption of the surrounding oxygen. Normal growth rates were accelerated in the <111> and suppressed in the <110> direction, and the growth habit changed from short prismatic to needle, whiskers or acicular form. Needle-shaped crystals with a square cross section were also obtained from the Cu_2O flux by Kunkle and Kohnke [106] and Shimada et al. [135, 136]. For instance, the growth rate ratio along the <111> and <110> directions increased from 1.08 for undoped SnO_2 to 4.2 in the case of doping the flux with 5 mol.% Cr_2O_3. The S face of (111) contains one PBC, while the K face contains no PBCs. The F face of (110) contains 2 PBCs. On the (111) face, oxygen atoms are alternately arranged in two- and three-fold coordination, while on the (110) face, oxygen atoms are arranged in three-fold coordination. When trivalent ion is adsorbed on the (111) surface, an oxygen atom in the neighborhood is desorbed to maintain charge neutrality and one PBC is broken. If the trivalent ion is adsorbed on the (110) surface, an oxygen atom is not destabilized in the neighborhood and not desorbed as in the case of the (111) surface. Flux-grown SnO_2 crystals reported by Fujiki and Suzuki [129] had different growth habit depending on flux type. $Na_2B_4O_7$ flux produced tubular crystals with developed {100}, {120}, and {101} faces. They were elongated in a direction inclined at 57° to the *c*-axis, similar to the morphology observed in natural cassiterite. Acicular crystals obtained from the $Na_2B_4O_7$–KF flux were surrounded by the {100}, {110}, and {101} faces. The crystals were elongated parallel to the *c*-axis, and some of them were twinned at the $(01\bar{1})$ plane. However, the acicular crystals obtained from the PbF_2 flux were bounded by a number of faces: {111}, {131}, {421}, {121}, and {411}.

6.4.4.2.3 *Examples*

(i) Kunkle and Kohnke [106] grew SnO_2 crystals from the Cu_2O flux. The mixture of Cu_2O and SnO_2 with the volume ratio of 2:1 was loaded into a 50 cm^3 Pt crucible that was heated up in a muffle resistance furnace. The temperature at the base and rim of the crucible was 1250°C and 1150°C, respectively. After a week of growth duration, the furnace was shut off and naturally cooled down to RT. The crystals were removed from the crucible by dissolving the flux in HCl. The obtained colorless crystals were needles through platelets to rods with depth in the crucible. The rods had a square cross section of 1 mm^2 and length of 5–10 mm. Long dimension was in the *c*-axis crystallographic direction, while the side surfaces were {100} planes. The crystals contained 0.0021 wt.% of Cu and 0.02 wt.% of Si.

(ii) Cu_2O flux for growing SnO_2 crystals was also used by Kawamura et al. [130, 131], who also investigated the impact of different dopants ranging from mono- to hexavalent ions. The amounts of Cu_2O, SnO_2, and dopants (typically as oxides) were 3.96 g, 1.04 g, and 0.05–0.1 g, respectively. The experiments were done in a 20 cm^3 Pt crucible placed within a resistively heated furnace. The furnace was heated up to 1280°C within 1 h, soaked for 3 h, and then cooled down to RT. The crystals were removed from the flux in a hot (80°C) HCl aqua solution for 2–3 days. The obtained crystals were of micrometer size only (about 10–20 μm). In liquid phase epitaxy (LPE) growth experiments, Kawamura et al. [132] were able to determine the solubility of SnO_2 in Cu_2O flux, which is between 2.5 and 3.5 mol.% in a temperature range of 1250–1320°C.

(iii) Fujiki and Suzuki [129] used 30 or 50 ml Pt crucibles and three different fluxes: $Na_2B_4O_7$ (92.3 wt.%), $Na_2B_4O_7$–KF (88.5–3.3, 74.7–6.8, and 43.5–43.5 wt.%), and PbF_2 (92.8 and 94.3 wt.%)

and SnO_2 powder in an amount 5.7–18.5 wt. % of the total weight of the starting material. The maximum temperature to which the flux with SnO_2 was heated up was 1200°C for $Na_2B_4O_7$, 1000–1200°C for $Na_2B_4O_7$–KF, and 990–1000°C for PbF_2. The cooling-down rate to the final temperature of 790–890°C was between 2 and 4 K/h. The highest flux losses of about 25–26.4 wt.% were noted for the PbF_2 flux, while for other fluxes below 10 wt.%. The $Na_2B_4O_7$ flux produced small tubular crystals below 1 mm in size; however, adding KF to $Na_2B_4O_7$ increased the crystal size up to $5 \times 0.2 \times 0.2$ mm^3 of acicular shape. Also, the PbF_2 flux produced relatively large crystals of $2 \times 0.2 \times 0.2$ mm^3 having an acicular shape.

(iv) Shimada et al. [135, 136] used the B_2O_3–V_2O_5 flux and Zn_2SnO_4 or SnO_2 as the Sn source. The molar ratio of the B_2O_3:V_2O_5 flux was 1:1. The starting material of 15–18 g containing the flux and SnO_2 powder was loaded into a 30 ml Pt crucible with a cover. The crucible was soaked at 1300°C for 20 h in air and then cooled down to 900–1000°C at the rate of 5 K/h. Next, the still molten material was poured on a stainless tray. SnO_2 crystals were then removed from the flux in hot 50% nitric acid. The obtained crystals were needle shaped of about 1×1 mm^2 in cross section with the long dimension along the *c*-direction. They were transparent and pale red in color. If the molar ratio of B_2O_3:V_2O_5 was 4:1, mainly $Bi_2Sn_2O_7$ was formed with a minor amount of SnO_2 crystals. In the case of Zn_2SnO_4 as the Sn source, the obtained needle-shaped crystals were pale brown or brown when grown at higher temperature or with a higher V_2O_5 content.

6.4.5 Conclusion

The best bulk SnO_2 crystals in terms of size and structural quality were obtained by the PVT method. Standard resistive furnaces produced relatively large crystals with a maximum volume of about 1 cm^3, but often the crystals were contaminated with elements from the furnace (mainly Al and Si). An inductively heated furnace with an Ir crucible produced yet larger crystals of 5–6 cm^3 with low level of residual contamination. The PVT-grown crystals also revealed quite low EPD and good rocking curve, enabling the preparation of wafers for epitaxy. CVT-grown SnO_2 crystals were smaller in size as compared with PVT-grown crystals but had relatively high purity. Often they had a twinned complex shape. On the other hand, the crystals obtained by the hydrothermal and flux methods were much smaller than those grown from the gas phase. Additionally, flux-grown crystals were contaminated with flux elements. Intentional doping was practiced mainly to control the electrical properties (Sb), optical properties (Eu, Tb), and the growth morphology (flux method). The summary of basic outcome on bulk SnO_2 crystals is provided in Table 6.1.

6.5 Thermal Stability of Bulk SnO_2 Crystals and Their Surfaces

6.5.1 Thermal Stability

Application of SnO_2 crystals for high-temperature operation (e.g., high-temperature gas sensors) or as substrates for epitaxy requires crystal stability at temperatures in quest. This can be roughly concluded from annealing experiments by checking mass loss as a result of thermal decomposition and determination of the surface damage. By evaluation of the mass loss of the SnO_2 crystals (Fig. 6.28) obtained by the PVT method (Galazka et al. [90, 114]), it can be substantially concluded that stoichiometric and high-purity SnO_2 crystals are thermally stable up to about 1200°C in non-reducing atmospheres (O_2, Ar) and 20 h annealing time, wherein the mass loss (i.e., decomposition) proceeds faster in the case of neutral as compared with the oxidizing atmosphere. As shown in Fig. 6.29a, originally brownish samples annealed in the presence in O_2 remain clear and transparent even after annealing at 1400°C for 20 h; however, there are signs of thermal decomposition on the surface. In the case of black samples, they usually do not become entirely colorless, but partly pink–purple. When highly reducing atmosphere containing H_2 is used, the mass loss starts above 600°C after 10 h annealing, and at or above 700°C, the mass loss proceeds very fast. This is due to the reducing action of H_2 leading to the formation of metallic Sn:

Table 6.1 Summary of bulk SnO_2 crystals obtained by different growth methods. EPD and FWHM indicate etch pit density and full width at half maximum of the rocking curve, respectively

Phase	Method	Crystal shape	Maximum crystal size [mm]	Dopants	Basic structural quality	Reference
Gas	PVT[(a)]	Needles/fibers, rods, plates	20×7×7, 30×4×2, 15×3×0.1	Sb, In, Ga	EPD=10^2–10^5 cm^{-2}	94, 95, 105, 109, 110, 111, 112, 113, 116, 117
	PVT[(b)]	Hemispheres	25×10	—	FWHM<50 arcsec, Low-angle grain boundaries	90, 114
	CVT	Rods, plates, prisms, granules, butterflies, leaves, dendrites	2×5×15	Sb, Eu, Tb	Twins, complex forms	119, 120, 121, 122, 123, 124, 125
Solution	Hydrothermal	Needles/rods	—	—	Twins	129
	Flux	Needles/fibers, rods, plates, prisms	10×1×1, 3×3×3	Mono- to hexavalent ions	Flux impurities, twins	106, 129, 130, 131, 132, 133, 134, 135, 136

[(a)]standard furnace, resistive heating; [(b)]Ir crucible, inductive heating, sapphire substrate

$$SnO_2(s) + 2H_2(g) \rightarrow Sn(l) + 2H_2O(g) \quad (6.13)$$

This reaction reaches the equilibrium coefficient of unity (K_{eq} = 1) already at 550°C, while at 800 and 1000°C, it is 16 and 61, respectively. Figure 6.29b shows the effect of annealing of SnO_2 samples in the presence of H_2 (5%) + Ar at 800°C and 1000°C for 10 and 1 h, respectively. After annealing at 800°C, the crystal surface is entirely covered with small metallic Sn droplets, while after annealing at 1000°C, the sample was almost entirely reduced to Sn with large metallic Sn droplets.

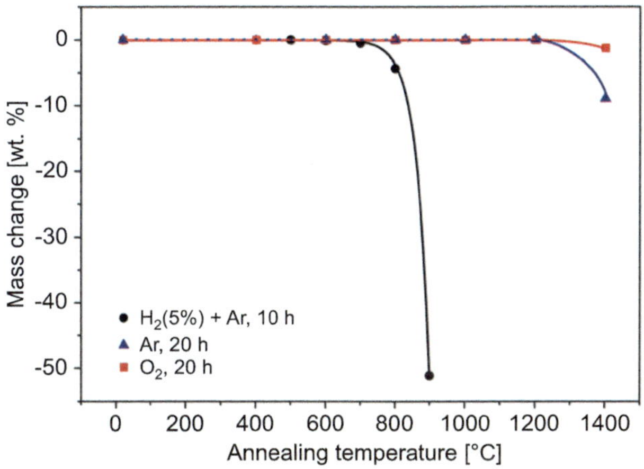

Figure 6.28 Thermal stability of PVT-grown SnO_2 crystals when subjected to annealing at different temperatures and atmospheres. Reprinted (with minor modifications) with permission from Ref. [90], Copyright 2014, John Wiley and Sons.

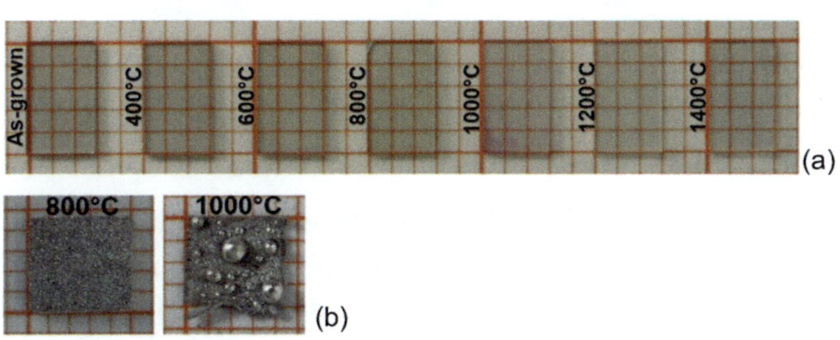

Figure 6.29 SnO_2 samples (PVT-grown) after annealing at 400–1400°C in O_2 for 20 h (a); and at 800°C/1000°C in H_2 (5%) + Ar for 10 h/1 h (b).

6.5.2 Surfaces of SnO_2 and Their Stability

The stability of SnO_2 surfaces and their selection are important for epitaxial growth, gas sensing, and catalytic properties. The knowledge of the surface composition and structure enables a proper surface selection for an intended application. A good analysis of low-index SnO_2 surfaces was carried out by Batzill and Diebold [137]. Experimentally studied bulk SnO_2 crystals were obtained by the flux method, while the study techniques involved low-energy electron diffraction (LEED), scanning tunneling microscopy (STM), and low-energy ion scattering (LEIS).

Different theoretical calculations [138–140] revealed the following order of the surface energy of a stoichiometric bulk SnO_2 crystal: (110) < (100), (010) < (101), (011) < (001). This energetic relation is often reflected in real crystals obtained from the flux and gas phase, as shown in Fig. 6.30. In this example, the surface coverage by low-index faces follows the relation of the surface energy, with the largest faces {110} as the most favorable and with no occurrence of the {001} faces.

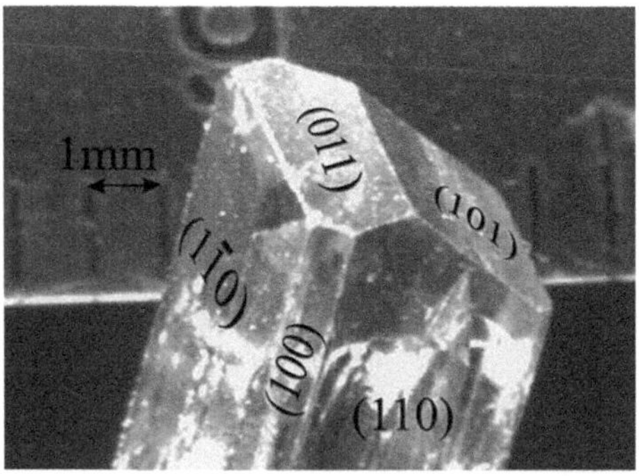

Figure 6.30 Bulk SnO_2 crystal obtained by the PVT method with indicated surfaces. Reprinted from Ref. [73], Copyright 2005, with permission from Elsevier.

A representation of low-index surface termination of bulk SnO_2 crystals, following Batzill and Diebold [137], is shown in Fig. 6.31 in both configurations: stoichiometric surface (left) and reduced surface (right). Experimental study of the (110) surface revealed its complex reconstruction depending on the preparation conditions. This is because of the stoichiometric (110) surface that is composed of five- and six-fold coordinated Sn atoms, three-fold coordinated O atoms, and two-fold coordinated bridging O atoms, while at the reduced surface, the six-fold coordinated Sn atoms become four-fold coordinated.

Vacuum-prepared (110) surface by ion sputtering is always oxygen deficient. The as-prepared and annealed at low temperatures (110) surfaces revealed a combination of 4×1 and (2×2) LEED domains. Annealing at high temperatures (about 1200 K) led to disappearance of the 4×1 domains with 1×1 reconstruction, and with 1×2 reconstruction after further annealing. Similar surface reconstruction, 1×2, was also observed in vacuum-annealed samples that were before annealed in the presence of oxygen. The change in the surface reconstruction from 4×1 to 1×2 upon vacuum annealing is likely related to the decomposition and creation of a reduced (110) surface. An oxygen loss from the (110) surface starts above 440 K in ultra-high vacuum conditions [141]. Sensato et al. [79] concluded that relaxation of the oxidized and reduced (110) surfaces requires displacement of in-plane O and Sn atoms. Moreover, O_2 adsorption does not fully reconstruct the reduced (110) surface.

A sputtered and annealed (600–1000 K) in vacuum (100) surface revealed 1×1 LEED reconstruction. STM studies revealed two types of steps on such surface: (i) straight and oriented along the [001] direction, which corresponds to a unit cell height, and (ii) mainly aligned with [010] direction, which consists of two steps with a half unit cell each [137, 142]. An oxygen depletion layer starts to form at and above 610 K.

Sputtered and vacuum-prepared (101) surfaces at the same temperature range revealed 1×1 LEED reconstruction. Prior exposure to oxygen substantially did not affect the surface reconstruction. The oxygen desorption from the (101) surface starts at about 550 K.

Both reduced (100) and (101) surfaces are formed by removing bridging oxygen from stoichiometric bulk SnO_2 and surface Sn atoms are reduced to Sn^{2+}. A reduced (110) surface is likely formed by removing a bridging oxygen and additionally every in-plane oxygen atom, which will form Sn^{2+} on the surface. Such reduced surface shows a surface SnO composition that is Sn-terminated. Indeed, Themlin et al. [143] showed in their resonant photoemission study that a part of Sn^{4+} changes their oxidation state to Sn^{2+} near the (110) surface.

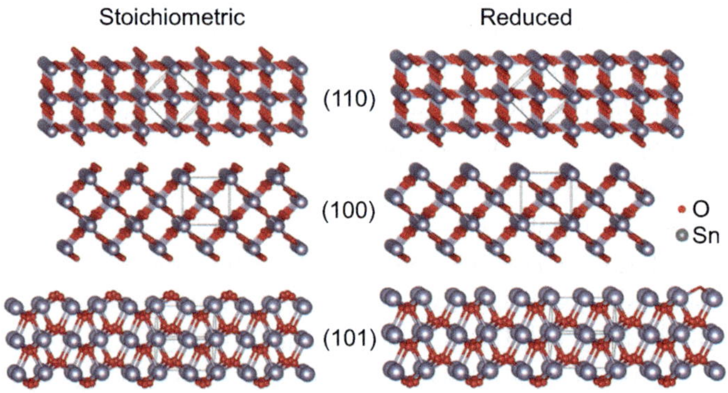

Figure 6.31 Ball-and-stick models of SnO_2 low-index surfaces (110), (100), and (101), which are stoichiometric and reduced. The reduced surfaces were obtained by removing two-fold coordinated bridging oxygen rows from the stoichiometric surfaces. It has been shown that the so-obtained reduced (100) and (101) surfaces are thermodynamically favored for low oxygen chemical potential. The reduced (110) surface exhibits a high surface energy and, thus, is unlikely to form. Drawing by R. Schewski (Leibniz-Institut für Kristallzüchtung, Berlin, Germany) based on Ref. [137].

6.6 Thermal and Mechanical Properties

6.6.1 Thermal Properties

The melting point of stoichiometric SnO_2 could not be properly determined so far. Galazka et al. [90] concluded in a series of experiments at high temperatures up to 2100°C and at high $p(O_2)$ up to 150 bar that it was not possible to melt SnO_2. Another report shows lower melting point of 1999°C [104]. This could be the result of non-stoichiometric composition containing an excess of elemental Sn, which acted as a flux. Therefore, the melting point of stoichiometric SnO_2 is rather above 2100°C.

Thermal expansion of solid SnO_2 was measured by Pannetier and Denes [76] in a temperature range of 273–593 K (0–320°C). A change in both lattice parameters a and c follows the following temperature relations:

$$a\ (\text{Å}) = 3.7984 + 3.31 \times 10^{-5}\ T + 7.9 \times 10^{-8}\ T^2$$

$$c\ (\text{Å}) = 4.8368 + 1.25 \times 10^{-4}\ T + 6.2 \times 10^{-8}\ T^2$$

where T is in Kelvin.

These relations are visualized in Fig. 6.32. The anisotropy in thermal expansion along the *a*- and *c*-axes is large, mainly due to strong Sn–O bonds, wherein the expansion in the *ab* plane comes from the O–O repulsive interaction. The thermal expansion coefficient in the temperature range of 0–320°C is about 2.66×10^{-5} and 3.66×10^{-5} K^{-1} for the *a* and *c* lattice parameters, respectively.

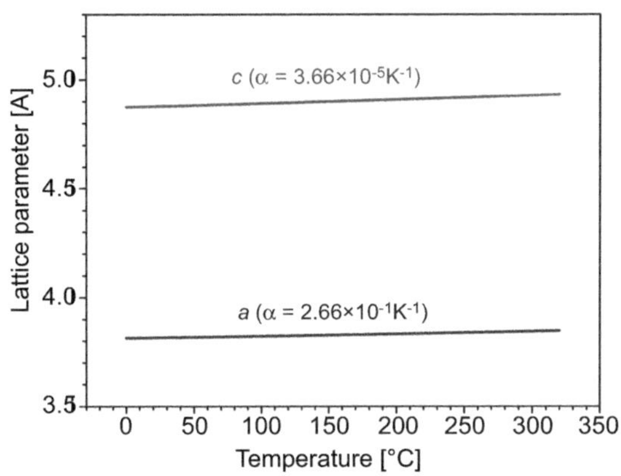

Figure 6.32 Thermal expansion of SnO_2 above RT, based on relations given in Ref. [76].

Thermal conductivity of SnO_2 measured by the laser flush method showed surprisingly high values, even at high temperatures, as illustrated in Fig. 6.33 (Klimm and Galazka [144]). At RT and 900°C, it has values of 51.4 and 11.6 W/mK, respectively. These values are much higher when compared not only with other TSOs (except ZnO, which has thermal conductivity of about 100 W/mK at RT), but also with other oxide single crystals. It is comparable with the thermal conductivity of MgO single crystals (50 W/mK) [145] and larger than that of sapphire (34 W/mK) [145].

Molar heat capacity (C_{pm}) was measured and theoretically predicted at a value of 52.6 J/molK at RT [146, 147]. It increases to a value of about 88.6 J/molK at about 1500°C according to FactSage database [148].

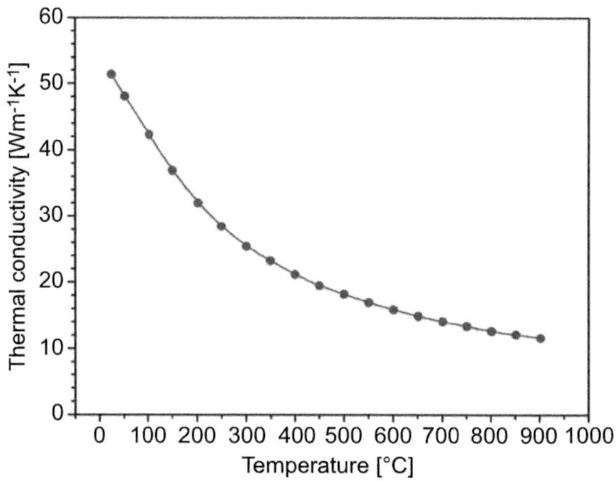

Figure 6.33 Thermal conductivity of PVT-grown SnO_2 crystals along the [001] direction [144].

6.6.2 Mechanical Properties

From measured vibrational frequencies of natural SnO_2 (cassiterite), Katiyar et al. [149] derived sound velocities of phonons propagating along [001], [100], [110], and [101] (between 1.37 and 7.54×10^3 m/s), which enabled calculation of six independent elastic constants (in GPa): $c_{11} = c_{22} = 216.7$, $c_{12} = 190.6$, $c_{13} = c_{23} = 134.5$, $c_{33} = 312.8$, $c_{44} = c_{55} = 55.5$, and $c_{66} = 194.2$. From the Voigt approach, the bulk modulus B_0 and shear modulus G are calculated from the following relations:

$$B_0 = \frac{1}{9}[(c_{11} + c_{22} + c_{33}) + 2(c_{12} + c_{13} + c_{23})] \tag{6.14}$$

$$G = \frac{1}{15}\left[(c_{11} + c_{22} + c_{33}) - (c_{12} + c_{13} + c_{23}) + 3(c_{44} + c_{55} + c_{66})\right] \tag{6.15}$$

And both Young modulus and Poisson ratio can be calculated as follows:

$$E = \frac{9B_0G}{3B_0 + G} \tag{6.16}$$

$$\nu = \frac{3B_0 - 2G}{2(3B_0 + G)} \tag{6.17}$$

The calculated moduli and Poisson ratio yield: B_0 = 185 GPa, G = 80 GPa, E = 210 GPa, and ν = 0.311. For comparison, bulk modulus of SnO_2 calculated by Camargo et al. [150] is 138.4 GPa.

SnO_2 is a moderately hard material having Mohs hardness of 6.5 [137]. The density of solid SnO_2 is 6.99 g/cm^3 [137].

6.6.3 Conclusion

Due to high thermal instability of SnO_2 at high temperatures, it was rather impossible to measure its melting point. It is likely over 2100°C, although lower value of about 2000°C was reported. There is some anisotropy in thermal expansion coefficient along a and c crystallographic axes (about 25%). Thermal conductivity measured along [001] is high, about 50 J/molK, which is one of the highest among TSOs. The Mohs harness is 6.5, indicating moderately hard material. Experimental determination of the elastic constants enabled estimation of bulk, Young's, and shear moduli, as well as the Poisson ratio. Basic physical properties of bulk SnO_2 single crystals are collected in Table 6.2.

Table 6.2 Thermal and mechanical properties of bulk SnO_2 single crystals

Property	Unit	Value	Reference
Solid density	g/cm^3	6.99	137
Melting point	°C	>2100	90
Thermal expansion coefficient of the lattice parameter (0–320°C)	K^{-1}	a: 2.66×10^{-5} c: 3.66×10^{-5}	76
Thermal conductivity at RT	W/mK	51.4	144
Molar heat capacity at RT	J/molK	52.6	146, 147
Elastic constants	GPa	$c_{11} = c_{22}$ = 216.7 c_{12} = 190.6 $c_{13} = c_{23}$ = 134.5 c_{33} = 312.8 $c_{44} = c_{55}$ = 55.5 c_{66} = 194.2	149

(Continued)

Table 6.2 (*Continued*)

Property	Unit	Value	Reference
Bulk modulus	GPa	185[(a)]	149
Shear modulus	GPa	80[(a)]	149
Young modulus	GPa	210[(a)]	149
Poisson ratio	—	0.311[(a)]	149
Mohs hardness	—	6.5	137
Sound velocity along [001], [100], [110], [101]	m/s	1370–7540	149

[(a)]Calculated from elastic constants given in Ref. [149] and Eqs. 6.14–6.17.

6.7 Optical Properties and Effect of Annealing

6.7.1 Transmittance, Absorption Edge, and Bandgap

SnO_2 crystals obtained from the gas phase had high structural quality and were large enough for basic optical characterization. Examples of transmittance spectra of as-grown and O_2-annealed SnO_2 crystals obtained by Galazka et al. [90, 151] are shown in Fig. 6.34. All the samples, whether as-grown or annealed, reveal a steep absorption edge at about 357 nm (3.47 eV) for the (001) orientation and good transparency in the visible spectrum. A similar value (3.5 eV) of the fundamental absorption edge was experimentally concluded by Nagasawa et al. [120]. High absorption in the near infrared (NIR) spectrum in the case of as-grown crystals results from high free electron concentration at the level of 10^{18} cm^{-3}. Summitt and Borrelli [152] gave an empirical relation for the absorption coefficient caused by the free carrier absorption in the spectral range of 0.5–6 μm at RT: $\alpha = 1.72\ n_e\lambda^3$, where α is the absorption coefficient in cm^{-1}, n_e is the free electron concentration in 10^{18} cm^{-3}, and λ is the wavelength in μm. The λ^3 dependance of the absorption coefficient in the spectral region of 600–2000 nm as the result of phonon-assisted absorption by the free carriers was confirmed theoretically by Peelaers et al. [153].

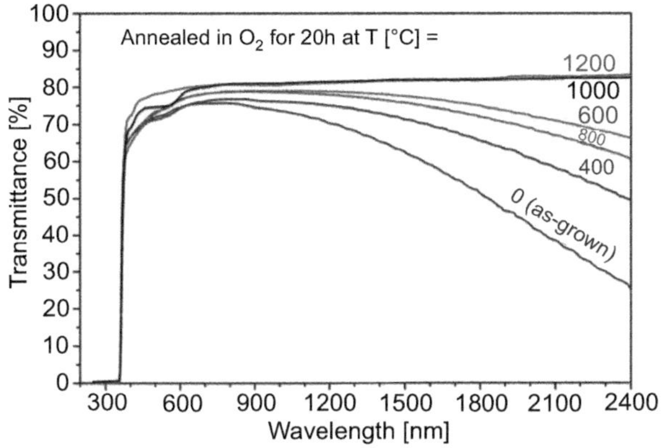

Figure 6.34 Transmittance spectra of as-grown and annealed SnO_2 crystals in O_2 for 20 h at different temperatures. The crystal wafer's thickness and orientation were 0.5 mm and (001), respectively. They were obtained by the PVT method [151].

With annealing temperature in O_2 for 20 h, the transmittance improves in the NIR spectrum as a result of the decrease in free electron concentration. After annealing at or above 1000°C, no free carrier absorption is visible, which indicates semi- or insulating state of the samples. At the same time, there is also a transmittance improvement in visible spectrum, as the brownish coloration was removed, and the crystals became colorless, in particular after annealing at 1200°C. Nagasawa and Shionoya [154] showed that electrically insulating SnO_2 crystals after annealing in O_2 at high temperature of 1300°C and pressure of 9.8 bar for 8.5 and 28 h are fully transparent in infrared spectral region up to 6 μm.

SnO_2 belongs to the tetragonal system; therefore, dichroism is expected. Indeed, the near-edge transmittance (Fig. 6.35) is polarization dependent, whether the polarization is parallel ($E \| c$) or perpendicular ($E \perp c$) to the optical axis c. The transmittance edge for $E \| c$ and $E \perp c$ is about 330 and about 355 nm, respectively [90], with a difference of 25 nm.

The near-edge absorption of PVT-grown SnO_2, as shown in Fig. 6.36 (Galazka et al. [151]), was best fitted with photon energy $E_{ph} \propto \alpha^2$ (where α is the absorption coefficient) indicating a direct transition in line with theoretical predictions. Estimated direct optical bandgaps, $E_{g,dir}$, for $E \| c$ and $E \perp c$, are 3.77 ± 0.03 and 3.485 ±

0.015 eV, respectively. Nagasawa and Shionoya [155, 156] concluded that the fundamental absorption edge is likely related with the direct dipole-forbidden excitation. Moreover, the authors revealed from low-temperature spectroscopic measurements (1.3 K) that the fundamental absorption begins at 3.5 and 3.3 eV for $E \| c$ and $E \perp c$, respectively, and fulfills a hydrogen-like series:

$$E_n = 3.5971 - \frac{0.033}{n_q^2}\,\text{eV} \tag{6.18}$$

for quantum numbers n_q = 2, 3, 4, and 5.

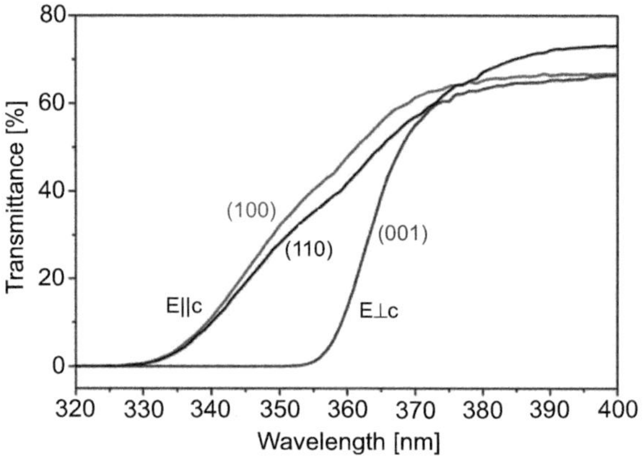

Figure 6.35 Near-edge transmittance of PVT-grown SnO_2 crystals depending on the crystal orientation. The wafer's thickness was 0.5 mm [151].

For PVT-grown crystals, Summitt et al. [157] concluded from the near-edge absorption coefficient the values for the direct transitions of 3.93 and 3.57 eV for $E \| c$ and $E \perp c$, respectively, while for indirect transitions ($E_{ph} \propto \alpha^{1/2}$) of 3.72 and 3.42 eV for $E \| c$ and $E \perp c$, respectively. Reddaway and Wright [158] gave the values for the direct transitions of 4.1 and 3.7 eV for $E \| c$ and $E \perp c$, respectively, while Fröhlich et al. [159] concluded a value of 3.56 eV. Feneberg et al. [160] reported from the ellipsometric measurements of thin films the fundamental, but dipole-forbidden bandgap at 3.596 ± 0.2 eV at RT. Examples of theoretical values for the fundamental bandgap are as follows: 3.65 eV [82], 3.6 eV [83], 3.07–3.36 eV [84], 3.5 eV [172], which in most cases are close to experimental values.

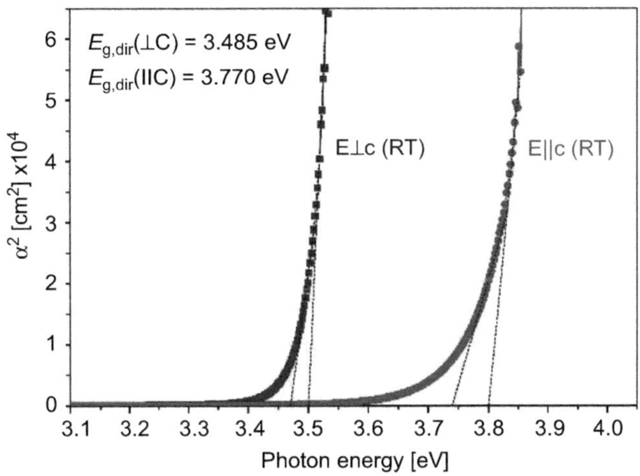

Figure 6.36 Polarization dependent absorption coefficient in power 2 versus photon energy at RT that indicates direct optical transitions. Dot black lines indicate extrapolations of nearly linear parts of the curves. The wafer's thickness was 0.5 mm [151].

Temperature dependence of the bandgap measured by Summitt and Borrelli [161] showed a change in the bandgap of 1.2 meV/K for both polarizations above RT (300–600 K), and a different shift below RT, which was 1.4 times larger for the $E||c$ as compared with $E\perp c$. This is the result of selection rules of electron–phonon interaction.

6.7.2 Dielectric Constant and Refractive Index

The rutile structure is birefringent with two independent components of the dielectric tensor $\varepsilon_\perp$ ($\varepsilon_{xx} = \varepsilon_{yy}$), which is ordinary, and $\varepsilon_{||}$ (ε_{zz}), which is extraordinary. High-frequency (ε_∞) and static (ε_r) dielectric constants for SnO_2 were determined both experimentally and theoretically. An average relative static dielectric constant can be calculated from the formula $\varepsilon_{AV} = (2\varepsilon_\perp + \varepsilon_{||})/3$.

Batzill and Diebold [137] reported the relative static dielectric constant $\varepsilon_{r\perp c}$ = 13.5 and $\varepsilon_{r||c}$ = 9.6 yielding ε_{rAV} = 12.2, while Van Daal [162, 163] reported for CVT-grown crystals (microwave measurements at 1–5 × 10^8 Hz) $\varepsilon_{r\perp c}$ = 14 ± 2 and $\varepsilon_{r||c}$ = 9.0 ± 0.5 with ε_{rAV} = 11.5. Theoretical values of the relative static dielectric constant given by Borges et al. [81] are $\varepsilon_{0\perp c}$ = 14.6 and $\varepsilon_{0||c}$ = 10.7

(ε_{0AV} = 13.3), by Dou and Persson [84] $\varepsilon_{0\perp c}$ = 12 and $\varepsilon_{0\|c}$ = 7.0 (ε_{0AV} = 14.3), and by Schleife et al. [82] ε_{0AV} = 12.3.

In terms of the high-frequency dielectric constant, Sunmitt [164] and Jarzębski and Morton [165] published values of $\varepsilon_{\infty\perp c}$ = 3.785 and $\varepsilon_{\infty\|c}$ = 4.175 giving $\varepsilon_{\infty AV}$ = 3.9, while Summitt and Borrelli [152] reported $\varepsilon_{\infty AV}$ = 4. Theoretical calculations by Dou and Persson [84] outputted values of $\varepsilon_{\infty\perp c}$ = 3.7 and $\varepsilon_{\infty\|c}$ = 3.9 (ε_{0AV} = 3.8), and by Schleife et al. [82] of $\varepsilon_{\infty AV}$ = 3.62. Taking into account the reported values, it can be concluded that ε_{rAV} = 12.2–14.6 and $\varepsilon_{\infty AV}$ = 3.62–4.18.

SnO_2 is also birefringent with two refractive indexes $n_{\perp c}$ (ordinary) and $n_{\|c}$ (extraordinary) for polarizations perpendicular and parallel to *c*-axis, respectively. A wavelength relation of both refractive indexes was published by Reddaway and Wright [158]. In the spectral region of 350–800 nm, $n_{\perp c}$ decreases from about 2.125 to about 1.97, while $n_{\|c}$ from about 2.18 to about 2.08. In the spectral region of 500–1000 nm, the birefringence $\Delta n = n_{\|c} - n_{\perp c}$ is about 0.1. The PVT-grown SnO_2 crystals measured by ellipsometry by Anooz and Galazka [166] in the spectral region of 250–800 nm (depicted in Fig. 6.37) showed a bit larger values of the refractive index, but with the same birefringence Δn = 0.1 (>400 nm). A large anisotropy is observed below 300 nm; however, at 300–330 nm, the refractive index is close to isotropic.

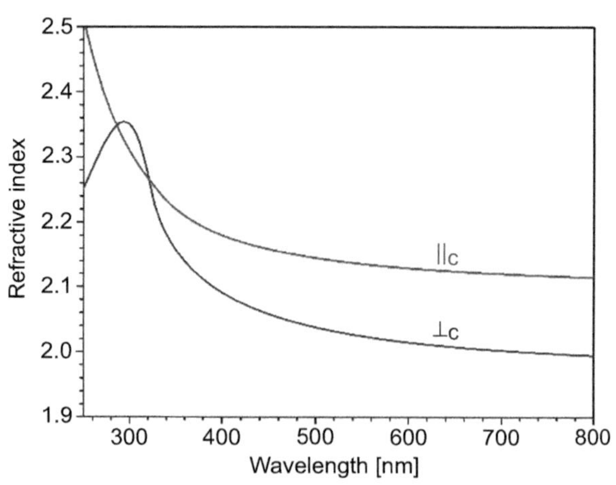

Figure 6.37 Wavelength-dependent refractive index of SnO_2: for polarization perpendicular ($n_{\perp c}$) and parallel to ($n_{\|c}$) *c*-axis at a spectral region of 250–800 nm for PVT-grown crystals (ellipsometry) [166].

Similar values of the refractive indexes at short wavelengths were also measured by Thiel and Helbig [110] on PVT-grown SnO_2 crystals. Values of $n_{\perp c}$ and $n_{\parallel c}$ at 350 nm are 2.09 and 2.17, respectively. The birefringence at 500–650 nm is also about 0.1. The isotropic point is located at 308 nm, i.e., beyond the absorption edge.

Detailed analysis of the dielectric function of SnO_2 thin films can be found in the works of Feneberg et al. [167–169].

6.7.3 Luminescence

In photoluminescence (PL) experiments, Fillard and de Murcia [170] and de Murcia et al. [116] reported two emissions, green and red, at 29–77 K under UV excitation (303 nm) of SnO_2 crystals obtained from the gas phase. PL output revealed a broad emission between 400 and 700 nm. The maximum of the green and red emissions was recorded in the ranges of 530–545 nm and 624–645 nm, respectively. At higher temperatures, the green emission disappears in favor of the red emission, which is likely due to energy transfer through impurities. Crabtree et al. [119] reported for undoped CVT-grown crystals PL emissions peaking at 443, 496, 539 eV, and 590 nm.

Cathodoluminescence (CL) of CVT-grown crystals reported by Crabtree [124, 125] shows at RT a broad emission between about 450 and 750 nm, while doping with Eu and Tb introduced to the CL spectra several sharp emission bands characteristic for rare earth ions corresponding to f $\rightarrow$ f transitions.

CL spectra measured by Schewski and Galazka [171] at RT of as-grown SnO_2 crystals obtained by the PVT method show a broad emission roughly between 400 and 700 nm (Fig. 6.38), similar to the CL spectra of the CVT-grown crystals. Here two local maxima at about 520 (green) and 564 nm (orange) are well distinguished. After annealing in O_2 at 1000°C for 20 h, the overall emission intensity increases and the absolute maximum of the emission shifts from the green to the orange emission. Also, an additional weak emission at 650 nm (red) appears. A sample annealed in the presence of H_2 (5%) at 600°C for 10 h revealed a similar CL spectrum structure to that annealed in the presence of O_2, but the overall emission significantly decreases as compared with the emission of the as-grown SnO_2. Basically, the structure of the CL spectra is similar to each other;

however, the intensities of particular emissions change. The broad emission in the visible spectrum of SnO_2 crystals under electron excitation is similar to other TSOs, such as β-Ga_2O_3 and In_2O_3, and it is assigned to self-trapped excitons (STEs). This has a confirmation in theory, as Varley et al. [172] calculated a value of a self-trapped hole (STH) emission of 2.27 eV for SnO_2, which corresponds to 546 nm, i.e., it is between both local maxima for experimentally observed green and orange emissions. STE is formed by a conduction electron that is coulombically attracted to the STH. The calculated self-trapping energy of 0.07 eV indicates the stability of the STH formed. STHs are localized mainly on single oxygen atoms [172].

Temperature-dependent radioluminescence spectra (X-ray excitation) of PVT-grown single crystals measured by Drozdowski et al. [173] showed at low temperatures (10–200 K) a weak emission peaking at about 490 nm, which vanishes at RT. The peak position was found to be temperature independent.

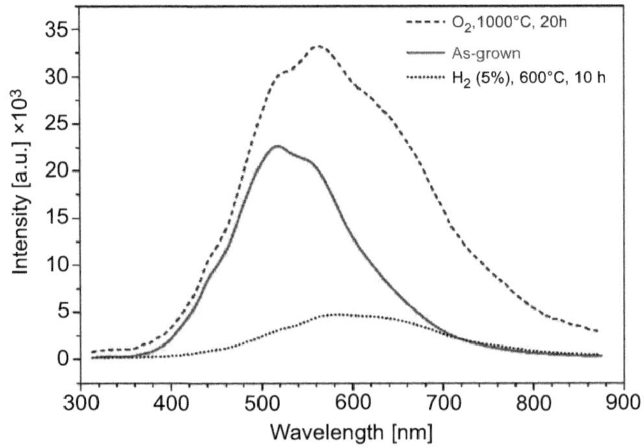

Figure 6.38 CL spectra at RT of PVT-grown SnO_2 crystals [171].

6.7.4 Phonon Modes

In the tetragonal rutile structure of SnO_2, there are 15 optical modes and 3 acoustic modes. Irreducible optical modes are described as follows [84, 170]:

$$A_{1g} + A_{2g} + B_{1g} + B_{2g} + E_g + A_{2u} + 2B_{1u} + 3E_u \quad (6.19)$$

where A_{2g} + B_{1u} are optically inactive, A_{2u} + E_u are infrared active, A_{1g}, B_{1g}, B_{2g}, and E_g are Raman active.

The three degenerate E_u modes and one A_{2u} mode correspond to the displacement of a cation and anion along [110] and [001] causing LO-TO splitting when a phonon wave at Γ-point of the BZ is parallel to the displacement [84].

Experimental values for phonon modes are given, e.g., by Summitt [164] (bulk crystals from vapor phase), Van Daal [163] (bulk CVT-grown crystals), Scott [174] (cassiterite), Katiyar et al. [149] (cassiterite), and Peercy and Morosin [175]. Examples of theoretical values are reported by Borges et al. [81] and Katiyar et al. [149]. Examples of both theoretical and experimental values of the phonon modes are included in Table 6.3.

Table 6.3 Optical modes in SnO_2 in [cm^{-1}]

Optical mode	Theoretical Ref. [81][a]	Experimental Ref. [164]	Theoretical Ref. [149]	Experimental Ref. [149]
B_{1g}	82.6/90.2	—	100	—
B_{1u}	138.4/133.7	—	140	—
E_u (TO)	200.3/196.5	243	236	244
E_u (LO)	251.9/237.1	273	268	276
E_u (TO)	270.5/253.3	284	297	293
E_u (LO)	306.9/294.0	368	377	366
A_{2g}	320.3/314.2	—	398	—
A_{2u} (TO)	457.4/424.2	465	512	477
E_g	462.4/432.6	—	441	476
B_{1u}	552.6/519.7	—	505	—
E_u (TO)	584.1/538.0	605	651	618
A_{1g}	617.3/577.6	—	646	638
A_{2u} (LO)	648.5/606.9	704	687	705
E_u (LO)	711.8/662.3	757	750	770
B_{2g}	733.8/688.2	—	752	782

[a]Two values depending on the calculation method

6.7.5 Conclusion

The optical properties of bulk SnO_2 single crystals were derived mainly for crystals obtained from the gas phase. SnO_2 crystals show anisotropy in optical properties depending on polarization vector whether it is parallel or perpendicular to the optical axis *c*. SnO_2 has a direct optical bandgap of about 3.77 eV for $E \| c$ and 3.485 eV for $E \perp c$, respectively. The fundamental bandgap is direct dipole-forbidden with a value of about 3.6 eV. The absorption edge originates at about 330 nm for $E \| c$ and 355 nm for $E \perp c$, respectively. In the spectral region of 350–800 nm, the refractive index is 2.16–1.99 for $\perp c$ and 2.22–2.11 for $\| c$, respectively, with a birefringence value of $\Delta n = 0.1$. The relative static dielectric constant is 9.0 ± 0.5 for $\| c$ and 14.2 ± 2 for $\perp c$, respectively. As-grown and undoped crystals show a broad emission band (photoluminescence and cathodoluminescence) roughly between 400 and 800 nm, which consists of three bands located in green, orange, and red regions of the visible spectrum, the intensity of which change upon annealing in the presence of oxygen (increases) and hydrogen (decreases). There are 15 optical modes in SnO_2 that were revealed both experimentally and theoretically.

Table 6.4 Basic empirical optical properties of SnO_2 crystals

Property	Unit	Value	Reference
Optical bandgap at RT	eV	3.77 ± 0.03 for $E \| c$ 3.485 ± 0.015 for $E \perp c$	151
Fundamental bandgap at RT	eV	3.596 ± 0.2 eV	160
Fundamental bandgap type		Direct dipole-forbidden	160
Bandgap shrinkage at 300–600 K	meV/K	1.2	161
Absorption edge	nm	330 for $E \| c$ 355 for $E \perp c$	90
Emission band peaks at RT (CL)	nm	520, 564, 650	171

Property	Unit	Value	Reference
Refractive index at 350–800 nm		2.125 to 1.97/2.16 to 1.99 for $\perp c$ 2.18 to 2.08/2.22 to 2.11 for $\|\|c$	158/166
Relative static dielectric constant		14.2 ± 2 for $\perp c$ 9.0 ± 0.5 for $\|\|c$	162
High-frequency dielectric constant		3.785 for $\perp c$ 4.175 for $\|\|c$	164, 165
Phonon modes	cm^{-1}	B_{1g}=100, B_{1u}=140, E_u(TO)=244, E_u(LO)=276, E_u(TO)=293, E_u(LO)=366, A_{2g}=398, A_{2u}(TO)=477, E_g=476, B_{1u}=505, E_u(TO)=618, A_{1g}=638, A_{2u} (LO)=705, E_u(LO)=770, B_{2g}=782	149

The basic experimental optical properties of bulk SnO_2 single crystals are shown in Table 6.4. More details on optical properties of SnO_2 can be found in the works of Jarzębski and Morton [165] and Jarzębski [176].

6.8 Electrical Properties and Effect of Annealing

The origin of the underlying *n*-type conductivity of SnO_2 crystals was the subject of intensive experimental and theoretical investigations. Although early experimental studies pointed to native point defects, i.e., oxygen vacancies and/or tin interstitials acting as shallow donors [120], recent computational evaluations rather excluded such possibility. The sources of the electrical conductivity have an extrinsic nature rather than intrinsic one. Theoretical results clearly indicate hydrogen acting as a shallow donor in SnO_2 crystals with no intentional doping. This is valid for both interstitial and substitutional (oxygen site) hydrogen. Both theory and experiments showed Sb as an effective *n*-type dopant. *p*-type conductivity has not been experimentally demonstrated for bulk crystals when doping with Ga and In [95], in line with the theory. In fact, MBE-grown thin films of In-doped SnO_2 were highly resistive [177], but not *p*-type. Although there are some reports on *p*-type conductivity in thin films with acceptor doping [101–103], such possibility needs further confirmation.

The sources of *n*-type conductivity can, therefore, be divided into three groups:

1. Interstitial and substitutional hydrogen H_i and H_O [83, 85, 86, 89]
2. Fluorine substituting oxygen, F_O [91, 92]
3. Some of the elements from group VB substituting tin: Sb_{Sn} [93, 94, 96–98], and possibly P_{Sn} and As_{Sn} [93]

As the *p*-type conductivity is rather not achievable, bulk SnO_2 crystals can be obtained as degenerate semiconductors, normal semiconductors, or electrical semi-insulators/insulators.

6.8.1 Free Electron Concentration, Hall Mobility, and Resistivity/Conductivity at RT: As-Grown Crystals and Effect of Annealing

For PVT-grown SnO_2 crystals obtained under the He + O_2 atmosphere, Marley and Dockerty [94] studied electrical properties (by Hall effect measurements) of three types of crystals: (i) as-grown and cooled down at a rate of 10 K/min, (ii) as-grown and quenched, and (iii) doped with Sb (about 300 wt. ppm). The cooled crystals (i) had brownish tint and a semi-insulating surface of 20–25 μm deep with the resistivity $\geq 10^6$ Ωcm, while the bulk was semiconducting with the resistivity (ρ) of 10^{-2} Ωcm, the free electron concentration (n_e) of about 10^{18} cm^{-3}, and the Hall mobility (μ) of about 150 cm^2/Vs. The quenched crystals (ii) had a brown coloration with $n_e = 2 \times 10^{18}$ cm^{-3} and μ = 150–200 cm^2/Vs. Annealing the cooled crystals (i) in O_2 at 1165–1390°C for 62–282 h decreased the free electron concentration to a level of 1.5×10^{16} – 5.7×10^{17} cm^{-3} and the Hall mobility to about 50–90 cm^2/Vs. Annealing the quenched crystals (ii) in dry argon at 1030–1060°C for 11.5–23 h decreased the free electron concentration by a factor of about 2 and also the Hall mobility to values of about 90–140 cm^2/Vs. Sb-doped SnO_2 crystals (iii) were blue and had the free electron concentration of about 8×10^{18} cm^{-3}, with no dependence on temperature (degenerate state). Further study of these crystals by Summitt and Borrelli [152] showed the following electrical properties at RT for slowly cooled

SnO_2 crystals (i): $n_e = 1 \times 10^{17}$–1.2×10^{18} cm^{-3}, μ = 91–152 cm^2/Vs, and ρ = 0.048–0.61 Ωcm, for quenched crystals (ii): n_e = 1.3–2.9 × 10^{18} cm^{-3}, μ = 166–178 cm^2/Vs, and ρ = 0.014–0.027 Ωcm, and for Sb-doped (ca. 500 wt. ppm) crystals (iii): $n_e = 5.2 \times 10^{18}$–1.9×10^{19} cm^{-3}, μ = 74–91 cm^2/Vs, and ρ = 0.0039–0.014 Ωcm.

With the use of elemental Sn as the starting material and elemental Sb as a dopant, Morgan and Wright [95] reported electrical properties of SnO_2 crystals grown by the PVT method with Hall effect measurements and In ohmic contacts. For Sb-doped crystals, the following properties were measured at RT: $n_e = 8 \times 10^{18}$– 8.8 × 10^{19} cm^{-3}, μ = 62–96 cm^2/Vs, and $\rho = 1.3 \times 10^{-2}$ and 7.5×10^{-4} Ωcm. These values were similar in both RT and at 77 K, indicating a degenerated state.

For undoped SnO_2 crystals obtained by the PVT method under Ar + O_2 and N_2 + H_2 atmospheres, de Murcia et al. [116, 117] and Thiel and Helbig [110], respectively, reported high resistivity of the obtained crystals, 10^{14} and 10–10^3 Ωcm, respectively. From annealing experiments of Marley and Dockerty [94], one can conclude a decrease in the free electron concentration of PVT-grown crystals by one to almost two orders of magnitude after annealing in O_2 at 1330–1390°C for 62–161 h. Annealing in Ar at 1030–1060°C for 11.5–23 h had, rather, a minor impact on the free electron concentration.

SnO_2 crystals obtained by Galazka et al. [90] with the use of CO_2 as the atmosphere in the PVT growth configuration were either electrically insulating or semiconducting depending on a relation between p(SnO) and $p(O_2)$ in the vicinity of the substrate (see Fig. 6.16). The as-grown crystals were semiconducting at p(SnO) ≥ $p(O_2)$; otherwise, they were electrical insulators. As-grown semiconducting SnO_2 crystals with no intentional doping had the following electrical properties at RT: $n_e = 3 \times 10^{17}$–1.9×10^{18} cm^{-3}, μ = 120–200 cm^2/Vs, and ρ = 0.03–0.1 Ωcm.

When subjecting semiconducting SnO_2 crystals obtained by the PVT method (Galazka et al. [90, 178] to annealing in O_2 for 20 h, after a gradual decrease in the free electron concentration by about one order of magnitude with temperature (Fig. 6.39a), they become electrical insulators (or semi-insulators) after annealing at 1000°C

(O_2) and 1200°C (Ar), as the Hall effect could not be measured. In the case of the oxidizing atmosphere (O_2), switching to the insulating state takes place after annealing time of 7.5 h (Fig. 6.40), while after annealing at 800°C, the samples remained semiconducting even after 20 h annealing time, but with lower free electron concentration. After annealing the as-grown samples in the presence of H_2(5%) for 10 h at temperatures up to 900°C, the free electron concentration substantially does not change; it scatters within a wide range around the level of 10^{17} cm^{-3}. If, however, electrically insulating samples (by previous annealing in O_2 at 1000°C for 10 h) were subjected to annealing in the presence of H_2, the free electron concentration gradually increases with temperature to the level of mid 10^{17} cm^{-3} after annealing at or above 600°C. The Hall effect could already be measured on samples after annealing at 400°C.

If we consider the Hall mobility of PVT-grown SnO_2 crystals (Fig. 6.39b), surprisingly it gradually decreases with annealing temperature after annealing in O_2 for 20 h by a factor of about 2. After annealing in the presence of H_2, the Hall mobility remains at the high level, between 170 and 240 cm^2/Vs. It drops to lower values after annealing at 900°C due to strong thermal decomposition. After annealing electrically insulating samples (previously annealed in O_2) in H_2, the Hall mobility remained at high level between 120 and 200 cm^2/Vs, i.e., in the same range as the Hall mobility of as-grown and H_2-annealed crystals. The resistivity shown in Fig. 6.39c is a function of the free electron concentration and electron mobility ($\rho = (e\mu n)^{-1}$). After annealing in O_2, the resistivity of SnO_2 crystal samples gradually increases with annealing temperature by about one order of magnitude, while after annealing those samples in the presence of H_2, which were previously annealed in O_2, the resistivity decreases by almost two orders of magnitude. For the as-grown crystal samples, their resistivities after annealing in the presence of H_2 remain scattered within one order of magnitude.

Annealing offers a very efficient tool for tuning electrical properties of PVT-grown SnO_2 crystals with several degrees of freedom arising from the annealing atmosphere, temperature, and time. The maximum of the free electron concentration was about 2×10^{18} cm^{-3}; the highest Hall mobility was 240 cm^2/Vs, and the lowest resistivity was 0.03 Ωcm.

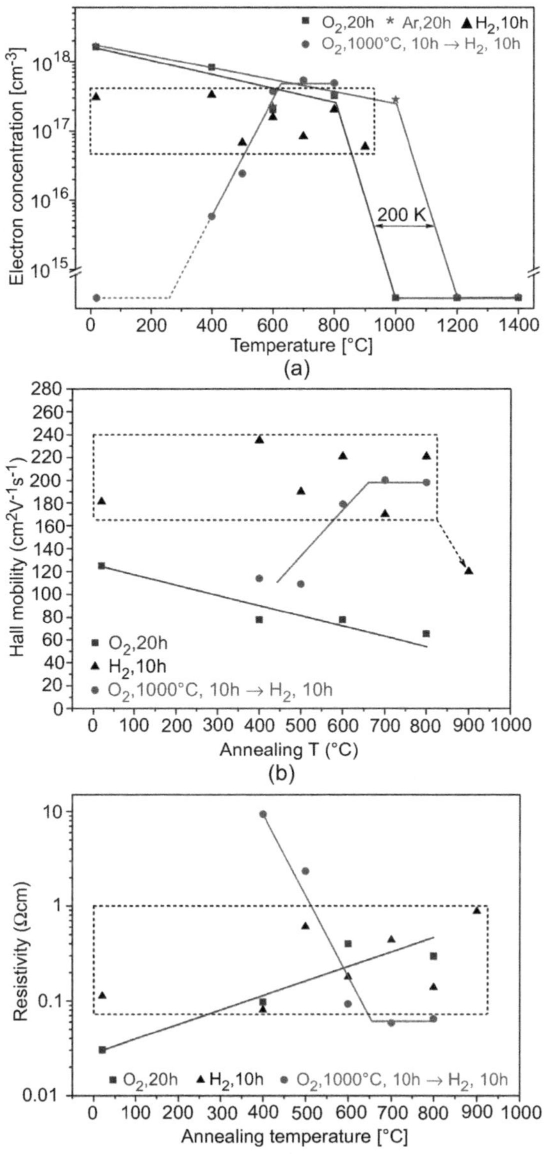

Figure 6.39 Effect of annealing of PVT-grown SnO_2 crystal samples in different atmospheres versus temperature on the (a) free electron concentration; (b) Hall mobility; and (c) resistivity. The set of samples used for annealing in the presence of H_2 had the initial free electron concentration lower than that used for annealing in the presence of O_2 and Ar [178].

Bierwagen and Galazka [179] concluded the anisotropy of Hall mobility of PVT-grown SnO_2 crystals basing on Hall effect measurements in the van der Pauw configuration of (100) and (110) oriented wafers. The anisotropy of the Hall mobility parallel to the *c*-direction and perpendicular to it is $\mu_{\|c}/\mu_{\perp c}$ = 1.26, and it corresponds to the anisotropy of the effective electron mass $m_{e\ \perp c}^{*} / m_{e\ \|c}^{*}$ = 1.28.

Basic electrical properties for undoped CVT-grown SnO_2 crystals reported by Crabtree et al. [119] at RT are as follows (by Hall effect measurements): $n_e = 10^{17}–10^{18}$ cm^{-3}, μ = 172–250 cm^2/Vs, and ρ = $3 \times 10^{-1}–3 \times 10^{-3}$ Ωcm, and by Nagasawa et al. [120]: n_e = 1.1–1.5 × 10^{17} cm^{-3}, μ = 180–240 cm^2/Vs, and ρ = 0.2–0.27 Ωcm. Annealing in O_2 at 1200°C for 5 h decreased the resistivity by about one order of magnitude, while annealing in vacuum (0.013 mbar) at 1200°C for 40 h increased the conductivity 500 times. Fonstad et al. [123] and Fonstad and Rediker [180] reported for undoped CVT-grown SnO_2 crystals either high resistivity of 6×10^5 Ωcm [123] or relatively low free electron concentration of 8.5×10^{15} cm^{-3} at RT [180]. In the latter case, a record high Hall mobility of 260 cm^2/Vs was measured. Sb-doped crystals showed the following values at RT: n_e = 8.6 × 10^{16}–2.2 × 10^{18} cm^{-3}, μ = 240–150 cm^2/Vs, and $\rho \approx$ 0.1 Ωcm. The anisotropy in electrical conductivity σ_a/σ_c was about 1.2 at RT.

In the case of flux-grown SnO_2 crystals, Kunkle and Kohnke [106] reported very high values of the resistivity at RT of $10^9–10^{10}$ Ωcm.

Basic electrical properties of bulk SnO_2 single crystals at RT are collected in Table 6.5.

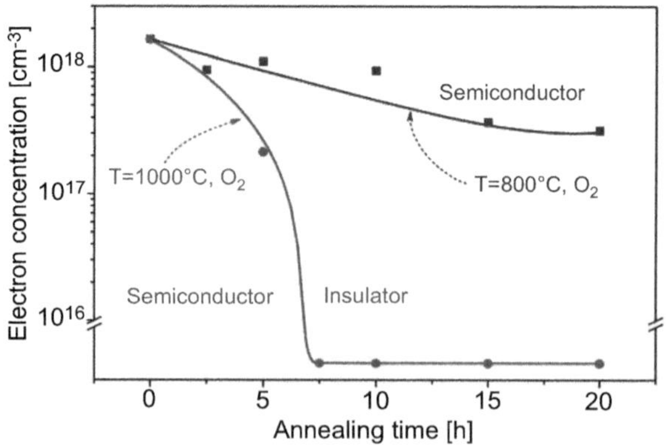

Figure 6.40 Impact of the annealing time and temperature on the free electron concentration of PVT-grown SnO_2 crystal samples. Reprinted (with minor modifications) with permission from Ref. [90], Copyright 2014, John Wiley and Sons.

Table 6.5 Basic electrical properties of as-grown bulk SnO_2 crystals at RT. n_e, μ, and ρ are the free electron concentration, electron mobility, and resistivity, respectively

Method	n_e [cm^{-3}]	μ [cm^2/Vs]	ρ [Ωcm]	Reference
PVT[(a)]: undoped	1×10^{17}–2.9×10^{18}	90–200	0.01–0.6	94, 152
PVT[(a)]: Sb-doped	2×10^{18}–8.8×10^{19}	60–200	0.00075–0.014	94, 95, 152
PVT[(b)]: undoped	3×10^{17}–1.9×10^{18}	120–200	0.03–0.1	90
CVT[(a)]: undoped	8.5×10^{15}–1.0×10^{18}	172–260	0.003–0.3	119, 120, 123, 180
CVT[(a)]: Sb-doped	8.6×10^{16}–2.2×10^{18}	150–240	0.1	180
Flux	—	—	10^9–10^{10}	106

[(a)]Standard furnace, resistive heating; [(b)] Ir crucible, inductive heating, sapphire substrate

6.8.2 Temperature-Dependent Electrical Properties

Analysis of temperature-dependent electrical properties of CVT- and PVT-grown bulk SnO_2 single crystals can be found, e.g., in the works of Fonstad et al. [123], Fonstad and Rediker [180], Marley and Dockerty [94], and Bierwagen and Galazka [181], respectively.

According to Fonstad and Rediker [180], temperature-dependent electrical properties of CVT-grown bulk SnO_2 crystals differ in between whether the crystals had initial high or low conductivity. The high-conductivity sample showed a little change in the free electron concentration and resistivity (by a factor of about 2 or less) indicating a degenerate state, while the low-conductivity sample revealed much stronger dependence by almost one order of magnitude in a temperature range of 80–325 K. The Hall mobility showed a maximum at 200 and 45 K for high and low-conductivity samples, respectively. The values of the Hall mobility for the high-conductivity sample at RT and 200 K (maximum value) were 150 and about 200 cm^2/Vs, respectively, while for low-conductivity sample, the values were 260 and 13500 cm^2/Vs. These results clearly show a strong dependence of the Hall mobility on the free electron concentration, much stronger as compared with other TSOs, such as ZnO, β-Ga_2O_3, and In_2O_3.

Marley and Dockerty [94] measured temperature-dependent electrical properties of a wide spectrum of PVT-grown bulk SnO_2 crystals: undoped as-grown (cooled down and quenched) and Sb-doped, annealed in Ar at 1030–1060°C for 11.5 and 23 h, annealed in O_2 at 1290–1390°C for 62–140 h at 1 atm and at 1165°C for 282 h at 0.06 atm. The results in a temperature range of 80–900 K are shown in Fig. 6.41. At RT, the resistivity, free electron concentration, and Hall mobility span in the range of 8×10^{-3}–1.7×10^{1} Ωcm, 7×10^{15}–7×10^{18} cm^{-3}, and 45–200 cm^2/Vs, respectively. Sb-doped samples (highly conducting) reveal a degenerate state as there is almost no change in the resistivity and the free electron concentration. Undoped samples, whether as-grown or annealed in Ar (1030–1060°C, 11.5–23 h), show a moderate dependence of ρ and n_e with temperature, both within one order of magnitude. The largest change in ρ and n_e as a function of temperature can be noted for samples annealed in O_2 at high temperatures (≥1290°C) for a long time (≥62 h) at atmospheric or low pressure (0.06 atm). The changes are as high as almost four orders of magnitude. In terms of the electron mobility, two different behaviors can be observed. Undoped as-grown samples show a moderate change in the mobility typical for other TSOs with a maximum of about 200–400 cm^2/Vs located at around 150 K. However, other annealed and Sb-doped samples reveal a maximum near RT and a strong decrease with lowering temperature ending with values below 25 cm^2/Vs at about 80 K. This study clearly shows the importance of the thermal history and doping of the crystals in electrical properties.

Other temperature-dependent free electron concentration and Hall mobility reported by Bierwagen and Galazka [181] on PVT-grown crystals of two orientations (100) and (110) are shown in Fig. 6.42. The results are qualitatively and quantitatively in line with moderately conducting SnO_2 crystals (n_e = 6–8 × 10^{17} cm^{-3}). The maximum Hall mobility of about 400 cm^2/Vs was measured at about 125 K. This temperature-dependent behavior is similar to the behavior of above-mentioned as-grown crystals studied by Marley and Dockerty [94].

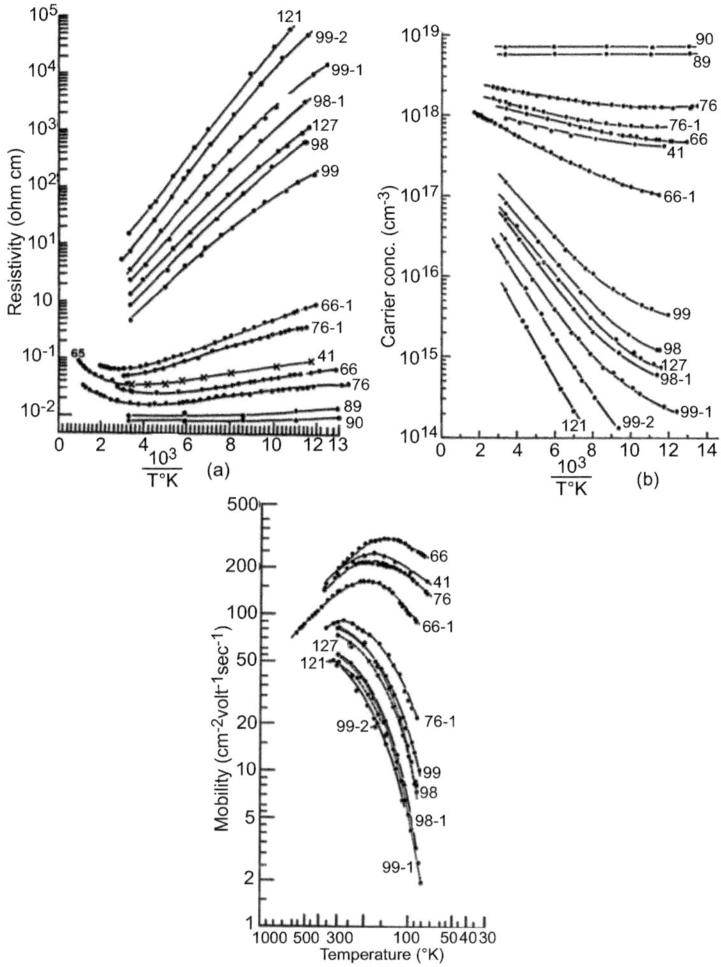

Figure 6.41 Temperature-dependent resistivity (a), free electron concentration (b), and Hall mobility (c) of PVT-grown SnO_2 crystals. Sample IDs are as follows: 76, 66 – as-grown, air quenched; 65, 41 – as-grown and cooled; 76-1, 66-1 samples 76 and 66 annealed in dry Ar at 1030°C, 11.5 h and at 1030–1060°C, 23 h, respectively; 99, 98, 127 – annealed in O_2 under 1 atm at 1390°C, 67 h, 1367°C, 75 h, and 1364°C, 62 h, respectively; 98-1, 99-1, 99-2 – samples 98 and 99 annealed in O_2 under 1 atm at 1360°C, 161 h, 1330°C, 140 h, and 1290°C, 71 h, respectively; 121 – annealed in O_2 under 0.06 atm at 1165°C for 282 h; 89, 90 – as-grown, Sb-doped. Reprinted with permission from Ref. [94], Copyright 1965, American Physical Society.

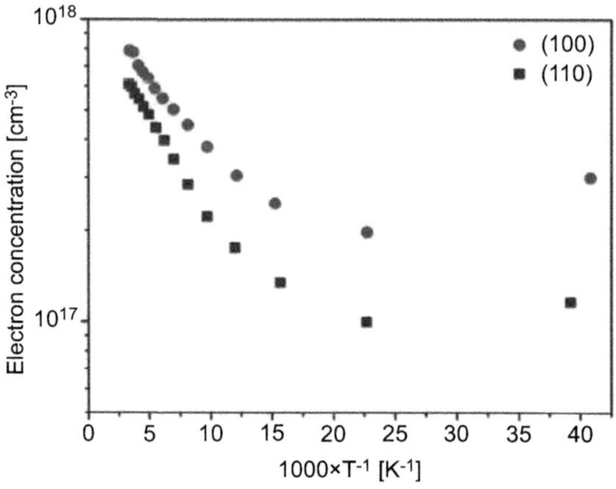

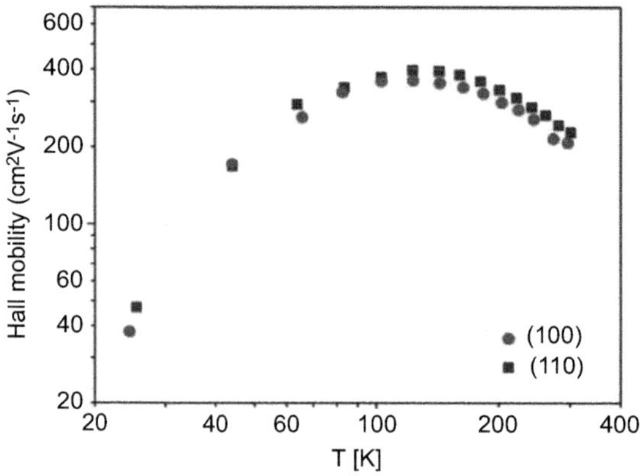

Figure 6.42 Temperature-dependent free electron concentration and Hall mobility of PVT-grown SnO_2 crystals having the free electron concentration of 6–8 × 10^{17} cm^{-3} at room temperature. Based on data extracted from Ref. [181].

6.8.3 Effective Electron Mass and Electron Mobility

In the course of experimental studies of electrical properties of bulk SnO_2 single crystals, a wide range of effective electron masses m_e^*/m_0 was reported, e.g., 0.14 [152] at $n_e \approx 10^{17}$–10^{18} cm^{-3}, 0.17 [119], 0.11–0.17 [95] at $n_e = 8.6 \times 10^{16}$ cm^{-3}, 0.22 [94] at $n_e \approx 10^{16}$–5.7×10^{18} cm^{-3}, 0.39 [180] at $n_e = 8.6 \times 10^{16}$ cm^{-3}, and 0.2–0.3 [163] at $n_e = 3 \times 10^{17}$ cm^{-3}.

Button et al. [182] showed experimental values of the effective electron mass by using cyclotron resonance technique and obtained the following values for directions perpendicular and parallel to the *c*-axis: $m_{e\perp c}^* = 0.299\ m_0$ and $m_{e\parallel c}^* = 0.234\ m_0$ at $n_e = 7 \times 10^{15}$ cm^{-3}, giving an arithmetical average $m_{e\,AV}^* = (2\ m_{e\perp c}^* + m_{e\parallel c}^*)/3 =$ $0.28\ m_0$. This also gives the anisotropy ratio of $m_{e\perp c}^*/m_{e\parallel c}^* = 1.28$. Similar anisotropy in electrical conductivity σ_a/σ_c a bit above 1.2 was measured at RT by Fonstad and Rediker [180]. Bierwagen and Galazka [181] demonstrated an almost constant Hall mobility ratio of $\mu_{\perp c}/\mu_{\parallel c} \approx 1.28$ for two different sample orientations in a temperature range of 100–300 K, which increases at lower temperatures (Fig. 6.43). Therefore, at that temperature range, $\mu_{\perp c}/\mu_{\parallel c} \approx$ $m_{e\perp c}^*/m_{e\parallel c}^*$.

For comparison, at high doping level of MBE-grown thin films ($n_e = 1.6 \times 10^{19}$–2.6×10^{20} cm^{-3}), Feneberg et al. [167, 169] derived from ellipsometric study the following experimental values of the effective electron masses: $m_{e\perp c}^* = 0.22$–$0.39\ m_0$ and $m_{e\parallel c}^* =$ 0.2–$0.26\ m_0$. The effective electron mass increases with the free electron concentration as a result of non-parabolic dispersion of the conduction band.

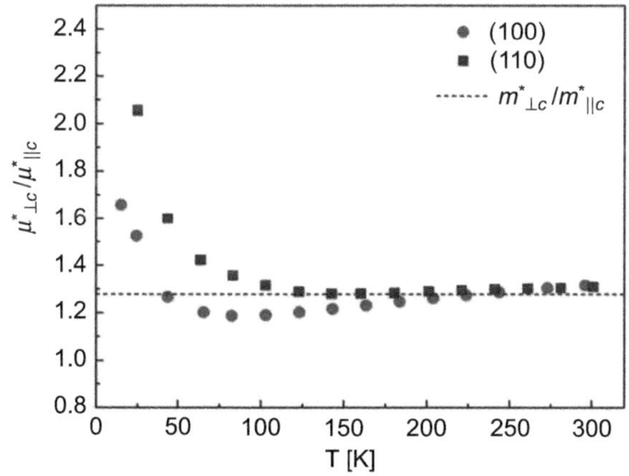

Figure 6.43 Ratio of the Hall mobility perpendicular ($\mu_{\perp c}$) and parallel ($\mu_{\parallel c}$) to the *c*-direction in PVT-grown SnO_2 crystals for two sample orientations (100) and (110). Based on data extracted from Ref. [181].

Theoretically predicted values of the effective electron mass in SnO_2 have a wide spread, but some of them are close to those measured experimentally on bulk crystals. Additionally, the theory could also predict effective hole masses m_h^*. For example, Borges et al. [81] predicted $m_{e\ \perp c}^* = 0.48\ m_0$ and $m_{e\ \|c}^* = 0.192\ m_0$ ($m_{e\ AV}^* = 0.38\ m_0$), as well as $m_{h\ \perp c}^* = 3.24–3.34\ m_0$ and $m_{h\ \|c}^* = 2.26\ m_0$ ($m_{h\ AV}^* = 2.95\ m_0$), Dou and Persson [84] calculated $m_{e\ \perp c}^* = 0.25\ m_0$ and $m_{e\ \|c}^* = 0.21\ m_0$ ($m_{e\ AV}^* = 0.24\ m_0$), and $m_{h\ \perp c}^* = 0.46\ m_0$ and $m_{h\ \|c}^* = 0.26\ m_0$ ($m_{h\ AV}^* = 0.39\ m_0$), while Schleife et al. [82] concluded theoretically the following values: $m_{e\ \perp c}^* = 0.26\ m_0$ and $m_{e\ \|c}^* = 0.21\ m_0$ ($m_{e\ AV}^* = 0.24\ m_0$), and $m_{h\ \perp c}^* = 1.21\ m_0$ and $m_{h\ \|c}^* = 1.47\ m_0$ ($m_{h\ AV}^* = 1.3\ m_0$).

Bulk SnO_2 single crystals have in average higher electron mobility as compared with previously discussed TSOs, i.e., ZnO, β-Ga_2O_3, In_2O_3. Several independent studies on bulk crystals obtained from the gas phase showed values of the Hall mobility at RT above 200 cm^2/Vs at the free electron concentration of about 10^{17} cm^{-3}, e.g., 240 cm^2/Vs by Galazka (Fig. 6.39b), Crabtree et al. [119], and Nagasawa et al. [120], and 260 cm^2/Vs by Fonstad and Rediker [180]. Moreover, Fonstad and Rediker [180] reported a record high value of 13,500 cm^2/Vs at 45 K at low level of the free electron concentration (1.9×10^{15} cm^{-3}). The maximum of the Hall mobility decreases with the free electron concentration and shifts toward higher temperatures. For example, at low $n_e = 8.5 \times 10^{15}$ cm^{-3}, the maximum Hall mobility of 13,500 cm^2/Vs was recorded at about 45 K, while at high $n_e = 2.2 \times 10^{18}$ cm^{-3}, the maximum Hall mobility of only about 200 cm^2/Vs was recorded at about 200 K. The Hall mobility versus temperature of CVT-grown crystals with different free electron concentrations shows behavior that, indeed, strongly depends on the free electron concentration [180].

The main limiting scattering mechanisms are the same as in other oxide and elemental semiconductors, i.e., phonon scattering (lattice) involving both optical and acoustical vibrational modes [94, 180], which is an intrinsic mechanism, and scattering at impurities at low temperatures, which is an extrinsic mechanism. Jarzębski and Marton [183] concluded that dominating scattering mechanism are optical modes above 300 K (RT), acoustic modes between 50 and 300 K, and ionized point defect scattering below 50 K. In the case of heavily doped crystals, scattering at ionized point defects dominates below RT.

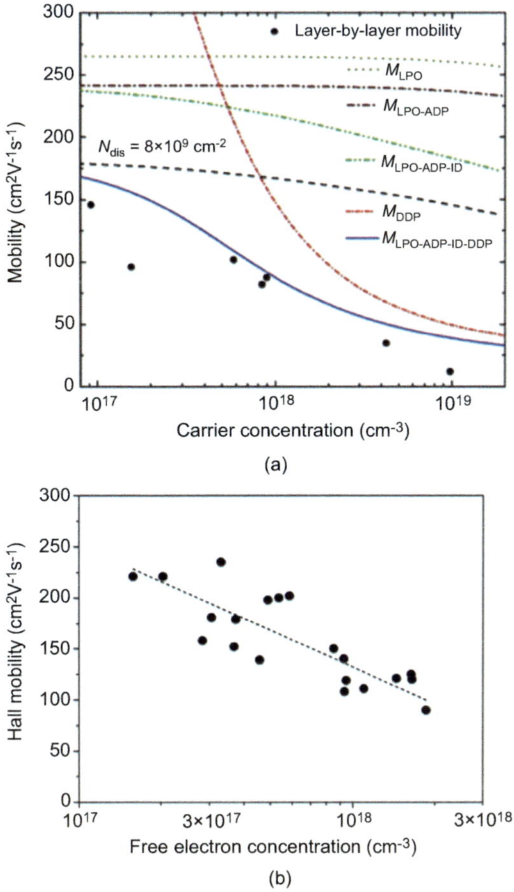

Figure 6.44 (a) Electron mobility versus experimental free electron concentration and simulation data. The experimental data were obtained by plasma-assisted molecular-beam epitaxy (PAMBE) on *r*-plane $(10\bar{1}2)$ sapphire in a layer-by-layer mode. LPO, ADP, ID, and DDP stand for longitudinal polar-optical mode, acoustic deformation potential, ionized defect, and dislocation deformation potential scattering, respectively. The dashed line $N_{dis} = 8 \times 10^9$ cm^{-2} shows the resultant mobility model curve regarding that constant value of dislocation concentration from the interface to the surface. Reprinted with permission from Ref. [184], Copyright 2012, American Physical Society. (b) Hall mobility versus free electron concentration measured on bulk SnO_2 single crystals obtained by the PVT method [185].

A detailed experimental and theoretical analysis of a relation between the electron mobility and the free electron concentration, including different scattering mechanisms, was reported by Farahani et al. [184] for heteroepitaxial films of SnO_2, as illustrated in Fig. 6.44a. From the simulation and analysis of layer-by-layer mobility versus carrier concentration, it was concluded that the strongest interaction with mobile electrons has longitudinal polar-optical phonons, and also acoustic phonons. Experimental electron mobility decreases with the free electron concentration (in the range of 10^{17}–10^{19} cm^{-3}) approximately linearly. However, the electron mobility of the heteroepitaxial films is greatly reduced by charged dislocations of high density.

Approximately a linear decrease in the Hall mobility versus free electron concentration is also observed for PVT-grown bulk SnO_2 single crystals, as shown in Fig. 6.44b (Galazka et al. [185]). An average Hall mobility of bulk SnO_2 crystals is higher than the electron mobility of heteroepitaxial SnO_2 films for the same free electron concentration by about 60%. This is likely due to much lower charged-dislocation density in bulk single crystals.

6.8.4 Ionization Energy of Shallow Donors

Bulk SnO_2 crystals have two donor levels with quite different ionization energies of $E_{CD1} \approx$ 4–50 meV and $E_{CD2} \approx$ 80–150 meV, respectively. Fonstad and Rediker [180] concluded the donor ionization energy of 10–15, 27.5 and 34, 140 meV at the net donor concentration $N_D - N_A > 2.2 \times 10^{18}$, 9.4×10^{16}, and 9.2×10^{15} cm^{-3}. In another study, Samson and Fonstad [87] measured two donor ionization energies of 35 ± 5 and 150 ± 10 meV at $N_D = 3.5 \times 10^{15}$ cm^{-3}, which were assigned to a doubly ionized oxygen vacancies, and 50 ± 5 meV, which was assigned to hydrogen. The ionization energy of shallow donors of 4–24 meV was reported by Nagasawa et al. [186]. The ionization energy of deeper donors of E_{CD2} = 81–138 meV was found by Marley and Dockerty [94] for a range of low net donor concentration of 1.5×10^{16}–5.7×10^{17} cm^{-3} with the ionization energy of isolated donors of $E_{CD2}{}^0$ = 150 meV. Jarzębski and Marton [183] stated that undoped bulk SnO_2 crystals have a shallow donor and deeper donor levels with ionization energies of 30 and 150 meV, respectively.

6.8.5 Deep-Level Traps

Deep-level traps in bulk SnO_2 crystals have not been studied in detail, and there is no clear picture thereof. Basing on photoconductivity and photoluminescence of SnO_2 crystals obtained by the PVT method, de Murcia et al. [116, 117] concluded a shallow electron trap at 0.14 eV below the bottom of the conduction band and three recombination levels at 0.7, 1.3, and about 2 eV above the top of the valence band. Thermally stimulated current (TSC) measurements of PVT-grown crystals by Fillard et al. [187] revealed deep traps at 0.22 eV and 0.44 eV below the bottom of the conduction band. Low-temperature thermoluminescence by Drozdowski et al. [173] of PVT-grown crystals (Galazka et al. [90]) did not show any electron traps. TSC measurements on flux-grown SnO_2 crystals by Houston and Kohnke [188] showed three trap levels at 0.23–0.29, 0.5–0.7, and 0.75–0.85 eV below the bottom of the conduction band. The latter trap could be associated with some of the flux impurities that are usually present in flux-grown crystals.

6.8.6 Surface Band Bending

To study band bending of PVT-grown bulk SnO_2 single crystals, Nagata et al. [189] applied a combined surface-sensitive soft- and bulk-sensitive hard-X-ray photoelectron spectroscopy, SXPES and HXPES, respectively. As illustrated in Fig. 6.45, (100)-oriented SnO_2 crystal samples show surface electron depletion layer (upward band bending) for both as-grown and O_2-annealed samples. The oxidization treatment reduced the in-gap states and free electron concentration by two orders of magnitude but did not change the surface Fermi level pinning approximately 0.4 eV below the E_{CBM}. The surface is depleted at a depth of several nm, similar to In_2O_3, where the electron accumulation layer is also confined to several nm.

In contrast to undoped bulk SnO_2 crystals, surface electron accumulation layer (SEAL) was observed in the layers upon gas adsorption [190] and in epitaxial thin films (101) grown by the plasma-assisted molecular-beam epitaxy (PAMBE) [191, 192]. The SEAL could be removed by an oxygen–plasma surface treatment [193].

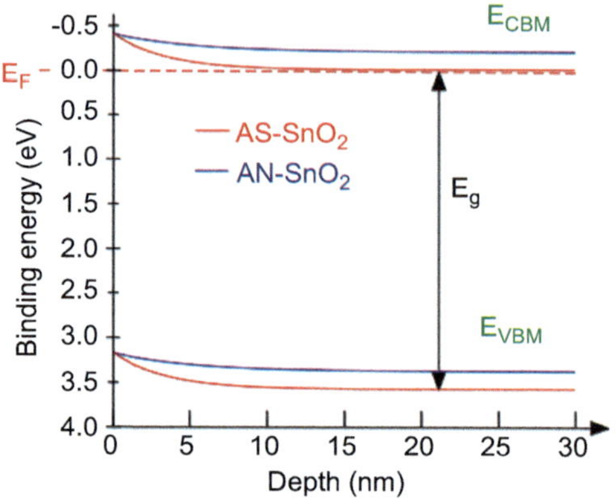

Figure 6.45 Schematic illustration of the band diagrams from surface (left) to bulk (right) estimated by XPES. E_{CBM}, E_{VBM}, E_g, and E_F are energies of the conduction band minimum, valence band maximum, bandgap, and Fermi level, respectively. AS and AN refer to as-grown and annealed (O_2, 1100°C, 20 h) SnO_2 crystals, respectively. Reprinted with permission from Ref. [189], Copyright 2019, The Japan Society of Applied Physics.

6.8.7 Conclusion

Among bulk SnO_2 single crystals, flux-grown crystals showed very high resistivity, while those obtained from the gas phase (PVT and CVT methods) revealed a typical semiconducting behavior with the free electron concentration, Hall mobility, and resistivity of about 10^{16}–3×10^{18} cm^{-3}, 90–260 cm^2/Vs, and 0.01–0.6 Ωcm at RT, respectively. The typical semiconducting behavior was confirmed by temperature-dependent measurements of the electrical properties. The highest low-temperature Hall mobility was measured as high as 13,500 cm^2/Vs. Doping with Sb increases the free electron concentration to a level close to 10^{20} cm^{-3} in bulk crystals obtained from the gas phase that can be assigned to a degenerate state. Annealing is an effective tool to tune electrical properties of bulk SnO_2 single crystals; particularly the crystals can be switched to electrically insulating or semi-insulating state by annealing either in an oxidizing or neutral atmosphere at high temperatures (≥1000°C). The semiconducting state of electrically insulating crystals can be recovered by annealing in the presence of hydrogen (≥400°C) for at least several hours. The electron mobility is anisotropic with the ratio of $\mu_{\perp c}/\mu_{\| c}$ = 1.28 around RT, which is equal to effective electron mass ratio $m_e^*{}_{\perp c}/m_e^*{}_{\| c}$. The effective electron mass is $m_e^*{}_{\perp c}$ = 0.22–0.39 m_0 and $m_e^*{}_{\| c}$ = 0.2–0.26 m_0 at a high free electron concentration ($n_e \approx 10^{19}$–3×10^{20} cm^{-3}), and about 0.17 at low and moderate free electron concentration ($n_e \approx 10^{16}$–10^{17} cm^{-3}). In undoped bulk SnO_2 crystals, there are two donor levels with ionization energies of 4–50 meV (shallow level) and 80–150 meV (deeper level). Bulk crystals show a surface electron depletion layer whether the crystal are as-grown or annealed in the presence of oxygen. Another analysis of electrical properties of SnO_2 crystals can be found in the work of Jarzębski and Morton [183].

A summary of basic electrical properties of bulk SnO_2 single crystals is included in Table 6.6.

Table 6.6 Basic electrical properties of bulk SnO_2 single crystals

Property	Unit	Value	Reference
Free electron concentration at RT, n_e	cm^{-3}	PVT: 1×10^{17}–2.9×10^{18}; PVT (Sb-doped): 2×10^{18}–8.8×10^{19} CVT: 8.5×10^{15}–1×10^{18} CVT (Sb-doped): 8.6×10^{16}–2.2×10^{18}	90, 94, 95, 152 119, 120, 123, 180
Maximum Hall mobility at RT	cm^2/Vs	PVT: 240[(a)] CVT: 260	178 180
Maximum Hall mobility at low temperature (45 K)	cm^2/Vs	13,500	180
Effective electron mass at $n_e = 7\times10^{15}$	—	0.299 for ⊥c 0.234 for ||c	182

(*Continued*)

Table 6.6 (*Continued*)

Property	Unit	Value	Reference
Ionization energy of shallow donors at n_e = 3.5×10^{15}–2.2×10^{18}	meV	E_{CD1}=4–50 E_{CD2}=80–150	87, 94, 180, 183, 186
Surface electron depletion layer	—	Present in ambient in both as-grown and O_2-annealed (100) crystals	189
Upward band bending	eV	0.4	189

[a]Annealed

6.9 Key Points

- Bulk SnO_2 single crystals cannot be grown directly from the melt. The best suited growth method for bulk SnO_2 single crystals in terms of size and structural quality seems to be physical vapor transport (PVT) method.
- It is a wide bandgap oxide semiconductor with anisotropic optical properties and optical bandgap of 3.77 ($\|c$) and 3.49 ($\perp c$) eV.
- Bulk crystals can be prepared as electrical insulator, normal semiconductor, or degenerate semiconductor. The highest conductivity can be obtained by Sb-doping.
- Bulk SnO_2 crystals show very high electron mobility up to 260 cm^2/Vs at RT. It is one of the highest electron mobilities among bulk TSO single crystals.
- SnO_2 has a high thermal conductivity (51.4 W/mK at RT), one of the highest among other TSOs.

6.10 Summary

Table 6.7 Main properties of bulk SnO_2 single crystals and growth techniques

SnO_2		
Property	**Value or description**	**Reference**
	Crystal system and structure	
Lattice parameter [Å]	a = 4.8382/4.7374 c = 3.8029/3.1864	76/77
System, structure	Tetragonal, rutile	
Space group	$P4_2/mnm$	
Cleavage plane	{110}, {011}	115, 120
	Bulk crystal preparation	
Maximum size: diameter × length [mm×mm], or width × length × thickness [mm×mm×mm]	PVT: 25×10 CVT: 2×5×15 Flux: 3×3×3	90 120 133, 134
	Thermal and mechanical properties	
Solid density [g/cm³]	6.99	137
Melting point [°C]	>2100	90
Thermal conductivity at RT [W/mK]	51.4	144
Thermal expansion coefficient for lattice parameter with respect to RT [K^{-1}]	a: 2.66×10^{-5} c: 3.66×10^{-5}	76
Elastic constants [GPa]	$c_{11} = c_{22}$ = 216.7 c_{12} = 190.6 $c_{13} = c_{23}$ = 134.5 c_{33} = 312.8 $c_{44} = c_{55}$ = 55.5 c_{66} = 194.2	149
Bulk modulus [GPa]	185[(a)]	149
Shear modulus	80[(a)]	149
Young modulus	210[(a)]	149
Poisson ratio	0.311[(a)]	149
Moss hardness	6.5	137

(*Continued*)

Table 6.7 (*Continued*)

Optical properties		
Optical bandgap [eV]	3.77±0.03 for $E\|\|c$ 3.485±0.015 for $E\perp c$	151
Fundamental bandgap [eV]	3.596±0.2 eV	160
Bandgap shrinkage at 300–600 K [meV/K]	-1.2	161
Emission band peaks at RT [nm]	520, 564, 650	171
Relative static dielectric constant	14.2±2 for $\perp c$ 9.0±0.5 for $\|\|c$	162
High-frequency dielectric constant	3.785 for $\perp c$ 4.175 for $\|\|c$	164, 165
Refractive index at 350–800 nm	2.125 to 1.97/2.16 to 1.99 for $\perp c$ 2.18 to 2.08/2.22 to 2.11 for $\|\|c$	158/166
Electrical properties		
Tunability of the free electron concentration at RT [cm^{-3}]	8.5×10^{15}–8.8×10^{19}	90, 94, 95, 152, 119, 120, 123, 180
Maximum experimental electron mobility at RT [cm^2/Vs]	240–260	178, 180
Ionization energy of shallow donors [meV]	4–50 80–150	87, 94, 180, 183, 186
Effective electron mass at low $n_e \approx 10^{16}$ cm^{-3}	0.299 for $\perp c$ 0.234 for $\|\|c$	182
Upward band bending [eV]	0.4	189

[a]Calculated from elastic constants given in Ref. [149].

References

1. H. Deng, J. Kong, P. Yang; *J. Mater. Sci: Mater. Electron.* **20** (2009), 1078–1082.
2. K. Nomura, J. Okabayashi, K. Okamura, Y. Yamada; *J. Appl. Phys.* **110** (2011), 083901.
3. A. A. Alsac, A. Yildiz, T. Serin, N. Serin; *J. Appl. Phys.* **113** (2013), 063701.

4. Q.-P. Tran, J.-S. Fang, T.-S. Chin; *Mater. Sci. Semicond. Process.* **40** (2015), 664–669.
5. S. Benkara, H. Ghamri, D. Rechem, M. Zaabat; *J. Mater. Res.* **32** (2017), 1594–1602.
6. G. E. Patil, D. D. Kajale, D. N. Chavan, N. K. Pawar, P. T. Ahire, S. D. Shinde, V. B. Gaikwad, G. H. Jain; *Bull. Mater. Sci.* **34** (2011), 1–9.
7. V. Consonni, G. Rey, H. Roussel, D. Bellet; *J. Appl. Phys.* **111** (2012), 033523.
8. A. R. Babar, S. S. Shinde, A. V. Moholkar, C. H. Bhosale, K. Y. Rajpure; *J. Semicond.* **32** (2011), 102001.
9. V. Gokulakrishnan, S. Parthiban, K. Jeganathan, K. Ramamurthi; *J. Mater. Sci.* **46** (2011), 5553–5558.
10. A. Lousa, S. Gimeno, J. Martí; *Vacuum* **45** (1994), 1143–1145.
11. V. Senthilkumar, P. Vickraman; *J. Mater. Sci.: Mater. Electron.* **21** (2010), 578–583.
12. M. Ghanaatshoar, M. Moradi, Z. Khodabandeh; *J. Supercond. Nov. Magn.* **26** (2013), 995–997.
13. H. W. Kim, S. H. Shim, C. Lee; *Ceram. Int.* **32** (2006), 943–946.
14. R. Rakesh Kumar, K. Narasimha Rao, K. Rajanna, A. R. Phani; *Mater. Res. Bull.* **48** (2013), 1545–1552.
15. A. F. Khan, M. Mehmooda, A. M. Rana, M. T. Bhatti; *Appl. Surf. Sci.* **255** (2009), 8562–8565.
16. H.-R. Kim, G.-H. Lee, D.-H. Kim; *J. Phys. D: Appl. Phys.* **44** (2011), 185203.
17. P. Wu, B. Zhou, W. Zhou; *Appl. Phys. Lett.* **100** (2012), 182405.
18. F. H. Aragón, J. C. R. Aquino, N. C. S. Gomes, J. D. Ardisson, S. W. da Silva, D. G. Pacheco-Salazar, J. A. H. Coaquira; *J. Eur. Ceram. Soc.* **37** (2017), 3375–3380.
19. B. L. Zhu, Y. T. Yang, W. C. Hu, J. Wu, Z. H. Gan, J. Liu, D. W. Zeng, C. S. Xie; *Appl. Phys.* **123 A** (2017), 217.
20. H. Kim, A. Piqué; *Appl. Phys. Lett* **82** (2004), 218–220.
21. H. Kim, R. C. Y. Auyeung, A. Piqué; *Thin Solid Films* **516** (2008), 5052–5056.
22. S. Nakao, N. Yamada, T. Hitosugi, Y. Hirose, T. Shimada, T. Hasegawa; *Thin Solid Films* **518** (2010), 3093–3096.
23. S. Yu, L. Ding, C. Xue, L. Chen, W. F. Zhang; *J. Non-Cryst. Solids* **358** (2012), 3137–3140.

24. M. E. White, M.-Y. Tsai, F. Wu, J. S. Speck; *J. Vac. Sci. Technol. A* **26** (2008), 1300–1307.

25. M.-Y. Tsai, M. E. White, J. S. Speck; *J. Cryst. Growth* **310** (2008), 4256–4261.

26. O. Bierwagen, T. Nagata, M. E. White, M.-Y. Tsai, J. S. Speck; *J. Mater. Res.* **27** (2012), 2232–2236.

27. R. F. Martinez-Gazoni, M. W. Allen, R. J. Reeves; *Phys. Rev. B* **98** (2018), 155308.

28. B. Vlahovic, M. Persin; *J. Phys. D: Appl. Phys.* **23** (1990), 1324–1326.

29. P. Rajaram, Y. C. Goswami, S. Rajagopalan, V. K. Gupta; *Mater. Lett.* **54** (2002), 158–163.

30. C.-Y. Wang, T.-W. Chen, C.-C. Lin, W.-J. Hsieh, K.-L. Chang, H. C. Shih; *J. Phys. D: Appl. Phys.* **40** (2007), 2787–2791.

31. S. Haireche, A. Boumeddiene, A. Guittoum, A. El Hdiy, A. Boufelfel; *Mater. Chem. Phys.* **139** (2013), 871–876.

32. D. Barreca, S. Garon, P. Zanella, E. Tondello; *J. Phys. IV France* **09** (1999), 667- 673.

33. X. Feng, J. Ma, F. Yang, F. Ji, F. Zong, C. Luan, H. Ma; *Mater. Lett.* **62** (2008), 1779–1781.

34. L. Kong, J. Ma, Z. Zhu, C. Luan, F. Ji; *J. Cryst. Growth* **312** (2010), 2931–2935.

35. Z. Zhu, J. Ma, L. Kong, C. Luan, Q. Yu; *J. Cryst. Growth* **324** (2011), 98–102.

36. R. Drevet, C. Legros, D. Bérardan, P. Ribot, D. Dragoé, C. Cannizzo, M. G. Barthés-Labrousse, A. Chaussé, M. Andrieux; *Surf. Coat. Technol.* **271** (2015), 234–241.

37. R. Khan, Zulfiqar, Z. S. Fashu, Y. Zaman; *J. Mater. Sci.: Mater. Electron.* **27** (2016), 5960–5966.

38. M. Kambe, M. Fukawa, N. Taneda, K. Sato; *Sol. Energy Mater. Sol. Cells* **90** (2006), 3014–3020.

39. N. S. Murty, S. R. Jawalekar; *Thin Solid Films***108** (1983), 277–283.

40. H. J. Snaith, C. Ducati; *Nano Lett.* **10** (2010), 1259–1265.

41. L. Xiong, Y. Guo, J. Wen, H. Liu, G. Yang, P. Qin, G. Fang; *Adv. Funct. Mater.* **28** (2018), 1802757.

42. J. Chantana, T. Hirayama, C. Ding, Y. Kawano, Q. Shen, K. Yoshino, S. Hayase, T. Minemoto; *Thin Solid Films* **669** (2019), 455–460.

43. M.-Y. Tsai, O. Bierwagen, J. S. Speck; *Thin Solid Films* **605** (2016), 186–192.

44. J.-H. Oh, T.-Y. Seong, H.-G. Hong, K.-K. Kim, S.-W. Yoon, J.-P. Ahn; *J Electroceram.* **27** (2011), 109–113.

45. Y. S. Liu, C. I. Hsieh, Y. J. Wu, Y. S. Wei, P. M. Lee, C. Y. Liu; *Appl. Phys. Lett.* **101** (2012), 122107.

46. G. Cannella, F. Principato, M. Foti, S. Di Marco, A. Grasso, S. Lombardo; *J. Appl. Phys.* **110** (2011), 024502.

47. J. Sun, A. Lu, L. Wang, Y. Hu, Q. Wan; *Nanotechnology* **20** (2009), 335204.

48. R. E. Presley, C. L. Munsee, C.-H. Park, D. Hong, J. F. Wager, D. A. Keszler; *J. Phys. D: Appl. Phys.* **37** (2004), 2810–2813.

49. L. Doohyeok, J. Youngin, K. Minsuk, K. Yoonjoong, K. Sangsig; *J. Nanosci. Nanotechnol.* **16** (2016), 11697–11700.

50. B. Jang, T. Kim, S. Lee, W.-Y. Lee, H. Kang, C. S. Cho, J. Jang; *IEEE Electron Device Lett.* **39** (2018), 1179–1182.

51. Y. Li, W. Yin, R. Deng, R. Chen, J. Chen, Q. Yan, B. Yao, H. Sun, S.-H. Wei, T. Wu; *NPG Asia Mater.* **4** (2012), e30.

52. T. Zhang, L.-H. Ding, W.-F. Zhang; *Chin. Phys. B* **21** (2012), 047301.

53. M. Alshehri, F. Al-Marzouki, A. Alshehrie, M. Hafez; *J. Alloy. Compd.* **757** (2018), 161–168.

54. F. H. Aragón, J. A. H. Coaquira, P. Hidalgo, S. L. M. Brito, D. Gouvêa, R. H. R. Castro; *J. Phys.: Condens. Matter* **22** (2010), 496003.

55. J. Okabayashi, S. Kono, Y. Yamada, K. Nomura; *J. Appl. Phys.* **112** (2012), 073917.

56. T. R. Cunha, I. M. Costa, R. J. S. Lima, J. G. S. Duque, C. T. Meneses; *J. Supercond. Nov. Magn.* **26** (2013), 2299–2302.

57. G. M. Stoian, P. A. Stampe, R. J. Kennedy, Y. Xin, E. Lochner, S. von Molnár; *J. Phys. D: Appl. Phys.* **47** (2014), 295002.

58. Y. Ziat, A. Benyoussef, A. El Kenz; *J. Phys. Chem. Solids* **75** (2014), 701–709.

59. J. Fan, E. Menéndez, M. Guerrero, A. Quintana, E. Weschke, E. Pellicer, J. Sort; *Nanomaterials* **7** (2017), 348.

60. S. Zhu, C. Chen, Z. Li; *J. Magn. Magn. Mater.* **471** (2019), 370–380.

61. K. D. Schierbaum, U. K. Kirner, J. F. Geiger, W. Göpel; *Sens. Actuator. B* **4** (1991), 87–94.

62. S. H. Saeedabad, G. S. Selopal, S. M. Rozati, Y. Tavakoli, G. Sberveglieri; *J. Electron. Mater.* **47** (2018), 5165–5173.

63. G. Korotcenkov, B. K. Cho; *IOP Conf. Series: Mater. Sci. Eng.* **18** (2011), 212008.

64. W. Chen, Q. Zhou, F. Wan, T. Gao; *J. Nanomater.* **2012** (2012), 612420.

65. A. Tischner, T. Maier, C. Stepper, A. Köck; *Sens. Actuator. B* **134** (2008), 796–802.

66. N. Barsan, J. Rebholz, U. Weimar; *Sens. Actuator. B* **207** (2015), 455–459.

67. C.-H. Lin, W.-C. Chang, X. Qi; *Procedia Eng.* **36** (2012), 476–481.

68. M. Kroneld, S. Novikov, S. Saukko, P. Kuivalainen, P. Kostamo, V. Lantto; *Sens. Actuator. B: Chem.* **118** (2006), 110–114.

69. Z. M. Jarzębski; *Phys. Status Solidi (a)* **71** (1982), 13–41.

70. K. L. Chopra, S. Major, D. K. Pandya; *Thin Solid Films* **102** (1983), 1–46.

71. E. Zorina, L. Poplevina, Y. Tsybulya; *High Energy Chem.* **20** (1986), 40–47.

72. S. Das, V. Jayaraman; *Prog. Mater. Sci.* **66** (2014), 112–255.

73. M. Batzill, U. Diebold; *Prog. Surf. Sci.* **79** (2005), 47–154.

74. S. C. Dixon, D. O. Scanlon, C. J. Carmalt, I. P. Parkin; *J. Mater. Chem. C* **4** (2016), 6946–6961.

75. V. G. Krishnan, P. Elango, A. Purushothaman, A. C. Bose; *Mater. Chem. Phys.* **199** (2017), 113–121.

76. J. Pannetier, G. Denes; *Acta Crystallogr. B* **36** (1980), 2763–2765.

77. A. Bolzan, C. Fong, B. J. Brendan, J. Kennedy, C. J. Howard; *Acta Crystallogr. B* **53** (1997), 373–380.

78. M. A. Mäki-Jaskari, T. T. Rantala; *Phys. Rev. B* **64** (2001), 075407.

79. F. R. Sensato, R. Custódio, M. Calatayud, A. Beltrán, J. Andrés, J. R. Sambrano, E. Longo; *Surf. Sci.* **511** (2002), 408–420.

80. Q.-J. Liu, Z.-T. Liu, L.-P. Feng; *Comput. Mater. Sci.* **47** (2010), 1016–1022.

81. P. D. Borges, L. M. R. Scolfaro, H. W. L. Alves, E. F. da Silva Jr.; *Theor. Chem. Acc.* **126** (2010), 39–44.

82. A. Schleife, J. B. Varley, F. Fuchs, C. Rödl, F. Bechstedt, P. Rinke, A. Janotti, C. G. Van de Walle; *Phys. Rev. B* **83** (2011), 035116.

83. D. O. Scanlon, G. W. Watson; *J. Mater. Chem.* **22** (2012), 25236.

84. M. Dou, C. Persson; *J. Appl. Phys.* **113** (2013), 083703.

85. A. K. Singh, A. Janotti, M. Scheffler, C. G. Van de Walle; *Phys. Rev. Lett.* **101** (2008), 055502.

86. P. D. C. King, R. L. Lichti, Y. G. Celebi, J. M. Gil, R. C. Vilão, H. V. Alberto, J. Piroto Duarte, D. J. Payne, R. G. Egdell, I. McKenzie, C. F. McConville, S. F. J. Cox, T. D. Veal; *Phys. Rev. B* **80** (2009), 081201.

87. S. Samson, C. G. Fonstad; *J. Appl. Phys.* **44** (1973), 4618–4621.

88. E. Korhonen, V. Prozheeva, F. Tuomisto, O. Bierwagen, J. S. Speck, M. E. White, Z. Galazka, H. Liu, N. Izyumskaya, V. Avrutin, Ü. Özgür, H. Morkoç; *Semicond. Sci. Technol.* **30** (2015), 024011.

89. W. M. H. Oo, S. Tabatabaei, M. D. McCluskey, J. B. Varley, A. Janotti, C. G. Van de Walle; *Phys. Rev. B* **82** (2010), 193201.

90. Z. Galazka, R. Uecker, D. Klimm, K. Irmscher, M. Pietsch, R. Schewski, M. Albrecht, A. Kwasniewski, S. Ganschow, D. Schulz, C. Guguschev, R. Bertram, M. Bickermann, R. Fornari; *Phys. Status Solidi* (*a*) **211** (2014), 66–73.

91. Z. Remes, M. Vanecek, H. M. Yates, P. Evans, D. W. Sheel; *Thin Solid Films* **517** (2009), 6287–6289.

92. F. Wu, X. Tong, Z. Zhao, J. Gao, Y. Zhou, P. Kelly; *J. Alloys Compd.* **695** (2017), 765–770.

93. J. B. Varley, A. Janotti, C. G. Van de Walle; *Phys. Rev. B* **81** (2010), 245216.

94. J. A. Marley, R. C. Dockerty; *Phys. Rev.* **140** (1965), A304–A310.

95. D. F. Morgan, D. A. Wright; *Brit. J. Appl. Phys.* **17** (1966), 337–340.

96. T. Nagata, O. Bierwagen, M. E. White, M.-Y. Tsai, Y. Yamashita, H. Yoshikawa, N. Ohashi, K. Kobayashi, T. Chikyow, J. S. Speck; *Appl. Phys. Lett.* **98** (2011), 232107.

97. O. Bierwagen, T. Nagata, M. E. White, M. Y. Tsai, J. S. Speck; *J. Mater. Res.* **27** (2012), 2232–2236.

98. M. E. White, O. Bierwagen, M. Y. Tsai, J. S. Speck; *J. Appl. Phys.* **106** (2009), 093704.

99. S. Lany, A. Zunger; *Phys. Rev. B* **80** (2009), 085202.

100. J. B. Varley, A. Janotti, C. G. Van de Walle; *Phys. Rev. B* **81** (2010), 245216.

101. A. Chaparadza, S. B. Rananavare; *Nanotechnology* **21** (2010), 035708.

102. J. M. Ni, X. J. Zhao, J. Zhao; *J. Inorg. Organomet. Polym.* **22** (2012), 21–26.

103. S. S. Pan, S. Wang, Y. X. Zhang, Y. Y. Luo, F. Y. Kong, S. C. Xu, J. M. Xu, G. H. Li; *Appl. Phys. A* **109** (2012), 267–271.

104. S. Cahen, N. David, J. M. Fiorani, A. Maître, M. Vilasi; *Thermochim. Acta.* **403** (2003), 275–285.

105. J. A. Marley, T. C. MacAvoy; *J. Appl. Phys.* **32** (1961), 2504–2505.

106. H. F. Kunkle, E. E. Kohnke; *J. Appl. Phys.* **36** (1965), 1489.

107. M. L. Harvill, R. Roy; *Am. Ceram. Soc. Bull.* **44** (1965), 297.

108. M. Nagasawa, S. Shionoya, S. Makishima; *Jpn. J. Appl. Phys.* **4** (1965), 195–202.

109. M. de Murcia, J. P. Fillard; *Mat. Res. Bull.* **11** (1976), 189–196.

110. B. Thiel, R. Helbig; *J. Cryst. Growth* **32** (1976), 259–264.

111. T. B. Reed, J. T. Roddy, A. N. Mariano; *J. Appl. Phys.* **33** (1962), 1014–1015.

112. D. B. Ghare; *J. Cryst. Growth* **23** (1974), 157–159.

113. T. Takizawa, T. Sakurai; *Jpn. J. Appl. Phys.* **12** (1973), 1323–1325.

114 Z. Galazka; "Growth measures to achieve bulk single crystals of transparent semiconducting and conducting oxides," *Handbook of Crystal Growth*, Vol. II, edited by T. Nishinaga and P. Rudolph, Elsevier, 2015, pp. 209–240.

115. F. P. Koffyberg; *J. Appl. Phys.* **36** (1965), 844–849.

116. M. de Murcia, M. Egee, J. P. Fillard; *J. Phys. Chem. Solids* **39** (1978), 629–635.

117. M. de Murcia, J. Bonnafe, J. C. Manifacier, J. P. Fillard; *J. Appl. Phys.* **49** (1978), 1177–1187.

118. www.icdd.com, 2019.

119. D. F. Crabtree, R. N. S. M. Mehdi, D. A. Wright; *Brit. J. Appl. Phys.* **2** (1969), 1503–1505.

120. M. Nagasawa, S. Shionoya, S. Makishima; *Jpn. J. Appl. Phys.* **4** (1965), 195–202.

121. M. Hirose, Y. Furuya; *Sci. Bull. Fac. Educ., Nagasaki Univ.* **32** (1981), 21–36.

122. Y. Furuya, M. Hirose, Y. Taneda; *Jpn. J. Appl. Phys.* **20** (1981), 785–786.

123. C. G. Fonstad, A. Linz, R. H. Rediker; *J. Electrochem. Soc.* **116** (1969), 1269–1271.

124. D. F. Crabtree; *J. Phys. D: Appl. Phys.* **7** (1974), L17–L21.

125. D. F. Crabtree; *J. Phys. D: Appl. Phys.* **7** (1974), L22–L26.

126. K. Matsumoto, S. Kaneko, K. Takagi; *J. Cryst. Growth* **40** (1977), 291–297.

127. I. P. Kuzmina, B. N. Litvin; *Kristallografiya* **8** (1963), 478–480.

128. M. L. Harvil, R. Roy; *Am. Ceram. Soc. Bull.* **44** (1965), 297.

129. Y. Fujiki, Y. Suzuki; *J. Japan. Assoc. Min. Petr. Econ. Geol.* **68** (1973), 277–283.

130. F. Kawamura, M. Kamei, I. Yasui, I. Sunagawa; *J. Am. Ceram. Soc.* **82** (1999), 774–776.
131. F. Kawamura, I. Yasui, M. Kamei, I. Sunagawa; *J. Am. Ceram. Soc.* **84** (2001), 1341–1346.
132. F. Kawamura, T. Takahashi, I. Yasui, I. Sunagawa; *J. Cryst. Growth* **233** (2001), 259–268.
133. S. Hidenori, U. Kenji; JP 60051699 A (1985).
134. S. Hidenori, U. Kenji; JP 60161387 A (1985).
135. S. Shimada, K. Kodaira, T. Matsushita; *Chem. Lett.* **5** (1976), 235–236.
136. S. Shimada, K. Kodaira, T. Matsushita; *J. Cryst. Growth* **59** (1982), 662–664.
137. M. Batzill, U. Diebold; *Prog. Surf. Sci.* **79** (2005), 47–154.
138. A. Beltrán, J. A. E. Longo, E. R. Leite; *Appl. Phys. Lett.* **83** (2003), 635–637.
139. J. Oviedo, M. J. Gillan; *Surf. Sci.* **463** (2000), 93–101.
140. P. A. Mulheran, J. H. Harding; *Modell. Simul. Mater. Sci. Eng.* **1** (1992), 39–43.
141. D. F. Cox, T. B. Fryberger, S. Semancik; *Phys. Rev. B* **38** (1988), 2072–2083.
142. K. Katsiev, M. Batzill, L. A. Boatner, U. Diebold; *Surf. Sci.* **602** (2008), 1699–1704.
143. J. M. Themlin, R. Sporken, J. Darville, R. Caudano, J. M. Gilles, R. L. Johnson; *Phys. Rev.* **42** (1990), 11914–11925.
144. D. Klimm, Z. Galazka; *Leibniz-Institut für Kristallzüchtung*, Berlin, Germany (2018), unpublished data.
145. A. M. Hofmeister; *Phys. Chem. Minerals* **41** (2014), 361–371.
146. O. Knacke, O. Kubaschewski, K. Hesselmann; *Thermochemical Properties of Inorganic Substances*, 2nd edn., Springer-Verlag, Berlin 1990.
147. J. Leitner, D. Sedmidubský, P. Chuchvalec; *Ceramics – Silikáty* **46** (2002), 29–32.
148. www.factsage.com.
149. R. S. Katiyar, P. Dawson, M. M. Hargreave, G. R. Wilkinson; *J. Phys. C* **4** (1971), 2421–2431.
150. A. C. Camargo, J. A. Igualada, A. Beltrán, R. Llusar, E. Longo, J. Andrés; *Chem. Phys.* **212** (1996), 381–391.
151. Z. Galazka, K. Irmscher, M. Pietsch; *Leibniz-Institut für Kristallzüchtung*, Berlin, Germany (2013), unpublished data.

152. R. Summitt, N. F. Borrelli; *J. Phys. Chem. Solids* **26** (1965), 921–925.

153. H. Peelaers, E. Kioupakis, C. G. Van de Walle; *Appl. Phys. Lett.* **100** (2012), 011914.

154. M. Nagasawa, S. Shionoya; *Jpn. J. Appl. Phys.* **10** (1971), 727–731.

155. M. Nagasawa, S. Shionoya; *Phys. Lett.* **22** (1966), 409–410.

156. M. Nagasawa, S. Shionoya; *J. Phys. Soc. Jpn.* **30** (1971), 158–167.

157. R. Summitt, J. A. Marley, N. F. Borrelli; *J. Phys. Chem. Solids* **25** (1964), 1465–1469.

158. S. F. Reddaway, D. A. Wright; *Brit. J. Appl. Phys.* **16** (1965), 195–198.

159. D. Fröhlich, R. Kenklies, R. Helbig; *Phys. Rev. Lett.* **41** (1978), 1750–1751.

160. M. Feneberg, C. Lidig, K. Lange, R. Goldhahn, M. D. Neumann, N. Esser, O. Bierwagen, M. E. White, M. Y. Tsai, J. S. Speck; *Appl. Phys. Lett.* **104** (2014), 231106.

161. R. Summitt, N. F. Borrelli; *J. Appl. Phys.* **37** (1966), 2200.

162. H. J. van Daal; *J. Appl. Phys.* **39** (1968), 4467–4469.

163. H. J. van Daal; *Solid State Commun.* **6** (1968), 5–9.

164. R. Sunmitt; *J. Appl. Phys.* **39** (1968), 3762–3767.

165. Z. M. Jarzębski, J. P. Morton; *J. Electrochem. Soc.* **123** (1976), 333C–346C.

166. S. Bin Anooz, Z. Galazka; *Leibniz-Institut für Kristallzüchtung*, Berlin, Germany (2019), unpublished data.

167. M. Feneberg, C. Lidig, K. Lange, M. E. White, M. Y. Tsai, J. S. Speck, O. Bierwagen, R. Goldhahn; *Phys. Status Solidi (a)* **211** (2014), 82–86.

168. M. Feneberg, C. Lidig, K. Lange, R. Goldhahn, M. D. Neumann, N. Esser, O. Bierwagen, M. E. White, M. Y. Tsai, J. S. Speck; *Appl. Phys. Lett.* **104** (2014), 231106.

169. M. Feneberg, C. Lidig, M. E. White, M. Y. Tsai, J. S. Speck, O. Bierwagen, Z. Galazka, R. Goldhahn; *APL Mater.* **7** (2019), 022508.

170. J. P. Fillard, M. de Murcia; *Phys. Status Solidi* **80** (1975), 279–287.

171 R. Schewski, Z. Galazka; *Leibniz-Institut für Kristallzüchtung*, Berlin, Germany (2013), unpublished data.

172. J. B. Varley, A. Janotti, C. Franchini, C. G. Van de Walle; *Phys. Rev. B* **85** (2012), 081109.

173. W. Drozdowski, W. Łachmański, K. Brylew, A. J. Wojtowicz; *Nicolaus Copernicus University*, Toruń, Poland (2015), unpublished data.

174. J. F. Scott; *J. Chem. Phys.* **53** (1970), 852–853.

175. P. S. Peercy, B. Morosin; *Phys. Rev. B* **7** (1973), 2779–2786.

176. Z. M. Jarzębski; *Phys. Status Solidi (a)* **71** (1982), 13–41.

177. M. E. White, O. Bierwagen, M. Y. Tsai, J. S. Speck; *Appl. Phys. Express* **3** (2010), 051101.

178. Z. Galazka, K. Irmscher, M. Pietsch; *Leibniz-Institut für Kristallzüchtung*, Berlin, Germany (2013), unpublished data.

179. O. Bierwagen, Z. Galazka; *Appl. Phys. Lett.* **112** (2018), 092105.

180. C. G. Fonstad, R. H. Rediker; *J. Appl. Phys.* **42** (1971), 2911–2918.

181. O. Bierwagen, Z. Galazka; *Appl. Phys. Lett.* **112** (2018), 092105.

182. K. J. Button, C. G. Fonstad, W. Dreybrodt; *Phys. Rev. B* **4** (1971), 4539–4542.

183. Z. M. Jarzębski, J. P. Morton; *J. Electrochem. Soc.* **123** (1976), 299C–310C.

184. S. K. V. Farahani, T. D. Veal, A. M. Sanchez, O. Bierwagen, M. E. White, S. Gorfman, P. A. Thomas, J. S. Speck, C. F. McConville; *Phys. Rev. B* **86** (2012), 245315.

185. Z. Galazka, K. Irmscher, M. Pietsch; *Leibniz-Institut für Kristallzüchtung*, Berlin, Germany (2015), unpublished data.

186. M. Nagasawa, S. Shionoya, S Makishima; *J. Phys. Soc. Jpn.* **20** (1965), 1093.

187. J. P. Fillard, J. Gasiot, M. de Murcia; *Phys. Status Solidi* **8** (1970), 893–898.

188. J. E. Houston, E. E. Kohnke; *J. Appl. Phys.* **36** (1965), 3931–3938.

189. T. Nagata, O. Bierwagen, Z. Galazka, S. Ueda, M. Imura, Y. Yamashita, T. Chikyow; *Jpn. J. Appl. Phys.* **58** (2019), 080903.

190. N. Barsan, J. Rebholz, U. Weimar; *Sens. Actuator. B* **207** (2015), 455–459.

191. T. Nagata, O. Bierwagen, M. E. White, M.-Y. Tsai, Y. Yamashita, H. Yoshikawa, N. Ohashi, K. Kobayashi, T. Chikyow, J. S. Speck; *Appl. Phys. Lett.* **98** (2011), 232107.

192. S. K. V. Farahani, T. D. Veal, J. J. Mudd, D. O. Scanlon, G. W. Watson, O. Bierwagen, M. E. White, J. S. Speck, C. F. McConville; *Phys. Rev. B* **90** (2014), 155413.

193. O. Bierwagen, M. E. White, M.-Y. Tsai, T. Nagata, J. S. Speck; *Appl. Phys. Express* **2** (2009), 106502.

Chapter 7

Barium Stannate ($BaSnO_3$)

7.1 Introduction

Ternary $BaSnO_3$ compound is a wide bandgap oxide showing *n*-type semiconducting behavior and transparency in the visible spectrum. It has the highest electron mobility among other transparent semiconducting oxides (TSOs) that triggered high-level interest in that material. Although $BaSnO_3$, as such, had occasionally been mentioned since 1946 [1] and 1958 [2], it was not a subject of detailed study in terms of electrical and optical properties. Such properties were reported from time to time for $BaSnO_3$ ceramics [3–7] and occasionally for thin films [8–10]. The situation drastically changed in 2012 when the first, but small, bulk single crystals were obtained [11] by the flux method, and a very high electron mobility was revealed (about 300 cm^2/Vs) [12]. That event was a turning point in obtaining $BaSnO_3$ thin films and bulk single crystals, as well as in investigating their physical properties enabling design and operation of novel devices utilizing two fundamental properties: perovskite structure and high electron mobility. These features make $BaSnO_3$ an ideal candidate for generating an effective two-dimensional electron gas (2DEG) at an oxide–oxide interface, e.g., $BaSnO_3$–$LaInO_3$ [13]. $BaSnO_3$ may, therefore, substitute a commonly used $SrTiO_3$ for that

Transparent Semiconducting Oxides: Bulk Crystal Growth and Fundamental Properties
Zbigniew Galazka

ISBN 978-981-4800-94-5 (Hardcover), 978-1-003-04520-5 (eBook)
www.jennystanford.com

purpose, which shows, however, very low electron mobility at room temperature (about 10 cm^2/Vs). Therefore, $BaSnO_3$ constitutes a perspective material for oxide electronics, e.g., high electron mobility transistors (HEMTs). Examples of thin-film field-effect transistors (TFTs) with a $BaSnO_3$ channel layer and $BaSnO_3$-based *p-n* diodes were demonstrated by Lee et al. [14].

The thin films of $BaSnO_3$ were heteroepitaxially grown mostly with the use of pulsed laser deposition (PLD) [15–23], and molecular-beam epitaxy (MBE) [24–27], but also other techniques were reported, such as chemical solution deposition (CSD) [28], sputtering [29–31], and sol-gel [32, 33].

The heteroepitaxial thin films, however, show very high dislocation density, which limits electron mobility [17]. Bulk $BaSnO_3$ single crystals of size enabling wafers for homoepitaxy are, therefore, in high demand. To meet that demand, melt growth and hydrothermal techniques rather than flux method would be required. But very high thermal instability and incongruent evaporation of that compound at high temperatures make such growth very challenging.

This chapter undertakes a discussion on the growth of bulk $BaSnO_3$ single crystals and shows a route for obtaining large-enough crystals for homoepitaxial purposes. It discusses the state of the art in growing bulk $BaSnO_3$ crystals (methods, size, quality), particularities of that compound from thermodynamic point of view, and physical properties revealed in the bulk crystals. Not all basic properties were shown by studying bulk crystals, and to achieve a larger picture of that compound, some findings for ceramics, thin films, and theory are also included.

7.2 Crystal Structure

Barium oxide (BaO) and tin(IV) oxide (SnO_2) form two intermediate compounds: orthostannate Ba_2SnO_4 and metastannate $BaSnO_3$ (see Section 7.4.1). Only $BaSnO_3$ (barium stannate) melts congruently and crystallizes in an undistorted ideal cubic, perovskite crystal structure with the space group $Pm\bar{3}m$. The crystal structure of $BaSnO_3$ is shown in Fig. 7.1.

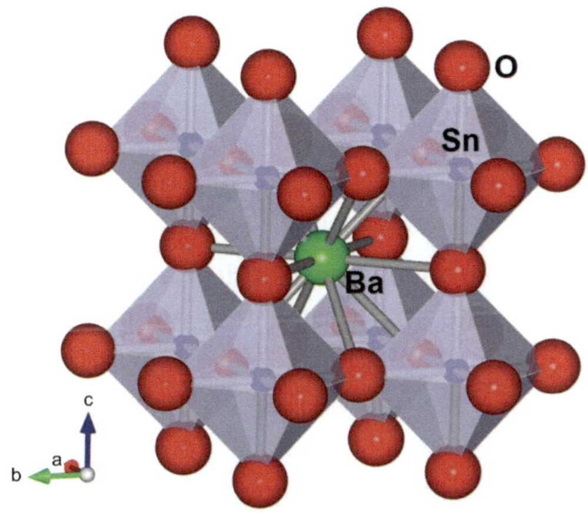

Figure 7.1 Perovskite crystal structure of $BaSnO_3$ consists of alternating Sn–O layers (Sn^{4+} in octahedral environment) and Ba–O layers with Ba^{2+} in dodecahedral environment.

The stability of the perovskite structure ABO_3 is defined by the Goldschmidt tolerance factor *t*:

$$t = \frac{(r_A + r_O)}{\sqrt{2}(r_B + r_O)} \tag{7.1}$$

where r_A, r_B, and r_O are ionic radii of the A-cation (Ba), B-cation (Sn), and O-anion (oxygen), respectively.

For $BaSnO_3$, the Goldschmidt tolerance factor is 1.024, indicating, indeed, an ideal and stable perovskite structure.

In the three-dimensional structure, $[SnO_6]^{8-}$ octahedra are corner-sharing and form layers, and the barium ions Ba^{2+} reside between these layers. Ba^{2+} can be substituted by several other ions, including La^{3+}. This introduces charge carriers into the crystal structure, which can move almost undisturbed along the Sn–O planes. This is the reason why high doping in $BaSnO_3$ does not necessarily lead to low carrier mobility, an aspect potentially interesting for devices (see Section 7.7).

The lattice constant of $BaSnO_3$ is $a = 4.117 \pm 0.002$ Å [34], while the Ba–O and Sn–O distances are 2.905 Å and 2.055 Å, respectively [35].

7.3 Band Structure and Native Point Defects

7.3.1 Band Structure

As in the case of other TSOs, $BaSnO_3$ was also the subject of an intensive theoretical study of the band structure, point defects, as well as structural, thermal, mechanical, optical, and electrical properties. For example, the band structure of this compound was calculated with different methods and approximations by Krishnaswamy et al. [36], Scanlon [37], Moreira et al. [38, 39], Dabaghmanesh et al. [40], Mizoguchi et al. [35], Liu et al. [41], Singh et al. [42], Stanislavchuk et al. [43], Kim et al. [44, 45], Kang et al. [46], Lochocki et al. [47], Monserrat et al. [48], Soleimanpour and Kanjouri [49], Li et al. [50], Bouhemadou and Haddadi [51], and Duarte et al [52].

The calculated band structure is shown in Fig. 7.2 [36]. The bottom of the conduction band (CB) originates from Sn *5s* (dominating), Sn *5p*, and O *2p* orbitals of high dispersion [36, 39, 43]. The top of the valence band (VB) is formed by O *2p* orbitals (dominating) with smaller contributions from Sn *5p,5s* and Ba *4d* orbitals [39, 43]. The top of the VB has a very small dispersion as compared with the dispersion of the CB. The bottom of the CB has a parabolic minimum at the Γ-point, while the top of the VB has a maximum at *R*-point of the Brillouin zone (BZ). The electric–dipole transition at the Γ-point from the valence band maximum (VBM) to the conduction band minimum (CBM) ($\Gamma \to \Gamma$) is allowed and leads to a direct optical bandgap $E_{g,dir}$. The transition from the VBM at the *R*-point to the CBM at Γ-point ($R \to \Gamma$) is also allowed and defines an indirect optical bandgap $E_{g,ind}$.

Exemplary calculated values of both bandgaps with density functional theory (DFT) and different approximations are: $E_{g,dir}$ = 2.88 and $E_{g,ind}$ = 2.4 eV (HSE06) [36], $E_{g,dir}$ = 3.68 and $E_{g,ind}$ = 3.2 eV (PBE0) [37], $E_{g,dir}$ = 3.6 and $E_{g,ind}$ = 2.6 eV (HSE06) [40], $E_{g,dir}$ = 3.21 and $E_{g,ind}$ = 2.73 eV (HSE) [48], $E_{g,ind}$ = 2.83 eV (GGA) [43], $E_{g,dir}$ = 3.0 eV (HSE06) [41], $E_{g,dir}$ = 3.46 eV and $E_{g,ind}$ = 2.98 eV (LDA) [12], $E_{g,ind}$ = 3.14 eV (PBE0) [18], $E_{g,dir}$ = 3.0 eV and $E_{g,ind}$ = 2.65 eV (MBJGGA) [15].

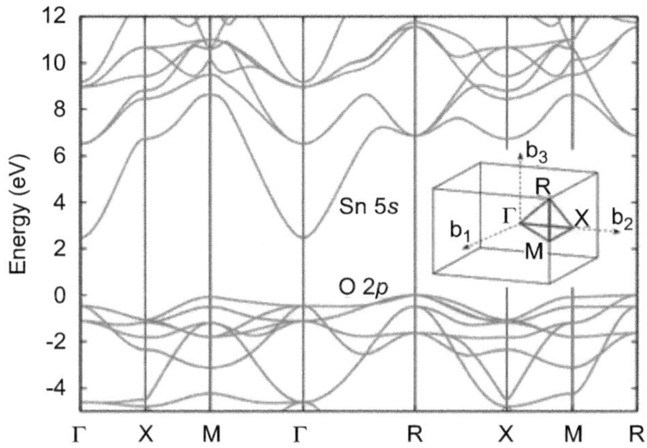

Figure 7.2 Calculated band structure of $BaSnO_3$ (first principles using the HSE06 hybrid functional). The inset indicates the high symmetry points in the Brillouin zone. The indirect gap R → Γ is 2.40 eV, and the direct gap at Γ → Γ is 2.88 eV. Reprinted with permission from Ref. [36], Copyright 2017, American Physical Society.

7.3.2 Native Point Defects

Native (intrinsic) point defects in $BaSnO_3$ were theoretically studied by Scanlon [37] and Dabaghmanesh et al. [40]. According to Scanlon [37], at metal-rich and oxygen-poor conditions (optimum for *n*-type defect formation), oxygen vacancies V_O are the lowest energy donors with the +2/0 transition level 0.37 eV below the CBM. This transition is too deep to provide sufficient electrical conductivity. Antisite intrinsic defect Sn_{Ba} is the next most stable defect and acts as an ultradeep donor with +1/0 and +2/+1 transitions levels of 0.73 eV and 0.41 eV above the VBM, respectively. Due to large difference between Sn- and Ba-site size in the $BaSnO_3$ structure, Sn can be stabilized in the Sn^{2+} charge state, avoiding this way a possible self-doping by substituting Ba^{2+} with Sn^{4+}. Interstitial Sn and Ba, Sn_i and Ba_i, both act as shallow donors but are of high energy; therefore, they will rather not play an important role in $BaSnO_3$ conductivity. Ba vacancy V_{Ba} has the lowest energy acting as an acceptor, but it is too high in energy to compensate V_O and Sn_{Ba}.

Under metal-poor and oxygen-rich conditions (optimum for *p*-type defect formation), the native acceptor defects are much lower in energy and can compensate the native donor defects. Interstitial oxygen O_i is the lowest energy acceptor defect; however, it relaxes from the ideal interstitial site to a lattice oxygen (O–O dumbbell). Its ionization levels are located in the CB, and they will not act as an effective charge compensator for deep acceptor V_{Ba}, V_{Sn}, and Ba_{Sn} native defects, despite much lower formation energy.

At all conditions, V_{Ba} is a dominating acceptor with the Fermi level being trapped in the bandgap that will compensate *n*-type conductivity, leading to an electrically insulating state. Therefore, intrinsic native defects will not enable to obtain *p*-type conductivity in $BaSnO_3$ at any conditions.

Calculations of Dabaghmanesh et al. [40] revealed that V_O^{2+} is a shallow donor, while V_O^{1+} is a deep donor, both with a relatively high formation energy.

In terms of extrinsic point defects theoretically studied by Scanlon [37], substitutional hydrogen H_O and interstitial hydrogens H_i^1 and H_i^2 all act as shallow donors in $BaSnO_3$ under metal-rich and oxygen-poor conditions. Moreover, H is very mobile in $BaSnO_3$. Other dopants, La substituting Ba (La_{Ba}), Sb substituting Sn (Sb_{Sn}), and F substituting O (F_O), are all shallow donors with F_O being the most soluble in $BaSnO_3$. The 0/+1 ionization levels are just below the CBM for Sb_{Sn}, and inside the CBM for La_{Ba} and F_O, respectively, favoring La_{Ba} and F_O extrinsic defects for high conductivity in $BaSnO_3$. At these conditions, Sb_{Ba} is a deep donor of high energy with little impact on the $BaSnO_3$ conductivity, as Sb will be stabilized in the +3 charge state by a large Ba-site. La_{Sn} is a deep acceptor with a high formation energy (4.32 eV), but it will not compensate La_{Ba}. F_i is too high in energy to compensate the low-energy F_O.

Under metal-poor and oxygen-rich conditions, the formation energy of H_O, F_O, H_i, and La_{Ba} are increased, as they all strongly depend on the oxygen partial pressure. The only stable shallow donor at these conditions is Sb_{Sn}. In other words, La_{Ba} can be compensated by La_{Sn} under these conditions. To achieve high-performance conductivity in $BaSnO_3$ by La-doping, the growth conditions should not be too metal poor and too oxygen rich to minimize the compensation by native and extrinsic La_{Sn} defects. F_O will not be compensated by F_i under oxygen-rich conditions, but it might be compensated by native defects under metal-poor conditions.

Dabaghmanesh et al. [40] also concluded that La_{Ba} forms a shallow donor with the +1/0 transition level in the CBM with no deep states in the bandgap. Additionally, they also found Y_{Ba} defect forming a shallow donor similar to La_{Ba} at both Sn-rich and O-rich conditions.

As a result of theoretical studies, it was experimentally confirmed in some instances that native defects cannot act either as shallow donors or as shallow acceptors due to self-compensation effect. High conductivity of $BaSnO_3$ can be easily achieved by doping with La, Y, H, and/or F under growth conditions that are not too metal poor and too oxygen rich.

7.4 Bulk Crystal Growth

7.4.1 Phase Diagram of the BaO–SnO_2 System

The first trials on melting the BaO–SnO_2 system were conducted by Wagner and Binder [2], who used an electric arc between carbon rods to melt sample sticks of different compositions. Other melting experiments were performed by inductive heating of a few milligrams of powdered samples in an Ir boat. These experiments enabled construction of a phase diagram for SnO_2 content 0–60 mol.%, as shown in Fig. 7.3. On the BaO-rich side, there is peritectically melting Ba_2SnO_4 and the congruently melting $BaSnO_3$. The melting point (MP) of the composition BaO:SnO_2 = 1:1, and hence congruent melting of $BaSnO_3$ derived from these experiments is 2057°C.

7.4.2 Historical Events

The first report on melting SnO_2–$BaCO_3$ system by Wagner and Binder [2] was published in 1958; however, no crystals were obtained. It seems that until 2012, bulk $BaSnO_3$ crystals have not been reported, when first very small $BaSnO_3$ crystals were obtained by Luo et al. [11] using the flux method. A breakthrough in obtaining bulk crystals was done by Galazka et al. [53] in 2017, who reported much larger crystals obtained directly from the melt for the first time. Surprisingly, the growth of crystals from the gas phase has not been reported. These main events are summarized in Fig. 7.4.

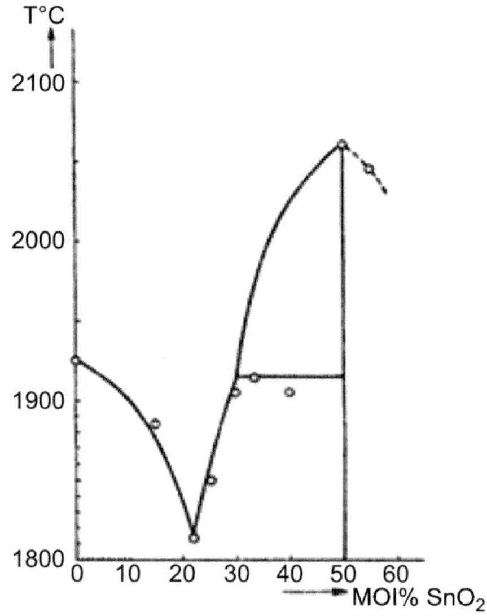

Figure 7.3 $BaSnO_3$–SnO_2 phase diagram. Reprinted with permission from Ref. [2], Copyright 2004, John Wiley and Sons.

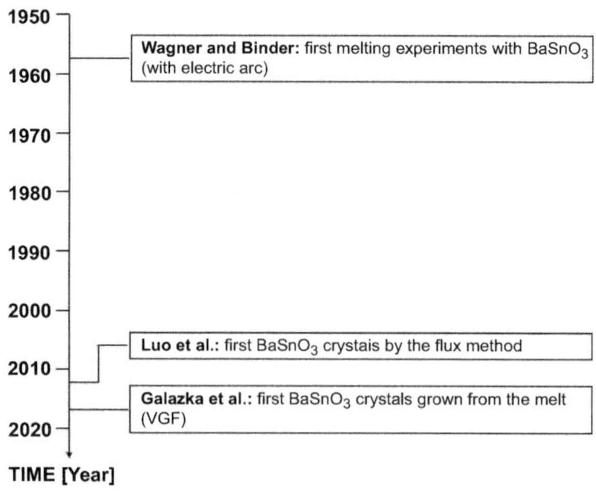

Figure 7.4 Main events in obtaining bulk $BaSnO_3$ crystals.

7.4.3 Flux

Due to the high MP of $BaSnO_3$ and very high thermal instability of SnO_2 at high temperatures (see Chapter 6), the flux method was found suitable for obtaining first, but very small bulk $BaSnO_3$ single crystals. The flux method operates at temperatures within the thermal stability of $BaSnO_3$. Flux-grown crystals were sufficient in size to reveal fundamental physical properties of $BaSnO_3$.

7.4.3.1 Crystal growth

Bulk $BaSnO_3$ crystals obtained by the flux method were reported by Luo et al. [11], Kim et al. [12, 44, 45], Stanislavchuk et al. [43], Lee et al. [14, 54, 55], and McCalla et al. [56].

The source material of $BaSnO_3$ whether undoped or doped was prepared by mixing SnO_2 and $BaCO_3$ powder in a stoichiometric composition. In the case of intentional doping, a powdered dopant oxide is added in a suitable proportion. First, the powders were mixed and fired at 1250°C for 6 h. Afterward, the fired powders were ground several times and pressed into a pellet, which was then sintered at 1400–1500°C for 20–48 h [11, 12, 44, 45]. During preparation, the source material was subjected to a high temperature of 1450–1500°C for solid-state synthesis. According to a study of solid-state synthesis in the SnO_2–$BaCO_3$ system, Huang et al. [57] concluded the formation of pure $BaSnO_3$ phase at temperatures at or above 1200°C. At lower temperatures, orthostannate Ba_2SnO_4 phase is formed due to an incomplete $BaCO_3$ decomposition. The obtained source material was then mixed with a PbO- [11, 43, 44], PbF_2- [14], Cu_2O- [12, 44, 45, 56], or Cu_2O-based [14, 54, 55] flux to form a starting material that was placed within a Pt crucible. The molar ratio of the $BaSnO_3$ source material (whether undoped or doped) to the PbF_2 flux was, e.g., 1:2.5 [14], while to the Cu_2O flux was, e.g., 1:15 [44, 45] or about 1:40 [56]. In the case of Cu_2O-based flux, the molar ratio of $BaSnO_3$ to CuO + Cu_2O, and $KClO_4$ was 1:10–80:0–3 [14, 54, 55].

Commonly used intentional dopants for the flux growth of $BaSnO_3$ crystals are La and Sb substituting Ba and Sn crystallographic sites, respectively. La or Sb molar concentration in the source material was up to 4% [11, 12, 44], and up to 10% [45], respectively. $KClO_4$ in the flux is also a source of doping with K, which acts as an acceptor compensating *n*-type conductivity. Other reported dopants were Pr and Nd with a molar concentration in the source material of 1–2% [56].

The Pt crucible with the starting material consisting of the $BaSnO_3$ source material and the Cu_2O- or Cu_2O-based flux was placed in a furnace, heated up in air to a maximum temperature of T_{MAX} = 1230–1250°C or above, slowly cooled down first to a final growth temperature of T_F = 1070–1210°C (e.g., at a rate of 1 K/h [14]), and then faster to RT [14, 44, 45]. The crystals were separated from the Cu_2O flux with dilute nitric acid [56]. When using the PbF_2 flux [14], the crucible with the charge was heated to T_{MAX} = 1200°C, slowly cooled down to T_F = 900°C with a rate of 5 K/h and then faster to RT. A typical heating time to the maximum temperature was about 5 h [14].

The obtained crystals were very small with exemplary maximum size of about 3 × 2 × 1 [11, 43], 2 × 2 × 1 mm^3 [44], 3 × 3 × 3 mm^3 [14], and 70 mg [56]. The maximum reported crystal size (in terms of edge dimension) is 5 × 3 × 1 mm^3 [14]. The maximum wafer size that can be fabricated from flux-grown crystals is about 3 × 3 mm^2 and occasionally 5 × 3 mm^2. Undoped $BaSnO_3$ crystals grown from the PbO flux were translucent orange or reddish platelets [11, 43], and gray when doped with La [11]. PbF_2 flux produced reddish crystals [14]. When using Cu_2O flux, the obtained crystals usually had a facetted cube-like shape with a size not exceeding 2 mm. As-grown and undoped crystals were usually translucent and colorless, while those doped with La and Sb were dark and substantially opaque (Fig. 7.5) [44, 45]. The same applies to crystals doped with Pr and Nd [7]. Cu_2O-based flux produced bluish to yellowish crystals with increasing $KClO_4$ concentration in the flux [14, 55]. The crystals grown from the PbO-based flux were contaminated with Pb, while those obtained from the Cu_2O flux did not show contamination with Cu [44].

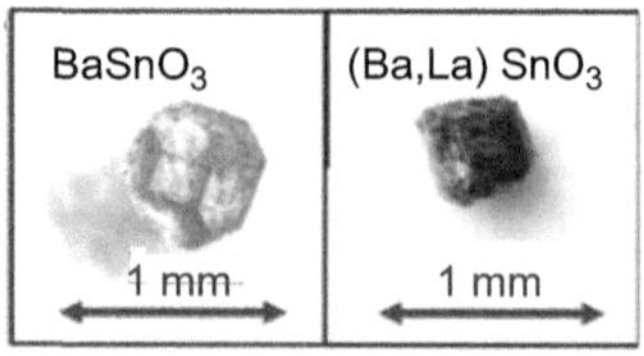

Figure 7.5 Undoped and La-doped $BaSnO_3$ crystals obtained from the Cu_2O flux. Reprinted with permission from Ref. [44], Copyright 2012, American Physical Society.

7.4.3.2 Intentional doping

La^{3+} and Sb^{5+} were found to be effective *n*-type dopants that substitute Ba^{2+} and Sn^{4+} ions in the $BaSnO_3$ crystal lattice, respectively. With the use of the flux method, a common practice is to add La_2O_3 or Sb_2O_5 to SnO_2 + $BaCO_3$ powder mixture during the preparation of the source material, resulting in the chemical formula $Ba_{1-x}La_xSnO_3$ and $BaSn_{1-y}Sb_yO_3$, respectively. When intentionally doped with La, there is some solubility limit above which a pylochlore $La_2Sn_2O_7$ phase is formed after synthesis at 1200°C in air, according to the study of Hadjarab et al. [7]. The La solubility limit, concluded from XRD measurements, was found at the level of 2 mol.% only, although Upadhyay et al. [58] reported a 5 mol.% limit for 1250°C sintering temperature. During the preparation of the source material for the flux experiments, Luo et al. [11] concluded the La solubility limit of about 3 mol.% for the synthesized powders. However, McCalla et al. [56] noticed the formation of the pyrochlore phase for La, Pr, and Nd doping at the molar concentration of 1–2%. Similar conclusion was drawn by Trari et al. [3] who doped $BaSnO_3$ ceramics with La. They also observed a pyrochlore phase $La_2Sn_2O_7$ at La concentration above 2 mol.%. Nonetheless, the obtained crystals doped with La with $x \leq 0.04$ did not show any secondary phase, only pure $Ba_{1-x}La_xSnO_3$ phase [11, 12, 44]. La incorporation into a growing crystal is approximately equivalent to the nominal doping of the sintered powder. For Sb doping, higher nominal concentrations were used in the source material, $y \leq 0.1$, with no formation of a secondary phase. Sb concentration in the crystals is smaller by a factor of 0.22–0.29 [45]. In fact, in the case of La-doping, a linear increase in the lattice parameter with the La content was observed [11, 44], but not for Sb doping, where a nonlinear relation was measured for both sintered powders and obtained crystals [45]. Yet another intentional dopant acting as the *n*-type compensator is K^+, which allows to obtain electrically insulating crystals [14, 54, 55]. The concentration of K impurities by electron probe microanalysis (EPMA) was found at the level of about 1.4×10^{19} cm^{-3}.

7.4.3.3 Examples

(i) Luo et al. [11] and Stanislavchuk et al. [43] grew crystals of a composition $Ba_{1-x}La_xSnO_3$ with $x = 0$–0.4. The source material was prepared from $BaCO_3$, SnO_2, and La_2O_3 at stoichiometric composition, which was mixed and sintered in an alumina crucible at 1500°C for 20 h in air. For $x = 0.04$ in the source material, however, a secondary pyrochlore phase ($La_2Sn_2O_7$) was detected in the XRD pattern. The sintered source material (with $x = 0$–0.03) was mixed with a PbO-based flux and loaded into a Pt crucible, which was heated up until the flux melted. Then the crucible was cooled down to RT. The obtained crystals were light reddish or orange ($x = 0$) or gray ($x = 0.01$, 0.02, and 0.03) translucent platelets with a typical size of about $3 \times 2 \times 1$ mm^3. Pb^{2+} may substitute Ba^{2+} in the $BaSnO_3$ crystal lattice due to not too large ionic radii difference in between. Indeed, the presence of Pb impurities in the crystal lattice, which make reddish or orange coloration, was identified.

(ii) The use of the Cu_2O flux for growing undoped and La- or Sb-doped $BaSnO_3$ crystals of a general formula $Ba_{1-x}La_xSnO_3$ and $BaSn_{1-y}Sb_yO_3$ ($x = 0$–0.04, $y = 0$–0.1), respectively, was reported by Kim et al. [12, 44, 45]. The source material was prepared from high-purity $BaCO_3$, SnO_2, and La_2O_3 or Sb_2O_5 powders at stoichiometric ratio, by mixing in an agate mortar, firing at 1250°C for 6 h, several grindings, pressing into pellets, and sintering at 1400–1450°C for 24–48 h. The sintered source material and the Cu_2O flux with a molar ratio of about 1:15 were loaded into a Pt crucible, which was heated up in a furnace in air above 1250°C, and then it was slowly cooled down first to 1210°C, then to RT. The obtained crystals were of cube-like shape with a size up to about 1–2 mm. Undoped crystals were substantially colorless, while those doped with La and Sb were dark. The crystals did not contain a secondary phase according to XRD measurements of powdered crystals, while EPMA did not reveal Cu impurities.

(iii) Lee et al. [14, 54, 55] used, for growing bulk $BaSnO_3$ crystals, Cu_2O-based flux consisting of CuO and Cu_2O with a molar ratio of $CuO:Cu_2O = 39:61$ to achieve eutectic between both oxides that lowers the melting point of Cu_2O from 1230°C to 1090°C. Additionally, $KClO_4$ was added as a flux constituent in order to compensate n-type conductivity by introducing

K^+ ions acting as compensating acceptors. The best molar ratio of the $CuO–Cu_2O$ flux to $BaSnO_3$ powder was found at about 50:1, while with the addition of $KClO_4$, the ratio of $CuO + Cu_2O:KClO_4:BaSnO_3$ = 50:2:1 was found to be the most optimal. The $BaSnO_3$ source material prepared in a similar way as above was mixed with the flux powder and loaded into a 400 ml Pt crucible that was placed in a furnace. The crucible was heated at the rate of about 240 K/h to 1230°C in air, then slowly cooled down at the rate of 1 K/h to 1070°C, and next to RT at a high rate. The obtained $BaSnO_3$ crystals were (001)-oriented plate-like crystals of about 2×2 to 3×3 mm^3 in a cross section. They were transparent with blue through bluish to yellowish coloration for $KClO_4$ concentration ranging from 0 to 3 mol.% in the flux. The polished (001) surfaces showed the full width at half maximum (FWHM) of the rocking curve of 0.022° (79 arcsec) and RMS roughness of about 0.3 nm, which was further smoothed to about 0.09 nm after annealing at 1000°C in air for 1 h. The (001) surface showed atomically flat terraces of about 4.1 Å, which corresponds to the unit cell parameter.

The summary of results for growing $BaSnO_3$ crystals with the flux method is provided in Table 7.1.

7.4.4 Growth from the Melt

7.4.4.1 Thermodynamics

Incongruent and strong evaporation of $BaSnO_3$ constituents at high temperatures makes the growth from the melt challenging. It requires consideration of the thermodynamic behavior of the $BaO–SnO_2$ system and application of technical solutions enabling to melt the system and obtain single crystals.

The starting material is composed of $BaCO_3$ and SnO_2. The use of $BaCO_3$ as the BaO source is due to the high hygroscopicity of BaO. $BaCO_3$ simply decomposes during heating up at about 1000°C into BaO(s) and CO_2(g). At higher temperatures (>1500°C), the released BaO reacts with SnO_2 in solid-state reaction to form $BaSnO_3$(s).

Table 7.1 Basic conditions for flux method and obtained $BaSnO_3$ single crystals. T_{MAX} and T_F are the maximum and final growth temperatures, respectively

Flux	T_{MAX} [°C]	T_F [°C]	$(T_{MAX}-T_F)/t$ [K/h]	Dopant	Maximum crystal size	Crystal coloration	Reference
PbF_2	1200	900	5	—	Surface area of 10 mm^2	Reddish	14
PbO	—	—	—	—	3×2×1 mm^3	Orange, Reddish	43, 11
PbO				La^{3+}		Gray	
Cu_2O	1250	1210	—	—	2 mm at edge	Colorless	44, 45, 12
Cu_2O				La^{3+}, Sb^{5+}		Dark	
$Cu_2O + CuO + KClO_4$	1230	1070	1	K^+	3×3×3 mm^3	Blue, bluish, yellowish	54, 55, 14

In the BaO–SnO_2 system forming $BaSnO_3$, both oxides are thermally unstable, and their volatile species have different partial pressures. Solid BaO becomes partially volatile at high temperatures:

$$BaO(s) \rightarrow BaO(s) + BaO(g) \quad (7.2)$$

while SnO_2 decomposes into tin suboxide and oxygen:

$$SnO_2(s) \rightarrow SnO(g) + \tfrac{1}{2}O_2(g) \quad (7.3)$$

Temperature-dependent partial pressures of the most volatile Sn- and Ba-containing species, SnO and BaO, respectively, at a total pressure of 10 bar are shown in Fig. 7.6. The partial pressure of BaO(g), p(BaO), increases with temperature and reaches a value of about 1 mbar at the MP of $BaSnO_3$, which is about 1855°C [53]. More importantly, p(BaO) is independent of oxygen partial pressure, $p(O_2)$, in a growth atmosphere. On the other hand, the partial pressure of SnO, p(SnO), is much higher than p(BaO) above 1300°C and it is $p(O_2)$-dependent. For more details on SnO_2 instability at high temperatures, see Chapter 6. At the MP of $BaSnO_3$, $p(SnO) = 1.9 \times 10^{-1}$, 1.0×10^{-1}, 4.8×10^{-2}, and 3.4×10^{-2} bar for O_2 concentrations in the growth atmosphere of about 1, 10, 50, and 100 vol.%, respectively, giving the ratio of $p(SnO)/p(BaO)$ = 158, 83, 40, and 28, respectively. At high O_2 concentrations, there is little gain in reducing p(SnO) as compared with low O_2 concentration region. Such incongruent evaporation of both volatile species will lead to a rapid composition shift in the starting material already during heating up prior melting, which is time dependent. High O_2 concentration in the growth atmosphere is a fundamental condition (although not sufficient) for minimizing composition shift in the BaO–SnO_2 system during the growth of $BaSnO_3$ single crystals from the melt.

Further insight into the behavior of the $BaCO_3$–SnO_2 system was achieved by thermal analysis (thermogravimetry analysis (TGA) and differential thermal analysis (DTA)) of the $BaCO_3$ + SnO_2 mixture resulting in a stoichiometric BaO–SnO_2 molar ratio after $BaCO_3$ decomposition, as presented in Fig. 7.7, which shows two subsequent heating runs of the same charge. A first heating led to the decomposition of $BaCO_3$ and BaO–SnO_2 solid-state synthesis, while the second heating showed the instability of $BaSnO_3$. The first heating revealed two distinct endothermic peaks (A) and (B) at 806 and 976°C in the DTA1 curve, which correspond to orthorhombic–hexagonal and hexagonal–cubic transitions of $BaCO_3$, respectively

[59]. The decomposition of $BaCO_3$ starts at about 1000°C, and it is completed at about 1300°C, as shown by the TG1 curve. This decomposition is also manifested in the broad DTA1 endothermic peak (C). Above 1300°C, there is also another mass loss (TG1 curve) but at a smaller rate. This is SnO_2(s) decomposition described by Eq. 7.3. Preparation of the starting material by sintering $BaCO_3$ + SnO_2 powders at temperatures above 1300°C will already cause a stoichiometry deviation toward BaO-rich composition before growth experiment.

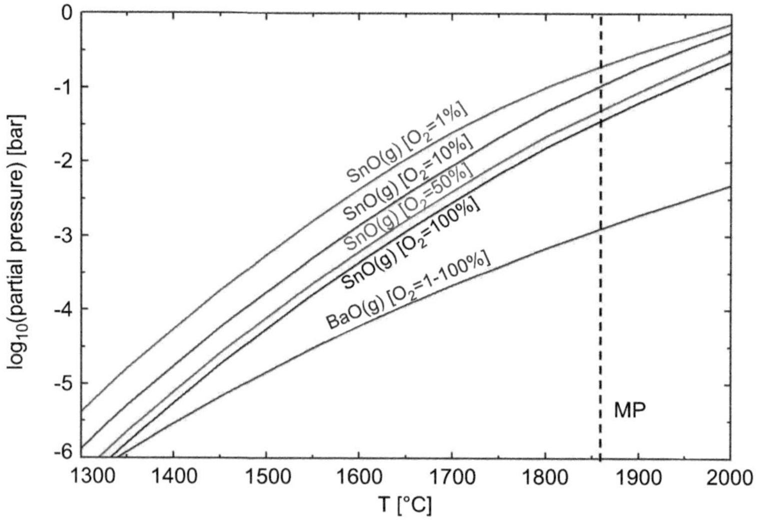

Figure 7.6 SnO(g) and BaO(g) partial pressures in the $BaCO_3$–SnO_2 system versus temperature at different oxygen partial pressures. Above 1300°C, all $BaCO_3$(s) is decomposed into BaO(s) and CO_2(g). For the calculations, the following input data were used: $BaCO_3$ = 1 mol, SnO_2 = 1 mol, O_2 + Ar = 1 mol with O_2 = 0.01, 0.1, 0.5, and 1 mol, and a total pressure of 10 bar.

During the first heating, a broad endothermic peak (D) develops at about 1500°C at the DTA1 curve, which could be assigned to solid-state synthesis of BaO(s) + SnO_2(s) $\rightarrow$ $BaSnO_3$(s). During the second heating of the same sample, the features manifested in the DTA1 curve are no longer present in the DTA2 curve, meaning a complete $BaCO_3$ decomposition and complete $BaSnO_3$ solid-state synthesis. In the TG2 curve, there is just a mass loss as a result of $BaSnO_3$ decomposition, which is identical to a mass loss from the SnO_2(s) decomposition. This mass loss is quite remarkable at 1600°C.

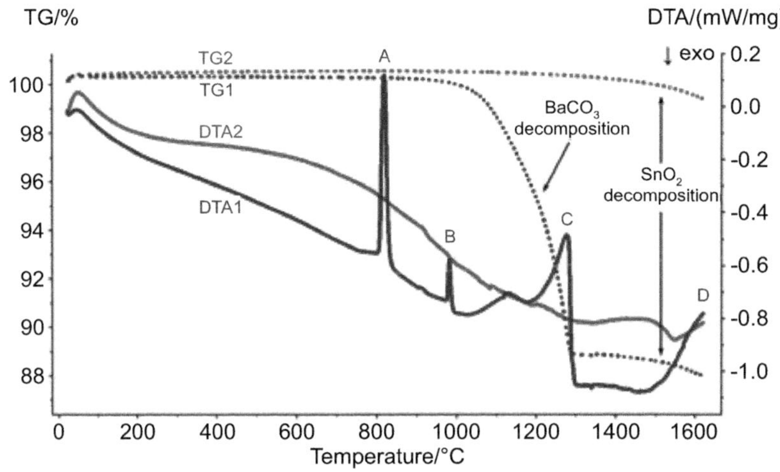

Figure 7.7 TG (dashed) and DTA (solid) curves of a $BaCO_3$–SnO_2 mixture versus temperature during first (DTA1, TG1) and second (DTA2, TG2) heating in pure oxygen. The heating rate was 10 K/min. Maximum temperature was 1620°C. Reprinted from Ref. [53], Copyright 2017, with permission from IOP Publishing.

7.4.4.2 Vertical gradient freeze (VGF)

7.4.4.2.1 *Growth conditions*

Melt growth experiments of bulk $BaSnO_3$ crystals were performed by Galazka et al. [53, 60] by utilizing a direct solidification method, which can be referred to as the vertical gradient freeze (VGF) method. To obtain bulk crystals of a reasonable size and quality, a number of growth conditions must be met, as listed below.

(i) The first step is a proper preparation of the starting material for melt growth experiments, which consists of $BaCO_3$ and SnO_2 powders. Both powders of 4N or 5N purity were first separately dried for 10–40 h at 150–200°C and 900–1200°C, respectively, then mixed, pressed at 2000 bar for 3 min, and sintered for 10–12 h at 750°C to avoid $BaCO_3$ decomposition. The starting composition was prepared either in the stoichiometric ratio of BaO/SnO_2 or with an excess of SnO_2 up to 10 mol.% to cover Sn losses at high temperatures that are unavoidable. Off-stoichiometric starting composition is an important growth condition minimizing the deviation from stoichiometry upon melting. For La-doped crystals, La_2O_3 powder was added to the starting material at a concentration up to 2 mol.% assuming a full Ba substitution by La: $Ba_{1-x}La_xSnO_3$, where $x = 0–0.02$.

(ii) A growth furnace constitutes an environment of growing crystals, so its design is one of the crucial factors in the overall growth process. As the decomposition rate of a thermally unstable oxide, including $BaSnO_3$, increases with temperature gradients, they should be as low as possible to decrease incongruent decomposition and evaporation of $BaSnO_3$ and to avoid its overheating. The growth furnace should also enable good communication of the $BaSnO_3$ starting material with oxygen in the growth atmosphere. At the level of the furnace design, these conditions were achieved by enclosing a metal crucible, made of Ir, with an Ir lid having defined openings for oxygen supply, and by using a good thermal insulation surrounding the Ir crucible with the Ir lid from all sides, with a window exposing a portion of the Ir lid surface for temperature monitoring with a two-color pyrometer. The crucible was heated up inductively by an RF coil powered by a 10–30 kHz generator. Before supplying a growth atmosphere to a growth chamber accommodating the growth furnace, it was evacuated with a vacuum pump to remove residual gases, particularly moisture. Low temperature gradients in the growth furnace constitute another important condition in growing $BaSnO_3$ crystals from the melt.

(iii) As discussed in the previous section, a crucial condition for minimizing the composition shift in the starting material during its heating up, melting, and cooling down is a high O_2 concentration in the growth atmosphere, pure O_2 in the best scenario. As evidenced by Galazka et al. [53] (see, e.g., Chapter 4), it is possible to use a very high O_2 concentration, including pure O_2, in a combination with Ir crucibles used for growing oxide single crystals from the melt. This technique enables to minimize not only evaporation of the main constituents but also the formation of elemental cations in the melt, which may form eutectic with Ir crucibles. In the case of $BaSnO_3$, very high O_2 concentration in the growth atmosphere is a *sine qua none* condition that decreases $p(SnO)$ and, therefore, the composition shift at high temperatures.

(iv) Another factor affecting the composition shift in the starting material through incongruent decomposition is heating and cooling times of the growth furnace, especially at a temperature range of about 1500°C to MP. Preferably, total heating time from RT to MP and cooling time from MP to RT are both not longer than 10 h. However, more important are the heating and cooling rates in the temperature range of about 1500°C to MP, where the decomposition rate is significant. In this temperature range, the preferred cooling and heating rates are about 200–300 K/h. The heating rate below about 1500°C can be smaller, e.g., 150–200 K/h, particularly to allow a complete $BaCO_3$ decomposition before the solid-state synthesis between BaO and SnO_2. Cooling rate below about 1500°C, particularly below about 1300°C, can be much smaller, e.g., 100–150 K/h. Moreover, at moderate temperatures of about 1000–1200°C, where $BaSnO_3$ is already thermally stable, the crucible with $BaSnO_3$ formed therein can be held for at least several hours to carry out in situ annealing in the presence of oxygen.

(v) Yet another very important growth condition minimizing the composition shift into the BaO-rich side is soaking time of the $BaSnO_3$ liquid phase. The decomposition rate of the liquid phase of unstable oxides, such as ZnO, β-Ga_2O_3, or In_2O_3, is much higher compared with that of the solid phase close to MP. The soaking time of the $BaSnO_3$ liquid should not be longer than about 1 h, otherwise, the volume of obtained crystals will drastically decrease.

(vi) Further preferential growth condition is an external pressure that partially overcomes $p(SnO)$ and $p(BaO)$ and improves the stoichiometry of the starting material upon melting, by reducing decomposition of SnO_2 to SnO (Eq. 7.3). A preferable overpressure is between 10 and 20 bar. At such high overpressure, the oxygen concentration should not be too high ($<$ 10 vol. %) to minimize incorporation of Ir particles through intensive oxidation of the Ir crucible.

A summary of crucial growth conditions for obtaining bulk $BaSnO_3$ single crystals from the melt is provided in Table 7.2.

Table 7.2 Important and preferential growth conditions for obtaining bulk $BaSnO_3$ single crystals from the melt

Condition	Unit	Value/description
Temperature of a starting material preparation	°C	<800
SnO_2 excess in a starting composition	mol.%	Up to 6
Low temperature gradients in a growth furnace	—	Ir crucible covered with an Ir lid and a good thermal insulation around the crucible
O_2 concentration in a growth atmosphere	vol.%	Up to 100
Heating/cooling rate of a growth furnace between about 1500°C and MP	K/h	200–300
Soak time of the $BaSnO_3$ liquid	min	<60
External pressure	bar	10–20

7.4.4.2.2 *Crystal growth*

As the starting material, different BaO–SnO_2 compositions were tested, from 1:1 to 1:1.1. Experimentally, it was found that excess SnO_2 higher than 10 mol.% leads to the formation of ceramic-like structure instead of single crystals. A preferred SnO_2 excess is between 2 and 6 mol.%, which depends on other growth conditions. For doping with La, La_2O_3 powder is added to the $BaCO_3$–SnO_2 starting material prior mixing, pressing, and sintering steps. The molar ratio of the La_2O_3 powder in the starting material was between 0.05 and 2 mol.%, assuming full substitutions of Ba-sites in the crystal structure.

After loading the as-prepared $BaCO_3$–SnO_2 starting material into the Ir crucible (either 40 or 60 mm diameter) in the growth furnace, the growth chamber is sealed and evacuated with a vacuum pump for 0.5–3 h to remove residual gases. In particular, moisture is critical, as it forms $Ba(OH)_2$ with BaO. Next, the growth chamber is filled up with a growth atmosphere containing a high concentration of O_2, up to 100 vol.%. At low and moderate temperatures, O_2 concentration can be low, such as 0.2–2 vol.% to minimize Ir oxidation in this temperature range. Here Ar (or N_2) + O_2, or CO_2 can be used as the growth atmosphere. Above that temperature, O_2 concentration must significantly increase to high values, preferably of 100 vol.% (pure O_2) before reaching MP. This procedure can be applied with a continued gas flow through the growth chamber. In the case of a "static" growth atmosphere (i.e., with no flow through the growth chamber), the growth atmosphere has a fixed composition containing a high O_2 concentration, preferably 40–100 vol.%, over the whole growth run. The growth of $BaSnO_3$ crystals from the melt can be conducted under the atmospheric pressure or under the external pressure up to 20 bar. In the case of the external pressure, the growth atmosphere has a fixed composition, e.g., O_2 = 1 bar and Ar = 19 bar. High O_2 concentration in the growth atmosphere combined high overpressure significantly decreases the composition shift toward the BaO-rich side. At high overpressure also CO_2 can be used as the growth atmosphere.

Then the growth furnace is heated up for 4–10 h until the MP of $BaSnO_3$, which can be detected by temperature measurements of the Ir lid enclosing the Ir crucible with a two-color pyrometer. Initial studies of melting behavior of the $BaSnO_3$ in a temperature range of 1800–1990°C revealed the MP of $BaSnO_3$ of 1855°C ± 25 K. With higher temperatures used for melting $BaSnO_3$, the volume of the obtained crystals drastically decreased due to fast and high deviation from the stoichiometry. Next experiments were conducted just around the MP of $BaSnO_3$. After melting, $BaSnO_3$ is soaked for about 5–10 min and then cooled down for 3–8 h with different cooling rates at different temperature ranges, faster at beginning of cooling down, and slower at lower temperatures. Such short soaking timing was necessary to minimize Sn losses from the starting material, but also Ir losses due to high O_2 concentration in the growth atmosphere (up to 100 vol.%). As explained in the previous section, at about 1000–1200°C during the cooling-down step, the crystals can be subjected to in situ annealing for at least several hours.

The crystals grown at ambient pressure are typically removed from the Ir crucible by immersing it in water for at least 1 day, because the excess of BaO(s) in the bottom part of the crucible, which is highly hygroscopic, reacts with water and forms barium hydroxide: $BaO + H_2O \rightarrow Ba(OH)_2$. On the other hand, $BaSnO_3$ single crystals are insoluble in H_2O or $Ba(OH)_2$ and remain intact. The crystallized material under high pressure combined with non-stoichiometric starting composition is more stoichiometric in the whole volume and more difficult to remove the crystals with water. In such cases, core drilling is utilized.

7.4.4.2.3 *Melt-grown $BaSnO_3$ crystals, their composition, and structural quality*

Due to a large composition shift in the starting material at high temperatures, molten and crystallized starting material within the Ir crucible at ambient pressure will contain an excess of BaO, the concentration of which decreases from the bottom toward the top of the crucible. The largest crystals are formed at the top part of the crucible due a high partial pressure of both O_2 (from atmosphere) and SnO (evaporation) in the gas reservoir just above the melt surface of $BaSnO_3$. On the other hand, at deeper levels, $BaSnO_3$ loses SnO(g) via decomposition and evaporation and is subjected to lower $p(O_2)$ due to limited O_2 diffusion inside the melt. The deviation from stoichiometry can be compensated by a non-stoichiometric starting composition (SnO_2 excess), better by an external overpressure, and the best by a combination of both.

Examples of bulk $BaSnO_3$ single crystals obtained in 40 and 60 mm diameter Ir crucibles are shown in Fig. 7.8. The coloration and size of the obtained crystals at ambient pressure vary from red through brown and dark brown to black depending on O_2 concentration in the growth atmosphere, doping (both SnO_2 excess and La_2O_3), temperature, and pressure. All crystals, except black ones, are substantially transparent. Typically, the undoped crystals are red to brown, while those La-doped were dark brown for low La_2O_3 doping level (<0.1 mol.%) to black for higher La_2O_3 doping level (>0.1 mol.%). Also, crystals obtained with a high SnO_2 excess are darker, usually dark red or brown. Crystals obtained under overpressure are brown to gray. Moreover, the size of the obtained single crystals decreases with temperature, i.e., the higher temperature above MP was used for melting the starting material, the smaller crystal grains were obtained. The same applies for doping with La_2O_3.

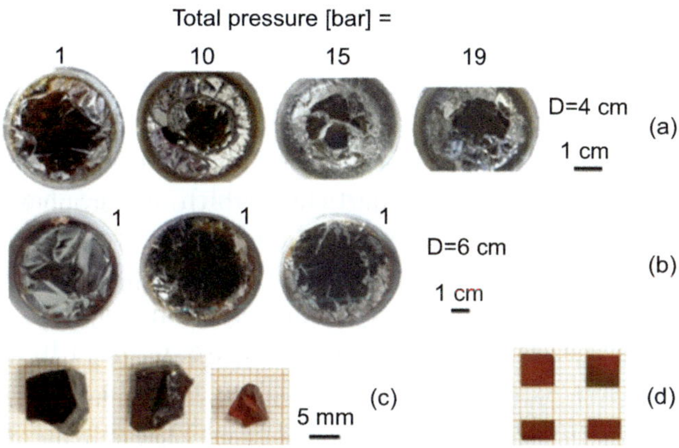

Figure 7.8 (a) Undoped $BaSnO_3$ obtained within a 4 cm diameter crucible at different external pressures; (b) undoped $BaSnO_3$ obtained within a 6 cm diameter crucible at ambient pressure; (c) $BaSnO_3$ single crystal grains extracted from both size crucibles with a volume of 0.1–1 cm^3 (the left-hand side and middle crystal grains are as-grown, while the right-hand reddish crystal grain is annealed at 1200°C in air for 40 h); and (d) as-cut wafers from annealed $BaSnO_3$ crystals.

Figure 7.8a shows a crystallized $BaSnO_3$ within 40 mm crucibles as a function of the external pressure up to 19 bar (O_2 = 1 bar + Ar = x_p – 1 bar, where x_p is the total external pressure) with SnO_2 excess in the starting material of 6 mol.%. The coloration of the crystals changes with pressure from reddish to brown with an increased volume of single crystalline $BaSnO_3$. At ambient pressure, single crystals of $BaSnO_3$ are confined to about 2–3 mm thick top layer, which increases with pressure to about 10 mm thick. A characteristic feature of crystallized $BaSnO_3$ under overpressure is the formation of a large single crystal grain at the central part, which is surrounded by small single crystals in the vicinity of the crucible wall. This is the result of a fast crystallization due to fast cooling down at high temperatures and additional cooling by gas convection under overpressure. With pressure, the MP of $BaSnO_3$ significantly increases to about 1970–2040°C at 19 bar depending on a growth furnace configuration.

Suitable growth conditions and larger crucibles (Fig. 7.8b) enabled to obtain as large as about 1 cm^3 single crystal grains after removing them from the crucibles (Fig. 7.8c), suitable for fabrication of 5 × 5 wafers (Fig. 7.8d) that can be used for homoepitaxy.

The as-grown crystals obtained in preferred conditions are usually dark red or brown (Fig. 7.8c, left and middle), while after annealing in the presence of O_2 at 1000–1400°C for at least several hours become light reddish and fully transparent (Fig. 7.8c, right). The brown and reddish coloration is likely associated with light extinction at metallic nanoparticles, which indeed have been identified in scanning transmission electron microscopy (STEM) image recorded with a high-angle annular dark field (HAADF) detector (Fig. 7.9a), indicating a higher atomic number in the probed volume. They are of metallic nature, perhaps Ir or Ir–Sn alloy particles as a result of Ir oxidation under a very high O_2 concentration in the growth atmosphere and formation of elemental Sn due to SnO_2 decomposition. Indeed, growth experiments under 19 bar overpressure with $p(O_2)$ = 4 bar revealed in obtained $BaSnO_3$ crystals Ir at a concentration of several hundred wt. ppm. So, at high pressure lower $p(O_2)$ is preferential (e.g., 1 bar). The nanoparticles, which were found in some areas of the crystals embedded in the $BaSnO_3$ matrix, are cuboid in shape with {100} facets. A transmission electron microscopy (TEM) image in the [001] projection of the crystal lattice (Fig. 7.9b) reveals mostly perfect crystalline material, with no dislocations, meaning their densities are not higher than 10^6 cm^{-2}. The bright spots correspond to Ba (Z = 56) and Sn (Z = 50) columns, while dark spots correspond to oxygen columns, which appear dark between the bright spots (low atomic number, Z = 8). The position of different columns are marked by differently colored balls in Fig. 7.9c, corresponding to Ba, Sn, and oxygen columns, respectively, which can be compared with an atomic structure model of $BaSnO_3$ in [001] perspective view.

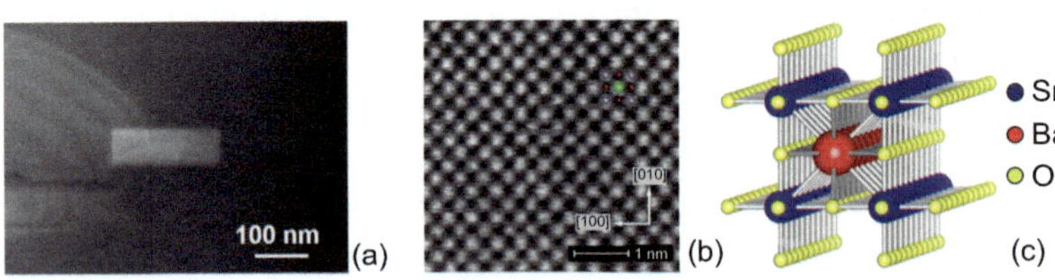

Figure 7.9 (a) STEM image of a defect in the $BaSnO_3$ matrix; (b) HAADF STEM image of $BaSnO_3$ single crystal; and (c) atomic structure model of $BaSnO_3$ in [001] perspective view. Reprinted from Ref. [53], Copyright 2017, with permission from IOP Publishing.

Rocking curve measurements of several samples from melt-grown $BaSnO_3$ crystals by high-resolution XRD (HR XRD) showed very low values at FWHM, such as 25 arcsec (Fig. 7.10).

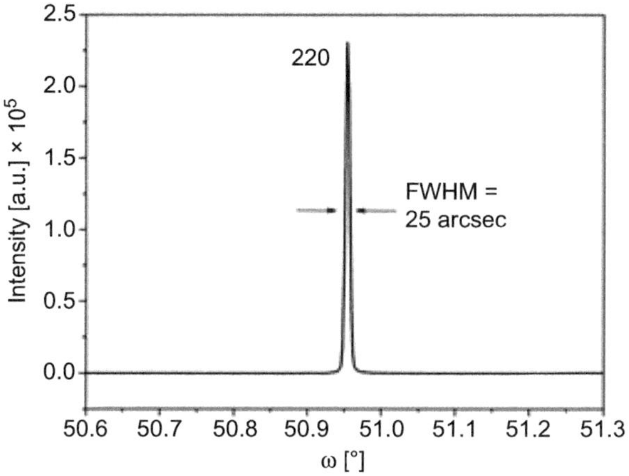

Figure 7.10 Rocking curve of a melt-grown $BaSnO_3$ wafer.

Etch pit analysis on the (100) surface after etching a crystal wafer in H_3PO_4 at 100°C for 30 min revealed a pit arrangement in groups mostly in a cross-like configuration (Fig. 7.11a,b) and around inclusions. Lines of pits propagate along orthogonal directions <010> and <001>, some of the lines also along <011> (Fig. 7.11b). All pits are substantially square in shape with crystallographically oriented edges that are aligned with the main crystallographic directions (Fig. 7.11c). The size of the pits is typically between 0.7 and 2 μm. The average etch pit density (EPD), which is associated with dislocation density, is at the level of 10^5–10^6 cm^{-2}, although there are also large, dislocation-free areas as well.

Despite the high instability of $BaSnO_3$ at high temperatures and high MP thereof, the obtained crystals revealed a pure perovskite phase $BaSnO_3$ according to the 2Θ scan of XRD of powdered crystals (Fig. 7.12) and PDF 01-074-1300 file. No other phases within the accuracy of the XRD, especially, no orthostannate Ba_2SnO_4 could be identified.

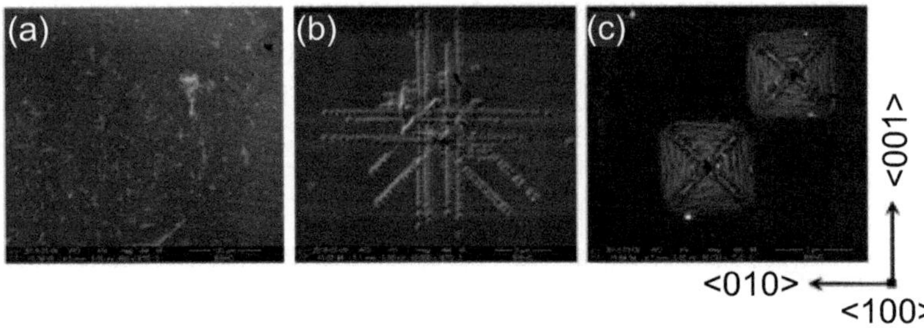

Figure 7.11 Images of etch pits on the (100) face of a wafer prepared from a melt-grown $BaSnO_3$ single crystal having the area of (in μm^2) of (a) 333 × 384, (b) 21 × 25, and (c) 6 × 7. Reprinted from Ref. [53], Copyright 2017, with permission from IOP Publishing.

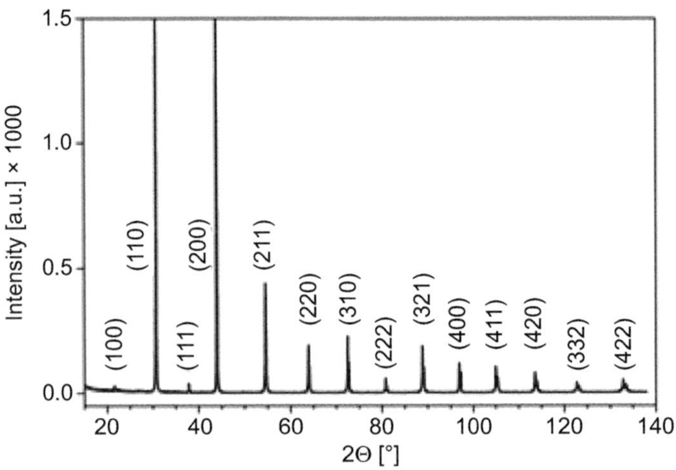

Figure 7.12 XRD pattern of powdered $BaSnO_3$ single crystals. Reprinted from Ref. [53], Copyright 2017, with permission from IOP Publishing.

Single perovskite phase of melt-grown $BaSnO_3$ crystals was confirmed by chemical analysis (ICP-OES) of main constituents forming the perovskite phase. As shown in Fig. 7.13, Sn and Ba concentrations in various $BaSnO_3$ crystals grown at different temperatures from the melt indicate stoichiometric composition within measurement errors independent of temperature, but simply less crystal volume was obtained at higher temperatures. $BaSnO_3$ heavily doped with La (1.24 wt.% in the crystal) showed distinguishable lower Ba concentration as compared with undoped crystals that indicate Ba substitution with La. It should be noted that the crystals "survived" from the treatment with water must have substantial stoichiometric composition; otherwise, they would be "dissolved" into tiny crystals upon $Ba(OH)_2$ action if any BaO excess was present in the crystals. The same effect would occur when the crystals are exposed to ambience by reaction with moisture within several days. In other words, exposure of the melt-grown $BaSnO_3$ crystals to water/moisture is a very simple and effective stoichiometry test involving no costs. Wafer preparation from crystals containing an excess of BaO (e.g., entrapped in defects) is not possible.

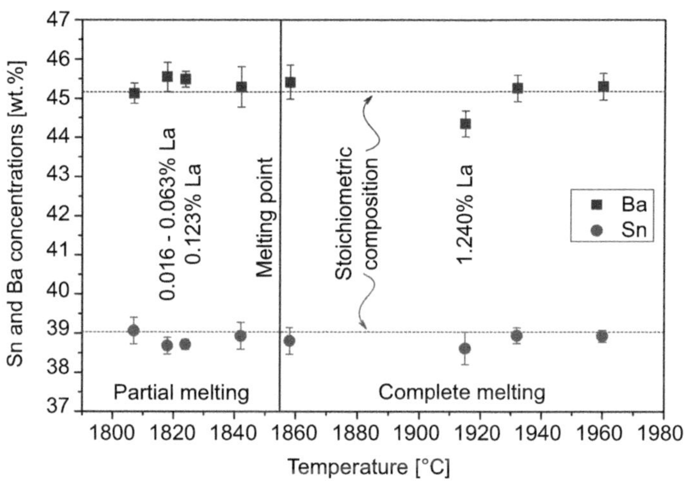

Figure 7.13 Sn and Ba concentrations measured by ICP-OES in various $BaSnO_3$ crystals grown at different temperatures. The samples obtained at temperatures lower than the complete melting of the starting material, indicated as the melting point, were extracted from molten and crystalized regions of the starting material. Reprinted from Ref. [53], Copyright 2017, with permission from IOP Publishing.

7.4.4.3 Examples

In each of the examples below, an Ir crucible was used, either of 40 or 60 mm in diameter, which was covered with an Ir lid having a central opening of 1–3 mm diameter for communication with the growth atmosphere. The crucible wall thickness was typically 1.5 mm. It was surrounded from all sides by a thermal insulation consisting of zirconia granules (just around the crucible) and low-density alumina elsewhere. Above a portion of the Ir lid (about 25% of the surface area and about 15–20 mm high), a free space was designed for temperature monitoring with a two-color pyrometer. Such limited free space exposed the lid for reliable temperature monitoring, on one hand, and ensured low temperature gradients inside the crucible, on the other hand. The crucible was inductively heated with an RF coil powered by an RF generator operating with either 10 or 30 kHz. The growth furnace with the coil was enclosed within a sealed growth chamber enabling the use of a suitable growth atmosphere and an overpressure of up to 20 bar. The starting material consisted of $BaCO_3$ and SnO_2, which were individually dried in air for 10–20 h at 130–200 and 800–1000°C, respectively, then mixed, and pressed at 2000 bar. In some of the experiments, the pressed powders were sintered in air for 10–40 h at temperatures below decomposition temperature of $BaCO_3$, typically at 700–800°C. In each case, the obtained $BaSnO_3$ crystals were removed from the crucible by immersing the crucible with the solidified material in water for 1–2 days. As mentioned above, an excess of BaO in the solidified starting material reacts with water and forms $Ba(OH)_2$, which releases $BaSnO_3$ crystals from the solid that remain intact. These experiments are described in detail in the following:

(i) The starting composition was stoichiometric, while the growth atmosphere was CO_2 under 2.4 bar overpressure. The crucible diameter and height were 40 mm, which accommodated 110 g of the pressed $BaCO_3$ and SnO_2 powders. The heating-up, soaking, and cooling-down times were 6 h, 10 min, and 4 h, respectively. The maximum recorded temperature was 1938°C. The obtained material consisted of many small brown $BaSnO_3$ crystals of about 2–3 mm in size at the uppermost part, and a mixture of BaO and small $BaSnO_3$ crystals below. The mass loss of the starting material due to SnO evaporation was about 8 wt.%. This example shows that too high temperature and not high-enough O_2 partial pressure lead to high losses of SnO_2 from the starting material and consequently small size and volume of the obtained $BaSnO_3$ crystals.

(ii) The starting composition contained 2 mol.% excess of SnO_2, while the growth atmosphere consisted of pure O_2. The crucible diameter and height were 40 mm, which accommodated 100 g of the pressed $BaCO_3$ and SnO_2 powders. The heating-up, soaking, and cooling-down times were 6 h, 15 min, and 4 h, respectively. The maximum recorded temperature was 1855°C. The solidified starting material consisted of several $BaSnO_3$ crystals at the uppermost part of a thickness of about 3 mm and a cross section of about 5×5 mm^2 in average. The remaining material below that $BaSnO_3$ layer consisted of a mixture of BaO and small $BaSnO_3$ crystals of 1–2 mm in size. The crystal coloration was red–brown. The mass loss of the starting material was about 6 wt.%. Lower operating temperature and high O_2 partial pressure enabled to obtain a larger size and volume of the $BaSnO_3$ crystals.

(iii) The Ir crucible dimensions and growth atmosphere were the same as in Example (ii). Here, the starting composition contained 3 mol.% excess of SnO_2 and 2 mol.% La_2O_3 as a dopant. The heating-up, soaking, and cooling-down times were 5 h, 20 min, and 3 h, respectively. The maximum recorded temperature was 1896°C. The solidified starting material consisted of several $BaSnO_3$ crystals at the uppermost part of a thickness of about 3 mm and a cross section of about 2×2 mm^2 in average. The remaining material below that crystalline $BaSnO_3$ layer consisted of a mixture of BaO and small $BaSnO_3$ crystals of 1–2 mm in size. The crystal coloration was black. The mass loss of the starting material was about 7 wt.%. High doping level with La and operating temperature much exceeding the MP of $BaSnO_3$ decreased the size and volume of the $BaSnO_3$ crystals.

(iv) The starting composition contained 10 mol.% excess of SnO_2, while the growth atmosphere consisted of 20 vol.% O_2 and 80 vol.% Ar. The crucible diameter and height were 40 and 60 mm, respectively, which accommodated 175 g of the pressed $BaCO_3$ and SnO_2 powders. Heating-up, soaking, and cooling-down times were 5 h, 15 min, and 5 h, respectively. The maximum recorded temperature was 1880°C. The solidified starting material consisted of several large, dark-brown $BaSnO_3$ crystals at the uppermost part of a thickness of about 5 mm with the size approaching 10×10 mm^2 in cross section. The remaining material below that $BaSnO_3$ layer consisted of a mixture of BaO and small $BaSnO_3$ crystals of 1–2 mm in size.

The mass loss was about 2 wt.%. Larger volume of the starting material combined with a suitable SnO_2 excess, operating temperature, and relatively high O_2 partial pressure enabled to obtain larger size and volume of the $BaSnO_3$ crystals.

(v) The Ir crucible was 60 mm in diameter and height, which accommodated 355 g of the $BaCO_3$ + SnO_2 starting material containing 3 mol.% excess of SnO_2. The growth atmosphere consisted of 20 vol.% O_2 and 80 vol.% Ar. The heating-up, soaking, and cooling-down times were 5 h, 10 min, and 4 h, respectively. The maximum recorded temperature was 1875°C. Similar to previous examples, the $BaSnO_3$ crystal layer was confined to about 3 mm thick top layer with a mixture of BaO and small $BaSnO_3$ crystals beneath that crystal layer. The mass loss was as high as 13 wt.%. The crystals were brown to dark brown with a size up to 8 mm. Despite the large volume of the starting material and SnO_2 excess, the volume of the obtained crystals was not larger than that obtained from small crucibles at optimized growth conditions. Larger cross section of the crucible and not high-enough O_2 partial pressure for the larger volume of the starting material led to a higher evaporation rate of SnO and consequently to higher losses of SnO_2 from the starting material.

(vi) For the experiment an Ir crucible of 40 mm in diameter and height was used. The starting composition contained 6 mol.% excess of SnO_2. The growth atmosphere consisted of O_2 at 1 bar and Ar at 18 bar. The heating-up, soaking, and cooling-down times were 3 h, 10 min, and 4 h, respectively. The maximum recorded temperature was 2040°C. The solidified material was basically pure $BaSnO_3$ (about 1 cm thick) consisting of several single crystal grains of red color.

(vii) Here, an Ir crucible of 60 mm in diameter and height and stoichiometric composition of the starting material were used. The growth atmosphere was composed of O_2 = 20 vol.% + N_2 = 80 vol.% under a total pressure of 19 bar. The heating-up, soaking, and cooling-down times were 5 h, 10 min, and 5 h, respectively. The obtained single crystal grains with a volume of up to 1 cm^3 each were black because of Ir contamination (several hundred wt. ppm). This example shows that at high overpressure, oxygen concentration in the growth atmosphere should be rather at low level (see Example (vi)).

7.4.5 Thermal Stability and Surfaces of Bulk Single Crystals

Studies of a surface preparation and its stability upon annealing were conducted by Lee et al. [55] on flux-grown $BaSnO_3$ crystals. (001) wafers were prepared first by mechanical polishing (with the use of colloidal silica suspension) and next by cleaning in different reagents (e.g., acetone, ethanol, isopropanol) and in de-ionized water for 10s. The polishing was performed at a variable out-of-plane and in-plane miscut angles of 0.05–1° and 0–45°, respectively, to create terraces on the (001)-oriented surfaces. The as-polished (001) surfaces did not show any terraces, while the RMS roughness was about 0.2–0.3 nm, which could be further improved by annealing at 1000°C or above in the presence of oxygen [54, 55].

In a series of annealing experiments at temperatures 1000–1300°C in oxygen at atmospheric pressure for 2–100 min with a high heating/cooling rate of about 600 K/h, there were no changes in surface morphology at annealing temperatures up to 1100°C, which consists of BaO and SnO_2 termination. However, annealing at 1200°C for 10–60 min revealed parallel terraces with the step height of about one unit cell at very small out-of-plane miscut angle of 0.05°. With higher miscut angle, 0.7°, the step height increases to 4–8 unit cells after annealing at 1250°C. For miscut angles of 0.05° to 1.0°, the terrace width decreased from about 500 to about 23 nm. X-ray photoemission spectroscopy (XPS) of polished and annealed atomically flat (001) surfaces showed that they consist mainly of SnO_2 termination, with a small amount of BaO_x. Indeed, theoretical study of the (001) surfaces by Slassi et al. [61] revealed that both terminations are stable and energetically favorable; therefore, they may coexist.

Annealing experiments on melt-grown $BaSnO_3$ crystals by Galazka et al. [53] in a temperature range of 1000–1400° in the presence of oxygen for 20–40 h lead to the following conclusions: (i) light-brown crystals remained light brown with a bit orange tint; (ii) dark brown became light brown; (iii) red-brown became red; (iv) very dark crystals (almost black) became either red or greenish; (v) crystals remained substantially transparent, with no major surface decomposition.

Annealing in the reducing atmosphere containing H_2 (5%) and Ar at 500°C for 10 h made the crystals a bit darker (reddish to brownish) with keeping very good transparency, while annealing at 600°C for 10 h produced black and opaque crystals.

The crystals annealed at 1200°C for 20 h in air did not show any mass loss, indicating good thermal stability at relatively high temperatures.

7.4.6 Conclusion

So far, bulk $BaSnO_3$ single crystals were grown only by the flux (solution) and VGF (melt) methods. Flux-grown crystals are small, rather not exceeding average dimensions of 3 × 3 × 3 mm^3 (volume of about 0.03 cm^3). They are of good structural quality and substantially free of flux impurities when grown from Cu_2O-based fluxes. From such crystals, small wafers of about 2 × 2 mm^2 could be prepared, although 3 × 3 or 5 × 3 mm^2 sized wafers are occasionally possible as well. The crystals were intentionally doped with La^{3+}, Nd^{3+}, Pr^{3+}, Sb^{5+}, and K^+ ions. The possibility of growing bulk $BaSnO_3$ single crystals directly from the melt opened a new route for obtaining much larger crystal volumes as compared with those obtained by the flux method. Despite high thermal instability of $BaSnO_3$ at high temperatures and high MP, as well as incongruent decomposition and evaporation, crystals as large as about 1 cm^3 were demonstrated. This enabled to fabricate 5 × 5 mm^2 wafers of epitaxial quality. Further development of the melt growth technique should spread the crystals to a wider community for epitaxy and device fabrication of unique performance. Also the hydrothermal technique may bring large $BaSnO_3$ crystals in the future once developed for this compound.

Similar to other TSOs, bulk $BaSnO_3$ crystals are thermally stable up to about 1200°C and about 500°C after annealing in the presence of oxygen (20–40 h) and hydrogen (10 h), respectively. Such thermal stability enables use of $BaSnO_3$ wafers for most, if not all, of the epitaxial techniques and high-temperature operation. The (001) surface allows for coexistence of both SnO_2 and BaO terminations; however, almost pure SnO_2 termination has been demonstrated.

A summary of bulk $BaSnO_3$ single crystals and wafers is given in Table 7.3.

Table 7.3 Outcome of bulk $BaSnO_3$ single crystals and wafers

Method	Crystal shape	Maximum crystal size [mm^3]	Dopants	Maximum wafer size [mm^2]	Basic structural quality of wafers	Feature	Reference
Flux	Polyhedrons, plates	3×3×3 5×3×1	La, Nd, Pr, Sb, K	5×3	FWHM=79 arcsec	Good quality, small volume	43, 44, 45, 11, 12, 54, 55, 14
VGF	Blocks	10×10×10	La	5×5	FWHM=25 arcsec EPD<10^6 cm^2	High quality, larger volume	53, 60

7.5 Thermal and Mechanical Properties

The melting point (MP) of $BaSnO_3$ reported by Wagner and Binder [2] and by Galazka et al. [53] are quite different, 2057°C and 1855°C ± 25 K, respectively. Calculated solid density of $BaSnO_3$ is 7.23 g/cm^3 with the use of lattice parameter of 4.117 Å [34].

Using stoichiometric $BaSnO_3$ ceramics with a bulk density of about 95% of the theoretical density, Maekawa et al. [6] measured a number of thermal and mechanical properties of that compound.

Within a temperature range of 27–1500°C, the average linear thermal expansion coefficient, determined by dilatometry, is 9.3×10^{-6} K^{-1}. The molar heat capacity C_p obtained by the DSC in a temperature range of 27–1000°C was expressed empirically by the following formula:

$$C_p = 112 + 2.74 \times 10^{-2} T - 1.55 \times 10^6 T^{-2} \text{ [J/Kmol]} \quad (7.4)$$

where T is the temperature in K.

With measured thermal diffusivity (α) by the laser flash method, the authors calculated thermal conductivity λ in the above-mentioned temperature range as $\lambda = \alpha \rho_s C_p$, which is 9.7 at RT and drops to 3.2 W/mK at about 1000°C.

Young's modulus E, shear modulus G, and Vickers hardness H_V measured at RT by Maekawa et al. [6] with the use of a diamond pyramid-type indenter are 244, 99.9, and 5 GPa, respectively. Basing on E and G, the bulk modulus $B_0 = GE/(9G - 3E) = 145.87$ GPa, while Poisson's ratio $v = (3B - E)/(6B) = 0.22$.

In a cubic system, there are only three independent constants that determine a complete elastic tensor: $c_{11} = c_{22} = c_{33}$, $c_{12} = c_{13} = c_{23}$, and $c_{44} = c_{55} = c_{66}$. Theoretical elastic constants calculated by Moreira et al. [38] are $c_{11} = 251$ GPa, $c_{12} = 82.4$ GPa, and $c_{44} = 98.1$ GPa; by Bouhemadou and Haddadi [51] are $c_{11} = 285.2$ GPa, $c_{12} = 68.5$ GPa, and $c_{44} = 84.3$ GPa; and by Duarte et al. [52] are $c_{11} = 296.3$ GPa, $c_{12} = 89.1$ GPa, and $c_{44} = 111.1$ GPa.

From the elastic constants, bulk modulus B_0, Young's modulus E, shear modulus G, and Poisson's ratio v can be calculated from the following Voigt relations:

$$B_0 = \frac{1}{3}(c_{11} + 2c_{12}) \quad (7.5)$$

$$E = \frac{(c_{11} - c_{12}) \times (c_{11} + 2c_{12})}{(c_{11} + c_{12})} \quad (7.6)$$

$$G = c_{44} \quad (7.7)$$

$$\nu = \frac{1}{2}\left(1 - \frac{c_{11} - c_{12}}{c_{11} + c_{12}}\right) \quad (7.8)$$

Basing on theoretical values of the elastic constants of Moreira et al. [38], Bouhemadou and Haddadi [51], and Duarte et al. [52] and Eqs. 7.5–7.8, the following quantities can be calculated: B_0 = 139/141/158 GPa, E = 210/259/255 GPa, G = 98/84/111 GPa, and ν = 0.247/0.194/0.231, respectively. Other reported values of theoretical bulk B_0, Young's E, and shear G moduli (in GPa) are B_0 = 139 [39], 132 [62], 146 [49], E = 207 [63], G = 84 [63].

Experimental values of longitudinal V_L and transverse V_T sound velocities measured by Maekawa et al. [6] on ceramic samples (ultrasonic pulse-echo method) are 6373 and 3814 m/s, respectively, with the corresponding Debye temperatures Θ_D = 522 K. At RT, Luo et al. [11] reported for flux-grown crystals V_L = 6150 m/s and V_T = 3333 m/s. Basing on these longitudinal and transverse sound velocities, the corresponding acoustic Debye temperatures are Θ_{DL} = 705 K and Θ_{DT} = 382 K. However, transverse and longitudinal thermal Debye temperatures derived from specific heat measurements are 330 K and 1365 K for two transverse and one longitudinal acoustic phonon modes [11]. Basing on specific heat and thermal expansion measurements, Kim et al. [44] also concluded two thermal Debye temperatures of 307 and 950 K with two-phonon contributions of different weights. The heat capacity below RT is mainly governed by the low-energy Debye phonons (307 K). Exemplary theoretical values of V_L, V_T, and Θ_D are 6220 and 3689 m/s, and 496 K [51] and 6674 and 3946 m/s, and 563 K [64]. McCalla et al. [56] concluded from the specific heat measurements the following values of Θ_D = 497, 516, and 485 ± 20 K for undoped, La-doped, and Pr-doped $BaSnO_3$ crystals obtained from the flux.

According to the study of Phelan et al. [65], $BaSnO_3$ is stable with its cubic structure under a pressure of up to 50 GPa as no splitting of Bragg reflections was observed. As shown in Fig. 7.14, the lattice parameter a decreases by 5.6% with increasing pressure to 50 GPa. The lattice parameter under both compression and decompression well fit a volume–pressure relation described by the third-order Birch–Murnaghan equation. It means that $BaSnO_3$ can accommodate strain arising from the lattice mismatch during heteroepitaxial growth and maintain the cubic system.

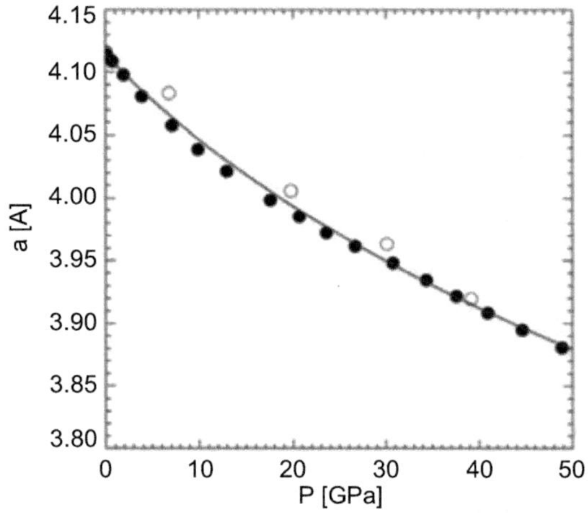

Figure 7.14 Pressure dependence of the lattice parameter *a* during compression (black circles) and decompression (open circles). The solid line is a relationship between volume and pressure (Birch–Murnaghan equation). Reprinted from Ref. [65], Copyright 2018, with permission from Elsevier.

Experimental and theoretical values of thermal and mechanical properties related to bulk $BaSnO_3$ ceramics and crystals are listed in Table 7.4.

Table 7.4 Thermal and mechanical properties of bulk $BaSnO_3$

Property	Unit	Value	Reference
Solid density	g/cm^3	7.23[(a)]	34
Melting point	°C	1855°C±25/2057	53/2
Thermal expansion coefficient of the lattice parameter (0–320°C)	K^{-1}	9.3×10^{-6} [(b)]	6

Property	Unit	Value	Reference
Thermal conductivity at RT	W/mK	9.7[b]	6
Molar heat capacity at RT	J/molK	137.4[b]	6
Elastic constants	GPa	c_{11} = 251[c]/285[c]/296[c] c_{12} = 82[c]/69[c]/89[c] c_{44} = 98[c]/84[c]/111[c]	38/51/52
Bulk modulus	GPa	146[b] 132[c]/139[c]/141[c]/146[c] /158[c]	6 62/38/51/49/52
Young's modulus	GPa	244[b] 207[c]/210[c]/255[c]/259[c]	6 63/38/52/51
Shear modulus	GPa	99.9[b] 84[c]/84[c]/98[c]/111[c]	6 51/63/38/52
Poisson's ratio	—	0.22[b] 0.19[c]/0.23[c]/0.25[c]	6 51/52/38
Vickers hardness	GPa	5[b]	6
Pressure stability of the cubic system	GPa	min. 50	65
Longitudinal sound velocity	m/s	6373[b] 6150 6220[c]	6 11 51
Transverse sound velocity	m/s	3814[b] 3333 3689[c]	6 11 51
Debye temperature	K	522[b]/497 705/382[d] 496[c]	6/56 11 51

[a]calculated value on the basis of lattice constant from Ref. [34]; [b]value for ceramic; [c]theoretical value; [d]longitudinal/transverse

7.6 Optical Properties

Optical properties of $BaSnO_3$ were intensively investigated both experimentally and theoretically to understand its fundamental properties and elicit its capability in a device design for specific fields of applications. The optical properties cover transmittance/absorption, bandgap, refractive index, reflectance, luminescence, and phonon modes. They constitute a basic package characterizing $BaSnO_3$ for interacting with electromagnetic radiation. As in previous chapters, this section provides measured and/or computed values and their relations without discussing the underlying physics. The properties relate mainly to bulk single crystals, which are compared with theoretical values.

7.6.1 Transmittance, Absorption Edge, and Optical Bandgap

Transmittance, absorption coefficient α, and optical bandgap versus photon energy E_{ph} for undoped, and Sb- and La-doped $BaSnO_3$ crystals obtained by the flux method were reported by Kim et al. [44, 45]. The undoped $BaSnO_3$ is fully transparent in the visible spectrum with the absorption edge originating at about 3.1 eV (400 nm). Sb doping induces high absorption in the visible spectrum and makes it opaque in the NIR spectral region as the results of a very high free carrier concentration n_e at the level of 10^{20} cm^{-3}. Very similar results were obtained for La-doped crystals. A linear approximation of the relation $E_{ph} \sim \alpha^{1/2}$, where E_{ph} and α indicate a photon energy and absorption coefficient, respectively, resulted in the indirect bandgap of about 2.95 eV. Sb- and La-doped $BaSnO_3$ with $n_e = 1.21 \times 10^{20}$, $n_e = 1.88 \times 10^{20}$, and $n_e = 2.39 \times 10^{20}$ cm^{-3} increased the bandgap by about $\Delta E = 0.1–0.2$ eV due to the Burstein–Moss shift described by

$$\Delta E(\text{eV}) = h^2\left(\frac{3n_e}{\pi}\right)^{\frac{2}{3}}/(8m_e^*) \tag{7.9}$$

where h is the Planck constant, n_e is the free electron concentration, and m_e* is the effective electron mass. ΔE can be partly compensated by the electron–electron and electron–impurity interactions that narrow the bandgap.

Transmittance spectra in the visible and NIR spectral regions of melt-grown undoped $BaSnO_3$ crystals by Galazka et al. [53] are shown in Fig. 7.15. Both almost colorless and red wafers show a steep absorption edge originating at 400 nm, wherein the red sample reveals a blue absorption, which is not present in the almost colorless sample. The red coloration and the associated blue absorption are likely due to scattering at nanoparticles of metallic nature as discussed in Section 7.4.4.2.3. Metallic particles embedded in a dielectric matrix give rise to light extinction by Mie scattering and absorption, as discussed in more detail in connection to In_2O_3 (Chapter 5). Both samples do not show any free carrier absorption; indeed, they were electrically insulating.

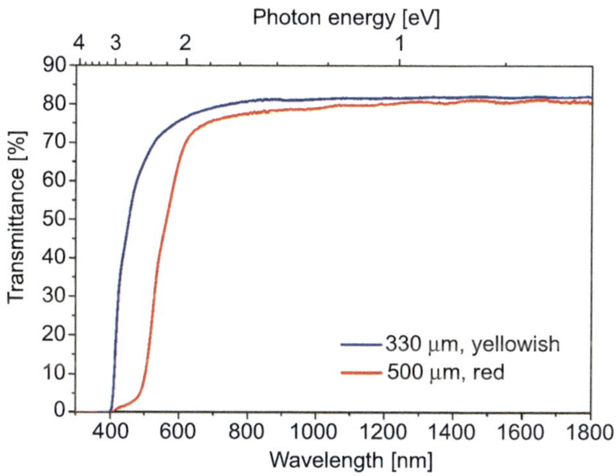

Figure 7.15 Optical transmittance spectra of two undoped and electrically insulating melt-grown $BaSnO_3$ crystal samples of different coloration, red (500 μm thick) and almost colorless with a yellowish tint (330 μm thick). Reprinted from Ref. [53], Copyright 2017, with permission from IOP Publishing.

The absorption coefficients α and $\alpha^{1/2}$ at 5 K and RT versus photon energy E_{ph} of a thin and nearly colorless melt-grown $BaSnO_3$ crystal sample are shown in Fig. 7.16a and Fig. 7.16b, respectively. Above about 3 eV, α increases very fast to about 700 cm^{-1} at about 3.1 eV. At 5 K, the relation $\alpha \propto (h\nu \pm E_{phonon} - E_{g,indir})^2$, where E_{phonon} and $E_{g,indir}$ are the energy of the momentum conserving phonon and

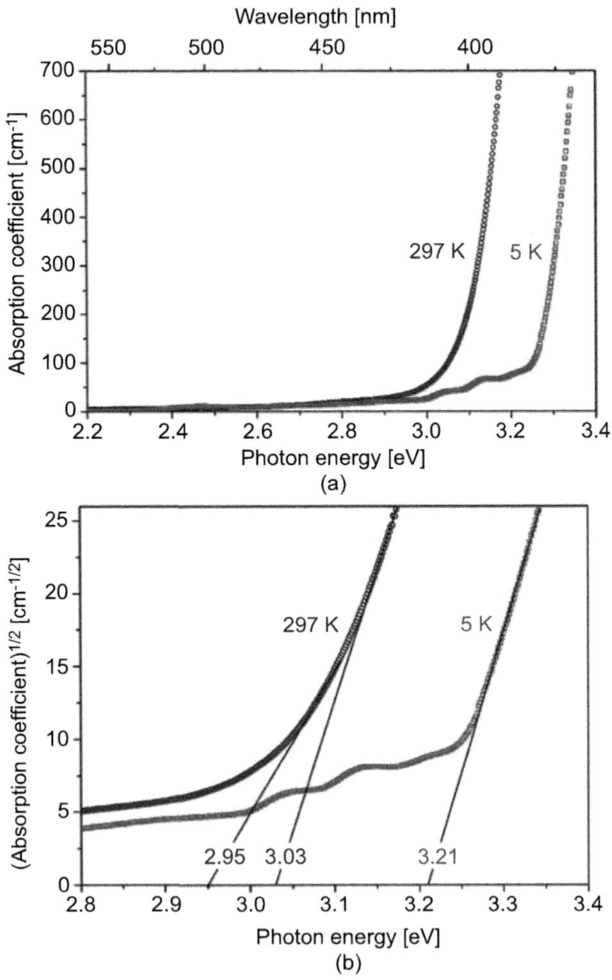

Figure 7.16 (a) Spectral dependence of the absorption coefficient measured at 297 K and 5 K for the nearly colorless transparent sample of 150 μm thickness. (b) Square root of the absorption coefficient versus photon energy. Thin black lines indicate extrapolations of nearly linear parts of the curves and the numbers give the extrapolated energy values. Reprinted from Ref. [53], Copyright 2017, with permission from IOP Publishing.

indirect bandgap, gives almost a perfect linearization that clearly points to the indirect bandgap of $E_{g,indir}$ = 3.17 ± 0.04 eV. At 5 K, the phonon absorption during indirect transition is strongly suppressed. On the other hand, the insufficient linearization of the data at RT might be due to the availability of several different phonon modes that can fulfil the momentum conservation and additionally due to disturbing weak defect related absorption. The roughly extrapolated data lead to the indirect gap $E_{g,indir}$ = 2.99 ± 0.04 eV at RT. Thus, the fundamental absorption edge is shifted to lower energies by about 180 meV from 5 K to RT. It can be concluded that the fundamental absorption edge is associated with phonon-assisted indirect-allowed optical transitions. As the phonon-assisted indirect-allowed optical transitions quickly reach very high absorption coefficients above 10^3 cm^{-1} for photon energies above the absorption onset, the direct transitions of much higher energies could not be observed for the present sample thickness of 150 μm.

Ellipsometric measurements of flux-grown undoped and La-doped $BaSnO_3$ crystals by Stanislavchuk et al. [43] revealed the onset of band-to-band absorption at about 3.5 eV, which corresponds to the direct bandgap at Γ-point of the BZ. Similar values of the direct bandgap of 3.56 and 3.36 eV were obtained for heteroepitaxial $BaSnO_3$ thin films [66, 67].

Theoretical values of the direct and indirect bandgaps show a large spread depending on the computational method. Examples of calculated (with DFT) direct $E_{g,dir}$ and indirect $E_{g,indir}$ bandgaps being close to experimental ones are $E_{g,dir}$ = 3.46 eV and $E_{g,indir}$ = 2.98 eV (Kang et al. [46], $E_{g,dir}$ = 3.68 eV and $E_{g,indir} \approx$ 3.2 eV (Scanlon [37], $E_{g,indir}$ = 3.14 eV (Duarte et al. [52], $E_{g,dir}$ = 3.21 eV and $E_{g,indir}$ ≈ 2.73 eV (Monserrat et al. [48], $E_{g,dir}$ = 3.0 eV and $E_{g,indir}$ = 2.65 eV (Soleimanpour and Kanjouri [49].

7.6.2 Dielectric Constant and Refractive Index

For flux-grown $BaSnO_3$ crystals, Luo et al. [11] and Stanislavchuk et al. [43] determined the dielectric function in the energy range of 0.75–5.9 eV at RT, which resulted in relative static and high-frequency dielectric constant values of ε_r = 17 ± 2 and ε_∞ = 3.3, respectively [43]. High-frequency dielectric constant from ellipsometric measurements of Anooz and Galazka [68] on melt-grown $BaSnO_3$ crystals is ε_∞ = 3.39 ± 0.005, which well matches the above-mentioned experimental values for flux-grown crystals.

For comparison, ε_r for heteroepitaxial films by hybrid MBE on Nb-doped STO was found by Nunn et al. [69] at a value of 14.8 ±

0.5, which remains substantially constant at the frequency range of 1 kHz to 1 MHz. Seo et al. [70] reported $\varepsilon_\infty = 3.79$ for heteroepitaxial $(Ba_{0.96}La_{0.04})SnO_3$ films on STO by pulsed electron deposition (PED) method. In the case of $BaSnO_3$ ceramics, Singh et al. [71] measured $\varepsilon_r = 20$. Examples of theoretical values of the high-frequency dielectric constant are $\varepsilon_\infty = 4.1$ [39] and $\varepsilon_\infty = 3.55$ [49]. Bévillon and Geneste [72] showed that the relative static dielectric constant decreases with pressure by a factor of about 2 at about 50–80 GPa and then increases to approximately the same value at 180 GPa as in the case of zero-pressure value.

Experimental refractive index *n* of melt-grown $BaSnO_3$ concluded from ellipsometric measurements is shown in Fig. 7.17 (Anooz and Galazka) [68]. It has a maximum of 2.35 at 355 nm (3.5 eV), which corresponds to the onset of band-to-band absorption, and it decreases to a value of 2.07 at 800 nm. In longer wavelengths, the refractive index is 1.94 ± 0.01 at 1 μm, and it decreases to 1.92 at 1.8 μm [53].

Examples of theoretical refractive index concluded from the dielectric function are 2.26 [49], 2.07 [64], and 2.2 [46].

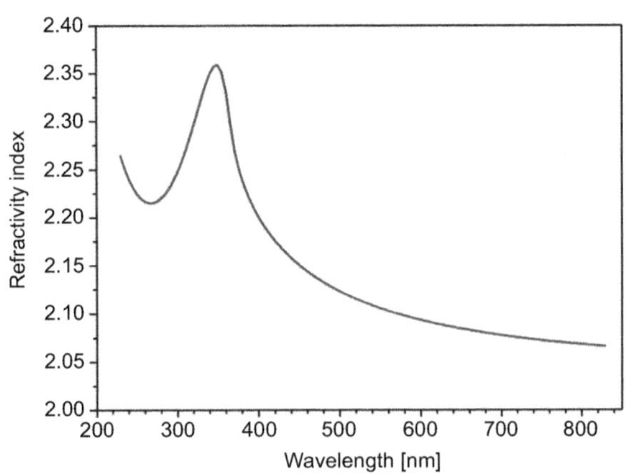

Figure 7.17 Refractive index of melt-grown bulk $BaSnO_3$ single crystal [68].

7.6.3 Luminescence

In the case of heteroepitaxial films by PLD on STO substrates as showed by Takashima and Inaguma [73], $BaSnO_3$ showed under

UV excitation (335 nm) a broad emission band peaking at 905 nm (NIR spectral region). Annealing decreased the peak intensity but did not change its peak position. In contrast to the films, cathodoluminescence measurements of melt-grown $BaSnO_3$ single crystals by Schewski and Galazka [74] did not show any luminescence at RT in a spectral region of 80–1270 nm.

7.6.4 Phonon Modes

In the cubic structure, there are phonon modes having the following irreducible representation: $3F_{1u} + F_{2u}$, where the three F_{u1} modes are IR-active modes and correspond to the acoustic phonon mode, while F_{2u} is an IR-inactive mode and corresponds to an optically silent mode [52]. In the ideal perovskite structure, there are no Raman-active modes. Calculated by Duarte et al. [52], IR-active modes can be summarized as follows: the first absorption peak that occurs at 146 cm^{-1} is assigned to translational movement of Ba atoms along *x*-, *y*-, and *z*-axes. The second absorption peak at 255 cm^{-1}, which has the highest intensity, is assigned to the angle bending of the Sn–O–Sn bonds along the *y*-axis with angular variation of the O–Sn–O bonds along the *z*-axis. The third absorption peak at 633 cm^{-1} corresponds to stretching and compressing vibrations of the Sn–O bonds along the *x*-, *y*-, and *z*-axes.

In an experimental study of flux-grown crystals by far-IR and Raman spectroscopies, Stanislavchuk et al. [43] found not only more than three IR-active modes of the LO and TO natures, but several Raman-active modes as well. The IR-absorption peaks were found at energies 135 (TO_1), 154 (LO_1), 244.5 (TO_2), 421 (LO_2), 628 (TO_3), and 723 (LO_3) cm^{-1}. The experimental Raman peaks were found at energies 115, 150–170 (LO_1 or M-point), 238, 408 (LO_2), 489 ($2TO_2$ or M-point), 543 ($TO_1 + LO_2$), and 724 (LO_3) cm^{-1}. Theoretical results at Γ-point revealed some discrepancies with measured values, which were assigned to lattice polarizability in polar perovskite crystals (LO modes) and their overtones. It should be noted that bulk crystals were grown from the PbO flux; therefore, relatively high concentration of Pb impurities is expected, which may influence the experimental IR and Raman spectra.

7.6.5 Conclusion

$BaSnO_3$ is an indirect wide bandgap semiconductor with an experimentally measured optical bandgap of about 3 eV, wherein the direct optical transition occurs at about 3.5 eV. Measured on bulk crystals, relative static and high-frequency dielectric constants are $\varepsilon_r = 17 \pm 2$ and $\varepsilon_\infty = 3.3$–3.39, respectively. The experimental refractive index in the visible spectrum (355–800 nm) is in the range of 2.35–2.07. In terms of luminescence, melt-grown $BaSnO_3$ crystals did not show any emission within 80–1270 nm, wherein the PL spectrum of heteroepitaxial films under UV excitations above the bandgap revealed a broad emission peaking at 905 nm. There are three IR-active modes in an ideal cubic perovskite with theoretical values of about 146, 255, and 633 cm^{-1}; however, experiment showed more active IR-absorption peaks and additionally Raman peaks, which are not present in the ideal cubic perovskite. A summary of basic optical properties is listed in Table 7.5.

Table 7.5 Basic experimental and theoretical optical properties of bulk $BaSnO_3$ crystals

Property	Unit	Value	Reference
Indirect bandgap [eV]	eV	≈3/2.98[(a)]	44, 53/46
Direct bandgap [eV]		≈3.5/3.46[(a)]	43
Fundamental bandgap		Indirect	44, 53/46
Absorption edge	nm	≈400	44, 45, 53
Emission band peak at RT	nm	None/905[(b)]	74/73
Static dielectric constant		17±2/14.8±0.5[(b)]/20[(c)]	43/69/71
High-frequency dielectric constant		3.3/3.39/3.79[(b)]/3.55[(a)]	43/68/70/49
Experimental refractive index at 355, 400, 600, 800, 1000, and 1800 nm		2.35, 2.2, 2.09, 2.07, 1.94, 1.92	68, 53
Theoretical refractive index		2.07–2.26	46, 49, 64
Experimental IR-active modes	cm^{-1}	135, 154, 244.5, 421, 628, 723	43
		146[(a)], 255[(a)], 633[(a)]	52

[(a)]theoretical value; [(b)]value for thin films; [(c)]value for ceramics

7.7 Electrical Properties

As already concluded from the theory (see Section 7.3.2), native point defects in $BaSnO_3$ cannot act either as shallow donors or as shallow acceptors due to self-compensation effect. As a result, extrinsic point defects were found to induce *n*-type conductivity in this compound. Among extrinsic factors, the following were experimentally and/or theoretically concluded as the source of the underlying electrical conductivity in $BaSnO_3$:

- La^{3+}, Y^{3+}, Pr^{3+}, and Nd^{3+} substituting Ba^{2+} site [11, 12, 37, 40, 44, 53, 56];
- Sb^{5+} substituting Sn^{4+} site [37, 45];
- H and F substituting O^{2-} (H_O and F_O) [37];
- Interstitial H and F (H_i and F_i) [37].

The most common shallow donor providing high free electron concentrations and very high electron mobility at high free electron concentrations in $BaSnO_3$ is La, which has intensively been practiced in the case of bulk crystals, thin films, and ceramics. This phenomenon can be explained by the following features [12, 14, 36, 45, 75, 79]:

i. High dispersion of the bottom part of the CB (enables high electron mobility);
ii. Low effective electron mass (enables high electron mobility);
iii. Low density of states (a reduction in the LO-phonon scattering rate);
iv. The conduction channel is formed by the Sn-subnetwork (Sn–O–Sn bonding of the SnO_6 octahedra), while La-doping occurs at Ba-subnetwork, which is physically separated from the Sn-subnetwork (minimizes scattering at ionized impurities);
v. Negligible interaction between La dopant and the conduction band (lack of significant orbital mixing between CBM and La states orbital);
vi. High dielectric constant that effectively screens scattering at charged structural defects, ionized impurities, and phonons.

7.7.1 Free Electron Concentration, Hall Mobility, and Resistivity/Conductivity at RT

Flux-grown $BaSnO_3$ crystals were intensively investigated for electrical properties whether undoped or doped with La and Sb. The electrical properties of undoped crystals depend on the flux composition. $BaSnO_3$ crystals grown by Lee et al. [25, 54, 55] from the Cu_2O–CuO–$KClO_4$ flux were electrical insulators.

The Cu_2O–CuO flux enabled Lee et al. [14, 55] to obtain undoped $BaSnO_{3-\delta}$ (supposed to be oxygen deficient) with electrically conducting behavior revealed by Hall effect measurements at RT showing the free electron concentration, resistivity, and Hall mobility of $n_e = 3 \times 10^{18}$ cm^{-3}, $\rho \approx 1 \times 10^{-3}$ Ωcm, and $\mu = 180$ cm^2/Vs. Undoped $BaSnO_3$ crystals obtained by McCalla et al. [56] from the Cu_2O flux showed at RT n_e = 2.15–3.56 × 10^{19} cm^{-3}, $\rho \approx$ 1.8–3.61 × 10^{-3} Ωcm, and μ = 80–106 cm^2/Vs. The results were similar to each other whether the crystals were as-grown or annealed in vacuum (1000°C, 1100°C) or in O_2 at 1 atm (1000°C) and 240 atm (550°C). At 5 K, the mobility increased to 100–145 cm^2/Vs.

For La-doped (2 mol.%) $BaSnO_3$ crystals grown from the PbO flux, Luo et al. [11] concluded from Hall effect measurements at RT the following electrical properties at RT: $n_e = 8.9 \times 10^{19}$ cm^{-3}, $\rho = 5.9 \times 10^{-4}$ Ωcm, and $\mu = 78$ cm^2/Vs. After annealing in O_2 at 700°C for 5 h, the electrical properties were similar to those of as-grown crystals.

La-doped $BaSnO_3$ crystals obtained by Kim et al. [12] from the Cu_2O flux revealed n_e = 8 × 10^{19}–6.8 × 10^{20} cm^{-3}, $\rho \approx 1.3 \times 10^{-4}$ Ωcm, and μ = 80–320 cm^2/Vs. The Hall mobility of 320 cm^2/Vs is the highest ever reported at RT not only for $BaSnO_3$, but also for all other TSOs. Further study by Kim et al. [44] of flux-grown La-doped (0–4 mol.%) crystals from the Cu_2O flux showed at RT (Hall effect measurements): n_e = 1 × 10^{19}–4 × 10^{20} cm^{-3}, ρ = 1–10 × 10^{-4} Ωcm, and $\mu \approx$ 200–300 cm^2/Vs.

Kim et al. [45] have also studied electrical properties of Sb-doped $BaSnO_3$ grown from Cu_2O flux. Here Sb^{5+} substitutes Sn^{4+} in the crystal lattice. The electrical properties of the crystals in a doping range of 0–10 mol.% in the flux were as follows (from Hall effect measurements): $n_e \approx$ 0.8–1.2 × 10^{20} cm^{-3}, ρ = 0.8–1.2 × 10^{-3} Ωcm, and μ = 30–80 cm^2/Vs. The Sb dopant can induce only half of the expected free electrons, with the other half remaining inactive. In other words, nearly half of the expected free carriers are likely to be trapped or localized.

Using the Cu_2O flux, McCalla et al. [56] obtained for La-doped $BaSnO_3$ crystal $n_e = 3.36 \times 10^{20}$ cm^{-3}, $\rho = 2.26 \times 10^{-4}$ Ωcm, and μ = 78 cm^2/Vs. Doping with Pr and Nd revealed the following electrical properties at RT: $n_e = 1.1 \times 10^{20}$ cm^{-3}, $\rho = 3.2 \times 10^{-4}$ Ωcm, and μ = 170–174 cm^2/Vs.

Figure 7.18a and Fig. 7.18b show a relation between resistivity ρ and Hall mobility μ versus free electron concentration n_e for La- and Sb-doped bulk $BaSnO_3$ single crystals grown from the Cu_2O flux and corresponding heteroepitaxial thin films obtained by the PLD on $SrTiO_3$ substrates. The resistivity of thin films doped with La or Sb is higher than that of bulk single crystals for the same free electron concentration by about one order of magnitude for $n_e < 10^{20}$ cm^{-3}. The Hall mobility of bulk single crystals shows very high and constant in average values of about 200–300 cm^2/Vs for $n_e \approx 0.3$–4×10^{20} cm^{-3}. Another series of La-doped crystals in Fig. 7.18 (squares) [12] shows a linear decrease in the Hall mobility with the free electron concentration in the relation $\mu \propto n_e^{-1}$, which corresponds to scattering at ionized impurities. In contrast, the Hall mobility of heteroepitaxial thin films is much lower as compared to that of bulk crystals and increases with the free electron concentration from about 20 cm^2/Vs to a saturation value of about 70–80 cm^2/Vs in a relation of $\mu \propto n_e^{1/2}$. This is due to a very high dislocation density in thin films at a level above 10^{10} cm^{-2} [19, 25, 76] and the presence of the rotational disorder of low-angle grain boundaries (the in-plane twist and out-of-plane tilt) [31], which all constitute the main source of electron scattering up to $n_e \approx 4 \times 10^{20}$ cm^{-3} above which scattering at ionized impurities starts to dominate.

When considering Sb-doped $BaSnO_3$ single crystals (Fig. 7.18b), they showed higher resistivity and lower Hall mobility (by a factor of about 4–7) as compared with La-doped crystals for the same free electron concentration, indicating a stronger scattering. Stronger scattering is also reflected in a steeper decrease in the Hall mobility with free electron concentration as compared with La-doped $BaSnO_3$ crystals. The reason for the μ versus n_e behavior is a different doping site: La substitutes a Ba-site that is physically separated from the transport channel formed by the Sn-subnetwork, while

Sb substitutes an Sn-site that forms at the same time the transport channel enhancing scattering at ionized impurities.

Undoped melt-grown $BaSnO_3$ crystals reported by Galazka et al. [53] showed either electrically insulating or semiconducting behavior according to Hall effect measurements, depending on crystal coloration, which in turn depend on growth conditions. The crystals of light brown and red color (melted at temperatures around MP) were electrical insulators, while dark-brown crystals (overheated and/or high Sn-excess in the starting composition) were semiconductors with $n_e \approx 5 \times 10^{17}$ cm^{-3}, $\rho = 0.2$ Ωcm, and $\mu = 50$ cm^2/Vs. Annealing undoped crystals in oxidizing atmospheres (air or O_2) at 1200–1400°C for 10–40 h resulted in insulating crystals independent of their original conductivity and coloration. Annealing insulating crystals in a reducing atmosphere (5% H_2 + Ar) at 600–700°C for 10 h yielded semiconducting crystals with $n_e = 10^{16}$–10^{18} cm^{-3} and $\mu < 10$cm^2/Vs, wherein annealing under reducing conditions at ≤500°C had no measurable effect on the electrical properties.

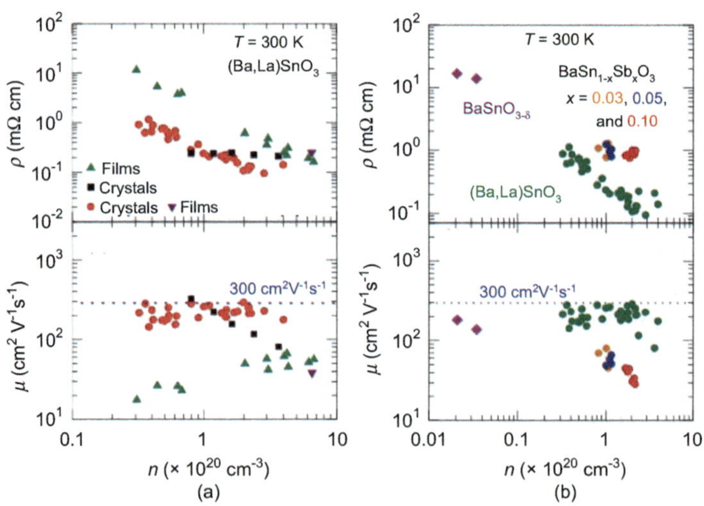

Figure 7.18 Resistivity ρ (upper panels) and Hall mobility μ (lower panels) of Sb-doped (a) and La-doped (b) $BaSnO_3$ single crystals and films as a function of the free electron concentration n. Figure (a) reprinted with permission from Ref. [44], Copyright 2012, and (b) reprinted with permission from Ref. [45], Copyright 2013, American Physical Society.

Melt-grown La-doped $BaSnO_3$ single crystals resulted in $n_e \approx$ 1–3.3 × 10^{19} cm^{-3}, ρ = 3 × 10^{-3}–8.6 × 10^{-4} Ωcm, and μ = 200–220 cm^2/Vs for La concentration of 0.016–0.123 wt.% La in the crystals. Comparing La concentration in the crystal (0.123 wt.% corresponds to 3.8 × 10^{19} La atoms cm^{-3}) with the free electron concentration (3.3 × 10^{19} cm^{-3}) suggests nearly complete electrical activation of the incorporated lanthanum and negligible compensation. It is to be noted that the metallic nanoparticles found in melt-grown crystals have only a marginal impact on the electrical transport properties (due to formation of Schottky barriers between the metallic nanoparticles and semiconducting host) as concluded from the analysis of the size and a distance between the nanoparticles (2 μm), which is five orders of magnitude larger than the depletion region (about 20 nm) arising from the electron concentration of about 10^{19} cm^{-3}.

The above-discussed basic electrical properties of as-grown undoped and doped bulk $BaSnO_3$ single crystals are listed in Table 7.6.

Doped bulk $BaSnO_3$ crystals are typically degenerate semiconductors with the Fermi level shifted to the conduction band. The ionization energy of shallow donors can be concluded only for normal semiconducting state with the free electron concentration rather not exceeding the level of about 10^{18} cm^{-3}. Such low doping level with the free electrons was obtained, e.g., by Ganguly et al. [30] with thin films grown by sputtering on MgO substrates, who concluded the activation energy of shallow donors of 68 meV at $n_e \approx 10^{17}$–10^{18} cm^{-3}. It should, however, be stressed that this value may not be representative as the films contained a high density of charged defects that lower the Fermi level below the band edge.

7.7.2 Temperature-Dependent Electrical Properties

Temperature-dependent (2–300 K) electrical properties were measured for undoped, La-, Sb-, Pr-, and Nd-doped flux-grown bulk $BaSnO_3$ single crystals by Kim et al. [44, 45] and McCalla et al. [56]. As shown in Fig. 7.19a [44], La-doped crystals revealed substantially temperature-independent, metal-like free electron concentration behavior indicating a degenerate semiconducting state. Indeed, in all cases, $n_e > 10^{19}$ cm^{-3}. Basing on Mott's criterion, the estimated metal-to-insulator transition takes place in a range of the electron

Table 7.6 Basic electrical properties of as-grown bulk $BaSnO_3$ crystals at RT

Method	Dopant	n_e [cm^{-3}]	μ [cm^2/Vs]	ρ [Ωcm]	Reference
Flux: Cu_2O–CuO–$KClO_4$	K	Electrical insulator			54, 55, 14
Flux: Cu_2O–CuO	—	3×10^{18}	180	1×10^{-3}	55, 14
Flux: Cu_2O	—	$2.15–3.56\times10^{19}$	80–106	$1.8–3.61\times10^{-3}$	56
Flux: PbO	La	8.9×10^{19}	78	5.9×10^{-4}	11
Flux: Cu_2O	La	$8\times10^{19}–6.8\times10^{20}$ $1\times10^{19}–4\times10^{20}$ 3.36×10^{20}	80–320 200–300 78	1.3×10^{-4} $1–10\times10^{-4}$ 2.26×10^{-4}	12 44 56
Flux: Cu_2O	Sb	$0.8–1.2\times10^{20}$	30–80	$0.8–1.2\times10^{-3}$	45
Flux: Cu_2O	Pr or Nd	1.1×10^{20}	170–174	3.2×10^{-4}	56
Melt: VGF	—	Electrical insulator			53
Melt: VGF	—	5×10^{17}	50	2×10^{-2}	53
Melt: VGF	La	$1–3.3\times10^{19}$	200–220	$8.6–30\times10^{-4}$	53

concentration of 5.5×10^{18} cm^{-3} to 1.5×10^{19} cm^{-3} (for electron effective mass of 0.2–0.4 m_0 and dielectric constant of 20) [27]. The resistivity decreases with temperature from RT down to 2 K by a factor of about 2. This results in a corresponding mobility increase. For comparison, La-doped $BaSnO_3$ thin films showed the same behavior of temperature-dependent electrical properties, although their residual resistivities are generally larger than those of the bulk crystals, indicating extra scattering at dislocations and grain boundaries.

Similarly, temperature-dependent electrical properties of Sb-doped $BaSnO_3$ crystals are depicted in Fig. 7.19b [45]. The crystals also showed basically temperature-independent free electron concentrations, indicating a degenerate semiconducting state. The resistivity decreases, and the Hall mobility correspondingly increases from RT to 2 K to a less extent (by about 20%) as compared with La-doped crystals (by about 50%). For the same free electron concentrations, Sb-doped bulk crystals revealed higher resistivities by five to ten times as compared with resistivities of La-doped bulk crystals, indicating much stronger electron scattering.

Undoped bulk $BaSnO_{3-\delta}$ crystals (Fig. 7.19b) [45] showed generally similar temperature-dependent behavior to those of La- and Sb-doped $BaSnO_3$ crystals as they are also degenerate semiconductor. Here, the electron Hall mobility has a small maximum and the resistivity a small minimum around 150 K. However, the resistivity is much higher as compared with doped crystals due to much lower free electron concentration by over one order of magnitude.

The highest Hall mobility of about 600 and 90 cm^2/Vs was measured at 2 K for La- and Sb-doped crystals, respectively. The undoped crystals showed the highest Hall mobility of about 160 cm^2/Vs between 100 and 200 K.

McCalla et al. [56] obtained a similar output of temperature-dependent electrical properties for flux-grown undated, La-, Pr-, and Nd-doped bulk $BaSnO_3$ crystals. Undoped and La-doped crystals showed only a small increase in the Hall mobility with a temperature decrease from RT to 5 K (by about 50%, from 78 to 120 cm^2/Vs), while Pr- and Nd-doped crystals revealed an increase by more than 100%, from 170 (Pr)/174 (Nd) at RT to 375 (Pr)/434 (Nd) cm^2/Vs at 5 K. Correspondingly, the resistivity decreased form about 3.2×10^{-4} (Pr and Nd) Ωcm at RT to 1.6×10^{-4} (Pr)/1.2×10^{-4} (Nd) Ωcm at 5 K.

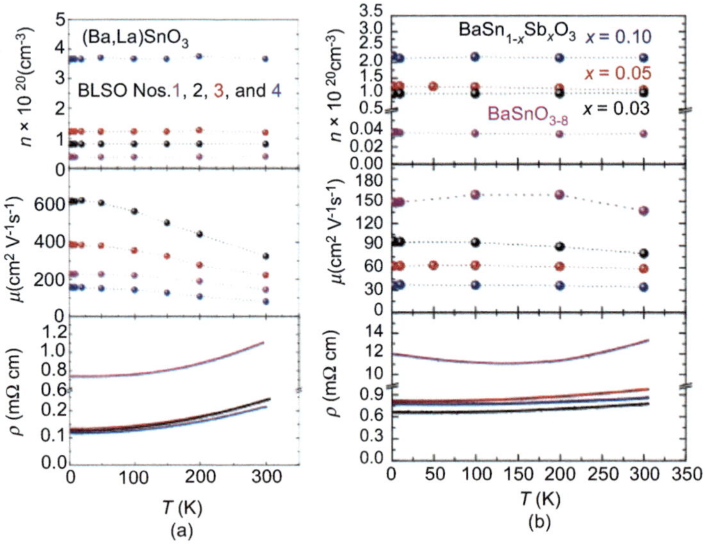

Figure 7.19 Temperature-dependent free electron concentration n (top panels), Hall mobility μ (middle panels), and resistivity ρ (bottom panels) for selected (Ba,La)SnO_3 (a) and $BaSn_{1-x}Sb_xO_3$ (b) single crystals having different free electron concentrations at RT labeled 1, 2, 3, and 4 for (Ba,La)SnO_3 and $x = 0.03$, 0.05, and 0.1 for $BaSn_{1-x}Sb_xO_3$. $BaSnO_{3-\delta}$ crystals are also included in (b). Figure (a) reprinted with permission from Ref. [44], Copyright 2012, and (b) reprinted with permission from Ref. [45], Copyright 2013, American Physical Society.

7.7.3 Effective Electron Mass and Electron Mobility

The effective electron mass m_e* is determined by the energy–momentum dispersion around the CBM. It is an important quantity in semiconductors as it is directly related to the carrier mobility:

$$\mu = \frac{e\tau}{m_e{}^*} \tag{7.10}$$

where μ is the electron mobility, e is the electron charge, and τ is the electron scattering rate.

Experimentally determined by Kim et al. [44, 45], the effective electron mass from the Burstein–Moss shift is $m_e^* \approx 0.6\ m_0$ for La- and Sb-doped bulk $BaSnO_3$ single crystals with $n_e > 1 \times 10^{20}$ cm^{-3}, where m_0 is the electron rest mass. McCalla et al. [56] estimated from the specific heat measurements $m_e^* \approx 0.4\ m_0$ in average (with high uncertainty) for undoped and Pr-, La-doped bulk crystals with $n_e > 1 \times 10^{20}$ cm^{-3}.

For comparison, exemplary values of m_e^* determined experimentally for heteroepitaxial La-doped $BaSnO_3$ films from IR spectroscopy are 0.35 m_0 for $n_e \approx 9 \times 10^{19}$ cm^{-3} (pulsed electron deposition on $SrTiO_3$ substrates) [70], and 0.2–0.43 m_0 for n_e = 2.3×10^{18}–1.8×10^{20} cm^{-3} (sol-gel on sapphire) [77]. Ellipsometry data obtained on heteroepitaxial La-doped $BaSnO_3$ films (by PLD on $SrTiO_3$ substrates) by Kim et al. [19] for n_e = 2.8–4 $\times 10^{20}$ cm^{-3} yielded m_e^* = 0.42–0.44 m_0.

Basing on the IR dielectric function obtained from transmittance/reflectivity, Fourier transform infrared spectroscopy (FTIR), and ellipsometry spectra of heteroepitaxial La-doped $BaSnO_3$ films (by PLD on $SrTiO_3$ substrates), Niedermeier et al. [78, 79] determined a relation between the effective electron mass and the free electron concentration, as shown in Fig. 7.20. Fitting the equation (dashed-dot line in Fig. 7.20)

$$m_e^* = m_0 \sqrt{1 + 2\beta \frac{\hbar^2 k^2}{m_0}} \tag{7.11}$$

where β is an empirical fitting parameter describing the degree of nonparabolicity of the CB, $k = (3n_e\pi^2)^{1/3}$ is the Fermi wave vector, and $\hbar$ is the reduced Planck constant, the electron effective mass data yields m_e^* = 0.19 ± 0.02 m_0 at the CBM. m_e^* remarkably increases from 0.19 to 0.36 m_0 for the free electron concentrations from 1.6 $\times 10^{19}$ to 7.6×10^{20} cm^{-3}, respectively. The resulted nonparabolicity parameter β is 0.37 ± 0.07 (eV)$^{-1}$, quite high as compared to other TSOs. However, the theoretical nonparabolicity parameter is about 0.25 [36].

Exemplary theoretical m_e^* values at the Γ-point of the BZ at CBM are close to that determined empirically (with respect to m_0): 0.2 [36], 0.2 [41], 0.22 [37], 0.21 [40], and 0.26 [50].

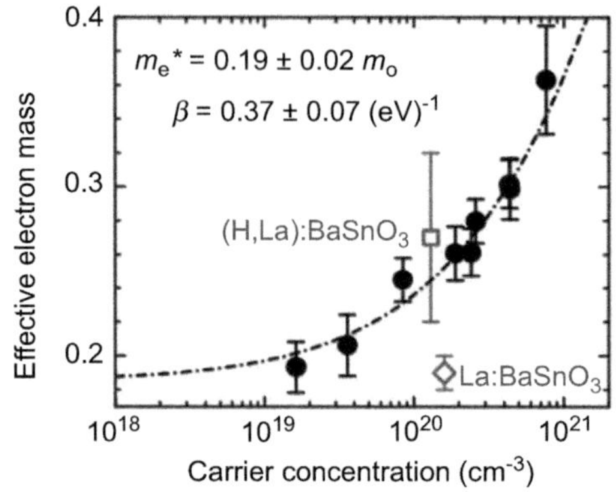

Figure 7.20 Relation between the effective electron mass and the electron concentration for heteroepitaxial La-doped $BaSnO_3$ films as determined by the plasma frequency in the Drude model for free carrier absorption (black filled circles). Data for La+H doped $BaSnO_3$ films (empty square) from Ref. [78] and La-doped $BaSnO_3$ (empty diamond) from Ref. [80] are given for comparison. Reprinted with permission from Ref. [79], Copyright 2017, American Physical Society.

As defined by Eq. 7.10, the electron mobility is inversely proportional to the effective electron mass m_e^* and directly proportional to the electron scattering rate τ. The effective electron mass arises from the CBM dispersion, and it constitutes the material's property. The remaining factors affecting the electron mobility are electron scattering sources [14, 45, 79]:

i. Acoustic and optical phonons;
ii. Ionized impurities;
iii. Neutral impurities;
iv. Oxygen vacancies;
v. Structural defects (grain boundaries, dislocations, stacking faults, including Ruddlesden-Popper type, etc.).

By fitting experimental μ data as a function of the free electron concentration n_e shown in Fig. 7.20 with the theoretical scattering sources (i)–(iv) summarized in Fig. 7.21, and taking into account considerations of Lee et al. [14] and Kim et al. [12, 44, 45], it can be concluded that:

i. Theoretical limit for the acoustic phonon scattering while neglecting other scattering sources yields a value of m_e* = 600 cm^2/Vs at RT;
ii. In undoped, oxygen-deficient $BaSnO_{3-\delta}$ single crystals, the dominating scattering source is oxygen vacancies;
iii. The electron mobility in La-doped crystals at the degenerate state is limited mainly by ionized impurities;
iv. In Sb-doped crystals at the degenerate state, the dominating scattering source is neutral and ionized impurities.

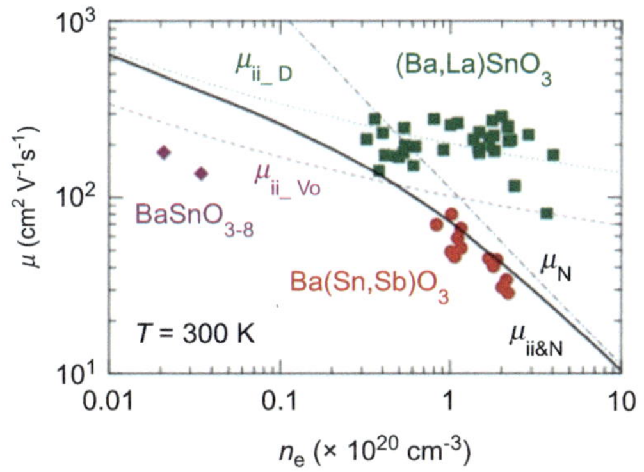

Figure 7.21 Experimental Hall mobility μ versus free electron concentration n_e for flux-grown $(Ba,La)SnO_3$ (green squares), $Ba(Sn,Sb)O_3$ (red circles), and $BaSnO_{3-\delta}$ (violet diamonds) single crystals, respectively. The dotted lines are the theoretically estimated mobility governed by ionized dopants (La or Sb) (μ_{ii_D}), oxygen vacancies (μ_{ii_Vo}), and neutral impurities (μ_N), respectively. The solid line shows a combined mobility governed by both ionized dopants and neutral impurities ($\mu_{ii\&N}$). Reprinted with permission from Ref. [45], Copyright 2013, American Physical Society.

Therefore, at normal semiconducting state ($n_e \leq 10^{18}$ cm^{-3}), high structural quality $BaSnO_3$ single crystals may show Hall mobility approaching about 400 cm^2/Vs. Indeed, Kim et al. [12, 44] reported for La-doped bulk $BaSnO_3$ crystals record high Hall mobility values at RT of 300–320 cm^2/Vs at still high $n_e \approx 10^{20}$ cm^{-3}. Reasons for high electron mobility in La-doped BaSn03 single crystals are outlined in items i.–vi. at the beginning of Section 7.7. Such high μ values will not, therefore, be available for doping the Sn-site by Sb, as it introduces an additional scattering source directly on the main conduction path.

As shown in Fig. 7.21, for bulk $BaSnO_3$ single crystals, the following relation $\mu \propto n_e^{-1}$ holds, which means the electron mobility decreases with the doping level. For comparison, in heteroepitaxial films, another relation was observed, $\mu \propto n_e^{1/2}$ (e.g., Kim et al. [12, 19]), indicating an increase in the electron mobility with the doping level, but with much lower overall mobility at similar free electron concentration. The reason for that is another dominating scattering mechanism, i.e., charged structural defects (scattering source v), mainly high density of threading dislocations that may create trap states for free electrons.

Niedermeier et al. [79] performed a detailed analysis of the electron mobility as a function of the free electron concentration for La-doped $BaSnO_3$ thin films obtained on MgO substrates by PLD and flux-grown bulk crystals from Ref. [44]. As shown in Fig. 7.22, indeed, the Hall mobility of the heteroepitaxial films at RT is governed by dislocation scattering of a density of about 10^{13} cm^{-2} at $6 \times 10^{18} < n_e < 2 \times 10^{20}$ cm^{-3} (degenerate state, region II), and by a combination of the scattering at optical phonons and ionized impurities at $n_e > 2 \times 10^{20}$ cm^{-3} (degenerate state, region III). Below $n_e < 6 \times 10^{18}$ cm^{-3} (region I), the Hall effect was not measurable due to the high density of trapping states that is higher than the doping level. This is the same effect as it was found for homoepitaxial films of β-Ga_2O_3 on (100)-oriented substrates, due to incoherent twin boundaries acting as electron traps (Schewski et al. [81], Fiedler et al. [82]). On the other hand, the Hall mobility of La-doped bulk $BaSnO_3$ single crystals is limited by optical phonon and ionized impurity scattering.

Krishnaswamy et al. [36] performed a theoretical analysis of the electron mobility in $BaSnO_3$ as a function of the free electron concentration and temperature taking into account multiple LO-phonon and ionized impurity scattering. From their results, shown in Fig. 7.23a,b, it can be concluded that:

i. At the non-degenerate semiconducting state of $BaSnO_3$ ($n_e \leq 10^{18}$ cm^{-3}), the resulted electron mobility at RT is limited by LO-phonon scattering (no screening) to a value of about 300 cm^2/Vs, which remains substantially constant;

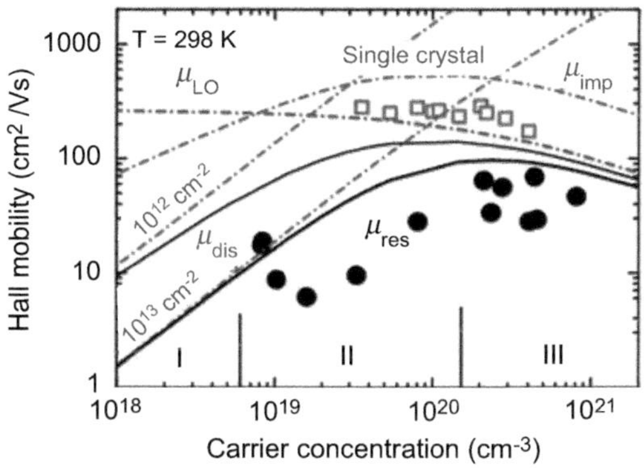

Figure 7.22 Room-temperature Hall mobility of heteroepitaxial La-doped $BaSnO_3$ films (filled circles) and bulk crystals (open squares, Ref. [44]) versus free electron concentration. In region I, the Hall effect was not measurable. The Hall mobility of the degenerate films is governed by dislocation scattering in the intermediate-doping regime (II) and LO optical phonon and ionized impurity scattering in the high-doping regime (III). μ_{imp}, μ_{LO}, μ_{dis} indicate mobilities due to scattering at ionized impurities, optical phonons, and dislocations, while μ_{res} is the resulted mobility when all these three scattering mechanisms are included. Reprinted with permission from Ref. [79], Copyright 2017, American Physical Society.

ii. At near or the degenerate state ($n_e > 10^{18}$ cm^{-3}), the resulted electron mobility at RT is affected by ionized impurities, which becomes dominant at $n_e > 10^{20}$ cm^{-3}. The maximum electron mobility is limited to a value of about 330 cm^2/Vs. An enhancement of the electron mobility at higher free electron concentrations, despite the strong contribution of ionized impurity scattering, is the result of the Fermi level shift to the CBM and significant screening;

iii. At RT, the electron mobility changes only slightly at a wide range of the free electron concentrations of about 10^{17}–10^{20} cm^{-3};

iv. At low temperatures, scattering at ionized impurities dominates, while at higher temperatures, scattering at LO-phonon reduces the mobility;

v. The maximum electron mobility occurs at about 50 K, which may reach a value above 3000 cm^2/Vs at normal semiconducting state ($n_e = 10^{17}$ cm^{-3}).

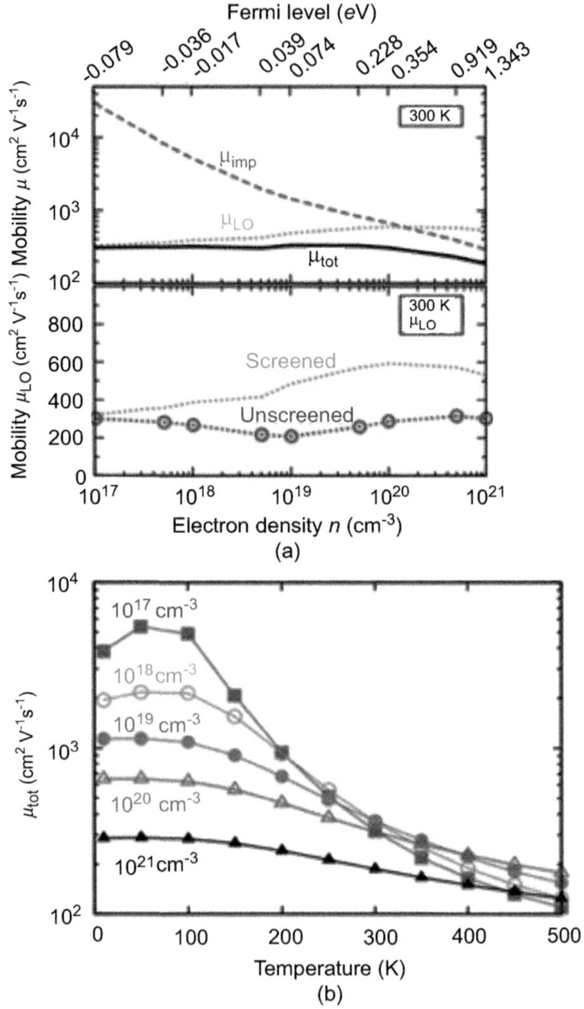

Figure 7.23 (a) Theoretical drift mobility versus free electron density (concentration) at RT for LO-phonon scattering μ_{LO}, ionized impurity scattering μ_{imp}, and total drift mobility μ_{tot} (top panel), and comparison of the screened and unscreened values for μ_{LO} versus electron density (bottom panel); (b) theoretical drift mobility μ_{tot} versus temperature as a result of a combination of LO-phonon and ionized impurity scattering for different electron densities of 10^{17}, 10^{18}, 10^{19}, 10^{20}, and 10^{21} cm^{-3}. Reprinted with permission from Ref. [36], Copyright 2017, American Physical Society.

7.7.4 Two-Dimensional Electron Gas

One of the features and challenges in the next generation of electronic devices, such as FETs, is the control of high-density currents at nanoscale, which can be generated at an interface of a heterostructure. This feature is called two-dimensional electron gas (2DEG). 2DEG systems were successfully achieved between (i) semiconductor (Si)–dielectric (oxide), (ii) nitrides (e.g., AlGaN/GaN), and (iii) dielectric (oxide)–dielectric (oxide). The system (i) enables to obtain the electron density of about 10^{12} cm^{-2}, system (ii) of about 10^{13} cm^{-2}, while the system (iii) of 3×10^{14} cm^{-2} (e.g., $SrTiO_3$–$GdTiO_3$) [83]. Therefore, the heteroepitaxial oxide systems, particularly consisting of perovskites, enable to obtain very high-density 2DEGs, which are not available for other systems so far. As a result, $BaSnO_3$, which has an ideal perovskite structure, has attracted much scientific attention to explore its capability in generating high-density 2DEGs. To generate 2DEG, confinement and a source of electrons are required. The source of electrons can be either extrinsic (eternal doping by polarization discontinuity) or intrinsic (self-doping by polar discontinuity, i.e., polar–nonpolar heterointerface). The criteria needed to generate and affect 2DEGs, following Bjaalie et al. [84], are as follows:

i. High structural quality of the system, including not too large lattice mismatch (low dislocation density), point defects (could be a source of electrons), and cation intermixing (causing not enough sharp interface);
ii. Minimized electron leak from the interface, by applying surfaces without surface states and/or symmetric interfaces;
iii. A suitable electronic structure, including a high density of state of the CB, a high effective electron mass perpendicular to interface, not too high dielectric constant, a high breakdown strength, and Mott insulator capabilities (CB comprised of *d*-states);
iv. Band alignment/offset;
v. High electron mobility, including low effective electron mass in the transport direction and low scattering rate.

$BaSnO_3$ possesses structural and electronic properties that may make it very attractive for generation 2DEGs when combined with another suitable perovskite. Particularly, $BaSnO_3$ is a nonpolar perovskite that can create a sharp interface with a lattice-matched polar perovskite (e.g., with $LaInO_3$). $BaSnO_3$ reveals a low effective electron mass and low scattering rate that enable a very high electron mobility. Further, the CB provides a high density of states, as it is formed by Sn *s*-states. Relatively wide bandgap of about 3 eV ensures a relatively large breakdown strength on one hand, and large band offsets on the other hand. Perovskites that may form with $BaSnO_3$ the 2DEG system include $LaInO_3$ [13, 85], $LaScO_3$ [86], $LaAlO_3$ [87], $SrTiO_3$ [87], and $KTaO_3$ or $KNbO_3$ [88]. The most promising 2DEG system seems to be $LaInO_3/BaSnO_3$, as it constitutes a polar/nonpolar system with a large CB offset of about 1.6 eV [85], not too high dielectric constant of $LaInO_3$ (39 [85]), and good lattice match between both materials (assuming pseudo-cubic structure). Indeed, Kim et al. [13] reported a polar interface state between nonpolar $BaSnO_3$ and polar $LaInO_3$ perovskites with the sheet conductance enhancement of more than 10^4, due to the existence of polarization in the polar $LaInO_3$. As shown in Fig. 7.24, the highest sheet conductance enhancement at the $LaInO_3$ (LIO)/ $BaSnO_3$ (BSO) takes place for $BaSnO_3$ doped with La (BLSO) at a concentration of 0.1–1 mol.%. For undoped $BaSnO_3$, there was almost no conductance enhancement, while for the La-doping level above 1 mol.%, the interface conductance likely merges with the film conductance due to very high film conductivity (degenerate state). In terms of the electronic reconstruction, doping $BaSnO_3$ with La moves the Fermi level close to the CBM and the downward band bending at the interface enables the presence of the mobile electrons therein. In this 2DEG system, $BaSnO_3$ must be electrically conducting and form the interface with a polar perovskite. For comparison, interfaces between nonpolar $BaSnO_3$ and nonpolar $BaHfO_3$ or $SrZrO_3$ did not show any conductance enhancement.

Alternatively, the 2DEG system can be generated by so-called δ-doping instead of forming a heterostructure.

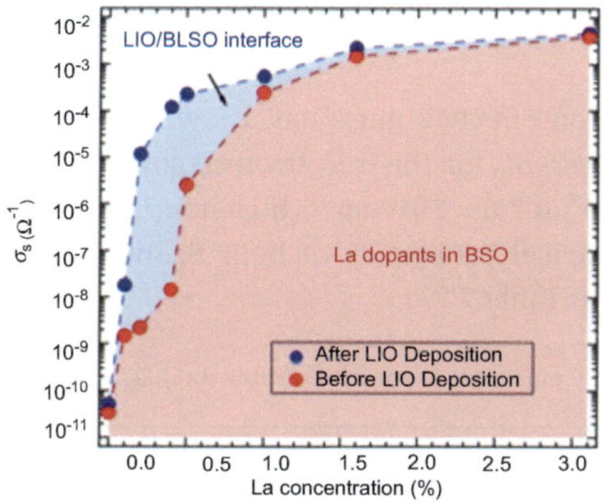

Figure 7.24 Sheet conductance changes for the $BaSnO_3$ film and $LaInO_3$/$BaSnO_3$ interface as a function of La-doping of the $BaSnO_3$ film (by PLD on $SrTiO_3$). LIO, BSO, and BLSO indicate $LaInO_3$, $BaSnO_3$, and La-doped $BaSnO_3$, respectively. Reprinted from Ref. [13] under Creative Commons license.

7.7.5 Conclusion

Bulk $BaSnO_3$ single crystals can be obtained either as electrical insulators, normal *n*-type semiconductors (free electron concentration of about mid-10^{17} cm^{-3}), or degenerate semiconductors (free electron concentration of about 3×10^{18}–6.8×10^{20} cm^{-3}). The crystals are highly conducting with typical resistivities of 10^{-3}–10^{-4} Ωcm. The most distinguished electrical property of bulk $BaSnO_3$ crystals is a very high electron (Hall) mobility approaching at RT values of 300–320 cm^2/Vs at very high free electron concentrations of about 10^{20} cm^{-3}. This is the result of doping the Ba-site with La^{3+} in the crystal lattice, which is physically separated from the conduction path through SnO_6 octahedra (reduced scattering at ionized impurities), low effective electron mass, high dispersion of the bottom part of the conduction band (enables high mobility), high dielectric constant (effective screening of charged defects, ionized impurities, and lower phonon scattering), and low density of states causing a reduction in the LO-phonon scattering rate. Doping the Sn-site with Sb^{5+} or the presence of high density of charged structural defects (films) decreases the electron mobility. In La-doped $BaSnO_3$ single crystals, the Hall mobility is limited by scattering at ionized impurities and LO-phonons.

The electron effective mass has a low value of 0.19 m_0, which increases to 0.36 m_0 for the free electron concentrations increasing from 1.6×10^{19} to 7.6×10^{20} cm^{-3} (high nonparabolicity parameter).

Basic electrical properties of bulk $BaSnO_3$ single crystals are summarized in Table 7.7.

Table 7.7 Basic electrical properties of bulk $BaSnO_3$ crystals

Property	Unit	Value	Reference
Range of free electron concentration	cm^{-3}	5×10^{17} – 6.8×10^{20}	11, 12, 53, 55, 14, 56
Experimental maximum electron mobility at RT	cm^2/Vs	320	12
Theoretical maximum electron mobility at RT	cm^2/Vs	≈330[(a)]	36
Experimental effective electron mass m_e^*/m_0 at free electron concentration $\leq 1.6\times10^{19}$ cm^{-3}		0.19±0.02[(b)]	79
Experimental effective electron mass m_e^*/m_0 at free electron concentration of 7.6×10^{20} cm^{-3}		0.36[(b)]	79
Theoretical effective electron mass m_e^*/m_0 at Γ-point of the CBM		0.2–0.26[(a)]	36, 37, 40, 41, 50
Sheet electron density of the 2DEG with $LaInO_3$	cm^{-2}	4.5×10^{13}	85

[(a)]theoretical value; [(b)]value for thin films

7.8 Key Points

- $BaSnO_3$ is an indirect wide bandgap oxide semiconductor with the optical bandgap of about 3 eV.

- It is an ideal perovskite with the tolerance factor close to unity.
- It can be heavily doped with the free electron concentration above 10^{20} cm^{-3}.
- $BaSnO_3$ shows the highest electron mobility among other TSOs, about 300–320 cm^2/Vs at high free electron concentration at the level of 10^{20} cm^{-3}.
- Bulk single crystals can be grown directly from the melt with the size and quality sufficient for homoepitaxial purposes.
- It is an ideal candidate for generating 2DEG.

7.9 Summary

Table 7.8 Basic properties of bulk $BaSnO_3$ single crystals

	$BaSnO_3$	
Property	**Value or description**	**Reference**
	Crystal system and structure	
System	Cubic, perovskite	
Lattice constants [Å]	a=4.117±0.002	34
Space group	$Pm\bar{3}m$	
	Bulk crystal preparation	
Maximum crystal volume [cm^3]	Flux: ~0.03 Melt (VGF): ~1	54, 55, 14 53
	Thermal properties	
Solid density [g/cm^3]	7.23$^{(a)}$	34
Melting point [°C]	1855±25	53
Thermal conductivity at RT [W/mK]	9.7$^{(b)}$	6
Thermal expansion coefficient (0–320°C) [K^{-1}]	9.3×10^{-6} $^{(b)}$	6
	Mechanical properties	
Elastic constants [GPa]	c_{11} = 251$^{(c)}$/285$^{(c)}$/296$^{(c)}$ c_{12} = 82$^{(c)}$/69$^{(c)}$/89$^{(c)}$ c_{44} = 98$^{(c)}$/84$^{(c)}$/111$^{(c)}$	38/51/52

Bulk modulus [GPa]	146[b]	6
Young's modulus [GPa]	244[b]	
Shear modulus [GPa]	99.9[b]	
Poisson's ratio	0.22[b]	
Vickers hardness [GPa]	5[b]	6
Longitudinal sound velocity [m/s]	6150	11
Transverse sound velocity [m/s]	3333	
Optical properties		
Indirect bandgap [eV]	≈3	44, 53
Direct bandgap [eV]	≈3.5	43
Fundamental bandgap	Indirect	44, 53, 46
Absorption edge [nm]	≈400	44, 45, 53
Luminescence bands at RT [eV]	None	74
Static dielectric constant at RT	17±2	43
High-frequency dielectric constant at RT	3.3	
Refractive index at 1.5–3.5 eV	2.07–2.35	68
Electrical properties		
Tunability of the free electron concentration at RT [cm^{-3}]	5×10^{17}–6.8×10^{20}	11, 12, 53, 55, 14, 56
Maximum experimental electron mobility at RT [cm^2/Vs]	320	12
Maximum theoretical electron mobility at RT [cm^2/Vs]	≈330	36
Effective electron mass at high free electron concentration (1.6×10^{19} to 7.6×10^{20} cm^{-3})	0.19–0.36[d]	79
Sheet electron density of 2DEG in $BaSnO_3$–$LaInO_3$ [cm^{-2}]	4.5×10^{13}	85

[a]calculated value on the basis of the lattice constant in Ref. [34]; [b]value for ceramic; [c]theoretical value; [d]value for thin films

References

1. H. D. Megaw; *Proc. Phys. Soc.* **58** (1946), 133–152.
2. G. Wagner, H. Binder; *Z. Anorg. Allg. Chem.* **297** (1958), 328–346.

3. M. Trari, J.-P. Doumerc, P. Dordor, M. Pouchard, G. Behr, G. Krabbes; *J. Phys. Chem. Solids* **55** (1994), 1239–1243.

4. B. Ostrick, M. Fleischer, H. Meixner; *J. Am. Ceram. Soc.* **80** (1997), 2153–2156.

5. S. Upadhyay, A. K. Sahu, D. Kumar, O. Parkash; *J. Appl. Phys.* **84** (1998), 828–832.

6. T. Maekawa, K. Kurosaki, S. Yamanaka; *J. Alloy. Compd.* **416** (2006), 214–217.

7. B. Hadjarab, A. Bouguelia, M. Trari; *J. Phys. D: Appl. Phys.* **40** (2007), 5833–5839.

8. H. F. Wang, Q. Z. Liu, F. Chen, G. Y. Gao, W. Wu, X. H. Chen; *J. Appl. Phys.* **101** (2007), 106105.

9. Q. Liu, J. Dai, Z. Liu, X. Zhang, G. Zhu, G. Ding; *J. Phys. D: Appl. Phys.* **43** (2010), 455401.

10. R. Takahashi, K. Valset, E. Folven, E. Eberg, J. K. Grepstad, T. Tybell; *Appl. Phys. Lett.* **97** (2010), 081906.

11. X. Luo, Y. S. Oh, A. Sirenko, P. Gao, T. A. Tyson, K. Char, S.-W. Cheong; *Appl. Phys. Lett.* **100** (2012), 172112.

12. H. J. Kim, U. Kim, H. M. Kim, T. H. Kim, H. S. Mun, B.-G. Jeon, K. T. Hong, W.-J. Lee, C. Ju, K. H. Kim, K. Char; *Appl. Phys. Express* **5** (2012), 061102.

13. U. Kim, C. Park, Y. M. Kim, J. Shin, K. Char; *APL Mater.* **4** (2016), 071102.

14. W.-J. Lee, H. J. Kim, J. Kang, D. H. Jang, T. H. Kim, J. H. Lee, and K. H. Kim; *Annu. Rev. Mater. Res.* **47** (2017), 391–423.

15. Q. Liu, J. Liu, B. Li, H. Li, G. Zhu, K. Dai, Z. Liu, P. Zhang, J. Dai; *Appl. Phys. Lett.* **101** (2012), 241901.

16. K. Balamurugan, E. S. Kumar, B. Ramachandran, S. Venkatesh, N. H. Kumar, M. S. R. Rao, P. N. Santhosh; *J. Appl. Phys.* **111** (2012), 074107 .

17. H. Mun, U. Kim, H. M. Kim, C. Park, T. H. Kim, H. J. Kim, K. H. Kim, K. Char; *Appl. Phys. Lett.* **102** (2013), 252105.

18. P. V. Wadekar, J. Alaria, M. O'Sullivan, N. L. O. Flack, T. D. Manning, L. J. Phillips, K. Durose, O. Lozano, S. Lucas, J. B. Claridge, M. J. Rosseinsky; *Appl. Phys. Lett.* **105** (2014), 052104.

19. U. Kim, C. Park, T. Ha, R. Kim, H. S. Mun, H. M. Kim, H. J. Kim, T. H. Kim, N. Kim, J. Yu, K. H. Kim, J. H. Kim, K. Char; *APL Mater.* **2** (2014), 056107.

20. J. Shiogai, K. Nishihara, K. Sato, A. Tsukazaki; *AIP Adv.* **6** (2016), 065305.

21. C. A. Niedermeier, S. Rhode, S. Fearn, K. Ide, M. A. Moram, H. Hiramatsu, H. Hosono, T. Kamiya; *Appl. Phys. Lett.* **108** (2016), 172101.

22. H. M. I. Jaim, S. Lee, X. Zhang, I. Takeuchi; *Appl. Phys. Lett.* **111** (2017), 172102.
23. B. S. Joo, Y. J. Chang, L. Moreschini, A. Bostwick, E. Rotenberg, M. Han; *Curr. Appl. Phys.* **17** (2017), 595–599.
24. S. Raghavan, T. Schumann, H. Kim, J. Y. Zhang, T. A. Cain, S. Stemmer; *APL Mater.* **4** (2016), 016106.
25. H. Paik, Z. Chen, E. Lochocki, A. H. Seidner, A. Verma, N. Tanen, J. Park, M. Uchida, S. L. Shang, B.-C. Zhou, M. Brützam, R. Uecker, Z.-K. Liu, D. Jena, K. M. Shen, D. A. Muller, D. G. Schlom; *APL Mater.* **5** (2017), 116107.
26. A. Prakash, P. Xu, X. Wu, G. Haugstad, X. Wang, B. Jalan; *J. Mater. Chem. C* **5** (2017), 5730–5736.
27. A. Prakash, P. Xu, A. Faghaninia, S. Shukla, J. W. Ager III, C. S. Lo, B. Jalan; *Nat. Commun.* **8** (2017), 15167.
28. R. H. Wei, X. W. Tang, Z. Z. Hui, X. Luo, J. M. Dai, J. Yang, W. H. Song, L. Chen, X. G. Zhu, X. B. Zhu, Y. P. Sun; *Appl. Phys. Lett.* **106** (2015), 101906.
29. K. Ganguly, P. Ambwani, P. Xu, J. S. Jeong, K. A. Mkhoyan, C. Leighton, B. Jalan; *APL Mater.* **3** (2015), 062509.
30. K. Ganguly, A. Prakash, B. Jalan, C. Leighton; *APL Mater.* **5** (2017), 056102.
31. H. Yun, K. Ganguly, W. Postiglione, B. Jalan, C. Leighton, K. A. Mkhoyan, J. S. Jeong; *Sci. Rep.* **8** (2018), 10245.
32. S. M. Xing, C. Shan, K. Jiang, J. J. Zhu, Y. W. Li, Z. G. Hu, J. H. Chu; *J. Appl. Phys.* **117** (2015), 103107.
33. C. Shan, T. Huang, J. Zhang, M. Han, Y. Li, Z. Hu, J. Chu; *J. Phys. Chem. C* **118** (2014), 6994–7001.
34. A. J. Smith, A. J. E. Welch; *Acta Crystallogr.* **13** (1960), 653–656.
35. H. Mizoguchi, H. W. Eng, P. M. Woodward; *Inorg. Chem.* **43** (2004), 1667–1680.
36. K. Krishnaswamy, B. Himmetoglu, Y. Kang, A. Janotti, C. G. Van de Walle; *Phys. Rev. B* **95** (2017), 205202.
37. D. O. Scanlon; *Phys. Rev. B* **87** (2013), 161201.
38. E. Moreira, J. M. Henriques, D. L. Azevedo, E. W. S. Caetano, V. N. Freire, E. L. Albuquerque; *J. Solid State Chem.* **187** (2012), 186–194.
39. E. Moreira, J. M. Henriques, D. L. Azevedo, E. W. S. Caetano, V. N. Freire, U. L. Fulco, E. L. Albuquerque; *J. Appl. Phys.* **112** (2012), 043703.

40. S. Dabaghmanesh, R. Saniz, M. N. Amini, D. Lamoen B. Partoens; *J. Phys.: Condens. Matter.* **25** (2013), 415503.

41. H.-R. Liu, J.-H. Yang, H. J. Xiang, X. G. Gong, S.-H. Wei; *Appl. Phys. Lett.* **102** (2013), 112109.

42. D. J. Singh, D. A. Papaconstantopoulos, J. P. Julien, F. Cyrot-Lackmann; *Phys. Rev. B* **44** (1991), 9519–9523.

43. T. N. Stanislavchuk, A. A. Sirenko, A. P. Litvinchuk, X. Luo, S.-W. Cheong; *J. Appl. Phys.* **112** (2012), 044108.

44. H. J. Kim, U. Kim, T. H. Kim, J. Kim, H. M. Kim, B.-G. Jeon, W.-J. Lee, H. S. Mun, K. T. Hong, J. Yu, K. Char, K. H. Kim; *Phys. Rev. B* **86** (2012), 165205.

45. H. J. Kim, J. Kim, T. H. Kim, W.-J. Lee, B-G. Jeon, J.-Y. Park, W. S. Choi, D. W. Jeong, S. H. Lee, J. Yu, T. W. Noh, K. H. Kim; *Phys. Rev. B* **88** (2013), 125204.

46. Y. Kang, H. Peelaers, K. Krishnaswamy, C. G. Van de Walle; *Appl. Phys. Lett.* **112** (2018), 062106.

47. E. B. Lochocki, H. Paik, M. Uchida, D. G. Schlom, K. M. Shen; *Appl. Phys. Lett.* **112** (2018), 181603.

48. B. Monserrat, C. E. Dreyer, K. M. Rabe; *Phys. Rev. B* **97** (2018), 104310.

49. S. Soleimanpour, F. Kanjouri; *Physica B* **432** (2014), 16–20.

50. J. Li, Z. Ma, R. Sa, K. Wu; *RSC Adv.* **7** (2017), 32703–32709.

51. A. Bouhemadou, K. Haddadi; *Solid State Sci.* **12** (2010), 630–636.

52. T. M. Duarte, P. G. C. Buzolin, I. M. G. Santos, E. Longo, J. R. Sambrano; *Theor. Chem. Acc.* **135** (2016), 151.

53. Z. Galazka, R. Uecker, K. Irmscher, D. Klimm, R. Bertram, A. Kwasniewski, M. Naumann, R. Schewski, M. Pietsch, U. Juda, A. Fiedler, M. Albrecht, S. Ganschow, T. Markurt, C. Guguschev, M. Bickermann; *J. Phys.: Condens. Matter.* **29** (2017), 075701.

54. W.-J. Lee, H. J. Kim, E. Sohn, T. H. Kim, J.-Y. Park, W. Park, H. Jeong, T. Lee, J. H. Kim, K.-Y. Choi, K. H. Kim; *Appl. Phys. Lett.* **108** (2016), 082105.

55. W.-J. Lee, H. Lee, K.-T. Ko, J. Kang, H. J. Kim, T. Lee, J.-H. Park, K. H. Kim; *Appl. Phys. Lett.* **111** (2017), 231604.

56. E. McCalla, D. Phelan, M. J. Krogstad, B. Dabrowski, C. Leighton; *Phys. Rev. Meter.* **2** (2018), 084601.

57. C. Huang, X. Wang, X. Liu, M. Tian, T. Zhang; *J. Eur. Ceram. Soc.* **36** (2016), 583–592.

58. S. Upadhyay, O. Parkash, D. Kumar; *Mater. Lett.* **49** (2001), 251–255.

59. I. Arvanitidis, D. Sichen, S. Seetharaman; *Metall. Mater. Trans. B* **27B** (1996), 409.

60. Z. Galazka, I. M. Hanke; *Leibniz-Institut für Kristallzüchtung*, Berlin, Germany (2018), unpublished data.

61. A. Slassi, M. Hammi, O. E. Rhazouani; *J. Electron. Mater.* **46** (2017), 4133–4139.

62. J. C. Farfán, J. A. Rodríguez, F. Fajardo, E. V. López, D. A. L. Téllez, J. Roa-Rojas; *Physica B* **404** (2009), 2720–2722.

63. E. Bévillon, A. Chesnaud, Y. Wang, G. Dezanneau, G. Geneste; *J. Phys.: Condens. Matter.* **20** (2008), 145217.

64. D. Cherrad, M. Maouche, M. Maamache, L. Krache; *Physica B* **406** (2011), 2714–2722.

65. D. Phelan, F. Han, A. Lopez-Bezanilla, M. J. Krogstad, Y. Gim, Y. Rong, J. Zhang, D. Parshall, H. Zheng, S. L. Cooper, M. Feygenson, W. Yang, Y.-S. Chen; *J. Solid State Chem.* **262** (2018), 142–148.

66. S. A. Chambers, T. C. Kaspar, A. Prakash, G. Haugstadand, B. Jalan; *Appl. Phys. Lett.* **108** (2016), 152104.

67. T. Schumann, S. Raghavan, K. Ahadi, H. Kim, S. Stemmer; *J. Vac. Sci. Technol. A* **34** (2016), 050601.

68. S. Bin Anooz, Z. Galazka; *Leibniz-Institut für Kristallzüchtung*, Berlin, Germany (2019), unpublished data.

69. W. Nunn, A. Prakash, A. Bhowmik, R. Haislmaier, J. Yue, J. Maria, G. Lastra, B. Jalan; *APL Mater.* **6** (2018), 066107.

70. D. Seo, K. Yu, Y. J. Chang, E. Sohn, K. H. Kim, E. J. Choi; *Appl. Phys. Lett.* **104** (2014), 022102.

71. P. Singh, B. J. Brandenburg, C. P. Sebastian, P. Singh, S. Singh, D. Kumar, O. Parkash; *Jpn. J. Appl. Phys.* **47** (2008), 3540–3545.

72. E. Bévillon, G. Geneste; *Phys. Rev. B* **75** (2007), 214106.

73. H. Takashima, Y. Inaguma; *Appl. Phys. Lett.* **111** (2017), 091903.

74. R. Schewski, Z. Galazka; *Leibniz-Institut für Kristallzüchtung*, Berlin, Germany (2018), unpublished data.

75. Z. Lebens-Higgins, D. O. Scanlon, H. Paik, S. Sallis, Y. Nie, M. Uchida, N. F. Quackenbush, M. J. Wahila, G. E. Sterbinsky, D. A. Arena, J. C. Woicik, D. G. Schlom, L. F. J. Piper; *Phys. Rev. Lett.* **116** (2016), 027602.

76. H. Mun, U. Kim, H. M. Kim, C. Park, T. H. Kim, H. J. Kim, K. H. Kim, K. Char; *Appl. Phys. Lett.* **102** (2013), 252105.

77. C. Shan, T. Huang, J. Zhang, M. Han, Y. Li, Z. Hu, J. Chu; *J. Phys. Chem. C* **118** (2014), 6994–7001.

78. C. A. Niedermeier, S. Rhode, S. Fearn, K. Ide, M. A. Moram, H. Hiramatsu, H. Hosono, T. Kamiya; *Appl. Phys. Lett.* **108** (2016), 172101.

79. C. A. Niedermeier, S. Rhode, K. Ide, H. Hiramatsu, H. Hosono, T. Kamiya, M. A. Moram; *Phys. Rev. B* **95** (2017), 161202.

80. S. J. Allen, S. Raghavan, T. Schumann, K.-M. Law, S. Stemmer; *Appl. Phys. Lett.* **108** (2016), 252107.

81. R. Schewski, M. Baldini, K. Irmscher, A. Fiedler, T. Markurt, B. Neuschulz, T. Remmele, T. Schulz, G. Wagner, Z. Galazka, M. Albrecht; *J. Appl. Phys.* **120** (2016), 225308.

82. A. Fiedler, R. Schewski, M. Baldini, Z. Galazka, G. Wagner, M. Albrecht, K. Irmscher; *J. Appl. Phys.* **122** (2017), 165701.

83. P. Moetakef, T. A. Cain, D. G. Ouellette, J. Y. Zhang, D. O. Klenov, A. Janotti, C. G. Van de Walle, S. Rajan, S. J. Allen, S. Stemmer; *Appl. Phys. Lett.* **99** (2011), 232116.

84. L. Bjaalie, B. Himmetoglu, L. Weston, A. Janotti, C. G. Van de Walle; *New J. Phys.* **16** (2014), 025005.

85. U. Kim, C. Park, T. Ha, Y. M. Kim, N. Kim, C. Ju, J. Park, J. Yu, J. H. Kim, K. Char; *APL Mater.* **3** (2015), 036101.

86. T. R. Paudel, E. Y. Tsymbal; *Phys. Rev. B* **96** (2017), 245423.

87. S. A. Chambers, T. C. Kaspar, A. Prakash, G. Haugstad, B. Jalan; *Appl. Phys. Lett.* **108** (2016), 152104.

88. X. Fan, W. Zheng, X. Chen, D. J. Singh; *PLoS ONE* **9** (2014), e91423.

Chapter 8

Ga-Based Spinels: $ZnGa_2O_4$, $MgGa_2O_4$, $(Zn,Mg)Ga_2O_4$, $CoGa_2O_4$

8.1 Introduction

Among different Ga-based spinel compounds, only $ZnGa_2O_4$ in the form of thin films [1] or ceramics [2] was shown as a transparent semiconducting oxide. Recently, Galazka et al. [3, 4] demonstrated semiconducting capabilities of $MgGa_2O_4$ and mixed $(Zn,Mg)Ga_2O_4$ single crystals. The advantage of Ga-based spinels is a cubic structure enabling more isotropic properties, easy wafer fabrication, and wide use as substrates thereof for homoepitaxy and heteroepitaxy. Furthermore, they have ultra-wide bandgaps (above 4.5 eV), which are suitable for UV electronics and optoelectronics. Importantly, as demonstrated by Galazka et al. [3–5], the growth of truly bulk single crystals of Ga-based spinels directly from the melt enables a suitable size of the crystals and wafers for a diversity of applications.

In fact, a number of devices have been demonstrated with the use of $ZnGa_2O_4$ ceramics, films, and nanostructures, including MOSFETs [6], photodetectors [7], electroluminescent devices (phosphors) [8–17], photocatalytic devices [18, 19], and gas sensors [20]. A demonstration of $MgGa_2O_4$-based devices includes phosphors [21] and laser host material [22]. On the other hand, $CoGa_2O_4$ revealed magnetic properties [23–25].

Transparent Semiconducting Oxides: Bulk Crystal Growth and Fundamental Properties
Zbigniew Galazka

ISBN 978-981-4800-94-5 (Hardcover), 978-1-003-04520-5 (eBook)
www.jennystanford.com

Among the discussed Ga-based spinel single crystals, $CoGa_2O_4$ crystals do not reveal semiconducting behavior; however, they show surface *p*-type conductivity after hydrogen annealing at elevated temperatures.

Besides semiconducting properties, Ga-based spinels were found as excellent substrates for magnetic films of $NiFe_2O_4$ and are suitable for other Fe-based magnetic films. Small lattice mismatch between $NiFe_2O_4$ thin films and $MgGa_2O_4$ and $CoGa_2O_4$ spinel substrates enabled to eliminate anti-phase boundaries in $NiFe_2O_4$ and explore its magnetic properties [26, 27]. $ZnGa_2O_4$ shows almost ideal lattice match with $NiFe_2O_4$ [5]. Moreover, $Zn_{1-x}Mg_xGa_2O_4$ offers tuning the lattice parameter between 8.281 and 8.334 Å. This seems to be another, very important field of application for bulk single crystals of Ga-based spinels.

The availability of bulk single crystals and exploration of their optical and electrical properties may lead not only to a better understanding of the underlying physics, but also to an expansion of the area of applications through utilizing specific properties characteristic for each of the materials.

8.2 Crystal Structure

$MeGa_2O_4$ (Me = Zn, Mg, Zn+Mg, Co) compounds are crystallizing in the spinel structure, which can be written as AB_2X_4, where X stands for the anion, which is oxygen, but other spinel-type compounds with X = F, S, Se can be found too. The anions form a cubic dense packing. The cations A and B are distributed over half of the octahedral and one-eighth of the tetrahedral sites, respectively (Fig. 8.1), resulting in space group $Fd\bar{3}m$ for the structure. A and B can be identical, and some mixed valence oxides such as $Fe(II,III)_3O_4$ and $Co(II,III)_3O_4$ are crystallizing as spinels [28, 29].

At very high temperature, the distribution of cations over the tetrahedral and octahedral sites is almost random, and this high degree of disorder results in a significant configuration entropy of 15.8 J/(mol·K) [30, 31] compared to the fully ordered state. This is one reason why many spinels are very stable compounds with high melting points. At lower temperatures, ordering of different A, B cations over the octahedral and tetrahedral sites is thermodynamically preferred.

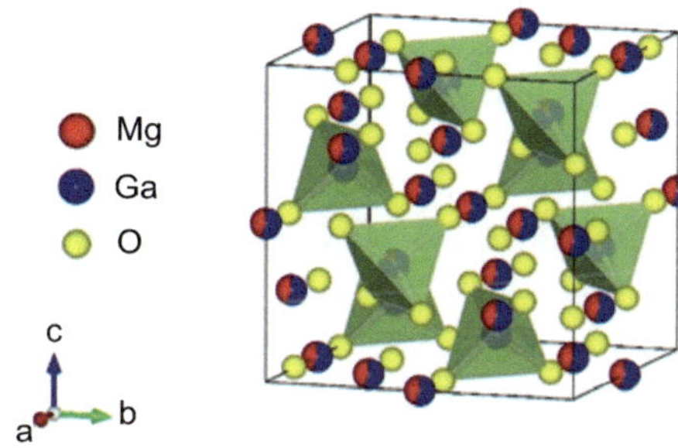

Figure 8.1 Spinel structure (mainly inverse) of $MgGa_2O_4$. Tetrahedral sites are occupied by 84% Ga (rest Mg), while octahedral sites by 55% Ga (rest Mg). Data from Ref. [31].

$MgAl_2O_4$ is the archetype of the spinel structure and found as a mineral in nature. There Mg^{2+} sits almost exclusively on tetrahedral sites, and Al^{3+} on tetrahedral sites. This type of site occupancy is the "normal spinel" structure. Alternatively, antisite cation disorder may occur where B occupies the available tetrahedral sites exclusively, and the octahedral sites are occupied by identical shares of A and B. This occupancy is called "inverse spinel." The normal and inverse spinel distributions represent extrema. The degree of inversion x characterizes the cation distribution in a more general formula $A_{1-x}B_x(A_xB_{2-x})O_4$. The parameter x can assume values between x = 0 (normal spinel) and x = 1 (inverse spinel). Inversion parameters x close to these extremes are called "mainly normal" or "mainly inverse," respectively. The random arrangement without site preference corresponds to $x = 2/3$.

An overview on the structural features of spinels is given by O'Neill and Navrotski [30], and according to their paper, $ZnGa_2O_4$ is a mainly normal spinel. The lattice parameter of ZGa_2O_4 measured by Phani et al. [32] for ceramics and Galazka et al. [5] for bulk single crystals are 8.3313 and 8.3336 Å, respectively. $MgGa_2O_4$, in contrast, is mainly inverse and has the lattice parameter of 8.281 Å as reported on bulk single crystals by Schwarz et al. [31]. It should be noted, however, that the lattice constant depends not only on the chemical composition but also on cation distribution over the tetrahedral and octahedral sites, respectively, and hence on the inversion degree. $ZnGa_2O_4$ forms a complete series of solid solution with spinel-type Zn_2SnO_4 [33]. And finally, the lattice parameter of $CoGa_2O_4$ is 8.3229 Å according to the work of Brik et al. [34].

8.3 Band Structure and Native Point Defects

Similarly to other TSOs, Ga-based spinels were also a subject of theoretical and experimental study for electronic structure and their properties, although not to such large extent. The outcome of the studies enables to design and fabricate devices with expected properties and functionality. Ga-based spinels with their crystal structure and physical properties may boost fundamental studies of this class of materials and concepts for new devices. Electronic structure is, therefore, a fundamental feature for next-step activities in material preparation, property exploration, and device fabrication.

8.3.1 Band Structure

The band structure of $ZnGa_2O_4$ was theoretically studied, e.g., by Sampath et al. [35], Pisani et al. [36], López et al. [37], Brik [38], Dixit et al. [39], Zerarga et al. [40], and Xia et al. [41]. An example of the calculated band structure of $ZnGa_2O_4$ is shown in Fig. 8.2. The valence band maximum (VBM) is primarily composed of O atoms (O *2p* orbitals), while the conduction band minimum (CBM) originates from Ga and O atoms, mainly Ga *4s* and O *2p* orbitals that form a dispersed *sp*-class band. The CBM is located in the Γ-point of the Brillouin zone (BZ), and the VBM is located in the K–Γ region, which indicates an indirect nature of this semiconductor in line with other theoretical calculations [37, 38, 40]. The calculated indirect and direct bandgaps are 4.36 eV and about 4.5 eV, respectively [41]. Other theoretical values of the indirect (K–Γ)/direct (Γ–Γ) bandgaps are: -/4.57–4.71 eV [39], 2.78/4.24 eV [37], and 2.69/2.74 eV [40]. However, due to a small energetic difference between the upper most valence bands of about 120 meV, $ZnGa_2O_4$ can be considered a pseudo-direct gap material.

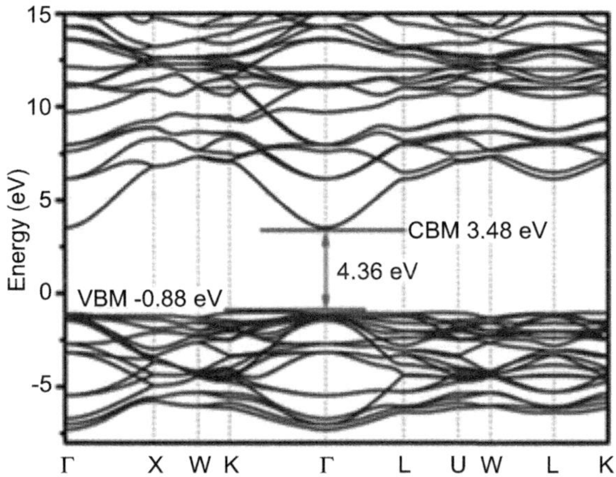

Figure 8.2 Calculated band structure of $ZnGa_2O_4$ (first principles using the GGA+U functional). The indirect gap $K \rightarrow \Gamma$ is 4.39 eV, and the direct gap at $\Gamma \rightarrow \Gamma$ is about 4.5 eV. Reprinted with permission from Ref. [41], Copyright 2018, American Chemical Society.

The band structure of $MgGa_2O_4$ was described only recently by Varley [42] and Thielert et al. [43], as shown in Fig. 8.3. Both theoretical (DFT, HSE06) and experimental (angle-resolver photoemission spectroscopy, ARPES) band structures reveal a very flat VBM with very little dispersion of about 0.6 eV [43]. The CBM is located at the Γ-point of the BZ, which in a combination of a local "maximum" of the VBM at Γ-point (Fig. 8.3b) points to a direct nature of the bandgap. Indeed, calculated values of the bandgap are about 5.013 and 5.014 eV [42] for the indirect and direct bandgaps, respectively, in a very good agreement with experimental bandgap of about 4.9 eV [3]. Similar to $ZnGa_2O_4$, the VBM of $MgGa_2O_4$ is composed by O *2p* states, while the CBM by Ga *4s* and O *2p* states.

The band structure of mixed $(Zn,Mg)Ga_2O_4$ has not been reported in the literature, although it is expected to be similar to the individual compounds of $ZnGa_2O_4$ and $MgGa_2O_4$, i.e., with a very flat VBM and CBM at the Γ-point of the BZ. The expected bandgap likely of a direct or pseudo-direct nature is between 4.6 and 4.9 eV.

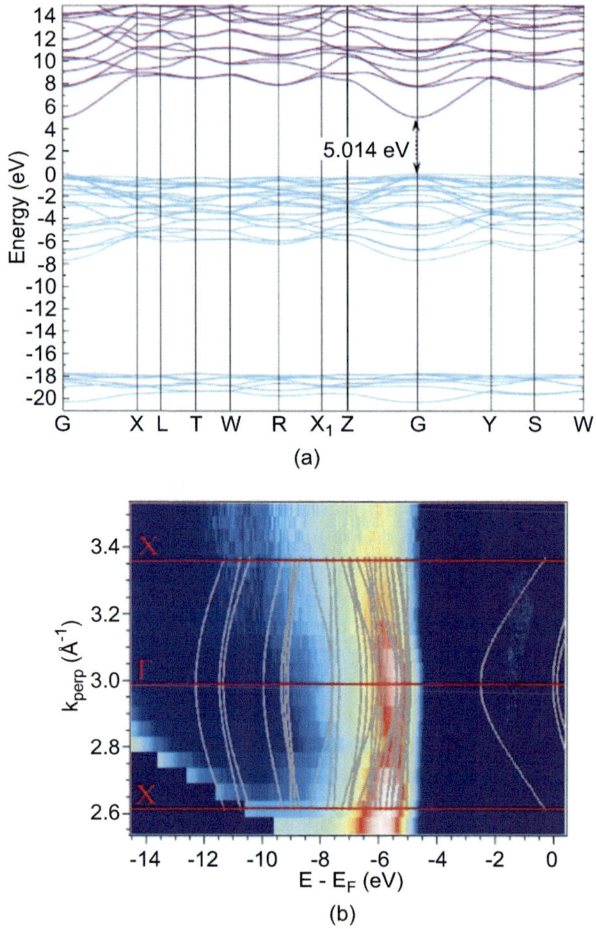

Figure 8.3 Calculated (HSE06) [42] (a) and experimental by ARPES (b) band structure of $MgGa_2O_4$. (b) shows k⊥ dispersion represented by energy distribution curves stacked as a color map and overlapped to the ab initio band structure calculations, represented by the gray dotted lines. The horizontal red lines indicate the Γ and *X* high-symmetry points of the Brillouin zone. Figure (b) reprinted with permission from Ref. [43], Copyright 2018, American Physical Society.

The band structure of $CoGa_2O_4$, calculated by Varley [44], is shown in Fig. 8.4. Here, unlike in the above-mentioned Ga-based

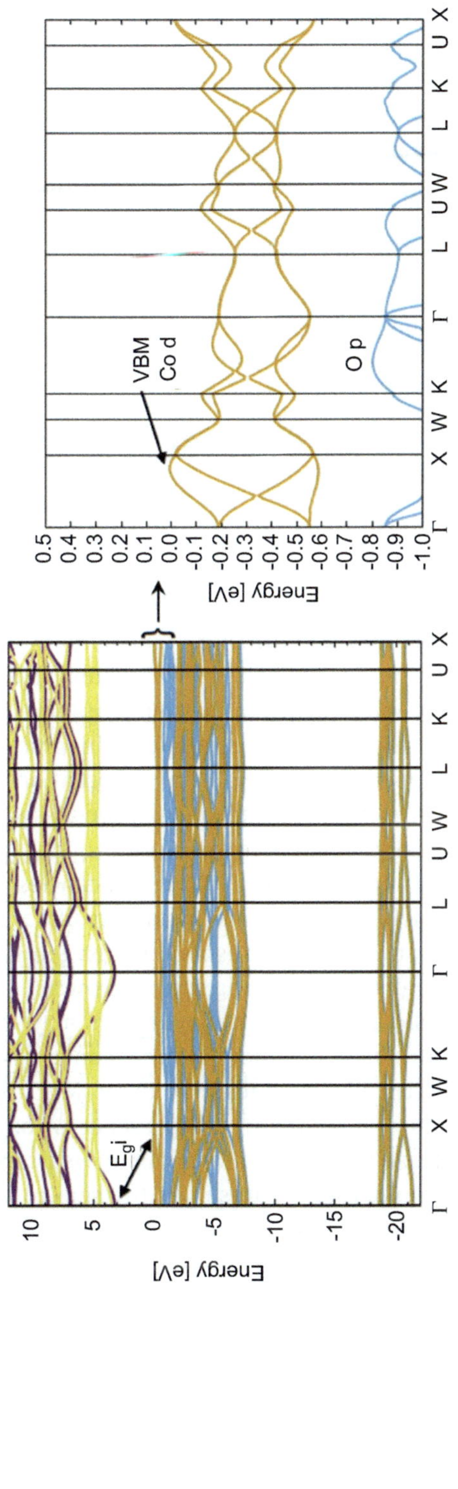

Figure 8.4 Calculated (HSE06) band structure of $CoGa_2O_4$ [44].

spinels, the VBM is predominantly composed by Co *d* and some O *p* states, while the CBM is dominated by Ga *d* states. The bandgap is spin dependent. The direct bandgaps (Γ – Γ) are 4.74 and 4.42 eV for spin up and down, respectively, and the indirect bandgaps (*X* – Γ) are 4.68 and 4.42 eV for spin up and down, respectively [44]. The energetic differences between the direct and indirect bandgaps are about 60 and 180 meV for spin up and down, respectively. Due to energetic difference between top valence bands at *X*- and Γ-points, although small, $CoGa_2O_4$ should rather be considered as the indirect bandgap material with the indirect bandgap of about 3.97 eV over both spins [44]. Spin-dependent bandgap structure may offer opportunities for $CoGa_2O_4$ in spintronics and optoelectronics.

8.3.2 Intrinsic and Extrinsic Point Defects versus Electrical Conductivity

The native point defects that may induce electrical conductivity in Ga-based spinels have not been theoretically investigated in detail. The possible native point defects in Ga-based spinels include atomic vacancies, interstitials, and antisites. Calculations by Xia et al. [41] showed that cation Ga-Zn exchange (antisites) in $ZnGa_2O_4$ is energetically unfavorable. However, oxygen vacancies at O^{2+} state (V_O^{2+}) are stable and act as deep donors. In practice, the cation exchange may appear during material preparation at high temperatures (e.g., when grown from the melt), when all spinels are rather random. Additionally, $MgGa_2O_4$ is the inverse spinel, in which the cations occupy intermixed position. Oxygen vacancies may be present in Ga-based spinels due to thermal instability of Ga_2O_3, ZnO, and CoO at high temperatures and corresponding oxygen loss if its partial pressure is not high enough during crystal growth.

In terms of extrinsic point defects, hydrogen may play an important role, as in the case of other TSOs. A conversion from semiconducting to electrically insulating state of $MgGa_2O_4$ and $ZnGa_2O_4$ takes place after annealing in the presence of oxygen at or above 700°C for at least 10 h, while annealing in the presence of hydrogen at the same temperatures preserves the semiconducting state [3, 5]. This may indirectly point to removal/incorporation of hydrogen from/to the crystals. Among other extrinsic point defects, nitrogen-substituting oxygen at 1-state (N_O^-) was found to be thermodynamically stable in $ZnGa_2O_4$ at all conditions.

8.4 Bulk Crystal Growth

8.4.1 Phase Diagrams

The pseudobinary MgO–Ga_2O_3, ZnO–Ga_2O_3, and CoO–Ga_2O_3 systems show some similarity with the prototype MgO–Al_2O_3 system, which is shown in Fig. 8.5. The spinel phase is the only intermediate phase that shows usually a more or less pronounced phase width. Unfortunately, all other spinel systems are not thoroughly investigated like this prototype, which is found as a mineral in nature and is used, e.g., as a precious stone: Actually, the "Black Prince's Ruby" of the Imperial Crown of England is a red spinel. The basic features of the Ga-based spinel phases will be given below in reference to Fig. 8.5.

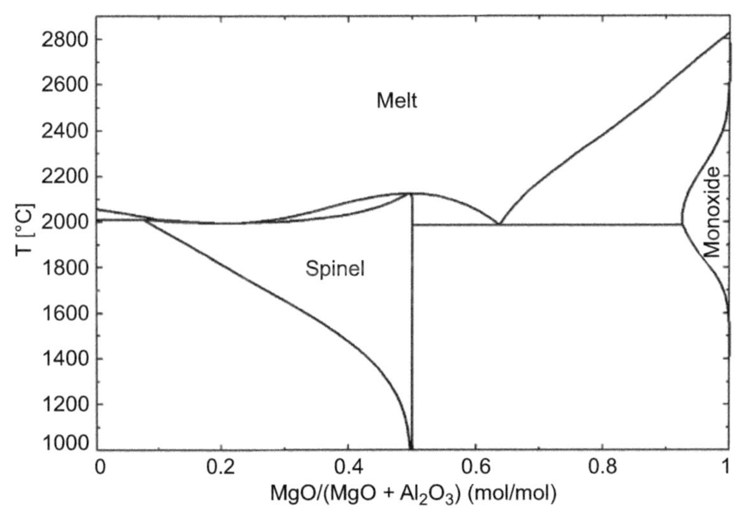

Figure 8.5 Calculated phase diagram of MgO–Al_2O_3 with the intermediate spinel phase $MgAl_2O_4$, which is the prototype of $MgGa_2O_4$, $ZnGa_2O_4$, and $CoGa_2O_4$.

Experimental data for the MgO–Ga_2O_3 system were compiled by Zinkevich et al. [45] and were used for a thermodynamic assessment of the whole system. The β-Ga_2O_3 phase has no significant solubility for Mg, while the solubility of the monoxide (MgO) phase for Ga_2O_3 is in a maximum of about 20%. Again, the $MgGa_2O_4$ spinel phase extends almost exclusively toward Ga_2O_3 with a maximum width from 30% to 50% MgO. In contrast to Fig. 8.5, the maximum of the spinel liquidus (eutectic melting point) coincides almost with the eutectic on the right-hand side. This means that with a slight excess of MgO, this phase will crystallize from the melt first. Thermodynamic data for the solid phases of this system were also measured by Jacob and Sivakumar [46]. The congruent melting point of $MgGa_2O_4$ was reported by Galazka et al. [3] during bulk crystal growth from the stoichiometric melt of 1930 ± 10°C.

For the ZnO–Ga_2O_3 system, no reliable data of the liquidus are available so far. This might result from the low thermal stability of both components at high temperature, particularly of pure ZnO that evaporates under ambient pressure at temperatures far below its melting point (1975°C). The congruent melting point of $ZnGa_2O_4$, 1900 ± 20°C, was reported recently by Galazka et al. [5] during the growth of bulk crystals from the stoichiometric melt. Neither in this work, nor elsewhere (Gribchenkova et al. [47]), there are hints of a significant phase width of the spinel phase. However, a series of hexagonal layered compounds $Zn_mGa_2O_{m+3}$ exist toward the ZnO side (Kimizuka et al. [48]).

A detailed study of the CoO–Ga_2O_3 system has not been published yet, but Kramer et al. [52] reported a congruent melting point of 1825 ± 5°C by pyrometric measurement during growth from the melt. A similar value (1810 ± 20°C) was also measured by Galazka [49]. Solid-state phase equilibria were investigated by Laqua et al. [50]. In analogy with Fig. 8.5, the solubility of CoO in Ga_2O_3 was measured. However, CoO (rocksalt structure, such as MgO) as well as $CoGa_2O_4$ spinel exhibits significant homogeneity width. It is remarkable that the spinel phase field obviously extends from the stoichiometric composition not only toward an excess of the 3-valent ion (like in Fig. 8.5), but also toward an excess of Co^{2+}. Solidus boundaries for the Ga_2O_3–CoO system are listed in Table 8.1.

Table 8.1 Experimental solidus boundaries of phase fields for the Ga_2O_3–CoO system in the scaling of Fig. 8.5 (calculated from data in Ref. [50])

Temperature [°C]	Spinel (Ga_2O_3-rich)	Spinel (CoO-rich)	CoO (Ga_2O_3-rich)
1250	0.445	0.539	0.970
1380	0.421	0.549	0.945
1464	0.399	0.555	0.934
1567	0.357	0.566	0.923

8.4.2 Historical Events

The first bulk single crystals of $ZnGa_2O_4$, obtained from flux, were reported in 1967 by Chase and Osmer [51]. The growth from the flux was continued by Boom et al. [52], van der Straten et al. [53], and Yan et al. [54, 55]. In 1995, Jia et al. [56] mentioned the growth of fiber $ZnGa_2O_4$ crystals from the melt with the use of laser heated pedestal growth (LHPD), while in 2018, Galazka et al. [5] reported for the first time truly bulk $ZnGa_2O_4$ single crystals obtained by the vertical gradient freeze (VGF) and Czochralski methods.

Bulk $MgGa_2O_4$ single crystals were first grown from flux in 1962 by Giess [57], by the Czochralski method in 1972 by Scott et al. [58], and by the optical floating zone (OFZ) method in 2010 by Suzuki et al. [59]. In 2014, Galazka et al. [3] discovered the semiconducting properties of $MgGa_2O_4$ and obtained bulk single crystals from the melt with the use of the Czochralski, Bridgman, and Kyropoulos-like methods. The LASSCGM, developed for In_2O_3 (see Chapter 5), also worked to some extent.

The mixed $(Zn,Mg)Ga_2O_4$ crystals were grown for the same time by the VGF method in 2018 by Galazka and Hanke [4].

The first reports on bulk $CoGa_2O_4$ single crystals obtained directly from the melt by the OFZ and Czochralski methods were published in 1985 by Kramer et al. [60]. The Czochralski- and Kyropoulos-like methods for the growth of bulk $CoGa_2O_4$ single crystals were used by Galazka [49] in 2014. In 1986, bulk $CoGa_2O_4$ crystals were obtained by Leccabue et al. [61] from the gas phase (CVT). In 2006, Nakatsuka et al. [62] reported the growth of $CoGa_2O_4$ single crystals by the flux method.

The summary of the main events in the bulk growth of Ga-based spinels is provided in Fig. 8.6.

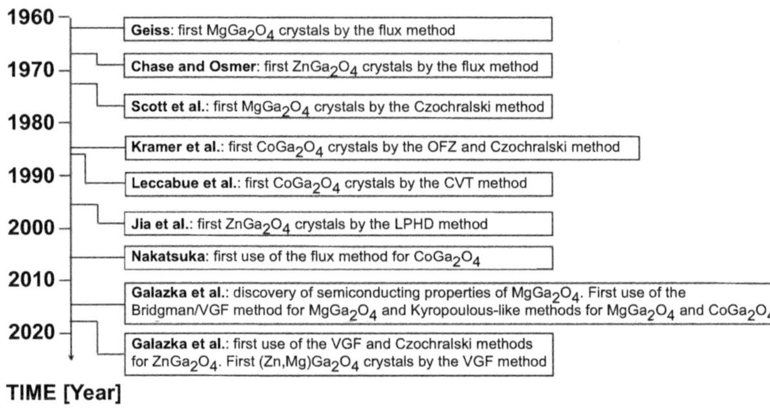

Figure 8.6 Main events in obtaining bulk $ZnGa_2O_4$, $MgGa_2O_4$, $(Zn,Mg)Ga_2O_4$, and $CoGa_2O_4$ single crystals.

8.4.3 Growth from the Gas Phase

Among different Ga-based spinels discussed in this chapter, only $CoGa_2O_4$ was grown from the gas phase with the use of the CVT method by Leccabue et al. [61]. The growth of crystals proceeded in the $CoGa_2O_4$ (source material)–Cl_2 (chemically active transport agent) system. The source material $CoGa_2O_4$ was prepared by solid-state reaction (sintering) of binary oxides CoO and Ga_2O_3 at 1200°C in air for 2 days. Prior use, the binary oxides were dried at 800°C in air for a day, and then ground and mixed. The growth environment was defined by a previously evacuated (10^{-6} Torr) sealed quartz ampoule containing the sintered powders and chlorine under a pressure of about 26.7 kPa at RT. The length and inner diameter of the quartz ampoule were 18 cm and 1.5 cm, respectively, with a distance between the source material and the growth zone of 9–10 cm. The ampoule was placed within a furnace with eight heating zones enabling different temperature profiles. An average temperature of the source material was 925°C, while the temperature difference between the source material and growth zone was 50–70°C.

Upon chemical reactions between $CoGa_2O_4$ and Cl_2, a number of chemical species are formed. Those having the highest partial pressure (ordered from the highest to the lowest) are $GaCl_3$, O_2, $CoCl_2$, $CoCl_3$, ClO_2, Cl_2O, Co_2Cl_4, and GaCl. High and similar partial pressures of Ga- and Co-containing species ($GaCl_3$ and $CoCl_2$) as well as O_2 enable a uniform transport of the species forming $CoGa_2O_4$ and substantially stoichiometric composition of the obtained crystals.

After the growth for 2 weeks, the obtained crystals were black and very small, about 1 mm in size. The inversion occupancy of the Co^{2+} ions at tetrahedral sites was determined at the level of 40%.

8.4.4 Growth from the Flux

8.4.4.1 Crystal growth

The flux method was used to grow bulk single crystals of $ZnGa_2O_4$ (Chase and Osmer [51], Boom et al. [52], van der Straten et al. [53], Yan et al. [54, 55]), $MgGa_2O_4$ (Giess [57]), and $CoGa_2O_4$ (Nakatsuka et al. [62]). Different fluxes were utilized, including Bi_2O_3 (or PbO) + PbF_2 + B_2O_3 [51], PbO + PbF_2 + B_2O_3, PbO + PbF_2 + B_2O_3 + SiO_2, Na_2O, and $Pb_2P_2O_7$ [53], and PbO + B_2O_3 [54, 55] for $ZnGa_2O_4$, PbO + PbF_2 [57] for $MgGa_2O_4$, and PbF_2 [62] for $CoGa_2O_4$. In all cases, a Pt crucible was used to accommodate the starting material consisting of a mixture of flux powders and source materials consisting of Ga_2O_3 and ZnO, MgO or CoO powders. The molar ratio of the source materials was either stoichiometric [51, 53–55, 62] or deviated from stoichiometry [53, 57]. The Pt crucible with the starting material was placed into a resistance furnace (e.g., a muffle furnace) and heated up to a maximum temperature T_{MAX} = 1200–1400°C [51–55, 57, 62] at which the flux melts. The molten starting material was soaked for several hours to 2 weeks [51, 52, 54, 55] and then slowly cooled down to a final growth temperature T_F = 700–1000°C depending on the flux composition at the rate of 0.5–5 K/h [51–55, 57, 62]. During that slow cooling down, the compound in quest nucleates at crystallization centers and slowly grows in the form of small bulk single crystals. Once the final growth temperature was reached, the molten starting material was cooled down to RT at a higher rate or quenched. Next, the crystals were released from the flux with a hot dilute acid solution, such as nitric acid [51, 52] or nitric–acetic acid [57] solution.

The obtained crystals had different shapes, such as octahedral, plates, or needles. The crystal size along an edge may reach 10 mm [53–55], with the largest dimensions of $10 \times 8 \times 7$ mm^3 [54, 55]. Some of the crystals, depending on growth environment, had inclusions and/or flux impurities [51, 57].

8.4.4.2 Examples

(i) For growing $ZnGa_2O_4$, Chase and Osmer [51] used a Bi_2O_3 (or PbO) + PbF_2 + B_2O_3 flux. The crystals were grown from the starting material consisting of a stoichiometric ratio of the source material ZnO and Ga_2O_3 (5 mol.% each) powders, and the flux containing 20 mol.% Bi_2O_3 (or PbO) + 68 mol.% PbF_2 + 2 mol.% B_2O_3 powders. The mixed powders of the starting material were loaded in a 100 ml Pt crucible, which was heated up in a resistance furnace that enabled vertical temperature gradients on the crucible wall of about 3 K/cm. For a furnace temperature of 1250°C, the temperature at the crucible bottom was 1100°C. Upon melting of the starting material, it was soaked for 1–2 weeks and then slowly cooled down at the rate of 4 K/h to 1000°C. After cooling down to RT, the crystals that were located mainly on the top of the flux were removed with hot 20% nitric acid. The obtained crystals contained some inclusions and had a size of 5–7 mm.

(ii) A similar flux for growing $ZnGa_2O_4$ crystals, i.e., PbO + PbF_2, was used by Boom et al. [52]. The starting material consisting of the source material and the flux of composition the same as in example (i) was loaded into a 50 ml Pt – 5% Rh crucible covered with a Pt lid, which was placed in a resistance furnace. The source material also contained Cr_2O_3 as a dopant. The crucible was heated up to T_{MAX} = 1250°C and kept at that temperature for 5 days. The following cooling rates were applied: 1250°C $\rightarrow$ 1100°C at 2 K/h, 1100°C $\rightarrow$ 900 at 6 K/h, and 900°C $\rightarrow$ RT at 20 K/h. The crystals were separated from the flux with hot 20% nitric acid. The obtained crystals of 0.06 cm^3 by volume were yellow to red-brown due to different Cr concentrations ranging between 0.05 and 0.5 wt.%. Also, other magnetic impurities were detected in the crystals, i.e., Fe, Mn, Ni, Pt, Rh, and Cu.

(iii) Van der Straten et al. [53] used several fluxes for growing bulk $ZnGa_2O_4$ crystals, including (a) $PbO + PbF_2 + B_2O_3$, (b) $PbO + PbF_2 + B_2O_3 + SiO_2$, (c) Na_2O, and (d) $Pb_2P_2O_7$. For the flux (a), the same composition as that of Chase and Osmer [51] was used (in mol.%: 5 ZnO + 5 Ga_2O_3 and 20 PbO + 68 PbF_2 + 2 B_2O_3). The composition of the starting material utilizing the flux (b) was as follows (in mol.%): 7.57 ZnO + 10.09 Ga_2O_3 and 29.34 PbO + 32.65 PbF_2 + 5.22 B_2O_3 + 15.13 SiO_2. With the use of the flux (c), the starting material had two compositions consisted of (in mol.%): 27.8/30 ZnO + 50/50 Ga_2O_3 and 22.2/20 Na_2O. Different molar ratios of ZnO/ Ga_2O_3 were used for the flux (d), with the starting material consisting of (in mol.%): 30/40/48/53.5 ZnO + 30/20/12/6.7 Ga_2O_3 and 40 $Pb_2P_2O_7$. The crystals were grown in 60 or 13 ml Pt crucibles that were heated up in a resistance furnace to a maximum temperature T_{MAX} with vertical temperature gradients on the crucible wall of 5 K/cm. After soaking, the molten starting material was slowly cooled down to the final growth temperature T_F at the rate of 0.5 K/h. T_{MAX}/T_F were as follows (in °C): 1250/1000, 1400/900, and 1300/900 for the fluxes (a, b), (c), and (d), respectively.

The crystal size along an edge was up to 5 mm, up to 10 mm (octahedral shape), less than 0.5 mm, and up to 4 mm (octahedral shape, plates, and needles) for the fluxes (a), (b), (c), and (d), respectively. The largest $ZnGa_2O_4$ crystals were obtained from the flux (b), indicating a positive impact of SiO_2 addition to the $PbO–PbF_2–B_2O_3$ system. The flux (c) appeared to be inefficient. In the case of the flux (d), which is nonvolatile, the best results were obtained for a molar ratio of ZnO/Ga_2O_3 = 4. A molar ratio of ZnO/Ga_2O_3 = 1 produced a mixture of $ZnGa_2O_4$ and Ga_2O_3.

(iv) Yan et al. [54, 55] used the $PbO + B_2O_3$ flux for growing $ZnGa_2O_4$ crystals. The starting material contained powders of binary oxides of a molar ratio of $ZnO:Ga_2O_3:PbO:B_2O_3$ = 1:1:1:2. The crystals were removed from the flux with a dilute nitric acid solution. The use of PbF_2 as the main flux constituent ($PbO:PbF_2:B_2O_3:ZnO:Ga_2O_3$ = 1:3:0.02:0.1:0.1) led to a massive mass loss (30 wt.%) due to high volatility of PbF_2. Moreover, PbF_2 has a high reactivity with ZnO and

Ga_2O_3 leading to a deviation from stoichiometry of resulted crystals. B_2O_3-rich flux composition provides a high solubility for $ZnGa_2O_4$. Mixed powders were loaded into a 50 ml Pt crucible, which was heated up in a SiC resistance furnace to 1250°C in air, soaked for 8 h, slowly cooled down to 850°C at the rate of 1 K/h, and then quenched to RT. The radial temperature difference between the center and furnace wall was 10 K. The crystals grew on the top surface of the flux, mainly at the central part, due to a lower $ZnGa_2O_4$ density as compared with the flux density, and radial temperature gradients. The obtained crystals were optically clear with average dimensions of $6 \times 5 \times 5$ mm^3, and the largest one of $10 \times 8 \times 7$ mm^3. The growth promoted the octahedral habit with well-developed (111) facets. The crystals had substantially stoichiometric composition. A large difference in ionic radii between Pb^{2+} (1.18) and Zn^{2+} (0.6) makes the incorporation of Pb into the crystal lattice rather unlikely.

(v) Bulk $MgGa_2O_4$ crystals were obtained by Giess [57] with the use of the PbO + PbF_2 flux. The starting material consisted of (in mol.%) 12 MgO, 8 Ga_2O_3, 40 PbO, and 40 PbF_2 ground powders, which were mixed and placed into a Pt crucible covered with a Pt lid to decrease the evaporation. The Pt crucible with the starting material was placed in a zirconia crucible, which was heated up in the resistance furnace to 1250°C in air. After melting the starting material, the furnace was cooled down to 850°C at the rate of about 0.5 K/h. Then the crucible was quenched. $MgGa_2O_4$ crystals were grown at the flux surface mainly due to a lower density as compared with the main solidified product Pb_2OF_2. The crystals were released from the solidified flux with dilute nitric–acetic acid solution.

The excess of MgO enabled to obtain relatively stoichiometric $MgGa_2O_4$ crystals. On the other hand, excess of Ga_2O_3 would lead to a larger composition change in the crystals, which had a tendency to twinning. The crystals obtained from the MgO-rich starting composition were octahedral with {111} faces, sometimes with dodecahedral {110} faces. The crystals contained about 200 wt. ppm of flux impurities.

(vi) To grow bulk $CoGa_2O_4$ crystals, Nakatsuka et al. [62] applied the PbF_2 flux. The starting composition consisted of CoO,

Ga_2O_3, and PbF_2 powders in the molar ratio of CoO:Ga_2O_3:PbF_2 = 1:1:9. The mixed powders were loaded into a 30 ml Pt crucible, which was heated up to 1200°C and next cooled down first to 700°C at the rate of 5 K/h, and second to 300°C at the rate of 10 K/h. Then the crucible was naturally cooled down to RT.

8.4.5 Growth from the Melt

8.4.5.1 Thermodynamics

Ga-based spinels have high melting points (MPs): $ZnGa_2O_4$ = 1900 ± 20°C [5], $MgGa_2O_4$ = 1950 ± 75°C [58] or 1930 ± 10°C [3], and $CoGa_2O_4$ = 1825 ± 5°C [60] or 1810 ± 20°C [49], and suffer from incongruent evaporation due to thermal decomposition of the constituents that leads to a shift in melt composition and the resulted crystal composition. Among the Ga-based spinels under discussion, the most unstable is the ZnO–Ga_2O_3 system. Therefore, growing Ga-based spinels from the melt requires a basic knowledge of thermodynamics to assess the melt behavior and growing crystals. The three spinels have similar melting points, so the evaporation rate will depend on the thermal stability of individual binary oxides forming the spinels, and oxygen partial pressure in a growth atmosphere. It should be stressed, however, that the thermal instability of resulted spinels might be lower than that of the individual binary oxides due to entropy stabilizing effect [63].

Figure 8.7 shows the most volatile species of different binary oxide ZnO–Ga_2O_3, MgO–Ga_2O_3, ZnO–MgO–Ga_2O_3, and CoO–Ga_2O_3 systems forming $ZnGa_2O_4$, $MgGa_2O_4$, (Zn,Mg)Ga_2O_4, and $CoGa_2O_4$ spinel compounds. The most volatile species of these compounds containing cations are Zn(g), Ga_2O(g), Mg(g) or MgO(g), and Co(g). In the ZnO–Ga_2O_3 system $p(Zn) \gg p(Ga_2O)$, which will lead to a fast composition shift toward Ga-rich composition. The ratio of $p(Zn)/p(Ga_2O)$ is very high, above 2000 at MP of $ZnGa_2O_4$ and in a wide range of $p(O_2)$. Therefore, it will require specific growth conditions to obtain stoichiometric $ZnGa_2O_4$ crystals. The MgO–Ga_2O_3 system shows an opposite direction, as $p(\text{MgO or Mg}) \ll p(Ga_2O)$. Here the $p(\text{MgO or Mg})/p(Ga_2O)$ ratio is very low, below 0.001 at MP of $MgGa_2O_4$ at low $p(O_2)$, and the resulted composition will have a tendency toward Mg-rich composition. The ZnO–MgO–Ga_2O_3 system to grow mixed (Zn,Mg)Ga_2O_4 crystals will result in the relation p(Zn) >> $p(Ga_2O)$ >> p(MgO) with the shift toward Mg-rich composition. In the case of the CoO–Ga_2O_3 system, p(Co) << $p(Ga_2O)$ with a p(Co)/$p(Ga_2O)$ ratio of about 0.05 at MP of $CoGa_2O_4$ and very low $p(O_2)$, which is required for the growth. Comparing the partial pressures of the most volatile species containing cations, we can conclude the following order of the thermal stability of the Gas-based spinels: $ZnGa_2O_4 < Mg_xZn_{1-x}Ga_2O_4$ (x = 0–0.5) < $CoGa_2O_4 < MgGa_2O_4$.

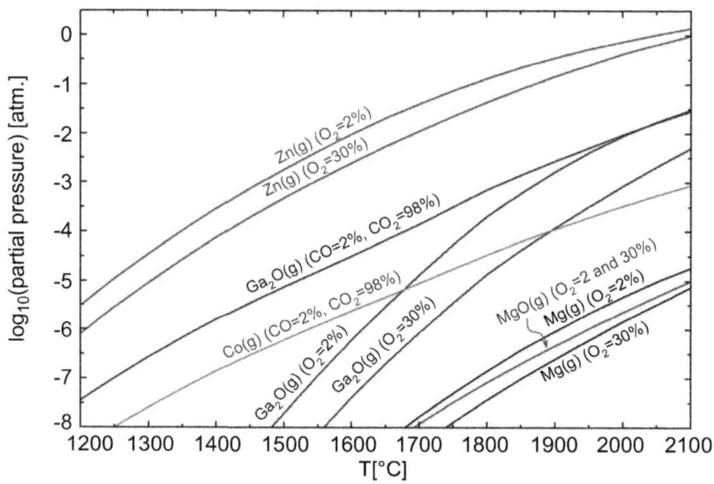

Figure 8.7 Partial pressures of the most volatile species Zn(g), MgO(g), CoO(g), and Ga_2O(g) in the ZnO–Ga_2O_3, MgO–Ga_2O_3, ZnO–MgO–Ga_2O_3, and CoO–Ga_2O_3 systems versus temperature at low (2%), and high (30%) oxygen concentration in a growth atmosphere ($ZnGa_2O_4$, $MgGa_2O_4$, (Zn,Mg)Ga_2O_4) and at slightly reducing atmosphere 2% CO + CO_2 ($CoGa_2O_4$).

To minimize the composition shift, several tools can be applied, including:

- High $p(O_2)$, which decreases the thermal decomposition of all the binary oxide systems;
- External overpressure, which minimizes the evaporation;
- Non-stoichiometric starting composition with suitable excess of a binary oxide having the highest partial pressure to compensate losses by evaporation;

- A proper growth method;
- Low temperature gradients achieved by a proper furnace design; and
- Short growth time.

These factors will be apparent further below when discussing different melt-growth techniques applied for Ga-based spinels.

8.4.5.2 Laser heated pedestal (LHPD) growth

The starting material for the LHPD growth of $ZnGa_2O_4$ crystal fibers as reported by Jia et al. [56] was prepared by sintering ZnO and Ga_2O_3 powders at 1100°C. To the starting material, MnO powder was also added at a concentration of 0.5 mol.%. Due to high and incongruent evaporation of Zn- and Ga-containing species (Fig. 8.7) at the MP of $ZnGa_2O_4$ and high surface-to-volume ratio of the grown fibers that enhances the evaporation, the applied pulling rate was very high, between 1 and 4 mm/min. The fiber crystals were grown in air. Growth experiments under nitrogen produced black particles on the crystal surface, which were of metallic nature, likely due to reduction of ZnO, MnO, and/or Ga_2O_3 to Zn, Mn, and/or Ga.

8.4.5.3 Optical floating zone (OFZ)

The OFZ was utilized to grow Ni-doped $MgGa_2O_4$ crystals by Suzuki et al. [59]. The starting material was prepared from $MgCO_3$, Ga_2O_3, and NiO powders at the molar ratio of 0.99:1.05:0.01 and 0.9:1.05:0.1. An excess of 5 mol.% of Ga_2O_3 in the starting material aimed to compensate losses of Ga-containing species due to Ga_2O_3 decomposition at high temperatures (see Chapter 4). The powders were mixed in an alumina mortar, calcined at 1300°C in air for 1 h, next ground, pressed at 100 MPa, formed in rod-shape, and sintered at 1200°C in air for 2 h. The rods of the starting material, functioned as seed and feed rods, were of 5–8 mm in diameter and 100–150 mm in length. The heat source was two 650 W halogen lamps that emitted light focused on a tip of the feed rod by double ellipsoidal mirrors. The growth atmosphere was dry oxygen flowing at the rate of 1–5 l/min. The growth and rotation rates (feed and seed rods) were 1–10 mm/h and 30 rpm, respectively.

The obtained Ni-doped $MgGa_2O_4$ crystals of about 4 mm diameter and 95 mm length were green and inclusion free. The optimum growth rate was about 5 mm/h or less, while the oxygen flow of at least 5 l/min. Higher growth rates produced cracks, while lower oxygen flow generated inclusions in the crystals.

The OFZ technique was also used by Kramer et al. [60] for growing bulk $CoGa_2O_4$ crystals. As a starting material for feed and seed rods a pressed and sintered powder of CoO and Ga_2O_3 at stoichiometric composition was used. Obtained crystals were used as seed rods for the next growth runs. The OFZ growth station was equipped with a halogen lamp and ellipsoidal mirrors. The growth atmosphere contained a few percent of oxygen to minimize the evaporation of Ga_2O, while the growth rate was in the range of 1–5 mm/h. The obtained crystals were of about 10 mm in diameter with a deep blue coloration.

8.4.5.4 Vertical gradient freeze (VGF)/vertical Bridgman (VB)

8.4.5.4.1 *Crystal growth*

The VGF or VB techniques were applied by Galazka et al. [3–5] to grow bulk $ZnGa_2O_4$, $MgGa_2O_4$, and mixed $(Zn,Mg)Ga_2O_4$ single crystals. In each case, Ir crucibles of either 40 or 60 mm diameter and cylinder length of 40 or 60 mm were used. The crucibles had either flat or conical bottom. The crucibles were covered with an Ir lid with a small central opening (1–3 mm) to lower temperature gradients inside the crucible and minimize the thermal decomposition of the constituents forming the starting material. The crucibles were surrounded from all sides by a thermal insulation with a small window exposing a portion of the lid for temperature monitoring with a pyrometer. The crucibles were heated inductively by an RF coil disposed around the thermal insulation and powered by an RF generator operating at 10 or 30 kHz.

Depending on the compound to be grown and its thermal stability (see Fig. 8.7), different growth atmospheres were used: N_2 for $MgGa_2O_4$, Ar + 8–40 vol.% O_2 for $ZnGa_2O_4$, and Ar + 10 vol.% O_2 for $(Zn,Mg)Ga_2O_4$. Typically, the crystals were grown under atmospheric pressure, although $ZnGa_2O_4$ crystals were also grown under overpressure of 2 and 10 bar to further decrease the evaporation of Zn(g).

The composition of the starting materials was as follows: $MgGa_{2+x}O_4$ with x = 0–0.02, $Zn_{1+y}Ga_2O_4$ with y = 0–0.04, and $Zn_{1-z}Mg_zGa_2O_4$ with z = 0.1, 0.25, and 0.5. A tiny excess of the Ga in $MgGa_2O_4$ and large excess of Zn in $ZnGa_2O_4$ aimed to compensate losses of those elements due to their high vapor pressures as compared with the other cation in the spinel structure. An advantage of the VGF or VB method is generally low losses of volatile species due to relatively short growth times and very low temperature gradients inside the crucible covered with the lid. The starting materials were typically prepared by firing the MgO, ZnO, and Ga_2O_3 powders individually in air for 10 h at 600–800°C, 600–700°C, and 800–1000°C, respectively, to remove moisture and other possible volatile impurities, such as carbonates. This step is referred to a drying step. Afterward, the powders were mixed at a suitable molar ratio and cold pressed, typically at 2000 bar for several minutes. The powders were then typically fired in air for 10–20 h at 800–1200°C to make the pressed powders more compact. This step will be called sintering step. Sintering temperature is too low for solid-state synthesis; however, it avoids thermal decomposition of any of the constituents that would lead to deviation from the pre-defined composition. For doping purposes to induce or increase the electrical conductivity, the starting materials contained additionally a suitable dopant oxide at a suitable concentration. For example, $MgGa_2O_4$ and $(Zn,Mg)Ga_2O_4$ crystals were doped with Si^{4+} by adding to the starting material SiO_2 at a concentration of 0.2–0.25 mol.%, while $ZnGa_2O_4$ was doped with Si^{4+} and Zr^{4+} by adding to the starting material SiO_2 and ZrO_2 at concentrations of 0.25 mol.%, each.

After detecting the liquid phase of the starting material within the crucible by temperature monitoring of the lid covering the crucible or visually when an opening in the lid was large enough for visual access to the starting material, the molten starting material was first homogenized at then cooled down at a rate of 14–50 K/h at the beginning, and at a higher rate of 100–200 K/h after melt solidification.

After growth, the crystallized starting material was core-drilled out of the crucible, which resulted in crystalline bulks of $ZnGa_2O_4$, $MgGa_2O_4$, and $(Zn,Mg)Ga_2O_4$, as shown in Fig. 8.8a, Fig. 8.8b, and

Fig. 8.8c, respectively. The bulks usually consist of several single crystal grains with a volume of 1–5 cm^3, as demonstrated in Fig. 8.8d. Such volume of single crystal grains enables fabrication of wafers of up to 10×10 mm^2.

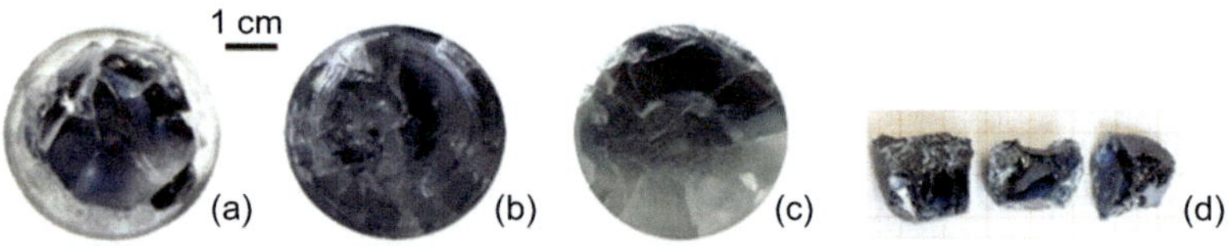

Figure 8.8 Crystallized $ZnGa_2O_4$ (a), $MgGa_2O_4$ (b), and $Zn_{0.5}Mg_{0.5}Ga_2O_4$ (c), within Ir crucibles (40 mm diameter) by the VGF/VB method with extracted single crystal grains of $ZnGa_2O_4$ (d) [4].

8.4.5.4.2 *Examples*

(i) Bulk $MgGa_2O_4$ single crystals were obtained by the Bridgman method by Galazka et al. [3]. An Ir conical crucible of the following dimensions was used: 60 mm diameter, 80 mm long cylindrical part, 24 mm long conical part, and 16 mm long seed portion. The crucible was covered with an Ir lid having a 2 mm central opening for communication with the growth atmosphere. As a starting material, a mixture of MgO (4N) and Ga_2O_3 (5N) powders at the molar ratio of 1:1 was used. The powders were separately dried at 800°C in air for 10 h, mixed, pressed at 2000 bar, and sintered at 1200°C in air for 10 h. In the conical crucible being embedded within a thermal insulation, 270 g of such prepared starting material was loaded. The crucible with the starting material was covered with the Ir lid and with a top thermal insulation. The growth atmosphere was N_2. The crucible was then heated up for 15 h to a maximum temperature T_{MAX} = 2050°C, soaked for 1 h, and then cooled down first to 1965°C at the rate of 14 K/h, next to 1450°C at the rate of about 50 K/h, and finally to RT at the rate of about 145 K/h. In this example, the melt was overheated by about 8.4% to induce electrical conductivity along with the neutral growth atmosphere. The solidified material was removed from the crucible by core-drilling from which several large crystal grains with dark-blue coloration were collected.

(ii) Bulk $ZnGa_2O_4$ single crystals were grown by the VGF method by Galazka et al. [5]. For the growth, two types of Ir crucibles were used: (a) cylindrical, 40 mm in diameter and height, and (b) conical, 40 mm in diameter, 60 mm long cylindrical part, and 24 mm long conical part ended with a cylindrical seed portion. ZnO and Ga_2O_3 powders of 5N purity, which constituted the starting material (200 g) loaded in the crucible, were first dried separately in air for 10 h at 600 and 800°C, respectively, then mixed at a molar ratio of 1:1, pressed at 2000 bar, and sintered at 700°C in air for 10 h. As a growth atmosphere, a mixture of 10 vol.% + Ar was used. The heating-up time was 8 h to minimize incongruent decomposition. After complete melting, the melt was soaked for a short time (about 10 min) and then cooled down to RT for 8–12 h, wherein the time at which the starting material was in the liquid phase was about 40 min. The solidified material after core-drilling consisted of three large single crystal grains having a volume of 2–3 cm^3 each. The crystal grains were blue with no cracks and visible inclusions.

(iii) Bulk $ZnGa_2O_4$ single crystals were grown by the VGF method by Galazka et al. [5]. Here a cylindrical Ir crucible was used, and the same powder preparation as in example (ii); however, the starting composition contained 0.2 mol.% excess of ZnO. The growth atmosphere contained a higher oxygen concentration, 21 vol.% mixed with Ar. The heating-up time, soaking time, and cooling-down time were 6 h, 5 min, and 10 h, respectively. The liquid phase of the starting material lasted for about 30 min. The solidified material after core-drilling consisted of several, dark-blue single crystal grains having a volume of 0.5–2 cm^3.

(iv) Bulk $Zn_{0.5}Mg_{0.5}Ga_2O_4$ single crystals were obtained by the VGF method by Galazka and Hanke [4]. A cylindrical Ir crucible of 40 mm diameter and height was used. The starting material of 200 g loaded in the crucible was prepared by individual drying MgO (4N), ZnO (5N), and Ga_2O_3 (5N) powders in air for 10 h at 600, 600, and 800°C, respectively. Then the powders were mixed together at the molar ratio of 0.5:0.5:1 and pressed at 2000 bar. The growth atmosphere consisted of 10 vol.%O_2 + Ar. The heating-up time, soaking time, and cooling-down time were 8 h, 30 min, and 8 h, respectively. The starting material was in liquid phase for about 60 min. The solidified material after core-drilling consisted of blue single crystal grains having a volume of 0.5–2 cm^3.

8.4.5.5 Czochralski and Kyropoulos-like methods

8.4.5.5.1 *Crystal growth*

The Czochralski method was used for growing bulk $MgGa_2O_4$ single crystals by Scott et al. [58] and by Galazka et al. [3], for $ZnGa_2O_4$ single crystals by Galazka et al. [5], and for $CoGa_2O_4$ single crystals by Kramer et al. [60] and Galazka [49]. Due to high MPs, inductively heated Ir crucibles were used in each case and a thermal insulation consisting of zirconia and alumina insulation surrounding the crucible from all sides with a free space over the crucible to accommodate a growing crystal. Among these spinels, $MgGa_2O_4$ is the most thermally stable and is transparent in the NIR region at RT, therefore, it enables easier heat dissipation from the growing crystal, facilitating a more stable growth and reasonable crystal size.

For $MgGa_2O_4$, Scott et al. [58] used a growth furnace consisting of an Ir crucible of 38 mm diameter and 2 mm thick wall, a 66 mm diameter zirconia crucible lined with zirconia felts and filled with zirconia powder in which Ir crucible was embedded. The crucible was heated up inductively by an RF coil. The starting material was prepared from MgO and Ga_2O_3 powders at the molar ratio of 1:1. The growth atmosphere consisted of 1 vol.% O_2 + 99 vol.% N_2. The growth and rotation rates were 12.5–25 mm/h and 20–45 rpm, respectively. The obtained crystals of about 12.5 mm diameter and 19–38 mm length were stoichiometric and clear.

To grow bulk $MgGa_2O_4$ single crystals by the Czochralski method, Galazka et al. [3, 64] used 40 mm diameter Ir crucible heated by an RF coiled powered by 10 or 30 kHz RF generator. The thermal insulation consisted of a combination of zirconia granules, zirconia felts, and low-density alumina. To further decrease temperature gradients in a pulling zone, an active (i.e., inductively heated) Ir afterheater was utilized above the crucible. The starting composition was typically stoichiometric or with a small excess of Ga_2O_3 (up to 0.02 mol.%) to cover losses of Ga-containing species at growth runs with low $p(O_2)$ in the growth atmosphere. Different growth atmospheres were used, including CO_2, 20% CO_2 + 80% N_2, 2–32 vol.% O_2 + Ar. The growth and rotation rates were 1–2 mm/h and 5–10 rpm, respectively. The crystals were grown on crystal seeds oriented along the <100> crystallographic direction, which was prepared from the first crystal grown on an Ir tube. The Czochralski-grown bulk $MgGa_2O_4$ single crystals were substantially colorless, clear, and transparent with the diameter and length of about 20 and 20–40 mm, respectively (Fig. 8.9a). The crystals had, however, a tendency to a foot formation and consequently to a spiral formation even at low crystallization ratio (below 30%). Although at RT the crystals were electrically insulating and transparent in the NIR spectrum, at high temperatures the crystals are much less transparent in that spectral region. Additionally the absorption at high temperatures can be further enhanced by free carrier absorption (semiconducting state), in a similar manner as Mg-doped Ga_2O_3, which shows semiconducting behavior at high temperatures while being electrically insulating at RT. Moreover, the foot/spiral formation may be enhanced if the melt viscosity is relatively high.

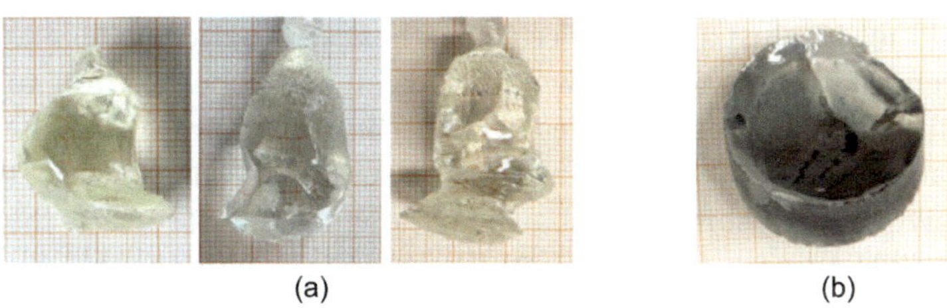

Figure 8.9 Bulk $MgGa_2O_4$ single crystals obtained by the Czochralski (a) [64] and Kyropoulos-like (b) methods. Figure (b) reprinted with permission from Ref. [3], Copyright 2015, John Wiley and Sons.

Although the Czochralski method produced a reasonable crystal volume of 6–12 cm^3, a yet larger volume could be obtained by the Kyropoulos-like method, which is a variant of the Czochralski method. Here the pulling rate was much smaller, below 0.5 mm/h, no crystal rotation, and a slow cooling of the crucible at the rate of 2–3 K/h until the whole melt crystallized on the crystal seed. The growth atmosphere was either CO_2 or Ar, which produced blue (Fig. 8.9b) and dark-blue crystals, respectively, whether undoped or doped with Si (0.2 mol.%). The blue coloration is the result of absorption in the red part of the visible spectrum, which is an extension of the absorption in the NIR region caused by the free carrier absorption, as the crystals showed semiconducting behavior. This technique produced single crystals with a size of 35 diameter and 20 mm length, respectively (volume of about 19 cm^3).

The Czochralski method was also tested by Galazka et al. [4, 5] to grow bulk $ZnGa_2O_4$ and $(Zn,Mg)Ga_2O_4$ single crystals. Both compounds undergo a strong incongruent evaporation leading to a fast composition shift; therefore, only short crystals (up to 10 mm) could be obtained. Additionally, they had a tendency to foot and spiral formation immediately after seeding, as the melt flow at the surface is additionally unstable due to the strong evaporation.

The Czochralski method was used by Kramer et al. [60] to grow bulk $CoGa_2O_4$ single crystals from the melt. For the growth, an inductively heated Ir crucible contained a $CoO–Ga_2O_3$ composition at the molar ratio of 1:1 as the starting material was used. An inductive RF coil was powered by a 10 kHz RF generator. The growth furnace located within a growth chamber consisted of zirconia thermal insulation surrounding the Ir crucible from the bottom and the side. Just above the crucible, there was a zirconia disk with a central opening for a growing crystal and an upper zirconia tube defining a space for the pulled crystal. The growth proceeded either at atmospheric pressure or at overpressure of 3.4 bar; however, the growth atmosphere was not given although due to the use of the Ir crucible, one can expect a neutral atmosphere with possible small concentration of oxygen. The growth and rotation rates were 1–6 mm/h and 20–50 rpm. The growth under atmospheric pressure was not successful due to very high evaporation rate of the starting material and highly unstable growth due to a concave interface toward the melt, which initiated a lateral growth over the melt surface and disconnection from the melt (melt-back). An additional reason was too high rotation rate (when the forced convection due to rotation overcomes the buoyancy) and high absorption in the NIR region causing difficulties in heat dissipation from the growing crystal. This mechanism is described in more detail in Chapter 4. An overpressure of 3.4 bar, combined with slower growth and rotation rates as well as water-cooled seed holder, enabled to minimize the evaporation and improved heat transfer through the growing crystal. As a result, the growth was more stable with a flat or slightly convex interface. The obtained crystals were of 10 mm diameter and 25 mm length.

Galazka [49] used for the growth of bulk $CoGa_2O_4$ single crystals by the Czochralski method an inductively heated Ir crucible of 40 mm diameter and height, and a thermal insulation similar to that used for $MgGa_2O_4$. However, the growth atmosphere for $CoGa_2O_4$ was different, slightly reducing to keep Co at oxidation state +2. It consisted of 2% CO + 98% CO_2. The starting composition was stoichiometric, while the growth and rotation rates were 1–2 mm/h and 5–10 rpm, respectively. Only short crystals (Fig. 8.10a) of about 20 mm with 15 mm diameter (volume of about 3 cm^3) could be obtained due to unstable growth arising from poor heat transport through the growing crystal. Larger crystal volume was obtained by the Kyropoulos-like method, similar to that used for bulk $MgGa_2O_4$ crystals. The growth atmosphere was the same as in the case of the Czochralski method, i.e., slightly reducing. The obtained crystals (Fig. 8.10b) were of 30 mm diameter and 25 mm length (total volume of about 18 cm^3 with a usable volume of about 10 cm^3). The crystals obtained by both the Czochalski and Kyropoulos-like methods were black and opaque in the visible spectrum (Fig. 8.10a,b). Moreover, the obtained crystals were slightly off-stoichiometric (by inductively coupled plasma optical emission spectroscopy (ICP-OES) with a deficiency of Co (by 0.5–1.5 wt.%) and a corresponding excess of Ga. This is likely due to the insufficient Co incorporation in a growing crystal rather than higher Co losses, as the Co losses are about eight times smaller than Ga losses (basing on the chemical analysis of deposition in a growth chamber due to evaporation). It cannot be ruled out that the melt contained not only Co^{2+}, but also Co^{3+} ions, which are not easily built into the crystal lattice, or Co sites are partly occupied by Ga^{3+} with associated oxygen vacancies or remain as vacancies. Additionally, it should be pointed out that the homogeneity range of the $CoGa_2O_4$ spinel phase is rather wide, at 1567°C from 35.7 to 56.6% CoO (Table 8.1). One can assume that preferably the congruently melting composition (= maximum of the liquidus curve) crystallizes from the melt; however, there is no thermodynamic condition that this is the 1:1 composition with 50% CoO. For the $MgAl_2O_4$ spinel, Mao et al. [65] found in a recent study the congruent maximum at 50.6% Al_2O_3, which is almost the same deviation from the 1:1 composition, which is measured for $CoGa_2O_4$.

(a) (b)

Figure 8.10 Bulk $CoGa_2O_4$ single crystals obtained by the Czochralski (a) and Kyropoulos-like (b) methods [49].

8.4.5.5.2 *Examples*

(i) Bulk $MgGa_2O_4$ single crystals were obtained by Galazka and Hanke [64] with the use of the Czochralski method. For that purpose, an Ir cylindrical crucible of 40 mm diameter and height was used. The starting material of stoichiometric composition and 180 g by weight was prepared by individually drying 5N purity Ga_2O_3 and MgO powders (800°C, air, 10 h), mixing together both powders, pressing the mixture at 2000 bar, and sintering at 1200°C in air for 10 h. For the growth a crystal seed (from previous $MgGa_2O_4$ crystal) oriented along the <100> direction was used. In one experiment, the growth atmosphere was pure CO_2 and the measured O_2 concentration (with a mass spectrometer) was 2 vol.%, while in another one, it consisted of 20% CO_2 + 80% N_2. The growth and rotation rates were 2 mm/h and 6 rpm, respectively, while the heating and cooling times were 15 h each. The obtained crystals of 18 mm diameter and 20–40 mm length were colorless. The crystals were transparent but had a rough surface due to insufficient oxygen partial pressure in the growth atmosphere. They grew in the form of a spiral, with a few small surface cracks.

(ii) Bulk $MgGa_2O_4$ single crystals were obtained by Galazka and Hanke [64] with the use of the Czochralski method. The starting material preparation (200 g) and composition, as well as crystal seed, were the same as in example (i). Here the growth atmosphere consisted of 32% O_2 + 68% Ar during crystal growth. During heating up, O_2 concentration was 10%, while during cooling down, it was switched off. The growth and rotation rates were 1.5 mm/h and 9 rpm, respectively, while the heating and cooling times were 10 and 16 h, respectively. The obtained crystal of 18 mm diameter and 25 mm length was substantially colorless with a straw tint and transparent. The crystal surface, in contrast to example (i), was smooth with no cracks or visible inclusions. The crystal had, however, a tendency to bend during growth (foot formation), but the growth was much more stable as compared with low oxygen concentration in the growth atmosphere.

(iii) Bulk $MgGa_2O_4$ single crystals were obtained by Galazka and Hanke [64] with the use of the Czochralski method. Starting material preparation and composition, as well as crystal seed, were the same as in example (i). Here the growth atmosphere consisted of 2–4 vol.% O_2 + Ar, which was unchanged during heating up, growth and cooling down. The growth and rotation rates were 1.5 mm/h and 9 rpm, respectively, while the heating and cooling times were 10 and 16 h, respectively. The obtained crystal of 20 mm diameter and 35 mm length was colorless with a smooth surface, and no cracks or visible inclusions. The bottom part of the crystal was fully facetted.

(iv) Bulk $CoGa_2O_4$ single crystals were obtained by Galazka [49] utilizing the Czochralski method. A used Ir cylindrical crucible was 40 mm in diameter and height. The starting material consisted of CoO (4N) and Ga_2O_3 (5N) powders at a molar ratio of 1:1. A special care was taken to prepare the starting material, as heating up of CoO (blue in color) in an oxidizing atmosphere at 800°C or above converts it to Co_3O_4 (i.e., CoO–Co_2O_3, which is almost black in color). Therefore, CoO was dried at 800°C in Ar for 10 h, while Ga_2O_3 was dried at 800°C in air for 10 h. Mixed and pressed (at 2000 bar) powders were then sintered at 850°C in Ar for 10 h. The blue coloration of the obtained starting material evidenced +2 oxidation state of Co. To avoid or minimize the change in Co oxidation state, the growth atmosphere consisted of 2% CO + 98% CO_2. As a seed, an Ir tube was used. The growth and rotation rates were 2 mm/h and 6 rpm, respectively. The obtained crystal of 15 mm diameter and 20 mm length was black and opaque. The crystal had a tendency to bend during growth. No cracks or visible inclusions were observed.

8.4.5.6 Composition and structural quality of bulk Ga-based spinel crystals obtained from the melt

Bulk Ga-based spinels obtained from the melt by the VGF/VB, Czochralski, and Kyropoulos-like methods had one single phase of $ZnGa_2O_4$ [5], $MgGa_2O_4$ [3], and $CoGa_2O_4$ [66], according to XRD measurements of powdered crystals, as shown in Fig. 8.11. The XRD patterns match database patterns labeled as PDF cards [67] 00-038-1240, 01-089-3082, and 01-075-4308 for $ZnGa_2O_4$, $MgGa_2O_4$, and $CoGa_2O_4$, respectively.

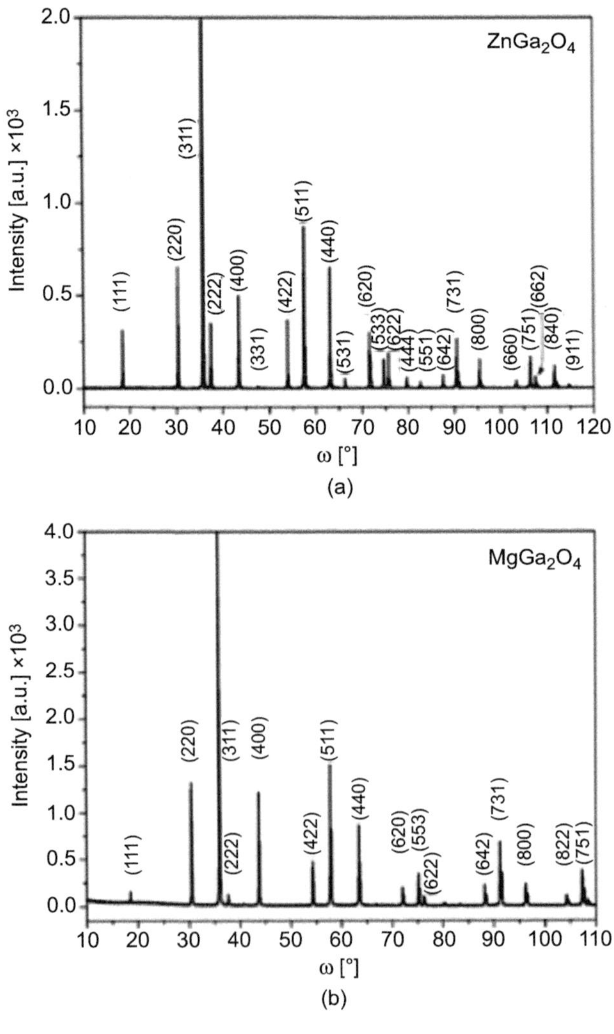

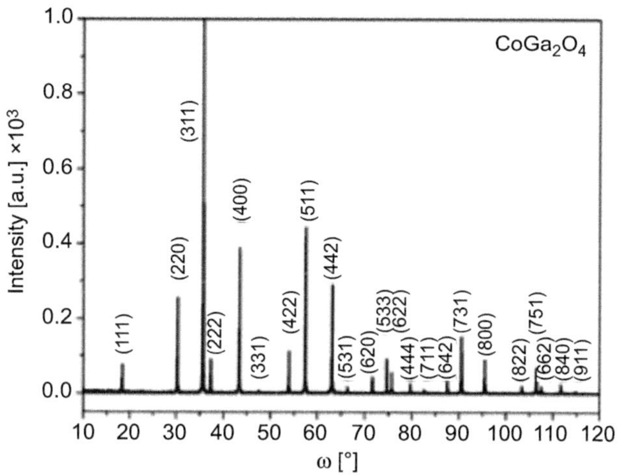

(c)

Figure 8.11 XRD diffraction patterns of powdered single crystals of (a) $ZnGa_2O_4$, reprinted from Ref. [5] under Creative Commons license; (b) $MgGa_2O_4$, reprinted with permission from Ref. [3], Copyright 2015, John Wiley and Sons; and (c) $CoGa_2O_4$ [66].

The melt-grown Ga-based spinel single crystals are substantially stoichiometric or very close to stoichiometric. A number of $MgGa_2O_4$ crystal samples (from four different crystals) measured by the ICP-OES (Bertram and Galazka [68]) showed Mg and Ga concentrations around stoichiometric composition within the measurement accuracy (Fig. 8.12a). This holds for semiconducting and electrically insulating crystals, for as-grown and annealed crystals in oxidizing atmospheres. X-ray fluoroscence (XRF) measurements of $ZnGa_2O_4$ crystal samples when grown at optimum conditions also showed substantially stoichiometric composition for colorless (electrical insulator) and light blue/blue (semiconducting) crystals; however, there was a small deviation from stoichiometry (excess of Ga and deficiency of Zn) for dark-blue crystals (degenerate semiconductor). $ZnGa_2O_4$ single crystals obtained at conditions deviated from optimum conditions, in particular when too long growth time was applied, revealed a much larger deviation from stoichiometry leading to the formation of another phase, β-Ga_2O_3. This happens for Ga/Zn ratio greater than 2.17.

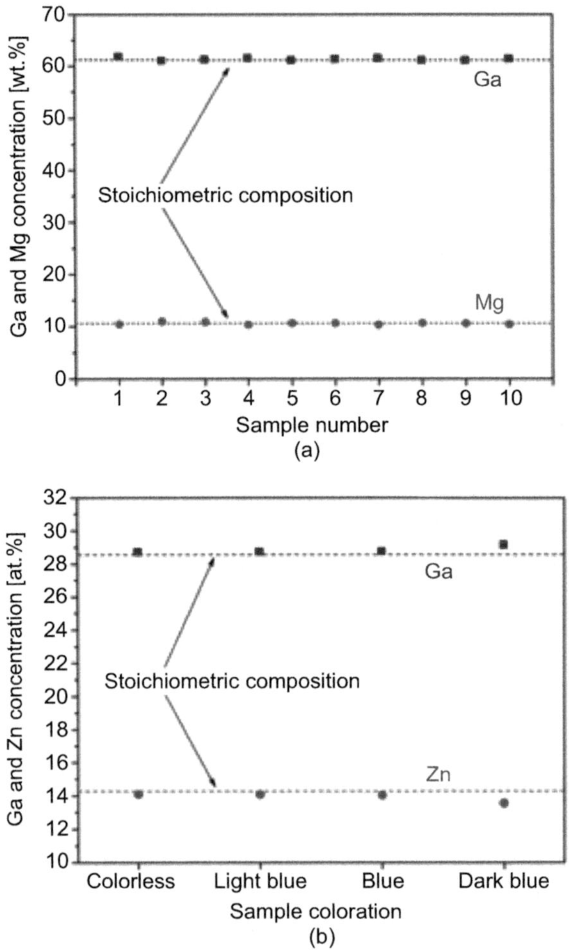

Figure 8.12 Composition of melt-grown bulk $MgGa_2O_4$ (a) [68] and $ZnGa_2O_4$ (b) single crystals measured by the ICP-OES (a) and XRF (b), respectively. The dashed lines indicate stoichiometric compositions. Figure (b) reprinted from Ref. [5] under Creative Commons license.

Bulk $CoGa_2O_4$ single crystals revealed a small deviation from stoichiometry. The molar ratio of $CoO:Ga_2O_3$ = 1:1 contains Co = 22.46 and Ga = 53.15 wt.%, wherein the obtained crystals measured by ICP-OES contained Co = 21–22 wt.% and Ga = 52.7–54.5 wt.% (Bertram and Galazka [69]).

$MgGa_2O_4$, $ZnGa_2O_4$, and $(Zn,Mg)Ga_2O_4$ obtained by the VGF/VB methods usually consisted of several, large single crystal grains having a volume of 1–5 cm^3. Areas contacting the crucible wall were often cracked due to differences in thermal expansion between the crystal and Ir, while those at the very bottom parts were often polycrystalline. In the case of the Czochralski and Kyropoulos-like methods, cylindrical bulk $MgGa_2O_4$, $ZnGa_2O_4$, and $CoGa_2O_4$ crystals were crack free in vast majority of growth runs.

The obtained crystal volume enabled fabrication of epi-ready wafers of size $5 \times 5 \times 0.5$ and $10 \times 10 \times 0.5$ mm^3 (Fig. 8.13) with the FWHM of the rocking curve of about 25, 50, and 72 arcsec for $ZnGa_2O_4$, $MgGa_2O_4$, and $CoGa_2O_4$, respectively.

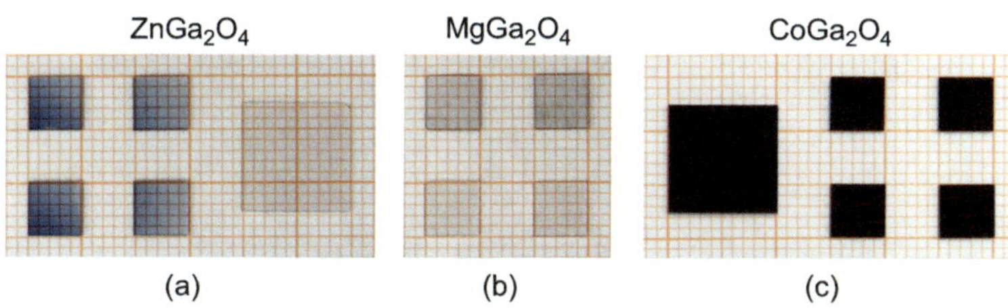

Figure 8.13 Epi-ready, (100)-oriented wafers of size $5 \times 5 \times 0.5$ and $10 \times 10 \times 0.5$ fabricated from melt-grown ZnG_2O_4, $MgGa_2O_4$, and $CoGa_2O_4$ single crystals, respectively. Blue and colorless $ZnGa_2O_4$ wafers are as-grown (semiconducting) and oxygen annealed (electrically insulating). Slightly bluish and colorless $MgGa_2O_4$ wafers are as-grown semiconducting and electrically insulating, respectively. Substantially black $CoGa_2O_4$ wafers are as-grown and electrically insulating.

All the crystals (including individual crystal grains) obtained by different melt-growth techniques had no visible macro-inclusions or voids. However, in some areas of the crystals, micro-voids or nano-inclusions were found, as a result of thermal decomposition of the binary oxides under applied growth atmosphere. The thermal decomposition may produce voids due the evaporation of Zn and Ga_2O, and metallic nanoparticles remaining in the crystals due to oxygen loss. $MgGa_2O_4$ and $CoGa_2O_4$ crystals occasionally revealed micro-voids, especially those grown at low $p(O_2)$, e.g., under Ar, N_2, or CO_2. This is likely due to losses of Ga_2O from the growing crystals. Crystals grown at higher $p(O_2)$ substantially did not show such voids, or at very low concentration.

$ZnGa_2O_4$ is the most unstable compound among other Ga-based spinels (see Fig. 8.7), and losses of Zn(g) are much higher as compared with Ga_2O(g) losses. Even when applying high $p(O_2)$, Zn losses are still noticeable. The micro-voids were observed only occasionally, but instead nanoparticles were observed under TEM (Fig. 8.14). The nanoparticles were oriented with respect to the $ZnGa_2O_4$ matrix and had a size and density of about 25 nm and 5×10^{14} cm^{-3}, respectively. The formation of the Moiré pattern at the particles indicates, indeed, a crystalline nature with a different lattice parameter or crystallographic orientation with respect to the matrix. The nature of the nanoparticles is not clear; however, they may have a metallic nature.

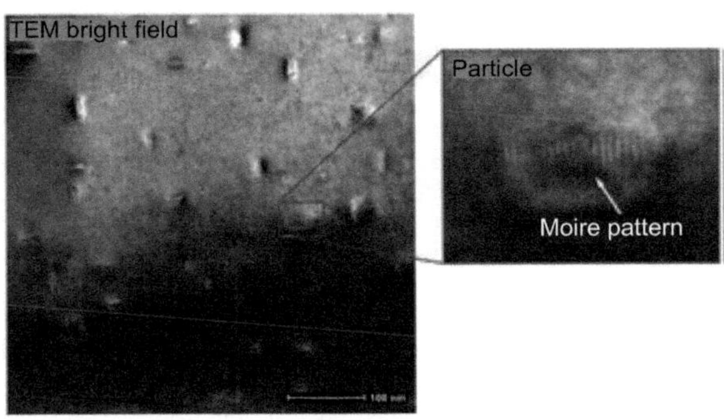

Figure 8.14 Nano-inclusions in $ZnGa_2O_4$ single crystals obtained from the melt. Reprinted from Ref. [5] under Creative Commons license.

8.4.5.7 Thermal stability

The $MeGa_2O_4$ spinel of a general formula AB_2O_4, where Me = Mg, Zn, Co, and Zn+Mg, should rather be described by the following formula: $Ga^{[4]}[MeGa]^{[6]}O_4$. The degree of inversion x may vary between $x = 0$ for a normal and $x = 1$ for an inverse spinel. For a fully random distribution of A and B atoms over the tetrahedral and octahedral sites, the inversion degree is $x = 2/3$, and typically this value is reached for every (normal and inverse) spinel at high temperatures T [70]. Distribution of cations A and B over octahedral and tetrahedral sites, basing on Hoffman and Brown [71], is summarized in Table 8.2.

Table 8.2 Cation distribution within a spinel structure of general formula AB_2O_4

Spinel structure	Octahedral site	Tetrahedral site	Inversion parameter *x*
Normal	B_2	A	0
Inverse	A + B	B	1
Intermediate	$A_\delta + B_{2-\delta}$	$A_{1-\delta} + B_\delta$	(0, 1)

The entropy contribution *S* of the spinel structure resulting from this disorder is minimum for $x = 0$, is larger ($2R\ln2$, where *R* is gas constant and "ln" is natural logarithm) for $x = 1$, and goes through a maximum for the random spinel with $x = 2/3$ [72]. The entropy change during transition between inverse and random spinel can be detected by changes in the specific heat capacity, e.g., by differential scanning calorimetry (DSC) measurements. An example of such change in the specific heat capacity for $MgGa_2O_4$, which is an inverse spinel at RT ($x \approx 1$), is shown in Fig. 8.15a. It summarizes four heating experiments from 800 to 1200°C at which randomization of cation distribution takes place. At 800°C, the sample was equilibrated for 16, 8, 4, and 2 h prior heating to 1200°C at 30 K/min. In each case, there is an increase in the specific heat capacity (indicated by endothermal downward bend of the DSC curve) at about 1000, 960, 880°C for equilibration times of 16, 8, and 4 h, respectively; however, shorter times (2 h) appear to be insufficient to reconstruct the low-*T* inverse spinel ordering from the high-*T* random ordering that was formed at 1200°C. The increase in the specific heat capacity becomes, at the same time, smaller with shorter equilibration times at 800°C. A similar situation was observed for $ZnGa_2O_4$, which is a normal spinel at RT ($x = 0$), where the sample was equilibrated at 600°C for 0, 0.5, 1, 2, and 4 h. In each case, the sample was first held at 1200°C for 1 h, cooled down to 600°C at the rate of 30 K/min, and then heated up again to 1200°C at the rate of 20 K/h. As shown in Fig. 8.15b, equilibration at 600°C (0 h) did not produce a detectable thermal effect; however, the equilibration times of 0.5 h and longer produced an endothermal reaction near 1000°C, which is associated with an increase in the specific heat capacity that becomes higher with the equilibration time at 600°C. The endothermal reaction at about 1000°C indicates a transition temperature between normal and random spinel. The thermal analysis of Ga-based spinels enables to draw the following conclusions:

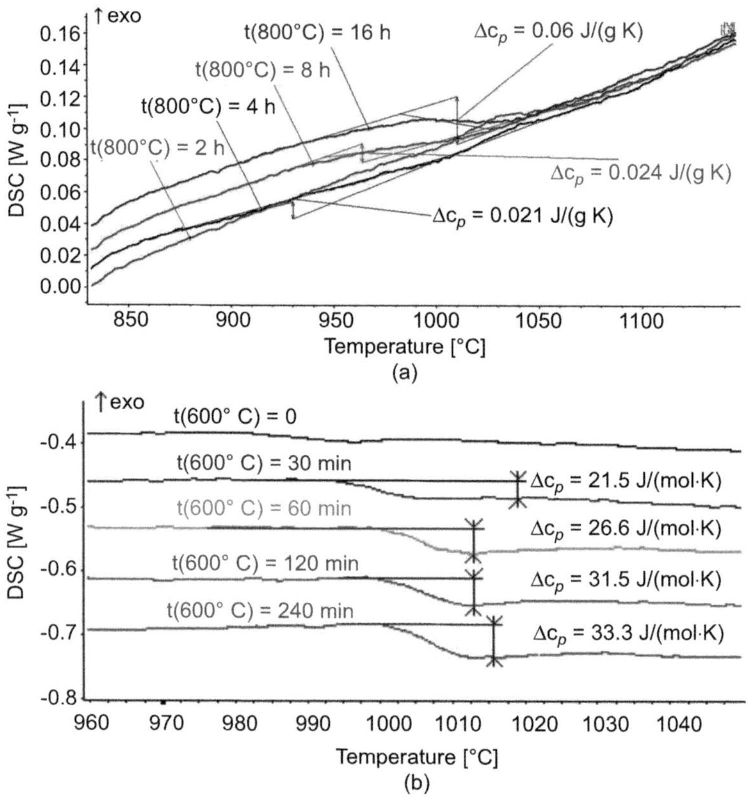

Figure 8.15 DSC measurements versus temperature of an $MgGa_2O_4$ (a) and $ZnGa_2O_4$ (b) single crystal showing specific heat changes indicating a transition from the inverse or normal to random spinel structure of the crystal. Prior reaching each of the heating curves, the sample was heated to 1200°C at the rate of 30 K/min and hold there for 1 h; and then cooled down to 800°C ($MgGa_2O_4$) or 600°C ($ZnGa_2O_4$) at the rate of 30 ($MgGa_2O_4$) or 20 K/min ($ZnGa_2O_4$). The parameters *t*(800°C) and *t*(600°C) at each of the curves refer to the equilibration times at 800°C and 600°C, respectively. Figure (a) reprinted with permission from Ref. [3], Copyright 2015, John Wiley and Sons, and (b) reprinted from Ref. [5] under Creative Commons license.

i. Whether the spinel is inverse ($MgGa_2O_4$) or normal ($ZnGa_2O_4$) at RT, it becomes random at high temperatures above 800°C, where Me^{2+} and Ga^{3+} cations are distributed randomly at tetrahedral and octahedral sites in the spinel structure;
ii. To approach an equilibrium cation distribution in the spinel structure, a sufficient annealing time is required. This also applies to cooling time of a crystal after growth from the melt at high temperatures. If the cooling time is not long enough, the obtained crystals may not be a fully inverse/normal spinel, but partly random, as not all ions have occupied own designated positions;
iii. When using the spinels as substrates for epitaxy, they should not be heated up to temperatures exceeding about 800°C for prolongated times (>2 h) as it may induce transition to the random spinel with different surface structural properties. The same relates to spinel-based devices operating at such high temperatures; and
iv. Reordering of Me^{2+} and Ga^{3+} cations has an impact on the electronic structure of the spinels and may lead, likely indirectly, to changes in electrical conductivity.

The inversion parameter $x = 0.575$ for flux-grown $CoGa_2O_4$ crystals was found by Nakatsuka et al. [62], which can be classified as a spinel between normal and inversed, although the authors call it a "largely normal spinel." Leccabue et al. [61] reported for CVT-grown $CoGa_2O_4$ crystals 40% inversion of Co on tetrahedral sites, and on the surface even 50%.

According to the theoretical work of Seko et al. [73], both the inverse $MgGa_2O_4$ and normal $ZnGa_2O_4$ spinels undergo tetrahedral–octahedral site exchange at 1300 and 1600 K, respectively, wherein the inverse $MgGa_2O_4$ spinel undergoes octahedral–octahedral site exchange at about RT (270 K). Tetrahedral–octahedral site exchange relates to inversion from inverse or normal to random spinel with the theoretical transition temperatures close to the experimental ones. However, the octahedral–octahedral transition, which is related to cubic-tetragonal phase transition, was not observed experimentally below RT, neither by ARPES [43] nor by thermal diffusivity measurements [31].

Another aspect relates to thermal decomposition of the Ga-based spinel single crystals at high temperatures at oxidizing and reducing atmospheres, which has an impact on the thermal stability of substrates used for epitaxy and device operation at high temperatures. A rough estimation of the thermal stability can be concluded from the mass loss after annealing melt-grown Ga-based spinel crystals at designated atmospheres, temperatures, and times (Galazka et al. [3–5, 49]).

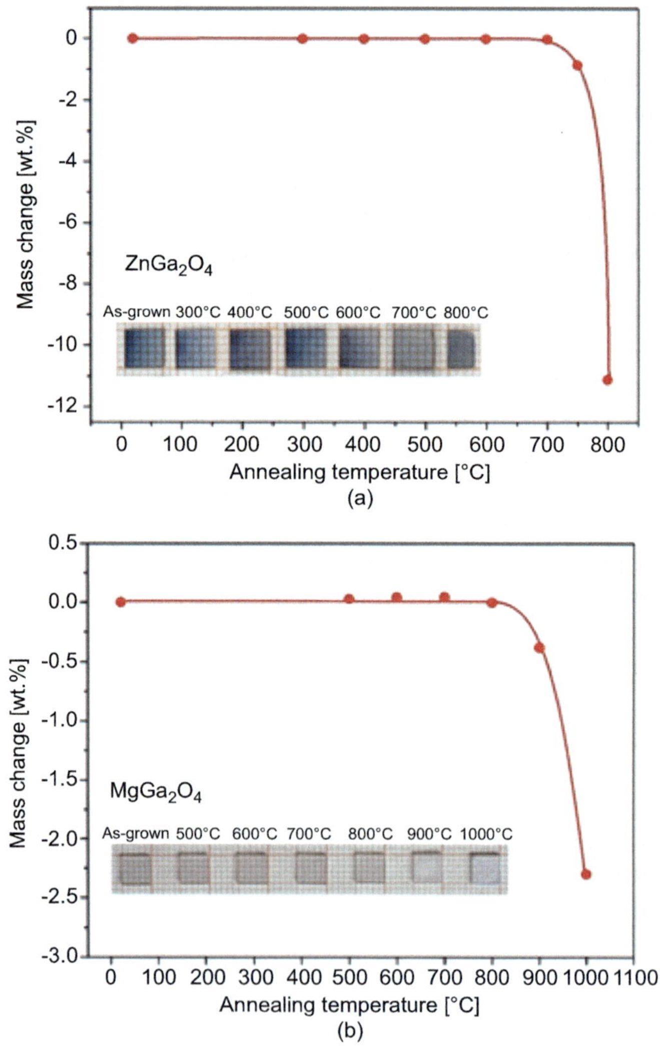

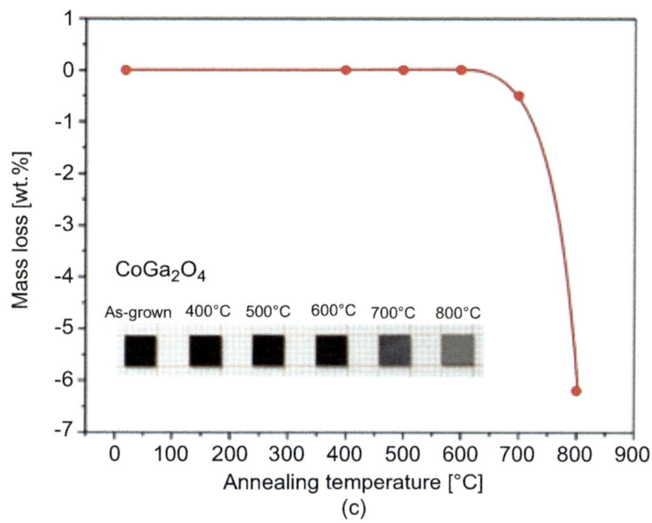

Figure 8.16 Mass change after annealing melt-grown bulk Ga-based spinel crystals in an H_2-containing atmosphere (5% H_2 + Ar) for 10 h along with the appearance change in the annealed samples: (a) $MgGa_2O_4$, (b) $ZnGa_2O_4$, and (c) $CoGa_2O_4$.

When considering an oxidizing atmosphere (O_2 or air, 10–20 h annealing time), all the spinels are substantially stable up to about 1100–1200°C when no mass loss was detected within measurement error (<0.05%). Above that temperature, the Ga-based spinels become thermally unstable and start to decompose with an increase in the mass loss with temperature. $MgGa_2O_4$ and $CoGa_2O_4$ are thermally stable up to about 1200°C, while $ZnGa_2O_4$ up to about 1100°C. This is not surprising as the thermal stability corresponds to the partial pressure of the most volatile species containing cations, which holds the following order at melting points of the compounds: $p(MgO) < p(CoO) < p(Zn)$ for $MgGa_2O_4$, $CoGa_2O_4$, $ZnGa_2O_4$, respectively. Mixed bulk $Zn_{0.5}Mg_{0.5}Ga_2O_4$ single crystals are stable also up to 1200°C.

The thermal stability of $ZnGa_2O_4$, $MgGa_2O_4$, and $CoGa_2O_4$ in reducing conditions (5% H_2 + 95% Ar, 10 h annealing time), as shown by temperature-dependent mass loss characteristics in Fig. 8.16, is much lower as compared to that at oxidizing conditions. Here, again, the most stable is $MgGa_2O_4$, up to 800°C, next are $ZnGa_2O_4$ and $Zn_{0.5}Mg_{0.5}Ga_2O_4$ up to about 700°C, and the least stable is $CoGa_2O_4$, up to about 600°C, above which a strong decomposition proceeds (high mass loss). In this process, the thermal stability of $CoGa_2O_4$ is enhanced by higher reactivity between the crystal surface and hydrogen. The decomposition of the compounds can be detected visually due to a change in the crystal sample appearance, as indicated by insets in Fig. 8.16, showing the crystal samples annealed in the presence of hydrogen at different temperatures. Simply, above a critical temperature, the crystal surface becomes rough and milky with a different color tint.

Compatible results on annealing flux-grown $ZnGa_2O_4$ crystals in pure hydrogen with longer times were reported by Yan et al. [55], who observed no mass loss after 100 h annealing at 600°C, tiny mass loss (1 wt.%) after 60 h annealing at 700°C, and full decomposition (mass loss 100 wt.%) after 30 h annealing at 800°C.

8.4.6 Conclusion

Bulk Ga-based spinel single crystals were grown either from flux or directly from the melt. In particular, VGF, Czochralski, and Kyropoulos-like methods enabled to obtain single crystals of a reasonable size (volume 10–24 cm^3) and high structural quality characterized by a single phase, substantially stoichiometric composition, lack of cracks, grains, visible inclusions, and a narrow FWHM of the rocking curve (<50 arcsec), which is sufficient to prepare substrates for epitaxial purposes, e.g., for electronic and optoelectronic devices or for magnetic Fe-based spinel films. On the other hand, the flux technique produced small crystal volumes, below 1 cm^3. The CVT method resulted in much smaller crystals.

Among $ZnGa_2O_4$, $MgGa_2O_4$, $(Zn,Mg)Ga_2O_4$, and $CoGa_2O_4$ spinels, the most thermally unstable compound is $ZnGa_2O_4$, which requires a proper design of a growth environment (starting composition, growth atmosphere, pressure, temperature gradients, and timing), wherein an addition of Mg improves the thermal stability of $ZnGa_2O_4$, however, in favor of poorer electrical properties (see Section 8.7). The most appropriate growth methods for these compounds are the VGF or VB. $CoGa_2O_4$ and $MgGa_2O_4$ compounds are thermally more stable than $ZnGa_2O_4$ and $(Zn,Mg)Ga_2O_4$ and can be grown by the Czochralski method and its modification (Kyropoulos-like method).

Table 8.3 Outcome of bulk Ga-based spinel crystals and their thermal stability

Phase	Method	Compound	Maximum crystal size in [cm] *D* – diameter, *L* – length, *V* – volume	Structural quality	Reference
Bulk crystals					
Gas	CVT	$CoGa_2O_4$	D=0.1	—	61
Solution	Flux	$ZnGa_2O_4$	1×0.8×0.7	Flux impurities,	51, 52, 53, 54, 55
		$MgGa_2O_4$	—	Inclusions, twins	57
		$CoGa_2O_4$	—		62
Melt	LHPG	$ZnGa_2O_4$	—	Inclusions	56
	OFZ	$MgGa_2O_4$	D=0.4, L=9.5	Inclusions, cracks	59
		$CoGa_2O_4$	D=1	—	60
	VGF/VB	$ZnGa_2O_4$	V=1–5	FWHM < 50 arcsec	5
		$MgGa_2O_4$		Nanoparticles,	3, 64
		$(Zn,Mg)Ga_2O_4$			4
	Czochralski	$ZnGa_2O_4$	D=1.5, L=1.5	FWHM<50 arcsec	5
		$MgGa_2O_4$	D=2, L=4/D=1.2, L=3.8		3/58
		$CoGa_2O_4$	D=1.5, L=2/D=1, L=2.5		49/60
	Kyropoulos-like	$MgGa_2O_4$	D=3.5, L=2	FWHM<50 arcsec	3
		$CoGa_2O_4$	D=3.5, L=2.5		49

Thermal stability

Compound	Transition temperature to random spinel [°C]	Stability temperature in O_2 (≥10 h) [°C]	Stability temperature in H_2 (10 h) [°C]	Reference
$ZnGa_2O_4$	1000	1100	700	5
$MgGa_2O_4$	800	1200	800	3
$CoGa_2O_4$	Not measured	1200	600	49

$ZnGa_2O_4$ and $(Zn,Mg)Ga_2O_4$ are normal; $CoGa_2O_4$ is largely normal, while $MgGa_2O_4$ is an inverse spinel at RT. Independent of the structure, all the spinels under consideration undergo a transition to a random structure at high temperatures, typically above 800°C.

Visible thermal decomposition of the spinels upon annealing takes place at or above 1200°C and at or above 600–800°C at oxidizing and reducing atmospheres, respectively, and at least 10 h annealing time.

The methods applied to grow bulk Ga-based spinel single crystals with a maximum crystal size and basic structural quality, as well as basic thermal stability of the crystals, are collected in Table 8.3.

8.5 Thermal and Mechanical Properties

8.5.1 Thermal Properties

The melting points of Ga-based spinels were measured for $ZnGa_2O_4$ by Galazka et al. [5] (1930 ± 10°C), for $MgGa_2O_4$ by Scott et all. [58] (1950 ± 75°C) and Galazka et al. [3] (1930 ± 10°C), and for $CoGa_2O_4$ by Kramer et al. [60] (1825 ± 5°C) and Galazka [49] (1810 ± 20°C).

The measured solid densities are 6.17 [74], 5.32–5.33 [75, 76], 6.01–6.03 [61, 62] g/cm^3 for $ZnGa_2O_4$, $MgGa_2O_4$, and $CoGa_2O_4$, respectively. An example of the calculated $ZnGa_2O_4$ solid density is 6.2 g/cm^3 [81], close to the measured value.

Thermal conductivities of the Ga-based spinels reported and measured by Galazka et al. [5] and Klimm and Galazka [77] are summarized in Fig. 8.17. Thermal conductivities for $ZnGa_2O_4$, $MgGa_2O_4$, $Zn_{0.5}Mg_{0.5}Ga_2O_4$, and $CoGa_2O_4$ at RT are 22.13, 16.26, 10.34, and 13.4 W/mK, respectively. At high temperatures region (above 700°C) the thermal conductivities of the Ga-based spinels show a small local maximum corresponding to endothermal bend of the DSC curves in Fig. 8.15. This indicates changes in the inversion degree, where there is a mutual exchange of cations between tetrahedrally and octahedrally coordinated sites. Such physical cation rearrangement at high temperatures contributes to heat transfer in a crystal in addition to phonon and electron transport according to Schwarz et al. [31] and Galazka et al. [5].

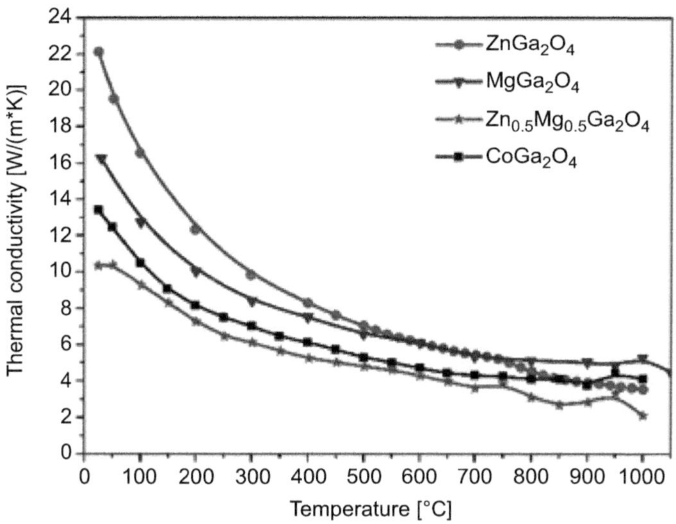

Figure 8.17 Thermal conductivities of $ZnGa_2O_4$, $MgGa_2O_4$, $Zn_{0.5}Mg_{0.5}Ga_2O_4$, and $CoGa_2O_4$ single crystals obtained from the melt [77], wherein data for $ZnGa_2O_4$ are extracted from Ref. [5].

Molar heat capacity of Ga-based spinels can be estimated with the FactSage package using Kopp's rule according to which a good approximation for heat capacity of a mixed oxide is the sum of the individual oxides forming the spinel: here $c_p(MeGa_2O_4) = c_p(MeO) + c_p(Ga_2O_3)$, where Me is a divalent metal. Calculated molar heat capacities for Ga-based spinels at RT: $c_p(ZnGa_2O_4)$ = 133 J/molK, $c_p(MgGa_2O_4)$ = 129 J/molK, and $c_p(CoGa_2O_4)$ = 147 J/molK.

The thermal expansion coefficient of melt-grown $MgGa_2O_4$ single crystals was measured by Hirschle et al. [75] with dilatometry at a temperature range of 103–1673 K. As shown in Fig. 8.18, the thermal expansion coefficient changes with temperature nonlinearly with a discontinuity at 860 K, which corresponds to the glass-like transition and it is associated with the above-discussed cation rearrangement with change in the inverse parameter. Here, the transition occurs at lower temperatures (near 600°C instead of about 800°C); however, this likely depends on the thermal history of the measured sample prepared from the same melt-grown $MgGa_2O_4$ single crystals as for DSC and thermal conductivity measurements. At RT, 500°C, 1000°C, and 1400°C, the thermal expansion coefficient is about 6×10^{-6}, 9×10^{-6}, 14×10^{-6}, and 18×10^{-6} K^{-1}, respectively.

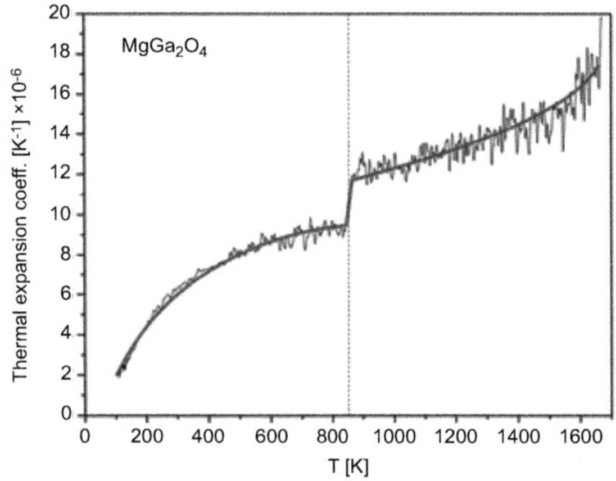

Figure 8.18 Thermal expansion coefficient of a melt-grown $MgGa_2O_4$ single crystal. Adapted from Ref. [75], Copyright 2018, with permission from AIP Publishing.

Thermal expansion of bulk $ZnGa_2O_4$ single crystals was measured by van der Straten et al. [53] (diffractometry) and expressed the thermal expansion in a temperature range of 20–1200°C by the formula: $a(\text{Å}) = 8.332 + 6.18 \times 10^{-5}\, T + 1.06 \times 10^{-8}\, T^2$, where a is the lattice parameter and T is temperature in °C. It results in the thermal expansion coefficient of about 6.2×10^{-5} K^{-1}.

For melt-grown $CoGa_2O_4$ single crystals, Kramer et al. [60] measured the thermal expansion (dilatometry) coefficient at a temperature range of 20–1000°C and concluded the thermal expansion coefficient of 9×10^{-6} K^{-1}.

8.5.2 Mechanical Properties

In a cubic system, a complete tensor of elastic stiffness coefficients is reduced to three independent components c_{11}, c_{12}, and c_{44}, in the same way as it was described in Chapter 7.

The coefficients c_{11}, c_{12}, and c_{44} measured by Hirschle et al. [75] on melt-grown $MgGa_2O_4$ single crystals by resonant ultrasound spectroscopy (RUS) decrease with temperature with a sudden change in c_{11} and c_{44} at 820–850 K, which corresponds to the discontinuity in thermal expansion. That is related to the transition from the inverse to the random spinel as concluded from the DSC (Section 8.4.5.7) and thermal conductivity (Section 8.6.1), although lower transition temperature (by about 200 K) was detected. The elastic stiffness coefficients and their change in a temperature range of 103–1673 K are collected in Table 8.4, which for comparison also contains theoretical values by Bouhemadou et al. [78] (DFT-LDA).

Table 8.4 Elastic stiffness coefficients of melt-grown $MgGa_2O_4$ single crystals and their change in a temperature range of 103–1673 K

c_{ij}	**Temperature range [K] Experimental, Ref. [75]**	**c_{ij} (T_0 = 295 K) [GPa] Experimental, Ref. [75]**	**dc_{ij}/dT [MPa/K] Experimental, Ref. [75]**	**c_{ij} (T_0 = 295 K) [GPa] Theoretical, Ref. [78]**
c_{11}	103–850	280.18(6)	−30.9(2)	262
	820–1673	290.9(1)	−50.3(1)	
c_{12}	103–1673	148.02(4)	−11.1(1)	145
c_{44}	103–850	129.87(3)	−10.3(1)	118
	820–1673	133.96(7)	−17.55(7)	

Having measured or theoretically predicted elastic stiffness coefficients c_{11} (longitudinal compression), c_{12} (transverse expansion), and c_{44} (shear modulus), it is possible to calculate other useful mechanical properties, i.e., bulk modulus B_0, Young's modulus E, shear modulus G, and Poisson's ratio ν from the following Voigt expressions:

$$B_0 = \frac{1}{3}(c_{11} + 2c_{12}) \tag{8.1}$$

$$E = \frac{(c_{11} - c_{12}) \times (c_{11} + 2c_{12})}{(c_{11} + c_{12})} \tag{8.2}$$

$$G = c_{44} \tag{8.3}$$

$$\nu = \frac{1}{2}\left(1 - \frac{c_{11} - c_{12}}{c_{11} + c_{12}}\right) \tag{8.4}$$

The corresponding bulk modulus B_0, Young's modulus E, shear modulus G, and Poisson's ratio v for experimental [75]/theoretical [78] data are 192/186 GPa, 178/158 GPa, 130/118 GPa, and 0.346/0.357, respectively.

Brik [38], Bouhemadou and Khenata [79], Kushwaha [80], Zeraga et al. [81], and Pandey et al. [82] reported for $ZnGa_2O_4$ the following theoretical values of elastic stiffness coefficients (in GPa): c_{11} = 249.5/228/348.6/390/288, c_{12} = 121.8/120/176.7/143/218, and c_{44} = 115.5/107/128/114/125. Basing on Eqs. 8.1, 8.2, and 8.4 and above-mentioned theoretical values of c_{ij}, the bulk modulus, Young's modulus, and Poisson's ratio are B_0 = 164/156/234/225/241 GPa, E = 170/145/230/313/100 GPa, and v = 0.328/0.345/0.336/0.268/0.431, respectively. For comparison, experimental bulk modulus for synthetized $ZnGa_2O_4$ is 233 GPa [83].

Longitudinal v_L, transverse v_T, and average v_{AV} sound velocities can be calculated from mechanical properties [38, 80], which in turn enable to calculate the Debye temperature Θ_D [80]:

$$v_L = \sqrt{\left[\frac{3B_0 + 4G}{3\rho_s}\right]}, \quad v_T = \sqrt{\frac{G}{\rho_s}}, \quad v_{AV} = \frac{1}{\sqrt[3]{\frac{1}{3}\left(\frac{1}{v_L^3} + \frac{2}{v_T^3}\right)}} \tag{8.5}$$

$$\Theta_D = \frac{h}{k_B}\sqrt[3]{\frac{3n_a N_A \rho_s}{4\pi M}}\, v_{AV} \tag{8.6}$$

where h and k_B are Planck's and Boltzmann's constants, N_A is Avogadro's number, ρ_s is solid density, while n_a is the number of atoms per unit cell [56].

Using experimental elastic stiffness coefficients for $MgGa_2O_4$ [75], v_L = 8287 m/s, v_T = 4943 m/s, v_{AV} = 5472 m/s, and the resulting Θ_D = 1540 K.

The theoretical values of sound velocities and the Debye temperature for $ZnGa_2O_4$ by Brik [38], Kushwaha [80], and Zeraga et al. [81] are as follows: v_L = 6342/7823/7849 m/s, v_T = 3728/4194/4356 m/s, v_{AV} = 4133/4684/4852 m/s, Θ_D = 944/591/603 K, respectively. These values are significantly smaller than those derived for $MgGa_2O_4$.

8.5.3 Conclusion

Melting points of Ga-based spinel compounds ($ZnGa_2O_4$, $MgGa_2O_4$, $CoGa_2O_4$) are high and span between 1810 and 1930°C. As they contain at least one thermally unstable binary oxide, such high melting points enhance incongruent decomposition and increase difficulties in obtaining bulk single crystals. Thermal conductivities at or near RT of $ZnGa_2O_4$, $MgGa_2O_4$, $Zn_{0.5}Mg_{0.5}Ga_2O_4$, and $CoGa_2O_4$ are 22.1, 16.26, 10.34, and 13.4 W/mK, respectively; however, at high temperatures (>800°C, when they become mainly random spinels), they are closer to each other. At high temperatures Ga-based spinels show a local rise in thermal conductivities indicating changes in the inversion degree associated with a mutual exchange of cations between tetrahedrally and octahedrally coordinated sites, which contributes to heat transfer in a crystal in addition to phonon and electron transport. Thermal expansion coefficient around RT decreases from 6.2×10^{-5} K^{-1} for $ZnGa_2O_4$ through 9×10^{-6} K^{-1} for $CoGa_2O_4$ to 6×10^{-6} for $MgGa_2O_4$.

Mechanical properties of $MgGa_2O_4$ were studied both experimentally and theoretically, while of $ZnGa_2O_4$ crystals, theoretically only. Mechanical properties of $CoGa_2O_4$ crystals were not reported so far. The tensor of elastic stiffness coefficients for these Ga-based spinels is reduced to three independent components: c_{11}, c_{12}, and c_{44}. For $MgGa_2O_4$, experimental values of the coefficients in a wide temperature range are (in GPa): c_{11} = 280–291, c_{12} = 148, and c_{44} = 130–134, while for $ZnGa_2O_4$ theoretical values of the coefficients are (in GPa): c_{11} = 228–390, c_{12} = 120–218, and c_{44} = 107–128, and show inconsistency. From the elastic stiffness coefficients, other mechanical parameters can be derived, such as bulk, Young's, and shear moduli, and Poisson's ratio. Knowing the values of mechanical properties, also sound velocities and Debye temperature can be concluded as well. The values of thermal and mechanical properties of Ga-based spinels are summarized in Table 8.5.

Table 8.5 Experimental and/or theoretical values of thermal and mechanical properties of bulk Ga-based spinel single crystals

Property	Unit	Compound	Value	Reference
Solid density	g/cm^3	$ZnGa_2O_4$	6.17	74
		$MgGa_2O_4$	5.32–5.33	75, 76
		$CoGa_2O_4$	6.01–6.03	61, 62
Melting point	°C	$ZnGa_2O_4$	1900±20	5
		$MgGa_2O_4$	1930±10/1950±75	3/58
		$CoGa_2O_4$	1810±20/1825±5	49/60
Thermal expansion coefficient of the lattice parameter	K^{-1}	$ZnGa_2O_4$	6.2×10^{-5} (20–1200°C)	53
		$MgGa_2O_4$	$6–18\times10^{-6}$ (20–1400°C)	75
		$CoGa_2O_4$	9×10^{-6} (20–1000°C)	60
Thermal conductivity at RT	W/mK	$ZnGa_2O_4$	22.1	5
		$MgGa_2O_4$	16.26	77
		$CoGa_2O_4$	13.4	77
Molar heat capacity at RT	J/molK	$ZnGa_2O_4$	133[a]	
		$MgGa_2O_4$	129[a]	
		$CoGa_2O_4$	147[a]	
Elastic stiffness coefficients at RT	GPa	$ZnGa_2O_4$	c_{11}=250[b]/228[b]/349[b]/390[b]/288[b] c_{12}=122[b]/120[b]/177[b]/143[b]/218[b] c_{44}=116[b]/107[b]/128[b]/114[b]/125[b]	38/79/80/81/82
		$MgGa_2O_4$	c_{11}=280.2/262[b] c_{12}=148/145[b] c_{44}=129.9/118[b]	75/78
		$CoGa_2O_4$	—	—

Property	Unit	Compound	Value	Reference
Young's modulus	GPa	$ZnGa_2O_4$	170[b,c]/145[b,c]/230[b,c]/313[b,c]/100[b,c]	38/79/80/81/82
$(c_{11}-c_{12})\times(c_{11}+2c_{12})/(c_{11}+c_{12})$		$MgGa_2O_4$	178[c]/158[b,c]	75/78
		$CoGa_2O_4$	—	—
Shear modulus	GPa	$ZnGa_2O_4$	116[b]/107[b]/128[b]/114[b]/125[b]	38/79/80/81/82
c_{44}		$MgGa_2O_4$	130[c]/118[b,c]	75/78
		$CoGa_2O_4$	—	—
Bulk modulus	GPa	$ZnGa_2O_4$	164[b,c]/156[b,c]/234[b,c]/225[b,c]/241[b,c]	38/79/80/81/82
$(c_{11}+2c_{12})/3$		$MgGa_2O_4$	192[c]/186[b,c]	75/78
		$CoGa_2O_4$	—	—
Poisson's ratio	—	$ZnGa_2O_4$	0.328[b,c]/0.345[b,c]/0.336[b,c]/0.268[b,c]/0.431[b,c]	38/79/80/81/82
$(1-(c_{11}-c_{12})/(c_{11}+c_{12}))/2$		$MgGa_2O_4$	0.346[c]/0.357[b,c]	75/78
		$CoGa_2O_4$	—	—
Longitudinal sound velocity	km/s	$ZnGa_2O_4$	6342[b]/7823[b]/7849[b]	38/80/81
		$MgGa_2O_4$	8287[c]	75
		$CoGa_2O_4$	—	—
Transverse sound velocity	km/s	$ZnGa_2O_4$	3728[b]/4194[b]/4356[b]	38/80/81
		$MgGa_2O_4$	4943[c]	75
		$CoGa_2O_4$	—	—
Debye temperature	K	$ZnGa_2O_4$	944[b]/591[b]/603[b]	38/80/81
		$MgGa_2O_4$	1540[c]	75
		$CoGa_2O_4$	—	—

[a]estimated with the FactSage package; [b]theoretical values; [c]calculated basing on data in the corresponding reference

8.6 Optical Properties

8.6.1 Transmittance, Absorption Edge, and Optical Bandgap

The transmittance spectra at RT of melt-grown (Czochralski, VGF) bulk $MgGa_2O_4$, $ZnGa_2O_4$, $Zn_{0.5}Mg_{0.5}Ga_2O_4$, and $CoGa_2O_4$ single crystals in the UV, visible, and NIR spectral regions by Galazka et al. [3, 5, 84] are shown in Fig. 8.19a, Fig. 8.19b, Fig. 8.19c, and Fig. 8.19d, respectively. In three former cases, the absorption edge is very steep and originates at 250, 275, and 268 nm for $MgGa_2O_4$ and $ZnGa_2O_4$, and $Zn_{0.5}Mg_{0.5}Ga_2O_4$, respectively. Similar absorption edge for $MgGa_2O_4$ and $ZnGa_2O_4$, 250 and 280 nm, was reported by Scott et al. [58] and Yan et al. [55] for Czochralski- and flux-grown crystals, respectively. The as-grown crystals reveal an absorption in the NIR and in red part of the visible spectrum, which is typical for the free

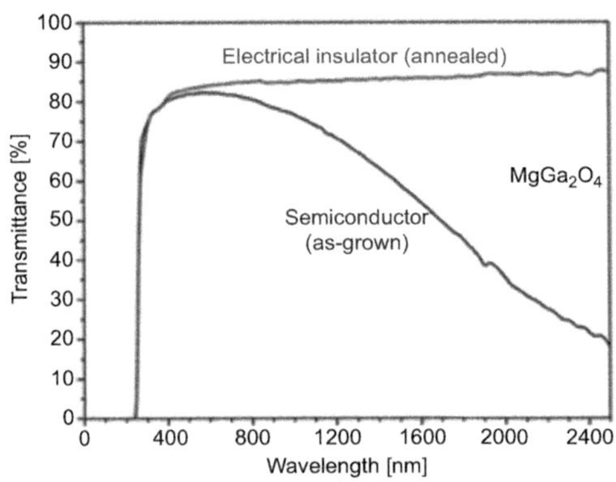

(a)

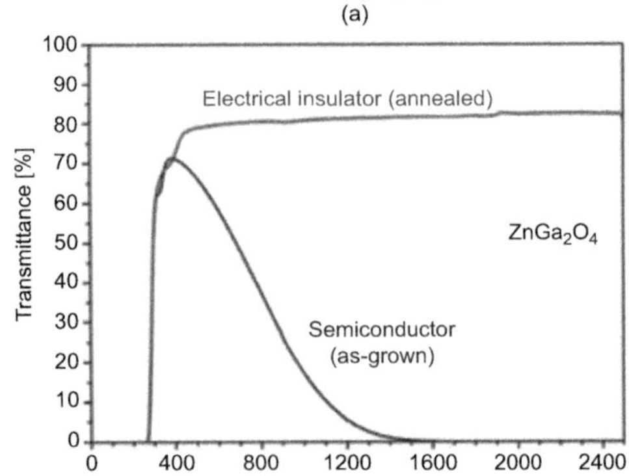

(b)

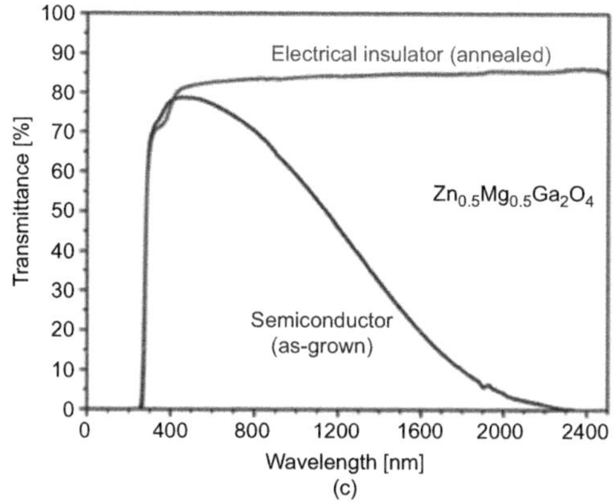

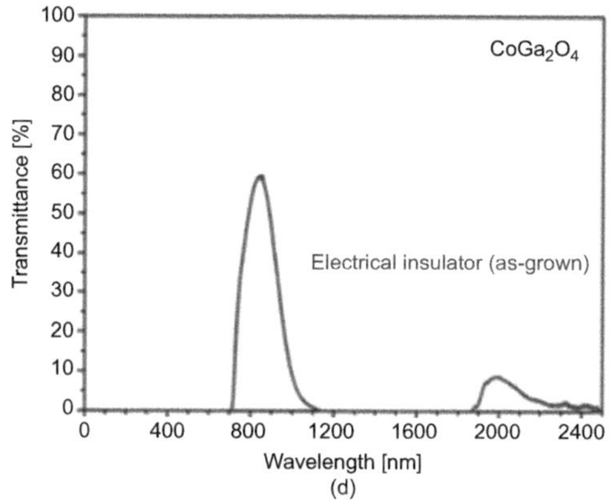

Figure 8.19 Optical transmittance spectra of melt-grown bulk $MgGa_2O_4$ (a), $ZnGa_2O_4$ (b), $Zn_{0.5}Mg_{0.5}Ga_2O_4$ (c), and $CoGa_2O_4$ (d) single crystal wafers (0.5 mm thick), which are either semiconducting (as-grown) or electrically insulating (annealed in air for 10 h at 1200°C for $MgGa_2O_4$, 800°C for $ZnGa_2O_4$, and 1000°C for $Zn_{0.5}Mg_{0.5}Ga_2O_4$. Figure (a) reprinted (with minor modifications) with permission from Ref. [3], Copyright 2015, John Wiley and Sons, and (b) reprinted from Ref. [5] under Creative Commons license.

carrier absorption as discussed in previous chapters in relation to other TSOs. Indeed, $MgGa_2O_4$, $ZnGa_2O_4$, and $Zn_{0.5}Mg_{0.5}Ga_2O_4$ showed free carrier concentration at the level of 10^{17}, 10^{19}, and 10^{19} cm^{-3}, respectively. Annealing these compounds in an oxidizing atmosphere for 10 h at 700°C or above turns $MgGa_2O_4$, $ZnGa_2O_4$, and $Zn_{0.5}Mg_{0.5}Ga_2O_4$ crystals into electrically insulating state, which is reflected in full transmittance in the visible and NIR spectral regions. Scott et al. [58] showed that Czochralski-grown $MgGa_2O_4$ crystals are transparent up to about 7 μm.

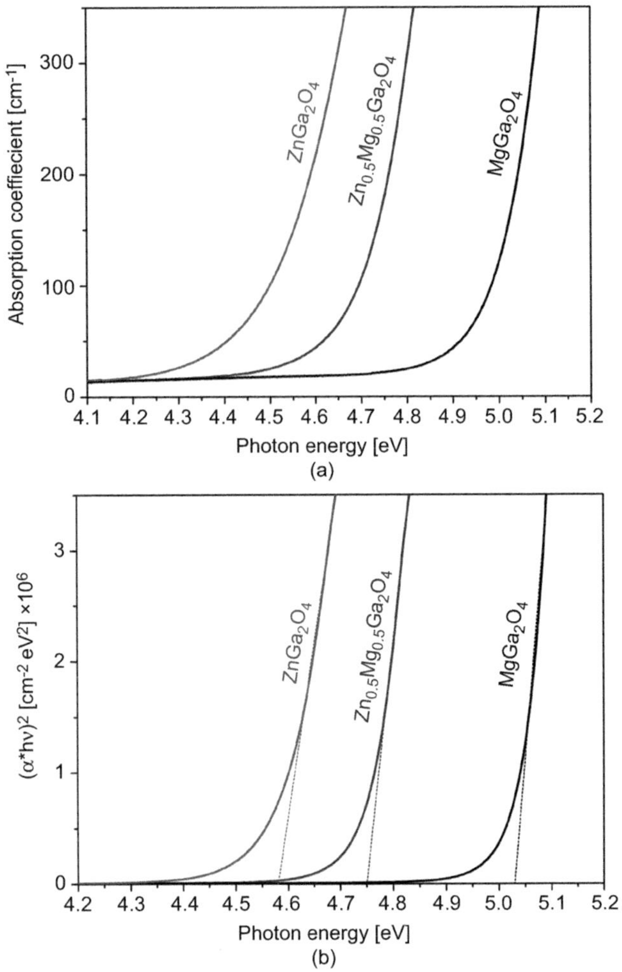

Figure 8.20 Spectral dependence of the absorption coefficient α (sample thickness of 150 μm) (a), and absorption coefficient α in power 2 versus photon energy (b) for melt-grown $ZnGa_2O_4$, $MgGa_2O_4$, and $Zn_{0.5}Mg_{0.5}Ga_2O_4$ crystal samples. Dot lines in (b) indicate an extrapolation of the linear part of the curve that corresponds to direct transitions [85].

On the other hand, $CoGa_2O_4$ is opaque in the visible and NIR spectral regions (Fig. 8.19d) with a transparency window between 800 and 950 nm. Kramer et al. [60] showed for melt-grown $CoGa_2O_4$ crystals transmittance that is basically opaque in the visible and NIR regions (up to 12 μm) with a transparent window between 3 and 5 μm.

To reveal optical transitions, whether direct or indirect, allowed or forbidden, the absorption coefficient α measured on thin samples (Fig. 8.20a) was plotted versus photon energy at different relations versus the photon energy E_{ph} to achieve the best linear fit. As shown in Fig. 8.20b, the best linear fit was found for the relation $\alpha^2 \sim E_{ph}$, which indicates a direct transition (Galazka et al. [5, 85]. The derived optical direct bandgaps at RT for $ZnGa_2O_4$, $MgGa_2O_4$, and $Zn_{0.5}Mg_{0.5}Ga_2O_4$ are 4.59 eV, 5.03 eV, and 4.75 eV, respectively. Basing on the above-mentioned absorption coefficient–photon energy relation measured for mixed $ZnGa_2O_4$–$MgGa_2O_4$ spinels at different molar ratios and having a general formula of $Zn_{1-x}Mg_xGa_2O_4$ (x = 0.1–0.5), the bandgap changes approximately linearly over the whole composition range.

For $MgGa_2O_4$, the bandgap decreases with temperature (up to 1400°C) at a rate of 1.35 meV/K [3]. Bulk $CoGa_2O_4$ single crystals are too opaque in the visible spectrum to derive the optical bandgap from absorbance measurements. On the other hand, ellipsometric measurements of melt-grown bulk $ZnGa_2O_4$, $MgGa_2O_4$, and $CoGa_2O_4$ single crystals showed the bandgaps of 4.84 ± 0.01 eV, 4.95 ± 0.017 eV, and 3.55 ± 0.05 eV, respectively [89].

For comparison, the theoretical bandgap of $MgGa_2O_4$ was found to be direct with a value of 5.01 eV [42], which matches the experimental value. Some of values of the theoretical direct bandgap of $ZnGa_2O_4$ that are close experimental ones are 4.24 [37], 4.57–4.71 [39], and about 4.5 eV [41]. A theoretical value of the indirect bandgap of $CoGa_2O_4$ is 3.97 eV (over both spins up and down) [44].

8.6.2 Dielectric Constant and Refractive Index

The relative static dielectric constant ε_r reported by Scott et al. [58] for melt-grown $MgGa_2O_4$ crystals measured along the ⟨111⟩ direction is ε_r = 11.7 over a frequency range of 1 kHz to 2 MHz, while by Kramer et al. [60] for melt-grown $CoGa_2O_4$ is ε_r = 9.2 over a frequency range of 0.2–9.2 GHz. It should be noted that these values are expected to be isotropic because ε_r is a second-rank tensor. $MgGa_2O_4$ ceramics revealed the relative static dielectric constant measured by Kan et al. [86] of ε_r = 9.3 (maximum value). The relative static dielectric constant for bulk $ZnGa_2O_4$ crystals was not measured so far; however, a value of ε_r = 9.8 at 9 GHz and ε_r = 10.2–10.4 at 13 GHz was reported for ceramics by Lu et al. [87] and Xue et al. [88], respectively. In this case, the values of the relative static dielectric constant depend on the ceramic density and sintering temperature.

High-frequency dielectric constant ε_∞ of melt-grown bulk $ZnGa_2O_4$, $MgGa_2O_4$, $Zn_{0.5}Mg_{0.5}Ga_2O_4$, and $CoGa_2O_4$ single crystals was concluded by Anooz and Galazka [89] from ellipsometric measurements, with the values of $\varepsilon_\infty = 2.6 \pm 0.04$, 2.36 ± 0.012, 2.6 ± 0.04, and 1.67 ± 0.17, respectively.

Examples of theoretical ε_∞ values for $ZnGa_2O_4$ are ~2.8 [38] and 3.4 [40]. Pandey et al. [82] calculated for $ZnGa_2O_4$ values of the static and high-frequency dielectric constants as $\varepsilon_0 = 13.72$ and $\varepsilon_\infty = 5.1$, respectively.

The refractive index n of melt-grown $ZnGa_2O_4$, $MgGa_2O_4$, $Zn_{0.5}Mg_{0.5}Ga_2O_4$, and $CoGa_2O_4$ derived from ellipsometric measurements by Anooz and Galazka [89] in a spectral region from 240 to 830 nm is shown in Fig. 8.21. The highest value of the refractive index in the measured region shows $CoGa_2O_4$, followed by $ZnGa_2O_4$, $Zn_{0.5}Mg_{0.5}Ga_2O_4$, and $MgGa_2O_4$, wherein at short wavelengths (<300 nm), the refractive index for $ZnGa_2O_4$ increases with the highest rate. Exemplary values of the refractive index at 300, 600, and 800 nm are n(300 nm) = 2.074/2.013/2.000/2.212, n(600 nm) = 1.948/1.846/1.880/1.999, and n(800 nm) = 1.936/1.935/1.869/1.980 for $ZnGa_2O_4$, $Zn_{0.5}Mg_{0.5}Ga_2O_4$, $MgGa_2O_4$, and $CoGa_2O_4$, respectively.

The refractive index reported by Suzuki et al. [22] for Ni-doped (0.143 at.%) melt-grown $MgGa_2O_4$ crystals is 1.88–1.85 for the spectral region 633–1544 nm. An order of magnitude higher Ni concentration in the crystals increases the refractive index very little, by about 0.6%.

For comparison, the refractive index of $ZnGa_2O_4$ thin films obtained by the mist-CVD decreases from about 2.13 to about 1.98 at short wavelengths of 320–600 nm [90], which is a bit smaller as compared with bulk crystals, while of thin films obtained by the RF sputtering in the same spectral region, the refractive index decreases from about 1.825 to 1.7–1.725 depending on annealing conditions [91], which is much smaller than that obtained for bulk crystals. An example of the theoretical refractive index for $ZnGa_2O_4$ is 2.1 at 600 nm [40], a bit higher than that measured for bulk crystals at the same wavelength.

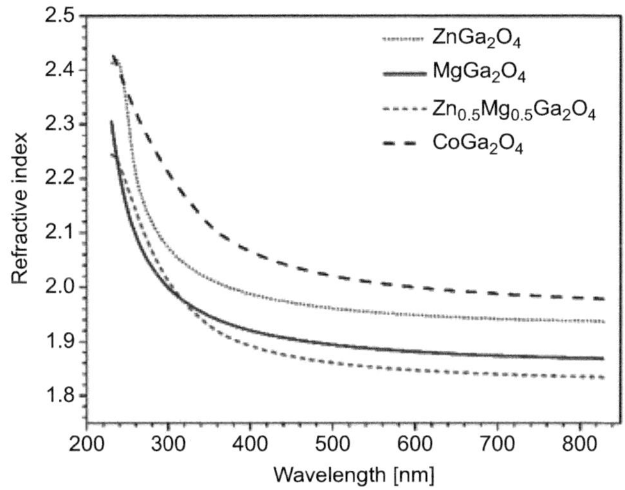

Figure 8.21 Refractive index of melt-grown bulk $ZnGa_2O_4$, $MgGa_2O_4$, $Zn_{0.5}Mg_{0.5}Ga_2O_4$, and $CoGa_2O_4$ single crystals derived from ellipsometric measurements [89].

8.6.3 Luminescence

The most intensively studied spinel in terms of luminescence is $ZnGa_2O_4$, as it was found as an efficient phosphor. Indeed, it reveals a strong intrinsic blue emission [92], and green to red extrinsic emissions when doped and co-doped with Mn [8], Eu [93], Cr [94], (Mn + Cr) [95], (Eu + Tb) [96], and (Ge + Li + Mn) [97]. $MgGa_2O_4$ was also claimed as a phosphor [98] and potential host for tunable infrared laser applications when doped with Ni [59].

The cathodoluminescence (CL) spectra of melt-grown bulk $ZnGa_2O_4$, $MgGa_2O_4$, and $Zn_{0.5}Mg_{0.5}Ga_2O_4$ single crystals by Schewski and Galazka [99] and Galazka et al. [3] show at RT a broad and structured emissions between 300 and 450, 300 and 550 nm, and 300 and 600 nm, respectively (Fig. 8.22). $CoGa_2O_4$ did not show any CL signal at the 250–1600 nm spectral range, likely due to a high absorption in the visible spectrum. The emissions are peaking in the blue region of the visible spectrum at about 350, 360, and 380–400 nm for as-grown $ZnGa_2O_4$, $MgGa_2O_4$, and $Zn_{0.5}Mg_{0.5}Ga_2O_4$, respectively.

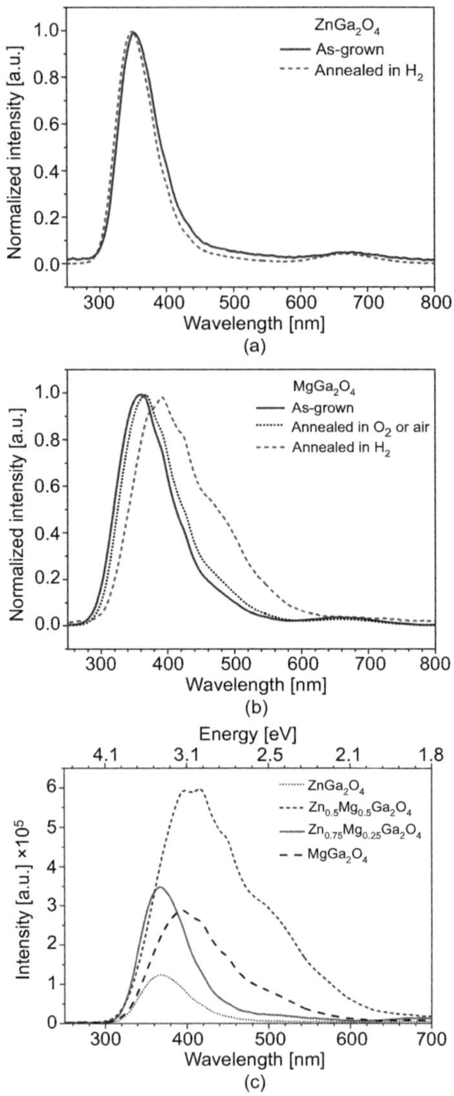

Figure 8.22 Normalized cathodoluminescence (CL) spectra of melt-grown bulk $ZnGa_2O_4$ (a) [99]; $MgGa_2O_4$ (b); and combined absolute spectra of $ZnGa_2O_4$, $MgGa_2O_4$, $Zn_{0.75}Mg_{0.25}Ga_2O_4$, and $Zn_{0.5}Mg_{0.5}Ga_2O_4$ (c) [99]. All spectra were taken at RT. Annealing of $ZnGa_2O_4$ in the presence of H_2 was done at 600°C for 10 h, while annealing of $MgGa_2O_4$ was carried out in the presence of O_2 at 1000°C for 20 h, and in the presence of hydrogen at 600°C for 10 h. Figure (b) reprinted with permission from Ref. [3], Copyright 2015, John Wiley and Sons.

The normalized CL spectra of $ZnGa_2O_4$ (Fig. 8.22a), whether as-grown or annealed in the presence of hydrogen, overlap with each other and show the structure consisting of several bonds. Very similar structure of the normalized spectra can be concluded for $MgGa_2O_4$ (Fig. 8.22b), where the structured emission is more pronounced. $MgGa_2O_4$, whether as-grown (semiconducting or insulating) or annealed in the oxidizing atmosphere, also revealed overlapped CL spectra consisting of five bands; however, annealing in the reducing atmosphere changed the relative intensities and makes the luminescence broader. And finally, $Zn_{0.5}Mg_{0.5}Ga_2O_4$ shows the broadest luminescence with a similar structure to those of $ZnGa_2O_4$ and $MgGa_2O_4$. The structure and energy positions of the CL spectra of bulk Ga-based spinels are very similar to those of the CL spectra of bulk β-Ga_2O_3 single crystals (see Chapter 4) and can be attributed to self-trapped excitons.

Figure 8.22c compares absolute intensities of CL spectra of as-grown bulk Ga-based spinel single crystals measured in parallel in one single run (the same measurement conditions). Interestingly, mixed bulk $Zn_xMg_{1-x}Ga_2O_4$ (x = 0.25 and 0.5) single crystals show much higher emission intensities than individual $ZnGa_2O_4$ and $MgGa_2O_4$ crystals; therefore, they might be good candidates for phosphor applications. For x = 0.25, the emission is very similar to $ZnGa_2O_4$, but with higher x = 0.5, it is more similar to the emission of $MgGa_2O_4$, but in both cases the emission is more intense and broader as compared with individual compounds.

For comparison, thin films of $ZnGa_2O_4$ showed different intensities of the emission structure. For example, films obtained by the mist-CVD had CL emission bands at 2.5 (496 nm), 2.9 (428 nm), and 3.4 (365 nm) eV [90], by RF magnetron sputtering at 420 (blue), 510 (green), and 705 (red) nm for undoped, Mn-doped, and Cr-doped [100] or at 490 nm (2.53 eV) [101], and by MOCVD at 332 nm (3.73 eV) [1]. The CL spectra of $ZnGa_2O_4$ ceramics showed peaks at 380, 360, and 505 eV for undoped, Ge-doped, and Mn-doped [12].

The photoluminescence (PL) spectra of Mn-doped bulk $ZnGa_2O_4$ single crystals grown by the LHPG showed a peak at about 505 nm under 266 nm excitation [56]. The PL spectra of undoped $ZnGa_2O_4$ thin films under UV excitation (energies higher than the bandgap) are generally similar to the CL spectra. However, emission spectra depend on the film stoichiometry, substrate type, growth and annealing conditions. For example, undoped PLD-grown films under 232–283 nm excitation revealed broad emissions peaking at about 450 nm [102], 460 [103, 104], 480–500 [105]. Mn-doped PLD-grown films under similar UV excitation showed the emission peaking at about 503–512 nm [106–109]. Emission at 502 nm under 267 nm excitation was also observed in Mn-doped $MgGa_2O_4$ ceramics [98].

8.6.4 Phonon Modes

Ga-based spinels of $MeGa_2O_4$ (Me = Zn, Mg, Co) show 42 normal modes, 3 acoustic modes, and 39 optical modes [80]:

$$A_{1g}(R) + E_g(R) + T_{1g} + 3T_{2g}(R) + 2A_{2u} + 2E_u + 5T_{1u}(IR) + 2T_{2u} \tag{8.7}$$

where R and IR are Raman- and infrared-active phonon modes. The modes can be either Raman or infrared active, but not both. There are five Raman active ($A_{1g} + E_g + 3T_{2g}$) and four IR active ($4T_{1u}$) modes.

Among Ga-based spinels, all phonon modes were experimentally and theoretically revealed for $ZnGa_2O_4$ only, by Gorkom et al. [112], and Kushwaha [80], López et al. [37, 111], and Lauwers and Herman [110] (Table 8.6). Low-frequency modes relate to vibrations of the GaO_6 octahedral units as Ga–O bonds are longer than Zn–O bonds, while high-frequency modes are related to vibrations of ZnO_4 tetrahedral units, where stretching is larger than bond bending.

Table 8.6 Calculated and experimental phonon modes of $ZnGa_2O_4$

	Ref. [80] (theoretical) [cm^{-1}]	**Ref. [37, 111] (theoretical) [cm^{-1}]**	**Ref. [110] (theoretical) [cm^{-1}]**	**Ref. [112] (experimental) [cm^{-1}]**
A_{1g} (R)	714	717	707	714
E_g (R)	638	395	402	638
T_{2g} (1)	614	618	631	611
T_{2g} (2)	469	488	492	467
T_{2g} (3)	183	186	180	—
T_{1u} (1) TO	574	580	566	570

	Ref. [80] (theoretical) [cm^{-1}]	Ref. [37, 111] (theoretical) [cm^{-1}]	Ref. [110] (theoretical) [cm^{-1}]	Ref. [112] (experi-mental) [cm^{-1}]
T_{1u} (2)	426	429	440	420
T_{1u} (3)	329	342	330	328
T_{1u} (4)	176	175	211	175
T_{1u} (1) LO	939	—	—	935
T_{1u} (2)	691	—	—	689
T_{1u} (3)	537	—	—	538
T_{1u} (4)	286	—	—	287
A_{2u} (1)	689	702	—	—
A_{2u} (2)	411	419	—	—
E_u (1)	567	563	—	—
E_u (2)	227	229	—	—
T_{1g}	638	366	—	—
T_{2u} (1)	461	450	—	—
T_{2u} (2)	138	135	—	—

8.6.5 Conclusion

Optical properties of Ga-based spinels in quest, except $ZnGa_2O_4$, have not been studied extensively, neither experimentally nor theoretically. However, the basic optical properties, as shown in Table 8.7, could be derived from existing data and extra measurements on bulk Ga-based spinel crystals. $ZnGa_2O_4$, $MgGa_2O_4$, and $Zn_{1-x}Mg_xGa_2O_4$ (x = 0.1–0.5) are transparent in the visible and NIR spectral regions, which is only affected by the free carrier absorption. On the other hand, $CoGa_2O_4$ is opaque in those spectral regions to a large extent. The absorption edge is located in the UV part of spectrum for $ZnGa_2O_4$ (275 nm), $MgGa_2O_4$ (250 nm), and $Zn_{0.5}Mg_{0.5}Ga_2O_4$ (268 nm). It could not be, however, estimated for $CoGa_2O_4$ because of its opacity in the visible spectrum. Due to the flatness of the VBM of the Ga-based spinels, their bandgaps can be considered as pseudo-direct with experimental values (based on optical measurements) of 4.59, 4.9–5.03, 4.75, and 3.55 eV for

$ZnGa_2O_4$, $MgGa_2O_4$, $Zn_{0.5}Mg_{0.5}Ga_2O_4$, and $CoGa_2O_4$, respectively, which are consistent with theoretical studies.

The experimental relative static dielectric constant is in the range of 9.2–11.7, while the high-frequency dielectric constant varies between 1.67 and 2.6. The refractive index in the visible spectrum holds the following relation: $n(CoGa_2O_4) > n(ZnGa_2O_4) > n(Zn_{0.5}Mg_{0.5}Ga_2O_4) > n(MgGa_2O_4)$, with the values of 2.212/2.074/2.013/2.000 at 300 nm, and 1.980/1.936/1.935/1.869 at 800 nm, respectively.

Table 8.7 Basic experimental and theoretical optical properties of bulk Ga-based spinel crystals

Property	Unit	Compound	Value/Description	Reference
Bandgap (direct or pseudo-direct or indirect)	eV	$ZnGa_2O_4$	4.59/4.5–4.7[a]	5/39, 41
		$MgGa_2O_4$	4.9–5.03/ 4.95/ 5.01[a]	3, 85/89/42
		$Zn_{0.5}Mg_{0.5}Ga_2O_4$	4.75	85
		$CoGa_2O_4$	3.55/3.97[a]	89/44
Absorption edge	nm	$ZnGa_2O_4$	275	5
		$MgGa_2O_4$	250	3
		$Zn_{0.5}Mg_{0.5}Ga_2O_4$	268	85
		$CoGa_2O_4$	-	-
Luminescence peak at RT	nm	$ZnGa_2O_4$	350	99
		$MgGa_2O_4$	360	3
		$Zn_{0.5}Mg_{0.5}Ga_2O_4$	380–400	99
		$CoGa_2O_4$	-	-
Relative static dielectric constant ε_r		$ZnGa_2O_4$	9.8[b]	87
		$MgGa_2O_4$	11.7	58
		$Zn_{0.5}Mg_{0.5}Ga_2O_4$	-	-
		$CoGa_2O_4$	9.2	60
High-frequency dielectric constant ε_∞		$ZnGa_2O_4$	2.6 ± 0.04	89
		$MgGa_2O_4$	2.36 ± 0.012	89
		$Zn_{0.5}Mg_{0.5}Ga_2O_4$	2.6 ± 0.04	89
		$CoGa_2O_4$	1.67 ± 0.17	89

Property	Unit	Compound	Value/Description	Reference
Experimental refractive index at 300/600/ 800 nm		$ZnGa_2O_4$	2.074/1.948/1.936	89
		$MgGa_2O_4$	2.000/1.880/1.869	89
		$Zn_{0.5}Mg_{0.5}Ga_2O_4$	2.013/1.846/1.935	89
		$CoGa_2O_4$	2.212/1.999/1.980	89
Phonon modes	cm^{-1}	$ZnGa_2O_4$	138[(a)]/-, 176[(a)]/175, 183[(a)]/-, 227[(a)]/-, 286[(a)]/287, 329[(a)]/328, 411[(a)]/-, 426[(a)]/420, 461[(a)]/-, 469[(a)]/467, 537[(a)]/538, 567[(a)]/-, 574[(a)]/570, 614[(a)]/611, 638[(a)]/638, 689[(a)]/-, 691[(a)]/689, 714[(a)]/714, 939[(a)]/935[(a)]	80/112
		$MgGa_2O_4$	-	-
		$Zn_{0.5}Mg_{0.5}Ga_2O_4$	-	-
		$CoGa_2O_4$	-	-

[(a)]theory; [(b)]ceramics

8.7 Electrical Properties

The origin of the electrical conductivity of Ga-based spinels was very little studied theoretically and experimentally. Intrinsic point defects rather do not form shallow donors [41], therefore cannot be responsible for the electrical conductivity. Rather, extrinsic point defects may form shallow donors, likely hydrogen as in the case of other TSOs. Indeed, annealing experiments of melt-grown bulk $MgGa_2O_4$, $ZnGa_2O_4$, and $(Zn,Mg)Ga_2O_4$ crystals by Galazka et al. [3, 5, 113] indirectly point to hydrogen as the source of the electrical conductivity. Annealing semiconducting crystals in an oxidizing atmosphere turns the crystals into electrically insulating state, which can be recovered by annealing in the presence of hydrogen. Annealing semiconducting crystals in the presence of hydrogen does not remove the electrical conductivity. Intentional doping of

$MgGa_2O_4$ with Si^{4+}, and $ZnGa_2O_4$ with Si^{4+} and Zr^{4+} that suppose to substitute Ga^{3+} in the crystal lattice, as well as with Y^{3+} substituting Zn^{2+} in $ZnGa_2O_4$, substantially did not affect the electrical properties as compared with undoped crystals [3, 5], likely due to very high free electron concentration in undoped crystals, which may overlap the electron concentration arising from the intentional dopant. Another possible extrinsic dopant that theoretically may induce the electrical conductivity in $ZnGa_2O_4$ is nitrogen, capable of forming shallow donors in the −1 charge state (N^{1-}) coexisted with oxygen vacancies in the +2 charge state (O^{2+}) [41].

8.7.1 Free Electron Concentration, Hall Mobility, and Resistivity at RT

Electrical properties at RT of as-grown and annealed bulk Ga-based spinel single crystals ($ZnGa_2O_4$, $MgGa_2O_4$, and $(Zn,Mg)Ga_2O_4$) obtained from the melt were reported so far only by Galazka et al. [3, 5, 113] (Table 8.8). As-grown bulk $ZnGa_2O_4$ single crystals showed very high electrical conductivity with the free electron concentration n_e, Hall mobility μ, and resistivity ρ of 3×10^{18}–9×10^{19} cm^{-3}, 20–100 cm^2/Vs, and 0.002–0.1 Ωcm, respectively [5].

As shown in Fig. 8.23, annealing $ZnGa_2O_4$ crystals in the presence of O_2 at or above 700°C for at least 10 h (here 20 h) makes the crystals electrically insulating (Hall effect not measurable) with a crystal color change from blue to colorless. Annealing in the presence of H_2 (5% H_2 + Ar) up to 800°C for 10 h substantially does not affect the electrical properties; the free electron concentration remains at a high level above 10^{19} cm^{-3}. On the other hand, annealing electrically insulating crystals in the presence of H_2 turned the crystals into semiconducting state with the electron concentration approaching 10^{19} cm^{-3}.

Also, Yan et al. [55] reported that electrically insulating flux-grown bulk $ZnGa_2O_4$ crystals turned into semiconductor-like state (resistivity of about 10 Ωcm at RT) when exposed to annealing in hydrogen at 600–800°C for 10–100 h. In this case, however, the semiconductor-like behavior was limited to a surface layer of a thickness of about 10 μm. The activation energy of shallow donors was found to be about 20 meV [55]. Hydrogen-annealed $ZnGa_2O_4$ ceramics (at 700°C for 5 h) also showed semiconducting behavior with the resistivity of about 0.03 Ωcm at RT [2]. For comparison, MOCVD-grown heteroepitaxial $ZnGa_2O_4$ films (on sapphire) showed the following electrical properties: $n_e = 1.9 \times 10^{14}$–6.7×10^{16} cm^{-3}, $\mu \approx 1.5$–5.5 cm^2/Vs, and $\rho = 68$–5730 Ωcm [1]. These examples clearly show that great electrical properties of bulk $ZnGa_2O_4$ single crystals obtained directly from the melt are not achievable either for ceramics and epitaxial films, or for flux-grown bulk crystals.

The semiconducting behavior of $MgGa_2O_4$ crystals was discovered by Galazka et al. [3], who reported $n_e = 3 \times 10^{17}$–2×10^{18} cm^{-3}, $\mu \approx 2$–7 cm^2/Vs, and $\rho = 0.1$–3.7 Ωcm. The Hall mobility is, however, quite low as compared with bulk $ZnGa_2O_4$ single crystals or other TSOs, which was substantially not affected either by annealing in the presence of oxygen or hydrogen, or by doping with Si. Annealing in the presence of hydrogen at 600–900°C for 10 h resulted in almost unchanged free electron concentration, which stays at a level of 10^{18} cm^{-3} (Fig. 8.23). Similar to $ZnGa_2O_4$, annealing $MgGa_2O_4$ single crystals in the presence of oxygen at or above 700°C for at least 10 h resulted in the electrically insulating state (Hall effect not measurable).

Melt-grown bulk $Zn_{1-x}Mg_xGa_2O_4$ ($x = 0.1$–0.5) single crystals obtained from the melt by Galazka and Hanke [4] revealed $n_e = 5 \times 10^{17}$–1×10^{18} cm^{-3}, $\mu \approx 10$–31 cm^2/Vs, and $\rho = 0.36$–0.61 Ωcm at RT [113]. Incorporation of Mg into $ZnGa_2O_4$ enabled to obtain lower free electron concentration, similar to that of $MgGa_2O_4$, but in favor of the Hall mobility that decreases from 31 to 10 cm^2/Vs with increasing Mg content, approaching that of $MgGa_2O_4$.

Melt-grown bulk $CoGa_2O_4$ single crystals showed electrically insulating state whether as-grown or annealed in the presence of oxygen (up to 1200°C) or hydrogen (below 600°C), but it showed the electrical p-type conductivity after annealing in the presence of hydrogen at 700°C. This is, however, related to surface conductivity as the result of thermal decomposition thereof. The samples showed electrically insulating state after removing the conductive layer. The sheet hole concentration, Hall hole mobility, and sheet resistance are as follows: $n_h = 5.2 \times 10^{17}$ cm^{-2}, $\mu_h \approx 2$ cm^2/Vs, and $\rho_h = 6.3$ Ωcm^2, respectively [113].

Table 8.8 Basic electrical properties of as-grown and annealed bulk Ga-based spinel crystals at RT

Compound	**State**	n_e **[cm^{-3}]**	μ **[cm^2/Vs]**	ρ **[Ωcm]**	**Reference**
$ZnGa_2O_4$	As-grown	3×10^{18}–9×10^{19}	20–100	0.002–0.1	5
	Annealed: (O_2, >700°C, >10 h)	Electrical insulator			
	Annealed: (H_2, 300–800°C, 10 h)	6×10^{18}–3×10^{19}	40–107	0.02–0.002	
$MgGa_2O_4$	As-grown	3×10^{17}–2×10^{18}	2–7	0.1–3.7	3
	Annealed: (O_2, >700°C, >10 h)	Electrical insulator			
	Annealed: (H_2, 600–900°C, 10 h)	1×10^{18}–2×10^{18}	4.5–6.5	0.6–1.5	
(Zn,Mg)Ga_2O_4	As-grown	5×10^{17}–1×10^{18}	10–31	0.36–0.61	113
$CoGa_2O_4$	As-grown	Electrical insulator			113
	Annealed: (O_2, >700°C, >10 h)	Electrical insulator			
	Annealed: (H_2, 700°C, 10 h)	5.2×10^{17} cm^{-2} (sheet, *p*-type)	2 (*p*-type)	6.3 Ωcm^2 (sheet, *p*-type)	

Note: O_2 means an atmosphere containing oxygen (O_2, air), while H_2 indicates 5% H_2 + 95% Ar.

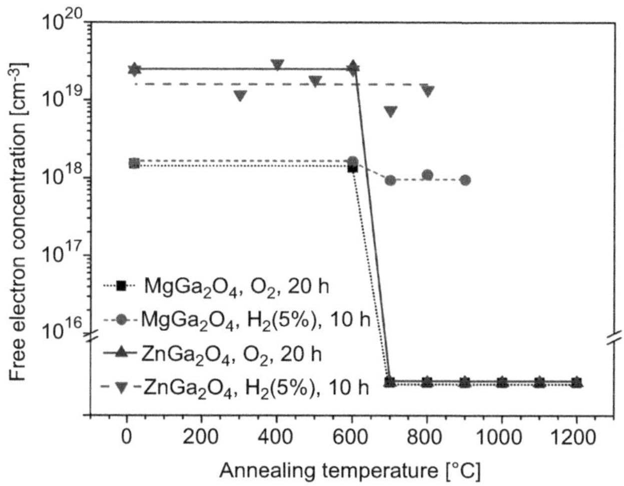

Figure 8.23 Free electron concentration versus annealing temperature, measured by the Hall effect after annealing. Values below break at 10^{16} cm^{-3} indicate not measurable Hall effect, or electrically insulating (or semi-insulating) state.

8.7.2 Effective Electron Mass and Electron Mobility

The effective electron mass in Ga-based spinels has not been experimentally determined so far. On the other hand, there are several reports on theoretical values of the effective electron and hole masses m_e* and m_h* at different transitions for $ZnGa_2O_4$. For example, Sampath et al. [35] calculated m_e^*/m_{e0} = 0.22, 0.21, and 0.23 for the $\Gamma \to X$, $\Gamma \to L$, and $\Gamma \to K$ transitions in the BZ, while Zeraga et al. [40] reported m_e^*/m_{e0} = 0.25 and 0.27 for the $\Gamma \to X$ and $\Gamma \to L$ transitions in the BZ, respectively. Xia et al. [41] provided the following values of the effective electron masses: m_e^*/m_{e0} = 0.66, 0.658, and 0.664 for the $\Gamma \to X$, $\Gamma \to L$, and $\Gamma \to K$ transitions, respectively, and of effective hole masses: m_h^*/m_{h0} = 5.91, 9.48, and 3.88 for the $K \to \Gamma$, $L \to \Gamma$, and $L \to U$ transitions, respectively. And finally, Dixit et al. [39] published values of m_e^*/m_{e0} = 0.23 and 0.25 along the [111] direction depending on the calculation method.

Theoretical effective electron masses for $MgGa_2O_4$ were calculated by Varley [42], which are 0.28, 0.31, and 0.29 for the $\Gamma \to X$, $\Gamma \to Y$, and $\Gamma \to Z$ transitions in the BZ, respectively.

In the case of $CoGa_2O_4$, Varley [44] concluded from his calculations that m_e^*/m_{e0} is fairly isotropic, but spin dependent: It is about 0.33 for one spin channel and about 0.38 for the other. The calculated effective hole mass m_h^*/m_{e0} for the indirect transition $\Gamma \rightarrow X$ is about 3.

This short overview of the theoretical effective electron and hole masses in $ZnGa_2O_4$, $MgGa_2O_4$, and $CoGa_2O_4$ showed a high isotropy for different transitions. m_e^*/m_{e0} seems to be in the range of 0.21–0.25 for $ZnGa_2O_4$, 0.28–0.31 for $MgGa_2O_4$, and 0.33–0.38 for $CoGa_2O_4$, respectively; therefore, the expected electron mobility is higher for $ZnGa_2O_4$ than for $MgGa_2O_4$ and $CoGa_2O_4$. Indeed, melt-grown bulk $MgGa_2O_4$ single crystals showed low Hall mobility of 2–7 cm^2/Vs at the free electron concentration in the range of about 10^{17}–10^{18} cm^{-3}, while $ZnGa_2O_4$ showed high Hall mobility of about 100 cm^2/Vs at a very high free electron concentration exceeding 10^{19} cm^{-3}. Interestingly, the Hall mobility of $ZnGa_2O_4$ increases with the free electron concentration approaching its maximum value at $n_e > 10^{19}$ cm^{-3} (Fig. 8.24a). Two possible reasons can be proposed for such behavior:

1. The presence of nanoparticles of metallic nature that form with the semiconducting host local Schottky contacts that affect current flow through the crystal at lower free electron concentrations, similar to that observed in the case of (100)-oriented homoepitaxial layers of β-Ga_2O_3, where the Schottky contacts are formed by charged incoherent twin boundaries [114, 115];
2. An enhancement of the electron mobility at high free electron concentrations (degenerate state) could be the result of the Fermi level shift to the CBM and significant screening, and limitation of the electron mobility by LO-phonon scattering (no screening) at low free electron concentration, similar to that observed in $BaSnO_3$ (Section 7.7.5).

When considering bulk $Zn_{1-x}Mg_xGa_2O_4$ single crystals, the Hall mobility drastically (Fig. 8.24b) decreases with Mg content, approaching low values (10 cm^2/Vs) at Mg = 50 mol.%, similar to that of pure $MgGa_2O_4$. To keep relatively high Hall mobility (>30 cm^2/Vs), Mg content in bulk $Mg_xZn_{1-x}Ga_2O_4$ crystals should not be higher than about $x = 0.1$.

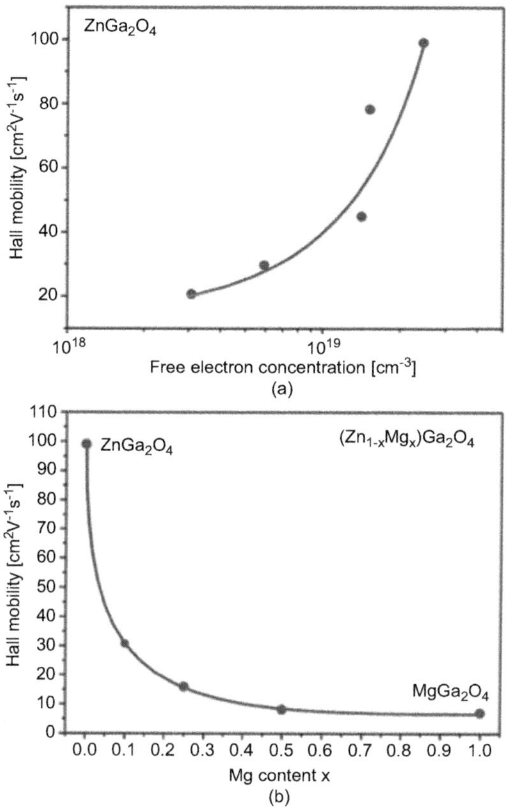

Figure 8.24 Hall mobility versus electron concentration for bulk $ZnGa_2O_4$ single crystals (a), and versus Mg content for mixed bulk $Zn_{1-x}Mg_xGa_2O_4$ single crystals (b). All the bulk crystals were obtained from the melt.

8.7.3 Conclusion

Among different bulk Ga-based spinel single crystals, $ZnGa_2O_4$, $MgGa_2O_4$, and $(Zn,Mg)Ga_2O_4$ revealed a semiconducting *n*-type behavior that converts into electrically insulating state after annealing in the presence of oxygen at or above 700°C for at least 10 h. On the other hand, annealing in the presence of hydrogen maintains the electrical properties even after annealing at high temperatures until a strong decomposition (up to 800–900°C). However, $CoGa_2O_4$ did not show semiconducting behavior either when as-grown, or after annealing in the presence of oxygen (≤1200°C) and hydrogen (<600°C). However, annealing $CoGa_2O_4$ in the presence of hydrogen at 700°C makes the surface *p*-type conductive with the hole mobility around 2 cm^2/Vs.

For melt-grown crystals, $MgGa_2O_4$ and $(Zn,Mg)Ga_2O_4$ have the free electron concentrations typical for normal semiconductors 3×10^{17}–2×10^{18} cm^{-3}), while $ZnGa_2O_4$ has the free electron concentration typical for a degenerate semiconductor (3×10^{18}–9×10^{19} cm^{-3}). The Hall mobility is the highest for $ZnGa_2O_4$ (about 100 cm^2/Vs), which decreases with Mg content in $(Zn,Mg)Ga_2O_4$ to the lowest values for $MgGa_2O_4$ (a few cm^2/Vs). Interestingly, the highest Hall mobility of $ZnGa_2O_4$ was measured at very high free electron concentrations ($>10^{19}$ cm^{-3}).

Theoretical values of effective electron mass of $ZnGa_2O_4$ and $MgGa_2O_4$ revealed their isotropy for direct and indirect transitions with values of 0.21–0.25 and 0.28–0.31, respectively. The summary of the basic electrical properties of $ZnGa_2O_4$, $MgGa_2O_4$, and $Zn_{1-x}Mg_xGa_2O_4$ ($x = 0.1$ and 0.5) is included in Table 8.9.

Table 8.9 Basic electrical properties of bulk Ga-based spinel single crystals

Property	Unit	Compound	Value	Reference
Tunability of free electron concentration n_e	cm^{-3}	$ZnGa_2O_4$	3×10^{18}–9×10^{19}	5
		$MgGa_2O_4$	3×10^{17}–2×10^{18}	3
		$Zn_{0.9}Mg_{0.1}Ga_2O_4$ $Zn_{0.5}Mg_{0.5}Ga_2O_4$	5×10^{17}–1×10^{18}	113
Experimental maximum electron mobility at RT	cm^2/Vs	$ZnGa_2O_4$	107 at $n_e = 1.2\times10^{19}$	5
		$MgGa_2O_4$	7 at $n_e = 2\times10^{18}$	3
		$Zn_{0.9}Mg_{0.1}Ga_2O_4$	31 at $n_e = 6\times10^{17}$	113
		$Zn_{0.5}Mg_{0.5}Ga_2O_4$	10 at $n_e = 1.5\times10^{19}$	
Theoretical effective electron mass		$ZnGa_2O_4$	0.21–0.25	35, 39, 40, 41
		$MgGa_2O_4$	0.28–0.31	42
		$Zn_{0.9}Mg_{0.1}Ga_2O_4$ $Zn_{0.5}Mg_{0.5}Ga_2O_4$	—	—

8.8 Key Points

- $ZnGa_2O_4$, $MgGa_2O_4$, and $Mg_xZn_{1-x}Ga_2O_4$ are ultra-wide bandgap semiconductors (bandgaps 4.6–5 eV) with a cubic structure.
- These Ga-based spinel crystals, including $CoGa_2O_4$, can be grown directly from melt as bulk single crystals with a reasonable volume and structural quality, sufficient for wafer fabrication dedicated for epitaxy.
- $ZnGa_2O_4$ showed very high electron mobility (100 cm^2/Vs) at a degenerate semiconducting state ($n_e > 10^{19}$ cm^{-3}), higher than that of many other TSOs at similar free electron concentrations.
- Two cation sites in these ternary systems offer two doping sites, similar to that of $BaSnO_3$.
- Bulk spinel single crystals of $ZnGa_2O_4$, $MgGa_2O_4$, $Mg_xZn_{1-x}Ga_2O_4$ (x = 0.1–0.5), and $CoGa_2O_4$ offer great lattice matching substrates for Fe-based magnetic spinel films.

8.9 Summary

Table 8.10 Main properties of bulk Ga-based spinel single crystals and growth techniques

$ZnGa_2O_4$, $MgGa_2O_4$, $Zn_{1-x}Mg_xGa_2O_4$ (x = 0.5), $CoGa_2O_4$			
Property	**Compound**	**Value or description**	**Reference**
	Crystal system and structure		
System	$ZnGa_2O_4$	Cubic	
	$MgGa_2O_4$	Cubic	
	$Zn_{0.5}Mg_{0.5}Ga_2O_4$	Cubic	
	$CoGa_2O_4$	Cubic	
Lattice constants [Å]	$ZnGa_2O_4$	8.3313	32
	$MgGa_2O_4$	8.281	31
	$Zn_{0.5}Mg_{0.5}Ga_2O_4$	—	—
	$CoGa_2O_4$	8.326	50

(*Continued*)

Table 8.10 (*Continued*)

Space group	$ZnGa_2O_4$	$Fd\bar{3}m$	30
	$MgGa_2O_4$	$Fd\bar{3}m$	30
	$Zn_{0.5}Mg_{0.5}Ga_2O_4$	—	—
	$CoGa_2O_4$	$Fd\bar{3}m$	30
		Bulk crystal preparation	
Growth methods	$ZnGa_2O_4$	Solution: Flux Melt: LHPG, VGF/VB, Czochralski	51, 52, 53, 54, 55, 56, 5
	$MgGa_2O_4$	Solution: Flux Melt: OFZ, VGF/VB, Czochralski, Kyropoulos-like	57, 59, 3, 64, 58
	$Zn_{0.5}Mg_{0.5}Ga_2O_4$	Melt: VGF/VB	4
	$CoGa_2O_4$	Gas phase: CVT Solution: Flux Melt: OFZ, Czochralski, Kyropoulos-like	61, 62, 60, 49
Maximum volume of a single crystal [cm^3]	$ZnGa_2O_4$	3	5
	$MgGa_2O_4$	19	3
	$Zn_{0.5}Mg_{0.5}Ga_2O_4$	2	4
	$CoGa_2O_4$	10	49
		Thermal and mechanical properties	
Solid density [g/cm^3]	$ZnGa_2O_4$	6.17	74
	$MgGa_2O_4$	5.32–5.33	75, 76
	$Zn_{0.5}Mg_{0.5}Ga_2O_4$	—	—
	$CoGa_2O_4$	6.01–6.03	61, 62
Melting point [°C]	$ZnGa_2O_4$	1900±20°C	5
	$MgGa_2O_4$	1930±10°C / / 1950±75	3 / 58
	$Zn_{0.5}Mg_{0.5}Ga_2O_4$	—	—
	$CoGa_2O_4$	1810±20°C/1825±5	49 / 60
Thermal conductivity at RT [W/mK]	$ZnGa_2O_4$	22.1	5
	$MgGa_2O_4$	16.26	77
	$Zn_{0.5}Mg_{0.5}Ga_2O_4$	10.34	77
	$CoGa_2O_4$	13.4	77

Thermal expansion at 20–1000°C [K^{-1}]	$ZnGa_2O_4$	6.2×10^{-5}	53
	$MgGa_2O_4$	$6–18\times10^{-6}$	75
	$Zn_{0.5}Mg_{0.5}Ga_2O_4$	—	—
	$CoGa_2O_4$	9×10^{-6}	60
Thermal stability at oxidizing/ reducing conditions [°C]	$ZnGa_2O_4$	1100/700	5
	$MgGa_2O_4$	1200/800	3
	$Zn_{0.5}Mg_{0.5}Ga_2O_4$	1200/700	4
	$CoGa_2O_4$	1200/600	49
Elastic stiffness coefficients at RT [GPa]	$ZnGa_2O_4$	c_{11}[(c)] = 250/228/349/390/288 c_{12}[(c)] = 122/120/177/143/218 c_{44}[(c)] = 116/107/128/114/125	38/ 79/ 80/ 81/ 82
	$MgGa_2O_4$	c_{11}=280.2/262[(c)] c_{12}=148/145[(c)] c_{44}=129.9/118[(c)]	75/ 78
	$Zn_{0.5}Mg_{0.5}Ga_2O_4$	—	—
	$CoGa_2O_4$	—	—
Young's modulus [GPa]	$ZnGa_2O_4$	170[(c,e)]/145[(c,e)]/230[(c,e)] /313[(c,e)]/100[(c,e)]	38/ 79/ 80/ 81/ 82
	$MgGa_2O_4$	178[(e)]/158[(c,e)]	75/ 78
	$Zn_{0.5}Mg_{0.5}Ga_2O_4$	—	—
	$CoGa_2O_4$	—	—
Bulk modulus [GPa]	$ZnGa_2O_4$	164[(c,e)]/156[(c,e)]/234[(c,e)] /225[(c,e)]/241[(c,e)]	38/ 79/ 80/ 81/ 82
	$MgGa_2O_4$	192[(e)]/186[(c,e)]	75/ 78
	$Zn_{0.5}Mg_{0.5}Ga_2O_4$	—	—
	$CoGa_2O_4$	—	—
Shear modulus [GPa]	$ZnGa_2O_4$	116[(c,e)]/107[(c,e)]/128[(c,e)] /114[(c,e)]/125[(c,e)]	38/ 79/ 80/ 81/ 82
	$MgGa_2O_4$	130[(e)]/118[(c,e)]	75/ 78
	$Zn_{0.5}Mg_{0.5}Ga_2O_4$	—	—
	$CoGa_2O_4$	—	—

(*Continued*)

Table 8.10 (*Continued*)

Poisson's ratio	$ZnGa_2O_4$	0.328[c,e]/0.345[c,e]/0.336[c,e]/0.268[c,e]/0.431[c,e]	38/ 79/ 80/ 81/ 82
	$MgGa_2O_4$	0.346[e]/0.357[c,e]	75/ 78
	$Zn_{0.5}Mg_{0.5}Ga_2O_4$	—	—
	$CoGa_2O_4$	—	—
		Optical properties	
Bandgap [eV]	$ZnGa_2O_4$	4.59/4.84/4.5–4.7[c]	5/ 89/ 39/ 41
	$MgGa_2O_4$	4.90/4.95/5.0[c]	3/ 89/ 42
	$Zn_{0.5}Mg_{0.5}Ga_2O_4$	4.75	85
	$CoGa_2O_4$	3.55/3.97[c]	89/ 44
Absorption edge [nm]	$ZnGa_2O_4$	275	5
	$MgGa_2O_4$	250	3
	$Zn_{0.5}Mg_{0.5}Ga_2O_4$	268	85
	$CoGa_2O_4$	—	—
Luminescence peaks at RT [eV]	$ZnGa_2O_4$	350	99
	$MgGa_2O_4$	360	3
	$Zn_{0.5}Mg_{0.5}Ga_2O_4$	380–400	99
	$CoGa_2O_4$	None	99
Relative static dielectric constant	$ZnGa_2O_4$	9.8[d]	87
	$MgGa_2O_4$	11.7	58
	$Zn_{0.5}Mg_{0.5}Ga_2O_4$	—	—
	$CoGa_2O_4$	9.2	60
High-frequency dielectric constant	$ZnGa_2O_4$	2.6±0.04	89
	$MgGa_2O_4$	2.36±0.012	89
	$Zn_{0.5}Mg_{0.5}Ga_2O_4$	2.6±0.04	89
	$CoGa_2O_4$	1.67±0.17	89
Refractive index at 300/600/800 nm	$ZnGa_2O_4$	2.074/1.948/1.936	89
	$MgGa_2O_4$	2.000/1.880/1.869	89
	$Zn_{0.5}Mg_{0.5}Ga_2O_4$	2.013/1.846/1.935	89
	$CoGa_2O_4$	2.212/1.999/1.980	89

Electrical properties			
Tunability of the free electron (or hole) concentration at RT [cm^{-3}]	$ZnGa_2O_4$	3×0^{18}–9×10^{19}	5
	$MgGa_2O_4$	3×10^{17}–2×10^{18}	3
	$Zn_{0.5}Mg_{0.5}Ga_2O_4$	5×10^{17}–1×10^{18}	113
	$CoGa_2O_4$	~5×10^{17} [(a)]	113
Max. experimental carrier mobility at RT [cm^2/Vs]	$ZnGa_2O_4$	107 at $n_e = 1.2\times10^{19}$	5
	$MgGa_2O_4$	7 at $n_e = 2\times10^{18}$	3
	$Zn_{0.5}Mg_{0.5}Ga_2O_4$	10 at $n_e = 1.5\times10^{19}$	113
	$CoGa_2O_4$	2[(b)]	113
Theoretical effective electron mass	$ZnGa_2O_4$	0.21–0.25	35, 39, 40, 41
	$MgGa_2O_4$	0.28–0.31	42
	$Zn_{0.5}Mg_{0.5}Ga_2O_4$	—	—
	$CoGa_2O_4$	0.33–0.38	44

[(a)]sheet hole concentration of a reduced surface; [(b)]hole mobility; [(c)]theoretical value; [(d)]ceramics; [(e)]calculated basing on data in the corresponding reference

References

1. R.-H. Horng, C.-Y. Huang, S.-L. Ou, T.-K. Juang, P.-L. Liu; *Cryst. Growth Des.* **17** (2017), 6071–6078.
2. T. Omata, N. Ueda, K. Ueda, H. Kawazoe; *Appl. Phys. Lett.* **64** (1994), 1077–1078.
3. Z. Galazka, D. Klimm, K. Irmscher, R. Uecker, M. Pietsch, R. Bertram, M. Naumann, M. Albrecht, A. Kwasniewski, R. Schewski, M. Bickermann; *Phys. Status Solidi* (*a*) **212** (2015), 1455–1460.
4. Z. Galazka, I. M. Hanke; *Leibniz-Institut für Kristallzüchtung*, Berlin, Germany (2018), unpublished data.
5. Z. Galazka, S. Ganschow, R. Schewski, K. Irmscher, D. Klimm, A. Kwasniewski, M. Pietsch, A. Fiedler, I. Schulze-Jonack, M. Albrecht, T. Schröder, M. Bickermann; *APL Mater.* **7** (2019), 022512.
6. L.-C. Cheng, C.-Y. Huang, R.-H. Horng; *J. Electron. Devices Soc.* **6** (2018), 432–437.
7. Z. Lou, L. Li, G. Shen; *Nano Res.* **8** (2015), 2162–2169.
8. T. Minami, T. Maeno, Y. Kuroi, S. Takata; *Jpn. J. Appl. Phys.* **34** (1995), L684–L687.

9. T. Minami, Y. Kuroi, T. Miyata, H. Yamada, S. Takata; *J. Lumin.* **72–74** (1997), 997–998.

10. G. Anoop, K. M. Krishna, M. K. Jayaraj; *J. Electrochem. Soc.* **158** (2011), J269–J272.

11. H. I. Kang, J. S. Kim, M. Lee, J. H. Bahng, J. C. Choi, H. L. Park, G. C. Kim, T. W. Kim, Y. H. Hwang, S. I. Mho, S. H. Eom, Y. S. Yu, H. J. Song, W. T. Kim; *Solid State Commun.* **122** (2002), 633–636.

12. J. S. Kim, H. L. Park, G. C. Kim, T. W. Kim, Y. H. Hwang, H. K. Kim, S. I. Mho, S. D. Han; *Solid State Commun.* **126** (2003), 515–518.

13. Y. S. Jeong, J. S. Kim, H. L. Park; *Solid State Commun.* **139** (2006), 157–160.

14. S. Itoh, H. Toki, Y. Sato, K. Morimoto, T. Kishino; *J. Electrochern. Soc.* **138** (1991), 1509–1512.

15. M. Flynn, A. H. Kitai; *J. Electrochem. Soc.* **148** (2001), H149–H153.

16. S.-H. Yang, T.-J. Hsueh, S.-J. Chang; *J. Electrochem. Soc.* **152** (2005), H191–H195.

17. A. Bessière, S. Jacquart, K. Priolkar, A. Lecointre, B. Viana, D. Gourier; *Opt. Express* **19** (2011), 10131–10137.

18. X. Xu, A. K. Azad, J. T. S. Irvine; *Catal. Today* **199** (2013), 22–26.

19. X. Li, X. Zhang, X. Zheng, Y. Shao, M. He, P. Wang, X. Fu, D. Li; *J. Mater. Chem. A* **2** (2014), 15796–15802.

20. C. Chen, G. Li, Y. Liu; *Powder Technol.* **281** (2015), 7–11.

21. B. Yasoda, R. P. S, Chakradhar, J. L. Rao, N. O. Gopal, C. N. Xu; *J. Appl. Phys.* **98** (2005), 053910.

22. T. Suzuki, M. Hughes, Y. Ohishi; *J. Lumin.* **130** (2010), 121–126.

23. D. Fiorani, S. Viticoli; *Solid State Commun.* **25** (1978), 155–157.

24. M. Ogata, H. Kozaka, N. Uryu; *Phys. Status Solidi (b)* **127** (1985), K33–K36.

25. J. L. Soubeyroux, D. Fiorani, E. Agostinelli; *J. Magn. Magn. Mater.* **54–57** (1986), 83–84.

26. A. V. Singh, B. Khodadadi, J. B. Mohammadi, S. Keshavarz, T. Mewes, D. S. Negi, R. Datta, Z. Galazka, R. Uecker, A. Gupta; *Adv. Mater.* **29** (2017), 1701222.

27. J. Shan, A. V. Singh, L. Liang, L. J. Cornelissen, Z. Galazka, A. Gupta, B. J. van Wees, T. Kuschel; *Appl. Phys. Lett.* **113** (2018), 162403.

28. H. J. van Hook, M. L. Keith; *Americ. Miner.* **43** (1958), 69–83.

29. K. Mocala, A. Navrotsky, D. M. Sherman; *Phys. Chem. Miner.* **19** (1992), 88–95.

30. H. S. C. O'Neill, A. Navrotsky; *Am. Mineral.* **68** (1983), 181–194.

31. L. Schwarz, Z. Galazka, T. M. Gesing, D. Klimm; *Cryst. Res. Technol.* **50** (2015), 961–966.

32. A. R. Phani, S. Santucci, S. Di Nardo, L. Lozzi, M. Passacantando, P. Picozzi, C. Cantalini; *J. Mater. Sci.* **33** (1998), 3969–3973.

33. G. Palmer, K. Poeppelmeier; *Solid State Sci.* **4** (2002), 317–322.

34. M. G. Brik, A. Suchocki, A. Kamińska; *Inorg. Chem.* **53** (2014), 5088–5099.

35. S. K. Sampath, D. G. Kanhere, R. Pandey; *J. Phys.: Condens. Matter.* **11** (1999), 3635–3644.

36. L. Pisani, T. Maitra, R. Valentí; *Phys. Rev. B* **73** (2006), 205204.

37. S. López, A. H. Romero, P. Rodríguez-Hernández, A. Muñoz; *Phys. Rev. B* **79** (2009), 214103.

38. M. G. Brik; *J. Phys. Chem. Solids* **71** (2010), 1435–1442.

39. H. Dixit, N. Tandon, S. Cottenier, R. Saniz, D. Lamoen, B. Partoens, V. Van Speybroeck, M. Waroquier; *New J. Phys.* **13** (2011), 063002.

40. F. Zerarga, A. Bouhemadou, R. Khenata, S. Bin-Omran; *Solid State Sci.* **13** (2011), 1638–1648.

41. Y. Xia, T. Wang, X. Zhao, X. Jiao, D. Chen; *J. Phys. Chem. C* **122** (2018), 5509–5517.

42. J. B. Varley; *Lawrence Livermore National Laboratory,* CA, USA (2017), unpublished data.

43. B. Thielert, C. Janowitz, Z. Galazka, M. Mulazzi; *Phys. Rev. B* **97** (2018), 235309.

44. J. B. Varley; *Lawrence Livermore National Laboratory,* CA, USA (2019), unpublished data.

45. M. Zinkevich, S. Geupel, F. Aldinger; *J. Alloy. Compd.* **393** (2005), 154–166.

46. K. T. Jacob, S. Sivakumar; *J. Alloy. Compd.* **775** (2019), 1357–1364.

47. N. A. Gribchenkova, A. V. Steblevsky, A. S. Alikhanyan; *J. Chem. Thermodyn.* **115** (2017), 1–6.

48. N. Kimizuka, M. Isobe, M. Nakamura; *J. Solid State Chem.* **116** (1995), 170–178.

49. Z. Galazka; *Leibniz-Institut für Kristallzüchtung*, Berlin, Germany (2014), unpublished data.

50. W. Laqua, B. Küter, B. Reuter; *J. Solid State Chem.* **17** (1976), 261–273.

51. A. B. Chase, J. A. Osmer; *J. Am. Ceram. Soc.* **50** (1967), 325–328.

52. H. van den Boom, J. C. M. Henning, J. P. M. Damen; *Solid State Comm.* **8** (1970), 717–719.

53. P. J. M. van der Straten, R. Metselaar, H. D. Jonker; *J. Cryst. Growth* **43** (1978), 270–272.

54. Z. Yan, H. Takei; *J. Cryst. Growth* **171** (1997), 131–135.

55. Z. Yan, H. Takei, H. Kawazoe; *J. Am. Ceram. Soc.* **81** (1998), 180–186.

56. W. Jia, H. Liu, S. Huang, X. Wu, L. Lu, W. M. Yen; *J. Electrochem. Soc.* **142** (1995), 1637–1640.

57. E. A. Giess; *J. Appl. Phys.* **33** (1962), 2143.

58. B. A. Scott, K. H. Nichols, R. M. Potemski, J. M. Woodall; *Appl. Phys. Lett.* **21** (1972), 121–122.

59. T. Suzuki, M. Hughes, Y. Ohishi; *J. Lumin.* **130** (2010), 121.

60. W. E. Kramer, A. M. Stewart, R. H. Hopkins; *J. Cryst. Growth* **73** (1985), 329–336.

61. F. Leccabue, C. Pelosi, E. Agostinelli, V. Fares, D. Fiorani, E. Paparazzo; *J. Cryst. Growth* **79** (1986), 410–416.

62. A. Nakatsuka, Y. Ikeda, N. Nakayama, T. Mizota; *Acta Crystallogr. E* **62** (2006), i109–i111.

63. C. M. Rost, E. Sachet, T. Borman, A. Moballegh, E. C. Dickey, D. Hou, J. L. Jones, S. Curtarolo, J.-P. Maria; *Nat. Commun.* **6** (2015), 8485.

64. Z. Galazka, I. M. Hanke; *Leibniz-Institute für Kristallzüchtung*, Berlin, Germany (2017), unpublished data.

65. H. Mao, M. Selleby, B. Sundman; *CALPHAD* **28** (2004), 307–312.

66. A. Kwasniewski, Z. Galazka; *Leibniz-Institute für Kristallzüchtung*, Berlin, Germany (2019), unpublished data.

67. www.icdd.com.

68. R. Bertram, Z. Galazka; *Leibniz-Institute für Kristallzüchtung*, Berlin, Germany (2014), unpublished data.

69. R. Bertram, Z. Galazka; *Leibniz-Institute für Kristallzüchtung*, Berlin, Germany (2015), unpublished data.

70. G. Andreozzi and F. Princivalle; *Am. Mineral.* **87** (2002), 838–844.

71. C. W. W. Hoffman, J. J. Brown; *J. Jnorg. Nucl. Chem.* **30** (1968), 63–79.

72. H. St. C. O'Neill, A. Navrotsky; *Am. Mineral.* **68** (1983), 181–182.

73. A. Seko, K. Yuge, F. Oba, A. Kuwabara, I. Tanaka; *Phys. Rev. B* **73** (2006), 184117.

74. PAULING FILE Multinaries Edition – 2012, Springer & Material Phases Data System (MPDS), Switzerland & National Institute for Materials Science (NIMS), Japan, 2016, Chief Editor Pierre Villars, Material Phases Data System (MPDS), CH-6354 Vitznau, Switzerland.

75. C. Hirschle, J. Schreuer, Z. Galazka; *J. Appl. Phys.* **124** (2018), 065111.

76. P. G. Casado, I. Rasines; *Z. Kristallogr.* **160** (1982), 33–37.

77. D. Klimm, Z. Galazka; *Leibniz-Institute für Kristallzüchtung*, Berlin, Germany (2018), unpublished data.

78. A. Bouhemadou, R. Khenata, F. Zerarga; *Eur. Phys. J. B* **56** (2007), 1–5.

79. A. Bouhemadou, R. Khenata; *Phys. Lett. A* **360** (2006), 339–343.

80. A. K. Kushwaha; *Comput. Mater. Sci.* **85** (2014), 259–263.

81. F. Zerarga, A. Bouhemadou, R. Khenata, S. Binomran; *Comput. Mater. Sci.* **50** (2011), 2651–2657.

82. R. Pandey, J. D. Gale, S. K. Sampath, J. M. Recio; *J. Am. Ceram. Soc.* **82** (1999), 3337–3341.

83. D. Errandonea, R. S. Kumar, F. J. Manjon, V. V. Ursaki, E. V. Rusu; *Phys. Rev. B* **79** (2009), 024103.

84. Z. Galazka, K. Irmscher, M. Pietsch; *Leibniz-Institute für Kristallzüchtung*, Berlin, Germany (2015), unpublished data.

85. Z. Galazka, K. Irmscher, M. Pietsch; *Leibniz-Institute für Kristallzüchtung*, Berlin, Germany (2019), unpublished data.

86. A. Kan, T. Moriyama, S. Takahashi, H. Ogawa; *Jpn. J. Appl. Phys.* **52** (2013), 09KH01.

87. X. Lu, W. Biana, C. Min, Z. Fu, Q. Zhang, H. Zhu; *Ceram. Int.* **44** (2018), 10028–10034.

88. J. Xue, S. Wu, J. Li; *J. Am. Ceram. Soc.* **96** (2013), 2481–2485.

89. S. Bin Anooz, Z. Galazka; *Leibniz-Institute für Kristallzüchtung*, Berlin, Germany (2019), unpublished data.

90. T. Oshima, M. Niwa, A. Mukai, T. Nagami, T. Suyama, A. Ohtomo; *J. Cryst. Growth* **386** (2014), 190–193.

91. O. M. Bordun, I. Y. Kukharskyy, V. G. Bihday; *J. Appl. Spectrosc.* **78** (2012), 922–926.

92. S. Itoh, H. Taki, Y. Sato, K. Morimoto, T. Kishino; *J. Electrochem. Soc.* **138** (1991), 1509–1512.

93. P. D. Rack, J. J. Peterson, M. D. Potter, W. Park; *J. Mater. Res.* **16** (2001), 1429–1433.

94. A. Bessière, S. Jacquart, K. Priolkar, A. Lecointre, B. Viana, D. Gourier; *Opt. Express* **19** (2011), 10131–10137.

95. J. S. Kim, J. S. Kim, T. W. Kim, H. L. Park, Y. G. Kim, S. K. Chang, S. D. Han; *Solid State Commun.* **131** (2004), 493–497.

96. E. Rusu, V. Ursaki, G. Novitschi, M. Vasile, P. Petrenco, L. Kulyuk; *Phys. Status Solidi (c)* **6** (2009), 1199–1202.

97. J. S. Kim, H. L. Park, G. C. Kim, T. W. Kim,Y. H. Hwang, H. K. Kim, S. I. Mho, S. D. Han; *Solid State Commun.* **126** (2003), 515–518.

98. B. Yasoda, R. P. S. Chakradhar, J. L. Rao, N. O. Gopal, C. N. Xu; *J. Appl. Phys.* **98** (2005), 053910.

99. R. Schewski, Z. Galazka; *Leibniz-Institute für Kristallzüchtung*, Berlin, Germany (2019), unpublished data.

100. H.-W. Choi, B.-J. Hong, S.-K. Lee, K.-H. Kim, Y.-S. Park; *J. Lumin.* **126** (2007), 359–364.

101. S.-H. Yang, T.-J. Hsueh, S.-J. Chang; *J. Electrochem. Soc.* **152** (2005), H191–H195.

102. S. S. Yi, I. W. Kim, H. L. Park, J. S. Bae, B. K. Moon, J. H. Jeong; *J. Cryst. Growth* **247** (2003), 213–218.

103. J. S. Bae, B. K. Moon, B. C. Choi, J. H. Jeong, S. S. Yi, I. W. Kim, J. S. Lee; *Thin Solid Films* **424** (2003), 291–295.

104. S. S. Yi, J. S. Bae, B. K. Moon, J. H. Jeong, I. W. Kim, H. L. Park; *Appl. Phys. A* **76** (2003), 433–437.

105. Y. E. Lee, D. P. Norton, C. Park, C. M. Rouleau; *J. Appl. Phys.* **89** (2001), 1653–1656.

106. I. Ahmad Md., M. Kottaisamy, N. Rama, M. S. R. Rao, S. S. Bhattacharya; *Scr. Mater.* **54** (2006), 237–240.

107. Y. E. Lee, D. P. Norton, J. D. Budai, M. Rouleau, J.-W. Park; *J. Electroceram.* **4** (200), 293–297.

108. P. Thiyagarajan, M. Kottaisamy, N. Rama, M. S. R. Rao; *J. Phys. D: Appl. Phys.* **42** (2009), 155301.

109. C. F. Yu, P. Lin; *Jpn. J. Appl. Phys.* **35** (1996), 5726–5729.

110. H. A. Lauwers, M. A. Herman; *J. Phys. Chem. Solids* **41** (1980), 223–230.

111. S. López-Moreno, P. Rodríguez-Hernández, A. Muñoz, A. H. Romero, F. J. Manjón, D. Errandonea, E. Rusu, V. V. Ursaki; *Ann. Phys.* **523** (2011) 157–167.

112. G. G. P. Van Gorkom, J. H. Haanstra, H. V. D. Boom; J. *Raman Spectrosc. 1* (1973), 513–519.

113. Z. Galazka, K. Irmscher, M. Pietsch; *Leibniz-Institute für Kristallzüchtung*, Berlin, Germany (2018), unpublished data.

114. R. Schewski, M. Baldini, K. Irmscher, A. Fiedler, T. Markurt, B. Neuschulz, T. Remmele, T. Schulz, G. Wagner, Z. Galazka, M. Albrecht; *J. Appl. Phys.* **120** (2016), 225308.

115. A. Fiedler, R. Schewski, M. Baldini, Z. Galazka, G. Wagner, M. Albrecht, K. Irmscher; *J. Appl. Phys.* **122** (2017), 165701.

Index

absorption 175, 176, 272–274, 312, 313, 315, 316, 323, 374, 431, 432, 596, 597, 599, 653, 654, 678, 683
- band-to-band 599, 600
- excitonic 176, 177, 182
- free carrier 174, 252, 272, 273, 275, 279, 291, 313, 394, 423, 431, 520, 521, 653, 654
- fundamental 522
- phonon-assisted 520

absorption coefficient 177, 273, 314, 425, 520–523, 596–599, 680, 681

absorption edge 174, 176, 182, 183, 315, 317, 423, 426, 431, 432, 525, 528, 596, 678, 687, 688
- fundamental 315, 316, 520, 522, 599
- near-edge 314, 323
- steep 313, 449, 520, 597

acceptor 179, 186, 188, 376, 468, 563, 564, 567
- compensating 198, 231, 331, 345, 376, 571
- deep 74, 339, 468, 564
- native 467
- neutral 177
- shallow 185, 467, 565, 603
- substitutional 468
- triple 375

activation energy 191, 200, 201, 351, 451

ampoule 5, 6, 88, 90–92, 98–100, 105, 107–110, 116, 117, 235, 236, 238, 383–385, 503, 504
- closed 107, 239, 493
- graphitized 108, 109
- integrated 90
- quartz 5, 90–92, 99, 101, 109, 235–237, 239, 383, 384, 493, 640
- standard 112

annealing temperature 141, 196–198, 304, 336, 338–340, 438–441, 513, 521, 532, 533, 666, 667, 693

annealing time 252, 303, 304, 332, 416, 417, 423, 439, 441–443, 532, 534, 667, 670

ARPES *see* spectroscopy, angle-resolved photoemission

autoclave 10, 11, 30, 60, 125, 126, 129–131, 139, 142–144, 146, 147, 149–159

bandgap 223, 224, 282, 425, 426, 449, 451, 523, 544, 546, 562, 564, 565, 596, 597, 632, 633, 681, 685, 687, 688
- dipole-forbidden 522
- direct 73, 177, 183, 203, 431, 432, 599, 602, 622, 632, 633, 636, 681
- fundamental 425, 431, 432, 522, 528, 548, 602, 622
- indirect 596, 598, 599, 602, 622, 636, 681
- ultra-wide 223, 349, 629, 697

bandgap shrinkage 432, 451, 528, 548

band structure 72, 73, 228, 229, 374, 465, 562, 632–635

$BaSnO_3$ 12, 14, 20, 21, 30, 31, 559–568, 570--588, 590, 592–594, 596, 600, 602–604, 606, 607, 612–614, 616–622

β-Ga_2O_3 20, 21, 30, 31, 223–234, 242, 243, 245, 248–250, 255, 263–265, 272, 275, 276, 279, 282, 283, 286, 288, 289, 290, 293–296, 298, 301–305, 307–316, 318, 320–326, 328–332, 334–336, 340, 342, 343, 344, 347–349
Bi_2O_3 12, 69, 242, 244, 641, 642
$Bi_4Ge_3O_{12}$ 20, 22, 23
binding energy 230, 376, 445, 544
  donor 191
  excitonic 177
Birch–Murnaghan equation 594
bixbyite 373, 374, 450
Bridgman configuration 120, 161, 305
Bridgman method 20, 75, 76, 162, 163, 174–177, 187, 193, 197, 203, 261, 271
Brillouin zone (BZ) 73, 229, 230, 315, 342, 374, 466, 527, 562, 563, 599, 632–634, 693
bulk modulus 169, 170, 173, 204, 309, 312, 350, 421, 422, 519, 520, 592, 673, 674, 677
Burton, Prim, and Slichter model 283
BZ *see* Brillouin zone

carrier concentration 194, 272–275, 279, 280, 288, 323, 541, 542, 596, 612, 615
carrier gas 64, 84–86, 94, 97–99, 106, 472, 479, 482, 492, 493, 496, 501, 502
CATA *see* chemically active transport agent
CB *see* conduction band
CBM *see* conduction band minimum
CCVT *see* contactless chemical vapor transport
chemically active transport agent (CATA) 2, 4–7, 76–78, 88–92, 234–236, 238, 241, 379, 383–387, 469, 471, 640
chemical solution deposition 560
chemical vapor deposition (CVD) 70, 223, 224, 226, 227, 372, 463
chemical vapor transport (CVT) 2–4, 7, 76, 98, 100, 112, 233–235, 378, 469, 471, 492, 496, 535, 545, 547
chemo-mechanical polishing (CMP) 140, 141, 149, 155, 156, 253, 256, 300, 301, 444
cleavage plane 227, 248, 251, 254, 260, 261, 275, 298, 310–312, 349, 450, 494
CL (cathodoluminescence) spectra 181, 321, 323, 326, 429, 430, 432, 525, 526, 685
  normalized 321, 685
  overlapped 685
CMP *see* chemo-mechanical polishing
$CoGa_2O_4$ 17, 18, 24, 31, 636, 637, 639–641, 645, 646, 654–661, 667, 668, 670–672, 675, 678–683, 687–689, 694–701
compound 3, 4, 14, 18, 20, 25, 30, 31, 44, 46, 48, 60, 64, 65, 630, 633, 667, 668
  complex 47
  endothermal 64
  non-metal 12
  spinel-type 630
  stable 46, 630
  unstable 21, 30, 469
  volatile 7, 9
conduction band (CB) 73, 191, 194, 195, 229, 315, 317, 466, 467, 539, 543, 562, 564, 603, 617–619

conduction band minimum (CBM) 199, 201, 202, 345, 346, 348, 374, 376, 466, 562–565, 610, 611, 615, 618, 620, 632, 633, 636
conductivity
  anisotropic heat 275
  electrical 174, 184–187, 230, 231, 252, 254, 282, 283, 293, 305, 329–332, 334, 336, 337, 339, 349, 371, 394, 395, 398, 399, 408, 433, 435, 467, 507, 529, 534, 539, 563, 603, 636, 649, 650, 665, 689, 690
  film 618
  high-performance 564
  hopping 341
  n-type 185, 203, 376, 433, 447, 467, 529, 530, 564, 567, 570, 603
  p-type 75, 158, 185, 186, 201, 203, 229, 329, 429, 467, 468, 529, 530, 564, 630, 691
  residual 434
  temperature-independent 307
  thermal 166, 167, 172, 173, 203, 204, 305–308, 311, 312, 350, 420, 422, 450, 479, 486, 517–519, 546, 547, 592, 595, 621, 671, 673, 676, 698
contactless chemical vapor transport (CCVT) 98, 100
crack 162, 171, 251, 407, 648, 651, 656, 657, 661, 668, 669
  peripheral 162
  radial 171
crystal coloration 108, 112, 117, 120, 134, 136, 145, 146, 148, 151, 160, 165, 241, 246, 248, 254, 291, 380, 506, 572, 587, 606
crystal diameter 16, 139, 247, 248, 251, 254, 260, 266, 269, 274, 296
crystal grain 121, 122, 247, 294, 380, 414, 580, 581, 588, 650–652, 661
crystal growth 2, 17, 57, 61, 95, 96, 99, 106, 126, 130, 223, 225, 242, 245, 247, 250, 258, 269, 294, 304, 310, 311, 329, 331, 372, 378, 379, 382, 393, 403, 471, 472, 479, 487, 489, 492, 494, 506, 567, 578, 636, 638, 641, 648, 652, 656
crystal lattice 44, 156, 185, 186, 198, 229, 231, 244, 276, 278, 288, 346
crystallization 4, 20, 21, 29, 109, 159, 164, 237, 272, 274, 397, 398, 401, 404
  fast 581
  parasitic 126
crystal nucleation 393, 500
crystal seeds 3, 125, 129, 147, 148, 251, 653
crystal structure 54, 71, 227, 276, 313, 372, 373, 465, 560, 561, 578, 630–632
  cubic bixbyite 373
  perovskite 560, 561
  wurtzite 71
crystal surface 99, 112, 259, 264, 265, 336, 513, 647, 657, 668
CVD *see* chemical vapor deposition
CVT *see* chemical vapor transport
CVT method 2, 4–7, 9, 30, 76–78, 82–88, 94, 96, 97, 115, 116, 232–234, 236, 379, 470, 494–498, 502, 503
Czochralski-grown β-Ga_2O_3 crystal 280, 281, 286, 287, 309, 334, 339
Czochralski method 22–26, 31, 232, 259–263, 265, 266, 269–272, 274–276, 278, 279, 281–283, 288, 290, 291, 314, 315, 317, 318, 327, 328, 330–332, 334, 335, 342–344, 639, 640, 652–658, 698

Debye temperature 44, 172, 173, 204, 307, 421, 422, 450, 595, 674, 675, 677
decomposition 51, 263, 264, 268, 302, 303, 410, 413, 415–417, 486, 488, 511, 515, 574, 575, 668
  controlled 28, 29
  incongruent 576, 577, 590, 651, 675
  temperature-dependent 409
decomposition rate 21, 164, 275, 288, 303, 399, 400, 406, 410, 472, 576, 577
deep-level traps 202, 205, 348, 543
defect 72, 74, 140, 148, 153, 295, 296, 376, 392, 393, 563, 564, 582, 585, 632, 636
  charged 300, 607, 619
  deep 182
  extended 21, 201
  extrinsic 74, 565
  grown-in 258
  in-gap state 377
  intrinsic 228, 231
  ionized 541
  line-shaped 296
  native 74, 230, 466, 564, 565, 633, 635
  native point 74, 177, 186, 187, 433, 467, 529, 603, 636
density functional theory (DFT) 73, 228, 562, 599, 633
DFT *see* density functional theory
differential scanning calorimetry (DSC) 53, 592, 663, 664, 673
differential thermal analysis (DTA) 573, 575
dislocation 23, 130, 148, 151, 152, 258, 393, 477, 479, 609, 612, 615
  screw 258, 392, 393
  threading 614
dislocation density 128, 130, 139, 140, 145, 152, 161, 477, 560, 583, 605, 617
dissolution zone 10, 11, 125, 131, 142, 147, 148, 150, 151, 155, 156, 160, 505, 506
DLTS *see* spectroscopy, deep-level transient
donor 179, 186, 188, 191, 542
  compensating 468
  deep 74, 186, 230, 329, 376, 466, 564, 636
  double 375
  intrinsic 376
  ionized 177
  isolated 191, 344, 348, 351, 447, 448, 451, 542
  lowest energy 563
  native 376
  neutral 177
  shallow 74, 75, 185–187, 191, 202, 231, 282, 283, 329–331, 333, 344, 376, 447, 449, 466–468, 529, 542, 546, 548, 563–565, 603, 607, 689, 690
  ultradeep 563
dopant 80, 131, 135, 137, 151, 176, 276–280, 283–285, 288, 289, 291, 313, 316, 319, 334, 499–501, 506
  acceptor 468
  active 499
  ionized 613
  stable 277
doping 129, 131, 132, 134, 135, 136, 138, 150, 151, 159, 161, 182, 185, 186, 198, 199, 228, 229, 248, 252–254, 279, 280, 288, 315, 321, 322, 333, 334, 434, 435, 500, 501, 580, 618, 619
  acceptor 530
  controlled 500
  eternal 617

double-bob Archimedean method 305
Drude model 612
DSC *see* differential scanning calorimetry
DTA *see* differential thermal analysis

edge-defined film-fed growth method (EFG method) 2, 26, 27, 254, 260, 271, 296, 325, 330, 335, 346, 350
effect
  capping 130, 154
  endothermal DSC 53
  entropic 47
  entropy-stabilizing 288, 645
  piezoelectric 191
  purifying 252
  self-compensation 187, 565, 603
  trapping 228
effective electron mass 348, 352, 446, 448, 451, 538, 539, 545, 548, 612, 620, 622, 693, 696
effective mass 194, 195, 202, 205, 446, 539, 609–611
effect of annealing 337, 513
EFG method *see* edge-defined film-fed growth method
elastic stiffness coefficient 309, 310, 421, 422, 450, 672–676, 699
electrical insulator 185, 187, 329, 333, 339, 531, 604, 606, 608, 678, 679, 692
electromagnetic field 28, 157, 397, 399, 400, 405
electron concentration 189, 191, 199, 201, 334–338, 340, 341, 343, 344, 348, 438–440, 442–444, 446, 447, 533, 534, 536, 690, 691, 694, 695
electron Hall mobility 342, 343, 609
electron mobility 188–190, 195, 198, 199, 201, 203, 341–344, 435, 436, 535, 536, 538, 540–542, 545, 546, 559, 560, 610, 612–615, 694
  experimental 205, 351, 451, 542, 548, 622
  theoretical 351, 451, 622
electron paramagnetic resonance 346
Ellingham-type diagram 49, 54, 57, 61, 62, 378
emission 179–183, 318, 319, 321, 430, 525, 526, 602, 683, 685, 686
  band-edge 319
  broad 428, 525, 526
  green 180–182, 319, 525
  near-band-edge 181
  orange 428, 525, 526
  red 525, 683
energy gap 179, 228, 345, 426, 428
  optical 314, 425
  quasi-direct 349
enthalpy 44, 45, 47
entropy 45, 47, 48
EPD *see* etch pit density
etching 84, 130, 139, 141, 143, 146, 148, 154, 155, 255, 257, 258, 295, 392
  chemical 130, 146
  dry 301
  high-temperature 476
  reactive ion 301
etch pit 139, 140, 154, 155, 257, 258, 296, 392, 393, 415, 477, 584
etch pit density (EPD) 108, 116, 139, 144, 145, 155, 160, 255, 259, 296, 414, 418, 476, 477, 510, 512
evaporation zone 79, 80, 471, 472, 480

exciton 176, 177, 179
ionized-donor-bound 179
neutral-donor-bound 179
self-trapped 313, 319, 326, 429, 432, 526, 685
expansion coefficient 167, 168, 308, 309, 419, 592

feed rod 14, 16, 17, 122, 248, 250–253, 647
Fermi level 466, 468, 544, 564, 607, 616, 618
Fermi level shift 321, 615, 694
Fermi wave vector 611
field-effect transistor 50, 70, 224, 560
filling factor 129, 131, 144, 154–159
flux 2, 3, 12–14, 60, 117, 119–124, 242–244, 388–394, 417, 418, 434–436, 448, 450, 506–510, 567, 568, 570–572, 608, 639, 641–644
flux-grown crystal 393, 394, 434, 435, 444, 445, 507, 508, 511, 543, 544, 567, 568, 590, 593, 599, 601
flux impurity 14, 242, 392, 512, 543, 590, 642, 669
flux method 3, 12, 13, 30, 377–379, 505, 507, 511, 514, 559–560, 565–567, 569, 571, 572, 639–641
formation energy 46, 74, 180, 226, 230, 231, 261, 321, 376, 444, 467, 468, 564
free electron concentration 188–199, 201–203, 281, 313–315, 326–328, 330–334, 336–344, 347, 348, 351, 352, 423, 433–449, 451, 520, 521, 530–545, 604–607, 609–611, 613–615, 619–622, 690, 691, 693–696
FTIR spectroscopy *see* spectroscopy, Fourier transform infrared
full width at half maximum (FWHM) 111, 112, 116, 117, 139–142, 144, 145, 151, 152, 156, 157, 159, 160, 162, 249, 251, 253, 255, 256, 414, 415, 477, 478, 512, 583, 591, 669
furnace 12, 13, 79, 84, 85, 94, 95, 97, 98, 114, 115, 119, 243, 244, 382, 383, 389–391, 471, 472, 482, 484, 485, 492, 493, 496–502, 509, 510, 570, 571
alumina 95
alumina-based 473
assembled growth 489
cold crucible growth 19
Czochralski growth 260
graphite growth 113
horizontal 79, 116, 387, 471, 492, 498, 503
infrared 252
multizone 107
resistively heated 499, 506, 509, 510
three-zone Kanthal 480
vertical 80, 81, 117
vertical Kanthal 92, 243
FWHM *see* full width at half maximum

Ga-based spinel 23, 25, 629, 630, 632, 636, 640, 645, 647, 662, 664, 666, 667, 670, 671, 675, 676, 685–690, 695–697
GDMS *see* glow discharge mass spectrometry
Gibbs free energy 45, 48, 94, 95, 102, 166
glow discharge mass spectrometry (GDMS) 133, 154, 259
Goldschmidt tolerance factor 561
grain 99, 147, 162, 226, 247, 251, 275, 294, 302, 418, 478, 581

grain boundary 121, 414, 491, 609, 612
  low-angle 23, 336, 477, 491, 605
growth atmosphere 8, 9, 24–26, 28, 29, 60, 61, 248–251, 253, 254, 259, 260, 262–266, 268–271, 279, 280, 286–288, 290–293, 331–336, 472, 488–490, 573, 576–580, 585–588, 645–648, 650–657
growth chamber 94, 159, 161, 248, 270, 400, 403, 489, 576, 578, 579, 654, 655
  high-pressure 157
growth condition 3, 4, 7, 9, 89, 90, 92, 142, 144, 145, 151, 152, 252, 254, 279, 473, 564, 565, 575, 578
growth duration 80, 84, 86, 87, 90, 92, 94, 96, 97, 115, 116, 123, 148, 155, 159, 505, 506, 509
growth environment 1, 135, 139, 144, 156, 176, 179, 180, 182, 184, 185, 187, 472, 473, 492, 493, 496, 640, 642
growth furnace 20, 23, 24, 95, 105, 106, 120, 121, 245, 247, 248, 259, 268, 269, 290–293, 403, 409, 479, 480, 485, 491, 492, 576–579
growth habit 86, 87, 119, 120, 135–137, 142, 149, 151, 384, 389, 394, 496, 500, 506, 508
growth interface 25, 87, 99–101, 107, 147, 273, 274, 286
growth method 2–4, 7, 29–31, 182, 184–187, 202, 296, 297, 319, 395, 397, 404, 405, 417, 418, 433, 435
growth morphology 135, 136, 137, 138, 139, 159, 180, 476, 494, 496, 507, 508, 511
growth rate 16, 24–26, 111–115, 126, 128, 131, 132, 135, 136, 138, 139, 143, 145–151, 153–156, 159, 161, 236, 237, 254, 392, 490, 491, 500, 648
growth temperature 5, 6, 8, 89–91, 96, 97, 99, 120, 122, 123, 184, 186, 235–238, 241, 242, 389, 474, 502, 503, 641
growth time 6, 9, 11, 12, 14, 16, 18, 19, 21, 28, 29, 31, 139, 388, 473, 502, 503, 647, 649
growth zone 4, 10, 11, 79–81, 83–85, 87–92, 94, 96, 97, 99, 107–110, 125, 126, 130–132, 135, 142–145, 149–153, 155, 156, 159, 233, 234, 383, 384, 386, 387, 471–474, 479–481

halide vapor phase epitaxy (HVPE) 224, 227, 346, 347, 372
Hall effect 184, 328, 446, 532, 614, 615, 690, 691, 693
Hall mobility 191–194, 196–199, 335–339, 341–343, 434–443, 447–449, 530, 532, 533, 535–538, 540–542, 544, 604–606, 609, 610, 613–615, 690, 691, 694–696
high-angle annular dark field 582
high electron mobility transistor 560
high-frequency dielectric constant 178, 184, 432, 451, 524, 529, 548, 599, 600, 602, 682, 688
high-temperature operation 415, 511, 590
HVPE *see* halide vapor phase epitaxy
hydrothermal condition 70, 134, 147, 388
hydrothermal method 2, 3, 10–12, 60, 75, 76, 123, 127, 134, 135, 138–140, 142, 143, 158, 174, 196, 198, 202, 203, 378, 379, 388, 470, 505

ICP *see* inductively coupled plasma
In_2O_3 31, 372, 377–380, 382, 383, 386–392, 394–398, 411, 413, 414, 416, 417, 419, 422–428, 432–437, 440, 441, 446–448, 450
In_2O_3 decomposition 380, 396, 397, 402, 406, 407, 409–411, 417, 431
inductive heating 16, 18, 20, 121, 399, 405, 485, 486, 512, 535, 565
inductively coupled plasma (ICP) 301, 655
insulation 8, 21, 23, 24, 26, 27, 289, 407, 414, 474, 486, 576, 578
  alumina thermal 290, 293
  ceramic thermal 471
  muffle thermal 481
  top thermal 407, 486, 650
  zirconia thermal 163, 654
intentional dopant 231, 246, 288, 289, 315, 349, 408, 434, 491, 567, 569, 690
intentional doping 131, 134, 198, 276, 330, 332, 347, 349, 433, 434, 447, 507, 529, 531, 567, 569
inverse spinel 631, 636, 662, 663, 670
ionization energy 229, 344, 346, 447, 466, 542, 545
ionized impurity 191, 341, 348, 438, 603, 605, 606, 612–615, 619, 620

Kopp's rule 671
Kyropoulos method 25, 26, 31

laser molecular beam epitaxy 227
LASSCGM *see* levitation-assisted self-seeding crystal growth method
lattice constant 203, 349, 561, 595, 621, 622, 631, 697
lattice parameter 225–227, 419, 420, 422, 450, 465, 516, 517, 519, 547, 592–594, 630–632, 672, 676
LEED *see* low-energy electron diffraction
levitation-assisted self-seeding crystal growth method (LASSCGM) 28, 29, 378, 379, 395, 397, 398, 400, 402, 406, 407, 411, 412, 414, 419, 420, 424, 427, 428, 430, 442, 443, 449, 450
liquid phase epitaxy (LPE) 387, 509
longitudinal sound velocity 171–173, 421, 422, 595, 622
low-energy electron diffraction (LEED) 299, 514
low-energy ion scattering 514
LPE *see* liquid phase epitaxy
luminescence 179, 180, 318, 320, 321, 323, 422, 428, 429, 525, 596, 600–602, 683, 685

mass loss 273, 303, 400, 401, 415–417, 487, 511, 574, 586–588, 590, 643, 666–668
mass transport 76, 81, 109, 237, 238
MBE *see* molecular beam epitaxy
melt growth method 3, 21, 30, 31, 187, 191, 345
metal-organic chemical vapor deposition (MOCVD) 70, 223, 227, 372, 463, 685
metal-organic vapor phase epitaxy 223
metal-oxide field-effect transistor (MOSFET) 224, 629

$MgGa_2O_4$ 17, 18, 31, 629–631, 633, 634, 636–641, 645, 646, 648–650, 652, 653, 656–659, 661, 663–668, 670–672, 674, 675, 678–685, 687–691, 693–701
microvoid 296, 483, 661, 662
Mie's theory 176
Miller index 145
MOCVD *see* metal-organic chemical vapor deposition
Moiré pattern 662
molecular beam epitaxy (MBE) 43, 70, 224, 227, 346, 347, 372, 463, 560, 599
MOSFET *see* metal-oxide field-effect transistor

near-infrared region (NIR region) 174, 272, 274, 313, 316, 323, 431, 432, 652, 654, 678, 680
NIR region *see* near-infrared region
NIR spectrum 423, 521, 653
nuclear reaction analysis 163
nucleation 1, 79, 107, 108, 147, 293, 392, 393, 500
  parasitic 105, 111
  spontaneous 7, 14, 94, 131, 132, 142, 143, 147, 162, 294

OFZ *see* optical floating zone
OFZ method 3, 14, 16, 17, 232, 247–249, 254, 325, 329, 333, 342
optical bandgap 176, 177, 182, 312, 324, 351, 422, 423, 425, 528, 548, 562, 596
optical floating zone (OFZ) 2, 16, 232, 247, 322, 325, 330, 335, 342, 639, 647, 698
optical phonon 167, 191, 194, 612, 614, 615
optical transition 228, 313, 315, 523, 599, 602, 681
overpressure 164, 166, 398, 401, 404, 406, 407, 409, 410, 413, 577, 580, 581, 646, 649, 654
oxidizing atmosphere 18, 19, 24, 303, 304, 431, 438, 440–442, 444, 483, 484, 657, 659, 679, 685, 689
oxygen fugacity 49, 52, 55, 61, 62, 65, 66
oxygen source 51, 55, 64, 77, 106, 281–283, 482, 488, 493, 496, 499

PAMBE *see* plasma-assisted molecular beam epitaxy
PBC model *see* periodic bond chain model
PED method *see* pulsed electron deposition method
periodic bond chain model (PBC model) 507, 508
perovskite 59, 602, 617, 618, 621
phase transition 4, 9, 10, 12, 14, 22, 25, 26, 43, 49, 272, 401, 479
phonon mode 183, 313, 322, 323, 430–432, 526, 527, 596, 599, 601, 686
  acoustic 601
  active 322
  experimental 686
  fundamental 430
  infrared-active 686
  longitudinal acoustic 593
  theoretical 182, 183
phonon scattering 192, 343, 540, 613
photon energy 177, 181, 314, 315, 317, 320, 425, 426, 428, 521, 523, 596–599, 680, 681
physical vapor transport (PVT) 2, 4, 7, 9, 76, 233, 234, 378–379, 434, 435, 469–471, 474, 476, 535, 545–547

Planck constant 446, 596, 611
plasma-assisted molecular beam epitaxy (PAMBE) 227, 541, 543
platelets 87, 107, 118, 235, 241, 245, 297, 474, 481, 507, 509
PLD *see* pulsed laser deposition
Poisson's ratio 169, 173, 204, 309, 312, 350, 518–520, 592, 595, 673–675, 677
polycrystal 16, 109, 111, 164, 255, 271, 293, 294
powder diffraction file 490
pulsed electron deposition method (PED method) 600, 611
pulsed laser deposition (PLD) 70, 224, 227, 372, 463, 560, 600, 605, 611, 614, 619
PVT *see* physical vapor transport
PVT-grown SnO_2 crystal 473, 483, 513, 518, 522, 524–526, 530, 532, 534, 537–539
PVT method 2, 3, 7, 8, 30, 75, 77, 79, 233, 379, 380, 470, 471, 475, 476, 478, 485–487, 489–493, 510, 511, 531

quartz tube 17, 79, 80, 84, 85, 87, 99, 100, 234, 252, 301, 492, 497–502

radiation 28, 272–274, 324, 398, 399, 408
  black body 272, 273
  electromagnetic 316, 596
  infrared 407
  nuclear 225, 324
Raman-active mode 182, 601
Raman spectrum 182, 322, 323, 601
reactive ion etching 301
reducing agent 7, 76, 79, 92–94, 98, 101, 105–107, 109, 116, 379–381, 435, 440
reducing atmosphere 16, 303, 304, 417, 429, 432, 438, 441, 443, 448, 666, 670
reducing conditions 50, 54, 65, 247, 313, 333, 606, 667, 699
refractive index 178, 179, 182, 317, 323, 324, 422, 425, 427, 428, 523–525, 528, 529, 596, 599, 600, 602, 682, 683
relative static dielectric constant 182, 204, 324, 432, 451, 523, 528, 529, 548, 681, 682, 688, 700
resistivity 184, 187–199, 329, 335, 337, 340–342, 434–443, 530, 532–537, 544, 604–606, 609, 610, 690, 691
RF (radiofrequency) coil 19, 21, 26, 27, 29, 159–161, 254, 259, 260, 292, 293, 398–400, 406, 648, 652, 654
RF generator 159, 160, 259, 586, 648
rotation rate 16, 18, 24, 25, 121, 250, 253, 254, 276, 290, 291, 647, 652–657

salt electrolysis 388, 394, 435
sapphire 14, 16, 20, 22, 23, 25, 27, 100, 226, 227, 380, 474, 476, 486, 491
SBD *see* Schottky barrier diode
scanning thermal microscopy 167
scanning transmission electron microscopy 582
scanning tunneling microscopy (STM) 299, 514, 515
scattering 341, 348, 438, 540, 541, 597, 603, 605, 614, 615, 620
  elastic Rayleigh 163
  inelastic neutron 182–184
  lattice vibration 191
  low-energy ion 514
Scheil–Gulliver equation 164, 283–289

Schottky barrier 445, 607
Schottky barrier diode (SBD) 224, 258
Schottky contact 445, 694
scintillator 70, 325, 326
SEAL *see* surface electron accumulation layer
seeds 10, 11, 107, 109, 116, 125, 126, 129–131, 133, 134, 135, 136, 137, 139, 140, 142, 145–157, 251, 253, 500
segregation coefficient 16, 245, 276, 282–289, 394
self-trapped exciton (STE) 313, 319, 323, 326, 429, 432, 526, 685
self-trapped hole (STH) 320, 429, 526
semiconductor 23, 27, 31, 327, 333, 334, 371, 433, 434, 530, 534, 606, 610, 678, 679, 696, 697
  classical 16, 22, 76, 274, 348
  compound 22, 23, 31, 274
  degenerate 187, 201, 347, 530, 546, 607, 609, 619, 659, 696
  elemental 23, 540
shear modulus 169–171, 173, 204, 309, 312, 421, 422, 518–520, 592, 673, 675, 677
SnO_2 383, 464, 468–477, 479–482, 485, 486, 489–492, 494, 495, 498–503, 505–511, 514, 519–521, 525, 526, 528–531, 540–547
solidification point (SP) 401, 402, 404, 410, 632
solution growth 60, 393
  classical 9
  high-pressure 10
  high-temperature 10, 12
  top-seeded 13, 121
SP *see* solidification point
spectral region 178, 313, 319, 520, 524, 528, 601, 653, 682, 687
  blue 282
  near-infrared 349, 449
  red 281
  visible 424, 431, 432
spectroscopy 611, 673
  angle-resolved photoemission (ARPES) 73, 229, 375, 377, 633, 634, 665
  bulk-sensitive hard-X-ray photoelectron 543
  deep-level optical 344
  deep-level transient (DLTS) 184, 199, 201, 344, 347
  Fourier transform infrared (FTIR) 176, 331, 611
  infrared 176, 331, 611
  photoinduced current transient 344
  positron annihilation 467
  Raman 183, 184, 323, 601
  muon-spin rotation and relaxation 331
  resonant ultrasonic 309
  scanning tunneling 377
  X-ray photoemission 73, 300, 589
STE *see* self-trapped exciton
Stepanov method 26, 271
STH *see* self-trapped hole
STM *see* scanning tunneling microscopy
stoichiometric composition 148, 567, 570, 584, 585, 588, 638, 641, 648, 656, 659, 660, 668
stoichiometric ratio 570, 575, 642
stoichiometry 64, 101, 113–115, 128, 150, 174, 176, 182, 186, 576, 577, 579, 580, 641, 644, 659, 660
sublimation 4, 7, 76, 77, 79, 115, 232, 233, 379
substrate 31, 97, 98, 113, 227, 256, 380, 416, 417, 477, 478, 485–488, 490, 492, 629, 630, 665, 666, 668

surface electron accumulation
layer (SEAL) 96, 444, 445,
448, 449, 543

technique 1–4, 7, 8, 14–16, 18, 20,
22, 23, 25, 28, 30, 70, 79, 81,
159, 161, 379, 380
Faraday rotation 194
epitaxial 223, 590
experimental 165, 347
floating zone 16
flux 10, 14, 123, 124, 246, 668
growth 1, 2, 20, 22, 26–28, 31,
117, 129, 343, 349, 433, 436
hydrothermal 30, 560, 590
low-temperature 396
melt-grown 76, 159
melt-growth 647, 661
plate-resonance 309
solution 30, 75, 123, 388, 505
sublimation 470
ultrasonic 169
temperature gradient 13, 14, 21,
23–26, 28, 29, 108, 109, 120,
260, 272, 274–276, 290–296,
404–408, 479, 480, 576,
642–644, 647–649
temperature monitoring 159, 576,
586, 648, 649
thermal analysis 57, 59, 573, 664
thermal decomposition 2, 3, 8,
378, 381, 464, 471, 511, 645,
646, 648, 649, 661, 666
thermal diffusivity 306, 350, 450,
592
thermal expansion 154, 162, 167,
172, 293, 305, 308, 311, 312,
350, 419, 516, 517, 672, 673
thermal expansion coefficient 173,
204, 308, 422, 517, 519, 594,
621, 671, 672, 675, 676
thermal history 242, 401, 536, 671
thermal instability 3, 14, 117, 281,
415, 417, 464, 560, 590, 645
thermal insulation 5, 8, 20, 21, 23,
24, 26, 27, 29, 114, 115, 161,
163, 186, 249, 259, 260, 291,
292, 406, 407, 484–487, 648,
650, 652
thermally stimulated current (TSC)
344, 543
thermal stability 59, 276–278,
281, 301, 303, 415, 590, 638,
645, 646, 648, 666–670
thermocouple 105, 115, 292, 305,
390, 400, 481, 496, 497, 499,
502, 504
top-seeded solution growth (TSSG)
13, 121, 124
transmittance 174–176, 312–316,
323, 422–425, 440, 520–522,
596, 597, 611, 678–680
transmittance spectrum 174, 313,
314, 423, 424, 520, 521, 597,
678
transparent semiconducting oxides
(TSOs) 1–7, 9, 12, 14, 20, 23,
24, 43, 44, 48–56, 60, 64–66,
69, 70, 75, 76, 463, 536, 559
transport agent 4, 5, 7, 89–92, 117,
187, 234, 239, 385–387, 503
transport rate 6, 89–92, 108, 383,
384, 504
transverse sound velocity 173,
422, 593, 595, 622
trap 199, 345–347, 468, 543
cold 501
deep 329, 344, 346, 347, 543
deep acceptor 201
deep defect 348
trap concentration 201, 202, 347
trap distribution 345, 347, 348
traveling-solvent floating zone
(TSFZ) 14, 121
TSC *see* thermally stimulated
current
TSFZ *see* traveling-solvent floating
zone

TSOs *see* transparent semiconducting oxides
TSSG *see* top-seeded solution growth
twin boundaries 256, 257, 266, 288, 393, 614, 694

undoped crystal 131, 134, 135, 250, 279, 281, 289–291, 334, 337, 338, 493, 500, 568, 570, 604, 606, 690

vacuum 90, 91, 105, 107, 108, 110, 114, 199, 235, 236, 415–417, 444, 445, 448, 449, 515
valence band maximum (VBM) 73, 201, 229, 345, 374, 376, 466, 544, 562, 563, 632, 633, 635, 636
vapor phase 75, 121, 122, 147, 194, 201–203, 378, 396, 464, 470, 527
vapor transport 7, 91, 107, 110, 387
  chemical 2, 4, 76, 233, 234, 378, 469, 471, 492
  physical 2, 4, 7, 76, 233, 379, 469–471, 546
VBM *see* valence band maximum
Verneuil method 14–16, 232, 245, 296, 333
vertical Bridgman method 291, 648
vertical gradient freeze (VGF) 2, 20, 21, 31, 291, 566, 575, 608, 639, 648, 649, 668, 678
VGF *see* vertical gradient freeze
VGF method 20, 21, 159, 271, 293, 295, 298, 639, 640
Vickers indentation method 171
volatile species 2, 7, 8, 233, 235, 237–240, 262, 265, 276–278, 395, 396, 573, 645, 646, 649
von Mieses stress 310

wafer fabrication 2, 27, 31, 310, 311, 650
wurtzite structure 71, 172, 202

X-ray excitation 323, 325, 432, 526
X-ray fluoroscence (XRF) 288, 659, 660
X-ray powder diffractometry 168
X-ray texture 414, 415, 477, 478
X-ray topography 139, 148, 153, 296
XRF *see* X-ray fluoroscence

Young's modulus 169–173, 204, 309, 312, 421, 422, 518, 520, 592, 595, 673, 674, 677

Zeeman splitting 194
zirconia 23, 260, 289–293, 407, 413, 446, 473, 501, 652
$ZnGa_2O_4$ 21, 24, 31, 629–633, 636–641, 643–652, 654, 658–668, 670–672, 674, 675, 678, 679, 681–691, 693–701
ZnO 75, 76, 78, 80, 83, 85–87, 105, 110, 111, 113–115, 117, 123, 125, 131–137, 139, 140, 142, 144, 146, 147, 150, 154, 155, 157, 158, 159, 160, 161, 162, 165, 167, 168, 169, 170–183, 187–189, 193–196, 199, 201–203
ZnO powder 92, 96, 97, 106, 109, 113, 126, 127, 142, 145, 147, 148, 151, 153, 156, 157

For Product Safety Concerns and Information please contact our EU representative GPSR@taylorandfrancis.com Taylor & Francis Verlag GmbH, Kaufingerstraße 24, 80331 München, Germany

Batch number: 10726505

Printed by Printforce, the Netherlands